Advances in Intelligent Systems and Computing

Volume 822

Series editor

Janusz Kacprzyk, Polish Academy of Sciences, Warsaw, Poland
e-mail: kacprzyk@ibspan.waw.pl

The series "Advances in Intelligent Systems and Computing" contains publications on theory, applications, and design methods of Intelligent Systems and Intelligent Computing. Virtually all disciplines such as engineering, natural sciences, computer and information science, ICT, economics, business, e-commerce, environment, healthcare, life science are covered. The list of topics spans all the areas of modern intelligent systems and computing such as: computational intelligence, soft computing including neural networks, fuzzy systems, evolutionary computing and the fusion of these paradigms, social intelligence, ambient intelligence, computational neuroscience, artificial life, virtual worlds and society, cognitive science and systems, Perception and Vision, DNA and immune based systems, self-organizing and adaptive systems, e-Learning and teaching, human-centered and human-centric computing, recommender systems, intelligent control, robotics and mechatronics including human-machine teaming, knowledge-based paradigms, learning paradigms, machine ethics, intelligent data analysis, knowledge management, intelligent agents, intelligent decision making and support, intelligent network security, trust management, interactive entertainment, Web intelligence and multimedia.

The publications within "Advances in Intelligent Systems and Computing" are primarily proceedings of important conferences, symposia and congresses. They cover significant recent developments in the field, both of a foundational and applicable character. An important characteristic feature of the series is the short publication time and world-wide distribution. This permits a rapid and broad dissemination of research results.

More information about this series at http://www.springer.com/series/11156

Sebastiano Bagnara · Riccardo Tartaglia
Sara Albolino · Thomas Alexander
Yushi Fujita
Editors

Proceedings of the 20th Congress of the International Ergonomics Association (IEA 2018)

Volume V: Human Simulation and Virtual Environments, Work With Computing Systems (WWCS), Process Control

 Springer

Editors
Sebastiano Bagnara
University of the Republic of San Marino
San Marino, San Marino

Riccardo Tartaglia
Centre for Clinical Risk Management
    and Patient Safety, Tuscany Region
Florence, Italy

Sara Albolino
Centre for Clinical Risk Management
    and Patient Safety, Tuscany Region
Florence, Italy

Thomas Alexander
Fraunhofer FKIE
Bonn, Nordrhein-Westfalen
Germany

Yushi Fujita
International Ergonomics Association
Tokyo, Japan

ISSN 2194-5357 ISSN 2194-5365 (electronic)
Advances in Intelligent Systems and Computing
ISBN 978-3-319-96076-0 ISBN 978-3-319-96077-7 (eBook)
https://doi.org/10.1007/978-3-319-96077-7

Library of Congress Control Number: 2018950646

This Springer imprint is published by the registered company Springer Nature Switzerland AG
The registered company address is: Gewerbestrasse 11, 6330 Cham, Switzerland

Preface

The Triennial Congress of the International Ergonomics Association is where and when a large community of scientists and practitioners interested in the fields of ergonomics/human factors meet to exchange research results and good practices, discuss them, raise questions about the state and the future of the community, and about the context where the community lives: the planet. The ergonomics/human factors community is concerned not only about its own conditions and perspectives, but also with those of people at large and the place we all live, as Neville Moray (Tatcher et al. 2018) taught us in a memorable address at the IEA Congress in Toronto more than twenty years, in 1994.

The Proceedings of an IEA Congress describes, then, the actual state of the art of the field of ergonomics/human factors and its context every three years.

In Florence, where the XX IEA Congress is taking place, there have been more than sixteen hundred (1643) abstract proposals from eighty countries from all the five continents. The accepted proposal has been about one thousand (1010), roughly, half from Europe and half from the other continents, being Asia the most numerous, followed by South America, North America, Oceania, and Africa. This Proceedings is indeed a very detailed and complete state of the art of human factors/ergonomics research and practice in about every place in the world.

All the accepted contributions are collected in the Congress Proceedings, distributed in ten volumes along with the themes in which ergonomics/human factors field is traditionally articulated and IEA Technical Committees are named:

I. Healthcare Ergonomics (ISBN 978-3-319-96097-5).
II. Safety and Health and Slips, Trips and Falls (ISBN 978-3-319-96088-3).
III. Musculoskeletal Disorders (ISBN 978-3-319-96082-1).
IV. Organizational Design and Management (ODAM), Professional Affairs, Forensic (ISBN 978-3-319-96079-1).
V. Human Simulation and Virtual Environments, Work with Computing Systems (WWCS), Process control (ISBN 978-3-319-96076-0).

VI. Transport Ergonomics and Human Factors (TEHF), Aerospace Human Factors and Ergonomics (ISBN 978-3-319-96073-9).

VII. Ergonomics in Design, Design for All, Activity Theories for Work Analysis and Design, Affective Design (ISBN 978-3-319-96070-8).

VIII. Ergonomics and Human Factors in Manufacturing, Agriculture, Building and Construction, Sustainable Development and Mining (ISBN 978-3-319-96067-8).

IX. Aging, Gender and Work, Anthropometry, Ergonomics for Children and Educational Environments (ISBN 978-3-319-96064-7).

X. Auditory and Vocal Ergonomics, Visual Ergonomics, Psychophysiology in Ergonomics, Ergonomics in Advanced Imaging (ISBN 978-3-319-96058-6).

Altogether, the contributions make apparent the diversities in culture and in the socioeconomic conditions the authors belong to. The notion of well-being, which the reference value for ergonomics/human factors is not monolithic, instead varies along with the cultural and societal differences each contributor share. Diversity is a necessary condition for a fruitful discussion and exchange of experiences, not to say for creativity, which is the "theme" of the congress.

In an era of profound transformation, called either digital (Zisman & Kenney, 2018) or the second machine age (Bnynjolfsson & McAfee, 2014), when the very notions of work, fatigue, and well-being are changing in depth, ergonomics/human factors need to be creative in order to meet the new, ever-encountered challenges. Not every contribution in the ten volumes of the Proceedings explicitly faces the problem: the need for creativity to be able to confront the new challenges. However, even the more traditional, classical papers are influenced by the new conditions.

The reader of whichever volume enters an atmosphere where there are not many well-established certainties, but instead an abundance of doubts and open questions: again, the conditions for creativity and innovative solutions.

We hope that, notwithstanding the titles of the volumes that mimic the IEA Technical Committees, some of them created about half a century ago, the XX Triennial IEA Congress Proceedings may bring readers into an atmosphere where doubts are more common than certainties, challenge to answer ever-heard questions is continuously present, and creative solutions can be often encountered.

Acknowledgment

A heartfelt thanks to Elena Beleffi, in charge of the organization committee. Her technical and scientific contribution to the organization of the conference was crucial to its success.

References

Brynjolfsson E., A, McAfee A. (2014) The second machine age. New York: Norton.

Tatcher A., Waterson P., Todd A., and Moray N. (2018) State of science: Ergonomics and global issues. Ergonomics, 61 (2), 197–213.

Zisman J., Kenney M. (2018) The next phase in digital revolution: Intelligent tools, platforms, growth, employment. Communications of ACM, 61 (2), 54–63.

Sebastiano Bagnara
Chair of the Scientific Committee, XX IEA Triennial World Congress

Riccardo Tartaglia
Chair XX IEA Triennial World Congress

Sara Albolino
Co-chair XX IEA Triennial World Congress

Organization

Organizing Committee

Riccardo Tartaglia (Chair IEA 2018)	Tuscany Region
Sara Albolino (Co-chair IEA 2018)	Tuscany Region
Giulio Arcangeli	University of Florence
Elena Beleffi	Tuscany Region
Tommaso Bellandi	Tuscany Region
Michele Bellani	Humanfactor[x]
Giuliano Benelli	University of Siena
Lina Bonapace	Macadamian Technologies, Canada
Sergio Bovenga	FNOMCeO
Antonio Chialastri	Alitalia
Vasco Giannotti	Fondazione Sicurezza in Sanità
Nicola Mucci	University of Florence
Enrico Occhipinti	University of Milan
Simone Pozzi	Deep Blue
Stavros Prineas	ErrorMed
Francesco Ranzani	Tuscany Region
Alessandra Rinaldi	University of Florence
Isabella Steffan	Design for all
Fabio Strambi	Etui Advisor for Ergonomics
Michela Tanzini	Tuscany Region
Giulio Toccafondi	Tuscany Region
Antonella Toffetti	CRF, Italy
Francesca Tosi	University of Florence
Andrea Vannucci	Agenzia Regionale di Sanità Toscana
Francesco Venneri	Azienda Sanitaria Centro Firenze

Scientific Committee

Sebastiano Bagnara (President of IEA2018 Scientific Committee)	University of San Marino, San Marino
Thomas Alexander (IEA STPC Chair)	Fraunhofer-FKIE, Germany
Walter Amado	Asociación de Ergonomía Argentina (ADEA), Argentina
Massimo Bergamasco	Scuola Superiore Sant'Anna di Pisa, Italy
Nancy Black	Association of Canadian Ergonomics (ACE), Canada
Guy André Boy	Human Systems Integration Working Group (INCOSE), France
Emilio Cadavid Guzmán	Sociedad Colombiana de Ergonomia (SCE), Colombia
Pascale Carayon	University of Wisconsin-Madison, USA
Daniela Colombini	EPM, Italy
Giovanni Costa	Clinica del Lavoro "L. Devoto," University of Milan, Italy
Teresa Cotrim	Associação Portuguesa de Ergonomia (APERGO), University of Lisbon, Portugal
Marco Depolo	University of Bologna, Italy
Takeshi Ebara	Japan Ergonomics Society (JES)/Nagoya City University Graduate School of Medical Sciences, Japan
Pierre Falzon	CNAM, France
Daniel Gopher	Israel Institute of Technology, Israel
Paulina Hernandez	ULAERGO, Chile/Sud America
Sue Hignett	Loughborough University, Design School, UK
Erik Hollnagel	University of Southern Denmark and Chief Consultant at the Centre for Quality Improvement, Denmark
Sergio Iavicoli	INAIL, Italy
Chiu-Siang Joe Lin	Ergonomics Society of Taiwan (EST), Taiwan
Waldemar Karwowski	University of Central Florida, USA
Peter Lachman	CEO ISQUA, UK
Javier Llaneza Álvarez	Asociación Española de Ergonomia (AEE), Spain
Francisco Octavio Lopez Millán	Sociedad de Ergonomistas de México, Mexico

Donald Norman   University of California, USA
José Orlando Gomes   Federal University of Rio de Janeiro, Brazil
Oronzo Parlangeli   University of Siena, Italy
Janusz Pokorski   Jagiellonian University, Cracovia, Poland
Gustavo Adolfo Rosal Lopez   Asociación Española de Ergonomia (AEE),
    Spain
John Rosecrance   State University of Colorado, USA
Davide Scotti   SAIPEM, Italy
Stefania Spada   EurErg, FCA, Italy
Helmut Strasser   University of Siegen, Germany
Gyula Szabò   Hungarian Ergonomics Society (MET),
    Hungary
Andrew Thatcher   University of Witwatersrand, South Africa
Andrew Todd   ERGO Africa, Rhodes University,
    South Africa
Francesca Tosi   Ergonomics Society of Italy (SIE);
    University of Florence, Italy
Charles Vincent   University of Oxford, UK
Aleksandar Zunjic   Ergonomics Society of Serbia (ESS),
    Serbia

Contents

Human Simulation and Virtual Environments

Accommodation Assessments for Vehicle Occupants Using Augmented Reality

Byoung-keon Daniel Park[(⊠)] and Matthew P. Reed

University of Michigan, Ann Arbor, MI 48109, USA
keonpark@umich.edu

Abstract. This paper presents a new accommodation assessment method for vehicle occupants using a statistical body shape model in an augmented reality (AR) environment. Vehicle occupant accommodation assessment is an important aspect of vehicle interior design. Variability in body dimensions of the target population is a key component in determining the overall user accommodation. Statistical body shape modeling enables quantitative representation and assessment a wide range of variability in anthropometry and posture. These statistical models provide a way to efficiently generate a realistic 3d body shape surface along with the standard body dimensions, anatomical landmark locations and joint locations. In the current study, an automotive posture body shape model based on data from 255 men and women ages 20 to 95 years old was used in a demonstration of AR technology. Typically, quantitative assessment of a physical vehicle requires time-consuming scanning to obtain a computer model that can be used with virtual assessment tools. We addressed this issue by using AR to enable assessment without explicit model building. Apple ARKit on an iPhone was employed in this study to implement the model in an augmented vehicle environment. The system allows the user to place a human model in a vehicle by detecting the seat surfaces. The user is able to manipulate the body shape to assess accommodation across the range of anthropometric variability. Interior accommodation was assessed by measuring the distances between the certain points from both the model and the augmented physical environment, in addition to a qualitative visual inspection. Opportunities and impacts of the proposed AR approach with digital human models in more applications are discussed.

Keywords: Augmented reality · Accommodation · Seat
Digital human models

1 Introduction

Accommodation assessment plays important role in automobile interior package design practices and vehicle occupant safety [1]. Typically, quantitative assessment of a physical vehicle in a virtual environment requires time-consuming scanning and manual processing to obtain a precise computer model of the vehicle that can then be analyzed using virtual assessment tools. In this paper, we bypass the generation of the vehicle model using augmented reality (AR). AR provides an interactive experience of a physical environment whose elements are *augmented* by computer-generated images.

© Springer Nature Switzerland AG 2019
S. Bagnara et al. (Eds.): IEA 2018, AISC 822, pp. 3–9, 2019.
https://doi.org/10.1007/978-3-319-96077-7_1

We implemented AR-based accommodation assessment using Apple ARKit to assess accommodation between a digital human model and a physical vehicle interior. An automotive posture body shape model based on data from 255 men and women ages 20 to 95 years old was utilized in this study. UMTRI has developed a range of statistical body shape models based on 3d anthropometric data for children and adult men and women with a wide range of body size and age. The models, some of which are available online (http://humanshape.org), provide a way to efficiently generate a realistic 3d body shape surface along with the standard body dimensions, anatomical landmark locations and joint locations. In the current proof of concept, a mobile app for iPhone was developed, allowing for generating a subject-specific avatar and positioning the avatar on a target seat position. Opportunities and impacts of the proposed AR approach with digital human models are discussed.

2 Methods

2.1 Statistical Body Shape Model

The current analysis was conducted using data from 255 men and women ages 20 to 95 years old. Standard anthropometric measures were also obtained and each participant was scanned minimally clad using a VITUS XXL laser scanner in a seated posture (Fig. 1). The scan data were fit using a homologous template mesh and procedures published previously [2]. The template-fitting procedure produces a watertight mesh with 14427 vertices and 28850 polygons. Following the fitting, the meshes were made symmetrical by averaging the corresponding left and right vertices.

Fig. 1. Body shape measurement data of seated adults.

The processed template fits were then analyzed along with the measured landmark and joint locations to develop a statistical body shape model (SBSM) using a method presented in our previous work [3]. In brief, a principal component (PC) analysis was used to represent the variability in the processed data with a few PC scores. A total of

60 PCs that together explain more than 99% of the data variance were retained from the PCA. A regression analysis was applied to associate the PC scores and the anthropometric predictors, such as stature, body mass index (BMI), age, and sitting height to stature ratio.

2.2 AR-Based Assessment Tool

An AR-based assessment tool for mobile devices was developed using Apple ARKit. ARKit recognizes notable features in the scene image and tracks differences in the positions of those features across video frames to create a correspondence between real and virtual spaces. Figure 2 shows feature points tracked by ARKit and an estimated floor plane based on the tracked feature points. These tracked feature points on the objects in the scene were used to place our body shape model and also to detect reference point to measure a distance between the model surface.

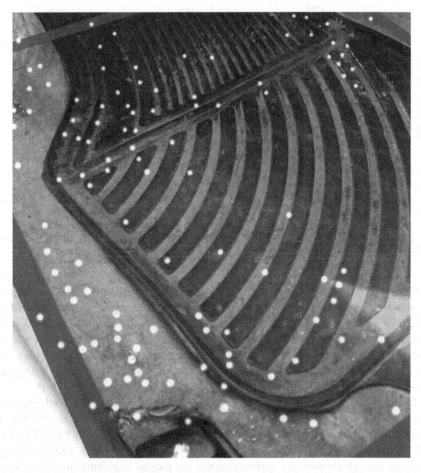

Fig. 2. Tracked feature points (yellow dots) and estimated horizontal plane (blue square) on a car floor. (Color figure online)

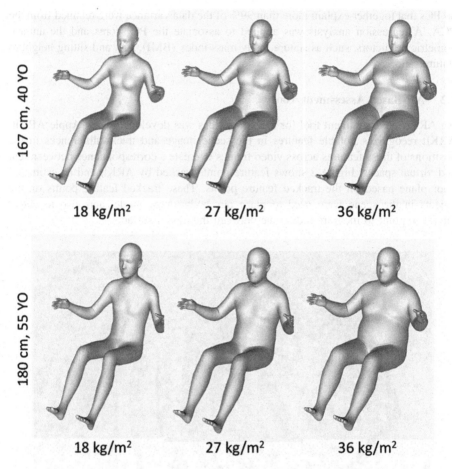

Fig. 3. A series of 3d avatars generated by varying the body mass index and gender while holding the stature and age.

One of the key factors for accurate accommodation measurement in AR is to place the virtual body shape model in a proper position in the augmented 3d space of the vehicle. We used a plane detection function of ARKit to detect the vehicle floor. An occupant posture prediction model developed in a previous study [4] was used with the estimated floor plane to roughly estimate the model position in the occupant space. The app interface allows the user to adjust the human model position and orientation manually.

Unity 3D, a cross-platform game engine developed by Unity Technologies, was used for visualizing and manipulating the seated body shape model in the AR. To measure the distance between a point on the vehicle interior surface and a point on the body shape model, a ray casting method was used. In brief, when a user touches the screen on the vehicle part, the system converts the picked two-dimensional coordinate into a 3d ray in the AR space and finds the interacting point between the ray and a surface fitted to neighbored feature points. When another point is selected on the model, the Euclidean distance is computed from the selected points.

3 Results

3.1 Representing Posture and Body Shape

Figure 3 shows a range of body shapes generated using the statistical body shape model (SBSM). The figures were obtained by manipulating the body mass index and gender parameters over a wide range while holding the stature and age parameters constant. The 3d body contours were generated by the SBSM along with 114 body landmarks and joint locations.

3.2 AR-Based Accommodation Assessment

Figure 4 shows an example of occupant accommodation assessment using the developed mobile app. A subject-specific female avatar was generated with the parameters of 167 cm of stature, 28 kg/m^2 of BMI and 40 years old for this assessment. The vehicle floor and the seat plane of a mid-size sedan were detected within a second. The model was positioned on a seat considering a normal sitting position. The model kept in the initial position in the augmented space while moving the phone to visually check the model from various viewpoints. The point-to-point distance measurements returned plausible results.

4 Discussion

To our knowledge, this is the first implementation of AR in vehicle occupant accommodation assessment. A statistical body shape model was utilized to test a wide range of anthropometric variability in body shapes in the AR-based assessment environment. The mobile app demonstrated good performance with minimal user interaction. Since this approach can be applied to any interaction analyses between a statistical human model and physical objects, this will open up various opportunities in ergonomics, anthropometry, product design and system assessment.

Major limitation of this approach is that the reliability of the measurements relies on the tracking accuracy of the device. Once the device lost the trackability of the scene, it is hard to reposition the model at the exact same previous position, and this would result in inconsistency between the measurements. The current system also requires manual adjustment of the model to find a proper sitting position. These limitations can be resolved by using a physical reference in the scene. For example, a statistical model of seat shape [5] could be used to estimate the seat H-point location and back angle, which could improve the posture and position of the model.

To further improve the performance, a depth detection method is needed. The current system does not consider the depths of surrounding objects, so the entire body shape is always shown even though the part of the body shape is occluded by vehicle

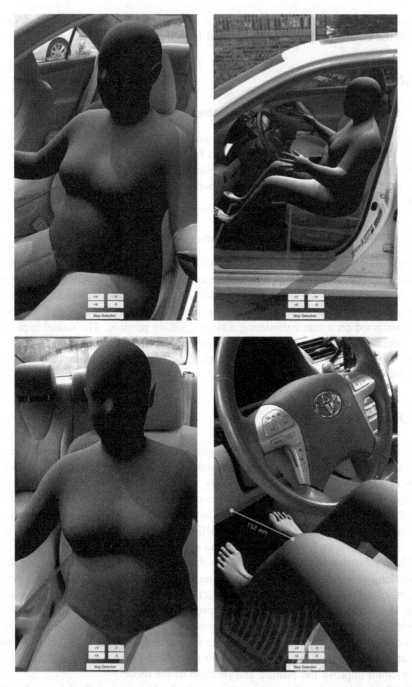

Fig. 4. Developed mobile app interface and a sample accommodation assessment of an avatar

parts. Using a fully poseable body shape model will allow further assessments, such as reach. We can also incorporate personal protective equipment and body borne gear to simulate occupational vehicle users [6].

References

1. Flannagan CAC, Manary MA, Schneider LW, Reed MP (1998) An improved seating accommodation model with application to different user populations. Technical Paper 980651. SAE Transactions: Journal of Passenger Cars, vol 107
2. Park B-K, Reed MP (2015) Parametric body shape model of standing children aged 3–11 years. Ergonomics 58(10):1714–1725
3. Park BKD, Ebert S, Reed MP (2017) A parametric model of child body shape in seated postures. Traffic Inj Prev 18(5):533–536
4. Park J, Ebert SM, Reed MP, Hallman JJ (2016) A statistical model including age to predict passenger postures in the rear seats of automobiles. Ergonomics 59(6):796–805. https://doi.org/10.1080/00140139.2015.1088076
5. Kim KH, Ebert SM, Reed MP (2016) Statistical modeling of automotive seat shapes. SAE Technical Paper 2016-01-1436. SAE International, Warrendale, PA
6. Zerehsaz Y, Jin J, Reed MP (2017) Development of seating accommodation models for soldiers in vehicles. Ergonomics 60(4):589–596

Who Should I Trust
(Human vs. Automation)? The Effects
of Pedigree in a Dual Advisor Context

Carl J. Pearson and Christopher B. Mayhorn[✉]

North Carolina State University, Raleigh, NC, USA
{Cjpearso, Chris_mayhorn}@ncsu.edu

Abstract. Source type bias (human vs automation) may influence the development of trust in decision aids. Situations involving two decision-aids may depend on the influence of pedigree (perceived expertise) such that decision making or reliance behavior is affected. In this task, the Convoy Leader decision-making paradigm developed by Lyons and Stokes (2012) was adapted to address advisor pedigree such that the human and automated information sources could be of high or low pedigree. Two hundred participants were asked to make eight decisions regarding the route taken by a military convoy based on intelligence (e.g., past insurgent attacks, Improvised Explosive Devices (IEDs) detected, etc.) provided by two information sources (human and automation) of varying degrees of pedigree. In two of these eight decisions, the decision-aids provided conflicting information. Results indicated that participants were likely to demonstrate a bias such that they were more likely to trust the information coming from the human advisor regardless of pedigree. This bias towards the human was only reversed when the automated decision aid was presented as having far greater pedigree. Measures of trust attitudes were highly indicative of decision making behaviors. The findings are addressed in terms of design within a dual-advisor context where human operators may receive conflicting information from advisors of different source types.

Keywords: Decision-making · Trust · Dual-advisor systems

1 Introduction

Automation is becoming more common in daily life (Parasuraman and Riley 1997), and we might find ourselves making decisions based on input received from different source types such as automated decision aids or other humans. One type of automation where trust and decision-making are particularly intertwined is within the context of an automated *decision support system* (DSS) where automation assists a human decision maker by organizing information and providing a recommendation on the best course of action (Sage 1991).

A review by Madhavan and Wiegmann (2007a) included a robust synthesis of research to create a model of how trust is similar and different between human decision aids and automated DSS. They found certain comparable characteristics of an automated DSS or human decision aid affect the development of trust: (1) source type

© Springer Nature Switzerland AG 2019
S. Bagnara et al. (Eds.): IEA 2018, AISC 822, pp. 10–17, 2019.
https://doi.org/10.1007/978-3-319-96077-7_2

indicates whether the adviser is human or automated, (2) pedigree is conceptualized as source credibility, or the degree to which an adviser is perceived as an expert, and (3) reliability indicates the rate at which the adviser makes a correct recommendation. These characteristics, or merely the perceptions of these characteristics, interact to form a basic level of trust in a DSS or decision aid. Madhavan and Wiegmann (2007b) validated these components with a single adviser supporting a participant. However, the impact of these characteristics with two advisers present requires further research and is the focus our experiment.

One of the few recent dual adviser context studies was conducted by Lyons and Stokes (2012), where they investigated perceived risk and decision-making (reliance on decision aids) between conflicting information supplied by human and automated advisers. They found that higher risk decreased reliance on the human DSS, which was unrelated to trust. However, a partial replication by Pearson et al. (2016) found different results: trust in human was higher in situations of high risk, but reliance was only marginally higher in the human. Both of these studies are limited by the uncontrolled third variable of pedigree, where the human was a high-ranking military official and the automated DSS was not distinguished in either direction.

The current experiment addressed these shortcomings from previous work by creating consistently controlled pedigree levels across advisers. Therefore, adviser pedigree's effect on trust and decision-making reliance can be clearly delineated when human operators are faced with a dual-advisor task where conflicting information is presented from automated and human sources.

2 Method

This study involved two differing decision aid sources (advisers) and pedigrees (perceived expertise) that gave information in conflict with one another. The task involved choosing the safest route for a military convoy, made to the likeness of the Convoy Leader software used in previous experiments (Lyons and Stokes 2012; Pearson et al. 2016). The design was a between groups 2×2 experimental manipulation where advisor source (human vs automated) and advisor pedigree (low vs high) were used to measure dependent measures including trust and behavioral reliance.

2.1 Participants

Amazon Mechanical Turk was used to recruit 200 participants within the United States. The average age was 39, with a range of 20 to 71 years, and women made up 42% of the sample.

2.2 Stimulus Materials

A set of eight maps in a satellite view (top down) were created that involve three possible route choices, along with past insurgent activity and past improvised explosive device (IED) locations (see Fig. 1). Participants were presented with the map and a key to the markings about past IED locations and past insurgent activity. Fifteen seconds into the map's appearance, a decision aid appeared. As the text visually appeared, it was accompanied by the auditory presentation of the decision aid recommendation text (a human voice for the human decision aid, and a robotic voice for the automation). The second decision aid appeared ten seconds after the first decision aid appears, along with its audio readout of the recommendation. Counterbalancing of recommendation presentation order was implemented through a blocked randomization between participants where the first 100 participants received the human recommendation first, and the second 100 participants received the automation recommendation first.

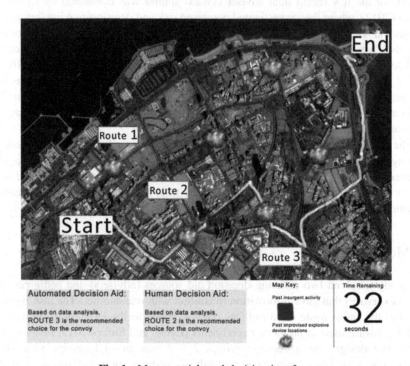

Fig. 1. Map materials and decision interface.

Once both decision aids appeared for ten seconds, participants had an option to click a button that said, "make a decision". This forwarded the browser to a decision page that no longer contained the map. If participants did not click the button, they were automatically forwarded to the next page after 60 total seconds. This decision page had a multiple-choice option to select the route of their choice (1, 2, or 3), along with information showing which options were recommended by a specific decision aid.

2.3   Pedigree Manipulation

A series of text-based profiles (see Fig. 2) were created to prime the participants with the level of pedigree involved with each adviser, similar to previous experimental pedigree protocols (Madhavan and Wiegmann 2007b). A measure of perceived

	High prestige	Low prestige
Human decision aid	Dr. Thomas Johnston is a US Army captain. He graduated with a bachelor's degree in engineering from West Point and a master of science degree in information sciences from MIT. Since finishing his education, Dr. Johnston has worked as an intelligence officer for 6 years. Dr. Johnston has a deep understanding of modern US military convoy capabilities, as well as current tactics used against the US by enemy forces. Dr. Johnston was recently appointed as the head of convoy logistics instruction at Fort Lee, Virginia.	John Thompson is a private in the US army. In his first year in the army, he has just begun to learn about the factors involved in military convoy operations. John Thompson has a partial understanding of current US military convoy tactics. However, he does not have any understanding of enemy countermeasures. He has not yet led a convoy in the field.
Automation decision aid	The Intelligent Route Choice (IRT) Program was created by computer scientists at MIT in 2004. It has a series of risk-calculating decision models capable of analyzing the best possible route using up-to-date enemy activity data and current satellite maps. The IRC Program possesses a vast amount of historical data to use in its situational analyses. Reviewed by many military computer scientists, the IRC is known to be an invaluable tool in military convoy operations in real world situations.	The Route Suggestion Program (RS) was created by researchers at Roxbury Community College in 2014 to assess military convoy route characteristics. The RS Program is at an experimental phase in its development process, going through early alpha stage testing. The RS Program possesses a limited set of past convoy data and uses unreliable algorithms to make recommendations. The RS Program is being considered by a military unit in the Middle East for a limited field test.

Fig. 2. Adviser profiles.

expertise was used as a manipulation check, adapted from Ohanian (1990) and involved rating if the decision aid was expert, experienced, knowledgeable, qualified, and skilled. Adviser profiles were piloted and iterated to ensure high pedigree advisers were perceived similarly as more expert than low pedigree advisers. The adviser pedigree treatment variable had four possible conditions: low pedigree human/low pedigree automation (AL HL, N = 50), low pedigree human/high pedigree automation (AH HL, N = 48), high pedigree human/low pedigree automation (AL, HH, N = 50), high pedigree human/high pedigree automation (AH HH, N = 52).

2.4   Dependent Measures

To assess the trust that participants felt for each decision aid, this study used a scale created by Merritt (2011) to measure trust ratings for automated and human decision support systems. This scale had two distinct aspects: trust and liking. As they were not intertwined, the liking section of the scale was removed without harming the validity of the trust section of the scale. The trust scale had six items (e.g.: "I can depend on the [human/system] or "I can rely on the [human/system] to do its best every time I take its advice"). Participant trust scores between advisers were ultimately made into an algebraic difference score, where positive values indicated human trust preference and negative values indicated automation trust preference.

2.5 Procedure

Participants completed a consent form before proceeding to the experiment. Before the experimental task, participants were given task instructions to pick the safest of three routes for their supply convoy illustrated on the map. After the instructions, participants were randomly assigned to an adviser pair, and the order of the pair was also randomly ordered (see Fig. 2). The adviser pedigree rating scale was given along with each adviser profile.

This map task (as shown previously in Fig. 1) occurred eight times total. In trials 1–3 and 5–7, advisers gave the same recommendation; in trials four and eight, advisers gave conflicting recommendations. The points of interest among these trials were on the reliance outcomes of the disagreement trials (trials four and eight). After the map trials, participants completed the trust scale for each decision aid, taken from Merritt (2011). Lastly, they provided basic demographic information (age, sex, and native language), before receiving their code to obtain payment via Mechanical Turk.

3 Results

An ANOVA was first conducted to examine the effects of adviser pedigree treatment group on adviser trust preference. Levene's test of equality of error variances was non-significant ($F(3,196)$, $p = .113$), so the analysis did not constrict degrees of freedom. There was a significant effect of pedigree treatment group on trust preference in decision aids, $F(3,196) = 25.241$, $p < .001$. A Games-Howell post hoc test was also conducted on the treatment groups to control unequal cell sizes (means listed below in Table 1). The matched pedigree groups (groups 1 and 4) were not significantly different from each other but were both significantly different from the non-matched pedigree groups (groups 2 and 3). Both unmatched pedigree groups (groups 2 and 3) were significantly different from each other and all matched pedigree groups (groups 1 and 4). All p values in the post hoc test were at .001 or lower. The significant ANOVA model supported that the pedigree manipulation treatments were effective in relation to trust scores. Further examination of confidence intervals showed that when pedigree levels were similarly high, there was significantly more trust in the human adviser.

Table 1. ANOVA treatment group trust preference estimated marginal means and confidence intervals.

Treatment group	Mean	Lower CI	Upper CI
1. AL HL	0.187	−0.135	0.509
2. AH HL	−0.747	−1.075	−0.418
3. AL HH	1.277	0.955	1.599
4. AH HH	0.349	0.034	0.665

Automated adviser trust was only significantly preferred when the automation pedigree far outweighed the human adviser's pedigree.

To determine how trust preferences predicted behavioral reliance outcomes, a series of multinomial logistic regressions were conducted to examine the effect of trust preference scores on the likelihood of behavioral reliance decisions (between self-reliance, automation, and human reliance, see Fig. 3). Among the eight trials each participant made a reliance decision, the trials where the advisers disagreed (trials four and eight) were considered the target trials.

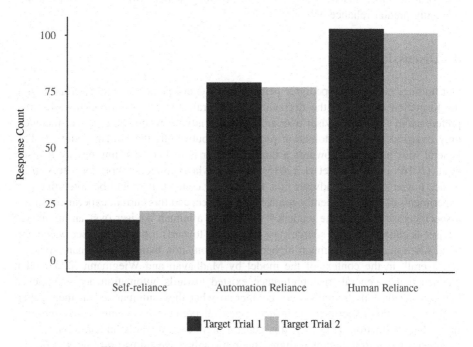

Fig. 3. Behavioral reliance decisions across target trials.

A multinomial logistic regression was conducted for the first disagreement trial with trust preference predicting the categories of behavioral reliance. A test of the full model against a constant only model was statistically significant, indicating that trust preferences reliably distinguished between route choices, ($\chi^2 = 110.07$, p < .001, df = 2). Nagelkerke's R2 of .502 indicated a moderately strong relationship between prediction and grouping. Participants were .27 times more likely to rely on the automated recommendation compared to the independent choice for every 1.3 steps towards preference of automation in the trust measure. Participants were 2.48 times more likely to rely on the human recommendation compared to the independent choice for every .9 steps towards preference of the human adviser in the trust measure.

A multinomial logistic regression was conducted for the second disagreement trial with trust preference predicting the categories of behavioral reliance (see Fig. 3). A test of the full model against a constant only model was statistically significant, indicating that trust preferences reliably distinguished between route choices, ($\chi^2 = 79.09$, $p < .001$, df = 2). Nagelkerke's R2 of .383 indicated a moderate relationship between prediction and grouping. Participants were .49 times more likely to rely on the automated recommendation compared to the independent choice for every .711 steps towards preference of automation in the trust measure. Participants were 2.26 times more likely to rely on the human recommendation compared to the independent choice for every .816 steps towards preference of the human adviser in the trust measure.

Considering both overall significant models and the expected predictor-outcome relationships individually, these results firmly suggest that trust preferences did significantly predict reliance behaviors.

4 Discussion

Our findings indicate that pedigree perceptions in decision aids are an important trait in the formation of trust in the decision aids. Overall, there was a trend towards trust preference in a human adviser over an automated adviser when pedigree was matched, only overridden when automation pedigree far outweighs the human pedigree. This general trust preference towards a human adviser is in line with findings by Pearson et al. (2016) and Merritt et al. (2015), both of whom found support for a trust preference towards a human adviser in a dual adviser context. It should be noted that past experiments did not sufficiently control for pedigree, and this current experiment shows a specific case wherein the tendency for trust in a human adviser over an automated adviser is countered. The source type preference towards a human adviser was countered when pedigree was much higher in the automation than in the human adviser.

Overall, in the context of the model by Madhavan and Wiegmann (2007a) that claimed that a general trust preference existed towards automation across matched pedigree levels, our findings were counter to what they summarized in their paper. Results from this experiment indicated a general preference for human advisers in a dual-adviser situation. However, this general preference towards human advisers was attenuated when automation pedigree far outweighed the human adviser's pedigree. This is similar to findings by Merritt et al. (2015) when manipulating adviser reliability, where they found a preference for human advisers compared to automated advisers in a dual-adviser setting. However, they theorized that these findings were due to participants weighing automation errors more heavily because of violating the perfection in automation schema (Dzindolet et al. 2003). Because our study did not involve performance feedback but still found a human trust preference, it appears the bias for human trust may go beyond just that of an automation perfection schema in a dual-adviser context.

Lastly, trust preferences were significantly related to behavioral reliance: more trust preference in a specific adviser type strongly predicted behavioral reliance on that adviser on target trials. These findings support the notion that trust is a valid construct as an attitude strongly related to behavioral reliance. This is partially consistent with past research findings. Merritt et al. (2015) found that trust was predictive of reliance, although only in the first half of their trial blocks.

In conclusion, this experiment provides a contribution to the literature by removing any unintentional pedigree confounds thereby allowing a clearer assessment of trust across source type. Ultimately, the significant prediction of reliance from trust preference supports the importance of understanding trust. Trust on its own is mainly valuable in its ability to predict behavioral outcomes. These experimental results showed that one could reasonably predict reliance between conflicting decision aids by assessing pre-existing self-reported trust measures.

References

Lyons JB, Stokes CK (2012) Human-human reliance in the context of automation. Hum Factors 54(1):112–121

Parasuraman R, Riley V (1997) Humans and automation: use, misuse, disuse, abuse. Hum Factors 39(2):230–253

Sage AP (1991) Decision support systems engineering. Wiley, New York

Madhavan P, Wiegmann DA (2007a) Similarities and differences between human-human and human-automation trust: an integrative science. Hum Factors 8:277–301

Madhavan P, Wiegmann DA (2007b) Effects of information source, pedigree, and reliability on operator interaction with decision support systems. Hum Factors 49(5):773–785

Pearson CJ, Welk AK, Boettcher WA, Mayer RC, Streck S, Simons-Rudolph JM, Mayhorn CB (2016) Differences in trust between human and automated decision aids. In: Proceedings of the 2016 symposium and bootcamp on the science of security. ACM, p 95

Ohanian R (1990) Construction and validation of a scale to measure celebrity endorsers' perceived expertise, trustworthiness, and attractiveness. J Advert 19(3):39–52

Merritt SM (2011) Affective processes in human-automation interactions. Hum Factors 53 (4):356–370

Merritt SM, Sinha R, Curran PG (2015) Attitudinal predictors of relative reliance on human vs automated advisors. Hum Factors 3(3–4):327–345

Dzindolet MT, Petersen SA, Pomranky RA, Pierce LG, Beck HP (2003) The role of trust in automation reliance. Int J Hum Comput Stud 58(6):697–718

Using 3D Statistical Shape Models for Designing Smart Clothing

Sofia Scataglini[1,2]([✉]), Femke Danckaers[3], Robby Haelterman[1],
Toon Huysmans[3,4], Jan Sijbers[3], and Giuseppe Andreoni[5]

[1] Department of Mathematics (MWMW), Royal Military Academy,
Renaissancelaan 30, 1000 Brussels, Belgium
sofia.scataglini@rma.ac.be
[2] Military Hospital Queen Astrid, Bruynstraat 1, 1120 Brussels, Belgium
[3] imec – Vision Lab, Department of Physics, University of Antwerp,
Universiteitsplein 1, 2610 Antwerp, Belgium
[4] Applied Ergonomics and Design, Department of Industrial Design, TU Delft,
Landbergstraat 15, 2628 CE Delft, Netherlands
[5] Department of Design, Politecnico di Milano, 20158 Milan, Italy

Abstract. In this paper we present an innovative approach to design smart clothing using statistical body shape modeling (SBSM) from the CAESAR™ dataset. A combination of different digital technologies and applications are used to create a common co-design workflow for garment design. User and apparel product design and developers can get personalized prediction of cloth sizing, fitting and aesthetics.

Keywords: Statistical body shape modeling (SBSM) · Anthropometry
Blender · Motion capture · Smart clothing

1 Introduction

Statistical shape modeling is a promising approach to map out the variability of body shapes, commonly used in 3D anthropometric analyses. With statistical shape models, a wide range of body shapes can be simulated. So, a wide range of 3D smart garment can be designed on it as well.

The design of smart clothing is crucial to obtain the best results. Identifying all the steps involved in the functional design workflow can prevent a decrease in the wearer's performance ensuring a successful design [1–5].

Nowadays, smart clothing provides a methodology to monitor mechanical, environmental, and physiological parameters in real time and in an ecological, non-intrusive approach. These parameters can be used to detect gesture or specific patterns in movement, design more efficient, specific training programs for performance optimization, and screen for a potential causes of injury.

Apparel sizing and fit impact every part of the apparel lifecycle, from design to manufacturing to the consumers. The main requirements that a smart clothing is called to achieve are functionality, usability, monitoring duration, wearability, maintainability, connectivity and washability [6]. On the other hand comfort is a fundamental and a

S. Bagnara et al. (Eds.): IEA 2018, AISC 822, pp. 18–27, 2019.
https://doi.org/10.1007/978-3-319-96077-7_3

transversal factor (also among the previous features) that has to be considered. Wearing an uncomfortable system comprises the user's ability to do his/her job.

Anthropometry is a key for the clothing design and the placement of smart textiles around the body. Volume, shape, weight, and adherence to the body of wearable devices must be designed to not affect or interfere with natural movements. Statistical body shape modeling (SBSM) can allow mapping out the variability of the shape of the clothes, thus contributing to more successfully sized and better fitting apparel (Fig. 1).

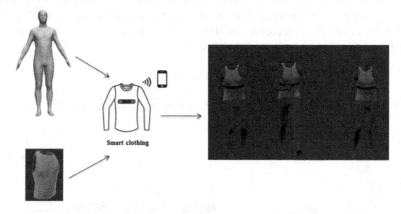

Fig. 1. SBSM for design smart clothing.

The innovative approach we propose, concerns a combination of different digital technologies and applications to create a common co-design workflow for the design of a garment [7]. Users or designers could upload the SBSM and get a personalized prediction of clothing size as well as personalized suggestions on how different products may fit their body enhancing performance and functionality [8].

In our case, the SBSM was built selecting 57 soldiers-like body shapes from the CAESAR™ database to design smart clothing for this specific military population.

2   Methods

In this section, the method for building an SBSM for designing smart clothing is described. First, a moving SBSM is built from 3D human body scans. Human activities can then be replicated based on body shape and a motion data collected on a subject by a mocap system. This provides a visualization of a digital human model based upon anthropometry and biomechanics of the subject [9]. Furthermore, an SBSM garment for a specific body shape can then be created. The rigging phase allows visualizing the garment fitting and aesthetics [10]. A clustering algorithm can finally be used to determine a sizing systems based on the biometric features of the subject [11, 12].

2.1 Building a Statistical Body Shape Model

First, a digitally modeled body shape with n uniformly distributed vertices is registered in a marker-less way to all input surfaces to obtain a homologous point-to-point correspondence. All input surfaces were corrected for posture, in a way that every shape was standing in the average posture. Then, a statistical shape model is built using principal component analysis on the corresponded surfaces [13, 14]. We selected 57 soldiers body shapes (male, height 1520 mm–2100 mm, age 18y–35y, BMI < 25) from the CAESAR™ database [15] to build our model. In this statistical shape model, the average shape $\bar{x}$ and the main shape modes, or the principal component (PC) modes of the shape model P, are incorporated (Fig. 2). This means that a new shape y can be formed by a linear combination of the PCs:

$$y = \bar{x} + Pb \tag{1}$$

with b the vector containing the shape parameters of a specific instance.

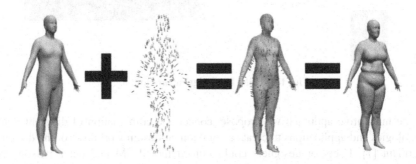

Fig. 2. Example of generating a new body shape with specific shape parameters. The average body shape $\bar{x}$ is added to the shape Pb. The result is the average body with a displacement vector for each vertex. When the displacement is applied to the shape, a new body shape is formed.

A mapping matrix M describing the relationship between the biometric features F (such as height, weight, gender,...) and the principal component weights matrix B of every input shape was calculated using multivariate regression, by $M = BF^+$ with F^+ the pseudoinverse of F. By multiplying M with a given feature vector $\mathbf{f}$, new PC weights can be generated: $b = Mf$. From these PC weights, a new body shape can be built.

2.2 Generating a Shirt

On this new body shape, a shirt is designed. This is done by calculating critical points. The root of the hips, neck, left shoulder and right shoulder were calculated. Then, a clipping plane was generated on every critical point with a pre-defined normal. For the hips, this normal was defined as $(0, 0, 1)$. The normal for the neck was $(0, 0.707, -1)$. The left shoulder plane had as normal $(-1, -0.2, 0)$, and the right shoulder had

$(1, -0.2, 0)$. After removing the arms, legs and head of the body shape, it was scaled by a scaling matrix S with the origin at $(0, 0, 0)$.

$$S = \begin{bmatrix} 1.1 & 0 & 0 \\ 0 & 1 & 0 \\ 0 & 0 & 1.1 \end{bmatrix} \qquad (2)$$

The shirt was uniformly resampled to 1000 points. The body mesh and the cloth were exported as object files.

2.3 Clustering

The surfaces are clustered by PC weights, using the k-means clustering algorithm of Hartigan and Wong [16]. K-means clustering aims to partition N observations into k clusters in which each observation belongs to the cluster with the nearest mean, serving as a prototype of the cluster. The goal is to minimize the within-cluster sum of squares:

$$\sum_{j=1}^{k} \sum_{i=1}^{N} \left\| x_i - \mu_j \right\| \qquad (3)$$

with x_i the vector holding the parameter weight vectors of a specific surface and, μ_j the average parameter weight vector of the set. In this case the N observations are the parameter weight vectors per surface. Finally, the characteristic shape per cluster was returned.

3 Experiment and Results

3.1 Combined Techniques (3D to 2D and 2D to 3D)

The co-design workflow (Fig. 3) was implemented in Blender [17]. Blender as an open source 3D computer graphic software can be used to rig the mesh cloth with an SBSM describing all the phases necessary to simulate garment fitting and aesthetics [18]. First, a moving SBSM was built from a population of 3D human body shapes (Fig. 2). Next, the garment shape and sizing system was created (Fig. 4). The following step involves importing the file in Blender as an object file (OBJ). After that, the UV mapping followed (Figs. 5 and 6). Finally there was the rigging phase where the cloth mesh was rigged to the humanoid mesh (Figs. 7 and 8).

3.1.1 Creating Mesh

A shirt was created for a specifically generated body shape, by cutting the surface on predefined locations (arms, legs, neck) and scaling the surface. The resulting shirt was shown in Fig. 4.

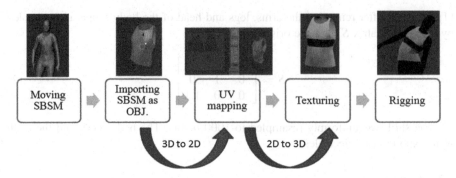

Fig. 3. Garment co-design workflow in Blender.

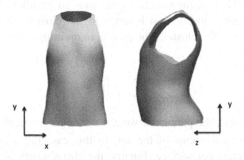

Fig. 4. The generated shirt (left-frontal view) and (right-sagittal view).

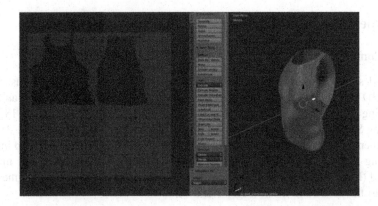

Fig. 5. UV's unwrapped (on the left) and UV mapping (on the right) in Blender.

3.1.2 UV-Mapping Your Meshes

UV mapping is the process of projecting the 3D surface onto a 2D texture surface [19]. Each face of the 3D model is mapped to a face of the UV map. Each face on the UV map corresponds to one face of the 3D model, and the UV preserves the edge relationships between faces of the 3D model.

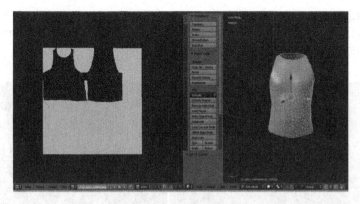

Fig. 6. UV's unwrapped (on the left) with the texture and UV mapping (on the right) in Blender.

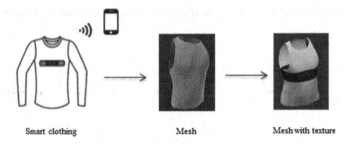

Smart clothing   Mesh   Mesh with texture

Fig. 7. Smart clothing texturing.

Fig. 8. The skeleton parented with the humanoid and clothing meshes.

The UV mapping process requires three steps: unwrapping the mesh (Fig. 5), creating the texture and applying the texture. Texturing is a process where an image is applied (mapped) to the surface of a shape or polygon (Fig. 6). In our case a texture representing the aesthetics of the cloth was mapped on the cloth mesh (Fig. 7).

3.1.3 Rigging

Rigging [17] refers to a process in which the mesh is attached to the skeleton (Fig. 8) associating a movement to a BVH file [20] (Fig. 9).

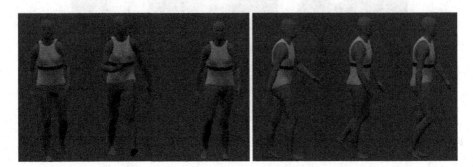

Fig. 9. Moving SBSM with clothing in Blender (left-frontal view and right-sagittal view).

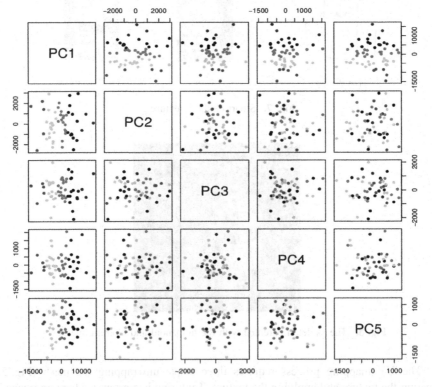

Fig. 10. Plot of the clusters in the PCA space. The clusters are shown by the same color. The red dots belong to cluster 1 (Small), the cyan dots to cluster 2 (Medium), the green dots to cluster 3 (Large), the black dots to cluster 4 (X-Large), and the blue dots to cluster 5 (XX-Large). The average shape per cluster is shown in Fig. 11. (Color figure online)

3.2 Clustering

The 57 body shapes were subjected to the clustering algorithm. In our experiments, the number of initial clusters was set to five, since that corresponds to the number of clothing sizes (Small, Medium, Large, X-Large and XX-Large) we were interested in identifying (Fig. 10). As can be seen from the results, the main difference between the clusters was the height. The body shapes that correspond to these measurements are shown in Fig. 11. Table 1 shows a subset of the body measure and standard deviation of the average shape of each cluster in mm.

For smart shirt design, the anthropometric cluster, gave an indication of the size constraints of an individual. While, moving SBSM was used to modify the aesthetics design and garment fitting.

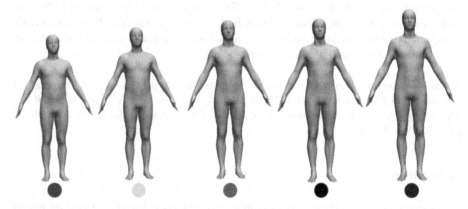

Fig. 11. Average shape of every cluster. Red: 1725 mm, 63.6 kg (Small), Cyan: 1787 mm, 72.4 kg (Medium), Green: 1856 mm, 77.1 kg (Large), Black: 1913 mm, 82.2 kg (X-Large) and Blue: 2000 mm, 91.3 kg (XX-Large). (Color figure online)

Table 1. Body measurements (mean values are given in mm along with the standard deviation in parenthesis).

Parameter	Small	Medium	Large	X-Large	XX-Large
Shoulder to wrist	603	631	652	689	721
	(26)	(16)	(20)	(21)	(51)
Chest circumference	902	942	940	979	1009
	(84)	(50)	(37)	(58)	(41)
Bust chest circumference under bust	902	942	940	972	1009
	(84)	(50)	(37)	(61)	(41)
Hip circumference	907	984	992	1007	1059
	(60)	(42)	(39)	(48)	(52)
Shoulder breadth	449	453	470	474	503
	(24)	(16)	(19)	(18)	(11)
Sitting height	915	927	958	977	1012
	(13)	(24)	(25)	(28)	(20)

(continued)

Table 1. (*continued*)

Parameter	Small	Medium	Large	X-Large	XX-Large
Stature	1725	1787	1855	1912	2000
	(31)	(19)	(15)	(21)	(34)
Waist circumference	793	841	828	880	889
	(55)	(46)	(53)	(66)	(47)

4 Conclusion

This study demonstrates a co-design approach to smart clothing development using moving statistical body shape models. This methodology can be applied to apparel design ensuring a more successful design. The advantage of using an SBSM is that our method can be applied to a whole range of body shapes, e.g. specific shape clusters that correspond to a specific size. Anthropometric soldier dimensions can be used to design ergonomic military equipment and functional clothing.

Acknowledgements. This work was supported by the Agency for Innovation by Science and Technology in Flanders (IWT-SB 141520). We acknowledge Alain Vanhove of the Royal Military Academy for his contribution in the 3D modeling. We would also like to thank all the participants in this study.

References

1. Scataglini S (2017) Ergonomics of gesture: effect of body posture and load on human performance. Ph.D. Politesi. https://www.politesi.polimi.it/handle/10589/136840
2. Scataglini S, Truyen E, Perego P, Gallant J, Tiggelen DV, Andreoni G (2017) Smart clothing for human performance evaluation: biomechanics and design concepts evolution. In: 5th International digital human modeling symposium, Germany, Bonn
3. Scataglini S, Andreoni G, Truyen E, Warnimont L Gallant J, Tiggelen DV (2016) Design of smart clothing for Belgian soldiers through a preliminary anthropometric approach. In: Proceedings 4th DHM digital human modeling, Montréal, Québec, Canada, 15–17 June
4. Andreoni G, Standoli CE, Perego P (2016) Defining requirements and related methods for designing sensorized garments. Sensors 16(6):769
5. Scataglini S, Andreoni G, Gallant J (2018) Smart clothing design issues in military applications. In: International conference on applied human factors and ergonomics (AHFE): advances in human factors in wearable technologies. Springer
6. Gemperle F, Kasaback C, Stivoric J, Bauer M, Martin R (1988) Design for wearability. In: Proceedings of the 2nd IEEE international symposium on wearable computers
7. Sanders EBN, Stappers PJ (2008) Co-creation and the new landscapes of design. Codesign 4(1):5–18
8. Gupta D (2011) Functional clothing-definition and classification. Indian J Fibre Text Res 36(4):321–326
9. Scataglini S, Danckaers F, Haelterman R, Van Tiggelen D, Huysmans T, Sijbers J (2018) Moving statistical body shape models using blender. In: International congress of ergonomics (IEA). Springer (in press)

10. Carulli M, Vitali A, Caruso G, Bordegoni M, Rizzi C, Cugini U (2017) ICT technology for innovating the garment design process in fashion industry. In: Chakrabarti A, Chakrabarti D (eds) Research into design for communities, vol 1. ICoRD 2017. Smart Innovation, System and Technologies, vol 65. Springer, Singapore
11. Vinué G, Simó A, Alemany S (2016) The k-means algorithm for 3D shapes with an application to apparel design. Adv Data Anal Classif 10:103
12. Viktor HL, Paquet E, Guo H (2006) Measuring to fit: virtual tailoring through cluster analysis and classification. In: Fürnkranz J, Scheffer T, Spiliopoulou M (eds) Knowledge discovery in databases: PKDD 2006. Lecture notes in computer science, vol 4213. Springer, Berlin, Heidelberg
13. Danckaers F, Huysmans T, Ledda A, Verwulgen S, Van Dogen S, Sijbers J (2014) Correspondance preserving elastic surface registration with shape model prior. In: International conference of pattern recognition, pp 2143–2148
14. Danckaers F, Huysmans T, Hallemans A, De Bruyne G, Truijen S, Sijbers J (2018) Full body statistical shape modeling with posture normalization. In: AHFE 2017: advances in human factors in simulation and modeling, pp 437–448
15. Robinette KM, Daanen HAM, Paquet E (1999) The CAESAR project: a 3-D surface anthropometry survey. In: Second international conference on 3-D digital imaging and modeling (Cat. No. PR00062), pp 380–386
16. Hartigan JA, Wong MA (1979) Algorithm AS 136: a K-means clustering algorithm. Appl Stat 28:100–108
17. Blender Online Community (2015) Blender-a 3D modeling and rendering package
18. Baran I, Popović J (2007) Automatic rigging and animation of 3D characters. ACM Trans Graph 26(3):72
19. Villar O (2014) Learning blender: a hands on guide to creating 3D animated characters, 2nd edn. Addison Wesley Professional, Boston
20. Dai H, Cai B, Song J, Zhang D (2010) Skeletal animation based on BVH motion data. In: 2nd International conference on information engineering and computer science, pp 1–4

Moving Statistical Body Shape Models Using Blender

Sofia Scataglini[1,2]([✉]), Femke Danckaers[3], Robby Haelterman[1],
Toon Huysmans[3,4], and Jan Sijbers[3]

[1] Department of Mathematics (MWMW), Royal Military Academy,
Renaissancelaan 30, 1000 Brussels, Belgium
sofia.scataglini@rma.ac.be
[2] Military Hospital Queen Astrid, Bruynstraat 1, 1120 Brussels, Belgium
[3] imec – Vision Lab, Department of Physics, University of Antwerp,
Universiteitsplein 1, 2610 Antwerp, Belgium
[4] Applied Ergonomics and Design, Department of Industrial Design, TU Delft,
Landbergstraat 15, 2628 CE Delft, Netherlands

Abstract. In this paper, we present a new framework to integrate movement acquired by a motion capture system to a statistical body shape model using Blender. This provides a visualization of a digital human model based upon anthropometry and biomechanics of the subject. A moving statistical body shape model helps to visualize physical tasks with inter-individual variability in body shapes as well as anthropometric dimensions. This parametric modeling approach is useful for reliable prediction and simulation of the body shape movement of a specific population with a few given predictors such as stature, body mass index and age.

Keywords: Statistical body shape modeling · Digital human modeling
Blender · Motion capture

1 Introduction

Statistical body shape modeling (SBSM) is an intuitive approach to map out body shapes variability of a 3D body anthropometric database. The shape variance is described by shape parameters, which can be adapted to form a new realistic shape. Furthermore, body shapes belonging to a specific percentile of a target group, can be visualized.

Body shape modeling can be classified as static or dynamic. Static represents the human body at one particular pose like standing or sitting. Dynamic shape modeling deals with shape variations due to pose changes or a subject moving during scanning. Body shape variation can be decomposed into rigid (associated with the orientations and positions of the segments) and non-rigid deformation (e.g. changes in shape because of soft tissues associated with the segment in motion) [1].

Nowadays, an inertial motion capture system (mocap) can capture the movement of a subject during a physical task [2]. This information can be translated as a skeletal animation as a Biovision Hierarchy (BVH) character animation file [3]. Commercial

© Springer Nature Switzerland AG 2019
S. Bagnara et al. (Eds.): IEA 2018, AISC 822, pp. 28–38, 2019.
https://doi.org/10.1007/978-3-319-96077-7_4

modeling tools (e.g. Autodesk Maya, Autodesk 3dsMax) and Open-Source software (Blender) can be used to integrate the body shape with the motion [4–6].

The 3D modeling software includes programs and tools for animation, simulation, rendering, video editing, motion tracking and game creation. Human activities can then be replicated based on body shape and a motion data collected on a subject. This provides a visualization of a digital human model based upon anthropometry and biomechanics of the subject [7].

In this paper, we propose a new framework to integrate the movement acquired by the inertial motion capture system with a statistical body shape model (SBSM) using Blender version 2.78 [4]. This allows product designers, ergonomists, and engineers to simulate realistic human behavior [8]. The data-driven human activities can be included into scenarios in a virtual environment to visualize whether particular tasks are supported or hindered by the design.

2   Methods

In this section, a framework to create moving statistical body shape models using Blender is described (Fig. 1). First, an SBSM is built from 3D human body scans. Next, the body mesh is rigged with a BVH file. As a result, motion is added to the SBSM.

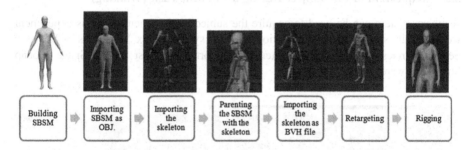

Fig. 1. A framework for moving SBSM using Blender.

2.1   Building a Statistical Body Shape Model

First, a reference surface is registered in a marker-less way to N target surfaces to obtain a homologous point-to-point correspondence [9]. Then, a statistical shape model, or principal component (PC) model, is built from the corresponded shapes [9, 10], as shown in Fig. 2. A specific feature of a person's shape, such as height, can be adapted by adding a linear combination of principal components to the person's shape vector. Furthermore, a body shape with specific features can be generated by multiplying a feature vector to the mapping matrix of the shape model [11, 12].

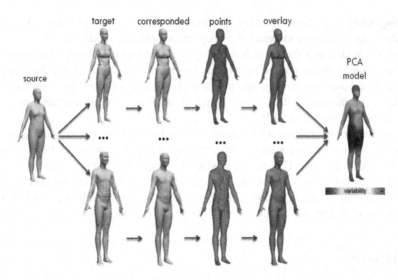

Fig. 2. Schematic overview of the building of an SBSM [9]. First, a source surface is registered to every target surface to obtain a point-to-point correspondence. From the corresponding surface, a statistical shape model is built.

2.2 Acquisition of the Subject During a Physical Task (Walking)

A mocap system can be used to acquire the subject's movement. For this experiment we asked the subject to wear a motion tracking system (17 x Yost Labs 3- Space™ Sensor Wireless Device). This system is supported by Yost Labs 3-Space Mocap Studio software in Fig. 3 [13].

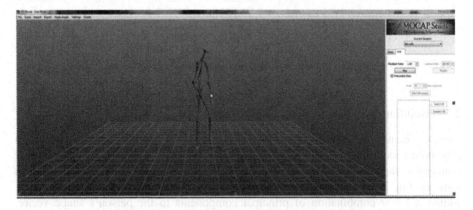

Fig. 3. YEI Mocap studio software data representation.

After an initial calibration on the T-pose, we asked the subject to walk. Motion data was recorded (number of frames: 285, frame time: 0.016 s) and exported as BVH file.

2.3   Blender Workflow

The Blender user interface is divided into three frames (Fig. 4): the toolbar, the 3D view and the tools region. The toolbar contains the following menus: File, Render, Window, and Help, the spin boxes and a short description of the Blender scene.

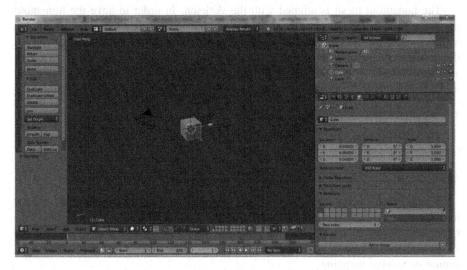

Fig. 4. Blender user interface.

The 3D view is the place where the DHM it is created. Finally, there is a tool region on the left of the panel that consists of tabs related to relations, animation, tool, physics, and grace pencil [14].

In this specific case, we defined a workflow that describes all the steps necessary to moving statistical body shape using Blender (Fig. 5):

(1) Importing the SBSM.

In this step the SBSM previously created is imported in Blender as an OBJ file from the Menu: Info Editor ▸ File ▸ Import/Export.

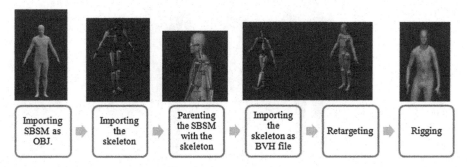

Fig. 5. Blender workflow.

(2) Importing the skeleton from the mocap system into Blender.

The skeleton from YEI Mocap is imported in Blender as an OBJ file from the Menu: Info Editor ▸ File ▸ Import/Export.

(3) Parenting the SBSM with the skeleton.

The humanoid mesh is aligned with the skeleton. In order to have the humanoid mesh assigned to the different bones that compose the skeleton we need to setup a Vertex Group. To do this, first we need to select the humanoid mesh, and shift select the armature (skeleton) and open the parent menu (Ctrl + P) to and select "with the automatic weights". This calculates the influence of a bone on vertices based on the distance from those vertices to a bone, the so-called bone heat algorithm [15].

(4) Re-importing the skeleton.

The skeleton from YEI Mocap is imported in Blender as a BVH file from the Menu: Info Editor ▸ File ▸ Import/Export.

(5) Retargeting

Using the retargeting tool, that is part of the Motion Capture Blender add on, we transfer the animation from the imported mocap skeleton (armature) to the SBSM. This is because we have two armatures: the SBSM parenting with one armature and the skeleton (armature) from the mocap system. At that phase you can use the "auto guess" feature or manual mapping.

(6) Final rigging

The armature from the motion tracking system is parented with the humanoid mesh.

3 Experiment and Results

In this section, the results of building an SBSM and adding motion to them is described (Fig. 6).

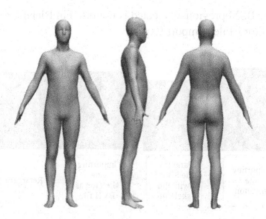

Fig. 6. Average soldier (male, height: 1840 mm, age: 27.9 years, BMI: 22.4).

3.1 Building a Statistical Shape Model

A statistical shape model was built from the CAESAR™ database [16]. We selected 57 soldier body shapes (male, height 1520 mm–2100 mm, age 18y–35y, BMI 18–25) to build our model (Fig. 6 and Table 1).

Table 1. Anthropometric measurements (mean values in mm).

Waist circumference	846
Chest circumference	951
Hip circumference	990
Arm length	654
Crotch height	873
knee height	570
Shoulder breath	466
Sitting height	953
Thigh circumference	568

From these meshes, we removed posture variances and finally built a statistical shape model (Fig. 7). Using this statistical shape model, a new body shape was calculated.

3.2 Importing the SBSM into Blender

The previously created SBSM (Fig. 7) was introduced in Blender as an object file (OBJ), (Fig. 8).

Next, the armature from the mocap system was imported into Blender and it was aligned with the humanoid mesh (Fig. 9).

The following step was the automatic weights parenting: calculating the influence of a bone on vertices based on the distance from those vertices to a bone, the so-called bone heat algorithm [15]. This influence was assigned as weights in the vertex groups. Weight painting mode was used to tweak which part of the mesh was affected by each group [17, 18], (Fig. 10).

Once it was verified that the armature was parented with the mesh, we imported the BVH file (number of frames: 285, frame time: 0.016 s) representing the walking of the subject into Blender (Fig. 11).

At that phase we linked each bone of the armature previously parented with the new one. This phase is called retargeting (Fig. 12). Using the retargeting tool, that is part of the Motion Capture Blender add on, we transferred the animation from the imported mocap skeleton (armature) to the SBSM. In this phase we changed the names of one armature and we manually linked each bone with the other.

The armature from the motion tracking system was parented with the humanoid mesh. When the rigging was completed, the armature (skeleton) could move and the DHM was animated accordingly (Fig. 13).

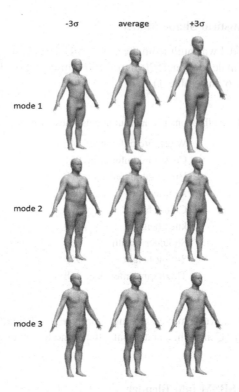

Fig. 7. The first three eigenmodes of the soldier SBSM plus (right) and minus (left) three standard deviations (σ) and the average body shape (center).

Fig. 8. The SBSM imported into Blender as OBJ file.

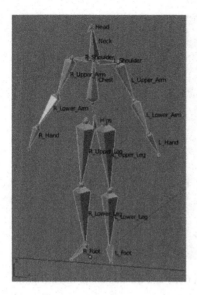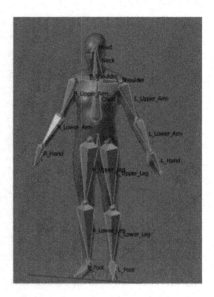

Fig. 9. Armature from the motion tracking system (left) and the humanoid mesh aligned with the armature (right).

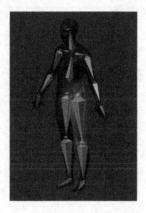

Fig. 10. Weight painting (RED = 100% weighted, BLUE = 0% weighted). (Color figure online)

4 Conclusion

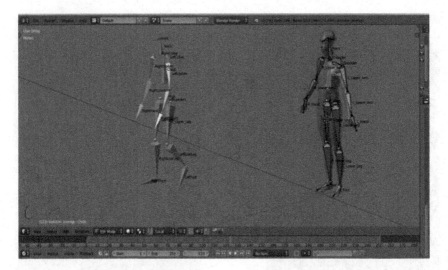

Fig. 11. Armature from the motion tracking system (left) and the humanoid mesh (right).

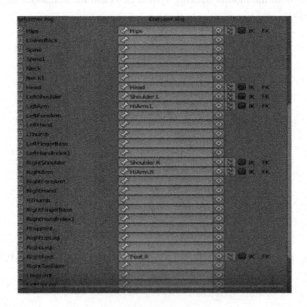

Fig. 12. Retargeting.

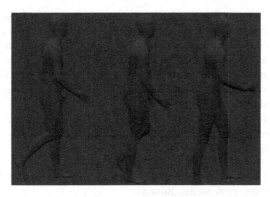

Fig. 13. Walking SBSM.

We proposed a framework to rig a statistical shape model in open source software. Results show that our framework leads to detailed, realistic body shapes, moving in a natural way. By using an SBSM, the same motion can be applied to a wide range of body shapes. Our proposed methodology allows users, designers, and ergonomists, to simulate realistic human movement.

Acknowledgement. This work was supported by the Agency for Innovation by Science and Technology in Flanders (IWT-SB 141520). We acknowledge Alain Vanhove of the Royal Military Academy for his contribution in the 3D modeling. We would also like to thank all the participants in this study.

References

1. Cherng Z, Mosher S, Camp J, Lochtefeld D (2012) Human activity modeling and simulation with high fidelity. In: Interservice/industry training, simulation and education conference (I/ITSEC)
2. Santos WR, Braatz D, Tonin D, Menegon LZ, Luiz N (2016) Analysis of the integrated use of a motion capture system with a digital human modeling and simulation software for incorporation of future activity. Gest Prod 23(3)
3. Dai H, CAI B, Song J, Zhang D (2010) Skeletal animation based on BVH motion data. In: 2nd International conference on information engineering and computer science, pp 1–4
4. Blender Online Community (2015) Blender-a 3D modelling and rendering package
5. https://www.autodesk.com/products/maya/overview
6. https://www.autodesk.com/products/3ds-max/overview
7. Scataglini S, Truyen E, Perego P, Gallant J, Tiggelen DV, Andreoni G (2017) Smart clothing for human performance evaluation: biomechanics and design concepts evolution. In: 5th International digital human modeling symposium, Germany, Bonn
8. Badler NI (1997) Virtual humans for animation, ergonomics, and simulation. In: Proceedings IEEE non rigid and articulated motion workshop
9. Danckaers F, Huysmans T, Ledda A, Verwulgen S, Van Dogen S, Sijbers J (2014) Correspondence preserving elastic surface registration with shape model prior. In: International conference of pattern recognition, pp 2143–2148

10. Danckaers F, Huysmans T, Hallemans A, De Bruyne G, Truijen S, Sijbers J (2018) Full body statistical shape modeling with posture normalization. In: Cassenti D (ed) Advances in human factors in simulation and modeling. AHFE 2017. Advances in intelligent systems and computing, vol 591. Springer, Cham (2017)
11. Danckaers F, Scataglini S, Haelterman R, Van Tiggelen D, Huysmans T, Sijbers J. Automatic generation of statistical shape models in motion. In: AHFE 2018: advances in human factors in simulation and modeling (in press)
12. Danckaers F, Huysmans T, Lacko D, Sijbers J (2015) Evaluation of 3D body shape predictions based on features. In: 6th International conference on 3D body scanning technologies, Lugano, Switzerland, pp 258–265
13. https://yostlabs.com/
14. Villar O (2014) Learning blender: a hands on guide to creating 3D animated characters, 2nd edn. Addison Wesley professional, Boston
15. Meyer M, Desbrun M, Schröder P, Barr AH (2003) Discrete differential-geometry operators for triangulated 2-manifolds. In: Hege HC, Polthier K (eds) Visualization and mathematics III. Mathematics and visualization. Springer, Berlin, Heidelberg
16. Robinette KM, Daanen HAM, Paquet E (1999) The CAESAR project: a 3D surface anthropometry survey. In: Second international conference on 3-D digital imaging and modeling (Cat. No. PR00062), pp 380–386
17. Baran I, Popović J (2007) Automatic rigging and animation of 3D characters. ACM Trans Graph 26(3):72
18. Scataglini S (2017) Ergonomics of gesture: effect of body posture and load on human performance. Ph.D., Politesi. https://www.politesi.polimi.it/handle/10589/136840

Lessons Learned from Crisis Situation Simulations for the Local Command Post (LCP) in Extreme Situation (ES)

Violaine Bringaud[✉]

R&D, EDF Lab Paris-Saclay,
7 boulevard Gaspard Monge, 91120 Palaiseau, France
violaine.bringaud@edf.fr

Abstract. Armed with feedback from the Fukushima accident, EDF's Nuclear Operations Division has launched an action programme that aims to implement additional crisis management means (added equipment and organisational methods) to meet the needs in case of extreme accident conditions. As part of this action programme, a multidisciplinary R&D team (ergonomics experts and human reliability engineers) was consulted to manage a trial campaign in collaboration with the Operator, the engineering department and the company's training department. Completed with the help of simulations, this test campaign enabled recreation of the individual and collective activity dynamic for the crisis organisation's team members and was analysed with help from the Model of resilience in situation.

With respect to the Local Command Post function, the goal of this document is to report how the implemented simulation system, coupled with the analysis of collected data, enabled the addition of activity understanding items from the Local Command Post. By relying on several examples, the goal was to question the reasoning of keeping this type of data in order to develop innovative models for extreme and unpredictable crisis preparation.

Keywords: Crisis management · Nuclear industry · Resilience

1 Introduction

For EDF, the Fukushima accident in 2011 was full of lessons about management of extreme situations (ES). Indeed, the notion of Extreme Situation emerged after the accident at Fukushima in order to pose the question of an organization's response to cumulative events beyond the design basis of the plant and of the organization itself. The characteristics of the situation result, a priori, from an initiating event such as an earthquake or a flood of unusual intensity:

- The degradation of the environment can go as far as an isolation of the site preventing the emergency teams from arriving.
- The conditions of intervention in the field can become very difficult.
- People on site may have been injured or contaminated by the initiating event or its consequences.

© Springer Nature Switzerland AG 2019
S. Bagnara et al. (Eds.): IEA 2018, AISC 822, pp. 39–49, 2019.
https://doi.org/10.1007/978-3-319-96077-7_5

- Internal and External means of communications may have been lost.
- An accidental situation can occur on several reactors simultaneously, involving the loss of internal and/or external electrical sources, and/or loss of cooling means.

We take as reference an Extreme Situation generated by an event resulting in the isolation of the site preventing the arrival of emergency teams and reinforcements in the immediate future, the loss of external power supply of the site, and where one or more reactors have lost their internal electrical sources and/or their source of cooling.

In these conditions, only the teams permanently present on site are present at the installation. There are ample communication means to signal an alert. However, in this specific situation, an emergency satellite phone is the only means of communication considered to be operational towards the outside. As such, the operating team is in relation with the National emergency teams, which are mobilised and available.

Armed with feedback from the Fukushima accident, EDF's Nuclear Operations Division has launched an action programme that aims to implement additional crisis management means (added equipment and organisational methods) to meet the needs in case of extreme accident conditions.

As part of this action programme, a multidisciplinary R&D team (ergonomics experts and human reliability engineers) was consulted to manage a trial campaign in collaboration with the Operator, the engineering department and the company's training department. These trials were implemented through extreme situation simulations and were meant to test the organisational options offered to improve management of a crisis in this context. Strengths to preserve and areas for improvement were identified [1], but organisational options were not called into question. This was the case for the introduction of a new distribution of missions taken into charge by a member of the operating team (generally the deputy operating chief) to apply the safety engineer procedure when the on-call safety engineer cannot access the site. This was also an opportunity to write messages while waiting for support to arrive. The alert mission was conducted by the Local Management Command Post (LCP), who was not reachable in this situation. It was recovered by the Shift Manager, who became LCP. The same goes for the transmission of information with National.

Also, the performance of extreme situation simulations enabled the dynamic recreation of the individual and collective activity of crisis organisation team members, who analysed the situation using the Model of resilience in situation (MRS) [2–4]. In this way, the document proposes focusing on the function of the Local Command Post (LCP) held by the Shift Manager of the operating team who, in the absence of the local crisis team (including the Local Management Command Post (LMCP)), ensures communication between operations in damaged units and the national crisis organisation. This makes the function a vital link in the management of extreme situation crisis management. The goal of this document is to discuss how the implemented simulation system, associated with the analysis of collected data, has made it possible to offer insight about LCP activity. By relying on several examples, the goal was to question the reasoning of keeping this type of data in order to develop innovative models for extreme and unpredictable crisis preparation.

The first part places Local Command Post (LCP) missions in relation with all extreme situation crisis organisation missions. The second part presents the simulation system and the methodology used in this test campaign. Following this, survey markers related to the Model of resilience in situation (MRS) are presented as a data analysis compilation. Lastly, by relying on this model, LCP activity examples are presented and it is possible to learn lessons from it to prepare crisis team members for extreme situation management.

2 The Local Command Post (LCP) Function in the Extreme Situation (ES) Crisis Organisation

Figure 1 represents the crisis organisation mobilised for the operation of damaged units in an extreme situation without the rigging of local on-call teams (in particular the Local Support Team and the Local Management Command Post), who cannot be present on-site.

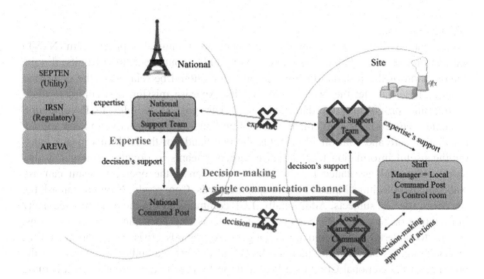

Fig. 1. Representation of the ES crisis organisation (according to Alengry) [5] (Color figure online)

In ES, the inaccessibility of a nuclear power plant prevents the rigging of on-call crisis team members, particularly the Local Support Team and the Local Management Command Post, who represent the expertise mission and decision-making on-site, respectively. The two communication channels associated with these missions are non-operational (cf. red crosses in Fig. 1). They are temporarily replaced by a single communication channel between the National Command Post and the Local Command Post. This specific configuration structures activities for those involved who occupy a position and, particularly, LCP at the time the alert is given and during crisis management.

- **At the time the alert is given.** In the given context whereby the installation's situation is deteriorating, operating instructions are driving the LCP to engage the In-house Emergency Plan, through which it is requested the Local Management Command Post be alerted. As the latter is not able to be reached, the procedure calls for an alert to be given at the national level, through a voice message. This message includes some basic general information (for example, the sensation of an earthquake or the first indications about installation status) to enable National emergency teams to mobilise and begin analysing the situation vis-à-vis unit operations. It is also requested that the time of first formal telephone contact (audio conference) between LCP and the National Command Post be noted. Also, if contact cannot be made with the on-call safety engineer in these early periods of installation deterioration, the operating team member responsible for the mission must apply the safety engineer procedure and redirect messages, as this mission begins.
- **After the alert, during crisis management.** LCP activity can be defined according to three dimensions.

1/Communication
On the one hand, this means supplying the National Technical Support Team (NTST) with as precise and reliable information as possible about the status of units, despite deterioration of the installation (little information should be relayed to the command room). These data let the NTST carry out its expertise mission and complete diagnostics and prognostics on damaged units. The goal is to offer LCPs operation strategies and actions to implement in a medium/long-term perspective to keep a given installation safe and controllable [5]. On the one hand, for the LCP this means transmitting useful information to National emergency teams to organise emergency dispatch and useful assistance for the site. In this sense, the operating team can ask National about technical questions and specific needs. Conversely, National can ask the site about specific subjects. Additionally, LCP communicates to the operating team any information originating with National that it deems important (answers to questions, actions to undertake, expected unit changes). As regards crisis organisation in ESs, exchange occurs between the national level and the site via audio conferences in the presence of the National Command Post and the LCP, who are autonomous regarding the planning of these communication points. In addition, un-programmed telephone calls are also possible between them. The team member who must write the message on-site has a hard copy that guides the entering of information for National to collect. These messages are adapted to extreme situations and can be complemented by LCP in preparation for an audio conference.

2/Represent the installation's status and its evolution.
It is expected that LCP take a step back regarding the installation's operations. To do so, the LCP seeks to maintain the most reliable and precise knowledge possible about the installation. It must be aware of the actions performed on units and associated effects on installation. In this context, LCP completes the data in the compendium sheet for the National emergency teams using its own information about the instrumentation and through consultation with the supervisor. As such, it builds a dynamic representation of the installation's status in a time frame of short/medium term operations.

3/Make decisions.
In real time, LCP makes decisions regarding the operations on each unit. It must share with the National Command Post all which relate to alternative procedure options. It is also charged with validating or invalidating recommendations made by the NTST which the National Command Post communicates and explains.

In order to recreate the management of ES crisis situations, the simulation system implemented for this trial campaign and described in the following paragraph, relied on the ES crisis organisation recently presented, and specifically with regard to the LCP's function.

3   The Simulation System and Methodology Implemented for the Trial Campaign

The ES trial campaign benefitted from a vast system, which mobilised for each trial up to 80 people belonging to various company departments (Operations, Engineering, Training, R&D). The preparation, creation and analysis of trial data are staggered over several months and the trial campaign lasted four years in total (2014–2017).

3.1   The ES Crisis Management Simulation System

During the trial campaign, several situations were simulated and four of these reproduced the conditions of an ES through a scenario with the following characteristics: An earthquake provoking a total loss of power, affecting two reactors simultaneously; this required that the operating team apply Emergency Operations to maintain the units in safe condition. The site is isolated, the on-call teams cannot find their command post. The shift operating team's chief of operations is the only function in the local emergency organisation likely to be operational at all times as an on-site LCP. As for the site, the simultaneous use of two full-scale simulators in the control room with two complete operating teams (operators in control rooms and agents on the ground) enables recreation of ES management, complete with the sharing of common equipment between two damaged units. At the national level, several posts of the National Command Post and National Technical Support Team are reproduced. The emergency satellite telephone link between the LCP and the National Command Post is also simulated. The creation of this scenario was guided by a search for representation and realism with regard to a very destructive ES from the viewpoint of installation operations.

3.2   Methodology Implemented for the Test Campaign

Through the progression of a 5-hour scenario, a multidisciplinary team (ergonomics experts and engineers specialised in human reliability [6]) collects simulation data and organises them for analysis using a thematic table dealing with team functioning, interactions between the various posts and ES management support tools.

Collection of Data Essentially by Observation
Observation of simulated work stations, on the spot debriefing after the completion of the scenario and consultation of technical log books (looked over by engineers specialised in human reliability) make it possible to collect data about crisis situation management. Six work places are observed: 2 command rooms, 2 field operator groups, the NTST and the National Command Post. Around 80 individuals are mobilised for a trial. In the command room, each observer follows a specific function and at least one observer is present in each other place. Throughout the scenario, observations are continually made in order to account for the time-sensitive dimension of crisis management. Compendium data modalities for this first step include note-taking, audio/video recording and recording of reactor status parameters.

After the scenario has reached an end, all team work documents are recovered (procedures, communication sheets, personal sheets). An on the spot debrief is organised for each of the six places observed, where observers and teams participating in the scenario are found. This debrief is led by an ergonomics expert and its goal is to survey the team's viewpoint regarding the scenario's result. How did each member react and what does this mean about the team's functioning as well as internal and external communication between the various simulated posts. It also involves reviewing the main key points that observers themselves identified.

During the day or days following, a debrief brings together observers in the six simulated places. On the one hand, it lets participants share what they observed and, on the other, sets out the first items necessary for an overall, multidisciplinary understanding of accident operation.

Organisation of Collected Data
Data are formatted as timetables and defining moments are also identified.

– *Timetable.*

As they are completed for each of the six observed locations, timetables make it possible to record moments observed for a specific team over a long period of time (actions/decisions and individual events such as equipment breakdowns or actions from another group). Placing timetables into perspective amongst themselves enables identification of synchronisation and de-synchronisation moments between the various groups.

– *Defining moments.*

Additionally, identification is performed for defining moments. A defining moment (technical, organisational or related to collective expertise) is a moment or repetition of a moment that establishes that an action (or lack thereof), decision, individual or collective initiative or observation indicates the resilience of an evaluated socio-technical system. Defining moments are elements which are central to the simulated situation analysis. They rely on the Model of resilience in situation (MRS).

Used to register LCP activity within a group player network, this model is presented in the next section.

4 Analysing Local Command Post (LCP) Activity Using the Model of Resilience in Situation (MRS)

4.1 Model of Resilience in Situation (MRS) Markers

The Model of resilience in situation (MRS) [1, 2] is an empirical model built on the basis of analysed real or simulated accidents and relies on the theoretical framework of Social Regulation [7]. It is based on a definition that considers resilience as the ability of an organisation to anticipate disturbances and resist them, all the while adapting where necessary, and returning to an acceptable state.

In this way, MRS offers the keys to explain how a socio-technical system ensures safe functioning by drawing from two processes, apparently contradictory, but which MRS brings together. These are anticipation and adaptation.

Applied to the socio-technical system related to crisis management, MRS is a system to describe the dynamic of operational and group management of a crisis situation by considering the alternation between stable rule execution phases and reconfiguration phases, defined to readjust operational rules which were judged inopportune. MRS applies to crisis organisation in its entirety as a dynamic network for workgroups (each of the six places observed during the trial campaign), which interact under the current organisational framework. Each group forms part of its own environment and is considered to be an interactive, distributed cognitive system. The system's overall socio-technical resilience considered (real organisation of crisis management), consequences result from interactions between individuals within the groups and between the groups themselves (within observed posts as part of related simulations).

Additionally, MRS takes from Hollnagel [3] the four macro functions of resilience presented as the "four cornerstones of resilience" (Fig. 2):

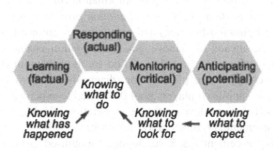

Fig. 2. The four cornerstones of resilience (Hollnagel)

MRS relies on this schema on the one hand to highlight the time constraint of the resilience dynamic and, on the other, to list the Hollnagel macro-functions as those which are most precise.

With respect to simulations of the evaluated crisis situations, only the short/medium term is considered (recent past/present/near future), which corresponds with two Response and Monitoring macro-functions. Organisational learning and Technical

Anticipation involve longer-term processes that largely surpass the real-time management of a crisis situation which returns, for instance, long loops from feedback and engineering.

Specific MRS functions relating to Response and Monitoring macro-functions are presented in the following table (Fig. 3).

	Recent past	Present	Near future
Monitoring	INFORMATION RETROSPECTIVE CONTROL DIAGNOSIS VERIFICATION		
Response		COOPERATION EXECUTION REAL-TIME CONTROL	COLLABORATION PROGNOSIS RECONFIGURATION ADAPTATION
	← SENSEMAKING →		

Fig. 3. Resilience functions (Bot) [8]

In the context of collective work, the dynamic interaction between the various functions oscillates between recent past/present/near future, and enables the organisation to be both robust and adaptable.

– *In robust phase, the execution process* involves implementing the decided strategy and is subject to an attribution of tasks and resources between team members (cooperation function). Robustness is obtained through execution (execution function) and the control of actions in progress to ensure that actions executed comply with operational rules (control function). The group continuously verifies that the strategy executed is appropriate (verification function). These actions are driven by the acquisition and sharing of information (information function).
– *During the reconfiguration phase, the adaptation process* involves ceasing irrelevant rules (reconfiguration function) and redefining operations; that includes goals, strategies and the means to achieve goals. Adaptation enables readjustment of operational rules (adaptation function). To do so, the behaviour of the installation and actions to complete are planned for (diagnostic and prognostic functions). During this phase, mobilised players pool the available expertise and communicate amongst themselves to co-develop the new operation to lead, which must be validated to be implemented (collaboration function).
– *Sensemaking* is the cement that holds together mobilised groups during situation management (sensemaking function). It is the cement of cooperation and collaboration and makes it possible to ensure coherence between the considered time frames (recent past/present/near future).

In summary, applied to data analysis for the previously presented simulations, MRS makes it possible to fully understand the socio-technical crisis management system in its entirety, by considering the interactions between the various posts that make it up. Within the group's network, the LCP plays a hinge role between the on-site operating team and the National emergency teams. By reconciling anticipation rationality and adaptation rationality, MRS also offers the keys to analysis, to better understand ES crisis situation management characterised by presenting a disadvantage and uncertain operating conditions in a reduced environment. Lastly, thanks to MRS, it is a benefit to address time-related specificities for the various posts involved in the crisis while also questioning the overall synchronisation of the socio-technical system. This offers an understanding of the crisis organisation's resilience.

The examples presented in the next paragraph rely on this analysis structure to report on LCP activity and its contribution to the resilience of the socio-technical system.

4.2   Two Examples to Question the Contribution of LCP to the Resilience of ES Crisis Organisation

Example 1: The Local Command Post Contributes to the Resilience of the Socio-Technical System
In the context of its expertise activity (using diagnostics/prognostics), the National Technical Support Team (NTST) is planning for the future by making a proposition to LCP in the form of an operational strategy enabling the recharging of equipment in the event that emergency methods fail (reconfiguration, adaptation). The LCP received the proposition as a response to a short-term question it posed previously. This example is demonstrative of a sensemaking problem, as NTST and LCP work on time spectra which are different, without this being expressed (collaboration difficulties). This situation creates a cognitive offset between the two parties. Nonetheless, by not validating the proposition, the LCP makes a decision adapted to a plant's "real-time" situation. Not applying the proposition does not further weaken the plant. Rather, it maintains it in a stable state. In this way, the LCP contributes to the socio-technical system's resilience.

Example 2: The Local Command Post Benefits from the Expertise of NTST for Overall System Resilience
Firstly, upon request from its team, the LCP validates in a circuit configuration in real time regarding equipment shared between two plants. Some time after this, while exchanging information by audio-conference between the site and National, the LCP indicates the position of the equipment. The National emergency teams do not understand the configuration presented and request the LCP explain it more thoroughly. This is done (collaboration). The National emergency teams question the relevance of this configuration and indicates that it will lead the subject's inquiry. Between two audio-conferences, the National emergency team group shares its concerns about the reported situation and forms a general vision of the situation (collaboration) through a sensemaking process. It comes to the conclusion that the presented configuration is unacceptable from an installation safety viewpoint, as it is causing the deterioration of two plants simultaneously. The NTST co-develops a new operating strategy which is disadvantageous for the installation with regard to long-term safety challenges. During the following audio-conference (90 min after the previous one), the National

emergency teams present to LCP their analysis of the situation (collaboration) and propose a change in strategy in order to preserve the two plants (reconfiguration, adaptation). The LCP validates this strategy change, which it communicates to its team as a priority operating goal (for execution).

In this example, using its ability to offer precise information about the configuration of installation equipment, the LCP enables the National emergency teams to fulfil their role as expert in on-site hindsight and reliance, as it has the means to call into question the operating goals to adapt to need and allow the crisis organisation to be resilient in its entirety.

5 Conclusion

Organised simulations which are part of the trial campaign for the EDF Post-Fukushima action programme were original in more than one way. On the one hand, they enabled recreation of the individual and collective activity dynamic for players involved in the management of ES crisis situations, meaning without local on-call teams and confronted with a deteriorating installation, which changes uncertainly and unpredictably. On the other, the model used to analyse the collected data during these simulations (MRS) offers an advancement in the knowledge used to manage this type of situation, by shedding light on resilience mechanisms. The keys to understanding how teams manage crisis situations are: firstly, considering the entirety of players involved in the crisis and their interactions, and secondly, taking into account antici-pation and adaptation to situations and understanding the different time strata for groups on-site and at the national level The presented examples involving the function of Local Command Post are an illustration of this.

Also, the knowledge produced by this action has led to a reflection about the evolution of preparation modes for crisis situations. This is done by offering the appropriate means to teams. The goal is also to understand the unpredictable dimension of some situations aside from situations covered by the reference base, to develop collective functioning and to be attentive to time-sensitive crisis management. In this perspective, the company is experimenting with innovative approaches to preparation, through reflective training systems [9] and peer-based experience sharing methods, while always considering the simulated or real life activity dynamic.

References

1. Baudard Q, De la Garza C, Le Bot P (2018) The model of resilience in situation and its contribution to crisis management. In: Proceeding of the European safety and reliability, ESREL 18 Conference, Trondheim, Norway
2. Le Bot P, Pesme H (2010) The model of resilience in situation (MRS) AS AN IDEALISTIC organization of at-risks systems to be ultrasafe. In: PSAM 10 – 10th international conference on probabilistic safety assessment & management, Seattle, Washington, USA
3. Hollnagel E (2011) Resilience engineering in practice. A guidebook. Ashgate (Ashgate studies in resilience engineering), Farnham, Surrey, England & Burlington, Vermont, USA

4. Massaiu S, Braarud P, Le Bot P (2013) Including organizational and teamwork factors in HRA: the EOS approach. In: IFE, coord.: EHPG 2013 Enlarged Halden Programm Group. Man-Technology Sessions. Storefjell Resort Hotel, Gol, Norway
5. Alengry J, Falzon P, De la Garza C, Le Bot P (2017) Diagnosis, forecast and sensemaking activities of a National Technical Support Team. In: Proceedings of the 13th international conference on naturalistic decision making, Bath, UK
6. De la Garza C, Pesme H, Le Bot P (2013) Interest of combining two human factor approaches for the evaluation of a future teamwork. In: Proceedings of the European safety and reliability, ESREL 13 conference, Amsterdam, Netherlands
7. Reynaud J-D (1997) Les règles du jeu. L'action collective et la regulation sociales, 3rd edn. A. Colin, Coll. "U", Paris
8. Le Bot P, Alengry J, De la Garza C (2016) Organising the operation of nuclear reactors in extreme situations: simulator based-test methodology. In: Proceedings of the 39th Enlarged Halden Programme Group Meeting, Fornebu, Norway
9. Alengry J, Falzon P, De la Garza C, Le Bot P (2018) What is "training to cope with crisis management situation"? A proposal of a reflexive training device for the National Technical Support Team. In: 20th international IEA conference, Firenze, Italy

Using Torrance Creative Thinking Criteria to Describe Complex Decision Making During an Outbreak Management by Public Health Experts

Liliane Pellegrin[1,3](✉), Leila Chassery[1,2], Nathalie Bonnardel[2],
Christelle Tong[1], Vincent Pommier de Santi[1,3], Gaëtan Texier[1,3],
and Hervé Chaudet[3]

[1] French Army Centre of Epidemiology and Public Health (CESPA), SSA,
Marseille, France
[2] Aix-Marseille Univ, Centre of Research in the Psychology of Cognition,
Language and Emotion (PsyCLE), EA 3273, Aix-en-Provence, France
[3] Aix Marseille Univ, IRD, SSA, AP-HM, VITROME,
IHU-Méditerranée Infection, Marseille, France
liliane.pellegrin_chaudet@univ-amu.fr

Abstract. This study focused on the application of Torrance framework about creative thinking in a complex professional context: the management and control of an outbreak by experts in epidemiology and public health. We argue that building accurate responses in this context depends on the complexity of situations, such as 'epidemiologic problems' experts have to deal with. Thus, depending on the problem's complexity, experts could possibly adapt their problem solving strategies, using either 'standard' strategies or more 'creative' ones.

Our goal was to characterize expert decision processes developed during critical situation (where rule-based strategies and usual procedures could be not satisfyingly applied) with regard to creativity criteria described by Torrance (fluency, flexibility, elaboration and originality). We carried out a simulated outbreak alert to study creative processes during experts problem-solving activities. This simulation was intended to put specialists in a context of epidemiological problem management, based on possible real practice but conducive to implement creative solutions. The analysis carried out on the observations allowed us to identify a total of 14 different themes, with 148 ideas expressed by the participants. The participants have therefore actively contributed to the elaboration of ideas as well as to the mutual enrichment and implementation of ideas. However, the number of evocated topics and ideas and their level of elaborations appears higher when epidemiologists are more experienced in their domain. Thus, creative thinking appears to be an important aspect of the epidemiological alert management and related to experience in this area.

Keywords: Creativity · Expert decision-making · Simulation · Epidemiology

© Springer Nature Switzerland AG 2019
S. Bagnara et al. (Eds.): IEA 2018, AISC 822, pp. 50–59, 2019.
https://doi.org/10.1007/978-3-319-96077-7_6

1   Introduction

We propose in this paper an applied study focusing on creative decision-making strategies used during a simulated critical outbreak management by experts. Epidemics, health threats are complex, time-pressure, uncertain and risky situations, which need to be understood and managed with a high-level of expert knowledge and skills. Epidemiology expertise could be described as based on both a deep understanding of an outbreak situation (identification and characterization of the accurate outbreak and its main features, anticipation of its plausible evolution with an acceptable confidence level) and a selection and elaboration of an accurate course of actions (medical countermeasures with respect to time constraints). These processes imply complex cognitive activities, such as problem solving, decision-making, anticipation and planning, dealing with uncertainty, time pressure and situation complexity. In this context, our objective was to study the experts' proficiency to find accurate solutions and responses to an outbreak spread in a given population, especially when confronted to a critical situation. Considering the point that building accurate responses depend on the epidemiologic problem's complexity, experts could adapt their problem solving strategies using either 'standard', 'typical' or 'creative' [1].

This study focused on the application of Torrance framework about creative thinking in a complex professional context consisting in the management and control of an outbreak by experts in epidemiology and public health. More precisely, our goal was to characterize expert decision processes developed during critical situation (where rule-based strategies and usual procedures could be not satisfyingly applied) with regard to creativity criteria described by Torrance [3, 4] (fluency, flexibility, elaboration and originality).

1.1   Context of the Study

Disease surveillance is defined as 'on-going systematic collection, analysis, and interpretation of data essential to the planning, implementation, and evaluation of public health practice, closely integrated with the timely dissemination of these data to those who need to know' [5]. These data are used to describe and to monitor the epidemic outcomes and public health emergencies in a given population. The epidemiological surveillance is based upon three linked basic activities, which are (1) the record, (2) the analysis of these health data, and (3) the dissemination of the obtained results (disease reporting) with the aim of undertaking control actions to limit infectious diseases spreading. French armed forces, which are deployed in different world areas, are exposed to various health threats, pathogens and risks related to local environment. A specific organization, the military Centre for Epidemiology and Public Health (CESPA) has in charge the health surveillance and epidemic alert management for the military population in metropolitan, overseas French territories and abroad deployments.

From a human perspective, outbreak detection and management is a high-level expert activity performed by specialists in epidemiology and public heath. Their first task is to assess the reality of an unusual epidemiological event, which could be considered as a potential health threat and risk for a given population. The experts usually described such event as an 'alarm', whose distinctive attributes are identified and analysed to confirm (or not) outbreak on a supervised population, seen as an 'epidemiological alert' [6]. Experts not only confirm the worst conditions of the health situation as well as the outbreak emergence, but also have o identify its main features, key elements for building a set of appropriate responses to limit its spread. These two tasks, the outbreak confirmation and the building of public health responses are based on expert activities, which are acquired through years of field experience. A team built by a minimum of 2 experts collectively performs these tasks in current situations, which could be extended if necessary to others members of the outbreak response team.

In this context, managing an outbreak belongs to professional situation where operators supervise and take decisions in dynamic and complex environments. Such environments are featured by multiple and interacting variables, temporal constrains, unclear or conflicting objectives, risks and uncertainty [7]. Dynamicity and complexity of processes involve that the operators have only a partial control of the situation processes [8]. In the context of outbreak management, they are faced to limited possibilities of control, due to a wide supervision field (biological, human, environmental attributes of the epidemics), remote actions to be performed dealing with response delays and disease spread, indirect access to crucial information to understand its causes and to project its evolution. Even if an outbreak alert can be considered as an abnormal, hazardous situation, the experts routinely manage it, applying procedures and operating rules, suited for 'usual' outbreaks such as seasonal flu or foodborne diseases, or to more complicated or less frequent ones. The public health surveillance actions are guided by both nationally acknowledged and internally established guidelines. Due to the complexity of epidemiology field, these procedures are quite flexible to ensure adaptable and accurate responses to field realities. But in some cases, they are not sufficiently suited for dealing with unexpected and highly complex outbreaks.

Such situations could engage the operators in a spiral out of control, handling major difficulties to contain them and to return to normalcy. Our study is focused on such critical situations, in which collaborative experts are confronted with difficulties to implement these guidelines. And our objective is to explore the team's capacities to overcome them by building new, unfamiliar strategies of understanding and action. For safety Human Factor approach, dealing with unforeseen and unexpected situations is managed by adaptive strategies based upon expertise, strong team coordination and responsive organisation evocated as 'managed safety' behaviours, complementary to ruled-based, 'controlled safety' ones [9–12]. The operators could readjust usual procedures to the current circumstances and even develop new strategies, 'invent new rules' [13, 14]. These skills devoted to anticipate, recognize and respond to adverse or unexpected events are required for an organisation to build resilience, defined by the ability to manage unexpected events or capacities to respond, monitor, learn and anticipate [15, 16]. Clinical psychologists have also been underlined the role of creativity in resilience as an effective tool to adaptation, adjustment and problem solving built by individuals confronted to adverse events [17].

Although creativity is a difficult notion to define, some authors have focused on the complementary factors that are mobilized in the creative act [18] or have studied creativity in open-ended problem solving and real-life situations [19–21]. In his initial work, Torrance [3, 4] defined creativity as the capacity to detect gaps, propose various solutions to solve problems, produce novel ideas, re-combine them, and intuit a novel relationship between ideas. Guilford and Torrance have defined criteria to characterize creativity: fluency (production of ideas), flexibility (production of different ideational categories), originality (production of unusual or novel ideas), and later elaboration (persistency on detailing an idea). Thus, in our study, we took into consideration these four criteria to analyse the activities of experts in epidemiology when building creative solutions to face the resolution of an epidemiological problem evolving to a critical situation. To explore creative processes implemented by the experts when solving such problem, we carried out a simulated outbreak alert. This simulation was intended to put operators in a context of epidemiological problem management similar to those possibly encountered in their real activity and conducive to implement creative solutions.

2   Materials and Methods

Two epidemiologists from the surveillance department, accepted to participate in this simulated outbreak alert. These two specialists were usually working together during weekly on-call duty. The first one was a confirmed expert (i.e. a domain specialist with more than 10 years of surveillance practice who can deal with complex cases) and the second one, was an intermediate expert (i.e., expert physician, with more than 3 years of surveillance practice who manage usual situations) referring to the description of usual levels of expertise in medical domains [22, 23]. The participants had three hours to achieve prescribed goals: to identify main features of the outbreak (time, location, concerned population), to build an outbreak situation diagnosis (aetiological causes, its plausible evolution with an acceptable confidence level), and to propose an adapted course of actions (public health countermeasures with the respect of time and organisational constraints).

During the simulation, they were locally assisted by the medical information staff, which routinely collects and organizes declared health related events (i.e. notifications of suspect cases) from local military medical units. They also called for expert advices a member of the medical intelligence department, as they usually do in real situations. They were located in a simulation room equipped with a video recording device, two telephones, phone books, an internet access, reference books, whiteboards, paperboard, papers and pencils). They could exchange with 5 others persons, who played the roles of different medical and military correspondents. These stakeholders were located in a separated room (the DIRection of Animation room) and they were also equipped with a computer connected to Internet, telephones and an audio recording device. They had a set of reference documents to supervise the scenario development (characters to play, information on the outbreak, simulation scenario, patient files, list of possible contacted persons).

The design of the scenario was done with the help of two confirmed experts (one from the public health military center, one from civilian university and infectious institute). We asked them to build a complex outbreak story with critical features in order to create an overflow of resources among the participants, thus encouraging them to build creative behaviors and decisions. The experts proposed a non-usual contamination of a rare disease of a military detachment deployed in Republic of Djibouti.

This outbreak context was complicated by a major organizational constraint for local French forces deployment (an international military exercise was organized as the first suspected cases were declared).

To collect data related to specialists' behaviours, two observers conducted a direct paper-and-pencil observation. Video and audio recording complemented these in situ observations in order to obtain a complete transcription of the management of the epidemiologic situation. We used EORCA, an activity-centred analysis method, devoted to collective medical activities in complex situations, to formalised observed 'events', which allowed us to take into consideration both actions, agents, tools and spatiotemporal features [24]. Then, some of these features were more precisely analysed and categorised with regard to the 4 criteria related to the Torrance Creative Thought Test (1976)- (see Table 1):

- *Flexibility*, defined as the individual capacity to express diverse ideas, belonging to different contextual fields or subjects. This analysis was based on the changes of themes addressed by the ideas.
- *Fluidity*, as the ability of an individual to produce numerous ideas, which are both unique and relevant to the task. Ideas were classified with regard to categories such as *information search about the outbreak, assumptions, to-do instructions countermeasures propositions*.
- *Elaboration*, as an individual's skill to develop, to enrich and to expand ideas. To analyse the elaboration of the ideas produced during the simulation, we noted all the actions and speech, individually or collectively, brought to explain, detail, expand, or implement an idea.
- *Originality*, which corresponds to the most important creative criterion, is associated with the concept of 'creative power' of an individual. It refers to a person's ability to produce unusual ideas, which move away from the obvious ones. To take into account this criterion, we asked a public health expert to evaluate the originality of all the ideas produced by specialists.

Table 1. Indices and scoring definitions for this study

Indices	Definitions	Scores
Flexibility	Change of topics	Number of discussed topics
Fluidity	Different ideas in the same topic	Number of produced ideas of a topic
Elaboration	Enrichment, expanding, development, achievement of an idea (*NB. Discourse elements considered as idea elaboration are only those, which complement this one*)	Number of added elements to an initial idea
Originality	Rarity of ideas	Expert evaluation (*NB. A simple identification of ideas considered as original by the expert instead of a planned 4 steps lickert scale*)

3   Results

Results based on the EORCA method showed a total amount of 1729 speech actions recorded during the 2h22mn of the outbreak scenario resolution. Creativity criteria score were used to analyse productions of the two specialists taken together as well as to characterize productions of the team composed of these two participants taken together (see Table 2). These results first indicated that this team was strongly active, building and implementing concepts, actions and decisions to deal with a difficult scenario. The analysis of the data we gathered shows that the two specialists expressed a total of 148 ideas (fluidity criteria), covering 14 different topics by the participants (flexibility criteria). 422 actions, and thus 24.4% of the speech acts, were categorized as elaborative elements of ideas.

Table 2. Observed scores relative to creativity criteria.

	Flexibility (topics)	Fluidity (ideas)	Elaboration	Originality
Epd1 (*Senior*)	14	87	241	1
Epd2 (*Intermediate*)	12	61	181	0
Total		148	422	

A senior, who was in charge of evaluating the originality of the full set of ideas, identified only one fully original concept (i.e. "We must re-examine the patients"). It corresponds to modifying the usual procedure by calling specifically contaminated patients, in order to get direct information from them about their exposition context. This expert explains his choice considering that this decision of action was innovative, not described in their current procedure of outbreak management. He also underlined the pertinence of this idea in context.

Concerning the level of expertise, we observe that the senior physician was more productive for all criteria than the less experienced one, i.e. with regard to the number of ideas, of evocated topics, and to the level of idea elaborations (see Table 2). Nevertheless, both participants have actively contributed to the enrichment and implementation of the ideas of the other one and to their elaboration. They elaborated a total of 422 elements to explain, enrich and implement the initial ideas (see Table 3). We observed that the ideas correspond to 14 different topics (see Table 3). Considering the numbers of ideas for each topic, 'biological analysis and sampling' involves the maximal number of ideas (16.2% of total evocated ideas), followed by 'questioning about hypothesis of food contamination' (9.5%). The three following issues, which correspond to identifying possible outbreak origins, patients' health status and their local management, are discussed with the same frequency. Team questioning about information sources, planning and outbreak spreading are a slight support to different ideas. In the same way, discussions about exposed population and outbreak time featuring result in less than 10 different ideas.

Thus, the process of idea elaboration appears to be an important feature of this outbreak management, strongly related to the key points of the situation. These focus of interest are related tasks and activities, which are particularly difficult to solve due to

Table 3. Corresponding total of elaboration elements to ideas and topics

Evocated main topics	Nb of ideas' topics	Nb of elaboration elements	Ratio elaboration/idea
Biological analysis and sampling	24	53	2.21
Food contamination	14	25	1.79
Outbreak possible origins (identification)	12	41	3.42
Patients health status (signs & symptoms, severity rate, clinical picture)	12	16	1.33
Patient management (hospitalisation, treatment, nursing..)	**12**	**69**	**5.75**
Military authorities demands, communication and local management	11	45	4.09
Geographical aspects of outbreaks	11	12	1.09
Diseases hypothesis, aetiology building	11	31	2.82
Points of contact (social network)	**11**	**62**	**5.64**
Time (events temporality, key dates)	9	9	1.00
Exposed population (patients, military units)	8	18	2.25
Outbreak spreading	6	21	3.50
Planning, strategies of decisions and actions	4	4	1.00
Information resources	**3**	**16**	**5.33**
Total	148	422	2.87

their complexity, criticality and uncertainty [25]. Consequently, individual and collaborative elaboration helps to explain, to detail or discuss the evocated ideas for building a relevant situation representation in the goal of efficient decision-making. Finally, we built for each topic, a ratio between the ideas and their corresponding elaborations, to assess the weight of the team cognitive effort to find accurate information or solutions. Three mains topics can be seen as cognitively engaging: first, patient medical care, then activities related to social network management, and at least, information resources. This last topic is associated with few ideas (3) but it requires a constant activity.

4 Conclusion

Although we have not yet precisely analysed the process of building creative solutions in an epidemiological context, the analysis derived from the Torrance Creative Thought Test allowed us to explore and characterize different aspects of the activities developed by specialists to manage a critical outbreak situation, occurring in a complex and critical-safety environment. The criteria we took into account (i.e., analysis of fluidity, flexibility and elaboration) allowed the identification of the main topics of interest for the specialists, with regard to the expressed ideas and their elaborations. The results we obtained illustrate the two main axes of team situation management: identifying the outbreak nature and managing the social context and the cognitive cost related to this complex situation.

Only one idea was judged as fully original. Because originality is an inseparable part of creativity, it is impossible to assert the presence of full creative processes by epidemiologists during this situation. We could interpret this result in two ways. First this lack of originality may be due to the epidemiologists' activities in relation to high flexibility of their procedure, as Okoli et al. [2] description of creative decisions for experienced fireground commanders. Secondly, the method we have chosen to assess the ideas' originality wouldn't be well suited. When applying the Torrance test, the assessment of the originality of participants' ideas is based on the frequency of occurrence of ideas with regard to the totality of ideas expressed by all participants, which corresponds to a 'statistical originality'.

In this study, we had only one team composed of two participants, which did not allowed us to apply the originality analysis classically used in creativity tests based on divergent thinking tasks. However, we have to point out the fact that epidemiological situations do not require only divergent thinking but also convergent thinking. Thus, although judgements of originality expressed by the senior expert can be partially subjective, it seems to us the best process to assess participants' ideas, extending to a panel of confirmed experts in the domain.

Finally, the results of this exploratory study open the questions about the process of building innovative or creative solutions in high-level expert activities, managing complex and critical situations. Dealing with epidemiological risks belongs to such work situations, classically featured by time-pressure, uncertainty, ill-structured problems and possible competing goals [26]. Identifying creativity also as the richness of possible courses of actions and strategies more than pure innovative concept, would be a pertinent approach, which need to be improved by furthers studies.

Acknowledgements. We are grateful to Dr. MA. Creach, Dr. F. Berger, Pr. JB Meynard from CESPA and C. Adamo from PsyCLe for their participation in this research as main Dirani facilitators. This research was granted by the PDH-1-SMO-4-17 funds (SSA/IRBA-Military Institute of Biomedical Research & DGA-Army General Direction).

References

1. Okoli JO, Weller G, Watt J, Wong BLW (2013) Decision making strategies used by experts and the potential for training intuitive skills: a preliminary study. In: Chaudet H, Pellegrin L, Bonnardel N (eds) Proceedings of the 11th international conference on naturalistic decision making (NDM 2013). Arpege Science Publishing, Paris, pp 227–232
2. Okoli JO, Weller G, Watt J (2016) Information filtering and intuitive decision-making model: towards a model of expert intuition. Cogn Technol Work 18(1):89–103
3. Torrance EP (1969) Creativity. What research says to the teacher?. National Education Association, Washington DC
4. Torrance EP (1965) Scientific views of creativity and factors affecting its growth. Daedalus Creat Learn 94(3):663–681

5. Thacker JB, Berkelman RL (1988) CDC surveillance update. Center for Disease Control and Prevention, Atlanta

6. Meynard JB, Chaudet H, Texier G, Queyriaux B, Deparis X, Boutin JP (2008) Real time epidemiological surveillance within the armed forces: concepts, realities and prospects in France. Rev Epidemiol Sante Publique 56(1):11–20

7. Klein GA (2008) Naturalistic decision-making. Hum Factors 50(3):456–460

8. Hoc JM, Amalberti R (2007) Cognitive control dynamics for reaching a satisficing performance in complex dynamic situations. J Cogn Eng Decis Mak 1(1):22–55

9. Amalberti R (2001) La conduite des systèmes à risques. PUF, Paris

10. Amalberti R (2013) Piloter la sécurité: théories et pratiques sur les compromis et les arbitrages nécessaires. Springer, Paris

11. Cuvelier L, Falzon P (2012) Sécurité réglée et/ou sécurité gérée? Quelles combinaisons possibles. In: Dessaigne MF, Pueyo V, Beguin P (eds) Innovation & Travail: Sens et Valeurs du changement, 47ème congrès de la SELF. Editions du Gerra, Lyon, pp 24–28

12. Morel G, Amalberti R, Chauvin C (2008) Articulating the differences between safety and resilience: the decision-making process of professional sea-fishing skippers. Hum Factors 50 (1):1–16

13. Nascimento A, Cuvelier L, Mollo V, Dicciocio A, Falzon P (2013) Constructing safety: from the normative to the adaptive view. In: Falzon P (ed) Constructive ergonomics. CRC Press, Boca Raton, pp 111–125

14. Dekker S (2017) The safety anarchist: relying on human expertise and innovation, reducing bureaucracy and compliance. Routledge, Abington

15. Hollnagel E, Woods DD (2006) Epilogue: resilience engineering precepts. In: Hollnagel E, Woods DD, Leveson N (eds) Resilience engineering: concepts and precepts. Ashgate, Aldershot, pp 326–337

16. Weick K, Sutcliffe K (2007) Managing the unexpected: resilient performance in an age of uncertainty. Jossey Bass, San Francisco

17. Metzl E, Morrel JM (2008) The role of creativity in models of resilience: theoretical exploration and practical applications. J Creat Mental Health 3(3):303–318

18. Lubart T, Mouchiroud C, Tordjman S, Zenasni F (2003) Psychologie de la créativité. Armand Colin, Paris

19. Bonnardel N (2006) Créativité et conception. Approches cognitives et ergonomiques. Solal Editions, Marseille

20. Bonnardel N, Bouchard C (2017) Creativity in design. In: Kaufman JC, Glaveanu VP, Baer J (eds) Cambridge handbook of creativity across different domains. Cambridge University, New York, pp 403–427

21. Bonnardel N, Wojtczuk A, Gilles PY, Mazon S (2017) The creative process in design. In: Lubart T (ed) The creative process: perspectives from multiple domains. Palgrave Macmillan, New York

22. Patel VL, Kaufman DR (2006) Cognitive science and biomedical informatics. In: Shortliffe EH, Cimino JJ (eds) Biomedical informatics. Health informatics. Springer, New York, pp 133–185

23. Hoffman RR, Ward P, Feltovich PJ, DiBello L, Fiore SM, Andrews H (2014) Accelerated expertise: training for high proficiency in a complex world. Psychology Press, Hove East

24. Pellegrin L, Bonnardel N, Antonini F, Albanese J, Martin C, Chaudet H (2007) Event oriented representation for collaborative activities (EORCA). A method for describing medical activities in severely-injured patient management. Methods Inf Med 46(5):506–515

25. Texier G, Pellegrin L, Vignal C, Meynard JB, Deparis X, Chaudet H (2017) Dealing with uncertainty when using a surveillance system. Int J Med Inform 104:65–73
26. Lipshitz R, Klein G, Orasanu J, Salas E (2001) Focus article: taking stock of naturalistic decision making. J Behav Decis Mak 14(5):331–352

What is "Training to Cope with Crisis Situations"? Developing a Reflexive Training Device for a Crisis Support Team

Julia Alengry[1,2(✉)], Pierre Falzon[1], Cecilia De La Garza[2], and Pierre Le Bot[2]

[1] Research Centre on Work and Development,
Conservatoire des Arts et Métiers, Paris, France
[2] EDF R&D, Pericles, Human Factors Group, Palaiseau, France
julia.alengry@edf.fr

Abstract. Support Teams functioning in crisis organization are challenged in the Post-Fukushima Context. Even though these teams are arranged by the company as a defined crisis structure, team members may work together only in crisis situations or crisis exercises: these teams are ephemeral. Training these teams in order to improve their teamwork inside a multi-level crisis organization is an important reliability stake. In this study, a reflexive training device for crisis management in the nuclear industry was designed and tested. Results show that it has allowed participants to debate some teamwork dimensions and helped them to build common references.

Keywords: Crisis management · Simulation-based training
Reflexive activities

1 Introduction: Training Support Teams in the Post-Fukushima Context

Since the Fukushima accident, High-Reliability Organizations are questioning their ability to cope with extreme and very rare accident situations, beyond those covered by the current design of technical and organisational emergency schemes.

The Fukushima accident happened on March 11, 2011: an earthquake followed by a tsunami led the Fukushima Daiichi facility into a black out, isolated from external backups. From March 11 to 15, on-site crisis cell had to cope with an extreme situation and try and manage 6 damaged units – at different level of severity and at different moments (Baudard and Le Bot 2017). Crisis Management was distributed between an on-site crisis cell, led by Masao Yoshida who had to manage 400 people; and two off-site crisis cells located in Tokyo. These off-site crisis cells were intended to assist the on-site crisis cell. The chain of command was disrupted during the emergency (National Diet of Japan 2012). Yoshida's testimony emphasized disagreements and the difficulty to share a same situation perception with off-site cells (Guarnieri et al. 2014). His testimony highlights the major issue of synchronisation between all the level of the

S. Bagnara et al. (Eds.): IEA 2018, AISC 822, pp. 60–67, 2019.
https://doi.org/10.1007/978-3-319-96077-7_7

crisis organization during the emergency and the risk of mismatch of the sensemaking between the stakeholders (Weick 1993).

In all High-Reliability Organizations, because of the potential severity of unforeseen situations, off-site crisis cells are implemented in order to focus on situation sensemaking and anticipate potential consequences and future states of the facility. During crisis management, these crisis cells are responsible for the anticipatory activity by remotely analyzing the situation and proposing possible action schemes. These support teams can build the overview of the situation that is necessary to adapt and adjust to the changing conditions of the situation (Hardy and Comfort 2015). They handle a second layer of activity (Artman and Waern 1999) by collecting data, assessing the situation and its evolution and allocating resources (Perry 2003, 1995; Artman and Waern 1999; Wybo and Kowalski 1998).

Thanks to their forecast activity, they can suggest actions or even changes to the current strategy, and review the goals so that they match the development of the situation and of the probable future facility states. The forecast activity is important for the revaluation of the actions to be implemented (Klein 1999). The anticipatory activity brings about a projection into the future. This activity is part of a different temporality than that of front-line teams. The forecast activity is critical in crisis management, especially in high-risk dynamic processes.

To fulfill their role, Support Teams have to manage two dimensions of teamwork: (1) teamwork inside the support team, who has to pool together different types of expertise (Crichton et al. 2005), (2) teamwork between the support team and the other stakeholders.

Teamwork is challenging, because of the characteristics of the crisis situation. The support team has to collaborate with distant stakeholders, in a dynamic and complex situation, under time pressure, high communication demands and finally too many or too few information to analyze (*ibid.*).

Another specificity characterizes these teams: they are *ad hoc* teams. They have generally trained periodically together, however teammates may work together only for the duration of the crisis (Salas et al. 2002). Teamwork processes have to be thought before, during and after crisis situation – real or simulated. In fact, teamwork is identified as one of the important incident command skills (Crichton et al. 2005).

In the organisation under study, the off-site crisis cell – i.e. the National Technical Support Team (NTST) - has to provide the Site with its expertise by proposing a diagnosis and a forecast of the development of the situation. It shall also provide the Site with information and proposal for action, even though the decision remains in charge of the actual management of the plant. Due to its off-site position, the NTST has more time to evaluate the situation and build an ad hoc proposal to cope with changing circumstances of the situation.

Two research questions arise: what is needed to train these crisis team members (NTST) to cope with unforeseen, unique, and high-risks situations? What collective skills need to be developed to prepare the crisis team to cope with and manage such situations?

Preparing teams cannot mean trying to expand the range of known-beforehand situations, since crisis situations are by definition unpredictable. It means enhancing deep reasoning abilities and developing appropriate group work competences. This can be done through simulation and debative debriefings.

The aim of this communication is to present the design of a reflexive training device for the NTST, tested in the nuclear industry.

Results show that it has allowed participants to debate some teamwork dimensions, such as the proposal process, its complexity, the roles and missions of each NTST member, and to identify new ways of functioning for others situations.

2   The Support Team Under Study: The NTST

Within the organization under study, different operational states of the facility are anticipated.

When a severe situation occurs, an in-house Emergency Plan is set off, leading to the mobilization of a multi-level crisis organization (Fig. 1):

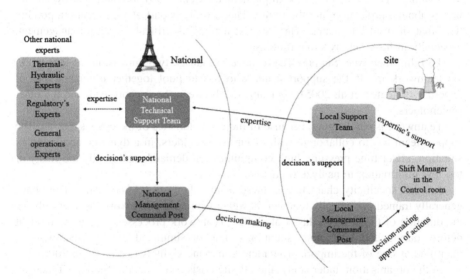

Fig. 1. The crisis management organisation under study

- Locally: Local Management Command Post, Local Support Team, and Shift Manager – in the control room.
- Nationwide: National Technical Support Team, National Management Command Post, DT (Technical Department) – a utility's team made of thermal-hydraulic experts, Framatome Crisis Team – general operations experts, Institute for Radiological and Nuclear Safety (IRSN) Crisis Team – regulator's experts.).

This study focuses on the National Technical Support Team (NTST). The NTST comprises twelve general operations and environmental high-level specialists. The NTST is not actually managing the plant. Its purpose is to make its expertise available to the Site by proposing a diagnosis and a forecast of the development of the situation. It also provides the Site with information, solutions, and proposals for action (through the Local Support Team or Shift Manager). The NTST develops a global picture through its technical skills that complement those of the Local Support Team. The NTST is an essential resource for the Site to anticipate problems to come and tasks to be undertaken. The proposals to the Site may concern changes to procedures, the implementation of preventive or mitigating actions. In this crisis organization, the NTST prognosis has an additional objective: to provide the National Management Command Post with population protection measures, while confronting their diagnosis/prognosis to the other national expert entities (DT, Framatome, IRSN).

Communications between the Local Support Team and the NTST are made through periodic messages received from the Site every 15 min describing the status of the facility and through audio-conferences with all expert entities held every 90 min approximately.

The NTST may contribute to the global resilience of the crisis organization by reevaluating goals and the course of action. The NTST's composition is stable, but teammates do not work together on a daily basis. Crisis exercises allow the testing of the crisis organization functioning during large scale simulations. They can also improve teamwork during simulation session.

3 Methodology

3.1 Reflexive Training Device Conception

Before the conception process, an analysis of the activity of the six NTST teams has been completed in order to understand their specificity and difficulties. 9 crisis simulation were observed and 15 interviews were conducted with NTST team members. Thanks to this data collection, a diagnosis of the current training methods has been made and shared with the NTST team members. This diagnosis emphasized specific needs to improve the NTST proposal activity and team debriefing after simulation scssion.

The Reflexive training device was conceived by a working group including Human Factors researchers, simulator instructors and NTST team members.

The device aims at developing the following collective skills (1) the efficiency of its proposals design process destined to the Site, its temporality, and deliberation about the alternatives; (2) the teamwork inside the NTST and with the simulated stakeholders (e.g. the site, the national command post).

To achieve these formative goals, the simulation technical device and the post-simulation activities were designed simultaneously. The Reflexive training device tested is divided into three parties: (1) Simulation, (2) Debriefing 1 and (3) Debriefing 2.

For instance, a simplified technical device was designed in order to focus on the NTST, its work environment and the requests of the other stakeholders. As Fig. 2 shows, requests from other stakeholders were simulated thanks to a serious game software.

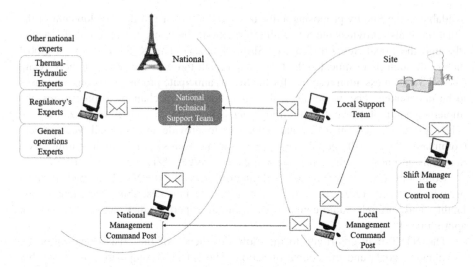

Fig. 2. Technical device used for the experimentation

Debriefings sessions were collective and focused on team functioning. Two debriefing sessions were proposed:

- Debriefing 1, which took place right after the simulation. Following the idea of self-debriefing, teammates were asked to identify their functioning modes regarding the simulation;
- Debriefing 2, which was a deeper debriefing session was conducted the same day, starting from a theme chosen by the team, and another theme chosen by the teammates and Human Factors observers.

Debriefings were conducted by a peer, who was trained to debriefing methods before the experimentation. On the experimentation day, the peer observed the simulation and then conducted debriefing sessions. He was asked to conduct debriefing with a non-normative point of view, and promote debate between experts and cross-learning.

3.2 Data Collection and Analysis

This reflexive training device and method was tested with one NTST team. The simulation lasted 3 h, and was audio and video recorded.

Debriefing 1 session lasted 1 h, and Debriefing 2 session, 2 h and 40 min. Both sessions were recorded and fully transcribed.

The analysis focus on the debriefing sessions in order to emphasize what has been discussed and lessons learned for teamwork. A thematic analysis of the debriefing sessions was developed targeting the teamwork points discussed.

4   Results

The results presented here focus on Debriefing 1. The instructions for this debriefing requested that half of the time should be spent on teamwork aspects. These teamwork aspects address different forms of synchronization: cognitive and operative, inside NTST and between NTST and the other stakeholders.

First, cognitive and operative synchronizations inside the NTST was raised throughout an analysis by teammates of good ways of functioning during the simulation. According to the teammates three team processes seem to reinforce operative and cognitive synchronizations. The three team processes judged to be positive are as follows:

- The centralization of information by the General Operation – a member of the NTST;
- The sharing and posting of proposals by the General Operation (same member has referred before);
- Synchronization moments when the whole team is gathered for the audio conferences.

These three team processes are not usual practices for the team. They appear to be positive and supported a reliable way to function. In fact, team members indicated during Debriefing 1 that these processes were favorable to team performance and needed to be perpetuated. They allow a better synchronization between the team members. Thanks to these team processes, team members shared a same picture of the situation and were able to provide better support to the Local Support Team by proposing a reprioritization of proposal during the simulation. This particular item is pointed out as a recurring difficulty.

Secondly, cognitive and operative synchronizations inside NTST issue were discussed through an elicitation and sharing of specific constraints of some team members. This elicitation permit to underline risk of desynchronization inside the NTST. The constraints and specific needs of some team members were shared during the debriefing. This sharing of constraints made it possible to emphasize issues of cognitive synchronization within the NTST, highlighting different temporalities and different forecasts to be articulated inside the NTST. Teammates were aware of the need to better understand the specificities of each of them to integrate these into the collective functioning. Awareness of these elements – e.g. specific constraints, risk of desynchronization, interdependence - can enhance team performance.

Finally, cognitive synchronization between NTST and other stakeholders was debated. This debate concerned the proposal process. In fact, team members did not agree on the way of involving stakeholders such as the National Crisis Director from the National Management Command Post (cf.2) in validating the proposal. This debate was not settled during this debriefing.

A thematic analysis of the second debriefing is in progress. The objective is to compare the themes that were discussed in the two debriefings, and the relationships between the two debriefings.

5 Discussion

The reflexive training device allow NTST team members to focus on teamwork dimensions during the simulation and during the two debriefing sessions. Team members' feedbacks about this experimentation is positive. During this experimentation, they were able to co-elaborate and produce four proposals to the Local Support Team. Considering the dynamics of a typical crisis situation that could occur, this work can be considered as efficient. The debriefing sessions helped to highlight positive teamwork dimensions and to look for ways to perpetuate them.

The debriefings also made it possible to clarify and make explicit the needs, the contributions and the constraints of each team member. The first debriefing session highlighted the risks of desynchronization and of lack of mutual awareness, which can affect team performance during crisis management.

The discussion of these teamwork dimensions is a first step towards building a common reference frame for an ephemeral team such as the NTST. A common reference frame is related to team rules, i.e. shared ways of functioning established before the situation. The common reference frame is made of meta-cooperative skills that guide teamwork and collaboration within the team, but also between the team and other teams. These skills can contribute to safe way of functioning by supporting a mutual awareness during the situation and a team awareness – which can be described as a continuous attention to the state of the team, in order to identify reconfigurations to operate in a situation to maintain safe operations.

The next steps are to compare the content of the two debriefings, identify the topics discussed and the new rules built by the team.

References

Artman H, Waern Y (1999) Distributed cognition in an emergency co-ordination center. Cogn Technol Work 1:237–246

Baudard Q, Le Bot P (2017) Modelling human operations during a nuclear accident: the Fukushima Daiichi accident in light of the MONACAS method. In: Cepin M, Bris R (eds) Safety and reliability—theory and applications. Taylor & Francis Group, London

Crichton MT, Lauche K, Flin R (2005) Incident command skills in the management of an oil industry drilling incident: a case study. J Conting Crisis Manag 13(3):116–128

Guarnieri F, Travadel S, Martin C, Portelli A, Afrouss A (2014) L'accident de Fukushima Dai Ichi, le récit du directeur de la centrale, vol 1. L'anéantissement. Libres opinions Presses des Mines, Paris

Hardy K, Comfort L (2015) Dynamic decision processes in complex, high-risk operations: the Yarnell Hill Fire, June 20, 2013. Saf Sci 71:39–47

Klein G (1999) Source of power. How people make decisions. The MIT Press, Boston

National Diet of Japan (2012) The official report of the Fukushima Nuclear Accident Independent Investigation Commission. http://www.nirs.org/fukushima/naiic_report.pdf

Perry R (1995) The structure and function of community emergency operations centres. Disaster Prev Manag 4(5):37–41

Perry R (2003) Emergency operations centres in Era of terrorism: policy and management functions. J Conting Crisis Manag 11(4):151–159

Salas E, Cannon-Bowers J, Weaver J (2002) Command and control teams: principles for training and assessment. In: Flin R, Arbuthnot K (eds) Incident command: tales from the hot seat. Ashgate, Aldershot

Weick K (1993) The collapse of sensemaking in organizations: the Mann Gulch disaster. Adm Sci Q 38(4):628–662

Wybo J-L, Kowalski K (1998) Command centers and emergency management support. Saf Sci 30:131–138

Innovations in Crowd Management: An Integration of Visual Closure, Anthropometry, and Computer Vision

J. W. Chin[1,2], T. W. Wong[1,2], and R. H. Y. So[1,2(✉)]

[1] HKUST-Shenzhen Research Institute, 9 Yuexing First Road, South Area,
Hi-tech Park, Nanshan, Shenzhen 518057, China
[2] Department of Industrial Engineering and Decision Analytics,
Hong Kong University of Science and Technology, Hong Kong, China
rhyso@ust.hk

Abstract. This paper reports the application of a bio-inspired computational artificial intelligent (A.I.) real-time crowd monitoring and management system that integrates ergonomics, anthropometric database, computer vision and decision analytics. The system matches and fits anthropometrically customized 3D human models into a 3D space that is dynamically constructed from videos captured by one or more surveillance cameras. This approach is consistent with the human visual closure effect when we estimate the number of people in moving crowds. Dynamic human movement data are optimally extracted from the video data and used to construct and train a crowd movement profile detector. Learning algorithms have been developed to detect deviations from the normal profile. Results of validations show that there remains a huge gap in the performance between a bio-inspired computational A.I. model and a normal human-being in the surveillance tasks in terms of reliability, but this is a notable first step of a reliable crowd management system not emphasizing on facial feature extraction.

Keywords: Visual closure · Anthropometry · Computer vision

1 Introduction

Anthropometry is the science of applying human dimension constraints analytically in the design of products and workplace (Pheasant and Haslegrave [1]). Although it has been integrated into 3D computer design programs like CATIA (e.g., Kayis and Iskander [2]), most of the applications had been restricted to human-in-the-loop processes. This is understandable as anthropometry is essential a design tool. With the advances in artificial intelligence algorithms, human-in-the-loop systems can be replaced by automated systems.

Automatic video analyses have been the subject of many studies. Feature extractions of human like figure have been used in many applications. Thanks to the advances in computer hardware, most of these applications can be conducted in real-time. In crowd management applications, however, feature extractions of human like figures can be expensive in computational cost and may not be applicable when each human

© Springer Nature Switzerland AG 2019
S. Bagnara et al. (Eds.): IEA 2018, AISC 822, pp. 68–74, 2019.
https://doi.org/10.1007/978-3-319-96077-7_8

figure consists only of a few pixels and with poor contrast. This is not uncommon when the video is captured by a normal surveillance camera with a wide field-of-view and under poor lighting. Our design and approach apply feature extraction not to individual human figure but to the static background information. In order words, each scene is analyzed to extract foreground that is moving and the background that remains static (Figs. 1 and 2). Once the areas are identified, the area will be filled with the optimal number of 3D anthropometric models scaled to the correct perspective (Fig. 3). This process is repeated for each frame. The approach is consistent with the visual closure effect that is associated with human intelligence (Thurstone and Thurstone [3], Botzum [4]). Results of validation studies show that the 3D anthropometric models moved in a smooth and natural way. This is remarkable as each frame was processed independently. This suggests that the procedure is robust even if there is no specific correlation processing across adjacent frames. Correlation analyses are then performed to match each model in adjacent frames so that their movement profiles can be extracted. Using these normal data as baseline, crowd or individual human movement deviates from the baseline can be detected.

Fig. 1. A snapshot of a video scene.

As the system separate the "moving" foreground and the "static" background, this system can easily detect an "unusual change" in the background such as an unattended luggage or fixture that "moves" into the scene and remain static for a period of time. Such data can potentially be developed into an anti-terrorist system for unmanned bomb detection. During the presentation, a demonstration of the system will be shown (Fig. 4).

Fig. 2. Areas (non-darken) dynamically estimated to contain changing visual content using artificial intelligence algorithm.

Fig. 3. Areas in red are the "Moving" foreground. Inspired by the visual closure effect that is associated with human intelligence, the artificial intelligence algorithm can estimate the number of the persons and their locations in the scene even without analyzing the features of humans. The green bounding boxes are inserted into the frame with correct perspective size whenever the A.I. "thinks" there should be a human (Color figure online).

Current Frame: 451, No. of Person: 11

Fig. 4. The processed frame: 3D anthropometric boundary boxes in corrected perspective sizes are projected into each video frame.

2 Human Model

The human model used in this project is based on the anthropometric estimated data by Pheasant [5]. In particular, Stature (1), Shoulder height (3), Hip height (5), Shoulder breadth (bideltoid) (17), Hip breadth (19), Head breadth (27) as shown in Fig. 5 were used to generate the human model. It can divided into three major parts (i.e. head and neck, body, and legs.). At a given height of a human in terms of pixels, the height and width of each part in terms of pixels can be calculated.

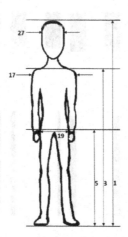

Fig. 5. Anthropometric data that is being used to generate human model (modified from [5])

The size of the head and neck (height and width) can be calculated as follows:

$$H^{pix}_{head\,and\,neck} = L^{pix}_{(1)} * \frac{L_{(1)-(3)}}{L_{(1)}} \qquad (1)$$

$$W^{pix}_{head\,and\,neck} = L^{pix}_{(17)} * \frac{L_{(27)}}{L_{(17)}} \qquad (2)$$

$H^{pix}_{head\,and\,neck}$ and $W^{pix}_{head\,and\,neck}$ denote the height and weight of the head and neck in terms of pixels respectively, $L_{(1)-(3)}$ denotes the length of head and neck in millimeter, which is the difference between the length of Stature (1) and the length of Shoulder height (3), obtained from [5].

The size of the body (height and width) can be calculated as follows:

$$H^{pix}_{body} = L^{pix}_{(1)} * \frac{L_{(3)-(5)}}{L_{(1)}} \qquad (3)$$

$$W^{pix}_{body} = L^{pix}_{(17)} \qquad (4)$$

Here we assumed the body width includes both arms, therefore the width of the body is equal to the Shoulder breadth (bideltoid).

Similarly, the size of the legs can be calculated as follows:

$$H^{pix}_{legs} = L^{pix}_{(1)} * \frac{L_{(5)}}{L_{(1)}} \qquad (5)$$

$$W^{pix}_{legs} = L^{pix}_{(17)} * \frac{L_{(19)}}{L_{(17)}} \qquad (6)$$

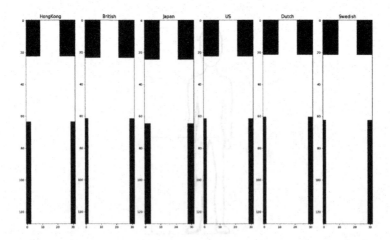

Fig. 6. Human model created using Pheasant's anthropometric data of human from different countries

Given same height of the human model in terms of pixels (i.e. 128 pixels), Fig. 6 shows the different human models generated using anthropometric data of humans from different countries.

3 Results

As shown in the Fig. 7, the accuracy of the location of every person is within 2 m and the accuracy of the location of most of the human models can be within 1 m, which is acceptable for crowd management in public areas. On the other hand, the accuracy of the estimated number of persons can be improved but the result is close to the actual number of persons in the scene (i.e. 14 persons). There are three interesting regions in the processed frame can be discussed.

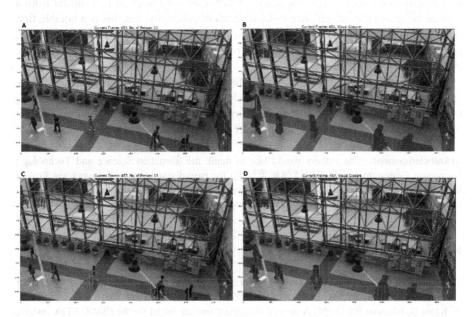

Fig. 7. **A** Processed frame 451, which is essentially Fig. 3. **B** How the A.I. "see" and "guess" the number of persons and location of them given the "red pixels" which are the "Moving" foreground of frame 451. **C** Processed frame 457. **D** "Moving" foreground of frame 457 (Color figure online).

First region is at the upper left corner of the frame in Fig. 7A. The A.I. failed to guess the correct number of persons due to heavy occlusion. However, when the humans are less being occluded by the green metal frame after few frames, it can successfully estimate the number of them.

Second region is at the middle of the frame in both Fig. 7A and C. The A.I. failed to recognize the human sitting at the bench. Interestingly, the human has less movements and he or she is being classified as the "static" background. New algorithm or engine can be developed to detect stationary humans in the scene to improve the reliability of the A.I. system.

Third region is at the bottom middle of the frame in Fig. 7C. The A.I. misjudges the shadow of the female as a person, and then two persons are being fitted into the region. Algorithms to remove the movements caused by the shadows can be implemented into the A.I. to reduce the false alarm.

4   Conclusion

This paper shows the proof of concept of a bio-inspired computational A.I. model, which combines anthropometric database, computer vision, decision analytics and artificial intelligence. It is consistent with the visual closure effect that is associated with human intelligence. From the results, it is noticeable that the A.I. is still far from a normal human in the surveillance tasks in terms of reliability, but this is a notable first step of the bio-inspired A.I. to grow into a much reliable and complex system. This approach is consistent with the approach of computational ergonomics (So and Lor [6]).

Human utilizes motion information, features from both foreground and background, and memory related to seen objects in the past to interpret the scene accurately. In the future, feature-based algorithm and memory module can be introduced into the A.I. system as part of the modules to better mimic human in the surveillance tasks.

Acknowledgement. The authors would like to thank the Shenzhen Science and Technology Innovation Committee (深圳市科技创新委员会) for partially supporting the work via Project JCYJ20170413173515472 (SZ-SZSTI1731). The study is also partially supported by the Innovation and Technology Commission via ITF Project ITS/170/15FP.

References

1. Pheasant S, Haslegrave CM (2005) Bodyspace: anthropometru, ergonomics and the design of work. CRC Press, 352 pp. ISBN 9780415285209
2. Kayis B, Iskander PA (1994) A three-dimensional human model for the IBM/CATIA system. Appl Ergon 25(6):395–397
3. Thurstone LL, Thurstone TG (1941) Factorial studies of intelligence. Psychrometric Monogr 2:94
4. Botzum WA (1951) A factorial study of the reasoning and closure factors. Psychometrika 16 (4):361–362
5. Pheasant S (2014) Bodyspace: anthropometry, ergonomics and the design of work: anthropometry, ergonomics and the design of work. CRC Press, Boca Raton
6. So RHY, Lor F (2004) Computational ergonomics? A possible extension of computational neuroscience? Definition, potential benefits, and a case study on cybersickness. In: McCabe PT (ed) Contemporary ergonomics 2004. CRC Press, Boca Raton, pp 405–409

A Unique Human Model to Simulate Various Types of Seat (Dis)comfort

Muriel Beaugonin[1], Zakariae Adel[2], and Caroline Borot[2(✉)]

[1] ESI Group, PARC d'Affaires SILIC,
99 rue des Solets, BP 80112, 94513 Rungis Cedex, France
[2] 5, Parc du Golf, 350 Rue Jean René Guillibert Gauthier de la Lauzière,
13090 Aix-en-Provence, France
Caroline.Borot@esi-group.com

Abstract. Human models combined with virtual prototypes is increasingly used by seat and interior engineers in both automotive and aeronautic industries. Testing virtually a seat design enables to anticipate potential discomfort issues, to compare seat variants through repeatable test procedures and this for a wide range of seat occupants' anthropometries. As comfort feeling is related to lots of factors, human models used for such virtual seat test must cover not only static comfort, through posture and pressure mapping evaluations, but also body vibration transmission related to external perturbations, and human thermal comfort sensation.

This paper will describe the different steps of development of these finite elements human models, from the initial collaboration with Hong-Ik University to select the targeted human anthropometries until the validation of their capabilities to predict the static (dis)comfort while seating in a virtual seat prototype. It will then detail the work done to capture also the physical phenomena related to the dynamic comfort, when a seat is submitted to some vibrations. This paper will also present the latest developments to include the prediction of thermal comfort.

Then, typical applications using those human models in interaction with virtual seat prototypes will be shown and will highlight the interest of a unique human model for prediction in diverse discomfort evaluations, such as static, dynamic and thermal comfort.

Finally, the last part will demonstrate the influence of human anthropometries for the static and dynamic comfort evaluation.

Keywords: Human model · Anthropometry · Static comfort
Dynamic comfort · Thermal comfort

1 Introduction

Within both automotive and aeronautic industries, seat engineers need to predict impact of design changes on seat (dis)comfort early in their conception process. Human models combined with virtual prototyping solutions allow to evaluate those seat designs before any real prototype is built, and with more repeatability than with volunteer tests on real seats. Indeed, first, as time goes by, volunteers anthropometry can

© Springer Nature Switzerland AG 2019
S. Bagnara et al. (Eds.): IEA 2018, AISC 822, pp. 75–85, 2019.
https://doi.org/10.1007/978-3-319-96077-7_9

evolve. Then, comfort is a subjective perception which can be affected by today's mood of volunteers and the seat fashion. Human models are a numerical solution to overcome those limitations.

2   Human Single Core Model for Various Seat Comfort Fields

The local level of compression of the seat due to the human seating has a strong influence on the dynamic performance of the occupied seat. Similarly, the interaction between the seat and its occupant has a primary impact on thermal performance. As a consequence, it is extremely important to have human models able to cover all aspects of comfort, i.e. static, dynamic and thermal.

2.1   Preliminary ESI Human Model for Seat Design Applications

Starting from 2003, ESI has initiated the development of human models dedicated to comfort prediction through its partnership with Hongik University. Using ESI's finite element (FE) solver capabilities, Pr. Hyung Yun Choi has developed a human model for static and riding comfort applications [1]. This human model anthropometry corresponds to a 50th percentile American male size, according to the UMTRI (University of Michigan Transportation Research Institute) [2] and AMVO (Anthropometry of Motor Vehicle Occupants) studies of the 1971–1974 HANES (Healthy And Nutrition Examination Survey) database. This model, initially developed for safety application, has been adapted to comfort requirements by comparison with experimental measurements on human volunteers for seating and dynamic load cases [3]. An additional work has been performed to further validate the comfort predictive capabilities of this human model for various postures [4].

2.2   New Generation of ESI Comfort Human Models

In order to be relevant, human models dedicated to comfort prediction need to be based on recent anthropometric information. Indeed, the overall size and shape of the population are changing continuously. For example, people are taller and larger than 30 years ago. Since the first comfort human model was based on an early 80's population, it was decided to develop a new generation of human models [5], in partnership with HYUNDAI Motor Company, based on a more recent anthropometric database.

Data Collection

Anthropometries Choice. To develop human model representatives of contemporary seat users, their anthropometry has to be identified. First, different percentiles anthropometries have been defined from a recent US national size survey, SizeUSA [6], which has collected dimensions of 10,800 volunteers, 18–65 years old from 2000 to 2003. Based on Caucasians between 36 and 45 years old, three percentiles, in terms of standing height and weight, have been identified: a 5%-tile female (called 'small female'), a 50%-tile and 95%-tile male (called 'mid-male' and 'large male' respectively) subjects. An equivalent work has been performed to identify anthropometry of small female, mid and large male representatives for Korean population, made with Korean national size survey, performed by Ergonomics Society of Korea, between 2003 and 2004. Table 1 provides obtained American and Korean percentiles definition.

Table 1. American percentiles definition

American percentile	5%tile Female	50%tile Male	95%tile Male
Height (cm)	153.7	177.8	190.5
Weight (kg)	49.9	85.7	117.5
Korean percentile	5%tile Female	50%tile Male	95%tile Male
Height (cm)	149	170.1	179.2
Weight (kg)	46.2	71.3	87.9

Volunteers Recruitment. Previously identified American and Korean percentiles anthropometries were used as target sizes to select volunteers for body shape acquisition. A first selection based on two primary parameters, height and weight, including possible standard deviation (one quarter of and half of standard deviation, for female and mid-male, and for large male, respectively) has been performed. In a second phase, 4 last primary dimensions (hip height, hip girth, bust, and back waist length) have defined the minimum and maximum of these body segments sizes. Selected subjects had to observe these limits.

3D Scanning Acquisition and CAD Modelling. The external geometry of selected volunteers in a driving posture was measured by a whole body 3D laser scanner (Model: Cyberware WB4). Undeformed shape of the buttocks and thighs were separately scanned from the volunteers propping on elbows while maintaining the knee angles the same as in a driving posture. The synthesis between the patches of the undeformed parts and the whole body scan image was carried out by using a three-dimensional scan modeling software, RapidForm.

Skeleton Location. The precise location of the skeleton to the skin surface in the model is important for an accurate prediction of the sitting pressure distributions in the finite element analysis. The 3D shape of the bony structures was obtained from Viewpoint CAD database and was scaled using X-ray images giving the distance between skin and bones of the trunk, while hip and thigh were measured by ultrasonic scanning. Joint locations were calculated by simple linear regression equations using surface landmarks.

Principles of Comfort-Specific Human Model Development

Bodies and Joints Modelling. The skeleton is modeled by a chain between rigid structures representing the main bones linked by kinematic joints. In particular, joints have been used to model the intervertebral disks of the cervical and lumbar parts of the spine, as well as all articulations of upper and lower limbs. The lumbar spine complex including the separated vertebrae has been calibrated to represent multi-directional motions, including springs and dampers derived from anatomical data.

Soft Tissues Modelling. For the head, the neck as well as for the arms, the lower legs and the feet, the soft tissues are not meshed. For these body segment, mass and inertia properties of body segments were obtained from GEBOD (GEnerator of BODY Data) with the user-supplied body dimensions option.

The skin is described with fabric membrane element with nonlinear fibers.

The abdominal organs are represented by an air-bag model with a membrane elements mesh, coarse at inner region but which becomes finer close to surface. The abdominal bag is then tied to the diaphragm but is free to slide on neighboring surfaces including the spine.

Trunk back, abdomen, pelvis and thigh regions are considered as deformable parts, meshed with tetraedric solid elements. Because these regions are in contact with the seat to support the body weight in a seating position, the mesh quality of those parts is closely related to the prediction of seating pressure distribution. Soft tissues such as flesh are generally considered to be incompressible materials due to their high proportion of water. They are considered to have a Poisson ratio nearly equal to 0.5. A distinction has been done between fat and muscle volumes in terms of material characterization, both modeled by an elastic Ogden model. Due to a paucity of material data for human tissues, material parameters have been obtained by parametric study based on previous study data [7].

New Generation of Human Models. Once a surface mesh and related skeleton-joints complex are created for a given individual, the above methodology is applied to develop this new generation of human models, for American and Korean percentiles, see Fig. 1.

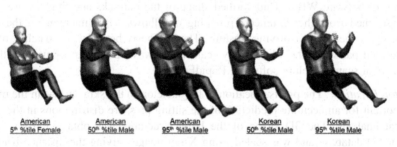

| American 5th %tile Female | American 50th %tile Male | American 95th %tile Male | Korean 50th %tile Male | Korean 95th %tile Male |

Fig. 1. New generation of ESI human models for comfort prediction

Validation in Seat Static and Dynamic Applications. Each developed human model has been validated against the measurement of a body pressure distribution on polyurethane foam block from a test performed with its corresponding volunteer. In order to achieve this, "Support Balance Diagram" of experiment and simulation were compared. Figure 2 shows that the patterns of support balance diagram for simulation and experiment are similar. Those validations confirmed that the FE models can predict seat pressure distribution patterns. Validation has also been performed for other seat pressure quantities, such as contact area, contact force ratio and sectional force ratio.

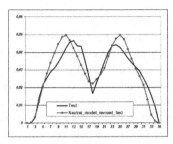 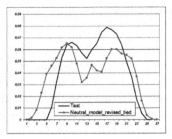

Fig. 2. Support balance diagram of seat cushion (left) and backrest (right) pressure distribution (Red lines; simulation results, Black lines: experiment results) (Color figure online)

For riding comfort requirements, the behavior of these new human models has been evaluated by comparison with experimental measurements on human volunteers, in dynamic load case during joint study with BMW [8]. New generation of comfort-specific human models provides equivalent results in terms of seat transfer function (STF) as the ones obtained during this study and compared to experimental measurements performed with volunteers at BMW group laboratories (see Fig. 3).

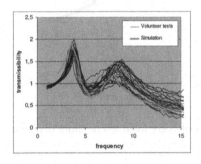

Fig. 3. Occupied seat transfer function: for volunteers and simulation

This STF exhibits a first peak around 4 Hz and a secondary peak around 8–10 Hz, which are characteristic of the interaction of the seat with a human occupant. Both peaks are confirmed by physical testing, and the STF predicted by numerical simulation compares well with the experimental STF measured on various human volunteers.

Adaption to Implicit Solving Schemes. For vibrations analysis, implicit simulation had an interest by reducing greatly the calculation CPU-time. Those human models were therefore adapted to be compliant for both explicit and implicit solving schemes with ESI FE solver. It was checked on the validation cases previously mentioned that the implicit scheme was increasing the accuracy for vibrations simulation.

Upgrade to Address Thermal Analysis. More recently, the modelling of the body thermal exchanges – internally and with its environment - has been added through a solving scheme similar to the one described by Fiala [9].

This model contains two main systems:

- A thermal passive system which models first the heat transfer inside the body through conduction and blood circulation. It takes also into account the heat generation coming from the human basal metabolism. Finally, it includes also the external heat exchanges through sweat evaporation, breath, convection with the air, conduction by contact with the seat and radiation. It takes into account clothing.
- A thermal active system. An active system representing the thermoregulation system of the human body (vasodilatation, vasoconstriction, shivering and sweating) is added to the human models. This thermoregulation is per nature transient and requires thus the direct coupling of the human model and the simulation of its environment: seat and air.

The thermal behavior of the digital human models has been validated using data from independent experiments in literature, as shown in Fig. 4.

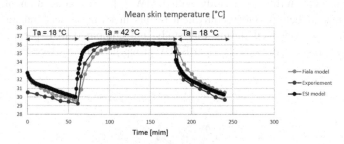

Fig. 4. Example of comparison between thermal behavior of ESI's human models, Fiala model and experimental tests and experimental results

Beyond the prediction of body local temperature distributions, this model converts those temperatures into local and global sensation (warm/cold) index, as well as local and global comfort index.

3   Performed Seat (Dis)comfort Prediction Based on a Human Single Core Model

To illustrate the interest of having a unique human model for prediction in diverse discomfort evaluations, such as static, dynamic and thermal comfort, the following scenario is used:

- Step 1: Sitting application – The Korean 50th percentile male model is seated to a seat with a frame considered as rigid.
- Step 2: Dynamic application - From seated HM50KR, vibrations have been imposed to the floor, while the seat frame is switched to a deformable model. Seat transmissibility function is obtained.
- Step 3: Thermal application - From seated HM50KR, thermal analysis is performed, with passive and active heating pads. The seat frame is kept as rigid part.

3.1   (Dis)comfort Prediction in Sitting Load Case

The selected human model is positioned geometrically close to the seat without intersection during the model setup, and is then seated during the solver computation by gravity with all the flesh and seat deformation occurring, see Fig. 5.

Fig. 5. Initial state (left) and Final static posture of seated HM50KR (right)

Several measurements are available to analyze the obtained static sitting posture. Amongst them, there are the H-Point location (not far from being the mid-point between two revolute joints representing the hips articulations), as well as the thighs angles, and the thoracic and lumbar spine angles, see Fig. 5.

Other quantities, such as proximity of the skin to the seat metallic frame, and flesh volume stress can be studied. But seating pressure distribution is one of the main factors for static comfort. It can be analyzed through sitting comfort criteria, which relate the objective pressure measurement to the subjective comfort feeling, such as:

- Mergl [10] criterion: The pressure map is divided into several areas and for each of them, four main parameters are extracted and related to the perceived discomfort: percentage of load on body part, gradient, maximum pressure and mean pressure.
- Zenk [11] criterion: A similar methodology is applied here with different areas and it is established that the maximum of comfort is provided in the ideal seat position as defined by an optimal load distribution.

Figure 6 shows an example of pressure distribution on cushion map, analyzed with the Mergl criterion. The maximum pressure zones are located at the level of HM50KR pelvis ischia. The analysis table indicates measures within and outside the comfort range, as well as its level of correlation with final comfort feeling.

3.2   (Dis)comfort Prediction in Dynamic Load Case

The evaluation of dynamic comfort consists in making the ratio between the applied motion on the seat rails and the final motion of the occupant at different locations. If the ratio is >1, the seat increases the vibrations; if the ratio is <1, the seat absorbs the vibrations.

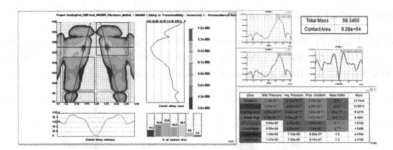

Fig. 6. Cushion map Pressure distribution with Mergl criterion (violet color: within/ outside: red color, from * weak to *** strong correlation) (Color figure online)

Once the human model is seated, the final state is used as starting state of dynamic load case. Vibrations are applied to the floor, while the seat frame is considered as deformable. The seat transfer function obtained for HM50KR is shown in Fig. 7. While the first peak located at frequency between 2 and 4 Hz, corresponds to the seat resonance and human first vibrations mode, the second one, between 6 and 12 Hz, is characteristic of the seat-occupant interaction.

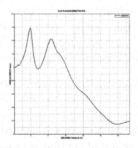

Fig. 7. Seat transfer function for HM50KR

3.3 (Dis)comfort Prediction in Thermal Load Case

The evaluation of thermal comfort helps dimensioning the thermal equipments such as heating pads in order to minimize their energy consumption, particularly in electric vehicles. Once the contact between seat and occupant is established through the seating simulation, this heating pad can be activated/deactivated according to a thermostat rule (see Fig. 8). The thermal properties of the seat, the convection with air and the heat transfer to the occupant through contact are taken into account. Different scenarios can be easily investigated by simulation with different heating pad patterns and different thermostat rules until finding the one which maximizes the comfort score.

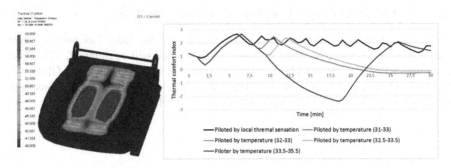

Fig. 8. Temperatures on seat and occupant comfort scores

4   Diverse Anthropometries and Population Percentiles

As indicated previously, human models, representatives of different population percentiles, have been developed. But in addition, scaling capabilities applicable to these human models have been developed to adapt them to other anthropometries. For illustration purpose, previous static and dynamic comfort load cases are performed with a small female model, 1.54 m and 50 kg, called 'HF05AM', while HM50KR represents an anthropometry of 1.70 m and 71 kg.

4.1   Effect of Anthropometry in Static Comfort Prediction

The effect of occupant anthropometry can be clearly seen on the cushion pressure map (see in Fig. 9). While HF05AM weight is smaller than the one of HM50KR, maximum peak pressure, located at pelvis ischia zone, is higher. This can be explained by different seated postures between HM50KR and HF05AM which has shorter legs and, thus, presents a higher pressure along thighs. Contrary to the Korean mid-male model, the American small female model doesn't respect the threshold of criteria concerning mass ratio at middle thigh zones, which is one of the comfort criteria strongly linked to comfort feeling.

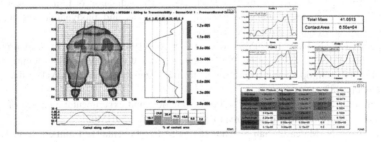

Fig. 9. Cushion pressure map for HF05AM

4.2 Effect of Anthropometry in Dynamic Comfort Prediction

When the weights of the two occupants are different, the foam has not the same compression level, and therefore the resonance of the seat will be different. This effect is highlighted on Fig. 10 of STF for vibratory load case with HM50KR and HF05AM. As expected, dynamic simulation of HF05AM, with a smaller weight, presents both displacement peak at higher frequencies than those of HM50KR run.

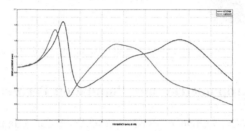

Fig. 10. Seat transmissibility function: HF05AM (in blue) versus HM50KR (in red) (Color figure online)

5   Conclusion

Thanks to ESI's Virtual Seat Solution human models library, it is possible to evaluate seat performance, precisely and accurately reflecting the interaction between seat and passengers for automotive and aeronautic seats. A unique single core model for the occupant as well as for the seat facilitates their usage for diverse (dis)comfort prediction, such as static, dynamic and thermal domains. Scaling capabilities enables to adapt available human models to specific anthropometries. Post-processing functionalities provides user-friendly ways to analyze the comfort performance.

References

1. Sah S, Hwang D, Choi H, Kim Y, Lee H (2003) The validation of finite element human model for occupant-seat interface pressure distributions and biodynamic characteristics under vertical vibration. SAE paper 20035431
2. Schneider LW, Robbins DH, Pflüg MA, Snyder RG (1983) Development of anthropometrically based design specifications for an advanced adult anthropomorphic dummy family, vols 1 and 2: anthropometric specifications for mid-sized male dummy, UMTRI final report, December 1983
3. Montmayeur N, Haug E, Marca C (2004) Numerical and experimental analyses of seating and riding comfort. In: Digital human modelling 2004, 04DHM03
4. Montmayeur N, Marca C, Sah S, Choi H (2005) Experimental and numerical analyses of seating pressure distribution patterns. In: Digital human modelling 2005, 05DHM16
5. Kim S-H, Hwang S-H, Lee K-N, Pyun J-K, Choi H, Kim K, Sah S, Montmayeur N (2007) New anthropometry of human body models for riding comfort simulation. In: Digital human modelling 2007, 07DHM26

6. http://www.tc2.com/what/sizeusa/index.html
7. Oomens CWJ, Bressers OFJT, Bosbomm EMH, Bouten C, Blader DL (2003) Can loaded interface characteristics influence strain distributions in muscle adjacent to bony prominences? Comput Methods Biomech Biomed Eng 6(3):80–171
8. Amann C, Huschenbeth A, Zenl R, Montmayeur N, Marca C, Michel C (2008) Virtual assessment of occupied seat vehicle vibration transmissibility. In: Digital human modelling 2008, 08DHM-12
9. Adel Z, Marca C, Borot C (2017) Development of a human thermal model to predict seat occupant thermal comfort. In: 1st international comfort congress, Salerno
10. Mergl C, Klendauer M, Mangen C, Bubb H (2005) Predicting long term riding comfort in cars by contact forces between human and seat. SAE 2005-01-2690
11. Zenk R, Franz M, Bubb H (2007) Spine load in the context of automotive seating. SAE 2007-01-2485

Simulation and Analysis of Human Neck Load and Injury During Sustaining Overload in Flight

Bao Jiayi[1,2], Zhou Qianxiang[1,2(✉)], Wang Xingwei[3], and Liu Zhongqi[1,2]

[1] School of Biological Science and Medical Engineering,
Beihang University, Beijing, China
zqxg@buaa.edu.cn
[2] Beijing Advanced Innovation Centre for Biomedical Engineering,
Beihang University, Beijing, China
[3] Air Force Aviation Medical Research Institute of China, Beijing, China

Abstract. Sustaining overload is the main factor of the pilots' neck injury during the arrested landing of carrier-based aircraft. Therefore, the finite element model simulation calculation method is used to analyze the biomechanical response of pilots' head and neck in the process of the carrier-based aircraft's arrested landing in this paper. We have established a finite element model of human head and neck with high biological fidelity, and verified the effectiveness using both static and dynamic methods. After that, the model is used to simulate the arrested landing process. Finally, the injury situation of the pilot's head and neck were analyzed and predicted based on the internationally recognized damage criteria. The results of this study may provide important theoretical basis and references for the design of the carrier-based aircraft landing training task as well as the design and improvement of the neck safeguard of the pilots.

Keywords: Sustaining overload · Arresting load · Neck injury
Simulation

1 Introduction

For a modern maritime power, the aircraft carrier is the core of naval operations, and the carrier-based aircraft is the wing of the navy [1]. The combination of them makes the navy have superior combat capabilities. In order to master the control of future wars, all major maritime powers are striving to develop aircraft carriers and carrier-based aircraft technologies.

The carrier-based aircraft needs to rely on deck-brake or other arresting gears to land on the deck of the aircraft carrier, and the deck-brake can make the aircraft stop at a speed of several hundred kilometers in about two seconds. In the short-term process of arrested landing, the pilot is subjected to a load in the horizontal direction for about 2 s. This load is called the landing arresting load. Arresting load can cause abnormal relative movement between the head and the limb, resulting in excessive cervical extension. Carrier-based aircraft pilots repeatedly exposed to the flight environment, can easily cause cervical spine whiplash injury. According to reports from the US military, carrier aircraft pilots have a high incidence of cervical pain.

During the process of the carrier-based aircraft arrested landing, the biomechanical response of the pilot's neck under sustaining overload was studied, and the stress and damage of the necks were simulated and analyzed, which can be used for the design and improvement of the pilot's necks protection device. It has great significance for minimizing the damage by sustaining overload to carrier-based aircraft pilots. This article will use the finite element model simulation method to calculate and analyze the damage of the head and neck of the carrier-based aircraft during the process of arrested landing.

2   Methods

For above research background, based on the human anatomy, this paper established a finite element model of human head and neck with higher biofidelity, and combined the experimental data in the literature to verify the effectiveness of the finite element model. Finally, the finite element model was used to simulate the force and damage of the carrier-based aircraft pilots during the process of arrested landing. Specific research content as follow:

2.1   Finite Element Model Establishment

In order to obtain more accurate simulation results, a finite element model of human head and neck with high biological fidelity was established in this paper, including the skull, vertebrae, intervertebral discs, ligaments, muscles, articular cartilage and other tissues.

Geometric Model Establishment
First of all, based on the basic human body data of China's carrier-based aircraft pilots in the 50th percentile, a healthy male volunteer aged 24 years and 172 cm in height and 62 kg in weight was selected to rule out head and neck skeletal malformations and lesions. Using GE's 16-slice spiral CT to scan him, could obtain human tissue scan tomography image from the head to the third thoracic spine (T3) and save the data as a DICOM file. Then the CT image data was imported into Mimics 17.0. There were 560 tomographic image with 512×512 pixels resolution. Threshold is used to extract the bone gray information, and the head and C1–T1 3D bone model are established respectively. The model is then imported into the RapidformXOR3 for smooth processing to obtain a smooth head and neck geometric model, as shown in Fig. 1.

After that, the geometric model was imported into Geomagic Studio 12.0. The model was first analyzed and checked by the grid doctor, and the smooth geometric features were further repaired. Then the surface is fitted and constructed to complete the conversion from point cloud data to a geometric surface model.

Between the vertebral bodies, the intervertebral discs are mainly used to bear pressure. The intervertebral disc is drawn in SolidWorks software. According to the outline of upper and lower vertebral bodies to draw the outline of the disc, and using

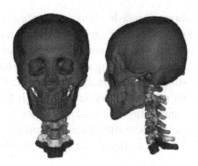

Fig. 1. Head and neck geometry model

scaling and other functions to divide the disc into the nucleus pulposus and the annulus fibrosus.

Finite Element Model Establishment

Meshing. Import the smooth optimized surface geometry model into Hypermesh 14.0 to divide the mesh. The dimensions of the grid cells are all 1 mm.

Soft Tissue Model Establishment. In ABAQUS, reference is made to the morphological characteristics of the human head and neck ligaments anatomy, and the starting and ending attachment points on the bone. Connecting the starting and ending points in the form of line units, and giving the corresponding cross-sectional area to complete the establishment of ligaments. The ligament finite element models established in this paper mainly include: anterior longitudinal ligament (ALL), posterior longitudinal ligament (PLL), ligamentum flavum (LF), joint capsular ligament (JCL), interspinous and supraspinous ligament (ISL & SSL). In addition, the connection between the head and cervical spine include an anterior occipital anterior membrane (AAOM), posterior atlantooccipital membrane (PPOM), transverse ligament of the atlas (TL), alar ligament (AIL), apical dental ligament (APL).

Muscles are constructed in the same way as ligaments. Refer to the head and neck anatomy, connect the muscles by attaching and affixing points, and give the cross-sectional area to simulate the major muscles of the head and neck [4], including the sternocleidomastoid muscle, musculus longus colli, musculus longus capitis, splenius capitis, scalenus anterior, scalenus medius, scalenus posterior, trapezius and so on.

Material Attribute. The selection of material properties is a crucial step in establishing a finite element model. Under the premise that the geometry of the model satisfies the accuracy of simulation, the material properties of each organization also have a direct impact on the simulation results. This article combines the data used in the literature [5] to study the material properties of the model. The vertebrae consisted of cortical bone and cancellous bone. Both the cortical bone and the cartilage endplates were homogenous in shell elements and the cancellous bone was selected as a homogeneous solid element. The annulus fibrosus and nucleus pulposus of the intervertebral disc are mainly subjected to pressure, so hexahedral solid elements are used, and material is modeled by isotropic linear elastic material. The ligament is mainly subjected to pulling force, and there is almost no mechanical response under compression. Therefore, the ligaments are modeled using isotropic linear elastic materials and are set to be incompressible. The material parameters of the articular cartilage given in the literature published by Yang et al. [6] are also modeled using the same linear elastic materials. The specific material properties of the finite element model established in this paper are shown in Table 1.

Table 1. Models' material parameter

Tissue name	Young modulus (Mpa)	Poisson ratio	Density (kg/m^3)	Element type
Cortical bone	10000	0.3	2.0E+03	Shell
Cancellous bone	450	0.3	1.0E+03	Solid
Vertebral endplate	1000	0.4	1.83E+03	Shell
Annulus fibrosus	3.4	0.4	1.2E+03	Solid
Nucleus pulposus	1.0	0.49	1.1E+03	Solid
ALL	28.2	0.4	1.1E+03	Truss
PLL	23	0.4	1.1E+03	Truss
LF	3.5	0.4	1.1E+03	Truss
JCL	5	0.4	1.1E+03	Truss
ISL & SSL	4.9	0.4	1.1E+03	Truss
TL	20	0.4	1.1E+03	Truss
APL	20	0.4	1.1E+03	Truss
AAOM	20	0.4	1.1E+03	Truss
PPOM	20	0.4	1.1E+03	Truss

Muscles have special mechanical properties: On the one hand, under the influence of external loads, muscles can passively bear the load. On the other hand, they can exert active contraction force through the control of nerves. This article refers to the study of Fan et al. [7], which only simulates the part of passive response of muscles. The Ogden model of superelastic materials is used as the muscle material. It is usually used to simulate soft tissue materials in the field of biomechanics.

Assembly of Model. After each part is assigned to the material properties, the assembly work is performed. The ligament is tied with its attachment area on the head and vertebral body so that relative movement does not occur. Similarly, the vertebral body and articular cartilage, muscles, annulus fibrosus and nucleus pulposus are also tied together [8]. Due to the sliding between the articular cartilage, a friction-free finite-slip surface contact method is established between each pair of articular cartilage. Since this article only analyzes the force and damage of the neck, the head is treated as a solid and the mass of 4.69 kg is added [9]. The finite element model after assembly is shown in Fig. 2. The model has a total of 617,648 nodes with 617,029 units, a total of 564,458 vertebrae, a ligament with 650 line units, and muscles with 1865 line units.

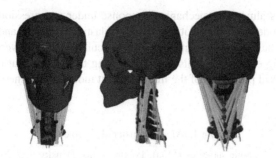

Fig. 2. Complete finite element mode

2.2   Verification of the Finite Element Model

The validity of the finite element model can be verified to ensure the accuracy of the simulation calculation. This article combines Panjabi et al.'s head and neck basic exercise experiment and Ewing et al.'s volunteer pre-collision experiment to verify the validity of the finite element model from both static and dynamic aspects.

Static Verification
Based on Panjabi et al.'s head and neck basic exercise experiment [10], the model was simulated and calculated. The role of muscle is not considered in the simulation, and the six freedom degrees of the lower surface of the thoracic spine T1 are constrained. A reference point is selected at the rotation center of the head, and the movement is coupled with other nodes of the head. At this reference point, pure torque of 1.5 Nm is loaded in ±X, ±Y, and ±Z direction respectively [11]. According to the right-hand rule, the anteflexion, rear extension, left flexion, right flexion and axial rotation are respectively generated. And the intervertebral range of motion between adjacent vertebral bodies is calculated.

Dynamic Verification
The dynamic response of the model was verified by the data of the Ewing et al. volunteer crash experiment [12].

Only releasing the translational freedom in the Y direction and the rotational freedom in the sagittal plane of the first thoracic spine T1. The acceleration in the Y direction is applied to the entire T1, and the forward rotation angle on the sagittal plane to simulate the pre-collision experiment. The entire model is in the normal gravity field. The entire simulation process lasts for 250 ms. The loaded T1 acceleration and rotation angle are shown in Fig. 3.

2.3   The Arrested Landing Finite Element Simulation of Cervical Biomechanics

According to the experimental data given in [13], the acceleration-time curve of carrier-based aircraft with a weight of 50,000 lb is selected as shown in Fig. 4.

T1 has the translational freedom in the fore-and-aft direction and the rotational freedom of the sagittal plane, all other degrees of freedom are constrained to apply the

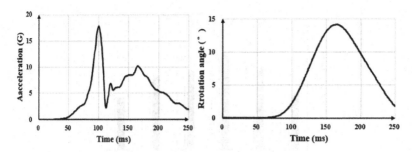

Fig. 3. T1 acceleration-time and rotation angle-time graph

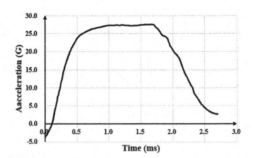

Fig. 4. Acceleration-time graph of arresting load

acceleration load to the entire T1 and give the entire model an initial velocity, 57 m/s. The entire model is in a 1G acceleration gravitational field, and considering the pilot wearing a helmet on the head, the mass of the head increases by 2.04 kg to 6.73 kg [14].

3   Results and Discussion

3.1   Verification Result

On the one hand, the static verification compares the calculated intervertebral motion with the experimental data and is divided into four processes: anteflexion, extension, lateral flexion and axial rotation (Fig. 5).

From the comparison of the model simulation data and the experimental data, we can see that the two data are basically consistent, and the motion of each vertebral body is basically within the error range of the experiment.

On the other hand, dynamic verification obtains the motion parameters of the front collision head from the results of the finite element model simulation. Select the acceleration in the Y direction of the center of gravity to compare with the experimental results, as shown in Fig. 6.

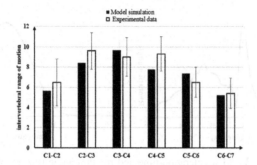

Fig. 5. Lateral flexion intervertebral range of motion comparison chart

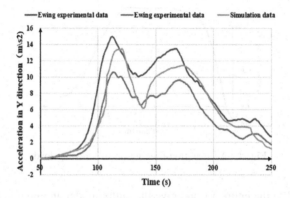

Fig. 6. Acceleration diagram in Y direction

From the comparison of simulation data and experimental data, the results obtained from the simulation in the head Y direction have good similarities with the experimental data, basically within the error range.

In summary, the human head and neck finite element model established in this paper has a good biological fidelity, and the dynamic response and the real experimental results have a high degree of similarity, it is considered that the finite element model is effective, can be used to dynamic simulation analysis of arrested landing process.

3.2 Analysis of Simulation Result

Stress Analysis of Vertebrae

After the simulation calculation, the stress cloud diagram at the time of maximum stress of each vertebrae is intercepted. As shown in Fig. 7, the vertebrae are not sufficiently stressed to directly cause damage to the vertebrae. However, there are stress concentration points in this process, and long-term repeated stress may cause fatigue hurt.

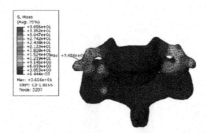

Fig. 7. Vertebrae stress cloud diagram

Stress Analysis of Intervertebral Discs

Select the stress of all nodes of the disc, find its mean, and finally output its stress-time curve, as shown in Fig. 8. The stress value of the C7–T1 intervertebral disc was the largest, and the stress of other intervertebral discs decreased gradually, but C4–C5 was larger than C5–C6. It can be seen that C4–C5 intervertebral discs are under greater stress during this exercise. Compared to other parts, it is also more likely to cause damage.

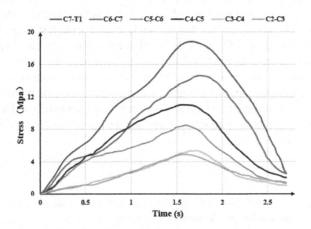

Fig. 8. Intervertebral discs stress-time graph

Maximum Tensile Strain of Ligaments

The maximum tensile strain of the main ligament can be obtained through simulation calculation. As shown in Fig. 9.

The ligament of the joint capsular stretches is the longest, and the spine supraspinous ligament takes the second place. It can be seen that excessive forward flexion of the neck can cause the joint capsular ligament to stretch for a long time, causing damage, affecting the stability of the articular cartilage, and eventually causing damage to the facet joints.

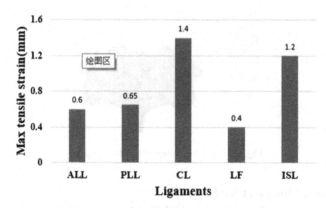

Fig. 9. Maximum tensile strain of ligaments

4   Conclusion

In summary, after finite element simulation and analysis, it can be known that the carrier's head and neck are subjected to continuous load during the process of blocking the carrier. This process will not cause damage to the vertebrae and spinal cord, but it may cause soft tissue damage. Over anteflexion head and neck, resulting in longer joint capsule ligament stretch, C4–C5 and C7–T1 intervertebral disc greater force, long-term repetitive motion is easy to form fatigue damage, affecting the stability of articular cartilage. Therefore, it may be considered to design a head restraint with a cushioning or fixing effect, fit the neck pillow of the neck curve more, and reduce the weight of the helmet to reduce the damage to the pilot. This study provides effective data support for the development of protective equipment and the design of the pilot training process.

References

1. Fengying Z (2007) Research on carrier-based carrier navigation technology. Nanjing University of Aeronautics and Astronautics, Nanjing
2. Zhang J, Wang Q, Li M et al (2012) Analysis of occupational characteristics of aircraft carrier pilots of foreign military carriers. J Navy Med 33(2):144–145
3. Zhang QH, Teo EC, Tan KW (2011) Effect of muscles activation on head-neck complex under simulated ejection. J Musculoskelet Res 40(40):155–165
4. Cai X, Yu Y, Liu Z et al (2014) Three-dimensional finite element analysis of occipitocervical fixation using an anterior occiput-to-axis locking plate system: a pilot study. Spine J Off J North Am Spine Soc 14(8):1399–1409
5. Yang KH, Hu J, White NA et al (2006) Development of numerical models for injury biomechanics research: a review of 50 years of publications in the stapp car crash conference. Stapp Car Crash J 50:429
6. Li F, Li H, Hu W et al (2016) Simulation of muscle activation with coupled nonlinear FE models. J Mech Med Biol 16(06):1–14
7. Raj PP (2008) Intervertebral disc: anatomy-physiology-pathophysiology-treatment 8(1):18–44

8. Meyer F, Bourdet N, Deck C et al (2004) Human neck finite element model development and validation against original experimental data. Stapp Car Crash J 48:177–206

9. Panjabi MM, Brand RA Jr, White AA 3rd (1976) Mechanical properties of the human thoracic spine as shown by three-dimensional load-displacement curves. Spine 58(5):42–52

10. Yoganandan N, Kumaresan S, Pintar FA (2001) Biomechanics of the cervical spine part 2. cervical spine soft tissue responses and biomechanical modeling. Clin Biomech 16(1):1–27

11. Ewing CL, Thomas DJ, Lustick L et al (1978) Dynamic response of human and primate head and neck to +Gy impact acceleration. Equipment 1978:549–586

12. Lv K, Zhu Q, Li X (2011) Modeling and simulation for arrested landing of carrier-based aircraft 2011:1928–1933

13. Parr MJC, Miller ME, Bridges NR et al (2012) Evaluation of the n_{ij} neck injury criteria with human response data for use in future research on helmet mounted display mass properties. Neuromodul Technol Neural Interface 56(1):2070–2074

Digital Human Model Simulation of Fatigue-Induced Movement Variability During a Repetitive Pointing Task

Jonathan Savin[1](✉), Clarisse Gaudez[1], Martine Gilles[1],
Vincent Padois[2], and Philippe Bidaud[2]

[1] INRS, rue du Morvan, CS 60027, 54500 Vandœuvre, France
jonathan.savin@inrs.fr
[2] Sorbonne Université, CNRS UMR 7222, Institut des Systèmes Intelligents
et de Robotique, ISIR, 75005 Paris, France

Abstract. Movement variability is an essential characteristic of human movement. It occurs in all kinds of activity including work-place tasks. However it is
almost ignored in workstation design, where expected movements are highly
standardized for productivity and quality considerations. Neglecting this variability may lead designers to omit parts of the future operator's movements, thus
leading to incomplete assessment of biomechanical risk factors.

This article describes a model-based virtual human controller intended to
simulate the movement variability induced by muscle fatigue during a repetitive
activity. It is built using a multibody dynamics framework and a 3-compartments
muscle fatigue model. The simulation of a repetitive pointing activity is described. Our demonstrator reproduces some of the adaptive behaviors described in
the literature. This demonstrator must still be validated by experimental human
data, but it opens interesting perspectives for DHM software improvements and
more reliable ergonomic assessments from the early stages of workstation design.

Keywords: DHM simulation · Movement variability · Muscle fatigue
Workstation design

1 Introduction

Movement variability is an intrinsic feature of human movement [1]: whether for a
given person at different times, or for different people, a prescribed movement is never
performed in exactly the same way twice. This seems to be related to the process of
control and regulation of movement in order to provide the essential adaptability and
flexibility to meet the characteristics related to the person, the task and the environmental constraints of the situation. Indeed, movement variability occurs in all kinds of
activity including occupational operations. However, despite its prevalence, workstation designers seem to be barely aware of this variability. They usually ignore it at the
workstation design stage, and tend more to define highly standardized operating modes
to satisfy productivity and quality considerations. However, neglecting this variability
may lead designers to omit parts of the future operator's movements, thus leading to
incomplete assessment of biomechanical risk factors.

© Springer Nature Switzerland AG 2019
S. Bagnara et al. (Eds.): IEA 2018, AISC 822, pp. 96–105, 2019.
https://doi.org/10.1007/978-3-319-96077-7_11

Since European workstation designers have the legal obligation to reduce occupational risks [2], providing them with tools accounting for this variability is therefore a great challenge. For instance, to date, commercial digital manikin software lack such features. As a preliminary step toward this ambitious goal, this paper describes a software demonstrator intended to simulate the effect of one source of variability during a repetitive activity, namely muscle fatigue. Section 2 recalls some basics on movement variability at the workplace. Section 3 describes our simulation framework and simulation results. Finally, Sect. 4 discusses these results and introduces the upcoming validation of our demonstrator.

2 Movement Variability and Workstation Design

2.1 Movement Variability at the Workplace

Movement variability is a highly complex and multifactorial phenomenon. It is manifested by more or less visible variations of the real activity: during the successive realizations of a prescribed task, the kinematic, kinetic and/or physiological characteristics of the movements of the operator change. When defining the geometry of a workstation or the location of protective devices, designers should therefore be able to assess how far their choices favor or impede future movement variability, and whether predictable variations in postures, velocities and exertions may lead to higher risks of musculoskeletal disorders.

Many parameters influence movement variability: the characteristics of the prescribed task (general posture of the operator, the dimensions and mass of the tools, the geometry of the workstation, pace) and the individual characteristics of the future operator (gender, anthropometry, physical performance). Furthermore, many other features, intrinsic to the operator, may intervene: the effects of age, expertise, learning phenomena, the presence of pain, muscle fatigue [1]. All these factors can influence movement variability at both the intra- and inter-operator levels, and are still currently ignored in tools used at the workstation design stage. As a preliminary step toward improved design tools, we developed a model-based demonstrator, as this approach has proved efficient for ergonomics studies [3, 4] and, unlike data-driven simulation, there is no need carry out new experimental acquisition for each new case of application.

2.2 Muscle Fatigue and Movement Variability

For the feasibility demonstrator, we chose to focus on one source of movement variability, namely, muscle fatigue. Indeed, muscle fatigue is likely to occur in most work situations, particularly in repetitive tasks (assembly, handling). It has been reported to modify general postures, ranges of motion, perceived posture and task precision [5–7]. These features may be critical in occupational activities where cycle-time and quality requirements must be fulfilled. Muscle fatigue issues are also a quickly developing field in the scientific community interested in virtual humans [8–11] and collaborative robots [12] although these applications are rather oriented towards task optimization in terms of fatigue than movement variability.

3 Simulation Framework

3.1 Muscle Fatigue Model

Among the different models of muscle fatigue identified in the literature, so-called "biophysical" compartment models appear more suitable for integration in digital manikin software [10] with respect to computation time, the number and complexity of their parameters and their ability to account for the time-history of muscle loading. Examples of implementation can be found in [13, 14]. The fatigue model chosen for our demonstrator was that proposed by Xia and Frey-Law [15, 16]. In this model, which we call XFL, muscle is composed of a constant number of fibers M_0. At each moment, under the effect of a command from the central nervous system, part of the fibers at rest (M_R) changes to active state (M_A), part of the fibers in active state changes to fatigue state (M_F) and part of the fatigued fibers returns to rest state (cf. Fig. 1). The distribution of the fibers at each moment validates the equation

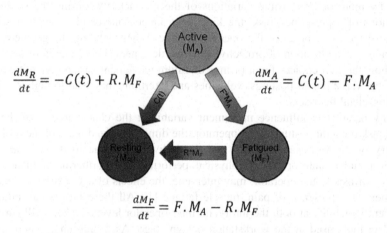

$$\frac{dM_R}{dt} = -C(t) + R.M_F \qquad\qquad \frac{dM_A}{dt} = C(t) - F.M_A$$

$$\frac{dM_F}{dt} = F.M_A - R.M_F$$

Fig. 1. Diagram of the principle of the XFL model from [16]

$$M_0 = M_A + M_R + M_F. \tag{1}$$

Only the active fibers participate in the effort so M_A also represents the current effort, expressed relatively to the maximum exertion. Changes from one state to another are defined by three coefficients denoted respectively C (command), F (fatigue) and R (recovery). Transition equations are detailed in Fig. 1. Command C is a function of the current activation M_A and the expected exertion TL (Target Load) expressed relatively to the maximum exertion. Two complementary descriptors of muscle fatigue can be defined (for more details, please refer to [15, 16]):

- the residual capacities RC, i.e. the effort potentially exerted if all the non-fatigued fibers $M_A + M_R$ were solicited together at the same instant;

- the central command *BE* (Brain Effort) defined as the ratio between the current exertion M_A and the residual capacities *RC*. It is also assumed to account for the perceived effort.

In addition, Xia and Frey-Law proposed generalizing their model to the case of an articulation, and not only to a muscle. To do this the authors proposed to take as set point *TL* the value of the expected torque, normalized in relation to the maximum torque Γ_{max} of the joint considered, instead of a muscular force. Figure 2 illustrates a simulation for the default shoulder fatigue parameters given in [16] and a static effort of 18% of the maximum torque for shoulder flexion (arm outstretched straight forward as in Fig. 4).

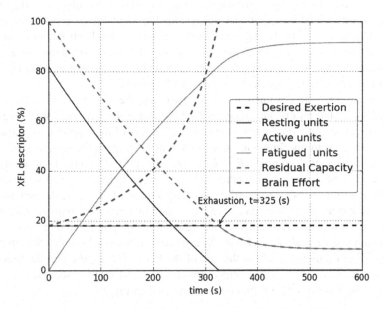

Fig. 2. Shoulder fatigue simulated for a constant exertion equal to 18% of the maximum exertion (keeping the arm outstretched straight forward)

3.2 Virtual Human Model and Control

The global framework considered for this study is XDE, a software developed by the French CEA, which has already been used to simulate activity at a workstation [3, 4, 17]. The virtual human is composed of 17 rigid body segments, actuated by 39 articular degrees of freedom assumed to be perfect hinges. The kinematics was defined according to the recommendations of the International Society of Biomechanics [18]. The control of the virtual human is based on optimizing elementary "tasks" used to describe the simulated activity (for instance, maintaining balance, reaching a point or a speed, exerting a force, maintaining a contact, keeping a posture) subjected to "constraints" (the fundamental equation of dynamics, bounded joints range of motion and actuating torques). Constraints are always fulfilled. On the contrary, tasks may not be perfectly achieved. Since tasks may be concurrent, a weight is associated to each of them so that significant actions or behaviors can be prioritized. At each simulation step, the multi-objective linear-quadratic programming (LQP) algorithm described in [19] computes the instantaneous expected torques. The XDE physical engine then computes the mathematical integration of the system's dynamics to solve the next state of the virtual human. The simulated motion is hence the trade-off between the constraints and the weighted tasks describing the activity.

3.3   Virtual Human Control and Fatigue: Additional Tasks and Constraints

In order to account for fatigue in virtual human simulation, we included XFL model instances to our virtual human model. Each fatigable joint was described by a pair of XFL models according to the semi-articulation method [20]. Each model of the pair was associated with a direction of the movement of the joint considered: for example one XFL model for flexion and one for extension. In the absence of co-contraction, each semi-joint "fatigues" while its counterpart "recovers").

The controller was designed to mimic two hypothesized behaviors: firstly, we assume that exerted efforts are limited with the offset of fatigue (the more fatigued a joint, the more its force production decreases). This loss of force production capacities is dealt with joint torque limitation constraints: at each instant, joints torques τ_k are limited to the residual capacity RC_k calculated by the XFL model. Secondly, we assume that the musculoskeletal control tries to preserve fatigued joints by transferring part of the exertions to other joints (the more fatigued a joint, the more the system seeks to lower its actuation). This behavior was translated into actuation redistribution tasks, namely torques minimization tasks whose weights ω_k are modulated depending on the central command BE_k calculated by the XFL model (cf. Sect. 4.1) so that the efforts are transferred from fatigued joints to the rest of the body. The LQP controller processes the set of these complementary fatigue-driven tasks and constraints added to those describing the activity without fatigue, as illustrated in Fig. 3.

4   Simulation of a Repetitive Pointing Activity

For the simulation case study, we considered the repetitive pointing activity with fatigue derived from that described in [5, 7, 21]. We simulated this experiment with a single virtual human, alternatively pointing a proximal and a distal target placed at 30% and 100%, respectively, of the length of an outstretched arm, with an expected cycle time of 2 s. During the movement, the elbow was to be kept at the altitude of the shoulder (an elliptic obstacle was placed under the elbow, cf. Fig. 4).

4.1   Simulation Parameters

The fatigable joints were the right shoulder and the right elbow. The associated R and F parameters were defined as in [16]. Joint torques were normalized according to the maximum effort values found in [22] for the shoulder and in [23] for the elbow. The

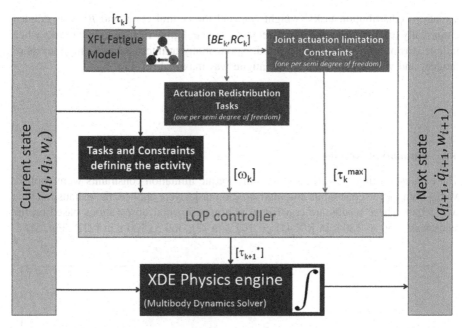

Fig. 3. Diagram of the principle of the controller: the physics engine solves the dynamics of the system at time $i + 1$ given its current state at time i and the actuation calculated by the LQP algorithm from the tasks and constraints defining the activity and fatigue-driven additional tasks and constraints.

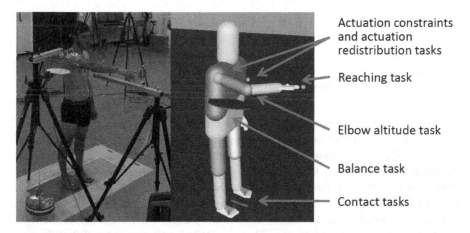

Fig. 4. Experimental set-up and simulation of the repetitive pointing activity. The main simulation tasks are balance, ground contact, elbow altitude and reaching targets; actuation redistribution tasks are set for the right shoulder and elbow.

main simulation tasks are balance ($\omega = 1$), general posture ($\omega = 10^{-2}$), ground contact ($\omega = 5$), elbow altitude ($\omega = 1$) and reaching targets ($\omega = 1$). The relation between the actuation redistribution tasks' weights ω_k and the central command BE was defined empirically (cf. Eq. 2) so that these weights were negligible relative to other tasks when fatigue was null or low, they increased exponentially at intermediate to high levels of fatigue, and were equal to 10 when fatigue was maximal (cf. Fig. 4).

$$\omega_k = 10^{-4} \cdot 2^{\frac{\ln(10^5)}{\ln(2)} \frac{BE}{100}}.$$ (2)

4.2   Simulation Results

The exhaustion time was 318 s when only torque limitation constraints were applied. When both torque constraints and minimization tasks were applied, the exhaustion time was 333 s. These results are consistent with experimental observations based on the same task (450 ± 180 s in [5], 413 ± 162 s in [7], 360 ± 120 s in [21]) (Fig. 5).

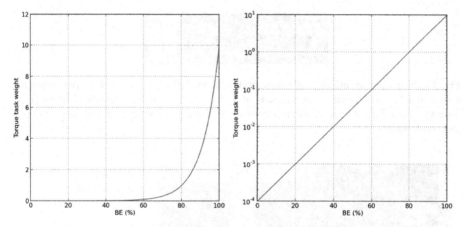

Fig. 5. Weight of the actuation redistribution tasks applied to the virtual human control

Torque limitation constraints induced changes in movement only at the very end of the simulation, shortly before exhaustion (pointing the last 4 targets, the virtual human hits the elliptic plate under the elbow). Otherwise, joint angles and trajectories remained identical. Torque limitation tasks induced more progressive variations in the movement on various joints. For instance, shoulder flexion and upper-trunk lateral inclination changed with the onset of fatigue by about 5° (cf. Fig. 6); fore-arm pronation and hand abduction angles decreased by about 3°; the altitude of the right shoulder increased by about 16 mm. These variations may appear small, but the pre-scribed task was very constrained. These results are partly consistent with the observations found in [5]: a shoulder elevation of about 12 mm, shoulder decreased by about 8. Some of these results need further comparison. For instance, the simulation also showed a rotation and a lateral inclination of the upper-trunk, which were not documented in the experiments cited, and a shift of the pelvis by about 15 mm towards the dominant side (cf. Fig. 6).

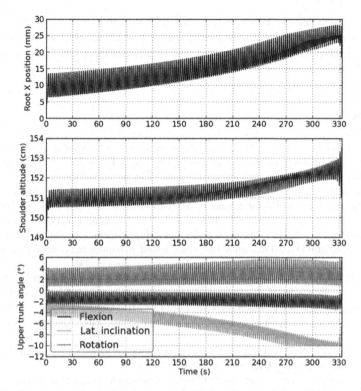

Fig. 6. Some adaptations of the movement simulated by the fatigue-driven controller

5 Discussion and Perspectives

Our virtual human controller implements joint torque limitation constraints and minimization tasks depending on the modeled state of muscle fatigue to account for intuitive behaviors induced by muscle fatigue (decrease of force production capacity and adaptive strategy through transfer of exertions). It was used to simulate a repetitive pointing task, but it may also be used for other activities since it can cope with other kinds of tasks, including external wrench tasks. For instance, the clip fitting activity described in [3] could be simulated 'as is' with our controller. The key points of simulation configuration were, in addition to the fatigue R and F parameters, the reference values for maximal joint actuation. Certain of the latter can be found in the literature, although they may sometimes be approximate or incomplete.

Using *a priori* parameterization and without any task-specific control derived from observed adaptation strategies, our demonstrator induced progressive variations of the movement prescribed, and partly consistent with the experimental results in terms of exhaustion time and kinematics variations between the beginning and the end of the experiment.

However, we had only reference data for the beginning/end comparison, which is why an experiment has recently been carried out and is currently being analyzed to compare intermediate variations as well (not only beginning/end variations), identify subject-specific strategies to compensate the effects of muscle fatigue during the performance of the pointing task considered, and implement these strategies to enrich the control of the demonstrator.

References

1. Gaudez C, Gilles MA, Savin J (2016) Intrinsic movement variability at work. How long is the path from motor control to design engineering? Appl Ergon 53(Part A):71–78
2. Directive of the European Parliament and of the council on machinery (2006)
3. De Magistris G, Micaelli A, Evrard P, Andriot C, Savin J, Gaudez C, Marsot J (2013) Dynamic control of DHM for ergonomic assessments. Int J Ind Ergon 43:170–180
4. Maurice P, Measson Y, Padois V, Bidaud P (2013) Assessment of physical exposure to musculoskeletal risks in collaborative robotics using dynamic simulation. In: Padois V, Bidaud P, Khatib O (eds) Romansy 19—robot design, dynamics and control: proceedings of the 19th CISM-Iftomm symposium. Springer, Vienna, pp 325–332
5. Fuller JR, Lomond KV, Fung J, Côté JN (2009) Posture-movement changes following repetitive motion-induced shoulder muscle fatigue. J Electromyogr Kinesiol 19:1043–1052
6. Srinivasan D, Sinden KE, Mathiassen SE, Côté JN (2016) Gender differences in fatigability and muscle activity responses to a short-cycle repetitive task. Eur J Appl Physiol 116:2357–2365
7. Emery K, Côté JN (2012) Repetitive arm motion-induced fatigue affects shoulder but not endpoint position sense. Exp Brain Res 216:553–564
8. Ma L, Chablat D, Bennis F, Zhang W (2009) A new simple dynamic muscle fatigue model and its validation. Int J Ind Ergon 39:211–220
9. Brouillette D, Thivierge G, Marchand D, Charland J (2012) Preparative study regarding the implementation of a muscular fatigue model in a virtual task simulator. Work J Prev Assess Rehabil 41:2216–2225
10. Rashedi E, Nussbaum MA (2015) A review of occupationally–relevant models of localised muscle fatigue. Int J Hum Factors Model Simul 5:61–80
11. Li Y, Zu X, Zhou Q (2013) Study on fatigue analysis and evaluation method of ergonomic virtual human. In: World congress on medical physics and biomedical engineering, 26–31 May 2012, Beijing, China. Springer, New York, pp 2011–2014
12. Peternel L, Tsagarakis N, Caldwell D, Ajoudani A (2016) Adaptation of robot physical behaviour to human fatigue in human-robot co-manipulation. In: 16th International conference on humanoid robots (humanoids), pp 489–494
13. Silva MT, Pereira AF, Martins JM (2011) An efficient muscle fatigue model for forward and inverse dynamic analysis of human movements. Iutam Symp Hum Body Dyn 2:262–274
14. Pereira AF, Silva MT, Martins JM, de Carvalho M (2011) Implementation of an efficient muscle fatigue model in the framework of multibody systems dynamics for analysis of human movements. Proc Inst Mech Eng Part K J Multi-Body Dyn 225:359–370
15. Xia T, Frey-Law LA (2008) A theoretical approach for modeling peripheral muscle fatigue and recovery. J Biomech 41:3046–3052
16. Frey-Law LA, Looft JM, Heitsman J (2012) A three-compartment muscle fatigue model accurately predicts joint-specific maximum endurance times for sustained isometric tasks. J Biomech 45:1803–1808

17. Savin J, Gilles M, Gaudez C, Padois V, Bidaud P (2017) Movement variability and digital human models: development of a demonstrator taking the effects of muscular fatigue into account. In: Advances in applied digital human modeling and simulation. Springer, New York, pp 169–179

18. Wu G, van der Helm FCT, Veeger HEJ, Makhsous M, Van Roy P, Anglin C, Nagels J, Karduna AR, McQuade K, Wang X, Werner FW, Buchholz B (2005) ISB recommendation on definitions of joint coordinate systems of various joints for the reporting of human joint motion—Part II: shoulder, elbow, wrist and hand. J Biomech 38:981–992

19. Salini J (2012) Dynamic control for the task/posture coordination of humanoids: toward synthesis of complex activities. https://tel.archives-ouvertes.fr/tel-00710013/document

20. Rodriguez I, Boulic R, Meziat D (2002) A joint-level model of fatigue for the postural control of virtual humans. In: Proceedings of the 5th international conference on human and computer, HC02

21. Fedorowich L, Emery K, Gervasi B, Côté JN (2013) Gender differences in neck/shoulder muscular patterns in response to repetitive motion induced fatigue. J Electromyogr Kinesiol 23:1183–1189

22. Lannersten L, Harms-Ringdahl K, Schüldt K, Ekholm J, Stockholm MUSIC 1 Study Group (1993) Isometric strength in flexors, abductors, and external rotators of the shoulder. Clin Biomech 8:235–242

23. Askew LJ, An K-N, Morrey BF, Chao EY (1987) Isometric elbow strength in normal individuals. Clin Orthop Relat Res 222:261–266

Towards Parametric Modelling of Skin Cancer Risk: Estimation of Body Surface Area Covered by Protective Clothing Using Base Mesh Modelling

Leyde Briceno[1,2], Simone Harrison[3,4], and Gunther Paul[1,2(✉)]

[1] AITHM, James Cook University, Townsville, QLD 4811, Australia
{leyde.briceno,gunther.paul}@jcu.edu.au
[2] Mackay Institute of Research and Innovation, Mackay, QLD 4740, Australia
[3] Skin Cancer Research Unit, College of Public Health, Medical and Veterinary
Sciences, James Cook University, Townsville, QLD 4811, Australia
simone.harrison@jcu.edu.au
[4] UV Radiation Group, School of Agricultural,
Computational and Environmental Sciences, University of Southern Queensland,
Toowoomba, QLD 4350, Australia

Abstract. The accumulated exposure to ultra-violet radiation creates an occupational and public health risk, and is carcinogenic to humans. The body surface area coverage by clothing (BSAC) contributes to skin cancer risk, and is a requirement in international standards on sun protective clothing, such as AS/NZS 4399:2017. BSAC is usually calculated utilising human subjects or physical mannequins using coating methods, indirect methods or direct measurements estimating the fraction of body covered. These methods are laborious and inflexible, and do not support computer based apparel design. To obtain a simpler, process integrated method, we determine the proportion of exposed body surface area using variable digital human models as virtual subjects, and image processing tools. Parametric, neutral posture human bodies of varying body stature, weight and age, including females and males, were generated in MakeHuman v1.1.1, and a protective clothing mesh, covering the minimum BSA specified in AS/NZS 4399:2017 was added. The MakeHuman definition of a human is based on fuzzy logic, with the main parameters normalised, and linked in a non-linear relation. The Whole Body Surface Area (WBSA) and the BSAC were obtained employing MeshLab, integrating elements on the respective surfaces, which were processed to improve precision. A procedure was developed to control geometric inconsistencies between the body base mesh and the clothing mesh. Thus different representative, generalized groups of subjects were analysed to explore BSAC. The method assists in the evaluation of exposed body areas in a wider spectrum of different occupations with their respective typical protective clothing conditions.

Keywords: Digital human modelling (DHM)
Body surface area coverage by clothing (BSAC)
Skin cancer · MakeHuman

© Springer Nature Switzerland AG 2019
S. Bagnara et al. (Eds.): IEA 2018, AISC 822, pp. 106–116, 2019.
https://doi.org/10.1007/978-3-319-96077-7_12

1   Introduction

Solar ultraviolet radiation (UVR) is a known carcinogen [1]. It is the main environmental risk-factor for cancer of the skin, of which there are three major types: basal cell carcinoma; squamous cell carcinoma; and lastly, malignant melanoma which develops in the pigment producing cells (melanocytes) and is more likely to metastasize than the other types of skin cancer [1, 2].

Skin cancer is the most common form of cancer in Caucasian populations worldwide [2], as well as being the most expensive in terms of direct costs to the health system [3]. However, theoretically, it is the easiest form of cancer to prevent and much easier than internal cancers to detect and treat early, because the warning signs manifest on the surface of the skin where they can easily be observed [2, 4]. It is estimated that about 80% of melanoma cases arising in Australia's predominantly fair-skinned, sun-loving population could be prevented by reducing sun-exposure, with prevention being most effective when multiple sun-protection strategies (such as avoiding peak UVR; applying high SPF sunscreen; wearing sun-protective clothing; seeking shade; wearing a hat and sunglasses when outdoors) are used in combination, from an early age [4–6].

Sun-protective clothing, such as broad-brim hats, sleeved shirts, Nylon Elastane rash-vests, and all-in-one protective swimsuits provide a physical barrier that reduces the amount of UVR reaching the skin [7]. Clothing is also less prone to incorrect use than sunscreen, where issues of using too little product, or not re-applying the product can significantly reduce its effectiveness [8].

Furthermore, the first RCT randomized controlled trial (RCT) to evaluate the efficacy of sun-protective clothing showed that children in the intervention group who routinely wore garments made of darker, tightly woven fabrics covering more body surface area (BSA) developed fewer pigmented moles than controls and therefore had a lower lifetime risk of developing melanoma [8, 9].

In 1996, Australia pioneered a reproducible measurement and classification protocol based on the relative ranking of UVR transmittance through fabric, known as the ultraviolet Protection Factor (UPF) [10]. This lead to the publication of AS/NZS 4399:1996, the joint Australian and New Zealand Standard for the evaluation and classification of sun-protective clothing [11]. The original Standard also provided specifications for UPF labels for garments and fabrics wishing to claim a sun-protective advantage [11]. Industry standards modelled on AS/NZS 4399:1996 have since been implemented in Britain, Europe, and the USA [12–15].

AS/NZ 4399:1996 and its associated UPF rating system were adopted almost universally by the textile industry [11]. However, the original standard only considered the UVR transmittance of the fabric, without taking into consideration the proportion of the BSA covered by the design of the garment [7]. In recent years, it became apparent that this was providing an avenue for manufacturers of brief swimwear and apparel to claim a sun-protective advantage and display a UPF label because most of these brief garments/swimwear were made of fabrics with low UVR-transmittance [7].

Minimum BSA coverage specifications have been incorporated into the European Standard for sun-protective clothing [13] as well as the recently revised joint Australian and New Zealand sun-protective clothing Standard published September 2017 (AS/NZS 4399:2017) [16].

Only those garments marketed in Australia and/or New Zealand that meet the minimum requirements for BSA coverage as specified in the revised standard AS/NZS 4399:2017 will now be permitted to display a UPF label or claim a sun-protective advantage [16]. A new index for sun-protective clothing called "the Garment Protection Factor (GPF)" which simultaneously considers BSA coverage of a garment and fabric UPF has been proposed since AS/NZS 4399:2017 was published in September 2017 [17].

This new index is compatible with current international standards for clothing labelled as sun-protective such that a GPF greater than or equal to zero can only be achieved by meeting the minimum requirements of these standards [17]. It extends the benefits of the present standards by providing an incentive for clothing manufacturers to design garments that cover a greater proportion of the BSA, yielding a higher GPF value indicative of a better sun-protective rating [17].

Current swing tags and labels displayed on clothing marketed as sun-protective do not routinely communicate the importance of garment coverage, although evidence suggests that this may be as important in preventing skin damage as the UVR-transmittance of the fabric [8, 9, 17].

The two components of the GPF index, namely fabric UPF and the BSA coverage ideally should be reported on the swing tag together with the overall GPF rating, making it easier for consumers to compare the sun-protective capabilities of different garments, leading to more informed purchases. Broad acceptance of the GPF would encourage manufacturers to design sun-protective garments that exceed the minimum standard for BSA-coverage as well as standardize the evaluation and labelling of sun-protective clothing across global markets, with positive implications for consumer awareness and skin cancer prevention world-wide. To achieve this goal, a method must be provided to enable the garment industry or retailers to efficiently calculate and assess BSA coverage of garment.

2   Materials and Methods

2.1   Virtual Human Dataset

MakeHuman v1.1.1 software [18] was used to generate a virtual human dataset from continuous ratio scale anthropometric data and [0, 1] normalised ratio scale inputs. The US National Health and Nutrition Examination Survey (NHANES) database was used to determine population anthropometric data [19].

For the MakeHuman definition of a human is based on fuzzy logic, absolute, ratio scaled body measures cannot be entered as input for modelling. With the purpose of generating human bodies from both absolute and normalized relative measurements, a MakeHuman plugin procedure was developed.

Input values for this calculation were obtained from the MakeHuman membership functions for age and height (body stature). For instance, making use of the minimum, average and maximum height functions (Fig. 1) and the height related membership functions (Fig. 2), a normalized value in the [0,1] space can be assigned to **Height** (Eq. 1). In addition to height, waist circumference was chosen as a measure that differentiates body types, as weight is not represented in absolute values in the software. Waist circumference is determined from weight and muscle indices in Make-Human, as shown in Table 1.[1]

$$(minI * minH) + (aveI * aveH) + (maxI * maxH) = \textbf{Height} \qquad (1)$$

A total of 144 virtual subjects were created, stratified by gender (female, male), age groups (<20; 20–<30; 30–<40; 40–<50; 50–<60; 60–<70; 70–<80; 80–<90 years), height (P5, P50, and P95), and waist circumference (P5, P50, and P95) based on the two relative indices, 'muscle' and weight (Table 1).

2.2 Sun Protective Clothing

Sun protective clothing was created using definitions from the Australian Standard AS/NZS 4399:2017 [20], which determines the amount of body surface that a piece of clothing needs to cover in order to claim a Ultraviolet Protection Factor (UPF) rating.

MakeHuman add-ons in Blender were used to make all-in-one clothing for both male and female gender. The boundaries in this garment were made following AS/NZS 4399 definitions, whereas upper garment sleeves shall cover three quarter of the upper arm, which is defined to go from a 'shoulder point' (per definition corresponding with the outermost extent of the A-C shoulder joint, at the junction of the clavicle and the scapula) to the elbow joint, and the lower body garment shall cover halfway between the crotch and knee [20].

Angles and lengths between the 'shoulder point' and elbow point were calculated on a T-pose, using the interactive ruler/protractor function in Blender (Fig. 3).

2.3 Body Surface Area

Whole Body Surface Area (WBSA) and Body Surface Area Not Covered by protective clothing (BSANC) were calculated using a script developed in Blender. This script has an intermediate step where the body mesh elements are processed.

The body mesh featured quadrilateral elements with 14311 vertices and 28348 faces for the full uncovered body. During the mesh processing these elements were divided into 112797 vertices and 225136 faces, and the whole surface was smoothed using the Blender function 'Subdivide Smooth' (number of cuts = 2 and smoothness factor = 0.5), which applies a Catmull-Clark Interpolation. WBSA was defined as the sum of the surfaces of all n elements $e_1 ... e_n$ (Eq. 2) on the whole body surface. BSANC was defined as the sum of the surfaces of all m elements $e_1 ... e_m$ (Eq. 3) on the body surface not covered by protective clothing (Fig. 4). Body Surface Area Covered by Protective Clothing (BSAC) was calculated as the difference between WBSA and BSANC.

---

[1] minI, aveI, maxI: minimum, average and maximum membership function
minH, aveH, maxH: minimum, average and maximum height.

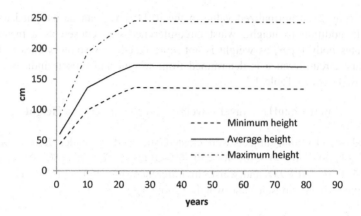

Fig. 1. Minimum, average and maximum height vs age for the male gender in MakeHuman

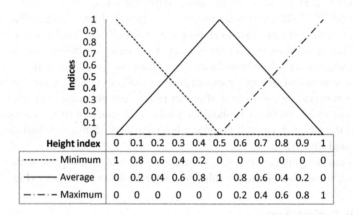

Fig. 2. Membership functions for the height parameter in MakeHuman

Table 1. Waist circumference in relation to weight and muscle indices

Waist circumference percentile	Muscle Index	Weight Index
P5	0.5	0.25–0.5
P50	0.33–0.5	0.5–1
P95	0.055–0.25	1

$$WBSA = \sum_{i=1}^{n} e_i \qquad (2)$$

$$BSANC = \sum_{i=1}^{m} e_i \qquad (3)$$

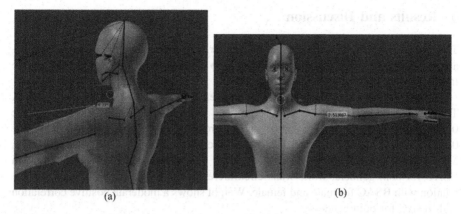

Fig. 3. Measurements on a T-pose (a) angle measurement on a woman (b) length measurement on a man

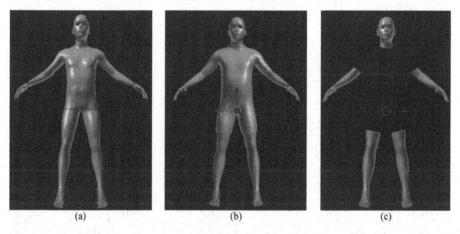

Fig. 4. Male subject of 16 years, height P50, and waist circumference P50. (a) Whole Body Surface Area (b) Body and protective clothing assembled (c) Body surface area not covered by protective clothing

During the body generation process, the set of vertices associated with body cavities were removed, and consequently body surface area reflects only the outer body shell.

2.4 Data Analysis

The scatterplot matrices and correlation analyses were performed using Spearman's Correlation Coefficient in R software [22].

3   Results and Discussion

This preliminary study was stratified by gender, age group, two anthropometric vari-ables (height and waist circumference) and two dependent inputs (weight and muscle indices).

BSAC was plotted against all study parameters, separated by gender (Figs. 5 and 6). These results suggest that BSAC is correlated with height (Female = 0.523 p-value 0.0097, Male = 0.543 p-value 0.0068), waist circumference (Female = −0.488 p-value 0.0157, Male = −0.602 p-value 0.0019) and weight (Female = 0.409 p-value 0.0470, Male = 0.439 p-value 0.0321). Height shows a moderate positive correlation with BSAC for both genders; while waist circumference exhibits a moderate negative cor-relation with BSAC for male and female. Weight shows a moderate positive correlation with BSAC for both genders.

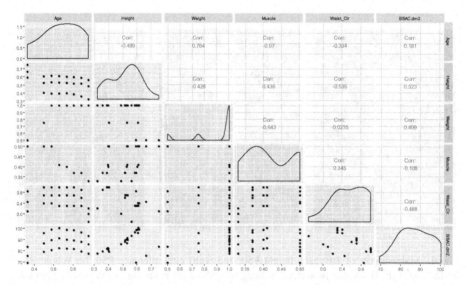

Fig. 5. Complete correlation matrix, Female (P5, 50 and P95 Height; P50 Waist Circumference)

Pearson's correlation was used to measure the relation between WBSA and BSAC by gender and percentile. As expected, Figs. 7 and 8 indicate a linear relationship between WBSA and BSAC for all-in-one clothing for both genders. In addition, the relative, percentiled body surface areas covered by this garment were 52.6% (Female P5), 49.3% (Male P5), 54.2% (Female P50), 50.7% (Male P50), 56.3% (Female P95) and 53.2% (Male P95).

The WBSA computed were compared with results from Lee and Choi [22], which were conducted using the traditional alginate method. The mean WBSA obtained by Lee and Choi were slightly smaller than ours for both genders; however the mean height and its standard deviation are higher in our study (Table 2).

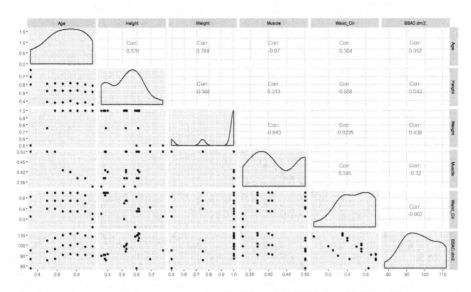

Fig. 6. Complete correlation matrix, Male (P5, P50 and P95 Height; P50 Waist Circumference)

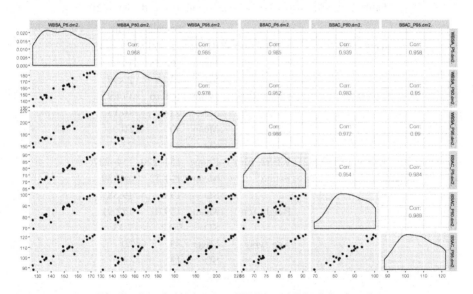

Fig. 7. Correlation between WBSA and BSAC for female

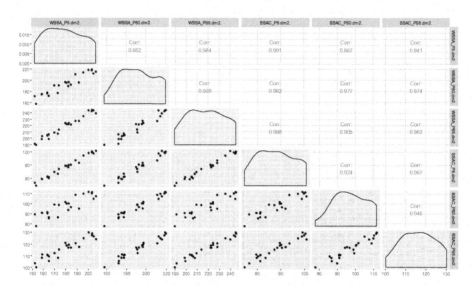

Fig. 8. Correlations between WBSA and BSAC for Male

Table 2. Comparison of study results with Lee and Choi [22].

			Female	Male
This study	WBSA (dm²)	Range	126–217	152–245
		Mean	166.38	196.74
		SD	23.14	23.99
	Height (cm)	Mean	161.16	174.9
		SD	9.69	10.13
	Age (yrs)	Mean	50	50
		SD	23	23
Lee and Choi [22]	WBSA (dm²)	Range	154–220	154–227
		Mean	164.52	183.39
		SD	19.6	19.38
	Height (cm)	Mean	159.01	172.91
		SD	7.52	7.43
	Age (yrs)	Mean	35	30
		SD	13	11

4   Conclusion

Based on the results presented, it can preliminarily concluded that the method provides a valid tool for determining BSAC. In the next stage however, more precise definitions of anthropometric measurements in the Australian Standard AS/NZS 4399 are required to calculate precisely BSAC, and thus improve reliability and repeatability for WBSA and BSAC calculations. Moreover, it is required to validate our model based approach against a database of physically scanned body measurements. Given the current limitations of the existing Australian Standard, this work will need to emanate outside the Standard, and may eventually lead to further Standard revision.

References

1. International Agency for Research on Cancer, World Health Organization (1992) IARC monograph on the evaluation of carcinogenic risks to human: Solar and ultraviolet Radiation, Lyon, vol 55
2. Lucas R, McMichael A, Smith W, Armstrong BK (2006) Solar ultraviolet radiation: global burden of disease from solar ultraviolet radiation. In: Prüss-Üstün A, Zeeb H, Mathers C, Repacholi M (eds) Environmental burden of disease series no. 13. World Health Organization, Geneva
3. Doran CM, Ling R, Byrnes J, Crane M, Searles A, Perez D, Shakeshaft A (2015) Estimating the economic costs of skin cancer in New South Wales, Australia. BMC Public Health 1 (15):952
4. Armstrong BK, Kricker A (1993) How much melanoma is caused by sun exposure? Melanoma Res 3:395–401
5. Holman CDJ, Armstrong BK (1984) Cutaneous malignant melanoma and indicators of total accumulated exposure to the sun: an analysis separating histogenic types. J Natl Cancer Inst 73:75–82
6. Harrison SL, MacLennan R, Speare R, Wronski I (1994) Sun exposure and melanocytic naevi in young Australian children. Lancet 344:1529–1532
7. Harrison SL, Downs N (2015) Development of a reproducible rating system for sun protective clothing that incorporates body surface coverage. World J Eng Technol 3:208–214
8. Harrison SL, Buettner PG, MacLennan R (2005) The North Queensland sun-safe clothing study: design and baseline results of a randomized trial to determine the effectiveness of sun-protective clothing in preventing melanocytic nevi. Am J Epidemiol 161:536–545
9. Harrison SL, Buettner PG, MacLennan R, Woosnam J, Hutton L, Nowak M (2010) Sun-safe clothing helps to prevent the development of pigmented moles—results of a randomised control trial in young Australian children. Ann ACTM 11:49–50
10. Gies HP, Roy CR, Elliott G, Zongli W (1994) Ultraviolet radiation protection factors for clothing. Health Phys 67:131–139
11. Standards Australia/Standards New Zealand (1996) AS/NZS 4399:1996 sun protective clothing—evaluation and classification. Standards Australia, Sydney and Standards New Zealand, Wellington
12. The British Standards Institution: BS 7949:1999 Children's clothing. Requirements for protection against erythemally weighted solar ultraviolet radiation (1999). https://shop.bsigroup.com/ProductDetail/?pid=000000000030159040. Accessed April 2018
13. European Committee for Standardization CSN EN 13758-2 + A1 Textiles—solar UV protective properties—Part 2: classification and marking of apparel (2003). https://www.en-standard.eu/csn-en-13758-2-a1-textiles-solar-uv-protective-properties-part-2-classification-and-marking-of-apparel/. Accessed April 2018
14. American Association of Textile Chemists and Colorists: AATCC TM183-2014 Transmittance or Blocking of Erythemally Weighted Ultraviolet Radiation through Fabrics (2014). https://members.aatcc.org/store/tm183/579/. Accessed April 2018

15. American Society for Testing and Materials: ASTM D6603-12 standard specification for labeling of UV-protective textiles (2012). https://www.astm.org/Standards/D6603.htm. Accessed April 2018
16. Standards Australia: AS/NZS 4399:2017 sun protective clothing—evaluation and classification (2017). http://www.standards.org.au/. Accessed April 2018
17. Downs NJ, Harrison SL (2018) A comprehensive approach to evaluating and classifying sun-protective clothing. Br J Dermatol 178:958–964
18. Bastioni M, Re S, Misra S (2008) Ideas and methods for modeling 3D human figures: The principal algorithms used by MakeHuman and their implementation in a new approach to parametric modeling. In: Proceedings of the 1st Bangalore annual compute conference. COMPUTE'08. ACM, New York, pp 10:1–10:6. https://dl.acm.org/citation.cfm?doid=1341771.1341782. Accessed April 2018
19. Fryar CD, Gu Q, Ogden CL, Flegal KM (2016) Anthropometric reference data for children and adults: United States, 2011–2014. National Center for Health Statistics. Vital Health Stat 3(39). https://www.cdc.gov/nchs/data/series/sr_03/sr03_039.pdf. Accessed April 2018
20. Cignoni P, Callieri M, Corsini M, Dellepiane M, Ganovelli F, Ranzuglia G (2008) MeshLab: an open-source mesh processing tool. In: Sixth eurographics Italian chapter conference, pp 129–136. http://vcg.isti.cnr.it/Publications/2008/CCCDGR08/MeshLabEGIT.final.pdf. Accessed April 2018
21. Venables WN, Smith DM (2018) R Core Team: an introduction to R. notes on R: a programming environment for data analysis and graphics. Version 3.4.4. R Foundation. https://cran.r-project.org/manuals.html. Accessed April 2018
22. Lee J, Choi J (2009) Estimation of regional body surface area covered by clothing. J Hum Environ Syst 12(1):35–45

Grasping Simulations Using Finite Element Digital Human Hand Model

Gregor Harih[1(✉)] and Mitsunori Tada[2]

[1] Laboratory for Intelligent CAD Systems, Faculty for Mechanical Engineering,
University of Maribor, Maribor, Slovenia
gregor.harih@um.si
[2] Digital Human Research Group, National Institute
of Advanced Industrial Science and Technology, Tokyo, Japan

Abstract. When developing new handheld products, engineers must consider ergonomics to increase the human-product performance, comfort, and lower the risk of cumulative trauma disorders. Extensive knowledge and lack of computer aided design software in terms of hand ergonomics prevents the improvement of handheld product ergonomics. The main research topic is therefore prehensile hand grasp with a handheld object. The nature of the human hand has prevented direct measurements of stresses, strains, forces, and contact pressure on the hand during movement and grasping. Therefore, several researchers tried to develop a feasible digital human hand model for hand biomechanics and product ergonomics. In this paper we present a viable method to determine realistic human hand movement and use this data to drive the developed finite element hand model for usage in hand biomechanics and product ergonomics. The model geometry has been acquired using medical imaging and appropriate numerical model definition inside finite element software has been defined. Grasping techniques and hand movement were then recorded using motion capture system and were input into the model. Based on numerical tests, the model has proven to be numerically feasible and stable. It shows reasonable biomechanical behaviour of movement and soft tissue deformation and corresponds well with experiments of contact area and pressure measurement and tendon/muscle force.

Keywords: Digital human hand model · Human grasp · Numerical simulation

1 Introduction

1.1 Human Grasping

The human hand is one of the most sophisticated bio-mechanical tools of a human. The main function of the hand is the interaction with the physical environment, where the most important is the prehensile hand grasp with a physical object. Thereby it can be effectively used as a tool for work, as well as an interface to use various powered and non-powered hand tools and products. Product designers have to consider ergonomics to increase the human-product performance, comfort, and lower the risk of cumulative trauma disorders (CTD) [1]. CTDs can be defined as a set of syndromes, which can be characterized as discomfort with persistent pain in the muscles, tendons, and other soft

© Springer Nature Switzerland AG 2019
S. Bagnara et al. (Eds.): IEA 2018, AISC 822, pp. 117–127, 2019.
https://doi.org/10.1007/978-3-319-96077-7_13

tissue; and joint movement inability or with reduced mobility. The most common CTDs as a result of usage of handheld products are bursitis, carpal tunnel syndrome, hand-arm vibration syndrome, ischemia, white finger syndrome, etc. It has been shown that CTDs result for major sick leaves of workers, which presents high costs for the company and high costs with diagnostics and treatment.

1.2   Digital Human (Hand) Models

Modern companies hugely rely on computer tools nowadays. They allow comprehensive and integrated approach to product development. Usually those computer tools cover the area of graphic representation, simulation, engineering analyses and animation of the developing product, however ergonomic views are poorly represented. In computer tools where they are present their use is usually limited to workplace ergonomics [2]. Smaller products, such as various hand tools, devices, etc. therefore cannot be considered within existing computer tools, which requires the use of classic methods: designing with anthropometric design tables and iterative design approach using physical prototypes. This approach usually requires longer development time which increases development costs and result in products with inferior ergonomics.

Increase in computer power has led to broader dissemination and use of digital human models (DHM) for animation, ergonomic analyses and simulations [3]. Computer tools allow analyses in an early stage of product development, therefore these tools contribute to lower development time and lower development costs and increased productivity of the company, which increases the competitiveness of the product on the market.

Extensive knowledge, which is necessary to comply with ergonomic principles and poor integration of ergonomics in modern CAD tools has led to low compliance with ergonomic principles within the product development [4]. Digital human hand models (DHHM), which are a part of a whole digital human model are used for evaluation of sight and simulations of hand reach. Newer DHMs, which are based on kinematics and biomechanics are usually being used in the field of workplace ergonomics for task evaluations such as lifting, pushing, etc. [5]. These analyses do not require anthropometrically and anatomically accurate DHHMs, which makes them inapplicable in the field of product ergonomics where grasp is the main ergonomic design attribute of the shape of the product. Therefore DHHM that consider only kinematics and biomechanics of the hand, but ignore the anatomical shape of the hand and its deformations of soft tissue cannot be used for accurate ergonomic analyses and product shape determination. Seo and Armstrong showed that soft tissue deformation is crucial for stable manipulation of different object in hands [6]. For maximum stability high contact area should be established between the object and human hand during grasping, therefore it is crucial to include soft tissue deformation in the ergonomic grasp analyses.

Several authors tried to develop DHHM with autonomous grasping using algorithms, however due to inaccurate hand models and complexity of human grasping ergonomic analyses can be also inaccurate. Endo et al. developed a DHHM called DhaibaWorks, which supports semi-autonomous grasps of the DHHM [7]. Due to the complexity of autonomous grasps using computer algorithms, this system uses utilizes additional contact points which have to be determined by the user to perform a feasible grasp. The developed DHHM considers the skin deformation during hand movement based on algorithm, however it does not consider mechanical behaviour of soft tissue during grasping of the object. Hence mechanical loads on the hand (stress, strain, contact pressure) cannot be predicted, which prevents the product shape optimisation.

1.3   Finite Element Method in Human Hand Biomechanics and Ergonomics

Several researchers have already utilised finite element method for modelling and simulating the biomechanical behaviour of the hand during different manual tasks and estimating the resulting loads. We have already shown that deformations and stresses that represent basic results of the structural analysis using FE method are also an important aspect within the field of biomechanics and ergonomics [8]. Researchers started approaching the problem using simplified 2D FE fingertip models during flat contact and analysed the mechanical responses to various loadings [9]. It has been shown that the soft tissue of the fingertip presents non-linear hyper-elastic material properties and experiences high local stresses and strains under dynamic loading and higher frequency produces higher stresses compared to lower frequency. It has also been shown that the surface curvature of the grasped object has a big influence on the resulting contact pressure on the fingertip [10]. Those objects that follow the shape of the hand and the skin's surface result in much lower contact pressures and local deformations of the skin and subcutaneous tissue, which can prevent discomfort and several disorders. A 2D FE model has also been used for predicting the responses of the soft tissue within different depths to vibration exposures and the correlation of the mechanical stimuli in mechanoreceptors [11]. It has been shown that dynamic strains induced by low frequency vibration will penetrate deeper in the soft tissue, whilst high frequency vibrations are concentrated within the skin layer. We used the same 2D FE fingertip model to analyse mechanical responses during flat contact using various hand-handle interface materials [8]. Simulations have shown that the proposed hyper-elastic foam materials with their distinctive mechanical behaviour can lower the contact pressure due to the deformations of the foam material, whilst still maintaining the stability of the product in the hands.

As shown, most authors used 2D FE fingertip models due to their simplicity and reduced computational costs. We have also shown that 3D FE fingertip models have an advantage over 2D FE models since they can provide more accurate results and additional insights into the third dimension despite being more intricate to construct and simulate [12].

Wu, Welcome and Dong [13] have already proposed a 3D FE fingertip model for simulating the dynamic response of a fingertip to dynamic loading, however the fingertip was geometrically simplified and symmetrical. Chamoret et al. [14] developed a 3D FE DHHM with multiple nonlinearities: geometrical, material, frictional contact and impact. Anatomical correctness of the 3D FE DHHM was achieved using CT scan of a hand. Authors presented a case study of an impact analysis of 3D FE DHHM pushing against a wall, however they did not present a possibility to use the 3D FE DHHM for grasping simulations.

The purpose of this paper was to explore the phenomena of human grasping and propose a viable method to determine realistic human hand movement and use this data to drive the developed FE hand model for usage in hand biomechanics and product ergonomics.

2   Materials and Methods

Methods of the development of the finite element digital human hand model mainly comprised of geometry acquisition using medical imaging and appropriate numerical model definition inside finite element software as outlined below. Grasping techniques and hand movement were then recorded using motion capture system and were input into the FE-DHHM.

2.1   Geometry Acquisition and 3D Reconstruction

For the purpose of this research, the FE-DHHM with basic FE definition has been already developed by previously [15]. Geometry acquisition of the human hand was based on medical imaging and reverse engineering afterwards. The surface model during the reverse engineering procedure was then converted into a solid model, afterwards all anatomical structures were imported into the Abaqus FE software in watertight IGES file format for the appropriate definition of the FE model (Fig. 1).

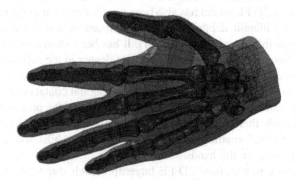

Fig. 1. A figure caption is always placed below the illustration. Short captions are centered, while long ones are justified. The macro button chooses the correct format automatically.

2.2   Finite Element Method

Material Parameter Determination. Skin and subcutaneous tissue are highly complex structures; therefore their material properties are hard to measure and define. It has been shown, that skin and subcutaneous tissue have non-linear viscoelastic properties, where the skin is stiffer than the subcutaneous tissue [16]. To get realistic response in terms of soft tissue deformation, stresses, and contact pressure of a numerical human hand model during movement and grasping, it is crucial to utilize a material model, which represents the actual material behaviour as accurate as possible. Therefore Ogden hyper-elastic material model has been used for the soft-tissue definition. Proposed material models and properties have been already used by us in our previous research and showed great correspondence to experiments [8, 17].

Boundary Conditions. The structure of human hand consists of 27 bones, numerous muscles, ligaments, tendons and other anatomical structures [18]. Bones with other structures are responsible for more than 20 degrees of freedom. Kinematics and dynamics of such complex system combined with unique material behaviour presents a great challenge for numerical simulations, which can be only tackled with eligible simplifications to make the finite element model numerically feasible and stable.

Grasping simulations can be only performed when appropriate input is given to the FE model. In case of our FE-DHHM input can be angle data of each joint or tendon/muscle forces as follows.

Angle driven joint model. Since we are interested in the loads inside the soft tissue, joints can be simplified to the extend they are numerically low-cost but still provide accurate movement of the bones. In this case the movement and deformation of soft tissue can be also considered as bio-mechanically accurate, since the material model has been verified and validated by us in previous papers.

In this manner, the joint definition was defined as follows on the example of finger joint between distal and proximal phalange bone (Fig. 2). Firstly, the center of the rotation in proximal phalange bone is identified where a new local coordinate system is created. The coordinate system is then oriented in such manner that the "y" axis corresponds with the axis of the rotation of the joint axis. In the center of the new local coordinate system a new reference point (RP-PIP) is created. Another reference point (RP-DIP) is also created on the surface of the distal phalange bone. Both reference points are then connected to bones using constraints, which fix their translations and rotations relative to the bone. Between both reference points a rigid wire is then created, which is used to define a connector that allows rotation only in "y" axis at the RP-PIP point. In this manner a simplified, numerically stable and bio-mechanically correct joint can be defined. Other joints in human hand are then defined analogously considering their local features.

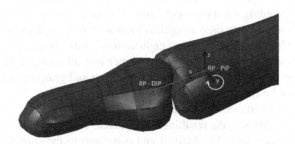

Fig. 2. Definition of an angle driven joint.

The proposed FE-DHHM was then coupled to an appropriate motion capture system where joint angles of hand movements were extracted and feed into our FE model from real life grasping scenarios. Hereby realistic hand posture and grasping pattern can be obtained, which can provide results in terms of soft tissue stresses, strains, and contact pressures when an object is grasped.

Tendon driven joint model. To obtain also the muscle/tendon forces, the angle driven joint definition has been upgraded with an additional reference point RP-T_1 at the location where the tendon is attached to the distal phalange bone (Fig. 3). This reference point has been attached to the bone and all translations and rotations have been fixed. Another reference point RP-T_2 has been defined in the direction where the tendon is placed inside the sheats. Both reference points have been connected with a hinge wire connector, which allows rotation in the "x" axis at both reference points.

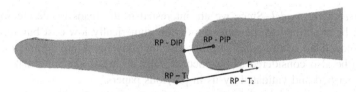

Fig. 3. Definition of the tendon driven finger joint model (upper: schematics, lower: FE definition).

To simulate the finger movement, concentrated force is applied at the RP-T_2 in the direction of the tendon, the "y" axis. This simulates the muscle force, which is transferred to the tendon and finally to the adjacent bone. Due to the moment arm, which is created based on the joint definition, a moment is created around the "x" axis of RP-PIP, which ultimately rotates the distal phalange bone around the center of rotation of the joint. The deformation of soft tissue is solely a result of material properties and joint definition.

Grasping – Numerical Simulations. Since human hand grasping is extremely complex phenomena, translation of all characteristics into computer simulation is also extremely complex task. Therefore, researchers have developed several techniques, such as video recording of movement, optical systems, magnetic systems, inertial systems, etc. In our case, we used an optical motion capture systems Smart-D from the company BTS. The system consisted of eight cameras with reflective markers attached to the hand. Cameras were placed in such manner that all markers were visible to at least three cameras all the time during hand movement and grasping. This ensured that joint angles could be extracted for any given time.

Subjects were instructed to grasp a pliance® cylinder sensor from the Novel company (Fig. 4). This way the resulting contact area, contact pressure, individual and combined finger force could be obtained and compared to the simulations performed later. Joint angles were extracted using the DhaibaWorks digital human hand model [19] and were fed into the FE-DHHM from this study.

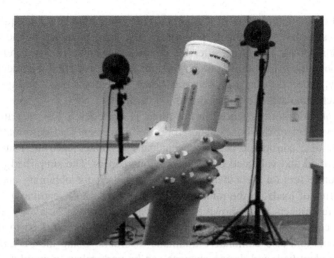

Fig. 4. Grasping the pliance® cylinder sensor with reflective markers attached to the hand.

3   Results and Discussion

Grasps generated by the mathematical models are usually evaluated by the operator visually or by calculating grasp quality using different methods within the software. This kind of evaluation can be unreliable, since real-world grasping is very complex and is also dependents on the subjective comfort rating of the user [20]. It has been shown that perceived subjective comfort is strongly correlated with user performance, therefore it is necessary to incorporate this aspect of product evaluation during the design phase. Comfort is affected by physical, physiological, and psychological factors; and is a subjectively-defined feeling that differs from person to person. Therefore it cannot be simply predicted neither by objective methods (such as grip-force and pressure measurement, electromyography, biomechanical hand-models, finite element analyses, etc.) nor by the resulting mathematical models that can only predict the physical aspects on the perceived comfort [20].

Whole hand FE models have not been developed yet due complexity of the human hand anatomy and resulting complex biomechanics. Therefore, we developed a numerically feasible and stable angle and muscle/tendon driven joint model, which allows finger movement based on joint angle rotation or muscle/tendon force. Successful development of such biomechanical system requires simplifications, which are a compromise between accuracy of the biomechanical behavior of the model and model complexity and calculation times. Based on the biomechanics of the joint, the joint rotation center is defined based on anatomical and topological features of the bones and does not consider the joint as two bones sliding on cartilage supported by ligaments, since such biomechanical system would be difficult to simulate using the finite element method. Used simplifications are also justified as our interest is in normal forces of the finger, tendon forces and contact pressures on the soft tissue. Based on our previous research, we have shown that such simplifications are reasonable and allow maintaining high level of accuracy of the simulated system.

The soft tissue deformation of the hand during simulation is a consequence of the bone link structure movement prescribed by the boundary conditions (i.e. muscle/tendon force). Therefore, after the simulation completed, we carefully investigated the bone movement during finger flexion. We observed that joint definition was set correctly, since the distance between bones was maintained during the finger flexion, which simulates the bones sliding on cartilage.

The simplified joint introduced in this paper is numerically low-cost and provides accurate movement of the bones. Therefore, the movement and deformation of soft tissue can be also considered as bio-mechanically accurate, since the material model has been verified and validated by us in previous papers. The simplified joint model using the connectors on the other hand cannot be used for obtaining and assessing results in terms of loads on the joints (contact pressures on the cartilage).

Several researchers have shown that a power grasp produces a very uneven distribution of contact pressure on the hand and fingers, which can lead to discomfort, pain, and acute and also cumulative traumatic disorders. Using the FE-DHHM these areas can be identified, and design changes can be undertaken to avoid such issues.

Therefore, we investigated the resulting contact area, contact pressure and also tendon forces from the experiments to the results from the simulation (Figs. 5 and 6).

Fig. 5. Numerical simulation of grasping a cylindrical handle (left) with the resulting contact area and contact pressure (middle) and contact area and contact pressure from the experiment (right).

Based on the results from the experiments and numerical simulations it is evident that results in terms of contact area correspond well (Fig. 6).

We additionally compared the tendon driven joint model to the angle driven joint model developed by us in previous research. Results have shown that the definition using the tendon driven model is comparably numerically low-cost, since one additional rigid wire representing the tendon, does not significantly increase the calculation time.

To validate the newly developed muscle/tendon driven joint model, we extracted the results of tendon force and reaction force on the cylinder from the simulation. We plotted the relationship between tendon force and resulting fingertip reaction force from our model and compared them to reproduced results from Kursa et al. [21].

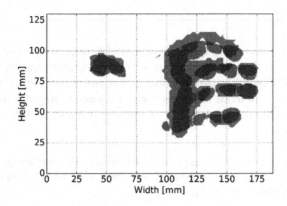

Fig. 6. Comparison of contact area between experiment (grey) and simulation (blue) (Color figure online)

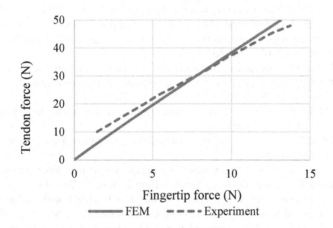

Fig. 7. Tendon force in comparison to the fingertip (reaction) force from the experiment and FEM.

Results from the simulations show good correspondence to the results from the experiment (Fig. 7).

In this regards future work should include throughout process of verification and validation of the developed FE-DHHM. Future FE-DHHM should also include more anatomical structures (skin, subcutaneous tissue, nail, capsules, synovial fluid...) and more realistic joint definitions to obtain even more accurate numerical results. Additionally, the FE-DHHM should be used for studying ergonomic and also bio-mechanic aspects of new product development ultimately leading to injury identification and also prevention.

4 Conclusion

In this paper we presented a newly developed FE finger joint model driven by a muscle/tendon force. The model has proven to be numerically feasible and stable and shows reasonable biomechanical behaviour of movement and soft tissue deformation. Results of the model in terms of tendon force compared to fingertip force correspond well to the experiments, which shows the proposed model is accurate and biomechanically correct.

Acknowledgements. The authors acknowledge the project (Development of a computational human hand model for ergonomic product design, Project ID: Z2-8185) was financially supported by the Slovenian Research Agency.

References

1. Salvendy G (2012) Handbook of human factors and ergonomics. Wiley, Hoboken
2. Shuxing D et al (2009) Study of method for computer aided ergonomics knowledge management and design aiming at product design. In: Computer-aided industrial design and conceptual design, 2009. CAID&CD 2009
3. Badler N (1997) Virtual humans for animation, ergonomics, and simulation. In: Proceedings of the 1997 IEEE workshop on motion of non-rigid and articulated objects (NAM'97). IEEE Computer Society, p 28
4. Karwowski W, Soares MM (2011) Human factors and ergonomics in consumer product design. CRC Press, Boca Raton
5. Chaffin DB (2005) Improving digital human modelling for proactive ergonomics in design. Ergonomics 48:478–491
6. Seo NJ, Armstrong TJ (2008) Investigation of grip force, normal force, contact area, hand size, and handle size for cylindrical handles. Hum Factors 50:734–744
7. Endo Y et al (2008) Virtual ergonomic assessment on handheld products based on virtual grasping by digital hand. SAE Int J Passeng Cars Electron Electr Syst 116:877–887
8. Harih G, Dolšak B (2014) Recommendations for tool-handle material choice based on finite element analysis. Appl Ergon 45:577–585
9. Wu JZ et al (2002) Simulation of mechanical responses of fingertip to dynamic loading. Med Eng Phys 24:253–264
10. Wu JZ, Dong RG (2005) Analysis of the contact interactions between fingertips and objects with different surface curvatures. Proc Inst Mech Eng H 219:89–103
11. Wu JZ et al (2006) Analysis of the dynamic strains in a fingertip exposed to vibrations: correlation to the mechanical stimuli on mechanoreceptors. J Biomech 39:2445–2456
12. Harih G, Tada M, Dolšak B (2016) Justification for a 2D versus 3D fingertip finite element model during static contact simulations. Comput Methods Biomech Biomed Eng 19:1409–1417
13. Wu JZ, Welcome DE, Dong RG (2006) Three-dimensional finite element simulations of the mechanical response of the fingertip to static and dynamic compressions. Comput Methods Biomech Biomed Eng 9:55–63
14. Chamoret D et al (2013) A novel approach to modelling and simulating the contact behaviour between a human hand model and a deformable object. Comput Methods Biomech Biomed Eng 16:130–140

15. Harih G, Tada M (2016) Development of a finite element digital human hand model
16. Wu JZ et al (2007) Simultaneous determination of the nonlinear-elastic properties of skin and subcutaneous tissue in unconfined compression tests. Skin Res Technol 13:34–42
17. Harih G, Tada M (2015) Finite element evaluation of the effect of fingertip geometry on contact pressure during flat contact. Int J Numer Methods Biomed Eng 31:1–13
18. Brand PW, Hollister A (1999) Clinical mechanics of the hand, 3rd edn. Mosby, St. Louis
19. Endo Y, Tada M, Mochimaru M (2014) Reconstructing individual hand models from motion capture data. J Comput Design Eng 1:1–12
20. De Looze M, Kuijt-Evers L, Van Dieën J (2003) Sitting comfort and discomfort and the relationships with objective measures. TERG Ergon 46:985–997
21. Kursa K et al (2005) In vivo forces generated by finger flexor muscles do not depend on the rate of fingertip loading during an isometric task. J Biomech 38:2288–2293

Gathering 3D Body Surface Scans and Anthropometric Data as Part of an Epidemiological Health Study – Method and Results

Dominik Bonin[1](✉), Dörte Radke[2], and Sascha Wischniewski[1]

[1] Federal Institute for Occupational Safety and Health, Dortmund, Germany
bonin.dominik@baua.bund.de
[2] Institute for Community Medicine - SHIP/KEF,
University Medicine Greifswald, Greifswald, Germany

Abstract. Gathering reliable and up-to-date full datasets of ISO-compliant anthropometric measures is time consuming and expensive. To use synergetic effects an explorative method was tested to use the environment of an existing population-based epidemiological study and extend the study design with additional three-dimensional body surface scans. Besides the standard algorithm based measure extraction, additional 34 ISO 7250-1 measures were extracted by manually identifying 44 anatomical landmarks on the 3D images. High priority was given to quality and reliability of the data. Therefore, standard operating procedures (SOPs) as well as a stringent quality assurance procedure were defined. In addition, three manually taken measures (height, waist-and hip-circumference) were compared to the extracted body scan data. The results of the manual reading process showed ambiguous results: some ISO measures showed a good intra- and interobserver reliability whereas others were difficult to identify on the scan images.

Keywords: 3D body surface scanning · Anthropometric database
Scanning methodology

1 Introduction

The use of anthropometric data offers manifold opportunities: for example for the human centered design of work places, appropriate scaling of digital human models for a subsequent individualized worksystem or enhanced ergonomic product design.

As the authors have shown in previous publications, it is possible to use publicly available data, for example from national surveys, and rework the data with a virtual synthesis, so that the data could be used without privacy concerns [1, 2]. The results were quite promising, but a synthetisized dataset still is dependent from its source data. So even if there are publicly available datasets, they may be outdated or not

---

D. Bonin and D. Radke—Equal contribution.

© Springer Nature Switzerland AG 2019
S. Bagnara et al. (Eds.): IEA 2018, AISC 822, pp. 128–140, 2019.
https://doi.org/10.1007/978-3-319-96077-7_14

comprehensive enough for a multivariate analysis. Therefore, a thorough anthropometric dataset is needed. Gathering such a dataset is time consuming and expensive. Furthermore, for manual measures, high qualified and experienced anthropologists are needed on site. The aim of this study was to use the environment of an existing epidemiological health study, to save time and curb the expenses.

The Study of Health in Pomerania (SHIP) is a population-based epidemiological health study in north-east Germany conducted by the University Medicine Greifswald [3]. As part of SHIP three body scans were performed within the standard study design. In cooperation with the Federal Institute for Occupational Safety and Health the existing measures were extended with two additional body scans to extract anthropometric measures according to ISO 7250-1:2010 [4]. The automatic measure from the original third scan was incorporated into one of the new scans. In total, four scans per participant were performed.

2 Methods

Between 2014 and 2016 3D surface scans from 1.644 participants were collected by trained and certified examiners using the Vitus smart XXL 3D body scanner (Human Solutions, Kaiserslautern, Germany). This type of body scanner was tested in earlier studies and passed the necessary accuracy tests according to [5]. Each participant was scanned in four positions; about 150 measurements calculated from a standard scan were obtained by the proprietary algorithms implemented in the AnthroScan software, version 3.0.7 (Human Solutions, Germany), during examination. In a later reading step additional 34 ISO 7250-1 measures were extracted by manually identifying 44 anatomical landmarks on the 3D images of two scan positions, sitting and standing, respectively.

Results obtained in the reading process were evaluated with SHIPs standard quality assurance procedures including extreme and missing value analysis and reader bias. Finally, 1.600 out of 1.644 datasets were used within the presented study. The others were excluded due to large numbers of missing values, e.g. as a result of compression stockings, medical appliances, or other assistive technology.

Each participant received a handout with 22 selected measurements and a black-and-white image of his or her scan as incentive.

2.1 Body Scans

The body scan examination was conducted in a predefined workflow:

- Scan 1: posture (a), ISO scan sitting
- Scan 2: posture (b), ISO scan standing
- Scan 3: posture (c), standard scan
- Manual somatometry, adding five reflective physical markers
- Scan 4: posture (d), marker scan

A visualization of the described scan postures is presented in Fig. 1. These drawings only serve as a simplified sketch within the AnthroScan wizard. The detailed

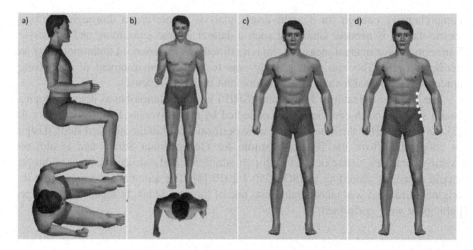

Fig. 1. Scan positions: a: ISO scan sitting, b: ISO scan standing, c: standard scan, d: marker scan (modified screenshots from the AnthroScan wizard)

posture information was provided in written form next to the drawings in the wizard, during examiner education and in the body scan SOP.

The two additional scan positions (a) + (b) were chosen on the basis of ISO 20685 recommendations [6]. The sitting ISO scan posture was slightly modified with respect to the parameter elbow-grip length and elbow-fingertip length. The subject sits erect with upper arms hanging downwards, the forearms horizontal, the left hand extended. Departing from the ISO 20685 standard (where both hands are extended), the right hand holds a measuring rod with grip axis vertical.

The standing ISO scan posture was modified from the ISO 20685 posture to include the parameter grip axis height. The participants need to stand fully erect without stiffness, feet closed and shoulders relaxed. The right upper arm is hanging downwards, forearm horizontal and parallel to the sagittal plane, with fist clenched. The left arm is hanging freely downwards, the hand holding a measuring rod in the sagittal plane, grip axis horizontal.

Posture (c) is the standard scan position for the automated, algorithm based measure extraction. Within the somatometry weight, height, waist and hip circumferences were measured manually. Weight was determined using a Soehnnle S20 scale. Height was taken with a Soehnle 5001 ultrasonic height measure employing a tubular spirit level for horizontal positioning (Soehnle Industrial Solutions, Backnang, Germany). The circumferences were taken with a standard measuring tape. Further, five reflective physical markers were attached at the left body side of the participant, identifying (1) trochanter major, (2) spina iliaca anterior superior, (3) 10th rib, and the positions of the manual (4) waist and (5) hip measurements. These markers were detected on the subsequent fourth scan (position d).

2.2   Anthropometric Parameters

The following table provides an overview of the examined anthropometric parameters. Small measures like hand, feet, and head measures (except head circumference) were excluded by definition. The data storage was conducted in agreement with ISO 15535 [7]. The first ten rows, containing information about the population, dates, location, etc. and ISO 7250-1 parameters that were not observed were skipped in this table for legibility (Table 1).

Table 1. Anthropometric measures. The column "field name" covers the numerical parameter names according to ISO 7250-1:2010, the column "field description" gives the full parameter name, and the column "origin" states, how the data were acquired (bs = body scan algorithm, bsiso = manual post processed reading values, soma = somatometry).

ID	Field name	Field description	Origin
11	4.1.01	Body mass (weight)	soma
12	4.1.02	Stature (body height)	bsiso
13	4.1.03	Eye height	bsiso
14	4.1.04	Shoulder height	bsiso
15	4.1.05	Elbow height	bsiso
16	4.1.06	Iliac spine height, standing	bsiso
17	4.1.07	Crotch height	bsiso
18	4.1.08	Tibial height	bsiso
19	4.1.09	Chest depth, standing	bsiso
20	4.1.10	Body depth, standing	bsiso
21	4.1.11	Chest breath, standing	bsiso
22	4.1.12	Hip breath, standing	bsiso
23	4.2.01	Sitting height	bsiso
24	4.2.02	Eye height, sitting	bsiso
25	4.2.03	Cervicale height, sitting	bsiso
26	4.2.04	Shoulder height, sitting	bsiso
27	4.2.05	Elbow height, sitting	bsiso
28	4.2.06	Shoulder-elbow length	bsiso
29	4.2.07	Elbow-wrist length, right	bsiso
30	4.2.08	Shoulder (biacromial) breath	bsiso
31	4.2.09	Shoulder (bideltiodal) breath	bsiso
32	4.2.10	Elbow-to elbow breath	bsiso
33	4.2.11	Hip breath, sitting	bsiso
34	4.2.12	Popliteal height, sitting	bsiso
35	4.2.13	Thigh clearance	bsiso
36	4.2.14	Knee height	bsiso
37	4.2.15	Abdominal depth, sitting	bsiso
39	4.2.17	Buttock-abdomen depth, sitting	bsiso

(continued)

Table 1. (*continued*)

ID	Field name	Field description	Origin
51	4.3.12	Head circumference	bs
56	4.4.03	Elbow-grip length, right	bsiso
57	4.4.04	Fist (grip axis) height	bsiso
58	4.4.05	Forearm-fingertip length, left	bsiso
59	4.4.06	Buttock-popliteal length (seat depth)	bsiso
60	4.4.07	Buttock-knee length	bsiso
62	4.4.08	Neck circumference	bsiso
64	4.4.10	Waist circumference	bs
65	4.4.11	Wrist circumference, right	bs
66	4.4.12	Thigh circumference, right (horizontal)	bs
67	4.4.13	Calf circumference, right	bs

2.3   Semi-automatic Reading Process

The reading of scans (a) and (b) (Fig. 1) is performed by trained and certified readers (see below) with a standardized software setup. A list of 44 preprogrammed landmarks (LM) is displayed (wizard by Human Solutions; order of landmarks adjusted by SHIP) and each LM is identified using a reticle on a three-dimensional view of the scan. The reader is presented with two views of one scan at a time showing the region for the landmark at different zoom factors and/or perspectives. At least one of the views is aligned with one of the spatial axes.

AnthroScan software supports the reader by several features: the reticle snaps to the body surface next to the mouse and reticle coordinates are displayed. In addition, individual scan views can be turned horizontally and zoomed using mouse and keyboard. With all landmarks labelled, AnthroScan calculates the values of 34 parameters. Potentially problematic parameters, e.g. neck circumference, are reviewed by the reader, landmarks adjusted and the calculation repeated, if necessary.

Each reading is documented in ShipBoss, a fully dynamic web application providing electronic case report forms (eCRF) with parameter dependencies. Readers report begin and end of the reading as well as quality-relevant parameters such as correct arm postures. A remark field is available to record all other peculiarities.

2.4   Quality Assurance

To ensure a high data quality and reliability standard operating procedures (SOPs) were defined for each part of the examination. To minimize rater-related effects a certification process for each involved examiner and reader was conducted. Quality assurance and data cleaning involved several tasks. Missings were coded, extreme values checked and observer variability evaluated. Further values were removed depending on the operational data recorded at examination and reading times. For example, body height was set to missing, if the participant refused to wear a bathing cap (excluding bald persons). Examiner and reader remarks were evaluated to exclude data of participants wearing medical appliances.

2.5 Examiner Certification

Five volunteers were examined twice by each examiner. Each examination comprised the full body scan and somatometric procedures as described above. The performance was observed and rated by a member of the SHIP quality assurance team responsible for these examinations (process certification). In addition, measurements of selected parameters were analyzed regarding their intra- and inter-observer reliabilities using Bland-Altman plots [8] (measurement certification).

2.6 Reader Certification

The certification was performed with 10 body scans. The series of readings was performed twice in random order with at least a reading-free day in-between both series. Mean bias and standard deviation were defined for a passing grade. As there were no standards available for this definition the recommendations from ISO20685 for the comparison between digitally extracted and manually taken measurements were taken as guidance. The reader with the best intra-reader reliability was chosen as reference for the inter-reader evaluation. Bland-Altman plots were used for visualization.

In addition to the landmark identification, each body scan was examined by the reader to rate the posture and arm position of the participant; the latter having an impact on the quality of several of the ISO parameters. Certification results for these categorical data were compared to a pre-defined gold standard result. The forearm parameters were binary coded (0 = correct posture, 1 = misaligned) whereas the upper arm parameters were classified in three categories: 1 = upper arm relaxed, 2 = slightly abducted, 3 = distinctly abducted. For any other peculiarities, the available remark field was used. Certification results were presented to the readers, problems discussed and calibration performed for conspicuous parameters. Depending on the number and severity of the problems the certification was repeated for some or all readers.

2.7 Comparison Manual vs. (Semi-)automatic Scan Values

Besides the distinct examiner and reader certification, three parameters were available for a direct comparison between somatometry and automatically determined AnthroScan values. For each manual measure from the somatometry, up to four parameters from the standard scan were available. Mean bias and standard deviation as well as Bland-Altman plots for visualization were evaluated to find the best matching parameter.

2.8 Statistical Analyses

All statistical analyses were performed and plots were generated using the programming language and free software environment for statistical computing and graphics that is supported by the R Foundation for Statistical Computing (R version 3.4.2) [9].

3 Results

3.1 Results Reader Certification

The results of the reader certification are summarized in Tables 2 and 3. The values show the absolute mean bias in cm of the intra- and interreader variability for each evaluated parameter. R1–R6 are the abbreviations for each certified reader, the column max ISO shows the maximum recommended differences according to ISO 20685.

Table 2. Intra-reader variability of the manual reading process. Values in cm.

Parameter	Ref	R1 MB	R2 MB	R3 MB	R4 MB	R5 MB	R6 MB	max ISO
stature_standing	Intra	0	−0.01	−0.03	−0.04	−0.05	−0.01	0.4
eye_height_standing	Intra	0	−0.22	0.12	1.26	−0.04	−0.09	0.4
shoulder_height_standing	Intra	−0.14	0	0.17	0.19	0.02	0.14	0.4
elbow_height_standing	Intra	0.05	−0.05	0.14	−0.05	−0.13	0	0.4
iliac_spine_height_standing	Intra	−0.03	−0.36	−0.45	0.15	−0.11	0.68	0.4
crotch_height_standing	Intra	0.14	0.10	0.04	0.18	0.07	0.04	0.4
tibial_height_standing	Intra	−0.64	0.93	−0.54	0.75	0.10	0.37	0.4
chest_depth_standing	Intra	−0.13	−0.06	−0.11	−0.02	−0.10	0.16	0.5
body_depth_standing	Intra	−0.06	0	−0.02	0.03	0	−0.05	0.5
chest_breadth_standing	Intra	−0.07	−0.17	−0.02	−0.16	0	−0.04	0.4
hip_breadth_standing	Intra	0.02	−0.04	−0.13	0	0.11	0.02	0.4
grip_axis_height_standing	Intra	−0.03	0.01	0.03	0	−0.07	−0.06	0.4
sitting_height	Intra	0.04	−0.09	0.06	−0.07	0.08	−0.55	0.4
eye_height_sitting	Intra	0.07	−0.13	0.28	−0.10	0.10	−0.57	0.4
cervicale_height_sitting	Intra	0.41	−0.06	0.24	0.01	−0.04	−0.28	0.4
shoulder_height_sitting	Intra	0.02	0.10	0.31	−0.09	−0.12	−0.39	0.4
elbow_height_sitting	Intra	−0.01	−0.08	−0.03	−0.10	0.06	−0.66	0.4
shoulder_elbow_length	Intra	0.04	0.17	0.37	0.04	−0.19	0.27	0.5
elbow_wrist_length	Intra	0.07	−0.37	−0.31	−0.23	0	0.14	0.5
biacromial_breadth	Intra	0.23	−1.63	−0.63	−0.12	0.80	−0.73	0.4
bideltoid_breadth	Intra	−0.28	0.40	0.16	0.20	0.79	−0.15	0.4
elbow_to_elbow_breadth	Intra	−0.48	−0.25	0.07	−0.15	−0.05	−0.30	0.4
hip_breadth_sitting	Intra	−0.32	0.09	−0.35	0.15	0.66	−0.16	0.4
popliteal_height_sitting	Intra	0.12	0.01	0.19	−0.09	−0.03	−0.08	0.4
thigh_clearance_sitting	Intra	−0.08	0.40	0.29	−0.01	−0.09	−0.64	0.5
knee_height_sitting	Intra	−0.10	−0.28	−0.55	0.49	−0.30	−0.19	0.4
abdominal_depth_sitting	Intra	−0.08	−0.05	0.03	0.04	−0.01	−0.10	0.5
thorax_depth_at_nipple	Intra	1.13	0.09	−2.18	0.41	−0.53	0.52	0.5
buttock_abdomen_depth_sit.	Intra	−0.08	0.03	0	−0.08	−0.05	−0.05	0.5
elbow_grip_length	Intra	−0.06	−0.15	0.31	−0.03	0.03	−0.29	0.5
forearem_fingertip_length	Intra	0.01	0	−0.02	−0.05	−0.29	−0.32	0.5
buttock_popliteal_length_sit.	Intra	−0.01	0.15	0.03	−0.08	0.05	−0.10	0.5
buttock_knee_length_sitting	Intra	−0.04	−0.02	−0.05	0	−0.02	−0.03	0.5

Table 3. Absolute inter-reader reliability. The column ref indicates the reader with the best intra-reader reliability that served as reference.

Parameter	Ref	R1 MB	R2 MB	R3 MB	R4 MB	R5 MB	R6 MB	max ISO
stature_standing	R1		0.02	0.02	0.02	−0.03	0.03	0.4
eye_height_standing	R1		0.21	−0.07	1.13	0.20	−0.22	0.4
shoulder_height_standing	R2	0.30		0.21	0.80	0.06	0.36	0.4
elbow_height_standing	R6	0.39	0.14	0.04	−0.12	0.05		0.4
iliac_spine_height_standing	R1		−2.46	−1.35	−1.54	2.11	−0.01	0.4
crotch_height_standing	R3	−0.01	−0.09		0.02	−0.22	0.01	0.4
tibial_height_standing	R5	−0.73	−0.27	0.94	0.90		0.11	0.4
chest_depth_standing	R4	1.18	0.21	0.86		0.18	−0.02	0.5
body_depth_standing	R2	−0.02		−0.02	0	0	0	0.5
chest_breadth_standing	R5	0.08	0.19	0.13	−0.34		−0.01	0.4
hip_breadth_standing	R4	−0.27	−0.16	0.39		−0.26	−0.31	0.4
grip_axis_height_standing	R4	−0.09	−0.05	−0.18		−0.18	−0.07	0.4
sitting_height	R1		0.18	−0.27	0.17	0.26	−0.02	0.4
eye_height_sitting	R1		0.39	−0.24	0.46	0.33	−0.28	0.4
cervicale_height_sitting	R4	−0.31	0.78	1.38		0.64	0.71	0.4
shoulder_height_sitting	R1		−0.14	−0.31	0.19	0.15	0.15	0.4
elbow_height_sitting	R1		0	−0.73	−0.42	−0.19	−0.49	0.4
shoulder_elbow_length	R1		−0.18	0.40	0.59	0.36	0.64	0.5
elbow_wrist_length	R5	0.57	0.02	0.82	0.53		−0.52	0.5
biacromial_breadth	R4	0.06	1.17	−0.23		0.54	−1.48	0.4
bideltoid_breadth	R6	0.32	0.82	0.68	1.01	−0.96		0.4
elbow_to_elbow_breadth	R5	0.82	−0.50	0.91	1.32		0.67	0.4
hip_breadth_sitting	R2	0.38		0.59	1.25	−0.34	0.41	0.4
popliteal_height_sitting	R2	0.09		−0.18	0	−0.17	−0.08	0.4
thigh_clearance_sitting	R4	−0.01	0.17	−0.45		0.43	−0.09	0.5
knee_height_sitting	R1		0.65	0.08	−4.46	0.39	−0.62	0.4
abdominal_depth_sitting	R5	−0.01	−0.02	0.11	0.06		0.02	0.5
thorax_depth_at_nipple	R2	−0.35		−1.82	−1.10	−1.04	−1.01	0.5
buttock_abdomen_depth_sit.	R3	−0.10	−0.04		0	0.01	−0.09	0.5
elbow_grip_length	R4	0.27	−0.20	−0.15		−0.47	−0.58	0.5
forearem_fingertip_length	R2	0.25		0.34	0.16	−0.40	−0.15	0.5
buttock_popliteal_length_sit.	R1		0.01	−0.21	0.39	−0.41	−0.58	0.5
buttock_knee_length_sit.	R4	0.02	0.04	−0.11		0.08	0.02	0.5

Due to reader independent problems with the algorithm based extraction of the parameter "4.2.16: thorax depth at nipple" this parameter was not included in the evaluation of the reader certification.

3.2 Results Categorical Evaluation

Weighted Cohen's kappa was used to assess the accordance of the categorical parameters to the pre-defined gold standard [10]. To circumvent the limitations known to exist for binary outcomes [11, 12] further statistical parameters (p_0, p_{first}, p_{second})

Table 4. Accordance of categorical parameters for all six readers (R1–R6) in relation to the gold standard. The values represent weighted Cohen's kappa for the upper arm parameters consisting of three categories each and p_0 for the forearm parameters consisting of two categories each.

Parameter	R1	R2	R3	R4	R5	R6
sitting_left_upper_arm	0.00*	0.72	0.50	0.64	1.00	1.00
sitting_right_upper_arm	0.52	0.80	0.77	0.60	0.69	0.69
standing_left_upper_arm	1.00	0.84	0.77	1.00	1.00	1.00
standing_right_upper_arm	0.60	0.85	0.26	0.69	0.69	0.69
sitting_left_forearm_front				0.95		0.95
sitting_left_forearm_horiz						0.95
sitting_right_forearm_front		0.90		0.95	0.90	0.95
sitting_right_forearm_horiz	0.85	1.00	0.90	1.00	0.95	1.00
standing_right_forearm_front	0.85	0.85	1.00	0.90	0.95	0.95
standing_right_forearm_horiz	0.95	0.90				

*Reader one only used a single category for the parameter sitting_left_upper_arm. Therefore it can be treated as a two-category comparison with $p_0 = 0.90$.

were calculated; p_0 representing the total accordance. Table 4 shows the values for weighted kappa and p_0 for the three-category and two-category parameters, respectively. Empty cells represent overall agreement with a single category only.

3.3   Results Comparison Manual vs. (Semi-)automatic Scan Values

Manual somatometric measurements were compared to the related (semi-)automatic measurements employing Bland-Altman plots and mean bias calculations; the results

Table 5. Mean bias and standard deviation of automated scan values compared to manual somatometry, * indicating the best matching parameter

Parameter	Sex	n	MB [%]	SD [%]	MB [cm]	SD [cm]
sta_stature*	Male	740	0.3	0.6	0.5	1.1
sta_stature*	Female	845	0.5	0.5	0.8	0.8
circ_waist	Male	737	1.5	2.1	1.6	2.1
circ_waist	Female	849	3.8	3.8	3.5	3.3
circ_high_waist*	Male	737	0.5	2.4	0.5	2.5
circ_high_waist*	Female	849	0.4	3.6	0.2	3.3
circ_hip	Male	738	4.7	3.3	4.8	3.4
circ_hip	Female	851	5.6	4.3	5.8	4.3
circ_high_hip	Male	738	3.4	4.8	3.8	5.2
circ_high_hip	Female	853	−4.0	4.1	−4.0	4.2
circ_middle_hip*	Male	737	3.0	4.0	3.3	4.4
circ_middle_hip*	Female	849	0.8	2.7	0.8	2.9
circ_buttock	Male	738	3.6	2.9	3.7	3.0
circ_buttock	Female	853	4.3	3.6	4.5	3.6

are shown in Table 5 and one example plot for one length and one circumference measurement is presented in Fig. 2. The manual body height measures (soma: som.-groe) were compared with parameter 4.1.02: stature, standing (bsiso_sta_stature).

To compare the waist circumferences, two different parameters from the automated scans were available in AnthroScan, waist (circ_waist) and high waist (high_waist). The latter was showing a better accordance with the manual somatometry. Regarding the hip measurements, the parameter middle hip showed the best accordance of the four different hip measures available in AnthroScan (see Table 5).

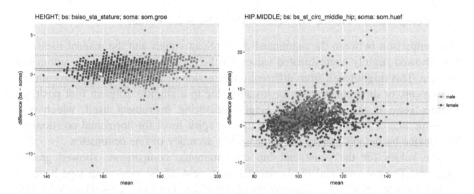

Fig. 2. Bland-Altman plots comparison manual vs. (semi-)automatic scan values. Left: body height as example for length measures. Right: hip middle as example for circumference measures. bs = default body scan, soma = manual somatometry

4 Discussion

The primary goal for the presented study was a broad collection of anthropometric measures as add-on to an existing population-based health study. For this kind of examination a narrow timeframe is probably one of the most challenging factors. The standard examination procedure cannot be disturbed arbitrarily as the participants' availability is dependent on the study workflow. The merging of posture b and c from ISO 20685 to the presented ISO standing scan posture and the minor changes of the arm posture in the ISO sitting scan offered the possibility to capture the targeted ISO measures with only two additional scans. As a tribute to the timeframe labelling landmarks with physical markers on the participant prior to scanning is not realizable for all positions as the process is very time consuming and thus not feasible for such a high number of participants. In addition, the labelling itself requires a highly experienced anthropometrist, as labelling errors also have a major influence on measurement quality [13]. To cope with these challenges a process was designed where the data collection is independent from the analysis, while still maintaining the highest possible quality standards. Thus, a thorough training of the examiners is an important prerequisite. An error occurring during reading still can be corrected at a later timepoint, but a bad scan image is not changeable after the participant has left the investigation center.

Therefore, the participant's posture is checked twice: first by the examiner prior to saving, a second time by the scan reader prior to the reading process. The mandatory categorical classification of the arm positions serves as a quick indicator for the posture quality, which worked well in practice. However, during analysis the requirement for additional optional fields to describe problems in the overall posture became apparent. As it is difficult to later separate such entries from the common remark field, the extra fields would simplify the identification of posture related measurement errors. For following studies, optional fields for the rating of the sitting and standing postures were added.

4.1   Comparison Manual vs. (Semi-)automatic Scan Values

The comparison between the automated AnthroScan values and the manual measurement showed slightly overestimated values for the parameter stature by the automated measure. This might result from different compression of the hair layer underneath the bathing cap, or different posture for instance due to a slightly different foot position. Another issue is the used manual height measuring instrument itself, which is a ultrasonic height measure employing a tubular spirit level for horizontal positioning. This device by default merely has a measuring accuracy of one centimeter.

The values for the waist and hip circumference comparisons showed greater variances. The pressure and the horizontal alignment of the measuring tape have an influence on the reliability, whereas the AnthroScan function "closedTapeMeasure" acts across the most outer points of the body shape in one horizontal plane. The mean biases were still in an acceptable range, but showing distinctly higher standard deviations than the height measures. This strengthens the assumption that the body shape had a major influence on the measurement of circumferences. This result is in line with previous comparison studies. For example, Han [14] found that the difference between the methods increase significantly with increasing BMI. However, it is noteworthy, that the results of the comparison based on automated value extraction throughout showed higher variances than recommended in ISO 20685.

4.2   Reader Certification

In this first part of the study, six persons were certified for the manual reading. These readers initially were non-experts who were trained before the certification. The intra-reader reliability showed quite promising results for the initial certification, except for a few parameters like tibial height, biacromial breath or thigh clearance (see Table 2). Without the possibility to palpate bony landmarks on the scan images, a few of the landmarks are difficult to identify. We therefore expected a higher variability in the measurements based on these landmarks.

The inter-reader variability showed ambiguous results (see Table 3). The fact that the reader with the best intra-reader reliability was chosen as the reference may lead to a possible misinterpretation, e.g. if one reader is consistently away from the desired landmark, the intra-reader reliability is good but will not represent the desired value. For the future, the definition of landmark standards for each scan image is in progress, so that every reader will be compared to a fix reference.

Higher intra- and inter-reader differences may be the result of the chosen method, i.e., due to the missing physical markers, as well as the selection of scans, some of which were intentionally more difficult to read. For example, the inclusion of scans with misalignments in posture is used to certify the categorical parameters in addition to the measurement parameters.

In summary, the presented study successfully implemented the extraction of ISO 7250-1 based measurements from body scans in an existing population-based epidemiological health study. The use of the existing infrastructure and patient management worked well, especially in due consideration of the narrow timeframe. The additional scans took about four to five minutes, which was feasible to implement within the standard workflow. The readers were comparably unexperienced at the beginning, but were consistently trained. A continuous quality management ensures high quality scan images and documents reader and time effects regarding the reading parameters. One advantage of this method is, that once a good scan image is stored, a related reading parameter still can be reevaluated, e.g. if better algorithms or identification methods will be developed or a relevant standard changes. Therefore, this method shows a reasonable potential for being used for the extensive collection of 3D scans for subsequent extraction of anthropometric measures.

5 Conclusion and Outlook

The use of existing structures from established epidemiological surveys seems to be a good starting point for a broad and effective collection of anthropometric datasets. There are several benefits like the shared participant recruitment, the availability of examiners and existing infrastructure. The presented method extends the automated algorithm based data extraction with a manual post processing. This reading process allows the extraction of ISO 7250-1 measures, independent from the location or locally available anthropologists. The categorical judgement of body scan postures previous to the reading process serves as a good indicator for quality and comparability of the data, as the body posture has an influence on several ISO 7250-1 measures.

Outlook: For the manual reading parameters, a validation study (n = 40) is currently in process comparing all ISO reading measures with classic manual measurements with anthropometer and beam caliper. The results will be available by the end of 2018. The data collection is still ongoing; a survey of the second SHIP cohort (SHIP-Trend-1) will deliver additional body scan data from up to 3000 participants in 2019. However, the results of the validation study will show if the collected datasets can be used in ISO 7250-1 conform databases.

Acknowledgement. We thank all SHIP participants, examiners, and readers. The Study of Health in Pomerania is part of the Community Medicine Research Network of the University Medicine Greifswald, which was supported by the German Federal State Mecklenburg-West Pomerania. The extension to the body scan examination and reading was funded by the Federal Institute for Occupational Safety and Health (BAuA), Germany.

References

1. Wischniewski S, Bonin D, Grötsch A (2015) Virtual anthropometry—synthesis and visualisation of virtual anthropometric populations for product and manufacturing engineering. In: The Proceedings of the 19th Triennial Congress of the International Ergonomics Association—Melbourne, 9–14 Aug 2015
2. Wischniewski S, Grötsch A, Bonin D, Parkinson M (2017) Synthesis and validation of a virtual anthropometric user population of German civilians based on an up-to-date representative dataset. May ed. BAuA: Focus. BAuA, Dortmund
3. John U, Hensel E, Lüdemann J, Piek M, Sauer S, Adam C, Born G, Alte D, Greiser E, Haertel U (2001) Study of Health in Pomerania (SHIP): a health examination survey in an east German region: objectives and design. Sozial-und Präventivmedizin 46(3):186–194
4. International Organization for Standardization, ISO 7250:2010: Basic human body measurements for technological design
5. Kouchi M, Mochimaru M, Bradtmiller B, Daanen H, Li P, Nacher B, Nam Y (2012) A protocol for evaluating the accuracy of 3D body scanners. Work 41(Supplement 1): 4010–4017
6. International Organization for Standardization, ISO 20685, 3-D scanning methodologies for internationally compatible anthropometric databases
7. International Organization for Standardization, ISO 15535, General requirements for establishing anthropometric databases
8. Martin Bland J, Altman D (1986) Statistical methods for assessing agreement between two methods of clinical measurement. Lancet 327(8476):307–310
9. R Core Team (2017) R: a language and environment for statistical computing. R Foundation for Statistical Computing, Vienna
10. Viera AJ, Garrett JM (2005) Understanding interobserver agreement: the kappa statistic. Fam Med 37(5):360–363
11. Feinstein AR, Cicchetti DV (1990) High agreement but low kappa: I. The problems of two paradoxes. J Clin Epidemiol 43(6):543–549
12. Cicchetti DV, Feinstein AR (1990) High agreement but low kappa: II. Resolving the paradoxes. J Clin Epidemiol 43(6):551–558
13. Kouchi M, Mochimaru M (2011) Errors in landmarking and the evaluation of the accuracy of traditional and 3D anthropometry. Appl Ergon 42(3):518–527
14. Han H, Nam Y, Choi K (2010) Comparative analysis of 3D body scan measurements and manual measurements of size Korea adult females. Int J Ind Ergon 40(5):530–540

Virtual Reality Simulation and Ergonomics Assessment in Aviation Maintainability

Fabien Bernard[1,2(✉)], Mohsen Zare[1], Jean-Claude Sagot[1],
and Raphael Paquin[2]

[1] ERCOS Group (Hub), Laboratory of ELLIADD-EA4661,
UTBM-University of Bourgogne Franche-Comté, Belfort, France
[2] Airbus Helicopters, Marignane, France
fabien.bernard@airbus.com

Abstract. This study aims to know better the potential of simulation tools used currently by maintainability engineers to analyse Human Factors/ergonomics (HFE). Non-ergonomics experts can use digital/physical simulation tools through virtual reality platforms and physical mock-ups to analyse whether the design is well adapted to future users, especially maintenance operators in the aviation field. Knowing the potential of these simulation tools would be the primary step in developing a new way of working for engineers to integrate HFE better in the design process of the aviation industry.

Keywords: Human factors/ergonomics · Simulation tools · Virtual reality
Physical mock-up · Maintainability · Aeronautics

1 Introduction

Human performance has been already studied in the aviation field, especially in regard to aircraft accidents [1]. At the beginning of the aviation story, the focus was to improve the technology and the reliability of machines [1]. During the Second World War, Human Factors/Ergonomics (HFE) were considered to anticipate the interaction between machine and pilot (particularly during the initial phase of pilot training in a simulator) and to design the flight deck [1, 2]. HFE was also used to improve safety and comfort in the cockpit and the passenger cabin during the following decades [3]. However, root-cause analyses of aviation accidents show that 12% of all accidents were due to HFE reasons during the maintenance activity [4]. Since design engineers in the maintainability department interface with other engineering activities (e.g., aerodynamic, hydraulic and electric integration, and architecture), HFE in maintainability could enhance the quality of maintenance activities and reduce the percentage of mistakes/errors [5]. HFE can be integrated into the early design process of maintainability and proactively evaluate the interaction between the future maintenance activity of the operator and the system component.

To consider HFE in the design process of maintainability, design engineers immerse the maintenance activities in a digital mock-up (DMU) and physical mock-up (PMU) in the early design phase [6]. They use different ergonomic tools to proactively analyse the future interaction between human activities and the system components

© Springer Nature Switzerland AG 2019
S. Bagnara et al. (Eds.): IEA 2018, AISC 822, pp. 141–154, 2019.
https://doi.org/10.1007/978-3-319-96077-7_15

either in DMU or PMU. Maintainability engineers often use their background/ experience and focus almost entirely on the physical dimension of HFE due to the lack of ergonomic skills and knowledge [7, 8]. They often underutilize the potential of DMU and PMU for performing ergonomics analysis on simulated maintenance tasks [9]. Virtual reality (VR) - as a typical DMU tool- is used to study the accessibility of the operator and his interaction with the architecture in maintainability [10]. Being easy to edit and cost-effective, VR is widely used by the aviation industries to evaluate the ergonomic features of a task, while PMU to simulate maintenance tasks is less present [11]. However, immersive interaction alone is not sufficient to represent the full interaction and assess the real operator's behavior. The aviation industries often use PMU to design space forms and architectural aesthetics. PMU is modified and optimized as the design of the product progresses. However, this design approach is more focused on technic (techno-centric design) rather than on the interactions between operators and the components of the architecture (anthropo-centered design) regarding the organization, mental processes (e.g., perception, reasoning), and motor responses [12–14]. Previous studies have shown the potential of PMU for performing ergonomic analyses, especially for the workstation design by directly involving the operators [15, 16]. PMU could enable the engineers to study the interactions between the operators and the future design, to collect feedback from the operators, and to investigate cognitive, organizational and physical workloads objectively [16].

Few studies have investigated the contribution of DMU and PMU to the design process. Pontonnier et al. [17] showed a significant difference between the results of VR compared to the real environment in various aspects such as performing a task efficiently, time management, natural movements, and access to different zones [17]. Aromaa et al. [18] showed that both simulations - virtual reality and augmented reality - are suitable for evaluating ergonomic factors for the designer in the maintenance platform of a rock crushing machine. The challenge in the maintainability department of the aeronautical industry is to better integrate the HFE during the design phases through the simulation tools (DMU and PMU). Engineers will use the simulation tools more efficiently when they know the differences and complementary of these tools, particularly for assessing ergonomic factors. The first step in this improvement is to study the existing gap between the simulation tools.

This study was, therefore, designed to investigate the gap between ergonomics analysis performed by DMU and that by PMU. We hypothesize that this investigation would help the designers to use the simulation tools by considering their limitations. First, we present the general context of study, maintainability and the most common simulation tools used to perform maintainability and ergonomics analysis. Subsequently, the experimentation, and the subjective and objective ergonomics assessments in both settings (DMU and PMU) are presented and compared.

2   The Context of the Study

According to the European standard EN 13306:2010 "Maintenance and terminology of maintenance" [19], maintainability defines the "ability of a product to be maintained or repaired easily during maintenance operations that include functions for which it was designed". Maintainability stakeholders must anticipate the future maintenance operator activities in three main ways: architecture/toolings/operator's way of work [20]. Taking into account the human activity in maintenance is a challenge in maintainability, particularly regarding customer feedback. The aeronautical industry has made a critical observation about maintenance because the tasks are expensive, reaching between 12–15% of the total airline costs per year [21]. Furthermore, the duration of each maintenance task (referred to as Mean Time To Repair; MTTR) must be respected [22]. The ratio of "hours of maintenance/flight hours" is a priority, particularly in the field of helicopters. Additionally, the French Association of Maintenance Engineers (AFIM) conducted a survey among a sample of 2500 maintenance professionals from various trades in industrial and service sectors which revealed that 62% of them consider their occupation causes pain [23]. Improving health and safety remain major issues of maintainability studies.

To perform their analysis, the maintainability stakeholders used simulation tools, digital or physical [24, 25] but very few HFE objective outcomes were extracted [7, 8]. However, simulation tools are also used by ergonomists to carry out activity analysis, proving their efficiency at covering all categories of indicators, organizational, economic and technical, integrating HFE in the early design process [26, 27].

3   Experimentation

3.1   Research Setting

The PMU used for our experiment represents the upper deck of a helicopter, where the engine is located as illustrated in Fig. 1.

Fig. 1. Helicopter representation with upper deck as the work area

To access the motors, one solution is to create scaffolding allowing the operator access to the right engine or the left engine. Figure 2 shows the top view of the upper deck. The maintenance operation will be carried out from the corridor of maintenance scaffolding. Depending on the arrangement of the technical elements, not all symmetrical, the manipulations are carried out on both sides, left and right. Some tasks must or can be performed through hatches requiring work to be done standing, kneeling or lying down.

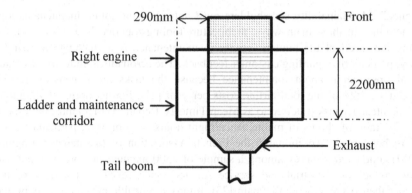

Fig. 2. Top view of upper deck layout with ladder and maintenance corridor

3.2 Specification of Simulation Tools

We used the VR technology in the factory being studied to simulate eleven maintenance tasks. The virtual and real worlds were synchronized in this experimentation so that we kept the simulation setting as close as possible to the usual process and the current industrial context. The VR platform is included in the design process and offers interaction between the user and the immersive environment [28, 29]. The VR platform, called RHEA (Realistic Human Ergonomics Analysis), is based on "Virtools" software ensuring a high level of realism. This software allows a quick transfer with the CAD system to reduce the loop in the design process. For this study, we integrated physical parts in VR: elements to simulate the contacts with the

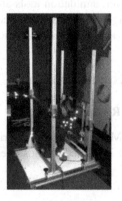

Fig. 3. Physical part integrated during the virtual reality simulation

architecture of the upper deck, components for assembling/disassembling to reproduce the real volume, weight and gravity center, and a tracker to carry a component virtually (Fig. 3).

The PMU designed for this study was composed of wooden parts and the real components of the upper deck such as a generator. Each operator performed the task in realistic conditions, with real architecture constraints, and the removable parts (having real weight and gravity center). The torque force and small removable components (screws, nuts) were also set up.

To analyze the HFE, we put two cameras on both simulation tools to record the maintenance activity without disturbing the operator. The position of the camera was ideally placed to record the front, back and sagittal view and all displacement into and around the work area.

3.3 Maintainability Tasks Selected

From this work situation, eleven maintenance tasks were evaluated. They were selected for their recurrence and their representativeness in relation to the whole helicopter in terms of postures, gestures or effort among other things. These tasks are included in the regular and mandatory maintenance, performed every week for the higher frequency (task 1) and every month for the least frequent (task 7) (Table 1).

We will describe the VR platform and PMU in the following paragraph.

Table 1. Description of each task and the main postures associated

Task	Description	Main posture
1	Accessibility between ladder and upper deck	Standing
2	Disassembly/assembly of the main generator	Standing and bending
3	Accessibility of the electric plug behind the fire wall	Kneeling
4	Manipulation of engine electric plug (floor)	Kneeling/lying down
5	Disassembly/assembly of the hydraulic pump	Kneeling/standing
6	Accessibility of the drain collector	Kneeling
7	Disassembly/assembly of the auxiliary generator	Standing and bending
8	Disassembly/assembly of engine support	Lying down
9	Tightness check of the RPM	Kneeling
10	Tightness check of the suspension support	Kneeling
11	Tightness check of the main rotor actuator	Kneeling

3.4 Participants

Six male operators were selected according to the standard NF EN ISO 15537 "Principles for selecting and using test persons for testing anthropometric aspects of industrial products and designs" [30]. Two main characteristics were considered in selecting the subjects: anthropometric parameters (height 176.3 ± 7.3 cm, weight 81.5 ± 13.4 kg) and work experience in terms of knowledge and expertise concerning maintenance of helicopter engine systems. The mean age and the length of work experience were 44.5 (± 7.6) years and 18.5 (± 8) years, respectively. We included only subjects without musculoskeletal disorders and all the people consented to participate in our study. One operator (out of six subjects) could not finish all of the eleven tasks in the VR platform because he had motion sickness preventing him from performing all the tasks (only 4 out of 11).

The experimentation started on the physical mockup, with each of six operators performing the eleven tasks individually over two hours. Between each operator, we took half an hour to recondition the simulation tool in its initial setup. The same protocol was deployed within the VR platform six weeks later. In total, 24 h of experimentation were analyzed based on direct observation and video recording.

3.5   The Investigation Instrument

Table 2 presents the indicators and the measurement tools used in studying them. The activity of the operators was analyzed through two indicators commonly used in maintainability: physical accessibility and both assembly and disassembly [26]. They are often associated with indicators dedicated to ergonomic analysis. We have thus selected five indicators that allow us to approach all the dimensions of HFE as defined by the IEA, namely the physical, cognitive and organizational dimensions [31]. Several measurement tools to quantify them have been used, namely:

Table 2. Ergonomic indicators related to maintenance tasks and measurement tools used to evaluate these indicators.

Indicators	Measurement tools	Metrics
Posture	Standard EN 1005-4:2008	Unacceptable posture
		Acceptable posture under conditions
		Acceptable posture
	RULA	Scale from 1 to 7
Muskuloskeletal symptoms	Nordic Questionnaire	Neck
		Shoulders
		Upper body
		Elbows
		Lumbar
		Wrists/hands
		Hips
		Knees
		Ankles/feet
Effort	Borg (perceived physical exertion)	Scale from 0 to 10
Safety at work area	APACT check-list (section 9)	Scale from 0 to 10
Time	Duration	Task duration
Mental Worklaod	NASA-TLX	Scale from 0 to 100

- The Standard EN 1005-4:2008 [32] (Safety of machinery - Human physical performance - Evaluation of working postures and movements in relation to machinery). It details the operator's activity depending on the posture (angle of main articulations), dynamic or static effort and time and divides the activity into three parts: Unacceptable posture; Acceptable under conditions; Acceptable.
- The RULA (Rapid Upper Limb Assessment) assesses the exposure of individual operators to ergonomic risks associated with upper extremity MSD. It takes into account three main data means: force, frequency and posture [33]. We used this tool to measure the posture of our subjects in a specific task.

- The Borg scale to measure the perception of effort during physical activity. Its recent version CR10 rated from 0 for no effort to 10 for an effort considered as "very, very hard" was retained in accordance with the work of the author [34].
- Safety work area by the APACT check-list (Association for Improvement of Working Conditions) [35] with 22 parameters evaluating the HFE of the workplace in terms of ergonomic standards and standards. Essentially, we used parameter 9, corresponding to "work area safety", to assess the risks associated with working at height, falls, burns or breaks by combining the level of personal protection and the inviolability of work protection thus corresponding for us to the organizational dimension of the task. The score is ranked from 0 to 10 and considered damaging below 6.5.
- NASA-TLX [36] is a subjective assessment of mental workload based on six dimensions: mental demand, physical demand, time requirement, personal performance, frustration and stress. It allows an overall score rated from 0 to 100 to be obtained, specifying if the operator is in underload (<30) or mental overload of work (>30).
- Nordic questionnaire [37]: we will use the body map of musculoskeletal regions based on the Standardized Nordic.

The results were statistically analyzed by the T-test to investigate the differences between the indicators.

4 Results

The mean percentage of exposure time to unacceptable posture for DMU was 49.8 ± 19.2, and for PMU 55.8 ± 21.1. The data for acceptable posture under conditions was 34.4 ± 13.4 and 28.3 ± 15.2 for DMU and PMU, respectively. The mean percentage of exposure time to unacceptable posture and acceptable posture under conditions were significantly different between DMU and PMU (Fig. 4).

The participants in the study reported different musculoskeletal symptoms when performing maintenance tasks on PMU compared to DMU. Musculoskeletal symptoms were significantly higher for neck on PMU (0.1 ± 0.02) than on DMU (0.02 ± 0.001). The time needed to perform tasks was significantly higher on DMU than on PMU. The mean of mental workload for PMU differed significantly between PMU and DMU (35.55 ± 1.1 vs 42.57 ± 1.3, respectively). For the other tools (Rula, Borg and APACT), we found no significant difference between DMU and PMU. Table 3 shows the results of the different ergonomic tools used to assess the maintenance tasks in DMU and PMU.

Table 4 presents the score of assessments for perceived physical exertion (Borg scale), duration and mental workload, task by task, between digital mock-up and physical mock-up. We observe, for each task, a difference between the digital and physical tool, except for Borg. Duration and NASA-TLX are clearly different, confirming the previous statistical observation. We also observe the highest scores for tasks 2 and 10 due to the long time and the complexity of the accessibility associated with a static effort to retain an alternator of 18 kg. A long static effort quickly causes pain; the

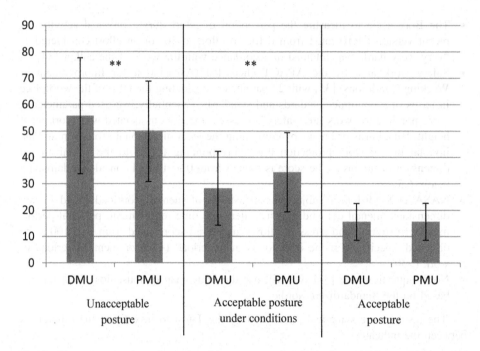

Fig. 4. The percentage time of unacceptable posture, acceptable under conditions and acceptable between DMU and PMU (**p < 0.01)

Table 3. Indicators assessed by both physical and digital tools (*p < 0.05; **p < 0.01)

Ergonomics tools	N	PMU mean (SE)	DMU mean (SE)
RULA	59	5.76 (0.22)	5.75 (0.22)
Musculoskeletal symptoms			
Neck*	59	0.1 (0.04)	0.02 (0.003)
Shoulders	59	0.05 (0.01)	0
Upper back*	59	0.13 (0.05)	0.03 (0.002)
Elbows	59	0.07 (0.02)	0.02 (0.003)
Lower back**	59	0.34 (0.04)	0.15 (0.05)
Wrists/hands	59	0.02 (0.006)	0.05 (0.005)
Hips	59	0.08 (0.007)	0.07 (0.07)
Knees	59	0.25 (0.05)	0.27 (0.09)
Ankles/feet	59	0.03 (0.004)	0.08 (0.005)
Perceived physical exertion (Borg scale)	59	2.64 (0.26)	3.01 (0.23)
Safety work area	59	5.20 (0.16)	5.20 (0.16)
Duration of task execution**	59	225.95 (31.04)	139.10 (6.4)
Mental workload (NASA TLX)**	59	35.55 (2.21)	42.57 (2.59)

Table 4. Task execution duration and mental workload for each task on digital mock-up (DMU) and physical mock-up (PMU)

	Duration of task execution (sec) Mean (SE)		Mental workload (score 0–100) Mean (SE)	
	DMU	PMU	DMU	PMU
Task 1	165 (9)	7.8 (0.2)	34.3 (6.1)	18.2 (5.9)
Task 2	84 (5.3)	393 (52)	61.3 (7.2)	48.8 (6.5)
Task 3	141 (11.2)	75 (11)	35.3 (6)	27.6 (3.7)
Task 4	184 (12.1)	221 (7.5)	37.6 (4.9)	30.1 (2.7)
Task 5	179 (9.3)	696 (140)	45.1 (7.1)	41.5 (5.3)
Task 6	56 (4.8)	12 (0.7)	32.4 (5.2)	28.1 (5.3)
Task 7	175 (8.6)	80 (4.3)	41.1 (5.6)	31.7 (6)
Task 8	114 (9.9)	239 (34)	43 (6.2)	37.8 (6)
Task 9	99 (8.1)	121 (21)	43 (4.6)	26.8 (3.9)
Task 10	168 (7.2)	393 (56)	51 (7.3)	56.1 (8.7)
Task 11	170 (6.9)	227 (48)	42 (5.5)	36.7 (5.3)

operator was clearly irritated and wanted to accomplish the task as soon as possible, reinforcing the mental workload. Tasks 1 and 9 were the easiest, respectively a displacement in the engine cowling and checking and tightening torque - very light effort - yet we observe the biggest gap in the mental workload, higher by an average 16 points for the DMU. Overall, mental workload is higher for the digital simulation than for the physical simulation in every maintenance situation studied.

5   Discussion

To help maintainability for stakeholders, simulation tools seem the most appropriate technology to perform reliable and understandable ergonomics studies for non-HFE experts. To better understand the efficiency of two main simulation tools, VR (in the digital category) and physical mock-up (in the physical category), this paper aims to compare both simulation tools through a specific industrial case study. Some relevant indicators have been studied through a deep ergonomics analysis - with the same operators and the same maintenance tasks - covering three dimensions defined by IEA (2000), physical, organizational and cognitive. Firstly, the category of posture is significantly different between the digital and physical simulation tool for unacceptable posture and acceptable posture under certain conditions. However, the overall proportion seems to be respected whatever the simulation tool; the longest part being performed with an unacceptable posture, the second being carried out with an acceptable posture under certain conditions and the last with an acceptable posture. Additionally, through all the maintenance tasks studied, we observed a significant difference between digital and physical simulation tools for three indicators out of six: musculoskeletal symptoms, duration of task execution, and mental workload. However, the ergonomic indicators - posture evaluated by RULA, perceived physical exertion assessed by Borg and safety at work area assessed by APACT - were not significantly different between the simulation tools.

Considering the physical mock-up as the reference, we can explain the difference by two main complementary reasons on the VR platform: physical contacts are not fully representative and the environment perception in DMU can cause a few disruptions. However, according to the Borg scale, the difficulty in perception assessed by the operator is similar to that in physical mock-up. We can justify it by the realism of contacts and forces with real parts and maintenance tools to track on the VR platform. Additionally, few factors impact the realism through the digital technology. Firstly, the space of work is limited by the volume of the room, to better perceive the real work space new technologies are emerging [38]. The environment perception can be altered by the quality of initial Computer Aided Design (CAD) transferred to VR technology and also the graphic displayed inside the helmet offering various characteristics like field of view [39]. Mainly, the spatial perception is not fully realistic [40] regarding the gap between the real environment perception and that already known by the maintenance operators. Even if the maintenance operator has been trained previously (one week before the experimentation), he can be unsettled and need a few minutes to find his bearings in a virtual environment compared to his knowledge of the real world. Motion sickness also appears as a far from negligible problem especially during a long virtual simulation [41]. It was necessary to have a long break between the tasks for an operator to continue the experimentation in safe conditions. In this case, the sequence is not fully realistic because an operator does not take a break between each task on the physical mock-up.

Despite these remarks justifying the gap for certain indicators between the real and virtual worlds, the safety, posture and force indicators did not present a significant difference. We can explain this investigation by the integration of the physical interaction directly into the VR platform [29], superimposing virtual and real contact. Figure 5 shows an operator performing the same task at the same time through both simulation tools. We observe the same physical behavior, the same contact with the environment.

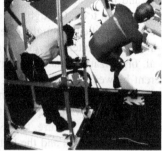

Fig. 5. Operator performing the same maintenance task between physical mock-up & virtual reality platform.

It appears that the main way to improve the efficiency of MMU will be the development and integration of physical parts into the VR platform to reduce the gap between real and digital simulation. Many studies work on the haptic system in various fields like aeronautic [42], medical [43] or automotive industries [44]. However, this technology is very expensive which justifies exploiting the current potential of tracking technology [45, 46]. It is possible to insert physical parts into the platform to represent the main contacts (hatch, table or wall) and add mobile parts with a tracker [47]. In our case study, the mobile part was, for example, a real torque wrench and an alternator reproducing real effort and biomechanical constraints. Using real parts with trackers is cheaper than the haptic system and can provide a deep benefit to HFE assessment. Additionally, in maintenance, torqueing bolts is very frequent but there is no system that allows this force to be reproduced in VR platforms. Another study has already detailed this specific need [48].

We can assume that new management and exploitation of the VR technology could have a significant impact on a better integration of HFE, first within the maintainability department but also in all departments around a single project with convergence points.

6 Conclusion

Through this industrial experimentation conducted on the maintenance activity of a helicopter manufacturer, we have confirmed the potential interest of using the digital simulation tool through the VR platform in specific conditions: inserting physical parts to simulate real contacts and forces with the participation of real operators. We found similar results for biomechanical indicators (such as posture and force) in DMU and PMU, but the mental indicator was different - showing the difficulty of assessing the cognitive aspect in a virtual environment. Improving the results of DMU for ergonomic purposes, and reducing the gap with the physical simulation tool, must focus on operator training in digital technology and enhancing the physical part.

Acknowledgement. I express my deepest and sincere gratitude to all the operators who participated in this experimentation: Jean-Michel, Franck F, Franck R, Charly, Eric, Philippe and Thierry.

References

1. Wiener EL, Nagel DC (eds) (1988) Human factors in aviation. Gulf Professional Publishing, Houston
2. Horeman T, Akhtar K, Tuijthof GJ (2015) Physical simulators. In: Effective training of arthroscopic skills. Springer, Berlin, pp 57–69
3. Spenser J (2008) The airplane. Collins, New York
4. Hobbs AN (2000) Maintenance 'error', lessons from the BASI survey. Flight Saf Aust 4:36–37
5. Gruber M, De Leon N, George G, Thompson P (2015) Managing by design. Acad Manag J 58(1):1–7

6. Stoffregen T, Bardy BG, Smart L, Pagulayan R (2003) Virtual and adaptive environments: applications, implications, and human performance issues, chap. On the nature and evaluation of fidelity in virtual environments, pp 111–128

7. Bernard F, Bazzaro F, Paquin R, Sagot JC (2017 January). Consideration of human factors in aeronautical maintainability. Annual reliability and maintainability symposium. (IEEE), Jan 2017, pp 37–43

8. Broberg O (2007) Integrating ergonomics into engineering: empirical evidence and implications for the ergonomists. Hum Factors Ergon Manuf Serv Ind 17(4):353–366

9. Sharma HK, Singhal P, Sonia P (2018) Computer-assisted industrial ergonomics: a review. In: ergonomic design of products and worksystems—21st century perspectives of Asia (pp 37–48). Springer, Singapore

10. Amundarain A, Borro D, Matey L, Alonso AG, de Guipúzcoa T (2003) Occlusion culling for the visualization of aeronautical engines digital mock-ups. In: Proceedings of virtual concept, Biarritz, France, pp 5–7

11. Yongsheng S, Yu L (2012) Application of DELMIA on maintainability design of aircraft. In Proceedings of the 2nd international conference on computer application and system modeling, 4p

12. Sagot, J. C. (1999). Ergonomie et conception anthropocentrée. Document pour l'Habilitation à diriger des recherches, Institut National Polytechnique de Lorraine (INPL), Nancy, 21

13. Czerniak JN, Brandl C, Mertens A (2017) Designing human-machine interaction concepts for machine tool controls regarding ergonomic requirements. In: IFAC-PapersOnLine, vol 50(1), pp 1378–1383

14. Bittencourt JM, Duarte F, Béguin P (2017) From the past to the future: integrating work experience into the design process. Work 57(3):379–387

15. Das B, Sengupta AK (1996) Industrial workstation design: a systematic ergonomics approach. Appl Ergon 27(3):157–163

16. Meister D (2014) Human factors testing and evaluation, vol 5. Elsevier, Amsterdam

17. Pontonnier C, Dumont G, Samani A, Madeleine P, Badawi M (2014) Designing and evaluating a workstation in real and virtual environment: toward virtual reality based ergonomic design sessions. J Multimodal User Interfaces 8(2):199–208

18. Aromaa S, Väänänen K (2016) Suitability of virtual prototypes to support human factors/ergonomics evaluation during the design. Appl Ergon 56:11–18

19. CEN/TC319.EN13306:2010 Maintenance–maintenance terminology. European Standard, Bruxelles

20. Lee SG, Ma YS, Thimm GL, Verstraeten J (2008) Product lifecycle management in aviation maintenance, repair and overhaul. Comput Ind 59(2):296–303

21. Čokorilo O (2011) Aircraft performance: the effects of the multi attribute decision making of non time dependant maintainability parameters. Int J Traffic Transp Eng 1(1):42–48

22. Chang Yu-Hern, Wang Ying-Chun (2010) Significant human risk factors in aircraft maintenance technicians. Saf Sci 48(1):54–62

23. AFIM (2004) Association française des ingénieurs et responsables de maintenance- santé et sécurité au travail: les métiers de la maintenance en première ligne. Guide nationale de la maintenance

24. Regazzoni D, Rizzi C (2014) Digital human models and virtual ergonomics to improve maintainability. Comput Aided Design Appl 11(1):10–19

25. De Sa AG, Zachmann G (1999) Virtual reality as a tool for verification of assembly and maintenance processes. Comput Graph 23(3):389–403

26. De Leon PM, Díaz VGP, Martínez LB, Marquez AC (2012) A practical method for the maintainability assessment in industrial devices using indicators and specific attributes. Reliab Eng Syst Saf 100:84–92

27. Perez J, Neumann WP (2010) The use of virtual human factors tools in industry—a workshop investigation. Ryerson University, Human Factors Engineering Lab Technical Report, 3

28. Garza LE, Pantoja G, Ramirez P, Ramirez H, Rodriguez N, Gonzalez E, Quintal R, Perez JA (2013) Augmented reality application for the maintenance of a flapper valve of a Fuller-Kynion Type M Pump. Procedia Comput Sci 25:154–160

29. Seth A, Vance JM, Oliver JH (2011) Virtual reality for assembly methods prototyping: a review. Virtual Real 15(1):5–20

30. ISO 15537:2004 Principles for selecting and using test persons for testing anthropometric aspects of industrial products and designs

31. International Ergonomics Association (2000) What is ergonomics. IEA members" and "study programs" (nd). Available at http://www.iea.cc/index.php

32. NF EN1005-4:2008 Human physical performance Part 4: evaluation of working postures and movements in relation to machinery

33. McAtamney L, Corlett EN (1993) RULA: a survey method for the investigation of work-related upper limb disorders. Appl Ergon 24(2):91–99

34. Borg G (1982) Psychophysical bases of perceived exertion. Med Sci Sport Exerc 14:377–381

35. Chitescu C, Sagot JC, Gomes S (2003) Favoriser l'articulation "Ergonomie/conception de produits" à l'aide de mannequins numériques. Dans les actes de la conférence 10eme Séminaire CONFERE (Collège d'Etudes et de Recherches en Design et Conception de Produits) sur l'Innovation et la Conception, Belfort, July 2003, pp 3–4

36. Hart SG, Staveland LE (1988) Development of NASA-TLX (Task Load Index): results of empirical and theoretical research. Adv Psychol 52:139–183

37. Kuorinka I, Jonsson B, Kilbom A, Vinterberg H, Biering-Sørensen F, Andersson G, Jørgensen K (1987) Standardised Nordic questionnaires for the analysis of musculoskeletal symptoms. Appl Ergon 18(3):233–237

38. Burns A, Salter T, Sugden B, Sutherland J (2018) U.S. Patent No. 9,865,089. U.S. Patent and Trademark Office, Washington, DC

39. Bowman DA, McMahan RP (2007) Virtual reality: how much immersion is enough? Computer 40(7):36–43

40. Loomis JM, Philbeck JW (2008) Measuring spatial perception with spatial updating and action. In: Carnegie symposium on cognition, 2006, Psychology Press, Pittsburgh, PA, US

41. Chen W, Chao JG, Zhang Y, Wang JK, Chen XW, Tan C (2017) Orientation preferences and motion sickness induced in a virtual reality environment. Aerosp Med Hum Perform 88 (10):903–910

42. Savall J, Borro D, Gil JJ, Matey L (2002) Description of a haptic system for virtual maintainability in aeronautics. In: IEEE/RSJ international conference on intelligent robots and systems, vol 3, pp 2887–2892

43. Wang R, Yao J, Wang L, Liu X, Wang H, Zheng L (2017). A surgical training system for four medical punctures based on virtual reality and haptic feedback. In: 2017 IEEE symposium on 3D user interfaces (3DUI), Mar 2017, pp 215–216

44. Langley A, Lawson G, Hermawati S, D'Cruz M, Apold J, Arlt F, Mura K (2016) Establishing the usability of a virtual training system for assembly operations within the automotive industry. Hum Factors Ergon Manuf Serv Ind 26(6):667–679

45. Riley S (2016) U.S. Patent No. 9,403,087. U.S. Patent and Trademark Office, Washington, DC

46. Meier P, Holzer S (2015) U.S. Patent No. 9,165,405. U.S. Patent and Trademark Office, Washington, DC
47. Menezes P, Gouveia N, Patrão B (2018) Touching is believing-adding real objects to virtual reality. In: Online engineering and internet of things. Springer, Cham, pp 681–688
48. Lawson G, Salanitri D, Waterfield B (2016) Future directions for the development of virtual reality within an automotive manufacturer. Appl Ergon 53:323–330

Applying Adaptive Instruction to Enhance Learning in Non-adaptive Virtual Training Environments

Robert A. Sottilare[✉][iD]

US Army Research Laboratory, Orlando, FL 32826, USA
robert.a.sottilare.civ@mail.mil

Abstract. This paper discusses the use of adaptive instruction to guide learning by stimulating non-adaptive virtual training environments. Adaptive instruction is sometimes referred to as differentiated instruction and is any learning experience tailored to meet the needs and preferences of an individual learner or team. An intelligent tutoring system (ITS) is the technology which delivers adaptive instruction. Adaptive instructional systems (AISs) use human variability and other learner/team attributes along with instructional conditions to develop/select appropriate strategies (domain-independent policies) and tactics (actions). The goal of adaptive instruction is to optimize learning, performance, retention, and the transfer of skills between the training environment and the work or operational environment where the skills learned during training are to be applied.

Keywords: Adaptive instruction · Intelligent tutoring systems (ITSs)
Virtual training environments (VTEs)

1 Introduction

Many training environments today use augmented, mixed or virtual reality technologies to provide each learner or team of learners with the visual and aural cues (and less often olfactory and tactile cues) to facilitate a realistic experience with the goal of enhancing their cognitive, psychomotor, or social skills. These cues are required to stimulate learner decisions and actions during the execution of a task (e.g. identify friend or foe) under a specified set of conditions (e.g., varying distance, lighting conditions, and weather phenomena) to a minimum set of standards (e.g., high accuracy (>95% correct)).

Adaptive instructional systems (AISs) are "computer-based systems that guide learning experiences by tailoring instruction and recommendations based on the goals, needs, and preferences of each learner in the context of domain learning objectives" [1]. AISs are intelligent and situationally aware of both the learner's attributes and the context of the instruction (e.g., conditions at some instance of the instructional path). However, most training environments today are non-adaptive or minimally adaptive where they tailor content based only on the learner's performance and tailor it the same way for every learner. Still other training environments use human-in-the-loop approaches to make decisions and take actions to change the path or outcomes of

© Springer Nature Switzerland AG 2019
S. Bagnara et al. (Eds.): IEA 2018, AISC 822, pp. 155–162, 2019.
https://doi.org/10.1007/978-3-319-96077-7_16

scenario-based training. Many of these systems are very good for training tasks and over time effectively guide learning and develop skills, but there is room for improvement. We propose an approach to reduce the human workload during simulation while simultaneously enhancing the tailoring and thereby the efficiency for both individual learners and teams.

The limitation of current training environments and their associated technologies is primarily tied to their inability to adapt to the needs and preferences of each individual learner. In other words, the learning environment reacts the same way for every user or at best usually only varies with changes in learner performance. AISs model the learner or team to provide tailored experiences based on their domain competence (prior knowledge), their goals, their preferences (e.g., personality factors) and their cognitive, affective, and physical states.

Today, military and civilian organizations have extensive inventories of non-adaptive or minimally adaptive training systems. If there was a methodology to stimulate these systems and their decision processes, their strategies and tactics would be tailored to each individual/team and this would enable the acceleration of learning by reducing previously mastered (redundant) content and by motivating and engaging learners to progress toward their personal goals. The Generalized Intelligent Framework for Tutoring (GIFT) is an augmentation technology which can be applied to many training simulations to guide training in an adaptive fashion.

GIFT is an open-source architecture for authoring, delivering, guiding, and evaluating tailored, computer-based instruction for individuals and teams of learners [2]. As of this writing, GIFT is used by over 1500 adaptive instructional researchers and developers in 76 countries. This paper reviews the modeling, actions, interactions, and specifications needed to drive adaptive instruction in currently non-adaptive systems. The ability of GIFT as an adaptive instructional architecture to interoperate with non-adaptive systems has the potential to significantly reduce the time and cost of instruction across many military and civilian training domains. We start by discussing the return-on-investment (ROI) for augmentation with adaptive instruction.

2 Return-On-Investment (ROI) for Adaptive Instruction

In this section, we attempt to build a case for the ROI for augmenting training systems with adaptive instruction. The typical cost of developing AISs today is significantly higher than non-adaptive systems based on the need to provide additional content for the many new paths created by tailoring instruction to each individual's (or teams) goals, needs, and preferences. Additional cost is due to the time to generate additional content and the expert skill level required for the various phases of the AIS authoring process (e.g., instruction system design, content curation, and content sequencing). The number of hours required to generate one hour of adaptive instruction is 200–300 h of authoring time compared to 40–60 h to generate one hour of non-adaptive instruction. So there is a significant difference in cost between adaptive and non-adaptive systems. Is it worth the investment? Yes.

While there is a higher cost for adaptive instruction, this cost may be recouped by a reduction in both the learner's contact hours and the human workload during instruction. Tailoring instruction reduces the time to reach competency (accelerated learning) by accounting for prior knowledge and identifying opportunities to skip or reduce instruction for previously-learned concepts. Adaptive instruction also reduces the amount of time lost to *off-task behaviors* (e.g., doodling, texting or talking instead of listening) usually encountered when learners are bored [3]. Boredom often occurs when there is a mismatch between the domain competency of the learner (prior knowledge) and the instructional content [4] and tailoring content to each learner is much more engaging.

There is also high potential to reduce the authoring costs of AISs through automation, but we will delay this discussion until Sect. 5 - *Next Steps*. Instead, we will focus next on opportunities to leverage existing non-adaptive systems for adaptive instruction by driving them with adaptive instructional engines like GIFT.

3   A Model of Interaction Design for AISs

Interaction design is "a user-oriented field of study that focuses on meaningful communication of media through cyclical and collaborative processes between people and technology" [5]. In this case we are discussing the interaction of learners immersed within virtual training environments where their learning experience is guided by an AIS (e.g., intelligent tutoring system).

In a typical non-adaptive training system (Fig. 1), learner act on the environment and then observe outcomes resulting from their actions. Based on this observation, learners decide what action to take next. The system is considered non-adaptive because the environment only changes or responds to learner stimuli. The environment is not intelligent enough to change on its own (e.g., goal-based behavior).

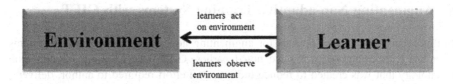

Fig. 1. A model of interaction design for non-adaptive instructional systems

An example of this non-adaptive training process follows: a learner, immersed in a virtual environment, has the goal to navigate from one point in the environment to another. Using a map, the learner estimates the distance and time to move from one point to another and acts on the environment by moving forward toward the south, the direction indicated on the map to get to the next point. When the learner believes he has arrived at the assigned point, he observes the virtual terrain and its nearby features (e.g., hills, roads or rivers) to see if they correspond to his position on the map. If they match up, the learner knows that he has reached the goal. If they don't, the learner attempts to reorient himself on the map and adjust his course. The environment in this case is static.

If we augment this non-adaptive system with an adaptive element (Fig. 2), we expand the capabilities to tailor instruction either by enabling the adaptive tutor to alter the challenge level (e.g., increase the difficulty of the scenario when the learner's performance is very high) or to scaffold the learner (e.g., support or encourage the learner through hints, prompts, or reflective dialogue).

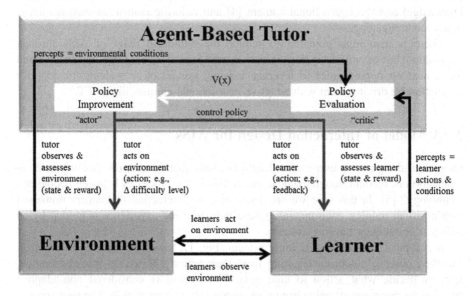

Fig. 2. A model of interaction design for adaptive instructional systems

4   Augmenting Non-adaptive Training Systems with GIFT

In this section we discuss options for augmenting non-adaptive instructional systems with GIFT to create AISs. As we noted earlier, GIFT is a learning augmentation technology that was created to reduce the time and effort required to author, delivery, manage, and evaluate AISs. *Learning augmentation technologies* enhance human productivity or capability [6] with respect to learning outcomes (e.g., knowledge and skill acquisition, performance, retention, and transfer of training). Examples of AISs in the role of learning augmentation technologies include intelligent tutoring systems, personal assistants, and intelligent media used to guide or support learners during training and educational experiences.

Using the model of interaction design in Fig. 2 as a basis, our next step is to focus on additional detail to illustrate how GIFT (or other tutoring architectures) might be used to augment instruction in currently non-adaptive training systems. Figure 3 takes the elements of the agent-based tutor illustrated in Fig. 2 and begins to define its various models and functions.

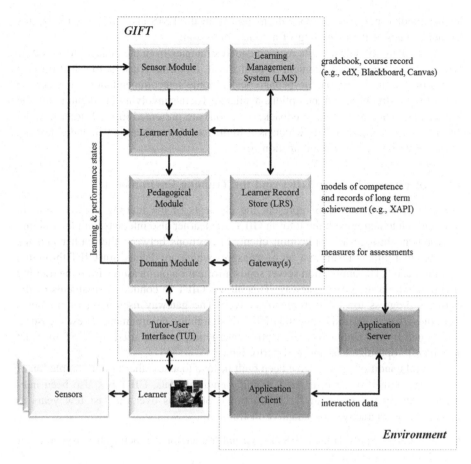

Fig. 3. Interaction between learner, GIFT, and application (environment)

4.1 Elements of GIFT

The elements of GIFT are shown as green boxes in Fig. 3. Starting at the top, the sensor module collects raw and processed state data from individual sensors (e.g., Microsoft Kinect or Zephyr Bioharness) and either processes the raw data or passes on the behavioral or physiological states to the learner module. The learner module tracks the real-time states of the learner (e.g., performance, learning, behavioral or physio-logical) during adaptive instruction. Learner state information is sent to the pedagogical module, the learning management system (LMS), and the learner record store (LRS. The pedagogical module uses learner states to select optimal domain-independent instructional strategies based on best practices found in the literature. The LMS logs achievements related to objectives (called concepts in GIFT) for all learners in a course of instruction and the LRS logs individual achievements across instructional experi-ences (e.g., courses, lessons, reading assignments) and constitutes the long term model of the learner based on their experiences and achievements. During instruction, the learner module may also accept information from the LMS and LRS to initialize the learner's states at the beginning of a course or lesson.

The domain module accepts recommended strategies from the pedagogical module and it either stores domain content or points to it. Its primary task is to select optimal tactics or actions based on the pedagogical module's recommendations, the current state of the instruction, and the options available. Tactics involving interaction with the learner (e.g., encouragement, feedback, direction) are presented to the learner through the tutor-user interface (TUI) which is a multi-modal interface and includes textual, verbal, and non-verbal communication modes.

4.2   Interactions Between GIFT and the Training Environment

The orange boxes in Fig. 3 represent the elements of the training environment. In addition to interacting with the tutor in GIFT, the learner also interacts with the training environment through an application client. Interactions between the learner and the training environment result in a set of measures which are defined for GIFT through a condition class. The application server sends information on measures from the training environment via the GIFT gateway. This allows GIFT to conduct assessments of the learner's progress toward learning objectives. The gateway uses a set of measures (variables) that are passed between GIFT and the training environment. These measures are defined by the author of the adaptive instruction and are used by GIFT to make decisions to optimize learning and performance.

Several condition classes have been built in Java to assess the start or completion of conditions as they occur in external training environments. GIFT has also been integrated with several non-adaptive instructional platforms over the last few years to support more interactive and adaptive instruction:

- Engagement Skills Trainer (EST) – virtual trainer for instructing the psychomotor task of marksmanship
- Virtual BattleSpace (VBS) – game-based, desktop training environment for instructing tactical tasks
- Virtual Medic (VMedic) – game-based, desktop training environment for instruction combat casualty care tasks
- Augmented Reality Sandtable (ARES) – projection of tactical maps on 3-D bed of sand
- Dynamic Environment Testbed (DE Testbed) – a virtual environment used to instruct the control and application of heavy construction equipment (e.g., excavator)
- Learning Tools Interoperability (LTI) – facilitates the integration of GIFT with learning management systems and their associated courses
- Unity Environment – provides an exemplar for integrating GIFT with Unity WebGL applications for training
- Distributed Interactive Simulation (DIS) – provides an exemplar for integrating GIFT with DIS-compliant simulations.

Concepts have been developed to ease the integration of previously non-adaptive training modes with GIFT, but have yet to be validated through experimentation:

- Land Navigation – use of adaptive instruction to drive planning and execution of land navigation in both virtual and live training contexts [7]
- Adaptive Triage – a mix of immersive displays and pressure sensors to facilitate live training of hemorrhage control using tourniquets and pressure bandages [8].

5 Next Steps

Based on the interaction design, GIFT in combination with a training environment is an augmentation to both the learner (guides and tailors learning) and the previously non-adaptive training system (expands instructional capabilities). We are currently examining commonalities in the measures of various tasks in order to identify measures that might cut across a variety of domains. The interaction within each task being instructed will influence which measures are needed to assess progress toward learning objectives. For example, location is likely a measure that can be used for a variety of tasks. Interaction and perception modes will also play a key role in the selection of measures. Tasks requiring visual, aural, and haptic perception will determine success criteria for psychomotor tasks.

Next steps are to demonstrate GIFT with a variety of instructional environment to support training and education in cognitive [9], affective [9], psychomotor [10], and especially team task domains [11–13]. Validation of new adaptive instructional concepts through a testbed functionality is a priority. Another objective is to understand how GIFT can drive embedded adaptive instruction in equipment and vehicles [14–16]. The goal for future integrations is to reduce the time and skill required to enable adaptive instruction in the broadest possible context at the lowest cost.

References

1. Sottilare R, Brawner K (2018) Exploring standardization opportunities by examining interaction between common adaptive instructional system components. In: Proceedings of the first adaptive instructional systems (AIS) standards workshop. US Army Research Laboratory, Orlando, Florida
2. Sottilare RA, Brawner KW, Goldberg BS, Holden HK The generalized intelligent framework for tutoring (GIFT). Concept paper released as part of GIFT software documentation. US Army Research Laboratory – Human Research & Engineering Directorate (ARL-HRED), Orlando, FL. https://gifttutoring.org/attachments/152/GIFTDescription_0.pdf
3. Baker RS, Corbett AT, Koedinger KR, Wagner AZ (2004) Off-task behavior in the cognitive tutor classroom: when students game the system. In: Proceedings of the SIGCHI conference on Human factors in computing systems. ACM, pp 383–390
4. Vygotsky L (1987) Zone of proximal development. In: Mind in society: the development of higher psychological processes, vol 5291, p 157
5. Graham L (1998) Principles of interactive design, 1st edn. Delmar Cengage Learning, Boston, p 240. ISBN 0827385579

6. Techopedia (2018) What is human augmentation? https://www.techopedia.com/definition/29306/human-augmentation. Accessed 20 May 2018

7. Sottilare R, LaViola J (2015) Extending intelligent tutoring beyond the desktop to the psychomotor domain: a survey of smart glass technologies. In: Proceedings of the interservice/industry training simulation & education conference, Orlando, Florida, December 2015

8. Sottilare R, Hackett M, Pike W, LaViola J (2016) Adaptive instruction for medical training in the psychomotor domain. J Def Model Simul Appl Methodol Technol. https://doi.org/10.1177/1548512916668680

9. Holden H, Sottilare R, Goldberg B, Brawner K (2012) Effective learner modeling for computer-based tutoring of cognitive and affective tasks. In: Proceedings of the interservice/industry training simulation & education conference, Orlando, Florida, December 2012

10. Sottilare RA, LaViola J (2016) A process for adaptive instruction of tasks in the psychomotor domain. In: Design recommendations for intelligent tutoring systems, p 185

11. Bonner D, Walton J, Dorneich MC, Gilbert SB, Winer E, Sottilare R (2015) The development of a testbed to assess an intelligent tutoring system for teams. In: Proceedings of the "developing a generalized intelligent framework for tutoring (gift): informing design through a community of practice" workshop at the 17th international conference on artificial intelligence in education (AIED 2015), Madrid, Spain, June 2015

12. Sottilare RA, Burke CS, Salas E, Sinatra AM, Johnston JH, Gilbert SB (2017) Designing adaptive instruction for teams: a meta-analysis. Int J Artif Intell Educ. https://doi.org/10.1007/s40593-017-0146-z

13. Fletcher JD, Sottilare RA (2017) Shared mental models in support of adaptive instruction for team tasks using the GIFT tutoring architecture. Int J Artif Intell Educ. https://doi.org/10.1007/s40593-017-0147-y

14. Sottilare R, Marshall L, Martin R, Morgan J (2007) Injecting realistic human models into the optical display of a future land warrior system for embedded training purposes. J Def Model Simul 4(2):97–126

15. Alexander T, Sottilare R, Goldberg S, Andrews D, Magee L, Roessingh J (2012) Enhancing human effectiveness through embedded virtual simulation. In: Proceedings of the interservice/industry training simulation & education conference, Orlando, Florida, December 2012

16. Sottilare R (2009) Making a case for machine perception of trainee affect to aid learning and performance in embedded virtual simulations. NATO research workshop (HFM-RWS-169) on human dimensions in embedded virtual simulations, Orlando, Florida, October 2009

Contact Pressure Analysis for Wearable Product Design

Wonsup Lee[1]([⊠]) [iD], Jin-Gyun Kim[2] [iD], Johan M. F. Molenbroek[3] [iD],
Richard H. M. Goossens[3], Hayoung Jung[4], and Heecheon You[4] [iD]

[1] Handong Global University, Pohang 37554, South Korea
W.Lee@Handong.edu
[2] Kyung Hee University, Yongin 17104, South Korea
[3] Delft University of Technology, 2628CE Delft, The Netherlands
[4] Pohang University of Science and Technology, Pohang 37673, South Korea

Abstract. 3D body scanning has been used broadly including digital human modeling, simulation, ergonomic product design, and so forth. This research used template-registered faces of 336 Koreans in order to use them to design an oxygen mask that provides good fit to Korean faces. The finite element analysis method is applied onto the template-registered faces to predict the contact pressure of a mask design onto different faces. The average and variation of the estimated contact pressure values among all the Korean faces were analyzed for evaluation of the appropriateness of a mask design for Koreans. The proposed method can be usefully applied to find an optimal shape of wearable products for a specific target population.

Keywords: 3D face scanning · Finite element analysis · Pressure estimation
Ergonomic product design

1 Introduction

3D body scanning and its analysis technologies have been usefully applied in studies on the ergonomic design of a form of a product. A 3D scanned image of a human body contains not only anthropometric dimensions (e.g., length, width, circumference) but also complex dimensions such as arcs, cross-sectional curves, surfaces, areas, and volumes. A product design based on those 3D shape characteristics of the human body has resulted in gaining a better fit and increased comfort, satisfaction, and safety for the users. Especially, a wearable product such as an oxygen mask, a full-face mask for snorkeling, or a VR headset needs to have an ergonomically designed shape, which can fit well to a certain amount of a target population.

An optimal shape of a design component can be determined by analyzing contact pressure occurring between the design component and 3D models. Lee et al. [1, 2] proposed a method for virtual fit analysis to find an optimal design of a pilot oxygen mask which accommodates the Korean Air Force pilots ($n = 336$). However, the contact pressure was not systematically considered with the virtual fit analysis method in their study. The finite element (FE) analysis techniques have been introduced for the

S. Bagnara et al. (Eds.): IEA 2018, AISC 822, pp. 163–169, 2019.
https://doi.org/10.1007/978-3-319-96077-7_17

contact pressure estimation of a product using 3D body scan images [3–5]. However, the previous FE studies used only a few body models (say, one or two) for the pressure estimation of a product, while various shape of the human body which can occur different contact pressure characteristics was not fully considered.

This study is aimed to develop a design method of an ergonomic wearable product based on numerous 3D body scan database and finite element (FE) analysis technology which is applied for estimation of contact pressure between a 3D body scan image and a design of a product.

2   Methods

2.1   3D Face Scans

3D face scan images of Korean Air Force pilots collected by Lee et al. [6] were used in this study. 3D faces of 336 pilots (male: 278, female: 58; age: 20 s to 40 s) were scanned using the Rexcan 560 (Solutionix Co., South Korea) 3D scanning system.

Landmarking and alignment were applied on 3D face images. 14 anthropometric landmarks were identified on the 3D face area to measure 22 facial dimensions related to designing the medical face mask. All 3D heads were aligned with the origin point at the sellion landmark, then aligned with two vectors (one vertical vector parallel to the Y axis and passing through sellion and supramentale and one horizontal vector parallel to the X axis and passing through left and right tragion) [2].

2.2   Template Model Registration

A template face model having appropriate mesh structure was prepared with considering the FE analysis, then all the 3D faces were template-registered using a non-rigid ICP registration method. Template registration, a computer graphic method, is one of useful technique that makes different 3D body scans having same vertex points and mesh structure. In this study, first, a symmetrized template face model having 1,340 vertex points was sophisticatedly generated by hand based on an average-sized face (see Fig. 1) using the RapidForm 2006 (INUS Technology, Inc., Korea) image processing software. Facial area related to the mask design (e.g., around nasal bridge, nasal side, chin) has denser vertex points than the other facial areas. Deformation and template model registration methods were applied to earn template-registered face images using the Matlab (MathWorks, Inc., Natick, MA, USA) programming. A hybrid registration approach proposed by Lee et al. [7, 8], which is using the bounded biharmonic weights (BBW) mesh deformational algorithm and non-rigid iterative closet point (ICP) registration to individual 3D head scans was used in this study. The hybrid registration approach which consider landmark's location by the mesh deformation algorithm and the registration methods could provide the accurate correspondence of mesh topology across all template-registered faces.

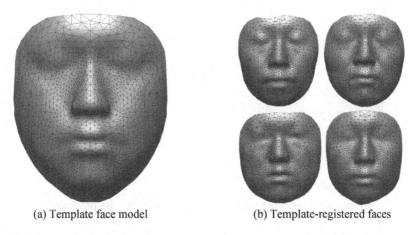

(a) Template face model   (b) Template-registered faces

Fig. 1. The template face and the template-registered faces (illustrated)

Fig. 2. The curvy shape of a mask's part that is contacting to the face

2.3   Contact Pressure Estimation Using Finite Element Analysis

From 3D scanned image of a product, a curvy shape that is contacting to a facial area was saved as a spline curvature. The curvature was defined as a spline curve consisted with 10 points as shown in Fig. 2. While the curvature is virtually aligned to all template-registered faces, the pressure of the curvature towards a 3D face was predicted using finite element analysis [9]. In this work, we applied triangular shell elements for the 3D facial modeling based on the 3D scanned image data as shown in Fig. 1. 2,624 elements and 1,340 nodes were then used. Since the contact pressure estimation of the facial skin with small deformation was mainly focused on, the simple linear elastic material properties were applied on the nodes except all the boundary nodes which are fixed and unmovable. For static analysis, the following equation was used:

$$f = Ku \tag{1}$$

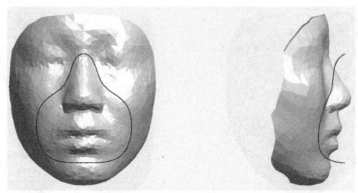

(a) face without contacting to the mask curvature: no pressure.

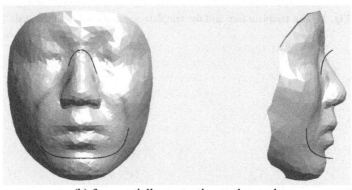

(b) face partially contacting to the mask

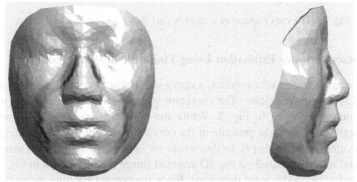

(c) face fully contacting to the mask

Fig. 3. The estimated contact pressure between the template-registered face and the mask defined as a curvy shape (illustrated) (Color figure online)

where K is a global stiffness matrix, u is a displacement vector for all nodes, and f is a force vector for all nodes, respectively. The elastic modulus and Poisson ratio which are necessary for calculation of the K are defined as 0.3 and 0.5 in this study. After the curvature was aligned at the required position by referring to Lee's study [10],

The displacement vector u was derived as the Euclidean distance between the curvature and each node. By solving Eq. (1), the contact pressure f was calculated. In this work, only z-translation at the nodal coordinates was simply considered with the stiffness matrix K to compute the force. The contact pressure is defined as the force exerted on a surface divided by the area (m^2) over which that force is applied. However, instead of the contact pressure, this study used the force which is proportionally equivalent to the contact pressure if the area is unconsidered. As illustrated in Fig. 3, the estimated contact pressure was shown as intensity of the red color. The pressure calculation process was implemented by MATLAB environment.

3   Results

As results, the average and variation of the predicted facial contact pressure between all the template-registered faces and a mask design were analyzed. Figure 4 is showing one average face with two different patterns of the average contact pressure calculated

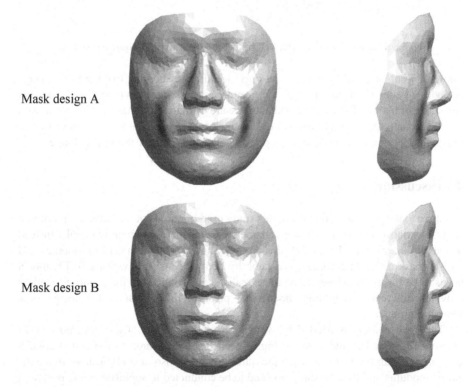

Fig. 4. Average contact pressure derived by two different mask designs (illustrated).

Mask design A

Mask design B

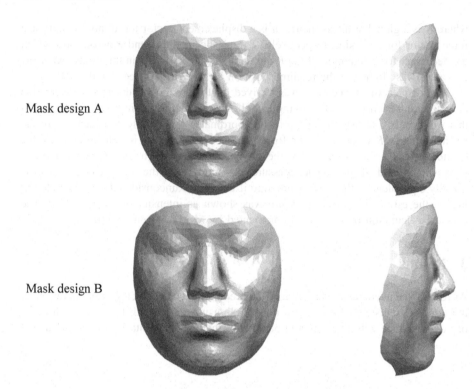

Fig. 5. Standard deviation analyzed by two different mask designs (illustrated).

using different shape of the mask. The Figs. 4 and 5 are showing that the lower image presents a better result in terms of the average and standard deviation of the contact pressure estimation. The average contact pressure of the lower image is smaller as well as more equally-distributed than that of the upper image. And the standard deviation of the contact pressure of the lower image is smaller than that of the upper image.

4   Discussion

Compare to the previous FE studies, this study used hundreds of face images in order to analyze variation of contact pressure characteristics that are different by people. Instead of making the realistic FE model, this study assumed that the material properties and face-to-mask contacting context are same by people. Since the validity of FE model needs to be further improved to be more realistic and robust, but the proposed method provides differences regarding the contact pressure that is useful for design of a wearable product.

The further study is needed for validation of the method. The estimated contact pressure illustrated as red color in Fig. 3 need to be compared to the actual contact pressure that can be measured by a pressure film [11]. Also, a study to investigate the elastic modulus and the Poisson ratio need to be conducted to sophisticate the proposed method.

The FEA-based contact pressure analysis method proposed in this study can be usefully applied to find an optimal shape of a product's part that contacts the human body. Different body parts and different product designs will be further considered in order to investigate an applicability and usefulness of the developed method.

Acknowledgement. This research was jointly supported by the National Research Foundation (NRF) of Korea funded by the Ministry of Science, ICT, and Future Planning (MSIP) (NRF-2018R1C1B5047805).

References

1. Lee W, Kim H, Jung D, Park S, You H (2013) Ergonomic design and evaluation of a pilot oxygen mask. In: Proceedings of the human factors and ergonomics society (HFES) 57th annual meeting, San Diego, CA, USA
2. Lee W, Yang X, Jung H, You H, Goto L, Molenbroek JFM, Goossens RHM (2016) Application of massive 3D head and facial scan datasets in ergonomic head-product design. Int J Digit Hum 1(4):344–360
3. Dai JC, Yang JZ, Zhuang ZQ (2011) Sensitivity analysis of important parameters affecting contact pressure between a respirator and a headform. Int J Ind Ergon 41(3):268–279
4. Lei Z, Yang J, Zhuang Z (2012) Headform and N95 filtering facepiece respirator interaction: contact pressure simulation and validation. J Occup Environ Hyg 9:46–58
5. Lei ZP, Yang J, Zhuang ZQ, Roberge R (2013) Simulation and evaluation of respirator faceseal leaks using computational fluid dynamics and infrared imaging. Ann Occup Hyg 57(4):493–506
6. Lee W, Jeong J, Park J, Jeon E, Kim H, Jung D, Park S, You H (2013) Analysis of the facial measurements of Korean air force pilots for oxygen mask design. Ergonomics 56(9):1451–1464
7. Lee W, Goto L, Molenbroek JFM, Goossens RHM (2017) Analysis methods of the variation of facial size and shape based on 3D face scan images. In: Proceedings of the human factors and ergonomics society (HFES) 61st annual meeting, Austin, TX, USA
8. Lee W, Goto L, Molenbroek JFM, Goossens RHM, Wang CCL (2017) A shape-based sizing system for facial wearable product design. In: Proceedings of the 5th international digital human modeling symposium, Bonn, Germany
9. Bathe KJ (2006) Finite element procedures. Klaus-Jürgen Bathe, Cambridge
10. Lee W (2013) Development of a design methodology of pilot oxygen mask using 3D facial scan data. Pohang University of Science and Technology, Pohang
11. Lee W, Yang X, Jung D, Park S, Kim H, You H (2018) Ergonomic evaluation of pilot oxygen mask designs. Appl Ergon 67:133–141

Simulation Techniques for Ergonomic Performance Evaluation of Manual Workplaces During Preliminary Design Phase

Francesco Caputo[1(✉)], Alessandro Greco[1], Marcello Fera[1],
Giovanni Caiazzo[2], and Stefania Spada[2]

[1] Department of Engineering, University of Campania Luigi Vanvitelli,
via Roma 29, 81031 Aversa, CE, Italy
{francesco.caputo,alessandro.greco,
marcello.fera}@unicampania.it
[2] FCA Italy – EMEA Manufacturing Planning and Control – Ergonomics,
Gate 16, Corso Settembrini 53, 10135 Turin, Italy
{giovanni.caiazzo,stefania.spada}@fcagroup.com

Abstract. Among the technologies included in Industry 4.0, the fourth industrial revolution, Digital Manufacturing (DM) represents a new approach to evaluate the performance of production processes in a virtual environment.

DM can be seen as the industrial declination of Virtual Reality (VR) that, by using an integrated computer-based system, allows creating simulation, 3D visualization and provides different tools to define the product and the manufacturing process simultaneously.

Virtualization and simulation of production processes generate benefits for companies in terms of time and costs, optimizing the assembly line and providing parameters for studying human-machine interaction.

Regarding this last topic, the aim of this paper is to propose an innovative procedure to support the workplaces design, based on simulation techniques that allow setting a virtual scenario in which a Digital Human Model (DHM) is able to carry on assembly tasks. Data from simulations can be analyzed and used to assess ergonomic indexes in a preventive and proactive approach.

As other automotive manufacturers, Fiat Chrysler Automobiles (FCA) applies EAWS (European Assessment Work Sheet), a first level screening, to assess the ergonomic biomechanical overload of workplaces in the design phase, according to international standards (ISO 11226 and ISO 11228-1, -2, -3).

The ergonomics risk assessment, since the design phase, allows identifying critical issues and to define and put in practice corrective actions in the earlier phase, being more successful and less expensive.

In order to support the procedure proposed in this research, a case study is described, based on the EAWS index evaluation of a workstation in a FCA plant assembly shop.

The simulation has been realized by using PLM software Tecnomatix Process Simulate by Siemens® and the EAWS analysis has been performed by using EAWSdigital by MTM®.

The procedure can be considered innovative to support human-centered design of production process in developing new products.

© Springer Nature Switzerland AG 2019
S. Bagnara et al. (Eds.): IEA 2018, AISC 822, pp. 170–180, 2019.
https://doi.org/10.1007/978-3-319-96077-7_18

Keywords: Digital manufacturing · Digital human models · Simulation EAWS

1 Introduction

Assembly lines, especially in the automotive field, are very complex to design, due to a large number of variables affecting the production process (technological, environmental, ergonomic), not all of them always taken into account. In particular, ergonomic aspects are often coarsely considered during the draft phase, first due to their complexity in the evaluation, even if their integration during the design phase could allow the opportunity to design workplaces, in particular the manual ones, with a human-centered approach, as leaded by the new industrial paradigm Industry 4.0.

Within the Industry 4.0 pillars, an important aspect, treated by this research, is represented by Digital Manufacturing (DM), the industrial declination of Virtual Reality (VR), that integrates a wide set of technologies to support the production, from the design to the product realization, monitoring and optimizing the production processes.

In particular, realizing a virtual copy of a real production line, the so called Digital Twins (DT), allows replicating the real behaviour of all the resources of the factory, from systems to humans, analysing and optimizing their performances in real operating conditions. Integrating Digital Human Models (DHM) in this virtual scenario, it is also possible dealing with ergonomic issues, evaluating ergonomic indexes scores with a numerical approach [1].

Regarding the preventive study of ergonomic issues, as described in [2, 3], Fiat Chrysler Automobiles (FCA) uses a preventive ergonomic method, called Ergo-UAS. This method is applied during both Process/Product Design and Process Industrialization and it is composed by EAWS (European Assembly Work Sheet) and UAS (Universal Analyzing System). EAWS [4] is a first level ergonomic screening for the evaluation of biomechanical overload risk, according to ISO 11226 [5] and ISO 11228-1, -2, -3 [6–8]. The UAS is a typical example of MTM (Method-Time-Measurement) system which is used for the definition of times and methods of work, describing the sequence of operations of a specific work task, assigning a predetermined standard time from the direct observation of the worker and the nature of the movements during the given task [9].

In order to achieve these results, a lot of information, principally related to human factors, are necessary to satisfy ergonomic standards, in particular those ones concerning postural aspects, the effort exerted by the workers and the manual handling of loads. During the development of a new product, the evaluation of these factors by means of standard procedures is complicated because, as there is no assembly line yet, it is necessary to physically simulate the activity in the laboratory and, due to the large number of workplaces (WPs) on an assembly line, this would require extremely high timescales.

For these reasons, the development of a numerical procedure for a proactive ergonomic evaluation could represent an important innovation for a human-centered and ergonomically safe design of the assembly lines.

This paper is aimed in proposing an innovative procedure, supported by a real case study about two workplaces of a FCA assembly line, for a proactive ergonomic assessment based on EAWS since during the design phase, in order to make ergonomics a fundamental variable for WPs design validation. The procedure is based on the simulation of the working activities carried out by a Digital Human Model (DHM), already validated in [10], implemented in the Siemens® Tecnomatix Process Simulate software environment.

2   Methods

2.1   Technical Standards

In the European Union, the protection system of health and safety in a working environment is set up by EU-Machinery Directive [11] and EU-Framework Directive [12], which demand ergonomic risk analysis to be carried out in several phases of the product/production process development. The main standards, mentioned also by Italian Law [13], are listed in Table 1.

Table 1. Technical standards.

European union directives		
Ergonomic factors	Machinery directive	Framework directive
Postures	EN 1005-4	ISO 11226
Forces	EN 1005-3	ISO 11228-2
Lifting (and carrying) Push and Pull	EN 1005-2	ISO 11228-1 ISO 11228-2
Upper limbs	EN 1005-5	ISO 11228-3

The application of these directives is mandatory during the production phase, during which it is only possible a corrective ergonomic approach.

Any ergonomic standard is not mandatory during the design phase, even if it could be a crucial step for ensuring a safe design of the working environment, preventively avoiding several corrections during the production phase.

European Assembly Work Sheet – EAWS. As already mentioned, FCA applies Ergo-UAS method during the WPs design phase in order to validate the design and to balance the line. In particular, EAWS is a first level screening for the evaluation of biomechanical overload risk. It has been developed for evaluating the physical human factors (postures, forces and manual handling), pointing out the main problems and offering an opportunity to find design change solutions to overcome them.

The method, developed in Germany by researchers of Darmstadt University of technology and researchers of MTM Foundation, is mentioned in ISO TR 12295; it consists in filling-in a checklist composed by four sections, in each of which the following tasks are evaluated:

- Working postures and movements with low additional physical effort – section 1;
- Action forces of the whole body or hand/finger system – section 2;
- Manual Material Handling – section 3;
- Repetitive loads of the upper limbs – section 4.

The overview of EAWS checklist is shown in Figs. 1 and 2.

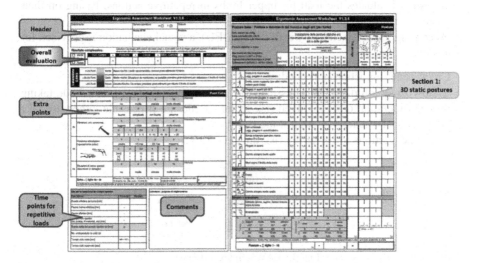

Fig. 1. EAWS checklist overview: headers and section 1.

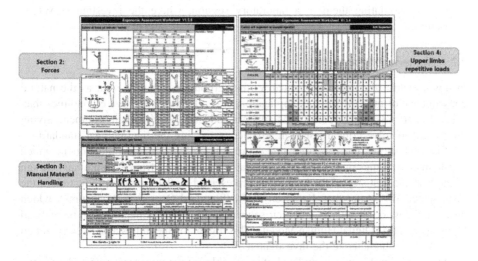

Fig. 2. EAWS checklist overview: section 2, section 3 and section 4.

Section 1 evaluates symmetric and asymmetric static working postures: postures are considered static if its duration is at least 4 s long. The score is assigned based on the range of values of assumed posture angles.

Section 2 evaluates the exerted forces by fingers, hands, arms and whole body. The minimum force value that contribute to the score is 30 N for fingers and hands and 40 N for arms and whole body.

Section 3 evaluates the Manual Material Handling, considered as carrying, repositioning, holding or pushing and pulling. The minimum mass value that contribute to the score is 3 kg.

Section 4 evaluates the load for upper limbs in repetitive actions, basing on their frequencies related to applied forces.

According to the Machine Directive (98/37/EU), a visual evaluation is provided for each section (Fig. 3):

- Green zone: no action is needed;
- Yellow zone: a further risk assessment, as well as an analysis, is carried out, taking into consideration other involved risk factors. Redesign if possible, otherwise take other recovery actions to control the risk;
- Red zone. Actions to lower the risk are necessary.

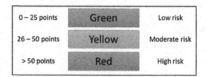

Fig. 3. EAWS index risk areas.

Currently, the checklist is filled-in by observing the whole working activities, physically simulating them in a laboratory session, where the investigated WP is reproduced.

2.2 Numerical Procedure

In this scenario, the approach named Virtual Ergonomics can be seen as the natural consequence of the DM strategy application. It is a kind of proactive ergonomics that, using digital tools, allows performing analysis and tests in a virtual environment, giving the chance to solve critical issues in advance, satisfying the international standards.

Figure 4 shows the proposed procedure for the WP design validation in which ergonomics becomes one of the design parameters. Starting from a preliminary design of the WP, the first step of the iteration is represented by the virtual reproduction of the investigated WP. After that, based on SOPs (Standard Operation Procedures), which describe the working activity and their sequence, the virtual simulation can be realized by using a DHM (Digital Human Model). This digital mannequin is able to carry out all the tasks that characterize the investigated WP.

Once the simulation is completed, it is possible extract numerical data and, thereby, assess the desired ergonomic index.

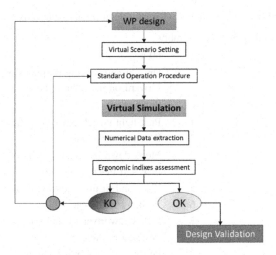

Fig. 4. WP design validation procedure based on ergonomic indexes assessment in virtual environment.

Depending on the score, it is possible validate the WP design or not. In this last case, there are two options: modifying the SOPs in case of medium or high risk score or modifying the design, if the previous solution was not effective, before restart the iteration.

3 Case Study

In order to prove the reliability of the proposed procedure, a case study has been investigated. In detail, it regards a real WP of a FCA plant assembly shop and the working activity is about the light switch cover and air nozzle assembly.

Table 2 shows a schematic description of the working activities. In this workstation, the task is characterized by two main activities: light switch cover assembly, in the lower zone of the steering, and foot air nozzle assembly, in the bottom of the gear shift zone.

The car is positioned on the Webb hook, with the reference frame (centered on the front axle) at 700 mm height (Fig. 5).

The devices used by the worker are two screwdrivers for tightening actions at 8 Nm torque.

About this working activity, the ergonomists provided the EAWS scores, evaluated by means of the traditional (observational) procedure (Table 3).

Table 2. Case study: Schematic description of the activity.

Light switch cover and air nozzle assembly		
Cycle time		60 s
Working conditions		Upright and moving posture; car in motion forward on the rotating hook
Main tasks		Sub-tasks
Light switch cover assembly	1	Pick light switch cover from the cart #1
	2	Walking to the car
	3	Move down the steering wheel lever
	4	Pick and Place ignition key on the car floor
	5	Place and fit in light switch cover
	6	Walking to the cart #2
	7	Pick screwdriver #1 and screws
	8	Walking to the car
	9	Perform 2 screwings to fix the cover
	10	Pick and place ignition key in ignition lock and move up the steering wheel lever
	11	Walking to the cart #2
Air nozzle assembly	12	Place screwdriver #1
	13	Pick screwdriver #2 and screws
	14	Walking to the cart #3
	15	Pick air nozzle from the cart #3
	16	Walking to the car
	17	Place air nozzle and perform 2 screwings
	18	Walking to the cart #2
	19	Place screwdriver #2

Fig. 5. Case study: virtual scenario.

Table 3. EAWS experimental score.

EAWS index	Whole body	Upper limbs
	121	44.5

4 Numerical Procedure and Evaluation

The case study has been implemented in a virtual scenario in order to numerically evaluate EAWS score, according to the procedure described in 2.2.

The simulation has been carried-out in TECNOMATIX PROCESS SIMULATE software environment by Siemens®, while the EAWS evaluation has been performed by using MTM® EAWSdigital code.

Within the virtual scenario, a DHM, named "Jack", is able to perform the working tasks. It is a digital cinematized mannequin, having realistic biomechanical properties, composed by 71 segments and 69 joints whit realistic degrees of freedom (dof's), based on NASA's researcher's studies [14].

5 Results

A high number of data can be extracted from simulations, as those ones regarding posture angles (Fig. 6).

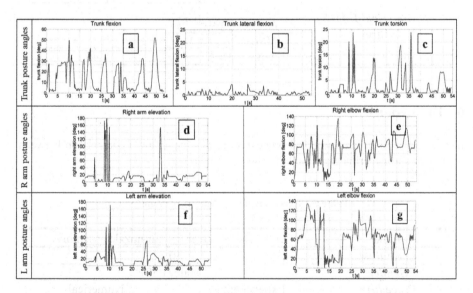

Fig. 6. Posture angles of trunk (a, b, c) and arms (d, e right arm; f, g left arm).

Posture angles are fundamental for evaluating the score about section 1. Regarding the trunk, the work task does not involve an excessive muscle load, since trunk flexion forward (Fig. 6a) in some intervals is within the range of 20°–60°, but never as static posture (i.e $\Delta t > 4$ s). The same goes for trunk torsion (Fig. 6c), which reaches values greater than 10° (the minimum value for asymmetric posture contribution), but with duration less than 4 s. Trunk lateral flexion (Fig. 6b) does not reach significant values, being ever less than 5°. About upper limbs, in Fig. 6d–g, the trends show that for both right and left arm, the elevation angles reach values greater than 90°, bringing the elbows over the should level and the hands over the head level. In this case, postures also cannot be considered as static and they do not contribute to the index.

About the section 2 of EAWS checklist, the index shows an extremely high value because of the use of screwdrivers with a tightening torque of 8 Nm, that, considering the screwdrivers geometry, means a counter-reaction force of about 80 N, exerted by the worker during the screwing tasks.

Section 3 does not contribute to the index because of the mass of handled parts is less than 3 kg.

Section 4, considering the number of repetitive actions related to exerted forces, shows a value included in medium risk area.

Table 4 shows the numerical results of EAWS scores.

Table 4. Numerical EAWS scores

Gender	Male (P50)		
	Whole body		Upper limbs
Numerical EAWS scores	Posture	2	
	Forces	100.5	
	MMH	0	
	Total	**102.5**	**45**

6   Discussions

In order to validate the numerical procedure, in this paragraph a comparison between numerical and experimental results is shown (Table 5).

Table 5. Experimental vs. Numerical results.

Task	Light switch cover and air nozzle assembly			
Cycle time	60 s			
Gender	Male (P50)			
Procedure	Experimental		Numerical	
Section 1	9		2	
Section 2	112		100.5	
Section 3	0		0	
Section 4	44.5		45	
EAWS index	Whole body	Upper limbs	Whole body	Upper limbs
	121	44.5	102.5	45

Numerical scores seem to be coherent with experimental ones. In particular, for what concerns section 1, the difference between experimental and numerical results is of 7 points, because numerical analysis allows achieving more accurate data than experimental ones, evaluated in observational way, without neglecting any kind of movements, even if they are not required to finalize the work task.

For what concerns section 2, the difference between experimental and numerical results is of 11.5 points, because of different postures assumed by the virtual mannequin and the workers in carrying out screwing tasks (Fig. 7). In particular, the real worker performs screwing in standing upright posture unlike the mannequin, who performs them in standing bent posture, contributing differently according to the Atlas of Forces [15]. Hence, despite the same exerted forces, numerical contribute to section 2 is less than experimental one because of the different assumed postures.

About section 3 and 4, the results are completely coherent.

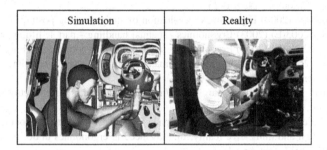

Fig. 7. Screwing task: simulation vs. reality.

7   Conclusions

A proactive ergonomic evaluation of workplaces is a formidable task to test the product feasibility since the design phase of a new product/production process, giving the opportunity to reduce time and costs and the possibility to change design parameters without risks.

Simulating operating tasks in a virtual environment provides a high number of data that allows evaluating the desired ergonomic index by means of a reliable numerical procedure.

In order to validate the proposed procedure, a real case study has been investigated. Numerical results have been compared with experimental ones, resulting coherent, despite some differences due to a more accurate analysis of numerical data with respect to the observational sensibility of the ergonomist.

This means that, in developing new products, the proposed method is ready to be applied and to give support to ergonomists and designers for human-centered WPs design, combining the expertise of ergonomics specialists with the advantages provided by virtual simulations and numerical data. This latter aspect would make it suitable for the design phase, with a significant reduction in costs for the implementation phase and an improvement in the working conditions of the operators, who will immediately have available a WP of which ergonomic efficiency has been previously validated.

References

1. Hovanec M, Korba P, Solc M (2015) Tecnomatix for successful application in the area of simulation manufacturing and ergonomics. In: 15th international SGEM geoconference on informatics, Albena
2. Spada S, Germanà D, Sessa F, Ghibaudo L (2015) FCA ergonomics approach in developing new cars: virtual simulation and physical validation. In: Proceedings 19th triennal congress of IEA, Melbourne
3. Vitiello M, Galante L, Capoccia M, Caragnano G (2012) Ergonomics and workplace design: application of ergo-UAS system in fiat group automobiles, work 41, pp 4445–4449
4. Schaub K, Caragnano G, Britzke B, Bruder R (2012) The European assembly worksheet. Theor Issues Ergon Sci 14(6):1–23
5. ISO 11226:2000 (2000) Ergonomics - evaluation of static working postures
6. ISO 11228-1:2003(E) (2003) Ergonomics - manual handling - part 1: lifting and carrying
7. ISO 11228-2:2006(E) (2006) Ergonomics - manual handling - part 2: pushing and pulling
8. ISO 11228-3:2007(E) (2007) Ergonomics - manual handling - part 3: handling of low loads at high frequency
9. International MTM Directorate. http://mtm-international.org/introduction-to-mtm-uas/ (online)
10. Caputo F, Greco A, D'Amato E, Notaro I, Spada S (2017) A preventive ergonomic approach based on virtual and immersive reality. In: Advanced in ergonomics in design: proceedings of the AHFE 2017, Los Angeles
11. Directive 2006/42/EC of the European parliament and of the council of 17 May 2006 on May 2006 on machinery and amending Directive 95/16/EC (recast.) (2006)
12. Council directive 89/39/EEC of June 1989 on the introduction of measures to encourage improvements in the safety and health of workers at work (1989)
13. D.lgs. 9 aprile 2008, n.81 - TESTO UNICO SULLA SALUTE E SICUREZZA SUL LAVORO
14. NASA (1987) Man-system integration standard (NASA-STD-3000)
15. Schaub K, Wakula J, Berg K, Kaiser B, Bruder R, Glitsch U, Ellegast RP (2014) The assembly specific force atlas. Hum Factors Ergon Manuf Serv Ind

Development and Evaluation of a Wearable Motion Tracking System for Sensorimotor Tasks in VR Environments

Andreas Mourelatos, Dimitris Nathanael$^{(\boxtimes)}$, Kostas Gkikas,
and Loizos Psarakis

School of Mechanical Engineering, National Technical University of Athens,
9 Iroon Polytechniou Street, 15780 Zografou, Athens, Greece
dnathan@central.ntua.gr

Abstract. The present communication reports on a series of experiments, aiming at exploring the effect of a virtual limb's visibility on task performance inside interactive Virtual Environments. To this end, a motion tracking system was developed, capable of tracking the movement of the human arm beginning from the shoulder up to and including the palm. Twenty two university students participated in a shooting task experiment, divided into two groups. One group first completed the experiment with full visibility of the arm, and then without, while the other completed the experiment in the opposite manner. Results show that while the arm's visibility had no significant effect on task performance, it does affect the subjects' subjective experience within the environment.

Keywords: Virtual reality · Motion-tracking · Shooting task

1 Introduction

Advances in computing and display technology have rendered Virtual Reality a rapidly evolving field of research, with a constantly growing number of applications and an increasingly commonplace presence in everyday life. VR environments and technologies are constantly becoming ever more inexpensive and readily accessible to humans, and seeing use for purposes exceeding entertainment.

The interest of the scientific community in Virtual Environments (VEs) has been both extensive and manifold, concerning issues ranging from the possible applications of such technologies in various aspects of life to the psychological effect that virtual stimuli can have on users.

The present communication reports on a series of experiments, which were designed and conducted with the aim of exploring the effect of limb visibility on task performance inside interactive VEs. The results of these experiments offer valuable insights regarding (i) the ability of a human using a VE to assimilate a virtual representation of their body with their body image, and subsequently incorporate this representation into their body schemas, (ii) the speed and ease with which that assimilation occurs, and (iii) the effect it has on the user's performance and subjective experience.

© Springer Nature Switzerland AG 2019
S. Bagnara et al. (Eds.): IEA 2018, AISC 822, pp. 181–188, 2019.
https://doi.org/10.1007/978-3-319-96077-7_19

For the completion of these experiments, the re-design and adaptation of an innovative motion tracking technology developed by the NTUA Lab of Cognitive Ergonomics was required, to facilitate its use within VR environments. This included the construction and programming of a functional prototype, as well as the development of a VE in which the prototype could be tested.

2   Motivation

Since the emergence of VR as a research tool, various papers have been released exploring the effect that the human body's visual representation, commonly referred to as a virtual or avatar body, has on the user's experience and performance within the VE. It has been demonstrated that a similarity in appearance or anatomy between the user's physical body and the avatar body's image is not a necessary prerequisite for the user to be able to experience an illusion of ownership towards the virtual body, as long as the user's own body schemas can be extended to control it.

For example, Slater et al. [1] showed that adult male subjects could identify with a virtual body representing a female child, while Kilteni et al. [2] found that users were capable of controlling a virtual arm up to four times the length of their physical one, as well as register it as their own limb when confronted with a threat to it. Won et al. [3] presented congruent results indicating that humans can learn to control bodies differing from their own in either (i) range of motion, such as a body with much more flexible lower limbs, or (ii) anatomy, such as one with a third arm.

Taking the above findings into consideration, exploring the assimilation of a fully controllable arm in a VE, which corresponds exactly to the user's physical arm, was deemed a significant avenue of research, as it provides both a new tool for interfacing with VEs (in the form of the aforementioned wearable system), and a gateway into research regarding the assimilation of dissimilar limb or body forms within such environments.

3   Development of IMU Wearable Tracking System

To this end, a motion tracking system was developed, capable of tracking the movement of the human arm beginning from the shoulder up to and including the palm.

The wearable system consists of three MPU 9250 IMUs connected to an Arduino Nano with 24AWG wires and mounted on a fingerless glove and elbow patch. Although various systems implementing arm tracking in VR already exist, the system in question retains certain benefits not found in existing systems. The main advantage it provides is the fact that it can function independently from any type of position tracking technology or display, rendering it entirely portable.

To explore the aforementioned questions, an environment was designed in the Unity Game engine, to create a Virtual Reality "shooting target practice" task. In this setting, a virtual arm within the environment could be entirely controlled by the movements of the user's arm in the real world with negligible lag. The shooting task was specifically selected as (ι) it imposes hand-eye coordination requirements on the user, allowing for the effective study the level of incorporation of an avatar body in VR environments and (ii) it is a sensorimotor task that by nature does not involve any tactile elements therefore rendering the test more realistic.

4 Experimental Design

The experiment was initially designed to test whether a correlation exists between the visibility of the virtual arm and the test subjects' performance in the task. Twenty two university students participated in the experiment, divided into two groups, one containing 12 and the other 10 participants, henceforth referred to as Group A & Group B, respectively. Group A consisted of 6 male and 6 female participants, while group B consisted of 5 male and 5 female participants. Participants were aged 20–29. Their experience in First Person Shooter games varied from none to very high.

The experiment involved two "trials" for each subject. Each trial was defined as a 1-min period during which the subjects use a laser beam fired from the end of the virtual arm to shoot down targets appearing in a 360° radius around them. If not shot down, each target would disappear after 10 s and another would appear at a different point in space. The subjects' goal was to shoot down the maximum amount of targets within the 1-min period. The task was completed from a standing position, ensuring that each subject had sufficient space to rotate around their axis, in order to reach targets appearing behind them, without being obstructed by the control and display devices. Between trials, the subjects were asked to fill out a questionnaire regarding their experience during the experiment.

Group A first completed the experiment with full visibility of the arm, filled in the questionnaire after this first trial, and subsequently ran the experiment again, this time having visibility of only the wrist and palm. To counterbalance learning effects, Group B completed the trials in the opposing order.

Both groups then completed the experiment for a third time, during which the subjects no longer had visibility of the beam. The visibility of the arm was determined accordingly to the groups' second trial, therefore Group A had visibility from the wrist down, whereas Group B had visibility of the full arm.

The aim of this third trial was to determine whether the usefulness of the arm's visibility is dependent on the nature of the task. It was theorised that, by removing the beam, the task of shooting targets was made not only harder, but qualitatively different from the original and more akin to real-world shooting, therefore forcing the subjects to adopt different sensorimotor routines for completing it. Evidently, no comparison of data between the first two trials and this third one could be deemed valid, as they originate from dissimilar tasks.

The addition of a fourth trial, where the arm's visibility would again be reversed with the beam remaining invisible, was rejected under the assumption that subjects were already trained in both conditions and thus learning effects were already compensated for.

5 Results

5.1 Average Scores per Trial

Firstly, as regards the average scores (targets hit) of the groups in each of the trials, the following data was acquired (Table 1):

Table 1. Average scores per trial

	Group 1			Group 2		
	Trial 1	Trial 2	Trial 3	Trial 1	Trial 2	Trial 3
Average	8.92	12.92	5.3	9.2	12.2	5.4
Std. Dev.	1.62	2.97	4.48	4.13	2.49	3.78

The average scores of Groups A and B in each of the trials were compared using a two-sample unequal variance t-test and found to not be statistically different.

5.2 Target Lifespan Charts

After these preliminary results were calculated, the following plots were produced. These show the time (in seconds) each target remained on screen before disappearing or being shot down on the Y axis, plotted during the total run-time of the experiment, displayed on the X-axis. The trend lines were obtaining by using linear regression on the data. Linear regression was chosen as a simple evaluation tool for observing trends appearing in the data, due to the lack of any sort of information as to what kind of formula would best represent the learning effect. The results are presented for both groups on the same axes, for the first two trials of the experiment. It is repeated at this point that Trials 1 & 2 were analysed separately from Trial 3, as they concern differentiated experimental tasks (Fig. 1).

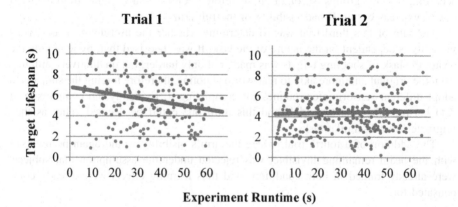

Fig. 1. Target lifespan charts, Trials 1 & 2

While the data is scattered, the linear regression trend lines indicate that during the first trial, targets stay on-screen for decreasing amounts of time as the experiments progresses.

During the second trial, that effect is no longer observed; targets remain on-screen for an amount of time that does not increase or decrease during the trial. That amount is approximately equal to the amount of time that targets stay on-screen at the end of the first trial (~ 4 s). A t-test was conducted, comparing the mean values of the second trial for each group with the estimated value reached at the end of the first trial (considered as a set value) by the corresponding group, demonstrating that these values can indeed be considered statistically identical.

These observations were further strengthened by evaluating the p-value of the regression curves. For all cases, the first trial showed a significant decreasing trend (p-value 0. 0004), while the second did not (p-value 0. 6743).

The same analysis was conducted for the third trial, wherein no significant trend was observed, as confirmed by the p-value of the regression trendline (0.6554).

5.3 Shots per Target Charts

Following this, charts displaying the number of shots fired by the players to shoot down each of the targets were produced. The charts were "normalised" by dividing the data points with the total number of shots fired during the trial, to compensate for the variance in total shots fired by each player. No significant trend was observed in any of the charts, the p-values of which are shown in Table 2.

Table 2. P-values of linear regression trend-lines of shots per target charts

	Trial 1	Trial 2	Trial 3
Groups A & B	0.9129	0.3743	0.0854

5.4 Length of Traced Curves

To supplement these results, the length of the curves traced by the virtual arm during the experiment was calculated. The average curve length for each group and trial can be seen in the following table (Table 3).

Table 3. Average length of traced curve per trial

	Group 1			Group 2		
	Trial 1	Trial 2	Trial 3	Trial 1	Trial 2	Trial 3
Average	80.40	82.45	63.04	72.71	74.02	57.81
Std. Dev	12.74	10.77	9.26	15.17	11.20	16.69

5.5   Questionnaire Results

The answers to the questionnaire indicate that the users did not experience large delays in their movements (17/22 reported below-average delay), or feel obstructed by the equipment used during the experiment (18/22 reported little or no obstruction), attesting to the fact that the experimental setup did not affect the results described above. One of the most interesting results obtained by the questionnaire is demonstrated by the answers to the following questions:

Question 4: How much harder or easier was the task made by the fact that your arm was invisible/visible?
    1-Much harder   *7-Much easier*

Question 5: How much was the naturalness of your interactions with the environment decreased or increased by the fact that your arm was invisible/visible?
    1-Severely decreased   *7-Severely increased*

In answering these questions, the two groups exhibited differences, with Group A reporting that the arm's visibility affected their effectiveness and experience in a positive way (10/12 and 9/12 above average in Questions 4 and 5, respectively), while Group B reported contrary results (just 2/10 and 1/10 above average in Questions 4 and 5, respectively).

6   Discussion

6.1   Average Scores

The fact that the two groups achieved the same average scores throughout trials, leads to the conclusion that the arm's visibility has no significant effect on the performance of the subjects. It is also obvious that familiarisation with the task improves performance, while the removal of the beam has an adverse effect on performance, as it forces the subjects to re-adapt their strategy, and increases the difficulty of the task. Still, no significant variance is observed in the groups' results on this trial, indicating that the arm's visibility is not of paramount importance even after modifying the task.

6.2   Target Lifespan

The decrease in target lifespan throughout the first trial and subsequent stabilisation during the second one clearly demonstrate the learning effect experienced by the experiment's participants – familiarisation with the task during the first trial leads to the targets being shot down faster, with a "plateau" value reached by the end of it and maintained throughout the second trial. The stability in target lifespan throughout the third trial can be attributed to the participants either (i) being unable to effectively re-adapt their strategy to the modified task within the allocated time or (ii) having already achieved a "maximum" level of familiarisation with the virtual arm, thus exhibiting similar "plateau" results to the second trial. It is worth noting that these two hypotheses are somewhat contradictory. Further experimentation on the modified task is necessary to determine which, if any, of the above hypotheses is true.

6.3 Shots per Target

It was theorised that the charts displaying the shots fired per target would mirror the trends displayed in the target lifespan charts, with familiarisation resulting in a smaller amount of shots required to hit each target however no such trends were observed in the results. The prevalent hypothesis for the interpretation of these results stems from the fact that the targets appear randomly in 3D space during the experiment. It is theorised that the location of the target, rather than the skill of the test subject, is the defining factor determining the amount of shots required to hit it. This conclusion was drawn based on the observation that nearby targets were much easier for subjects to hit than faraway ones, especially in the third trial were the beam was removed. Further testing is deemed necessary to support this hypothesis.

6.4 Lengths of Traced Curves

Similarly, it was initially theorised that an increase in curve length during the third trial of the experiment would be observed, owing to the hypothesis that test subjects would change the way they moved during the third trial. Namely, during the first two trials, due to the beam being visible, many subjects would retain their upper arm and forearm in an approximately fixed position, instead using the rotation of their wrist to aim, and recalibrating their shots according to the point in space that the beam would reach. During the third trial, subjects were unable to use that mechanism, as the beam was rendered invisible. Therefore, they would attempt to aim at targets by extending their physical arm in an attempt to align the virtual arm with the target in the VE. This movement has a much larger range-of-motion than the one employed during previous trials, leading to the assumption that the corresponding curves would exhibit longer length. The results did not confirm this hypothesis, however. Contrarily, the third trial exhibited the shortest average curve length of all.

The proposed explanation for this result is that the largest contribution to the overall curve length is not actually provided by aiming at any one target, but rather by the large movements of the arm when moving from one target to the next. As a result, the second trial, which usually exhibits the highest number of targets hit, contains more of these movements than any of the other two, especially the third one (where the scores are usually lowest), thus resulting in a larger overall curve length.

6.5 Questionnaire Results

The analysis of the questionnaire answers makes evident the fact that subjects belonging to Group A found that the visibility of their arms increased both their performance and the naturalness of their interactions, whereas those in Group B report the opposite effect occuring owing to the fact that their arms were invisible. This leads to the conclusion that, while the visibility of the arm seems to have had no significant effect on the users' objective measures of performance, it does affect satisfaction—users prefer being able to see their whole arm in the VE, rather than just the wrist and palm.

References

1. Slater M, Spanlang B, Sanchez-Vives MV, Blanke O (2010) First person experience of body transfer in virtual reality. PLoS ONE 5(5):e10564
2. Kilteni K, Normand JM, Sanchez-Vives MV, Slater M (2012) Extending body space in immersive virtual reality: a very long arm illusion. PLoS ONE 7(7):e40867
3. Won AS, Bailenson J, Lee J, Lanier J (2015) Homuncular flexibility in virtual reality. J Comput Mediat Commun 20(3):241–259

The Construction of High-Rise Building Fire Escape Scene

Hua Qin[✉] and Xiao-Tong Gao

Department of Industrial Engineering,
Beijing University of Civil Engineering and Architecture,
Beijing 100044, People's Republic of China
qinhua@mails.tsinghua.edu.cn

Abstract. In view of the needs of personnel behavior research in fire scenarios, a virtual experimental platform is set up to conduct research on personnel behavior in fire sites. The fire scenario is divided into two aspects: static and dynamic. In static aspect, the artistry and repeatability of the building are constructed by modeling software. In the aspect of dynamics, the change of building structure is realized by unity3D software, and the influence of structural change on Wayfinding Behavior is shown. It is intended to pave the way for the experimental platform for the study of personnel behavior in the subsequent fire field.

Keywords: Path seeking behavior · Complex building · Scene simulation

1 Introduction

With the development of economy and the increasing density of urban resources, the number of high-rise building fires has been increasing year by year. The issue of fire escapes caused widely concern by researchers. During the process of escaping from high-rise building, people need accurate information to action. But, because of high-rise building's characteristic which let high-rise building have a great impediment to people's access to spatial information. Whenever a fire happens, this situation is more obvious.

People in high-rise building will face more challenge to escape. The challenge divided into two aspects: one is static and the other is dynamic. Static characteristics are the inherent attributes of tall buildings: hierarchy, complexity, verticality, multichannelity. it is the inherent structural property of the building itself. The dynamic feature is the dynamic changes of tall buildings after a fire: the randomness of fire points, the spread of smoke, the randomness of collapsing, the flame spread on the same floor and the flame spread on different floors. The occurrence of high-level fire under real conditions is the process of alternating static and dynamic characteristics of high-rise building fire [1–5]. Personnel in the process of escape will continue to be affected by these characteristics.

In this paper, unity3D is used to design and construct a real fire near the scene of building fire escape. The purpose is to improve the accuracy of the virtual environment

© Springer Nature Switzerland AG 2019
S. Bagnara et al. (Eds.): IEA 2018, AISC 822, pp. 189–194, 2019.
https://doi.org/10.1007/978-3-319-96077-7_20

through the study of static dynamic characteristics and scene simulation, and help to set the scene of virtual environment fire research in the future.

2 Methodology

In order to reproduced the real dynamic fire scene. The spread model of fire and smoke and collapse are added to in dynamic fire scene.

In the static fire scene construction considering the artistic characteristics and structural characteristics of the high-rise building.

First, use AUOTOCAD to draw architectural drawings.

Second, using 3DMAX building the model and the collapses model.

Third, using unity3D import the model, then rendering the model, and then programming the game mode of the fire scene.

2.1 Static Fire Scene

Spatial perception process is affected by architectural characteristics. In the absence of fire, architectural artistry (the artistry makes people difficult to understand the structure of the building quickly), architectural repetition (the hierarchy of high buildings and the similarity of escape paths) affects the spatial perception process.

2.1.1 CAD Diagram of Static Scene

① Building Artistic

Architecture seeks artistic expression, such as circles, triangles, squares, and special shapes. Based on the actual building situation, this paper has modeled. The architectural drawings are as follows: (Figs. 1, 2, 3, 4).

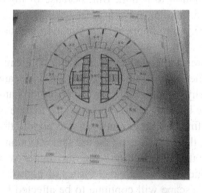   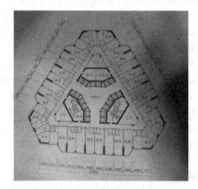

Fig. 1. Circular high-rise building   **Fig. 2.** Triangular high-rise building

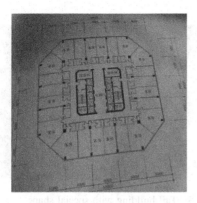

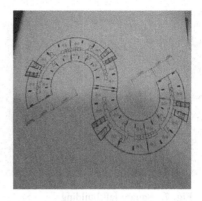

Fig. 3. Square tall building   **Fig. 4.** Tall building with special shape

In terms of angle, circular buildings have more ring paths, which have more obtuse angles. The square building has more right angle rotation. The path of a triangular building has more sharp angles. The passageways of special buildings are more complicated, which is a combination of the above situations.

② Repeatability

There is a high degree of similarity between the floors of high-rise buildings, and the passages between the same floors have very high similarity. In the process of building road finding, these similarities are mainly reflected in the rotation of angles.

2.1.2 Construction of Static Scene
See Figs. 5, 6, 7, 8.

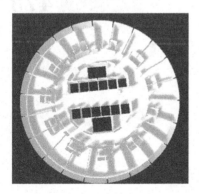

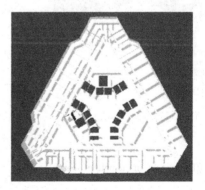

Fig. 5. Circular high-rise building   **Fig. 6.** Triangular high-rise building

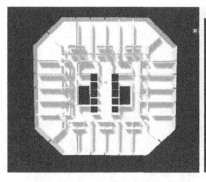
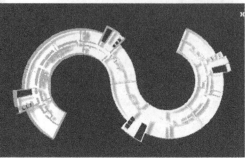

Fig. 7. Square tall building   **Fig. 8.** Tall building with special shape

2.2   Dynamic Fire Scene

In the case of fire, the structural mutagenicity (Architectural artistry, architectural reproducibility will be mutated by the change of fire), the randomness of the fire source, and the smog cover make the building a labyrinth that affects the space perception process. The decrease of efficiency of spatial information perception will increase the time of action and increase the risk.

2.2.1   The Construction of a Dynamic Scene (as an Example of a Common Building)

①fire   ②obstacle

③The impact on the path of search

The appearance of fire and obstacles affects the behavior of the road finding, as shown in the figure (Figs. 9 and 10):

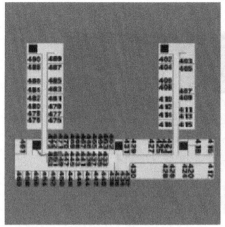

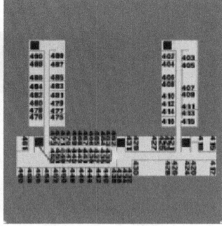

Fig. 9. Before the obstacle appears   **Fig. 10.** After the obstacle appeared

3 Implementation

3.1 Visual Angle

3.2 Find a Way to Escape

4 Discussion

This paper constructs a model of the influencing factors of the fire scene. By building a player can operate the game platform which simulation degree higher than the existing basis. Participants can experience and interact with the virtual fire scene. This makes it possible to study the personal behavior in virtual fire scenes.

References

1. Mileti DS (1974) A normative causal model analysis of disaster warning response
2. Watts JM (1998) Human behavior in fire. Springer, New York
3. Fraser JN (1999) Wiley online library modelling human behaviour within the fire risk assessment tool CRISP. Fire Mater 23:349–355
4. Kuligowski ED, Hoskins BL (2010) Occupant behavior in a high-rise office building fire
5. Bryan JL (1977) Smoke as a determinant of human behavior in fire situations (project people)

An Augmented Reality Application to Support Deployed Emergency Teams

Isabel L. Nunes[1,2(✉)], Raquel Lucas[1,3], Mário Simões-Marques[4],
and Nuno Correia[1,3]

[1] Faculdade de Ciencias e Tecnologia, Universidade NOVA de Lisboa,
2829-516 Caparica, Portugal
{imn, nmc}@fct.unl.pt, rv.lucas@campus.fct.unl.pt
[2] UNIDEMI, Faculdade de Ciencias e Tecnologia,
Universidade NOVA de Lisboa, 2829-516 Caparica, Portugal
[3] NOVA-LINCS, Faculdade de Ciencias e Tecnologia,
Universidade NOVA de Lisboa, 2829-516 Caparica, Portugal
[4] CINAV - Portuguese Navy, Alfeite, 2810-001 Almada, Portugal
mj.simoes.marques@gmail.com

Abstract. THEMIS-AR is an augmented reality (AR) app designed for mobile devices that was developed to assist first responders on disaster relief operations, which a very demanding context for users, regarding both the physical and the cognitive and emotional demands. In fact, the specificity of the usage environment imposes special care in addressing the interaction of the users with the system, regarding both the mobile devices technical characteristics and the workload. The paper provides a brief description of the THEMIS-AR architecture and features, focusing on the assessment of a concept demonstrator developed based on the UCD cycle (i.e., context of use characterization, requirements definition, solution implementation, and testing and validation). The usability testing setup and script adopted while performing field tests are described and the results obtained, namely regarding the application of the User Experience Questionnaire to benchmark the app evaluation in comparison with a set of previous evaluations made with this methodology. These results were compared with the user perceptions collected using a final questionnaire namely based on the System Usability Scale. The main findings resulting from this initial set of users are summarized and presented the way ahead for this component of the THEMIS project.

Keywords: Augmented reality · Emergency teams · User centered design
THEMIS · UEQ · SUS

1 Introduction

The *disTributed Emergency Management Intelligent System* (THEMIS) project is an undergoing R&D initiative launched to design and implement an intelligent system to assist and support real time Disaster Management activities during on disaster relief operations (DRO), namely considering the complex context of international interagency response to major disasters. As an intelligent system it will gathers information

© Springer Nature Switzerland AG 2019
S. Bagnara et al. (Eds.): IEA 2018, AISC 822, pp. 195–204, 2019.
https://doi.org/10.1007/978-3-319-96077-7_21

from multiple sources (e.g., users, sensors, crowdsourcing), provides situational awareness based on a georeferenced common picture which is shared among system users and offers advice and guidance on response priorities, resource assignment and procedures [1]. Figure 1 offers a high level conceptual perspective of THEMIS considering the scenario of a major disaster requiring international humanitarian assistance, provided by multiple agencies.

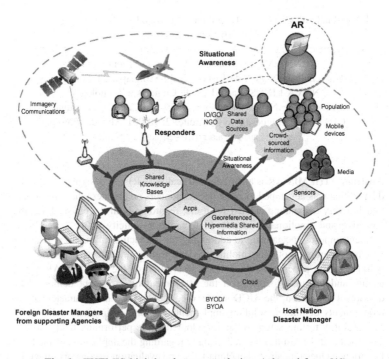

Fig. 1. THEMIS high level conceptual view (adapted from [1])

The Response phase of the Disaster Management cycle[1] is a complex process, usually dealing with a large amount of uncertain, incomplete and vague information, predominantly focused on immediate and short-term needs, which normally requires the coordination and collaboration among a variety of actors. There is no standard terminology and definitions regarding Disaster Management; nevertheless, the United Nations International Strategy for Disaster Reduction (UNISDR) organization proposes a terminology [2], which is used to identify some of the basic terms and definitions presented in Table 1. In general terms, a disaster is a serious disruption of the functioning of a community involving widespread human, material, economic or environmental losses and impacts, which exceeds the ability of the affected community or society to cope using its own resources.

---

[1] It is common to considers the Disaster Management cycle as having 4 phases: Preparedness (before the onset of a disaster), Response, Reconstruction, Mitigation.

Table 1. Disaster risk terminology adopted by the UN General Assembly [2].

Term	Description
Affected	People who are affected, either directly or indirectly, by a hazardous event. Directly affected are those who have suffered injury, illness or other health effects; who were evacuated, displaced, relocated or have suffered direct damage to their livelihoods, economic, physical, social, cultural and environmental assets
Disaster	A serious disruption of the functioning of a community or a society at any scale due to hazardous events interacting with conditions of exposure, vulnerability and capacity, leading to one or more of the following: human, material, economic and environmental losses and impacts
Disaster damage	Occurs during and immediately after the disaster. This is usually measured in physical units (e.g., square meters of housing, kilometers of roads) and describes the total or partial destruction of physical assets, the disruption of basic services and damages to sources of livelihood in the affected area
Disaster impact	The total effect (negative and positive) of a hazardous event or a disaster. The term includes economic, human and environmental impacts, and may include death, injuries, disease and other negative effects on human physical, mental and social well-being
Disaster management	The organization, planning and application of measures preparing for, responding to and recovering from disasters
Response	Actions taken directly before, during or immediately after a disaster in order to save lives, reduce health impacts, ensure public safety and meet the basic subsistence needs of the people affected

About a decade ago, in the aftermath of Hurricane Katrina, a report produced by US National Academy of Sciences [3] stressed the importance of disaster management and highlighted the role of information technology (IT) in disaster management. The same report stated that *"IT provides capabilities that can help people grasp the dynamic realities of a disaster more clearly and help them formulate better decisions more quickly"*; and concluded that *"IT has as-yet-unrealized potential to improve how communities, the nation, and the global community handle disasters"* and *"disaster management organizations have not fully exploited many of today's technology opportunities."* The report authors produced recommendations for improving disaster management, which included the pursuit of six key IT-based capabilities [3]:

- More robust, interoperable, and priority-sensitive communications;
- Improved situational awareness and a common operating picture;
- Improved decision support and resource tracking and allocation;
- Greater organizational agility for disaster management;
- Better engagement of the public;
- Enhanced infrastructure survivability and continuity of societal functions.

This report points the development of such capabilities as enabler of a flexible strategy of resilience and adaptability to the dynamics and inherent complexities of disasters. This requires the use of credentialing and identification tools for managing the flow of personnel in and out of incident areas to improve the efficiency of response and recovery operations; the evolution of information sharing towards distributed processing of continual messaging-streams fed by pervasive sensors, providing real time situational awareness data; coping with a "managed ad-hoc-racy" of disaster management and responder organizations that can evolve seamlessly and continuously over the entire course of a disaster; and leveraging the public as providers of information and sources of valuable technology tools [3].

The THEMIS project was proposed to contribute to improve the effective, efficient and timely engagement of the response teams, by offering increased situational understanding, and advice to the Disaster Management process. The THEMIS-AR app was developed to exploit emergent interaction technologies namely to assist first responders by increasing their perception of the disaster environment and incidents, support facilities and assets, as well as operational guidance, based on augmented reality[2].

In fact, nowadays there is a generalized use of mobile devices, and people are increasingly more familiarized with their functionalities (often based on sophisticated sensory capabilities), which turns these devices into very powerful and quite accessible tools for emergency team elements, involved in DRO, faced with multiple informational challenges.

The THEMIS-AR app initial development process was discussed in [4] describing the UCD approach used during the implementation of a demonstrator to validate the concept based on the assessment of the level of improvement achieved regarding, for instance, the guidance provided to response teams throughout their work, the situational awareness regarding their surroundings, or the navigation guidance towards the locations where their actions are needed.

Following the current introduction, Sect. 2 offers a brief overview of the UCD process applied in the development of the app. Section 3 presents the tests performed for assessing the app and discusses the results. Finally, the Conclusions section summarizes the work presented and offers a perspective of the future work.

2   THEMIS-AR

The THEMIS-AR is a demonstrator that was developed within the THEMIS project as a way of experimenting and validating the concept of using AR in the Preparedness and Response phases of the Disaster Management cycle. The development was based on the UCD cycle – encompassing the context of use characterization, the requirements definition, the solution implementation, and the testing and validation.

Regarding the context of use, the scenario of employment is the one of major disasters requiring international humanitarian assistance, provided by multiple agencies (as shown in the high level conceptual perspective presented in Fig. 1), offering an AR solution to equip emergency responders, as highlighted on the mentioned figure. It is

---

[2] Augmented Reality allows combining the vision of the real world, with virtual objects superimposed upon or somehow related with real world features.

worth noting that this context is very demanding for users, regarding the physical, cognitive and emotional dimensions. In fact, the specificity of the usage environment imposes special care in addressing the interaction of the users with the system, regarding both the mobile devices technical characteristics and the workload.

The app requirements are to augment the user perception of the situational context, by overlaying synchronized and context relevant information and guidance (e.g., symbols, pictures, text) to the image captured through the camera of the device. The set of app features includes the creation, edition, and visualization of geo-referenced information (e.g., location, characterization) of incidents, points of interest, and other response teams; exchange of operational instructions and reports; as well as advice and support regarding intervention priority or procedures. The THEMIS-AR app runs in smartphones or tablets held by the leaders of the response teams involved in DRO.

Regarding the solution implementation, the described demonstrator (illustrated in Fig. 2) was designed for tablets running Android, and implemented in Java, except for the AR functionality which used JavaScript. The Augmented Reality API was developed using Wikitude [5], which offers callJavascript methods used to interact with a Server database created using SQLite [6], a relational database management system. The Augmented Reality API assists the users by overlaying georeferenced information to the image captured by the mobile device camera. The application also offers functionalities to support the navigation towards incident or other point of interest sites. The demonstrator is currently being used as testbed to support future developments.

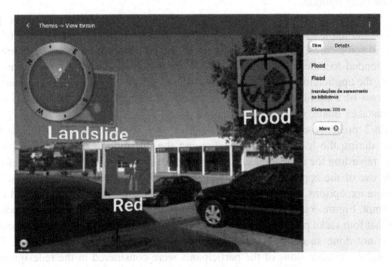

Fig. 2. Example of the THEMIS-AR API

The following section describes the demonstrator usability testing setup adopted for performing field tests, and the results obtained, namely regarding the application of the post-test User Experience Questionnaire (UEQ) [7] as methodology to assess six User Experience parameters (Attractiveness, Perspicuity, Efficiency, Dependability, Stimulation, and Novelty) and to benchmark the THEMIS-AR evaluation in comparison with a set of previous evaluations made with the UEQ method.

3 Demonstrator Usability Testing Methodology

The usability tests of the demonstrator were carried out in the Campus of the Faculty of Sciences and Technology of the New University of Lisbon, involving an initial group of 10 participants representing two groups: people with some experience in emergency response and people with no experience in emergency response. The participants' ages ranged from 23 to 46, with an average age of 26.6 years (standard deviation = 6.6). Eight were male and two were female. All participants were regular users of smartphones, eight were regular users of PC and only two of tablets.

For the field-tests a script was designed to test specific use cases (e.g., select the user team, understand the surrounding environment, enter data, insert team report, explore the map view, view and edit details of an incident, explore the AR view) focused on the app core functionality. The script guides users on performing twenty tasks considering a simulated scenario that involves the response to affected people and disaster damage. Before the main tests, a pilot test was conducted which resulted in some modifications to the script.

A database was created with the necessary data for carrying out the tests, matching several incidents and points of interest that were simulated in the Campus. Each test session involving a user was conducted individually and lasted approximately 1 h. After a brief introduction to the THEMIS-AR app, the users were prompted to perform the twenty tasks in the script. During the test session, users were invited to think aloud and share their opinion.

As the users performed the tasks, questions were asked about the conceptual aspects of the application. The answers were categorized using a Likert scale (one to five) or, in some cases, were only yes or no. At specific test times, questions were also asked intended to assess whether the users were understanding the information conveyed by the application (e.g., "How many injured people are there in the scenario?"). The success of a performing task also included an appreciation of the success of the same, mistakes made, opinions and suggestions.

Table 2 presents examples of tasks performed, shows THEMIS-AR interfaces captured during the tests, and graphs showing the distribution of the answers to the question regarding the corresponding task. Generically the answer reflected satisfaction with the use of the app and the tasks were performed with success. There was, however, some exceptions, with tasks where some participants were not successful at the first attempt. Figure 3 presents the error rate considering the twenty tasks. It is possible to note that four tasks present some challenge to a small number of participants and one task was not done successfully at the first attempt by any of the participants. These results and the observations of the participants were considered in the redesign of the functionality tested by the task.

On completion of the tasks, participants were asked to complete a post-test questionnaire, divided into two parts. The first part was intended to evaluate the performance of the application in six parameters (Attractiveness, Perspicuity, Efficiency, Dependability, Stimulation, and Novelty) using a questionnaire consisting of twenty-six items through which the participants expressed their opinion regarding the application, based on the UEQ scale. The second part was intended to evaluate user

Table 2. Example of THEMIS-AR scripted test tasks, interaction environment and results

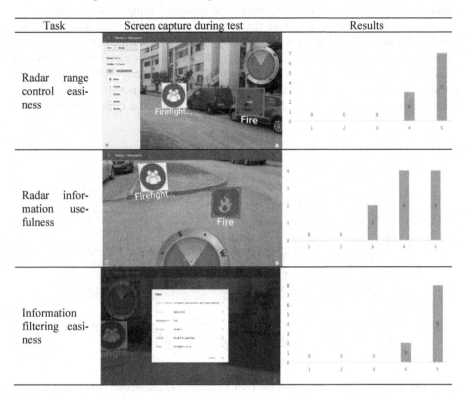

Task	Screen capture during test	Results
Radar range control easiness		
Radar information usefulness		
Information filtering easiness		

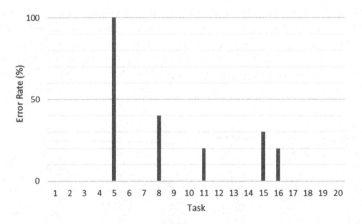

Fig. 3. Error rate of the tasks performed by the users

satisfaction based on the SUS scale [8] and also using questions that focused on the application domain. These final questions were evaluated using a Likert scale.

The analysis of the first part of the post-test questionnaire was done using the template Microsoft Excel file downloaded from the UEQ site, which groups the results to assess the Pragmatic Quality (Perspicuity, Efficiency, Dependability) and Hedonic Quality (Stimulation, Novelty). The results of the questionnaires were globally very positive. Figure 4 shows the evaluation results which very positive. Although Perspicuity was the parameter with the lowest result, it also had a quite positive result.

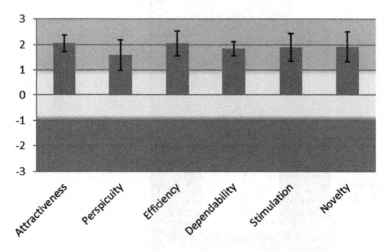

Fig. 4. Test results considering the User Experience Questionnaire (UEQ) scales

The UEQ also allows benchmarking the evaluated product regarding a set of evaluations made with the same method. The measured scale means are defined in relation to the values existing from a set of reference data. This dataset contains the results of 9905 participants from 246 studies related to different products.

Figure 5 shows the THEMIS-AR results reflecting the average of the assessments, corresponding to the line in the graph. Compared with other products evaluated with

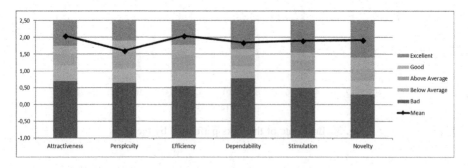

Fig. 5. Benchmark Chart based on the User Experience Questionnaire (UEQ) results

the UEQ method, THEMIS-AR results are considered "excellent" (in the range of the 10% best results) in all parameters except Perspicuity, which was assessed as "good" (10–25% best results). Therefore, the results are consistently above average.

The THEMIS-AR app results are summarized considering the three aspects of the pragmatic and hedonic quality: Attractiveness - 2.05; Pragmatic Quality - 1.83; and Hedonic Quality - 1.91. These results (illustrated in Fig. 6) are also quite good.

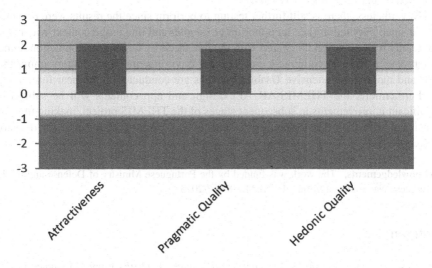

Fig. 6. Summary of the three pragmatic and hedonic quality aspects results

Considering the second part of the post-test questionnaire, the assessment was based on five statements. The first three statements are to evaluate the user satisfaction (based on the SUS scale) and the last two were focused on the application domain (based on the Likert scale). The results were once again very positive:

- Integration   assessed if the various functions of the application were well inte grated. The average result was 4.2;
- Consistency – assessed if the application was inconsistent. The average result was 1.3, indicating that users disagree that the application is inconsistent;
- Confidence – evaluated users' confidence in using the application. The average result was 4;
- Usefulness – evaluated the statement "I think the concept of this application will be useful during emergency responses." The average result was 4.8;
- Stimulation – evaluated the statement "I was able to have a good perception of the environment in which I am operating (regarding other teams, incidents and points of interest)." The average result was 4.4.

4 Conclusions

The goal with the THEMIS-AR Demonstrator was to proof the concept that offering response teams engaged on Disaster Relief Operations Augmented Reality functionality in mobile devices used by would be helpful and an added-value to the response teams. The initial results achieved with the Demonstrator are very positive and it is safe to conclude that the concept is valid.

Nevertheless, there is still further testing to perform since the results were obtained with a small population (i.e., an opportunity sample) and in a usage context which was not totally realistic when compared with the one faced by target user population, either in terms of individual equipment and working environment. These factors imply that new and more comprehensive Usability studies are conducted in the near future.

Furthermore, the THEMIS-AR API design and app contents will be subject of significant improvements in subsequent stages of the THEMIS project, making this app consistent with the other system environments (e.g, desktop application for decision makers).

Acknowledgements. The work was funded by the Portuguese Ministry of Defense and by the Portuguese Navy and by project UID/EMS/00667/2013.

References

1. Simões-Marques M, Correia A, Teodoro MF, Nunes IL (2018) Empirical studies in user experience of an emergency management system. In: Nunes I (ed) Advances in human factors and systems interaction(AHFE 2017), AISC, vol 592. Springer, Cham, pp 97–108
2. UNISDR. Terminology on Disaster Risk Reduction. United Nations International Strategy for Disaster Reduction. https://www.unisdr.org/we/inform/terminology. Accessed 27 May 2018
3. NAS (2007) Improving disaster management: the role of IT in mitigation, preparedness, response, and recovery. In: Rao RR, Eisenberg J, Schmitt T (eds) National academy of sciences. National Academies Press, Washington
4. Nunes IL, Lucas R, Simões-Marques M, Correia N (2018) Augmented reality in support of disaster response. In: Nunes I (ed) Advances in human factors and systems interaction(AHFE 2017), AISC, vol 592. Springer, Cham, pp 155–167
5. Wikitude Home page. https://www.wikitude.com. Accessed 27 May 2018
6. SQLite Home page. https://www.sqlite.org. Accessed 27 May 2018
7. UEQ. User Experience Questionnaire. http://www.ueq-online.org/. Accessed 27 May 2018
8. SUS. Measuring Usability with the System Usability Scale (SUS). https://measuringu.com/sus/. Accessed 27 May 2018

Evaluation of Repetitive Lifting Tasks Performed in Brick and Concrete Block Factories in Pakistan

Zafar Ullah[1(✉)], Shahid Maqsood[1], Rashid Nawaz[1], and Imran Ahmad[1,2]

[1] Department of Industrial Engineering,
University of Engineering and Technology, Peshawar 2500, Pakistan
zafarullah631@yahoo.com,
{smaqsood, rnawaz}@uetpeshawar.edu.pk
[2] Hanyang University Erica Campus,
Ansan, Gyeonggi-do Sangnoksu, South Korea
Imran86@hanyang.ac.kr

Abstract. Repetitive manual lifting and lowering tasks exposes workers to a high-risk musculoskeletal disorder. This paper aims at the assessment of postures of workers during the piling up process of a Concrete Block and Bricks (CB&B) at CB&B manufacturing factories in Pakistan and to study the effect of different risk factors such as contact force, stresses, and repetition of jobs that put muscles under redundant physical forces, which causes musculoskeletal disorders. The lifting indices (LI) of concrete blocks are 1.57 at the origin and 2.04 at the destination similarly LI for the bricks were 0.60 at the origin and 0.77 at the destination using the NOISH equation. For this purpose, the anthropometric data of 103 workers, working in 33 different factories, was collected. The postures were simulated and analyzed, using a human modelling solution HumanCAD software, with an objective to minimize the risk of work-related injuries, and stresses on the different parts of body was calculated. The results showed in current work environment, 38.83% of workers have lower back musculoskeletal disorder, followed by 31.06% with upper back, 29.12% with thorax and 24.27% with neck. Therefore, re-design the lifting methods of CB&B factories are essential in order to reduce the work-related injuries. The reports were shared with the CB&B industries.

Keywords: Musculoskeletal disorders · Lifting task · Lower back pain
Muscle stresses · Digital human modelling

1 Introduction

The International Labor Organization (ILO) current index shows that every year 317 million workers face work-related injuries and 2.02 million die every year due to work-related diseases and accidents, and further about 337 million serious work-related accidents each year [1]. The poor condition of organizational health and safety in the developing countries like Pakistan is due to many reason which includes lack of

S. Bagnara et al. (Eds.): IEA 2018, AISC 822, pp. 205–223, 2019.
https://doi.org/10.1007/978-3-319-96077-7_22

education and untrained workers. There are more than 7000 brick kilns operating in Pakistan which employed more than 100,000 workers [2]. Well trained workers are the asset to both un-interruptive production and economical gain [2, 3].

The high rate of work-related back injuries is a serious issue. Heavy lifting has been pointed out as one of the high risk factor of back injuries [4–7]. Ergonomists searching technique to quantify objectively the risk factors related to the back injuries [8]. As a result, different lifting assessment methods have developed with each one having different inputs, outputs, and succeeding interpretive capacities.

Manual lifting and lowering tasks is the common reason for Low Back Pain (LBP). Consequently, it is imperative to design tasks ergonomically and according to safety consideration. In 1981 NIOSH equation for health and safety practitioners for evaluating lifting related tasks was developed, which was revised for 1991 [9].

Work-Related Musculoskeletal Disorders (WMSDs) is a global issues in industrially oriented countries [10, 11]. WMSDs have been involved in many risk factors including force applied, working posture, vibration, biomechanical stressors and different other factors [12, 13]. The magnitude of the problem of WMSDs is now widely documented and a variety of diseases resulting from heavy work load, exertion, bending, repetitiveness, and prolong duration of jobs resulting in pain and may cause a permanent disabilities [14, 15]. There are numerous CB&B industries in Pakistan, where millions of workers are employed. In which workers perform repetitive work such as lifting and lowering of heavy load un-ergonomically that may cause WMSDs in different body parts such as neck, lower and upper back. Musculoskeletal problem caused a delay and heavy financial cost in the form of worker absenteeism and medical cost [16–18]. WMSDs is directly related to productivity and no proper attention has been given to the working environment and related workers in the industries [19–21].

Digital Human Modelling (DHM) tools are recently becoming a handy analysis tools for ergonomic design, workplace assessment in artificial digital space and for lowering or preventing WMSDs [22–24]. By integrating simulation, ergonomics solution methods, and evaluation, DHM enables the system's designer to improve and visualize workplace in digital space [25, 26]. One of the important benefit of DHM is the ability to perform ergonomic analysis alternatively through human mannequins in the workplace assessment and in the early design process [27, 28]. Comparative studies made through DHM tool show a precise postures fit [29, 30]. The simulation based studies though save time and money, however DHM has yet to be used for assessment of the risk factors in lifting and lowering tasks postures. The purpose of this study is to analyze risk factors of WMSDs within the handmade CB&B manufacturing factories and provide the companies an advice on improvements.

2 Methodology

WMSDs originate normally from performing repetitive tasks. If complete avoidance is impossible, high exposure to injury may be minimize by actions that include remodeling workplace for minimum movements, making lighter or safe for grip as well as workers training on ergonomic handling techniques. To achieve the objective, mimization of WMSDs, the study is designed in three steps. Firstly, the CB&B workers anthropometric data were obtained. Secondly, a questionnaire base study was conducted in over 33 factories, followed by the development of 3D Mannequin using HumanCAD software for stress analysis of postures [9]. The HumanCAD uses anthropometric data obtained from workers as input, as shown in Sect. 4, for modelling of 3D mannequines. The mannequin is postured according to the real life workers postures for analysis with actual loading conditions.

3 Anthropometric Data of the Workers

The larger and smaller bone calipers and other gauges available at Ergonomics and Biomechanics Research Lab at University of Engineering and Technology, Peshawar Pakistan, were used to record anthropometric data onto over 103 workers in the form of mean, standard deviation, maximum and minimum range are given in Table 1. The participants' different body parts were measures with maximum and minimum range. Every possible error in the measurement process was removed, so that the mannequin accurately mimics the real life workers.

Table 1. Anthropometric data of workers

S.No	ID	Descriptions	Mean	SD	Maximum	Minimum
1	S	Stature	166.13	6.63	178.01	156.41
2	HL	Hand length	18.57	1.08	20.21	16.32
3	SP	Span	167.77	5.62	176.52	158.22
4	SE	Shoulder Elbow Length	13.71	1.78	37.36	31.76
5	K	Knee height	52.79	2.56	57.67	48.79
6	F	Forearm Length	34.76	2.52	49.63	41.67
7	HB	Hand Breadth	7.94	0.59	8.86	07.09
8	FS	Fingertip to Shoulder Length	71.9	4.33	79.67	64.45
9	H	Hip Breadth	34.04	2.86	40.01	30.24
10	W	Weight	66.5	2.50	73.00	60.00

3.1 Questionnaire Survey

From Literature, the questions about WMSDs were identified. The questionnaire targeted the following relevant information:

(a) Personal information about Employees including height, age, weight, working experience, etc.
(b) Job-related information including working hour, posture, size and weight of objects handled, handling frequency, etc.
(c) WMSDs related information including, injury, pain nature and causes of WMSDS.

Due to the illiteracy high rate of the labor class working in brick and block factories, the questionnaires were completed on the spot from 103 people in 33 block factories in Pakistan. These focused workers were involved in work that required frequent bending, lifting, pushing and the use of excessive force. The permission was granted to our research group by various factories management and the university acknowledged the factory owners. The participants helped us in gathering the anthropometric data and all other information on voluntarily basis.

3.2 Results of Questionare Survey

The participant personal information's was recorded such as age, height, weight and working experience given in Table 2. It was essential parts of the study to collect data onto workers regarding discomforts and pain in any part of the body. Table 3 shows different body parts of frequencies and levels of discomforts. To calculate the body parts that had the most of the workers complaining of, we observed frequency tables and percentages of the range of the discomfort on responses of feeling the pain on that body part.

Table 2. Physical info and working experience

Age (years)	20–25	26–30	31–35	36–40	>40
	25	30	25	15	8
Height (cm)	<165	165–175	>175		
	55	30	18		
Weight (Kg)	<60	61–70	70–75	>80	
	35	45	20	3	
Work experience	<1 year	1–5 years	5–8 years	>8 years	
	20	60	20	3	

Table 3. Body parts associated with WMSDs

	No pain		Mild pain		Moderate pain		Severe pain	
	N	%age	N	%age	N	%age	N	%age
Neck	31	30.09709	23	22.3301	24	23.30097	25	24.27184
Shoulder	20	19.41748	32	31.06796	36	34.95146	15	14.56311
Elbow	29	28.15534	36	34.95146	30	29.12621	10	9.708738
Hip Thigh	33	32.03883	40	38.83495	24	23.30097	6	5.825243
Upper Back	11	10.67961	24	23.30097	28	27.18447	32	31.06796
Lower Back	11	10.67961	20	19.41748	24	23.30097	40	38.83495
Arm	15	14.56311	36	34.95146	24	23.30097	16	15.53398
Palm	41	39.80583	24	23.30097	30	29.12621	8	7.76699
Shank	65	63.1068	16	15.53398	12	11.65049	10	9.708738
Pelvis	33	32.03883	24	23.30097	26	25.24272	20	19.41748
Thorax	23	22.3301	24	23.30097	26	25.24272	30	29.12621

Table 3 shows that the body parts and discomfort values along with the percentage recorded during survey. It shows that lower back (38.83%) upper back (31.06%), thorax (29.12%) and neck (24.27%) were the affected body parts which workers complained of discomfort. Remaining body parts such as elbow (9.70%), hip thigh (5.82%), arm (15.53%), palm (7.76%), shank (9.70%), pelvis (19.41%) and shoulder (14.56%) were also noted to be affected. The minimum number of severe pain cases noted were on the hip thigh as only 5.82% severe pain cases were noted. Hence it can, be said that in the piling up processing the upper back, lower back as well as neck and thorax were the body parts where participants complain of the severe pain.

4 Digital Human Modelling

Digital human models are computer-generated representations of human beings used for biomechanical analysis. The mannequin was designing through HumanCAD software to mimic the real life industries workers posture with real loading conditions. The Ergo tool module available in the software provided the static biomechanical stress on the different body parts. In Sect. 4.1 the results of biomechanical stress on workers of bricks piling up process followed by the results of concrete blocks piling up process in Sect. 4.2.

4.1 Discussion of Biomechanical Stress in Brick Lifting

Four different postures of workers in lifting and piling up process are shown in Fig. 1. Mannequin in Fig. 1a picking the 7 kg loads with sitting position on knee above the hip, in Fig. 1b picking the same load in semi sitting position with align knee and hip position with hand extended and neck bending down 30° from the frontal plane. Similarly the mannequin in Fig. 1(c) unloading the load above the stature height of standing position with hand not extend, the mannequin in Fig. 1d unloading the load in standing position by hand extends.

The detail biomechanical stresses on the mannequin while lifting 7 kg load is given in the Tables 4, 5, 6, 7 and 8.

Biomechanical stresses in Tables 4, 5, 6, 7 and 8 are the Forces and Torque applied to the different body parts during lifting weights. Each table consists of 6 columns. The body parts are listed in column 1, columns 2, 3 and 4 shows the axis of forces applied. Column 5 shows the forces applied on the body parts and column 6 shows the torque on the body parts. Figure 2 shows the similar result in the graphical form.

Table 4 shows the static biomechanical stresses on different body parts, the highest force applied to pelvis (359.049 N) followed by thorax (268.708 N). Similarly the highest positive torque act on the thorax (47.12 NM) followed by 39 NM positive torque act on the pelvis (Fig. 3).

Table 5 shows the static biomechanical stresses on different body parts, the highest force applied on pelvis (359.049 N) followed by thorax (268.708 N). Similarly the highest positive torque act on the thorax (121.935 NM) followed by 127.191 NM positive torque act on the pelvis (Fig. 4).

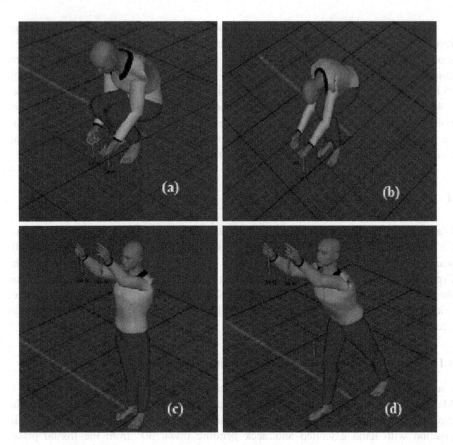

Fig. 1. Mannequin with brick load (a) sitting (b) semi sitting (c) standing and (d) reach posture

Table 4. Static biomechanical forces

Body parts	X	Y	Z	Force (N)	Torque (N.m)
Head	0	0	−90	65.629	0
Left Arm	180	60	−180	24.356	20.33
Left Foot	103.232	90	166.768	17.682	1.145
Left Forearm	−180	37.047	0	10.518	8.619
Left Palm	0	52.953	90	7.317	2.18
Left Shank	0	0	−90	49.872	1.145
Left Thigh	0	0	−90	121.998	2.777
Pelvis	0	0	−90	359.049	39.318
Right Arm	179.941	64.974	−0.016	25.267	22.609
Left Foot	159.587	90	110.413	17.682	1.094
Right Forearm	−0.072	−42.021	0.01	11.429	10.227
Right Palm	−0.093	47.98	90.016	7.317	3.369
Right Shank	0	0	−90	49.872	1.094
Right Thigh	0	0	−90	121.998	2.626
Thorax	0	0	−90	268.708	47.12

Table 5. Static biomechanical forces

	X	Y	Z	Force (N)	Torque (N.m)
Head	179.9998	−88.622	89.998	65.629	0
Left Arm	0	29.721	0	24.356	20.625
Left Foot	0	89.565	−90	17.682	1.141
Left Forearm	0	62.242	−180	10.518	13.855
Left Palm	0	−27.758	90	7.317	4.32
Left Shank	0	−20	−90	49.872	3.716
Left Thigh	−79.459	90	−10.541	121.998	32.043
Pelvis	0	0	−90	359.049	127.191
Right Arm	0.014	24.748	179.912	25.267	17.9
Left Foot	0	89.565	−90	17.682	1.09
Right Forearm	−179.9	−57.268	−179.55	11.429	12.132
Right Palm	−0.073	−32.732	89.907	41.317	2.984
Right Shank	0	−20	−90	49.872	3.7
Right Thigh	−79.373	90	−10.627	121.998	32.03
Thorax	0	−81.378	−90	268.708	121.935

Table 6. Static biomechanical forces

	X	Y	Z	Force (N)	Torque (N.m)
Head	0	−48	−90	65.629	0
Left Arm	0	10	0	24.356	2.482
Left Foot	0	78.317	−90	17.682	1.023
Left Forearm	0	42.52	180	10.518	1.82
Left Palm	0	−47.48	90	7.317	0.341
Left Shank	0	−33	−90	49.872	6.64
Left Thigh	−180	68	90	121.998	26.871
Pelvis	0	0	−90	359.049	85.635
Right Arm	−0.057	5.026	179.949	25.267	10.52
Left Foot	0	78.317	−90	51.682	5.33
Right Forearm	179.979	−37.547	−179.934	11.429	9.33
Right Palm	−0.131	−52.453	89.913	41.317	2.698
Right Shank	0	−33	−90	49.872	10.777
Right Thigh	−180	68	90	121.998	37.05
Thorax	0	−48	−90	268.708	71.677

Table 6 shows the static biomechanical stresses on different body parts, the highest force applied to pelvis (359.049 N) followed by thorax (268.708 N). Similarly the highest positive torque act on the pelvis (71.677 NM) and secondly 85.635 NM positive torque act on the pelvis (Fig. 5).

Table 7. Static biomechanical forces

	X	Y	Z	Force (N)	Torque (N.m)
Head	0	10.845	−90	65.629	0
Left Arm	180	63	−180	24.356	21.621
Left Foot	0	88.451	−90	17.682	1.131
Left Forearm	−180	40.047	0	10.518	9.578
Left Palm	0	49.953	90	7.317	2.718
Left Shank	0	−23	−90	49.872	4.397
Left Thigh	0	−23	−90	121.998	19.31
Pelvis	0	0	−90	359.049	54.23
Right Arm	179.847	67.973	0.086	25.267	23.648
Left Foot	−0.022	89.711	−89.978	17.682	1.092
Right Forearm	−0.107	−45.02	−0.43	11.429	10.923
Right Palm	−0.044	44.98	89.955	7.317	3.503
Right Shank	0	9	−90	49.872	3.23
Right Thigh	0	26	−90	121.998	18.213
Thorax	0	−30	−90	268.708	84.727

Table 8. Static biomechanical forces

	X	Y	Z	Force (N)	Torque (N.m)
Head	0	0	−90	65.629	0
Left Arm	180	60	−180	24.356	45.807
Left Foot	103.232	90	166.7688	17.682	1.145
Left Forearm	−180	37.047	0	10.518	20.55
Left Palm	0	52.953	90	7.317	5.828
Left Shank	0	0	−90	49.872	1.145
Left Thigh	0	0	−90	121.998	2.777
Pelvis	0	0	−90	359.049	94.509
Right Arm	179.587	64.974	−0.016	25.267	50.635
Left Foot	159.587	90	110.413	17.682	1.094
Right Forearm	−0.072	−42.021	0.01	11.429	24.291
Right Palm	−0.093	47.98	90.016	7.317	8.619
Right Shank	0	0	−90	49.872	1.094
Right Thigh	0	0	−90	121.998	2.626
Thorax	0	0	−90	268.708	100.534

Table 7 shows the static biomechanical stresses on different body parts, the highest force applied to pelvis (359.049 N) followed by thorax (268.708 N). Similarly the highest positive torque act on the thorax (84.727 NM) followed by 54.23 NM positive torque act on the pelvis (Fig. 6).

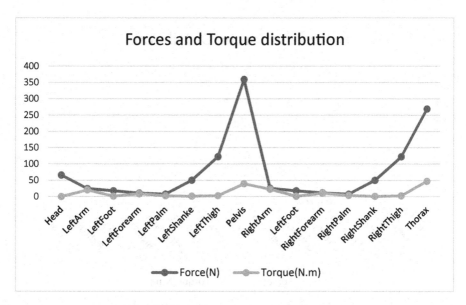

Fig. 2. Static biomechanical graph (SBG-1)

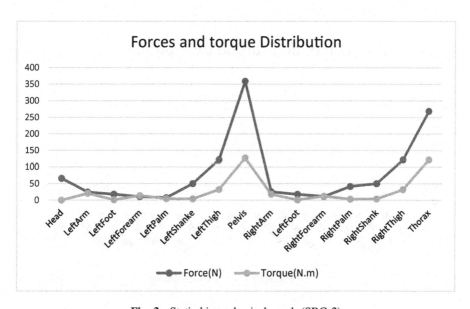

Fig. 3. Static biomechanical graph (SBG-2)

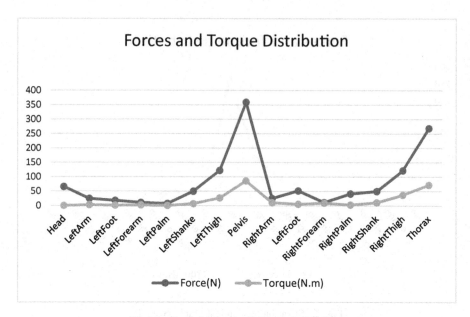

Fig. 4. Static biomechanical graph (SBG-3)

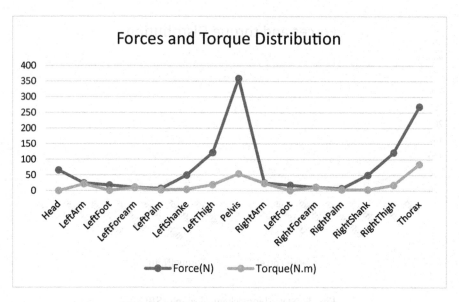

Fig. 5. Static biomechanical graph (SBG-4)

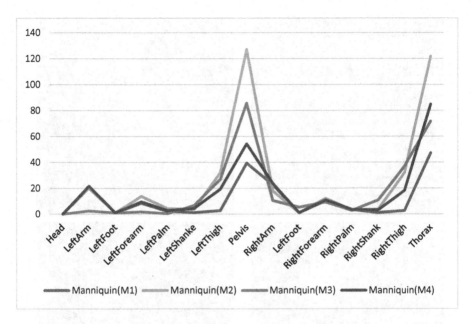

Fig. 6. Torque distributions at different body organs

The most stresses are concentrated on the pelvis region which is the considerable reason for Lumbosacral spine disorders and lower back pain. Due to posture in Fig. 1a the torque on pelvis is 39.318 NM and on thorax the torque is 47.120 NM while torque due to posture on Fig. 1b the torque is 127.191 N.m on pelvis and 121.935 N.m on thorax. Similarly in Fig. 1c the torque on pelvis is 85.637 NM and 71.677 NM on thorax and in Fig. 1d the torque on pelvis is 54.230 NM and on thorax 84.727 N.m.

4.2   Discussion of Biomechanical Stress in Concrete Block Lifting

A Mannequin was created in four different postures, these postures were assigned to pick of weight (18 kg) at a time in piling up process and the concrete blocks act as a weights due to gravity. According the workers lifting bricks in the piling up processing the stresses on the different body parts were calculated, the stresses distributions on different body parts are given in the tables. In the HumanCAD the Ergo tool of Static Biomechanics Tool was applied and all the forces and torque are displayed on the window screen. The detail of static biomechanical stress was given in the tables.

Mannequin in Fig. 7a picking the 18 kg loads in sitting position of knee above the hip, in Fig. 7b picking the same load in semi sitting position with align knee and hip position with hand more extended and neck bending 20° from frontal plane. Similarly the mannequin in Fig. 7c unloading the load above the stature height of standing position by hand not extended, the mannequin in Fig. 7d unloading the load in standing position with hand extended and neck bending 10° backward from the frontal plane.

Biomechanical stresses in Tables 4, 5, 6, 7 and 8 are the Forces and Torque applied on the different body parts in lifting weights.

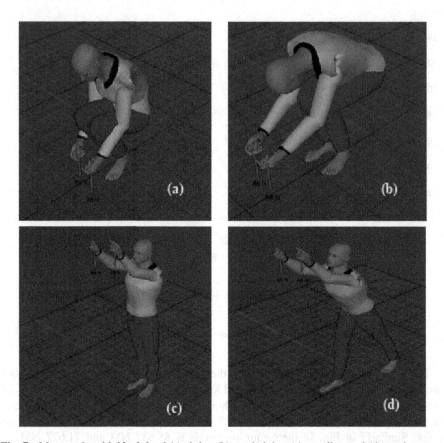

Fig. 7. Mannequin with block load (a) sitting (b) semi sitting (c) standing and (d) reach posture

Table 8 shows the static biomechanical stresses on different body parts, the highest force applied to pelvis (359.049 N) followed by thorax (268.708 N). Similarly the highest positive torque act on the thorax (100.534 NM) followed by 94.509 NM positive torque act on the pelvis (Fig. 8).

Table 9 shows the static biomechanical stresses on different body parts, the highest force applied to pelvis (359.049 N) followed by thorax (268.708 N). Similarly the highest positive torque act on the thorax (200.023 NM) followed by 221.478 NM positive torque act on the pelvis (Fig. 9).

Table 10 shows the static biomechanical stresses on different body parts, the highest force applied to pelvis (359.049 N) followed by thorax (268.708 N). Similarly the highest positive torque act on the thorax (142.684 NM) followed by 147.94 NM positive torque act on the pelvis (Fig. 10).

Table 11 shows the static biomechanical stresses on different body parts, the highest force applied to pelvis (359.049 N) and followed by thorax (268.708 N). Similarly the highest positive torque act on the thorax (100.534 NM) followed by 94.509 NM positive torque act on the pelvis (Fig. 11).

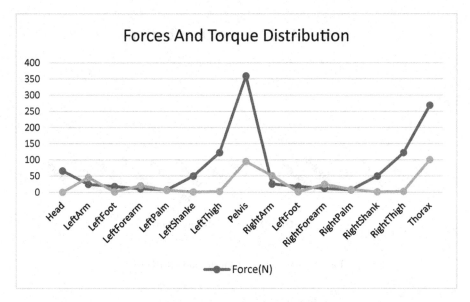

Fig. 8. Static biomechanical graph (SBG-5)

Table 9. Static biomechanical forces

	X	Y	Z	Force (N)	Torque (N.m)
Head	179.998	−88.622	89.998	65.629	0
Left Arm	0	29.721	0	24.356	47
Left Foot	0	89.565	−90	17.682	1.141
Left Forearm	0	62.242	−180	10.518	32.456
Left Palm	0	−27.758	90	7.317	10.864
Left Shank	0	−20	−90	49.872	3.716
Left Thigh	−79.459	90	−10	121.998	32.043
Pelvis	0	0	−90	359.049	221.478
Right Arm	0.014	24.748	179.912	25.267	42.146
Left Foot	0	89.565	−90	17.682	1.09
Right Forearm	−179.9	−57.268	−179.8	11.429	29.926
Right Palm	−0.073	−32.732	89.907	7.317	9.474
Right Shank	0	−20	−90	49.872	3.7
Right Thigh	−79.373	90	−10.627	121.998	32.03
Thorax	0	−81.378	−90	268.708	200.023

Similarly the torque distribution due to 17 kg of blocks the most stresses are concentrated on the pelvis region which is the considerable reason for Lumbosacral spine disorders and lower back pain. Due to posture in Fig. 7a the torque on pelvis is

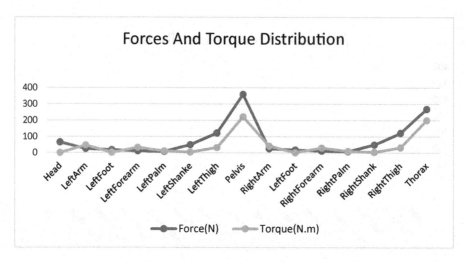

Fig. 9. Static biomechanical graph (SBG-6)

Table 10. Static biomechanical forces

	X	Y	Z	Force (N)	Torque (N.m)
Head	0	−48	−90	65.629	0
Left Arm	0	10	0	24.356	30.709
Left Foot	0	78.317	−90	17.682	1.023
Left Forearm	0	42.52	−180	10.518	25.556
Left Palm	0	−47.48	90	7.317	9.037
Left Shank	0	−33	−90	49.872	6.64
Left Thigh	−180	68	90	121.998	26.871
Pelvis	0	0	−90	359.049	147.94
Right Arm	−0.057	5.026	179.949	25.267	26.984
Left Foot	0	78.317	−90	17.682	0.965
Right Forearm	179.979	−37.547	−179.9	11.429	24.487
Right Palm	−0.131	−52.453	89.913	7.317	10.239
Right Shank	0	−33	−90	49.872	6.631
Right Thigh	−180	68	90	121.998	26.856
Thorax	0	−48	−90	268.708	142.686

94.509 NM and on thorax the torque is 100.534 NM while torque due to posture on Fig. 7b the torque is 221.478 NM on pelvis and 200.023 NM on thorax. Similarly in Fig. 7c the torque on pelvis is 147.940 NM and 142.686 NM on thorax and in Fig. 7d the torque on pelvis is 74.509 NM and on thorax 100.534 N.m (Figure 12).

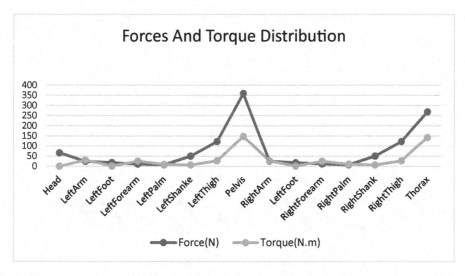

Fig. 10. Static biomechanical graph (SBG-7)

Table 11. Static biomechanical forces

	X	Y	Z	Force (N)	Torque (N.m)
Head	0	0	−90	65.629	0
Left Arm	180	60	−180	24.356	45.807
Left Foot	103.232	90	166.768	17.682	1.145
Left Forearm	−180	37.047	0	10.518	20.55
Left Palm	0	52.953	90	7.317	5.828
Left Shank	0	0	−90	49.872	1.145
Left Thigh	0	0	−90	121.998	2.777
Pelvis	0	0	−90	359.049	94.509
Right Arm	179.941	64.974	−0.016	25.267	50.639
Left Foot	159.941	90	110.413	17.682	1.094
Right Forearm	−0.072	−42.021	0.01	11.429	24.291
Right Palm	−0.093	47.98	90.016	7.317	8.619
Right Shank	0	0	−90	49.872	1.094
Right Thigh	0	0	−90	121.998	2.626
Thorax	0	0	−90	268.708	100.534

5 NIOSH Calculations

According to NIOSH 1991 equation the safe value for manual lifting are lifting Indices ≤ 1. Lifting Indices of the concrete blocks of weight 17 kg is 1.57, which is considerably large while LI of bricks of weight 7 kg is 0.60 at the origin and 0.77 at the destination, which is within safety limits.

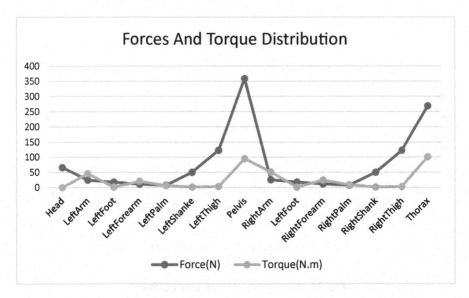

Fig. 11. Static biomechanical graph (SBG-8)

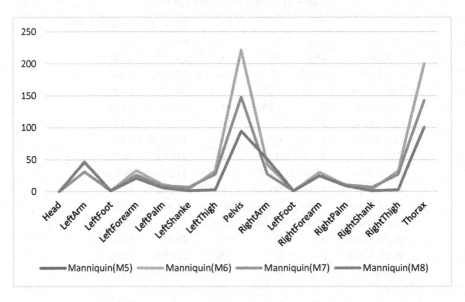

Fig. 12. Torque distributions at different body regions

6 Result and Discussion

According to NIOSH recommendations, the lifting tasks with Lifting index (LI) ≥ 1.0 shows a high risk of lifting-related low back pains [31–33]. Therefore, such tasks should be reshaping, resize, modify and redesigned to achieve an LI $\leq$ of 1.0. In the proposed research study the LI for bricks is 0.60 at origin and 0.77 at the origin which is satisfactory but in the concrete blocks case, the LI is 1.57 at the origin and 2.04 at the destination and which is significantly hazardous for workers health. Due to posture in Fig. 1a the torque on pelvis is 39.318 NM and on thorax the torque is 47.120 NM while torque due to posture on Fig. 1b is 127.191 N.m on pelvis and 121.935 N.m on thorax. Similarly in Fig. 1c the torque on pelvis is 85.637 NM and 71.677 NM on thorax and in Fig. 1d the torque on pelvis is 54.230 NM and on thorax 84.727 N.m. Similarly the torque distribution due to 17 kg of blocks the most stresses are concentrated on the pelvis region which is the considerable reason for Lumbosacral spine disorders and lower back pain. Due to posture in Fig. 7a the torque on pelvis is 94.509 NM and on thorax the torque is 100.534 NM, while torque due to posture on Fig. 7b the torque is 221.478 NM on pelvis and 200.023 NM on thorax. Similarly in Fig. 7c the torque on pelvis is 147.940 NM and 142.686 NM on thorax and in Fig. 7d the torque on pelvis is 74.509 NM and on thorax 100.534 N.m. WMSDs are noted as a result of the presence of different risk factors, including contact stress, force, vibrations, repetition,, and jobs that put muscles under redundant physical forces. No doubt that most of tasks performed at the blocks and bricks industries are posing a high risk of the workers. In our research, we noted that most of the workers had WMSDs symptoms of little pain to worst pain specifically back pain, lower back pain and upper back were the most frequent WMSDs complaints of blocks piling up workers. In addition, thorax pain and neck pain remains the second most common WMSDs complaint amongst workers. Improper lifting/lowering and stretch out bending are often indicated as causes.

7 Conclusion

Lifting and lowering are very prone to injuries. The most stresses are concentrated on the pelvis region which is the considerable reason for Lumbosacral spine disorders and lower back pain. In an industrially developing countries like Pakistan the source of exposure to WMSDs risks seems to be severe mainly because of the untrained workforce and due to the absence of the labor laws implementation. Though many studies have shown a significant relation between manual labor and WMSDs, in an industrially developing countries, people are exposed to work without knowing the new job physical demand. In this regard, there is a dire need for medical and physical examination as a prerequisite for a new jobs. In addition, workers should be trained on ergonomics basis before they are exposed to manual material handling. The conclusions drawn in this study as for a small population are more convincing to train the workers as workforce and implement the safety and labor laws in the country.

Acknowledgements. The authors are grateful to all Brick and Concrete Block industries administrations and workers who participated in this study. The Author(s) also like to thank Board of Studies and Advanced Research (BOSAR), University of Engineering and Technology, Peshawar for the research funding.

References

1. International Labor Standards on Occupational Safety and Health (2017) http://www.ilo.org/global/standards/subjects-covered-by-international-labour-standards/occupational-safety-and-health/lang–en/index.htm
2. Jamali AA et al (2016) Assessment of Occupational Health, Safety and Environment in Brick Kiln Industries at Tando Hyder, Pakistan
3. Dussault G, Franceschini MC (2006) Not enough there, too many here: understanding geographical imbalances in the distribution of the health workforce. Hum Resources Health 4(1):12. https://doi.org/10.1186/1478-4491-4-12
4. Russell SJ et al (2007) Comparing the results of five lifting analysis tools. Appl Ergon 38 (1):91–97. https://doi.org/10.1016/j.apergo.2005.12.006
5. Harreby M et al (1999) Risk factors for low back pain in a cohort of 1389 Danish school children: an epidemiologic study. Eur Spine J 8(6):444–450
6. Engels JA et al (1996) Work related risk factors for musculoskeletal complaints in the nursing profession: results of a questionnaire survey. Occup Environ Med 53(9):636–641
7. Marras W (2000) Occupational low back disorder causation and control. Ergonomics 43 (7):880–902
8. Dagli CH Artificial neural networks for intelligent manufacturing
9. Waters TR et al (1993) Revised NIOSH equation for the design and evaluation of manual lifting tasks. Ergonomics 36(7):749–776
10. Choobineh A, Tabatabaee SH, Behzadi M (2009) Musculoskeletal problems among workers of an iranian sugar-producing factory. Int J Occup Saf Ergon 15(4):419–424. https://doi.org/10.1080/10803548.2009.11076820
11. Choobineh A et al (2007) Musculoskeletal problems in Iranian hand-woven carpet industry: Guidelines for workstation design. Appl Ergon 38(5):617–624
12. Keyserling WM (2000) Workplace risk factors and occupational musculoskeletal disorders, Part 2: a review of biomechanical and psychophysical research on risk factors associated with upper extremity disorders. AIHAJ-Am Ind Hyg Assoc 61(2):231–243
13. Takala E-P et al (2010) Systematic evaluation of observational methods assessing biomechanical exposures at work. Scand J Work Environ Health 3–24
14. Stock SR (1991) Workplace ergonomic factors and the development of musculoskeletal disorders of the neck and upper limbs: a meta-analysis. Am J Ind Med 19(1):87–107
15. Pavlović-Veselinović S Repetition as a risk factor for the development of musculoskeletal disorders
16. Loisel P et al (2005) Prevention of work disability due to musculoskeletal disorders: the challenge of implementing evidence. J Occup Rehabil 15(4):507–524
17. Dall TM et al (2013) Modeling the indirect economic implications of musculoskeletal disorders and treatment. Cost Eff Resource Alloc 11(1):5
18. Dul J, Neumann WP (2009) Ergonomics contributions to company strategies. Appl Ergon 40 (4):745–752
19. Rost K, Smith JL, Dickinson M (2004) The effect of improving primary care depression management on employee absenteeism and productivity a randomized trial. Med Care 42 (12):1202
20. Braakman-Jansen LM et al (2011) Productivity loss due to absenteeism and presenteeism by different instruments in patients with RA and subjects without RA. Rheumatology 51 (2):354–361
21. Johns G (2011) Attendance dynamics at work: the antecedents and correlates of presenteeism, absenteeism, and productivity loss. J Occup Health Psychol 16(4):483

22. David G (2005) Ergonomic methods for assessing exposure to risk factors for work-related musculoskeletal disorders. Occup Med 55(3):190–199

23. Schaub KG et al (2012) Ergonomic assessment of automotive assembly tasks with digital human modelling and the 'ergonomics assessment worksheet'(EAWS). Int J Hum Factors Model Simul 3(3–4):398–426

24. Ding Z (2013) Manual assembly modelling and simulation for ergonomics analysis. Citeseer

25. Chang SW, Wang MJJ (2007) Digital human modeling and workplace evaluation: using an automobile assembly task as an example. Hum Factors Ergon Manuf Serv Ind 17(5):445–455

26. Duffy VG (2008) Handbook of digital human modeling: research for applied ergonomics and human factors engineering. CRC Press, Cambridge

27. Chaffin DB (2008) Digital human modeling for workspace design. Rev Hum Factors Ergon 4(1):41–74

28. Colombo G, Regazzoni D, Rizzi C (2013) Ergonomic design through virtual Humans. Comput Aided Des Appl 10(5):745–755

29. Álvarez-Casado E et al (2016) Using ergonomic digital human modeling in evaluation of workplace design and prevention of work-related musculoskeletal disorders aboard small fishing vessels. Hum Factors Ergon Manuf Serv Ind 26(4):463–472

30. Lämkull D, Hanson L, Örtengren R (2009) A comparative study of digital human modelling simulation results and their outcomes in reality: a case study within manual assembly of automobiles. Int J Ind Ergon 39(2):428–441

31. Waters TR (2007) When is it safe to manually lift a patient? AJN Am J Nurs 107(8):53–58

32. Chung MK, Kee D (2000) Evaluation of lifting tasks frequently performed during fire brick manufacturing processes using NIOSH lifting equations. Int J Ind Ergon 25(4):423–433

33. Waters T, Putz-Anderson V, Garg A (2003) Revised NIOSH lifting equation. Occup Ergon Eng Adm Controls

MakeHuman: A Review
of the Modelling Framework

Leyde Briceno[1,2] and Gunther Paul[1,2(✉)]

[1] AITHM, James Cook University, Townsville, QLD 4811, Australia
{leyde.briceno,gunther.paul}@jcu.edu.au
[2] Mackay Institute of Research and Innovation, Mackay, QLD 4740, Australia

Abstract. MakeHuman is an open source software rarely used in Ergonomic studies. Developed on open source Python code, the program creates realistic appearance 3D virtual human models, primarily focusing on morphing details. An intuitive graphical user interface working with sliders controls input parameters on normalized scales for the main parameters gender, age, muscle mass, weight, height, proportion and ethnicity. These input parameters govern associated output values, which mostly remain normalized. Height and age however are on an interval scale. MakeHuman Blender tools connect the MakeHuman and Blender programs, allowing users to modify a base mesh shape, create clothes, apply static poses or generate animations. In recent research work, MakeHuman was employed mostly to generate sets of virtual subjects. MakeHuman is a design (gaming) oriented, parametric virtual human modelling tool based on templates. A template model is transformed by means of scaling factors, resizing its segments and proportions, to create a set of human bodies compatible with the original base mesh. The template model is divided into 'areas of influence', and form factors are calculated to detect contraction or expansion, improving the use of targets in these areas. Fuzzy logic rules are employed in order to process inputs, which are linked directly to membership functions of fuzzy sets. With one morphing target file for each parameters' extreme values, multifactorial input change is amalgamated into a character, using an inference engine that produces a diversity of human bodies. The study aspires to assess the practicability of using the software in a Human Factors framework.

Keywords: Digital human modelling (DHM) · MakeHuman
Parametric modelling · Fuzzy logic · Blender

1 Introduction

Recent advances in computer graphics have allowed the emergence of new digital human modelling (DHM) tools with a more realistic human appearance [1]. Such tools have been increasingly used in various industries over the last two decades [2] and as consequence, a large number of human modeling tools have been developed in different platforms including open source software [1].

MakeHuman is an open source software developed on Python code, which creates male and female models with detailed morphological characteristics. MakeHuman is a

© Springer Nature Switzerland AG 2019
S. Bagnara et al. (Eds.): IEA 2018, AISC 822, pp. 224–232, 2019.
https://doi.org/10.1007/978-3-319-96077-7_23

parametric model [3], commonly used for generic human modeling, but rarely used in anthropometric or ergonomic (Human Factors) studies. However despite an increasing use of this software [4–8], limited information was found about its modelling conceptual framework. For this reason, in this study we applied an explorative analysis method to review the MakeHuman software based on the official source repository of the MakeHuman project [9], the application [10] and the documentation of code which are freely available [11]. We examined the parametric modelling approach, and the procedure to apply real human poses to the underlying biomechanical model (skeleton), with the objective to assess the practicability of using this open source software in a Human Factors/Ergonomics framework.

2   MakeHuman Model Approach

2.1   Parametric Model Description

'MakeHuman is a character creation suite, designed for making anatomically correct humans' [10]. The model combines a parameterized base mesh for human representation and a master skeleton or 'rig' that is associated with the base mesh and includes separate geometries for topology, clothes, tongue, eyes, eyelashes, eyebrows, teeth, and hair. Moreover, skin, hair, eyes and clothing allow the selection of material properties, and facial expressions can be selected from a database. The main normalized and macro-assigned model parameters are: gender, age, ethnicity, muscle, weight, proportions and height. Gender, face, torso, arms and legs, including hands and feet, have separate normalized detail settings to modify characters. Thirteen additional rigs are available in the rigs/skeleton library, however they are not endorsed by the Make-Human project. Furthermore, a group of alternative topologies are designed for special purposes and can be used to replace the base mesh for various applications such as car crash simulation. MakeHuman reports 21 human dimensions: height; chest, waist and hip circumference; neck circumference and height; upper arm circumference and length; lower arm length and wrist circumference; five other torso measurements; upper leg height and thigh circumference; knee circumference; lower leg height and calf circumference; and ankle circumference.

The MakeHuman model is based on templates. The body model is divided into segments or zones (Table 1) corresponding to specific templates/targets, which are applied on the human with the help of modifiers [3, 9, 11]. The master or default rig has 59 segments or bones in the head, 17 segments for the torso, 48 segments for the upper limbs and 38 segments for the lower limbs, 162 bones altogether. The spine is represented by three neck segments (neck01-neck03), two thoracic segments (*spine01-spine02*), two lumbar segments (*spine03-spine04*), and one sacral segment (*spine05*). The sacrum *spine05* is inverted towards anterior and ends in a node with the two pelvic segments *pelvis.R* and *pelvis.L*, and the anterior-posterior *root* segment which ends in the coccyx. The shoulder model is flexible enough to represent a complex shoulder rhythm, with a shoulder (*shoulder01.R*, *shoulder01.L*) and clavicle (*clavicle.R*, *clavicle. L*) segment on each side, connecting virtually to the *neck01/spine01* node. Chest circumference modelling is supported by two segments, *breast.R* and *breast.L*. The legs

Table 1. Body zones, templates/targets and modifiers

Body zones	Targets	Modifiers by target
Head	Left and right eye	(features) (size)
	Nose	(features) (size)
	Left and right cheek	(features) (volume)
	Left and right ear	(features) (size) (translation) (rotation)
	Mouth	(features) (size) (translation)
	Chin/jaw	(features) (size)
	Head	(features) (size)
	Neck	(features) (size) (translation)
Torso	Torso	(shape) (size) (translation) (muscle)
	Hip	(scale) (translation) (waist-hip position)
	Stomach	(shape) (tone)
	Buttocks	(volume)
	Pelvis	(size) (tone)
Upper body	Left and right shoulder	(muscle)
	Left and right upper arm	(scale) (fat) (muscle)
	Left and right lower arm	(scale) (fat) (muscle)
	Left and right hand	(scale) (translation)
	Fingers	(diameter) (distance) (length)
Lower body	Left and right upper leg	(scale) (fat) (muscle)
	Left and right lower leg	(scale) (fat) (muscle)
	Left and right foot	(scale) (translation)
Gender	Breast	(volume) (translation) (muscle) (weight)
	Genitals	(features) (size)

Source: Repository of the MakeHuman project

feature an upper leg segment (*upperleg01.R*, *upperleg01.L*) anatomically representing the femur neck, extending from the greater trochanter to the femur head; and an upper leg segment (*upperleg02.R*, *upperleg02.L*) representing the femur. The lower leg model from the knee to the talocrural joint consists of two approximately same length segments, *lowerleg01.R/lowerleg01.L* and *lowerleg02.R/lowerleg02.L*. The knee joint is represented as a simple node, and does not epitomise the anatomy of a tibiofemoral joint and a patellofemoral joint.

Macro-dependencies among the main parameters [9] are organized by groups, and when one parameter is modified a specific group of modifiers is applied (Table 2). Neither main, nor auxiliary model parameters in MakeHuman can be set numerically, and many model parameters are interdependent.

MakeHuman exports model geometry as Collada (.dae), Filmbox (.fbx), Stereolithography (.stl), Ogre 3D (.mesh.xml), Wavefront (.obj) or MakeHuman exchange (.mhx2) files, which allows for further processing of models in other software, such as Blender, OpenSim, Unity, Google Sketchup or 3DS MAX. The rig can be exported in Biovision Hierachy (.bvh) format for animations.

Table 2. Main parameters: modifier dependencies

Parameter	Modifier group
Age	Breast, macro details-universal/Universal (gender) (weight) (muscle), height, proportions
Gender	Breast, macro details-universal/Universal (age) (muscle) (weight), height, proportions
Muscle	Breast, height, proportions
Weight	Breast, height, proportions

Source: Repository of the MakeHuman project

2.2 Deformation Model

The body deformation is achieved essentially through morphing (makehuman/apps/warpmodifier.py) [3, 8]; it is applied by transforming a set of source vertices into a set of target vertices through interpolation. This transformation is limited to its area of influence [3] for calculating form factors and weights.

The form factor detects the contraction or expansion of the segment of the template [3]. The template model is deformed by applying a weight, which is the result of multiplying all the form factor values, according to the factor dependency group. MakeHuman uses its default skeleton as the basis of mesh vertex weighting for creating the poses available on the pose tab.

2.3 Fuzzy Set and Rules

MakeHuman uses fuzzy set theory [12] to describe the main parameters. A fuzzy set contains all the possible elements in the set under consideration, which is defined by a fuzzy rule-set; these rules assign to each element a value {0, 1} and there are fuzzy sets for each parameter (Table 3).

Table 3. Fuzzy sets and fuzzy rule-set for age and gender parameters

Parameter	Fuzzy set	Fuzzy rule-set
Age	Baby, child, young, old	if age < 25 　　Baby = max(0,1-age*5.333) 　　Child = max(0,min(1,5.33-age)-youngIndex) 　　Young = max(0,OldIndex*5.333-0.1875) 　　Old = 0 else 　　Baby = max(0,1-age*5.333) 　　Child = max(0,min(1,5.33-age)-youngIndex) 　　Young = max(0,OldIndex*5.333-0.1875) 　　Old = 0
Gender	Female, Neutral, Male	if Gender < 0.5 Female 　　else if Gender > 0.5 Male else Neutral

Source: Repository of the MakeHuman project

2.4   Procedure to Apply Human Poses

Skeleton driven deformation or body posing is modelled as a linear combination of rotation and translation morphing [3], where the skeleton is represented as segments linked by joints (nodes). The base mesh vertices are attached to the skeleton joints through weights, which is the amount of influence of individual joints on each vertex [9].

The MakeHuman skeleton in the global coordinate system is divided into body zones (see Table 1), and each zone into bones in their local coordinate systems, connected by joints which reference only to their master bone, following the reference mapping system [9–11] (Table 4).

Table 4. Torso body zone: MakeHuman bone group and joints

Body zone	Bone groups	Joints	Master bone
Torso	Breast (left) (right)	Head–tail	Spine 02
	Clavicle (left) (right)	Head–tail	Spine 01
	Shoulder (left) (right)	Head–tail	Clavicle (left) (right)
	Pelvis (left) (right)	Head–tail	Root
	Root	Head–tail	–
	Spine 01, Spine 02, Spine 03, Spine 04, Spine 05	Head–tail	Spine 02, Spine 03, Spine 04, Spine 05, Root
	Neck 01, Neck 02, Neck 03	Head–tail	Spine 01, Neck 01, Neck 02

Source: Repository of the MakeHuman project

Each joint has a number of degrees of freedom (translational and rotational) in a Cartesian coordinate system (x, y and z coordinates).

The MakeHuman software allows exporting base mesh human models with a skeleton as a .MHX file. Such files can be imported into Blender [13] to make use of the MakeHuman Blender tools, which allow to manually change a pose by specifying joint translational and rotational ranges of motion, using bone constraints and setting rotation limits. For instance, Table 5 shows the spine joint constraints for spinal motion based on the process of physical examination of spine and extremities [14].

Table 5. Spine joint constraints for spinal motions using MakeHuman Blender Tools

Pose	Joints	Limit rotation [7]	Limit rotation[a]	Master bone
Right bending	Spine 02	30°	X = 0 Y = -30° Z = 0	Spine 03
Left bending	Spine 02	30°	X = 0 Y = 30° Z = 0	Spine 03 (Fig. 1a)
Flexion	Spine 04	45°	X = 135° Y = 0 Z = 0	Spine 05 (Fig. 1b)
Extension	Spine 02	35°	X = 55° Y = 0 Z = 0	Spine 03 (Fig. 1c)
Right rotation	Spine 03	45°	X = 0 Y = 0 Z = 45°	Spine 04
Left rotation	Spine 03	45°	X = 0 Y = 0 Z = −45°	Spine 04 (Fig. 1d)

[a]Axial plane: x − y; coronal plane: x − z

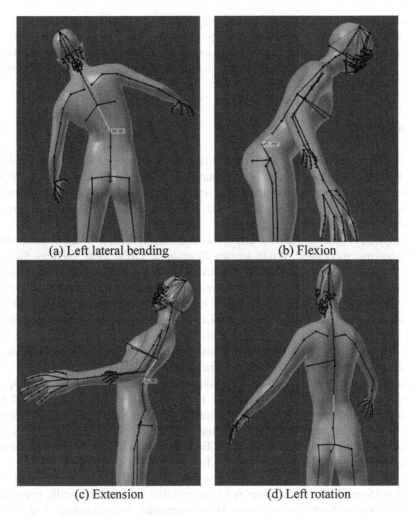

(a) Left lateral bending (b) Flexion

(c) Extension (d) Left rotation

Fig. 1. Spinal motion generation using MakeHuman Blender Tools

Human motion in reality however is much more complex. Trunk flexion occurs mainly in the cervical and lumbar spine. Moreover, trunk extension is generated predominantly in the lower cervical spine, between the 11th thoracic and the 2nd lumbar vertebrae, and between the lower lumbar vertebrae. Lateral flexion is strongest in the thoracic spine, and less significant in the cervical and lumbar spine. Trunk rotation is only possible in the thoracic, and mostly the cervical spine. Trunk rotation in the lumbar spine is marginal and can amount to 3–7 degrees between two vertebrae [15].

The Automatic IK solver in Blender is able to overcome this limitation, by translating bone chains using both forward and inverse kinematics. In order to operate correctly, Spline IK requires bone constraints, such as ROM and joint stiffness values around all rotational axes, which need to be determined from literature. Using Spline IK constraints in Blender, forward kinematics is used to position bones (spinal segments) along a control curve. Again, this control curve must be determined by the user. Similarly scaling, stretching and spline fitting parameters can be arbitrarily selected by the user, without any biomechanical relevance. It is therefore possible to create any aesthetically preferred, rather than anatomically correct human posture in MakeHuman and its related Blender tools.

3 Discussion

MakeHuman remains an interest group driven, open-source modelling environment with significant limitations in its documentation and accessibility of resources. While not suitable for commercial use, it is based on a mathematically complex human model with significant potential for scientific work. The tool has been vastly used for recreational purposes however.

Compared to other Ergonomic DHM models, MakeHuman is based on uncommon anthropometric modelling. While known Ergonomic DHM tools such as Siemens JACK, Human Solutions RAMSIS, imk EMA or others are building human models based on published and standardized international anthropometric databases, such as DIN, NHANES or ANSUR, where a user typically selects population percentiles of body proportions, MakeHuman does not offer reference to population anthropometry or direct numerical input of body proportions. Body proportions can however be controlled numerically through Python programming and extraction of numerical values from the fuzzy sets underlying the MakeHuman model.

MakeHuman is based on a limited biomechanical model with only eight spinal segments, compared to 24 human vertebrae. Hand, feet and head however are modelled in great detail, and the shoulder model permits a realistic shoulder rhythm. While this model may not be appropriate for the study of motion or spinal loads in manual materials handling, it can be applied in Ergonomic research, such as the detailed investigation of manual assembly tasks or foot operated control design.

Mobility of MakeHuman models must be controlled in Blender. Beyond predefined postures, MakeHuman does not limit mobility to biomechanically reasonable range of motion. Such controls must be implemented from literature, and neither MakeHuman nor Blender include such reference values for single joint or combined postural mobility.

The strength of the model lies in its high fidelity body mesh model. The model is well suited to problems such as body surface determination in medical and occupational health applications; head modelling in the design and assessment of personal protective equipment; or hand modelling for human machine interface design. The body mesh model is decidedly versatile and allows for custom-specific modifications; a range of mesh models for various applications is currently available in a library.

The practicability of using the software in a Human Factors framework yet needs to be established in a proper study. From a simple case-study perspective however, it appears that MakeHuman can be a useful ergonomic tool when used by experts with a significant ergonomic, mathematical and engineering background.

4 Conclusion

This study shows that the MakeHuman software may be used in ergonomic applications such as assembly tasks or control design, based on its great detail in hands and feet; however it has a limited biomechanical spinal model that may not be appropriate for the study of full body motion or spinal loads.

The main advantages of the MakeHuman software include its high fidelity body mesh model and the option of specific modifications using the MakeHuman Blender tools such as: changing, adding or removing elements affected on the body mesh by joint constraints, which allow adjusting the connection between the skeleton and the mesh (skinning process).

Future work includes examining postures in more detail for the application in ergonomic studies, and reviewing the skinning process in these cases.

References

1. Poirson E, Delangle M (2013) Comparative analysis of human modeling tools. In: 2nd international digital human modeling symposium, Ann Arbor, Michigan, United States. http://mreed.umtri.umich.edu/DHM2013Proceedings/Individual_Papers/dhm2013_submission_24.pdf. Last Accessed May 2018
2. Paul G, Wischniewski S (2012) Standardization of Digital Human Models. Ergonomics 55(9):1115–1118. https://doi.org/10.1080/00140139.2012.690454
3. Bastioni M, Re S, Misra S (2008) Ideas and methods for modeling 3D human figures: the principal algorithms used by makeHuman and their implementation in a new approach to parametric modeling. In: Proceedings of the 1st Bangalore annual compute conference. COMPUTE'08. ACM, New York, pp 10:1–10:6. https://dl.acm.org/citation.cfm?doid=1341771.1341782. Last Accessed April 2018
4. Buys K, Van Deun D, De Laet T, Bruyninckx H (2011) On-line generation of customized human models based on camera measurements. In: International symposium on digital human modeling, Lyon, France. http://citeseerx.ist.psu.edu/viewdoc/download?doi=10.1.1.473.6120&rep=rep1&type=pdf. Last Accessed May 2018
5. Piccirilli M, Doretto G, Adjeroh D (2017) A framework for analyzing the whole body surface area from a single view. PLoS ONE 12(1):e0166749
6. Wang Q, Jagadeesh V, Ressler B, Piramuthu R (2016) Im2Fit: fast 3D model fitting and anthropometrics using single consumer depth camera and synthetic data. In: IS&T international symposium on electronic imaging, 3D image processing, measurement (3DIPM), and applications, pp 3DIPM-045.1-3DIPM-045.7, Society for Imaging Science and Technology. https://doi.org/10.2352/ISSN.2470-1173.2016.21.3DIPM-045
7. Piérard S, Van Droogenbroeck M (2009) A technique for building databases of annotated and realistic human silhouettes based on an avatar. In: 20th annual workshop on circuits, systems and signal processing (ProRISC). Veldhoven, The Netherlands, pp 243–246
8. Van Deun D, Verhaert V, Buys K, Haexand B, Van der Sloten J (2011) Automatic generation of personalized human models based on body measurements. In: First international symposium on digital human modeling, 14–16 June 2011, Lyon, France
9. MakeHuman project source repository. https://bitbucket.org/MakeHuman/makehuman. Last Accessed May 2018

10. MakeHuman Community: Open source tool for making 3D characters. http://www. makehumancommunity.org. Last Accessed May 2018
11. MakeHuman Community: Wiki. http://www.makehumancommunity.org/wiki. Last Accessed May 2018
12. Zadeh L (1999) From computing with numbers to computing with words—from manipulation of measurements to manipulation of perceptions. IEEE Trans Circuits Syst Fundam Theory Appl 45(1):105–119
13. Blender Foundation: Blender. https://www.blender.org. Last Accessed May 2018
14. LeBlond RF, Brown DD, Suneja M, Szot JF (eds) (2014) DeGowin's diagnostic examination, 10th edn. McGraw-Hill Education - Europe, New York
15. Platzer W (1991) Taschenatlas der Anatomie. Band 1: Bewegungsapparat. Thieme, Stuttgart

Virtual and Augmented Reality: Innovation or Old Wine in New Bottles?

Thomas Alexander[⊠]

Fraunhofer FKIE, 53177 Bonn, Germany
Thomas.Alexander@fkie.fraunhofer.de

Abstract. Virtual Reality and Augmented Reality (VR/AR) have experienced a tremendous increase in attraction and application. Nowadays, VR-headsets are commercially available. They provide a much better resolution and much wider field-of view for a fraction of the original cost. The same development can be observed for AR, including smartphone or hand-held AR solutions. But beyond the fun of new technology, VR and AR are often lacking of a professional use case. They still fail to show their benefits and provide data for a detailed cost/benefit analysis. And with the growing dispersion and implementation into professional applications in modern industrial processes, various training and education applications, or advanced distant learning, there is an increasing need for concrete results and recommendations for usability and human-technology interaction.

They can be taken and applied from early studies in the area of Human Factors and Ergonomics. Although some of them are not important because of higher technological capabilities and performance (e.g. latency, resolution or field-of-view), most of them are still valid. They begin with cue conflicts, visual and multimodal perception and end with system performance, workload and simulator sickness.

This paper presents the relevance of HF/E for a meaningful, effective and efficient application of this technology. It shows also that the topic itself is not totally new and many requirements and results can be transferred from existing studies.

Keywords: VR/AR · Usability · Training and education

1 Introduction

1.1 Terminology of Virtual and Augmented Reality

Although the two terms *Virtual Reality (VR)* and *Augmented Reality (AR)* are widely used as buzzwords today, few of the people speaking about these technologies refer to the original sense of the terms. In a very general sense, *reality* is described as "the state of things as they actually exist, as opposed to an idealistic or notional idea of them." [1]. The term *virtual* origins from the terms *virtuel* (French: able to act) and *virtus* (Latin: virtue, valor, power, manhood). Today, virtual describes something that is almost as nearly as described, but not completely [2]. Therefore, Virtual Reality describes an experience that is close to, but not completely reality. VR today refers to a special

© Springer Nature Switzerland AG 2019
S. Bagnara et al. (Eds.): IEA 2018, AISC 822, pp. 233–239, 2019.
https://doi.org/10.1007/978-3-319-96077-7_24

technology that facilitates such an experience by special controls, computer processing hardware, and displays.

There are multiple definitions of VR [3–5]. Several of them are technology-oriented, but a vast amount of them refers to human factors and to the experience of being physically present in a VR. According to one of them, VR "is the experience of being in a synthetic environment and the perceiving and interacting through sensors and effectors, actively and passively, with it and the objects in it, as if they were real. Virtual Reality technology allows the user to perceive and experience sensory contact and interact dynamically with such contact in any or all modalities". [6]

In terms of Augmented Reality, Milgram et al. introduced the Mixed Reality (MR) continuum, which extends from full reality via augmented reality and augmented virtuality to full virtuality (i.e. practically a VR) [7]. This way it is possible to position the different technologies and approaches in a spectrum. Although it allows a comprehensive understanding, it has not replaced the broad use of the terms VR and AR.

According to [4], VR/AR-technology includes sensors (for capturing user actions, as controls), capable and powerful processing computing hardware (for calculating system responses to user actions) and output devices (for presenting the updated virtual reality). If the system is well-designed for the special application, it provides the feeling of *presence* or *immersion*. This is the subjective impression of being-there instead of being in reality [8].

1.2   History of VR/AR

The idea to present synthetic, computer-generated content in a realistic way is not new and has been around for decades. Simulators follow the same approach and simulate environments in a realistic way for a safe, effective and efficient education and training. Consequently, first VR-approaches derived from simulators. Ivan Sutherland presented his idea of a first "ultimate display" in the 60's [9]. He did not call it VR, but it was similar to today's solutions for head-mounted displays (HMDs). But technology was cumbersome and computing power limited so that the idea of VR was postponed. Instead, computer-based flight or driving simulators with a larger amount of hardware became more applicable.

Increasing computing-power and miniaturization of displays led to a revival of VR and AR in the late 80's, early 90's. But still the resolution and the field-of-view of the HMDs were limited, computing-power was too small and latency times were too large for a realistic, interactive experience. Instead, projection-based VR as virtual work-benches, large-scale projection walls or projection rooms were introduced and installed at many institutions or computing centers of universities [10–12]. They required new techniques and technologies for interacting, but they provided a good feeling of presence and interaction with the synthetic content. Moreover, they allowed a group of users to work cooperatively. But VR was a very cost-intensive solution then.

Recently, powerful graphical computers and capable HMDs became commercial available for a broad public. Everybody can buy a VR-system for a fraction of the costs of an original system. This is because of two reasons: A cheap, affordable but yet powerful computers (technology push) on the one hand and a growing market for VR-games (market pull) on the other. Today, technology is not an issue any more.

Instead, content and a useful application have become more important. While games (primarily those with a first-person perspective) have been one of the driving factors and main applications for leisure, meaningful professional applications are still limited.

A similar development can be observed for AR. The main difference is, that it is a few years behind. AR provides a mobile, natural presentation and interaction with computer-generated content. An ideal AR is seamless and merges synthetic and real environment. This requires special sensors, computing capabilities and displays. Although miniaturization, technological development and new sensors and displays are progressing, most solutions of today require corrections in terms of speed, registration and consistency of synthetic and real content. But they are already many demonstrations of smartphone- or tablet-based AR or novel AR-glasses showing future applications in many domains.

1.3 Applicability of VR and AR

VR and AR have been research topics for quite a while now. The interest in this topic has grown in the beginning, retarded for a while afterwards and returned lately. This is, because the idea of a synthetic, yet realistic environment is thrilling in the beginning, limited technological fidelity leads to disappointment and a further, more reasonable development of interest afterwards – if a topic is successful.

This development has been considered in the *Gardner Hype Cycle* for Emerging Technologies 2017 [13]. It shows that both, VR and AR are close to productivity – closer than, in facts, block chains ore AR-algorithms. With the growing need for mobility it is expected that AR will become even more useful in future.

The technological challenges made VR/AR a topic of interest for computer science, primarily for computer graphics. But because of the increasing applicability, it has become a topic for other disciplines as well. The close interaction between user and computer made it an important topic for Human Factors and Ergonomics. Knowledge and results from this domain prevented valuable requirements for the development, design and optimization of modern VR/AR systems. This refers to an application of VR/AR for various applications in education and training, system design, operational assistance etc. as well as basic requirements for VR/AR systems resulting from human perception, cognition and motor response.

2 Input/Output Technologies and HF/E

A reasonable application of VR and AR requires the inclusion of basic human characteristics and capabilities. There are multiple facets of this which are based on human information processing, including perception, cognition and motor response.

2.1 Visualization and Human Perception

Perception is the beginning of human information processing. With reference to VR and the natural experience, it is important to analyze each modality of perception first separately and then combined. For instance, human visual perception requires a minimum resolution of the display, spatial depth, colors, and, finally, low latency times between movement and visualization of the environment. Some of these characteristics hinder a realistic perception or may lead to unwanted side-effects. One example for this is simulator or cyber sickness. Its symptoms vary and extend from simple disorientation to serious dizziness or nausea. One of the main contributing factors is a sensorial conflict between visual and motion cuing. Other factors which have been identified as contributors to simulator sickness in virtual environment systems are divided into [14]:

- Characteristics of the user (workload, emotions etc.).
- Technology of the VR-system (missing modalities, poor resolution, latency etc.).
- Characteristics of the task (highly-dynamic movements etc.).

Several of the contributing technical factors, e.g. latency and image resolution, have been be reduced with increasing computing power, but several of them will remain in future. One of them is the accommodation/fixation error with VR/AR displays: In reality the focal length is adjusted so that a fixated point appears sharp on the retina. This is not the case in a VR/AR display, when the (virtual) projection plane is at a fixed distance.

This is only one example where there will be no easy technical solutions in the mid-term future. If this occurs, the task or content of the virtual environment has to be adjusted appropriately.

2.2   Interaction and Manipulation of Virtual Objects

Interaction and manipulation of virtual objects requires hardware devices for the control. The main control operations are: Generation of new objects, editing attribute values, erasing, placement, orientation, movement, and editing general object data.

Generation, editing and erasing of objects can easily be implemented. But placement, orientation, movement and editing of terrain data are more difficult, because interrelationships between different objects of the virtual environment have to be considered. This can lead to errors in the structure of the baseline visualization databank.

For correct placement and orientation of objects in the three-dimensional information about the object is necessary. Movement of objects is often critical, especially when it hits other objects. In this case a coalition detection and a physics engine are required in order to simulation realistic reactions. Although today's simulation software platforms provide such features in real-time, but there is no sufficient displays to present haptic feedback for a multi-modal cuing of the effects.

2.3   Spatial Orientation and Navigation in a Virtual Environment

Spatial orientation describes aligning or positioning in a space with respect to a specific direction or reference system. Navigation refers to change of position or orientation of the user. Both are based on a number of different sources of information and perceived through different sensory modalities. Successful spatial orientation and navigation involve different processes: sensing the environment, building up a mental spatial representation, and using mental spatial representation (e.g. walk, plan the next steps).

Sensing the environment refers to detecting landmarks, which define an area and control points which are used to determine directions. They are a base for building a mental spatial representation. This does not necessary need normal sensory, i.e. visual input. A mental spatial representation can also be based on synthetic representations, such as maps and texts. During navigation, mental representation of the current position and orientation in the environment is constantly updated. This refers to position, velocity and acceleration.

All of these information have to be well integrated in a VR for a correct representation of the environment. If there is a mismatch a wrong mental model, disorientation, breaks in presence and cybersickness are likely to occur.

3 Applications of VR and AR

3.1 Competency-Oriented Education and Training

Similar to conventional simulators, VR allows trainees to experience synthetic, computer-generated training environments in a realistic way. This prepares them for their future tasks and duties in a controlled, safe and realistic way. AR/VR subsume new types of paradigms and technological media that can provide a very realistic training environment with a natural interaction. By combining such controllable and safe training environments with a high degree of realism, it is possible to develop and enhance individual competencies.

In this case, trainees are able to experience full actions in a realistic environment. VR is also highly interactive by definition, which supports a successful, competency-oriented approach and a learning by experiencing. Training-rehearsal capabilities support individual reflection and extended possibilities for a specific feedback. Another benefits of VR technologies are their relatively low costs ("serious gaming") and their variability in terms of training applications. But it also has to be stressed that there is a vast need for intelligent and suitable training content, including environments, tasks and training rehearsal capabilities.

However, VR is only a technology and the effectiveness of VR technologies strongly depends on the task on which they are applied. Although VR is capable enough to accomplish real time applications and part task training, research is still needed in other areas. This is true for the integration into current education and training concepts.

3.2 Realistic System Design and Virtual Walkthroughs

VR allows a realistic presentation of synthetic designs and future products. Therefore, it allows for getting a first impression and identifying potential shortcomings of designs at an early system design phase. This application field of VR is to use the advanced visualization capabilities for designing, styling and presenting synthetic products in a virtual showroom. This way, future products (tools, objects, cars, busses, airplanes, buildings etc.) can be presented to a potential customer and experienced in a realistic way and a natural environment. Another topic is virtual manufacturing or cooperation.

It subsumes the application of VR in the production process in order to visualize working tasks or to program robots offline in a virtual work-cell. Virtual prototyping is a further topic in this area. It has been frequently used to design new production lines and car interiors. Moreover, simulation data, which is data representing the physical behavior of objects like driving dynamic data, can be visualized.

3.3   Operational Assistance

Operational assistance, or, to be more precisely, an application of VR/AR as a work aid has always been a topic of interest. This aspect refers to the capability of VR to explore complex data in an intuitive, naturally understandable way. This makes it an important tool for visualization of complex and abstract data (e.g. massive data analyses, wind turbulences). However, yet none of these applications has been successful. Although applying VR instead of an interactive visualization on a PC-display has clear advantages, it does not hide the fact that VR still requires several additional actions. This includes data preprocessing for the visualization, transfer to the VR system, setting up the system and, finally, wearing VR gear.

Another use of VR is tele-application. This refers to tele-communication, tele-collaboration, tele-presence, tele-manipulation, tele-medicine and tele-maintenance. By a combination of sensors, communication networks and displays, VR allows a user to be virtually present at another location. This also allows scaling of the environment or of user's actions, very important for large- or small-scale operations. Several successful implementations exist already, e.g. unmanned ground or air system control or minimal-invasive medical interventions. Others are likely to be developed in future.

The same is true for AR. AR has a potential to assist daily mobile life by adding relevant, geo-specific information into our sight. Unlike VR which requires additional hardware, AR can be used to get rid of mobile devices like smartphones or tablet computers. Thus, supporting or assisting functions become invisible and there is no need to carry or operate additional mobile systems – except for glasses. However, at this point the focus of the development is on visualization, i.e. output of information. The following step will be the input of information. There are a lot of open questions in this field and the current solutions, gestures or speech, might not be practical in every day's application. But this stresses the importance of HF/E to the development of the topic.

4   Conclusions

To answer the question in the title: VR/AR is clearly not a new technology. It has been around for a while by now. But VR/AR-technology has been struggling with technical shortcomings so far, so that the basic idea could not be realized. It also prevented a broad usage. However, technological development changed this making the systems capable, powerful and usable. But apart from gaming, just isolated success stories in education and training and for several tele-applications have been found. The reason is that content and the application-oriented optimization of the systems are still a challenge. This is especially true for AR, where the type of information input remains unclear. But future development of both, VR and AR, will have to rely heavily on HF/E for a practical, effective and efficient, successful use and application.

References

1. Oxford Living Dictionaries: Reality. https://en.oxforddictionaries.com/definitions/reality. Last Accessed 01 May 2018
2. Oxford Living Dictionaries: Virtual. https://en.oxforddictionaries.com/definitions/virtual. Last Accessed 01 May 2018
3. Burdea G, Coiffet P (1994) Virtual reality technology. Wiley, New York
4. Ellis SR, Kaiser MR, Grunwald AJ (eds) (1991) Pictorial communication in virtual and real environments. Taylor & Francis, London
5. Stanney KM (ed) (2002) Handbook of virtual environments. Erlbaum, Mahwah
6. Alexander T, Goldberg S (2005) Virtual environments for intuitive human-system interaction. RTO-TR-HFM-121-Part-I. NATO RTO, Neuilly-sur-Seine, France
7. MR-Continuum (Milgram)
8. Slater M, Pertaub D, Steed A (1999) Public speaking in virtual reality: facing and audience of avatars. IEEE Comput Graph Appl 19(2):6–9
9. Sutherland I (1965) The ultimate display. In: International federation of information processing, vol 2, p 506ff
10. Bullinger H-J, Brauer W, Braun M (1997) Virtual environments. In: Salvendy G (ed) Human factors and ergonomics. Wiley, New York, pp 1725–1759
11. Crux-Neira C, Sandin D, DeFanti T (1993) Surround-screen projection-based virtual reality: the design and implementation of the CAVE. In: Computer graphics (SIGGRAPH'93 proceedings), pp 135–142
12. Krüger W, Fröhlich B (1994) The responsive workbench. IEEE Comput Graph Appl 1215
13. Gartner Hype Cycle. https://blogs.gartner.com/smarterwithgartner/files/2017/08/Emerging-Technology-Hype-Cycle-for-2017_Infographic_R6A.jpg. Last Accessed 01 May 2018
14. Kennedy RS, Lane NE, Berbaum KS, Lilienthal MG (1993) Simulator sickness questionnaire: an enhanced method for quantifying simulator sickness. Int J Aviat Psychol 3:203–220

Design of Experiment Comparing Users of Virtual Reality Head-Mounted Displays and Desktop Computers

Steven C. Mallam[1(✉)], Salman Nazir[1],
Sathiya Kumar Renganayagalu[1,2], Jørgen Ernstsen[1], Sunniva Veie[1],
and Anders Emil Edwinson[1]

[1] Training and Assessment Research Group, Department of Maritime
Operations, University of South-Eastern Norway, Borre, Norway
{steven.mallam, salman.nazir, sathiya.k.renganayagalu,
jorgen.ernstsen, 144643, 04540}@usn.no
[2] Department of Virtual and Augmented Reality, Institute for Energy
Technology, Halden, Norway

Abstract. The use of computer-generated simulations have been standard practice in a wide range of tertiary and vocational education and training applications for decades. The growing ubiquity, relative affordability, increasing computing power and functionality of Virtual Reality headsets are creating new opportunities for personalized, immersive simulation experiences that can be used anywhere and anytime. For Virtual Reality headset experiences to be sustainable and appropriate for long-term usage in education and training programs, it is critical to investigate the practicalities of implementing such a technology. Thus, the investigation of emerging Virtual Reality technologies against conventional training systems can provide a better understanding of their impact. This paper presents an experimental design used to compare user performance, user motivation and user experience of searching tasks in identical virtual environments between two system configurations: (i) Virtual Reality Head-Mounted Display and a (ii) traditional desktop computer. A pilot study (participants N = 5) was performed with a between-group experimental design, using objective and subjective measures. The outcomes of this study and the lessons learned from developing, testing and refining the experimental design contribute to the broader knowledge of investigating and validating Virtual Reality Head-Mounted Displays for education and training applications.

Keywords: Immersive environments · Simulation · Experimental design
Computer based training · Objective performance assessment

1 Introduction

Recent technological advancements have enabled the development and proliferation of Virtual Reality Head-Mounted Display (VR HMD) systems into the consumer marketplace. Although VR HMD applications have been directed primarily towards entertainment industries, these technologies can be used as a tool for education and

© Springer Nature Switzerland AG 2019
S. Bagnara et al. (Eds.): IEA 2018, AISC 822, pp. 240–249, 2019.
https://doi.org/10.1007/978-3-319-96077-7_25

training (Huber et al. 2017). VR HMD offers the potential for cost effective, immersive, customized and on-demand training opportunities for students and professionals. The use of computer-generated simulations of virtual environments have been standard practice in a wide range of tertiary and vocational education and training applications for decades (Buttussi and Chittaro 2018; Castells et al. 2015). Traditional simulators generally have large physical footprints, high purchase and lifecycle costs, and require dedicated onsite technical and teaching personnel. This makes them expensive, and in most cases, exclusive to training facilities. Thus, the rapid development and proliferation of emerging VR HMD systems provide new and differing pedagogical tools. However, it is critical to investigate and better understand the effects, characteristics and potential added value of these emerging technologies in comparison to current training solutions.

1.1 Purpose

This pilot study is part of a broader research program investigating VR HMD systems as a tool in professional education training in the maritime domain. The purpose of this paper is therefore to present the experimental design, methods and tools used to compare VR HMD and conventional desktop-based computer systems to investigate: (i) user motivation, (ii) user performance and (iii) user experience.

2 Background

2.1 Simulations and Simulators in Education and Training

The modern perception of "simulators" are predominantly of computer-generated simulations imitating real-life systems and processes. However, the technologies can vary greatly, ranging from low-fidelity, rudimentary simulations to large, high-cost, full mission simulator systems (Berg and Vance 2017). Simulation-based training offers safe, comparably cost-effective and realistic training which would otherwise be challenging to expose trainees to due to ethical or economic considerations. Well-designed simulations enhance learning outcomes and performance (Nazir et al. 2012; Salas et al. 2012), improving trainees' perception and assessment of dangerous situations in real-life applications (Sanfilippo 2017). The latest VR HMDs are potentially market disrupting technologies, and facilitating the development of new possibilities for education and training (Jensen and Konradsen 2017). Preliminary research of VR HMD have found the enhanced immersion levels increases user engagement and performance (Buttussi and Chittaro 2018; Janssen et al. 2016).

2.2 User Motivation

The question of what motivates students and trainees is a critical aspect of creating sustainable and effective learning ecosystems and outcomes (Lai 2011). Gredler et al. (2004) describe motivation as an attribute triggering or preventing actions. Thus, motivation can be expressed as an aspect that makes us do what we do. Motivation can be divided into two categories: (i) intrinsic motivation relates to personal interest and ambition; (ii) extrinsic motivation relates to external rewards or pressure (Deci and Ryan 1985). In educational contexts, intrinsic motivation is more desirable, yielding better results than extrinsic motivation (Gredler et al. 2004). A trainee is more likely to experience intrinsic motivation if given the possibility to succeed through their own efforts and have a level of control over the process and outcome (Eccles and Wigfield 2002).

2.3 Skill Acquisition and User Performance

Skill acquisition is comparable to the process of learning, in which information is obtained and used to aid actions or processes (Garris et al. 2016). Skill acquisition can be separated into the three main categories: (i) cognitive, (ii) psychomotor/autonomous, and (iii) affective (Ackerman 1988; Langan-Fox et al. 2002). The measure of performance is an indication of skill acquisition. Performance is an aspect of skill acquisition which is affected by individual abilities (Langan-Fox et al. 2002). As the skill acquisition process evolves and subjects acquire a particular skill, the process becomes more automated and less flawed.

2.4 User Experience

User Experience (UX) is defined as *"a person's perceptions and responses that results from the use or anticipated use of a product, system of service"* (ISO 2010). Enhancing UX has positive implications for users, including increased enjoyment, satisfaction and productivity (ISO 2010). UX is a multifaceted concept (Saffer 2010), however, unlike design and usability-centered methods, UX focuses on the experience, while physical design is of secondary importance (Soegaard and Dam 2013).

3 Methods

3.1 Participants

Five participants were recruited for the pilot study (AGE AVG = 41.2 yrs; SD = 15.6 yrs; MIN = 28.0 yrs; MAX = 64.0 yrs). Participants were randomly assigned to either the control group (desktop computer, n = 3) or the experimental group (VR HMD, n = 2). Convenience and snowball sampling were utilized to recruit the participant pool. All participants completed an informed consent form prior to participation. This research project was approved by the Norwegian Centre for Research Data (project no. 57860).

3.2 Materials and Tools

Laboratory and System Configuration
Data collection was performed at InnoTraining's Virtual Reality Lab (http://innotraining.targlab.com/). The HTC Vive HMD and hand controllers (1st gen, 2016 model; Resolution: 1080 × 1200 per eye; Refresh Rate: 90 Hz; FOV: 110°) connected

to an ASUS computer tower (Graphics Card: GTX1080; Processor: Intel i7-6700 K @ 4.00 gHz; RAM: 32 GB). The desktop system was connected to a Dell U2717D monitor (Size: 68.47 cm; Resolution: 2560 × 1440; Response Time: 8.0 ms G2G; Refresh Rate: 60 Hz) with standard keyboard and mouse (see Fig. 1). Both VR and desktop participants used AKG K518 headphones.

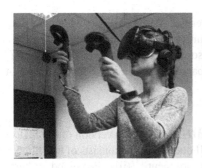

Fig. 1. Configuration of VR HMD (*left*) and, desktop computer (*right*) systems

Virtual Environment and Experimental Scenarios
The platform used for the virtual scenarios was *Fallout 4* (Bethesda Game Studios, v.1.10.40). *Fallout 4* was chosen for several reasons: (i) it is a well-established, highly-rated entertainment game series with high-quality graphics and reliability; (ii) it allows for customizable virtual environments, providing flexibility in developing virtual scenarios; (iii) its commercial availability for desktop and VR HMD configurations allow for a more valid cross-comparison between the two systems.

A customized virtual environment was built within *Fallout 4* by the authors. A simple maze was constructed and populated with various items of furniture (e.g. cabinets, chairs, etc.) (see Fig. 2). Throughout the maze, 10 objects (miniature statues) were hidden. Participants were instructed to search and retrieve the objects.

Fig. 2. Aerial perspective of the maze (*left*) and, first-person view of entrance to maze (*right*)

3.3 Measurements

Objective Performance Indicators

For both the control (desktop) and experimental (VR HMD) groups two performance indicators were established to measure proficiency of the task: (i) number of figures collected, and (ii) total time of completion (with a max time of 300 s allocated).

Physiological Measures

Participant Heart Rate (HR) and Electrodermal Activity (EDA) were collected in pre-experiment baseline measurements and during the experimental scenarios for analysis post-hoc. HR was collected via a HR sensor chest strap (Polar H10; Sampling Rate: 1 FPS) and EDA was collected via a wristband (Empatica E4, Sampling Rate: 4 FPS).

Post-Test Questionnaires

Intrinsic Motivation Questionnaire (IMI): An adapted version of the IMI was used (Monteiro et al. 2015). The adapted IMI questionnaire consists of five dimensions of intrinsic motivation: (i) Interest/Enjoyment, (ii) Perceived Competence, (iii) Effort/Importance, (iv) Pressure/Tension and (v) Value/Usefulness. Additional questions on task understanding and perceived immersion were added by the researchers.

User Experience Questionnaire (UEQ): The UEQ consists of 26 items measuring six dimensions of both usability ([i] efficiency, [ii] perspicuity, [iii] dependability), user experience aspects ([iv] stimulation, [v] novelty), and the main dimension: [vi] attractiveness (Laugwitz et al. 2008).

Simulator Sickness Questionnaire (SSQ): The SSQ (Kennedy et al. 1993) was used as both a monitoring instrument and post-hoc analysis of participants' occurrence and severity level of simulator sickness in the desktop and VR HMD scenarios.

3.4 Procedure

Participants were randomly assigned to the control group (desktop computer) or experimental group (VR HMD) in a between-group experimental design. Both groups were exposed to identical scenarios. The experimental procedure took approximately 90 min for each participant and was sub-divided into five phases (see Fig. 3).

(i) *Introduction:* Upon arriving at the lab, participants were provided an explanation of the experiment, tasks and expectations before filling out an informed consent form and introduction questionnaire consisting of demographic information and details regarding their technology utilization and gaming experience.

(ii) *Baseline Measurements:* Participants donned the Polar chest strap and Empatica E4 wristband with assistance from the researchers. They were then instructed to be seated in an erect position and as motionless as possible for a 5-minute period to establish their individual physiological baselines.

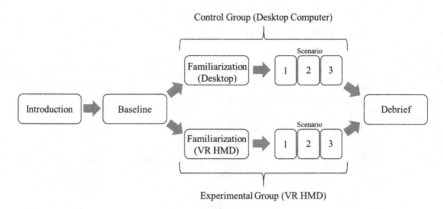

Fig. 3. Experimental Procedure

(iii) ***Familiarization:*** Participants were given a standardized 10-minute period with their assigned system to familiarize with the controls, navigation and interactions with the virtual environment. The familiarization period was scripted. The researchers verbally instructed participants to carry out various tasks to ensure all participants had similar experiences and basic skills required to interact with the virtual environment.

(iv) ***Experimental Scenarios (1, 2 and 3):*** The experimental scenarios and tasks were explained to the participants. They were instructed to find the ten objects within the maze as quickly as possible, with a maximum cut-off time of 300 s. They were given 3 separate attempts, with a 2-min rest period between each attempt. If a participant found the ten objects before the maximum time was reached, the round was ended.

(v) ***Debrief:*** Upon completion of the 3 experimental rounds, participants removed the HR and EDA devices, before completing a series of post-test questionnaires: (a) Simulator Sickness Questionnaire, (b) Intrinsic Motivation Questionnaire and (c) User Experience Questionnaire, and a brief exit interview about their experience.

4 Preliminary Results

4.1 VR vs Desktop Performance Score

User performance is expressed as a function of time used and round score. Thus, performance score (P) was calculated by dividing time in seconds (t) with number of objects successfully collected each round (n):

$$P = t/n \tag{1}$$

The performance score indicates the efficiency of collecting the objects within the maze; lower scores indicate higher efficiency and performance (see Fig. 4).

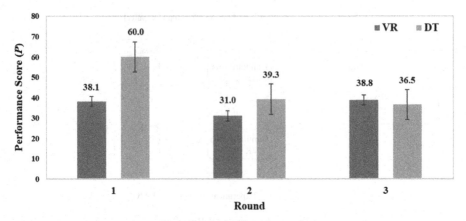

Fig. 4. Average performance scores of the five participants for each of round

5   Discussion

5.1   Implications for Education and Training

In order to keep users engaged in virtual learning environments and enhance the overall pedagogic experience, it is essential to better understand user motivation, user experience and learning outcomes specifically with VR HMD systems. There are two major aspects of VR HMD which are of interest to educators addressed within the current research: does VR HMD allow for (i) sustainable use, and (ii) enhanced learning opportunities and outcomes? In applying VR HMD for education and training, the technology must be accepted by not only the educators, but also by the students and trainees themselves. Thus, understanding aspects of user motivation and user experience can contribute to sustainable VR HMD usage. Additional technology must provide added benefit (e.g. by enhancing student learning outcomes or increasing accessibility). This points to the importance of empirically validating training and assessment methods, particularly when introducing emerging or alternative technologies.

VR HMD systems generally have a "wow factor" and novelty when introduced to new users. However, when implementing VR HMD simulations in training for professional education, these systems will require being repeatedly used over relatively long periods of time (e.g. ranging from individual training scenarios lasting minutes, to full courses with scenarios lasting many training hours over months/years). If the technology itself impacts user interest, willingness or ability to participate, it can have adverse effects on technology acceptance and diffusion. For example, cybersickness in VR HMD users is still a pervasive issue with existing technologies (Davis et al. 2014). Currently, VR HMD manufacturers warn of overuse, with Oculus proposing users take a 10–15 min break once every 30 min (Oculus 2018). This conflicts with current training practices in maritime simulator training exercises, which can sometimes last hours, with ideal lengths suggested between 40 and 60 min (Kobayashi 2005). Furthermore, the long-term usage of VR HMD on eyesight and general health has not been widely researched, and may be revealed as another limiting factor for the sustainable use of this technology for education applications.

The increased proliferation and affordability of VR HMD technologies is lowering the barrier to ownership. VR HMD systems are a relatively affordable technology for individuals to purchase (e.g. students), as they would a personal computer or mobile phone. This alters the traditional education ecosystem, where students would have to physically travel to centralized training facilities for simulator access, to a more personalized, distributed, and potentially on-demand simulation training.

However, the added value of VR HMD systems in delivering education and training must be better understood. The ultimate goal of utilizing simulators in specific educational programs is to enhance student competences. Thus, with the increased immersion that VR HMD users experience (Buttussi and Chittaro 2018; Janssen et al. 2016), coupled with more flexible education options and simulation opportunities, VR HMD may contribute positively to learning outcomes. How they are integrated into current educational programs and organized as a tool with traditional in-class lectures, online learning platforms and/or typical training simulators is not yet known.

5.2   Moving Forward: Developing VR HMD Experiments

The preliminary results indicate that technological novelty may have impacted the results (i.e. systematic error) from the self-reported measures. Considering that the novelty should wear off over prolonged and repeated usage, further study is necessary to reduce this systematic error. Thus, we propose increased attention toward understanding participants past gaming and VR HMD experience, as well as the amount and type of familiarization provided in subsequent experiments.

The VR HMD hardware had an impact on participant interaction with the virtual environment. Experimental group participants used hand controllers designed for VR applications with 6 Degrees of Freedom. The interaction with the virtual environment with hand controllers was unfamiliar for novices. While a familiarization period was conducted, preliminary results suggest the VR HMD hand controllers likely negatively impacted user performance, due to their novelty, in comparison to the control group (desktop keyboard and mouse configuration). It is believed that with repeated exposure that this effect will diminish. The researchers also underestimated the time and extent of the familiarization period required. For inexperienced participants, it was necessary to provide dedicated tasks to ensure a proper understanding of the basic skills required to interact with the virtual environment. The solution was to provide a longer and more detailed scripted familiarization period.

The current study focused on understanding differences between VR HMD and traditional desktop computer configurations for a relatively short-duration scenario. This is only one of many areas that needs to be investigated concerning the use of VR HMD in education and training. Another important aspect to study is the impact of prolonged exposure to VR HMD systems with regards to motivation, skill acquisition and health. Many training programs require long duration scenarios to improve trainees' skills and knowledge, and there is a lack of dedicated extended exposure and longitudinal VR HMD research. Understanding these effects in subsequent experiments can generate insights for optimizing education programs using VR HMD. For instance,

VR HMD is superior to full-scale simulators with regards to mobility and accessibility, so frequent and short-duration sessions may provide training methods which do not impede learning outcomes. Inputs and results from ongoing VR experiments and subsequent longitudinal studies are expected to provide valuable insight for not only training and education applications, but also entertainment and gaming stakeholders.

6 Conclusions

This pilot study provided insight into the design, evaluation and tools utilized for investigating user performance, motivation and experience between VR HMD and traditional desktop computer configurations. This pilot study intended to test and refine the experimental design and procedures for the project's full data collection. Therefore, this paper presents a generic, transferrable between-group experimental design comparing VR HMD systems to the differing "traditional" simulator configurations used across domains. This type of research, and experimental design, are valuable for the identifying strengths, weaknesses, and ultimately the value of utilizing emerging VR HMD technologies in education and training programs.

Acknowledgements. The authors would like to thank the Research Council of Norway for financial support of this research program (project number: 269424).

References

Ackerman PL (1988) Determinants of individual differences during skill acquisition: cognitive abilities and information processing. J Exp Psychol Gen 117(3):288–318

Berg LP, Vance JM (2017) Industry use of virtual reality in product design and manufacturing: a survey. Virtual Reality 21(1):1–17

Buttussi F, Chittaro L (2018) Effects of different types of virtual reality display on presence and learning in a safety training scenario. IEEE Trans Visual Comput Graph 24(2):1063–1076

Castells MI, Ordás S, Barahona C, Moncunill J, Muyskens C, Hofman W, Cross S, Kondratiev A, Boran-Keshishyan A, Popov A, Skorokhodov S (2015) Model course to revalidate deck officers' competences using simulators. WMU J Marit Aff 15(1):163–185

Davis S, Nesbitt K, Nalivaiko E (2014) A systematic review of cybersickness. In: Proceedings of the 2014 conference on interactive entertainment. ACM, New York, pp 1–9

Deci E, Ryan R (1985) Intrinsic motivation and self-determination in human behavior, vol 3. Plenum Press, New York

Eccles JS, Wigfield A (2002) Motivational beliefs, values, and goals. Annu Rev Psychol 53:109–132

Garris R, Ahlers R, Driskell JE (2016) Games, motivation, and learning: a research and practice model. Simul Gaming 33(4):441–467

Gredler ME, Broussard SC, Garrison MEB (2004) The relationship between classroom motivation and academic achievement in elementary school aged children. Fam Consum Sci Res J 33(2):106–120

Huber T, Paschold M, Hansen C, Wunderling T, Lang H, Kneist W (2017) New dimensions in surgical training: immersive virtual reality laparoscopic simulation exhilarates surgical staff. Surg Endosc 31(11):4472–4477

International Organization for Standardization (ISO) (2010) ISO 9241 Ergonomics of human-system interaction—part 210: human-centred design for interactive systems. ISO, Geneva

Janssen D, Tummel C, Richert AS, Isenhardt I (2016) Towards measuring user experience, activation and task performance in immersive virtual learning environments for students in immersive learning research network. In: Proceedings of international conference on immersive learning. Springer, Berlin, pp 45–58

Jensen L, Konradsen F (2017) A review of the use of virtual reality head-mounted displays in education and training. Educ Inf Technol 1–15

Kennedy RS, Lane NE, Berbaum KS, Lilenthal MG (1993) Simulator sickness questionnaire: an enhanced method for quantifying simulator sickness. Int J Aviat Psychol 3(3):203–220

Kobayashi H (2005) Use of simulators in assessment, learning and teaching of mariners. WMU J Marit Aff 4(1):57–75

Lai ER (2011) Motivation: a literature review. Pearson, London

Langan-Fox J, Armstrong K, Salvin N, Anglim J (2002) Process in skill acquisition: motivation, interruptions, memory, affective states, and metacognition. Aust Psychol 37(2):104–117

Laugwitz B, Held T, Schrepp M (2008) Construction and evaluation of a user experience questionnaire. In: USAB 2008. Lecture notes in computer science, vol 5298. Springer, Berlin, pp 63–76

Monteiro V, Mata L, Peixoto F (2015) Intrinsic motivation inventory: psychometric properties in the context of first language and mathematics learning. Psicologia: Reflexão e Crítica 28 (3):434–443

Nazir S, Totaro R, Brambilla S, Colombo S, Manca D (2012) Virtual reality and augmented-virtual reality as tools to train industrial operators. Comput Aided Chem Eng 30:1397–1401

Oculus VR (2018) https://scontent-arn2-1.xx.fbcdn.net/v/t39.23656/19896829_7716600130136 43_4087250127671001088_n.pdf?_nc_cat=0&oh=f27a10c0d20ccd00bae8c19c2e6ce8db& oe=5B7584EA. Accessed 02 May 2018

Saffer D (2010) Designing for interaction: creating innovative applications and devices, 2nd edn. New Riders, Berkeley

Salas E, Tannenbaum SI, Kraiger K, Smith-Jentsch KA (2012) The science of training and development in organizations: what matters in practice. Psychol Sci Public Interest 13(2):74–101

Sanfilippo F (2017) A multi-sensor fusion framework for improving situational awareness in demanding maritime training. Reliab Eng Syst Saf 161:12–24

Soegaard M, Dam RF (2013) The encyclopedia of human–computer interaction, 2nd edn. The Interaction Design Foundation, Århus

Virtual Simulation of Human-Robot Collaboration Workstations

Pamela Ruiz Castro[1]([✉]), Dan Högberg[1] [iD], Håkan Ramsen[2],
Jenny Bjursten[2], and Lars Hanson[1,3] [iD]

[1] School of Engineering Science, University of Skövde, Skövde, Sweden
pamela.ruiz.castro@his.se
[2] Volvo Trucks, Gothenburg, Sweden
[3] Industrial Development, Scania, Södertälje, Sweden

Abstract. The constant call in manufacturing for higher quality, efficiency, flexibility and cost effective solutions has been supported by technology developments and revised legislations in the area of collaborative robots. This allows for new types of workstations in industry where robots and humans co-operate in performing tasks. In addition to safety, the design of such collaborative workstations needs to consider the areas of ergonomics and task allocation to ensure appropriate work conditions for the operators, while providing overall system efficiency. The aim of this study is to illustrate the development and use of an integrated robot simulation and digital human modelling (DHM) tool, which is aimed to be a tool for engineers to create and confirm successful collaborative workstations. An assembly scenario from the vehicle industry was selected for its redesign into a collaborative workstation. The existing scenario as well as potential collaborative concepts are simulated and assessed using a version of the simulation tool IPS IMMA. The assembly use case illustrates the capabilities of the tool to represent and evaluate collaborative workstations in terms of ergonomics and efficiency assessments.

Keywords: Human-robot collaboration · Digital human modelling
Ergonomics · Virtual simulation · Collaborative workstations

1 Introduction

1.1 Collaborative Robots

Robots have become a vital element for numerous sectors of today´s manufacturing industry. However, there are still challenges to solve for the industry to efficiently react to the constant changing consumer behavior and the global competitiveness [1] and to gain from the introduction of new types of human-robot collaboration workstations in manufacturing. One challenge is that new types of human-robot collaboration workstations will demand the fulfillment of new regulations.

Currently, industrial robots are used for several processes at the assembly and production areas of large-volume manufacturing. However, safety specifications require

© Springer Nature Switzerland AG 2019
S. Bagnara et al. (Eds.): IEA 2018, AISC 822, pp. 250–261, 2019.
https://doi.org/10.1007/978-3-319-96077-7_26

most robots to operate behind secure barriers and when interaction zones are allowed the norms prevent the robot from operating at its full capacity, by limiting speed and force. Even though there are clear benefits on implementing robot operations in the manufacturing process it has not been easy to design and apply modifications or adapt new workstations. This since most robots are not easily programmed, and once a task is defined it is generally too specific to allow flexibility for different processes in the production line. This has led to high maintenance costs and ineffective floor usage [2].

Considering the continuous change in demands from the consumer and the market, manufacturing industries require to have more flexible workstations. Preferably enhanced flexibility also leads to quality improvements, or at least the new solutions must not negatively affect the quality during the manufacturing of the products. This has led to promoting human-robot collaborations, since robots can efficiently conduct tasks that require both repetitive and precise characteristics, while humans, on the other hand, are excellent at detecting inconsistencies and adapting to unexpected changes that could arise during the production [3, 4]. The two latter aptitudes challenges today's robotic solutions. Hence the focus on having closer interactions between the workers and the robots in the manufacturing processes.

Since technology has been evolving in robotics, there has been a particular focus for industrial applications since that may offer manufacturing companies competitive advantages. There has been continuous development of robots designed to collaborate in an open environment with the operators. One important area is to include built-in sensors that can ensure the safety of the workers [5]. Due to these developments, adjustments of the safety regulations have been made in order to allow closer collaborations. The supplement standard ISO 15066 [6] focuses on collaborative robot applications and dictate the implementation of enabling devices and emergency stops, as well as the type of collisions that could happen depending on the exposed area of the body of the operator.

Operators and robots have different skills and capabilities. Therefore a design objective is to find optimal solutions for the collaborations, in order to offer the highest efficiency of such hybrid workstations [7], while fulfilling the standards of the work environment and the appropriate measures.

1.2   Virtual Simulations

Recently there has been a tendency to postpone physical evaluations of design concepts for later stages in the development process of workstations. This to save time, cost and advance quality of eventual design solutions. This has led to an increase of design and evaluation tasks being carried out in digital (virtual) environments, e.g. using virtual scenarios to evaluate human work.

Several advantages have resulted from the virtual environments where virtual humans can be simulated [8]. From an ergonomic perspective it is possible to have detailed biomechanical measurements as well as testing concepts with a variety of human anthropometries simultaneously [9]. Correspondingly, there are models used to simulate the robotic tools for a specific environment. These possibilities have enabled evaluations of future workstations already in the virtual world, and to make risk assessments to improve safety. Such work has traditionally been done after acquiring the robot and setting the workstation [10].

Virtual simulations of robots and virtual humans can be made with multiple commercial software, with the limitation that for most cases the simulations need to be made separately. Only few existing software is capable to simultaneously simulate a human-robot collaboration workstation [11]. Therefore, simulation tools that can aid in the design of collaborative systems are being investigated for several manufacturing environments [12, 13], which can allow evaluating design alternatives and foreseeing human-robot interactions at early stages of the workstation design process. Conclusively, virtual simulation tools can assist to design optimized workstations for the human and the robot, by enabling testing several workflows and task allocation scenarios, and evaluating the most efficient design solutions, while proactively ensuring proper ergonomic conditions for operators [14].

1.3   Aim of the Study

The aim of this study is to verify the use of virtual simulations to evaluate human-robot collaborations from an early stage in the development process. Another aim is to validate the results of the predicted ergonomic assessments of the virtual operator with corresponding ergonomic assessment being based on recording of motions of real humans.

2   Method

2.1   Use Case

A use case from the manufacturing industry was selected as a base for implementing an industrial human-robot collaboration. The use case is based on an existing manual, power tool assisted, workstation in the production line of an engine assembly.

The existing workstation has one operator performing several tasks within a time frame. The tasks include the placement of all parts in a specific order and use of a power tool for tightening screws. There is a risk that the manual tasks cause musculoskeletal disorders since the operators perform highly repetitive movements, and have to reach for the tools in uncomfortable positions. These are the major reasons for investigating the possibly to modify this workstation into a human-robot collaboration workstation.

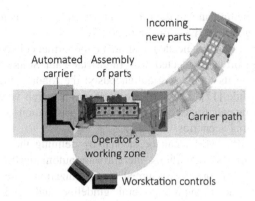

Fig. 1. Overview from the existing workstation for the use case.

The existing workstation is shown in Fig. 1. It consists of the following parts:

- A fixed section for the controls with two stands
- An automated carrier that arrives through the path with the engine
- A movable trolley with the new parts always arrive from the same side.

Furthermore, two different types of power tools (a single nut runner and a double nut runner) are fixed in the ceiling with pneumatic systems at a reachable distance for the operator to use when needed.

Since the automated carrier requires guidance throughout the assembly line, there is a path for the carriers that goes in between all workstations. Therefore it should be free from permanent tools to maintain the flow of the assembly.

The assembly starts when a new carrier guides itself into place inside the workstation area with an engine that comes from the previous workstation. Once in position, the operator can move the trolley with new parts to facilitate the placement of the ladder cover and the rest of the components that are to be assembled.

When the assembly has finished for that workstation the carrier continues into the next workstation, leaving room for a new carrier to arrive. The operator gives a signal in the control system when a new carrier arrives and the tasks begin, and gives another signal when the workstation has been cleared.

2.2 Simulation Software

Considering the constant change in technology and its applications, Digital Human Modelling (DHM) tools have taken an important role for the industry's development processes by allowing visualization and evaluation of concepts during early stages of the development process [10].

DHM tools are used for analyzing design scenarios with simulations, where the interaction of digital humans, referred to as manikins, can be evaluated in virtual environments. The use of these types of tools has increased for the design of collaborative workstations since early evaluations can give feedback of how the task allocation between the human and the robot can be arranged to increase efficiency and prevent safety or ergonomic risks for the operators.

Therefore, a DHM tool was needed to simulate the virtual collaborative workstation for this study. The software selected was Industrial Path Solutions [15] since it is possible to use two of its modules simultaneously: the robotics module [16] and the IMMA manikin module [17]. The software used is a research version, which is currently under development, focused on enabling the simulation of human-robot collaborations in industrial scenarios.

The manikin in IPS IMMA can be steered by defining the tasks that it should perform. Based on the task definitions, the software automatically computes motions based on mathematical algorithms designed for optimization, where the manikin can avoid collisions, move according to ergonomic guidelines and stay in balance. With the same task definitions, manikins with different anthropometry, referred to as a manikin family in the software, can be instructed to accomplish the same tasks. However, unique motions will be computed for each manikin [18], giving a more complete view of the tasks and observe the design limitations that may occur with some anthropometries.

The software has an ergonomic assessment functionality and a timeline that allows to evaluate the efficiency level of the workstation. For this study, the research version of the software has an ergonomic evaluation method based on RULA [19]. The final scores given from this method are evaluated as:

1-2 = acceptable posture if it is not maintained or repeated for long periods.
3-4 = further investigation needed, and changes may be required.
5-6 = investigation and change are required soon.
7 = immediate investigation and change.

An overview of the evaluations are given in IMMA, where an average of the RULA scores of the entire simulation is given per body zone and in total. For more detailed results, colour coded graphs are available for each body zone evaluation, with a general overview in which: green is considered risk-free for most operators, yellow ought to be investigated further, and red needs immediate attention and should be analysed in detail.

For the purpose of this study, following aspects were the focus of the collaboration: the location of the robot; and the sequence of tasks allocated between the robot and the operator. The design process followed for the collaborative workstation was as follows:

– Concept generation of the different aspects, resulting in 3 concepts per aspect. The concepts should consider the safety restrictions and working standards.
– Simulations of the concepts will be used as concept evaluations to determine the safety and efficiency of the concepts. This together with the input from the industry partners will determine the optimal concept.
– An optimization of the optimal concept will be used to provide a proposal for a collaborative workstation.

2.3   Verification Method

The design process using simulations to validate the concepts of the human-robot collaboration will serve as a verification that the DHM software can be used as a tool to evaluate human-robot collaboration workstation concepts.

In order to validate the results from the ergonomic assessments in the simulations, a physical prototype of the collaborative workstation was developed where operator motions were recorded with the mobile motion capture system, Xsens [20], and the data obtained were exported to IPS IMMA. Once in the software it was possible to compare the ergonomic assessments and time results of the simulated collaborative workstation with the physical prototype of the same concept. This verification only focused on the operator.

3   Results

3.1   Design Process with Simulations

Collaboration sequence and robot location were the aspects being addressed in the study. Several concepts were generated for each variable and tested in the software through simulations.

The main concepts of the aspects were simulated in the software. The concepts were evaluated with several criteria for the location of the robot, to ensure: robot reach; ease of space in the workstation; efficiency in the tasks performed; operators' safety; as well as operator variability.

The operator sequence for concept A is shown in Fig. 2. The sequence can be followed by the tags, where the first task is H1 followed by task H2, etc. This concept also evaluates the robot location, which is proposed to be mounted on a pedestal in the floor.

Fig. 2. Simulation of concept A's operator sequence, operator tasks are ordered starting with H1.

Fig. 3. Evaluation of concept B robot location, simulation performed with a manikin family.

In Fig. 3 concept B is being evaluated, where the robot is mounted in the ceiling and uses the double nut runner to tighten the screws. A family of manikins is used to test the workstation. The sequence proposed for this concept focuses on the robot tasks, in order to allow more time for the operator to place the parts, the robot sequence is designed to tighten only the first three sets of screws on one end and then move to the opposite end and tighten the rest of the screws. This type of sequence, although good for the operator, was evaluated by the industry and had to be modified since the tightening of the screws for the ladder frame have to follow a specific sequence to ensure the quality of the product.

Fig. 4. Comparison simulation of two different mounting concepts for the robot.

Concept C is evaluated in Fig. 4, where a single nut runner is used to tighten the screws. A comparison had to be made to see the robot sequence between 2 different mounting techniques to the ceiling. From the image it can be observed that while following the same path for tightening the screws, one robot (pink one) seems to be too close to the operator, especially taller operators. This represents a safety hazard since any unpredicted movement from the operator could result in a collision.

3.2   Simulation Results

After evaluating the different concepts and their characteristics, the sequence was optimized and a robot location was selected. The improved workstation is presented in Fig. 5, where the proposed robot location is mounted in the ceiling and secured on the side opposite to the operator. For the workstation, a single nut runner is suggested, which has to follow a predesigned tightening pattern.

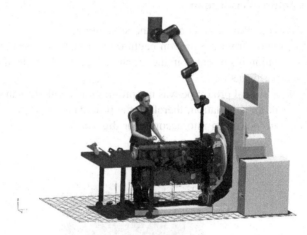

Fig. 5. Screen shot of simulation for optimized concept for the collaborative workstation.

The time distribution from the simulation of the human-robot collaboration is presented in Fig. 6 where the operator starts and finishes its tasks before the robot. The tasks of the operator are divided into seven sections and each represent the placement of certain components to be assembled. The robot time is divided into three sections which represent the tightening sequences. It is suggested that the robot has some evaluation of the placement of components before continuing with the next tightening sequence, to avoid collisions with the operator and ensure the quality of the product.

An ergonomic assessment was obtained from the simulation. The RULA scores obtained from the software's data for the cycle time of the workstation ranged between 2 and 4, which results in an acceptable workstation.

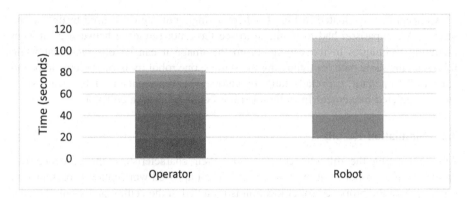

Fig. 6. Graph for the time distribution of each actor in the simulation

3.3   Motion Capture Verification

The physical prototype allowed to evaluate the simulation.

The time distribution from the motion capture recordings of a sample operator at the prototype workstation is presented in the graph of Fig. 7, where the different colors represent the tasks performed.

The robot in the physical prototype was not programmed with the same sequence or location as the one in the simulation, therefore the time data and interactions between operator and robot are not taken into account for this study.

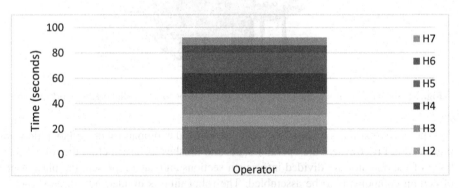

Fig. 7. Time distribution of sample operator at the prototype collaborative workstation

The RULA scores throughout time to evaluate the operator, were given by the IPS IMMA software after importing the motion capture recording data. It can be noted that most of the tasks performed in the workstation have a RULA score of 4 or less, except during 2 sections where the scores are of 6, but in both cases that score is given for less than 2 s each time, which represent less than 5% of the total cycle time.

4 Discussion

The results from the simulations and the motion capture verification were analyzed separately in order to give a better overview. From the results observed for the simulated collaboration, the time distribution between the actors in the workstation does not start at same time since the operator has to give clearance to the robot to begin its tasks. This causes the robot to start and end after the operator, leaving time for adjustment and possibly allowing to have more tasks performed within the workstation.

In order to have a frame of reference from the original use-case, the workstation was evaluated manually, by the author of this study, using a RULA spreadsheet. The assessment resulted in continuous RULA scores of 6 while handling the power tool to tighten the screws, and some peaks with a score of 7 between changing tools. By replacing the tasks related to the double nut runner the highest RULA scores could be eliminated and vibrations from the nut runner were also eliminated.

For the RULA analysis comparison, the scores were calculated through IPS IMMA software for both the simulation of the collaborative workstation and for the motion capture recording at the physical prototype.

The comparison between RULA scores of the simulation and the recorded motions of the prototype workstation shows that the overall scores throughout time for both cases ranges between scores 2 and 4 (Fig. 8). However, there are slight differences between both cases. The results of the simulation do not present any extreme scores, while the results of the motion capture data have two occasions in which the motions were evaluated with a score of 6.

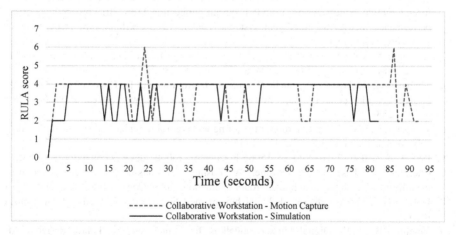

Fig. 8. Ergonomic evaluation comparison between simulated workstation and motion capture recording

From Fig. 8a comparison of time for the operator tasks can also be observed, where the cycle time for the operator according to the simulation is of 82 s, while the cycle time from the motion capture recording is of 92 s. Although there is a difference of 10 s between results, the obtained values from the simulation can provide an approximation of the cycle time expected of the future workstation.

From this study, it was noted that the simulations with IPS IMMA had a positive outcome throughout the design process of the workstation, since modifications could be visualized and evaluated early on. Both the robot locations and sequence for the collaboration could be modified and the effects of the changes could be observed with manikins of different anthropometrics. However, the time spent setting up the simulations was considerable and adjustments had to be made for every change applied in the workstation.

In conclusion, the software provided a good base when redesigning a collaborative workstation and the physical prototype proved that the results were quite accurate, although there were certain discrepancies. This could also happen in an existing workstation where operators may interact differently and external factors may affect the cycle times.

Acknowledgment. This work has been carried out within the Virtual Verification of Human-Robot Collaboration project, supported by VINNOVA/FFI and by the participating organizations. This support is gratefully acknowledged. Special thanks goes to Victoria Gonzalez, Volvo Trucks and FCC (Fraunhofer-Chalmers Centre).

References

1. Siciliano B, Khatib O (2016) Springer handbook of robotics, 2nd edn. Springer, Berlin
2. Jimmerson G, Menassa R, Pearson T, Shi J (2012) Levels of human and robot collaboration for automotive manufacturing. In: NIST special publication 1136, 2012 proceedings of the performance metrics for intelligent systems (PerMI–'12) workshop, U.S. Department of Commerce, pp 95–100
3. Krüger J, Nickolay B, Heyer P, Seliger G (2005) Image based 3D surveillance for flexible man-robot-cooperation. CIRP Ann Manuf Technol 54:19–22
4. Michalos G, Makris S, Tsarouchi P, Guasch T, Kontovrakis D, Chryssolouris G (2015) Design considerations for safe human-robot collaborative workplaces. Procedia CIRP 37:248–253
5. Wang L (2015) Collaborative robot monitoring and control for enhanced sustainability. Int J Adv Manuf Technol 81(9–12):1433–1445
6. ISO, T. S. 15066 (2016) ISO TS 15066-Robots and robotic devices-collaborative robots
7. Cherubini A, Passama R, Crosnier A, Lasnier A, Fraisse P (2016) Collaborative manufacturing with physical human–robot interaction. Robot Comput Integr Manuf 40:1–13
8. Duffy VG (2009) Handbook of digital human modeling. Taylor & Francis Group, Boca Raton
9. Chaffin DB (2001) Digital human modelling for vehicle and workplace design. SAE International, Warrendale
10. Ore F, Hanson L, Delfs N, Wiktorsson M (2015) Human-industrial robot collaboration—development and application of simulation software. Int J Hum Factors Model 5:164–185
11. Maurice P, Padois V, Measson Y, Bidaud P (2017) Human-oriented design of collaborative robots. Int J Ind Ergon 57:88–102

12. Ore F, Reddy Vemula B, Hanson L, Wiktorsson M (2016) Human-industrial robot collaboration—application of simulation software for workstation optimization. In: Proceedings of the 6th CIRP conference on assembly technologies and systems, pp 181–186

13. Zanchettin A, Ceriani N, Rocco P, Ding H, Matthias B (2016) Safety in human-robot collaborative manufacturing environments: metrics and control. IEEE Trans Autom Sci Eng 13(2):882–893

14. Ruiz Castro P, Mahdavian N, Brolin E, Högberg D, Hanson L (2017) IPS IMMA for designing human-robot collaboration workstations. In 5th international digital human modeling symposium, Bonn/Germany, June 26–28

15. IPS homepage. http://industrialpathsolutions.se/. Accessed 10 May 2018

16. Bohlin R, Delfs N, Mårdberg P, Carlson JS (2014) A framework for combining digital human simulation with robots and other objects, Tokyo, Japan

17. Högberg D, Hanson L, Bohlin R, Carlson JS (2016) Creating and shaping the DHM tool IMMA for ergonomic product and production design. Int J Digit Hum 1:132–152

18. Bohlin R, Delfs N, Hanson L, Högberg D, Carlson JS (2012) Automatic creation of virtual manikin motions maximizing comfort in manual assembly processes. In: 4th CIRP conference on assembly technologies and systems. USA Conference on Assembly Technologies & Systems, pp 209–212

19. McAtamney L, Corlett EN (1993) RULA: a survey method for the investigation of work-related upper limb disorders. Appl Ergon 24(2):91–99

20. Xsens homepage. https://www.xsens.com/. Accessed 26 May 2018

Avatar-Based Human Posture Analysis and Workplace Design

Soomin Hyun and Woojin Park(✉)

Seoul National University, Seoul 08266, South Korea
{xsmh, woojinpark}@snu.ac.kr

Abstract. Stressful working postures are associated with increased risks of work-related musculoskeletal disorders (WMSD). The reduction of stressful working postures is necessary for the safety and health of workers and is also related to increased productivity and worker satisfaction. The evaluation and control of work-related postural stress is currently achieved through a process that requires observation, analysis, and redesign (OAR approach). As an alternative overcoming the limitations of this approach, the authors have previously proposed an APAD approach [1]. The objective of the current paper is to report the initial results of the APAD research focusing on the modelling of individual PSM functions. 3 participants were asked to perform short-period static posture holding for 108 combinations of posture and hand load weights. Immediately after each task trial, the participants performed subjective ratings using the Borg CR10 scale to quantify the level of perceived postural stress. Individual PSM functions were derived through regression, and were evaluated by calculating the average absolute prediction error. The study results may serve as a basis for creating a large set of avatars that perceive postural stresses in a way similar to their actual human counterparts. Such avatars could be used to simulate various work situations in the digital world to aid in the design of less stressful and ergonomic workstations and work tasks.

Keywords: Working postures · Postural stress mapping

1 Introduction

1.1 Avatar-Based Posture Analysis and Workplace Design

Avatar-based posture analysis and workplace design (APAD) is a research project, which is currently being conducted by the researchers at the Seoul University Life Enhancing Technology Lab [1]. The research project is aimed at addressing the limitations of existing postural analysis tools. The main limitations are as follows:

1) Existing postural analysis tools do not consider the inter-individual variability of perceived stress. A lack of consideration for such variability is problematic because this may result in an erroneous assumption of perceived stress, assuming that all workers experience work-related postural stresses in a similar manner, in spite of anthropometric, anatomical, strength, and psychophysical differences.

S. Bagnara et al. (Eds.): IEA 2018, AISC 822, pp. 262–269, 2019.
https://doi.org/10.1007/978-3-319-96077-7_27

2) Current postural analysis tools require analysts to use the observation, analysis, and redesign approach to evaluate a work task. Essentially, this requirement of personnel to complete the task limits the number of workers and tasks that can be analyzed, due to financial, space, and time constraints.

3) The observation, analysis, and redesign approach that is currently in use can only be used after a certain work task has been enacted. This approach is reactive. A proactive approach is needed as it is more desirable to prevent musculoskeletal diseases as opposed to making after-the-fact corrections.

In the APAD project, these limitations are to be addressed by developing an avatar-based posture analysis system through the following four steps.

In this paper, a pilot study carried out for the first step of the process shown in Fig. 1 is described.

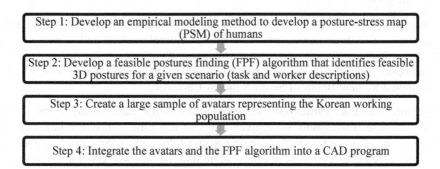

Fig. 1. Steps for avatar-based posture analysis and workplace design

1.2   PSM Function

Posture analysis tools have been widely used and researched since their introduction, due to their utility in evaluating postural stresses of manual tasks. Posture analysis tools are scoring tools that help determine the discomfort of a posture based on information such as joint angles and task frequency. Tools such as the Ovako Working Posture Analysis System (OWAS), Rapid Upper Limb Assessment (RULA), and Rapid Entire Body Assessment (REBA), etc. [2–7] have been used as methods of carrying out the process of observation-analysis-redesign (OAR), to accurately evaluate the postural stress of a designed work task. However, current posture analysis tools only provide one method for the whole population in evaluating postural stresses, which makes it difficult to take into account the distinctive physical and psychophysical characteristics of individuals, such as muscle mass or personal perception of pain. One way to compensate for this limitation is to understand the inter-individual differences in personal perception of pain, with the use of individual-specific postural stress functions. A past study by Ogutu and Park [8] has demonstrated the feasibility of developing individual-specific postural stress functions. However, this study only considered a single base function in modeling postural stress perception. Different forms of base functions need to be explored and their utility for predicting postural stress need to be empirically evaluated.

As a response to this knowledge gap, this pilot study evaluated various based functions for predicting postural stresses; the study was aimed at gaining insights into the optimal form(s) of individual postural stress mapping (PSM) functions. Posture-stress prediction functions were generated for each participant based on pilot experiment data. Then, the prediction functions were validated with test data, and compared in terms of adjusted R^2, average absolute prediction error, and maximum and minimum prediction error values.

2   Methods

2.1   Subjects

Three participants in their 20's (1 male and 2 females) participated in the pilot experiment. The three participants were free of musculoskeletal and/or neurological disorders. Each participant completed a consent form after being notified of the experiment's procedure and goals, and the data collection protocol was approved by the Institutional Review Board of Seoul National University.

Experimental Task. Participants performed a total of 108 static posture holding trials, each 10 s long. Participants were given 100 s of rest time in between task trials to minimize the effect of muscle fatigue on perceived discomfort. An extra 5 min of rest time was given after every ten postures. More time for rest was given if requested by the participant. The 108 posture holding trials were combinations of three load weights and 36 postures. The postures were recorded by attaching spherical reflective markers on the participants' body landmarks, which were recorded by an 8-camera motion capture system.

Procedure. Each participant was provided with a set of fitted clothes to wear at every data collection, and participants then had spherical reflective markers attached at body landmarks. Basic anthropometric measures were collected before the data collection. The participants were then instructed about the use of the Borg CR10 scale for rating perceived exertion [9], for which a visual aid was provided in plain view. The participants assumed a given posture for 10 s, and then rested for 100 s (or more, if requested by the participant). During each resting period, participants were asked to perform postural stress ratings for three parts of the upper body, which were the neck/shoulder, upper and lower arm, and lower back.

Data Analysis. This paper explored four simple base function forms consisting of various combinations of weighted local biomechanical loadings. Four regression models were developed for each participant as an attempt at predicting the individual's perceived postural stress. The four equations were selected based on published works related to postural stress prediction [10, 11], and are shown in Eqs. 1–4. A variable called the midpoint was utilized in some of the base functions, and represents the middle point between the upper and lower limits of the range of motion for a certain body joint. This decision was made based on the knowledge that perceived discomfort increases as a certain posture approaches the limits of a person's range of motion [3–6]. The formula for the midpoint is shown in Eq. 1.

$$\sum W_i |M_i| + C \tag{1}$$

$$\sum W_i |M_i| + \sum L_i * (M_i)^2 + W_{TW}(TW) + C \tag{2}$$

$$\sum W_i (MP_i) + W_{TW}(TW) + C \tag{3}$$

$$\sum W_i (MP_i) + \sum L_i (MP_i)^2 + W_{TW}(TW) + C \tag{4}$$

$$MP = \left| Theta_i - \frac{U_i + L_i}{2} \right|, \ L = 0 \tag{5}$$

M_i: absolute moment,
i: index of body joint (i = waist, shoulder, elbow)
W_i, L_i: weighting factor
TW: test weight
C: constant
MP_e: Midpoint (w: waist, s: shoulder, e: elbow)

The moment values were calculated in respect to the limb lengths of each individual, as well as their weight. For each participant, the collected data was separated into training (28) and testing (80) data. To create a model that best predicts the perceived stress at the extreme stress points, data points with a large discomfort score variability were selected as the training data set. Regression models were generated with the input being variables in the base function form, and the output as each participant's perceived stress. The resulting regression model with individual-specific weights was then validated by predicting the remaining 80 perceived stress scores. The accuracy of the prediction function was measured by adjusted R^2, average absolute prediction error, and maximum and minimum prediction error. The formula used to calculate average absolute prediction error is presented in Eq. 6.

$$\text{Average absolute prediction error} = \frac{\sum |Predicted \ tress \ Score - Actual \ Stress \ Score)|}{number \ of \ data} \tag{6}$$

3   Results

The three participants' stress rating distributions of the three participants is shown in Fig. 2. The discomfort score range and median scores for participants 1, 2, and 3 were 0 to 7, 0 to 10, and 0 to 5; and 4, 7, and 2, respectively. No outliers were detected.

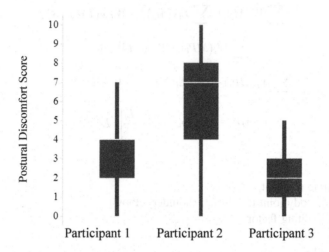

Fig. 2. Perceived stress ratings distributions of the three study participants

Regression Modelling. The best performing regression models for each participant are shown in Table 1. The adjusted R^2, average absolute prediction error (AAPE), and maximum and minimum error are shown in Tables 1, 2, 3 and 4. The lowest average absolute prediction error for each participant is marked in blue, and the highest, in red.

Table 1. Base function and formula with best performance in predicting perceived stress.

Participant	Base function form with best performance	Formula for predicting perceived stress
1	$W_iM_w + W_{TW}TW + C$	$0.020M_w + 0.643TW + 0.718$
2	$W_iM_w + C$	$0.021M_w + 3.252$
3	$W_iM_w + W_{TW}TW + C$	$0.00003823M_w + 0.498TW + 0.520$

Figures 3, 4 and 5 are scatterplots of the 80 testing data points (x-axis), compared with the postural stress predicted for the participant with the best fitting base function form, or lowest average absolute prediction error (y-axis). The line represents a perfect prediction model with a slope of 1. The darkness of the points represents the data density.

Table 2. Participant 1: adjusted R^2, AAPE, and max/min error for prediction formulas.

Formula	R^2	AAPE*	Max Error	Min Error
1	0.68	0.876	2.724	0.003
2	0.76	**0.861**	3.167	0.009
3	0.62	**1.385**	5.153	0.006
4	0.77	1.295	4.749	0.028

*Average absolute prediction error

Table 3. Participant 2: adjusted R^2, AAPE, and max/min error for prediction formulas.

Formula	R^2	AAPE*	Max error	Min error
1	0.52	1.484	5.570	0.007
2	0.53	**1.575**	5.159	0.022
3	0.07	**2.833**	7.574	0.152
4	0.15	2.655	6.940	0.226

*Average absolute prediction error

Table 4. Participant 3: adjusted R^2, AAPE, and max/min error for prediction formulas.

Formula	R^2	AAPE*	Max error	Min error
1	0.55	0.847	2.021	0.005
2	0.67	**0.783**	2.461	0.008
3	0.22	**1.202**	4.544	0.027
4	0.39	1.224	4.187	0.013

*Average absolute prediction error

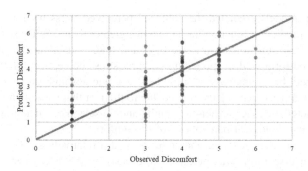

Fig. 3. Best fitting prediction function for Participant 1.

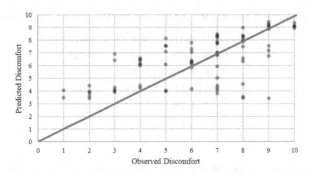

Fig. 4. Best fitting prediction function for Participant 2.

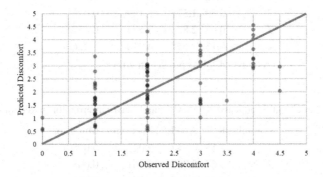

Fig. 5. Best fitting prediction function for Participant 3.

4 Discussion

The regression models predicted the subjects' perceived stress with acceptable accuracy - the best predicting function had R^2 values greater than 0.5. Two of the subjects' R^2 values for base function form 2 are 0.67 and 0.86, indicating that this particular function may perform well in modelling postural stress perception for some subjects. As hypothesized, base functions with the least average absolute prediction error were different for each participant. This may be due to the fact that participants had different physical characteristics such as height, weight, and muscle mass, in addition to possible differences in psychological perceptions of discomfort.

Another observation was that the subjects with a relatively smaller range of perceived postural discomfort was predicted (the best) by the same regression function form. The postural stress responses of participants 1 and 3 were best predicted by a prediction formula based on Eq. 1, and participant 2's postural stress was best predicted by a formula based on Eq. 2; participants 1 and 3 had a discomfort score range of 7 and 5, while participant 2 had a discomfort score range of 10. This may provide insight about the characteristics of participants whose perceived stress may be predicted accurately with base function 1.

On the other hand, the two participants whose perceived discomfort scores were best predicted with the same base function form showed a considerable difference in terms of the value of the constant and weights within the base function, which also supports the notion of an individual-specific postural stress function. It may be beneficial to explore individual characteristics such as muscle mass and personal tolerance to discomfort, in order to discover the reasons for these differences; such knowledge may aid in a more accurate diagnosis of an individual's PSM function.

This pilot study about the individual's PSM function showed positive results and lends support to the feasibility and future success of the APAD system. The authors are currently working on an experiment with additional participants, as well as planning for the next step in the APAD project. Future works may consider comparing the PSM functions of participants with similar characteristics such as gender, muscle mass, and height, in order to gain insight about the reason for the difference in perceived postural stress.

Acknowledgements. This research was supported by the National Research Foundation of Korea.

References

1. Hyun S, Park W et al (2017) Posture analysis and workplace design utilizing avatars. In: Proceedings of the Asia-design engineering workshop 2017. A-DEWS, Seoul, pp 7–71
2. Karhu O, Kansi P, Kuorinka I (1977) Correcting working postures in industry: a practical method for analysis. Appl Ergon 8:199–201
3. Mcatamney L, Corlett EN (1993) RULA: a survey method for the investigation of work-related upper limb disorders. Appl Ergon 24(2):91–99
4. Hignett S, Mcatamney L (2000) Rapid entire body assessment (REBA). Appl Ergon 31:201–205
5. Chung MK, Lee I, Kee D (2005) Quantitative postural load assessment for whole body manual tasks based on perceived discomfort. Ergonomics 48(5):492–505
6. Kee D, Karwowski W (2001) LUBA: an assessment technique for postural loading on the upper body based on joint motion discomfort and maximum holding time. Appl Ergon 32:357–366
7. David G, Woods V, Li G, Buckle P (2008) The development of the Quick Exposure Check (QEC) for assessing exposure to risk factors for work-related musculoskeletal disorders. Appl Ergon 39:57–69
8. Ogutu J, Park W (2011) An efficient method for modeling an individual's perception of postural stress. In: Proceedings of the human factors and ergonomics society annual meeting, vol 55, no. 1. HFES, Las Vegas, pp 625–629
9. Borg G (1982) Psychophysical bases of perceived exertion. Med Sci Sports Med 14:377–381
10. Marler RT, Rahmatalla S, Shanahan MK, Abdel-malek K (2005) A new discomfort function for optimization-based posture prediction. SAE Tech
11. Dysart MJ, Woldstad JC (1996) A new discomfort function for optimization-based posture prediction. J Biomech 29:1393–1397

Interactive Tools for Safety 4.0: Virtual Ergonomics and Serious Games in Tower Automotive

Antonio Lanzotti[1(✉)], Andrea Tarallo[1], Francesco Carbone[1],
Domenico Coccorese[2], Raffaele D'Angelo[3], Giuseppe Di Gironimo[1],
Corrado Grasso[4], Valerio Minopoli[1], and Stefano Papa[1]

[1] Fraunhofer Joint Lab IDEAS, University of Naples "Federico II", Naples, Italy
{antonio.lanzotti,andrea.tarallo,fcarbone,
giuseppe.digironimo,stefpapa}@unina.it
[2] Consorzio CREATE, Naples, Italy
domenico.coccorese@consorziocreate.it
[3] Direzione Regionale INAIL – CONTARP, Naples, Italy
r.dangelo@inail.it
[4] Tower Automotive Italy Srl, Pignataro Maggiore, CE, Italy
grasso.corrado@towerinternational.com

Abstract. This work focuses on an innovative training methodology based on the use of Virtual ergonomics and "serious games" in the field of occupational safety. Virtual Ergonomics was chosen as an effective and convincing tool for disseminating the culture of safety among the workers, while a "serious game" was developed to train operators on specific safety procedures and to verify their skills. The results of the experimentation in a real industrial case study showed that, compared to the traditional training methodology, multimedia contents and quantitative ergonomic analyses improve the level of attention and the awareness of the operators about their safety. On the other hand, Serious games turned out as promising tools to train the workers about safe operating procedures that are difficult to implement in a real working environment.

Keywords: Virtual reality · Ergonomics · Virtual humans
Occupational safety · Serious game

1 Introduction

The success of interactive design as a tool for improving human-machine interaction concerns not only web-based applications and mobile technologies, but also the industrial sector. The manufacturing industry 4.0, indeed, focuses on workers rather than just on industrial products and uses new technologies to facilitate their activities [1, 19].

To be truly effective, these innovations must involve every company department, in order to improve not only productivity in the strict sense, but also the occupational safety. It is indeed well-known that the prevention of injuries and the improvement of working conditions has a significant influence on productivity and company results [2].

© Springer Nature Switzerland AG 2019
S. Bagnara et al. (Eds.): IEA 2018, AISC 822, pp. 270–280, 2019.
https://doi.org/10.1007/978-3-319-96077-7_28

Therefore, improving workplace safety should not be considered merely an obligation of the law, but also a concrete economic opportunity for the stakeholders.

In particular, Italian *consolidated law on occupational safety* gives a special attention to **information, education** and **training** of the personnel as key factors to prevent injuries, since, regardless of the working context, each worker is responsible not only for his own safety, but also for the safety of his colleagues and, often, of external workers or even his fellow citizens.

The so-called *One Point Lessons* (OPL) about *Standard Operating Procedures* (SOP) are customary used for information purposes.

Frontal lessons are surely useful for the continuous improvement of occupational safety, but they are not always fully effective, especially when they are carried out at the end of a work shift. Moreover, the learning process is quite passive, since workers are asked to read documents or listen to the site safety manager.

Several companies also implement *Safety Talks* and *Safety Walks*. Safety Talks are short briefings used to create awareness among the workers of selected hazards in the workplace, while Safety Walks are periodic inspections around the site to identify safety issues. The latter often gives the workers the opportunity to highlight or disclose some lacks in the safety procedures or in the working environment.

Nevertheless, spreading *safety culture* among the workers is still a challenge. It is well-known that the involvement plays a key role in transmitting information effectively. Digital technologies that allow workers to "live" a simulated risk scenario help increase the concentration and speed up the learning process [3].

In particular, this work explores the application of interactive design techniques [27, 28] combined with *Virtual Ergonomics* (VErg) and the development of a computer-based *Serious Game* (SG) can improve the information flow about occupational safety and thus help prevent accidents and protect workers' health.

Virtual ergonomics uses *Digital Human Models* (DHMs) [4–6] and virtual reality technologies to take into account the human factor since the earliest stages of the design. The use of DHM tools have shown their power not only for product design and manufacturing [7], but also for workplace and process design [18, 26] in order to improve occupational health and safety [8–10].

Serious Games (SGs) have the same structure as videogames but with *an explicit and carefully thought-out educational purpose and are not intended to be played primarily for amusement* [11].

Several examples of SGs for workers training about safety do exist in the published literature, both desktop-based [12] and "immersive" [13–15].

Simulation through virtual reality technologies, in particular head-mounted displays, are more realistic and involving, but this does not imply necessarily a better usability or a higher educational effectiveness. Desktop-based SG are surely less emotionally involving and realistic, but they are more viable in any working context.

For this, the present works focuses on desktop-based applications.

2 Methodological Approach

As mentioned, virtual ergonomics simulations and serious videogames are here proposed as "communication tools" to improve the information flow about the company safety rules. This involve to different aspects of learning process, namely education and training (See Fig. 1).

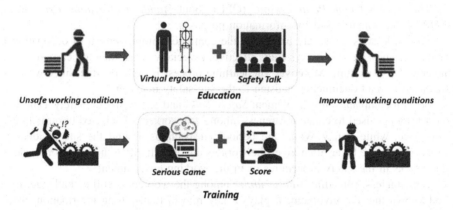

Fig. 1. Methodological approach

3 Virtual Ergonomics for Safety Education

DHM simulations are used for educational purposes. In principle, also real-life videos showing the workers how to correctly perform a certain operation could be used. However, digital simulations have a higher communicative power, since objective ergonomic indices related to correct as well incorrect working postures can be shown in real-time. This makes more awareness among the workers about the importance of observing proper working procedures.

More generally, multimedia digital content stimulates users' attention and increases his emotional involvement. This, in turn, can have a positive impact on the long-term memorisation of the information provided [16].

The software used for DHM simulations is Jack® by Siemens [17]. The anthropometrical characteristics of the digital humans are selected in order to comply with the actual target population of workers.

PEI index is used to evaluate the ergonomics performance (lower the better):

$$PEI = \frac{LBA}{3400N} + \frac{OWAS}{3} + \frac{RULA}{5}$$

where LBA, RULA and OWAS are synthetic ergonomics indices related to *Lower Back Analysis*, *Rapid Upper Limb Assessment* and *Ovako Working Posture Analysing System* [19], respectively.

3.1 Serious Games for Virtual Training

Just like videogames, serious games can convey information effectively, through emotionally-involving interactive tools. SG for virtual training simulates a working scenario in which the user is asked to choose among several options to complete a planned task (e.g. personal protection equipment to wear, tools to use, etc.).

The choices are organized in steps that cover all aspects of the training and are independent of each other: any mistake does not affect the subsequent step, nor implies any change in the options available or in the simulated scenario.

In order not to influence the player's performance during the simulation, the results are shown only upon completion of the SG. Users who have failed or have scored low are invited to repeat the entire simulation.

Usability of the SG [20] shall be particularly taken care of, since a good user experience encourage in the use of this digital tool.

From this point of view, questionnaires are a useful tool to evaluate the usability of the SG [21, 22] and to improve its characteristics.

Unity [23] is the game development platform chosen for the SG, due to its steep learning curve and its multiplatform framework that enable the deployment of the application across desktop, mobile, VR/AR devices, etc.

4 Case Studies

The methodology developed was applied to two industrial case studies provided by *Tower Automotive Italy Srl* [24], located in Pignataro Maggiore (CE), Italy. The trials were carried out in collaboration with *Istituto Nazionale per l'Assicurazione contro gli Infortuni sul Lavoro* (INAIL), which is the Italian public non-profit entity safeguarding workers against physical injuries and occupational diseases [25].

Tower Automotive Italy Srl manufactures steel components for automotive industry: steel sheets are processed in presses and the semi-finished products are subsequently assembled by welding (both automated and manual). The production process involves more than 200 employees, divided into 3 shifts of 8 h each, that work in various departments: sheet metal works, assembly and welding, tooling area, maintenance, warehouse and offices. Among other things, Tower Automotive Italy implements both *Safety Talks* and *Safety Walks* on a weekly basis since many years.

After 30 days of observation of the production process and the study of the risk conditions and near misses recorded in the company database between January and June 2017, two safety-related scenarios were selected.

The first case concerns the manual handling of an industrial cart. The second one concerns a possible dangerous condition during the steel sheet press processing.

4.1 Case 1 – Handling of Industrial Carts

As mentioned, the first case studied concerns the posture to be taken during the handling of industrial carts. In fact, some workers, occasionally, prefer to pull those carts rather than to push them (which would be the correct way according to UNI ISO

11228-2). Since this is an issue that can be solved with a convincing information, for this case we chose to use a virtual ergonomics approach based on digital human models.

The elements of the simulated scenario were modelled with *Catia V5* and *Solid-Works 2016* by *Dassault Systemes* and then imported into *Jack®* by *Siemens*.

Three possible postures have been simulated and analysed.

The first simulation concerns the incorrect posture that most workers often assume when towing the cart (Fig. 2).

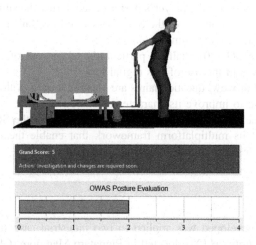

Fig. 2. First scenario: wrong posture for cart pulling

Similarly, the two correct postures were simulated: one for towing, to be used only for short distances when it is not possible to push the cart from a certain location (Fig. 3), and one for pushing (Fig. 4). The digital humans sample considered for the simulations consisted of 6 digital humans, 3 males and 3 females belonging to the 5th, 50th and 95th respectively.

Constant and equal forces were applied to the hands of the manikin to simulate the actual effort in handling the cart. Although the applied load is realistic, the value itself is not particularly important since the objective of the analysis was to compare different postures and not provide an objective reference value.

The LBA, OWAS and RULA [21] indices were evaluated separately and then the PEI synthetic index.

As expected, the results of the ergonomic analyses show that the PEI index do not vary much with sex and anthropometric characteristics, but rather with the scenario and, thus, with the posture considered (Table 1).

As expected, PEI values relating to scenario nr. 2, on average, are about half of those relating to scenario 1 (incorrect posture). However, they are higher than those of scenario 1 (ideal posture). This confirms the posture simulated in scenario nr. 2 is not the best one and must be taken only when it is unavoidable.

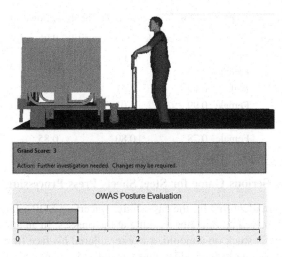

Fig. 3. Second scenario: correct posture for cart pulling

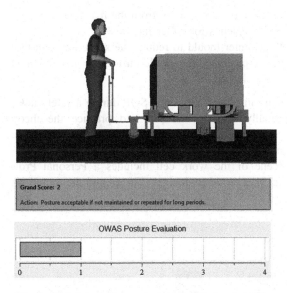

Fig. 4. Third scenario: correct posture for cart pushing

Such results were discussed during a Safety Talk in Tower Automotive, which involved approximately 150 workers, by means of multimedia presentations and virtual simulations. Eventually, a simple oral quiz was administered to the trainees to assess their level of attention.

Table 1. PEI index resulting from ergonomics analysis (lower the better)

Scenario	Sex	5th percentile	50th percentile	95th percentile
1	Male	1.98	2.03	2.12
	Female	2.07	2.11	2.18
2	Male	0.99	1.01	1.05
	Female	0.99	1.01	1.03
3	Male	0.77	0.79	0.84
	Female	0.78	0.80	0.85

4.2 Case 2 – A Serious Game for Steel Sheets Press Processing

The second problem identified for the case study concerns a SOP established following an accident that involved an operator working at the presses. In an attempt to remove a semi-finished product stuck on a mould, a worker injured his face (fortunately, with no major consequences) due to the sudden and unexpected movement of the metal sheet. Therefore, the safety procedure provides:

- Delimitation of a safety zone at 3 m from the machine,
- Closure of the side hatches for collecting waste,
- Lowering of the counter-mould to reduce the space between the two moulds,
- Prying between the piece and the mould through an iron rod at least 2.5 m long till the metal sheet is released

The operators are informed about this SOP during a safety talk.

However, normally nobody can put into practice the theoretical knowledge received, mainly due to the difficulty in arranging such simulated scenario in the real working environment.

The virtual scene of the work cell includes a Personal Protective Equipment (PPE) station (Fig. 5), a crowbars hanger (Fig. 6) and the mould pressing machine.

Fig. 5. Personal protective equipment selection

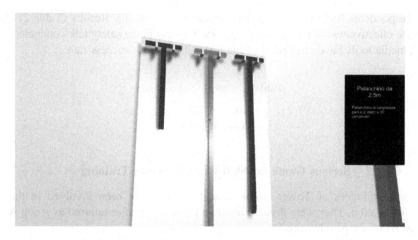

Fig. 6. Crowbars hanger

To pass the test, the user has to:

(a) Take the right PPEs from the table,
(b) Choose a crowbar of adequate length,
(c) Push on the right control buttons to put the press in a safe state

It worth noticing that, concerning step c, the correct sequence of buttons is crucial to achieve the goal.

Most PPEs shown are mandatory to do the task, but some of them are optional or even conflicting. The game is designed in such a way that the player cannot wear more than one PPE that protect the same part of the body at the same time, for example a bump cap and a helmet.

The final score indicates if the player has succeeded or failed, providing information about each choice made. A "game over" was also provided because some elements are mandatory, and their non-use shall lead to failure in any case.

For a better user's experience, the workers involved in the experimentation were briefly trained about the game and their goal. Moreover, a questionnaire was administered to evaluate the usability of the SG. Users could express their satisfaction in a five levels scale (Very unsatisfactory, Unsatisfactory, Indifferent, Satisfactory, Very satisfying).

5   Results

5.1   Case 1 – Virtual Ergonomics Simulations and Safety Talk

To evaluate the efficiency Virtual ergonomics and safety talks in the long term, three months after the meeting, a questionnaire was administered to ten workers. Just five of them had participated in the Safety Talk with Virtual Ergonomics, while the other five had attended the standard Safety Talk on the same topic.

Respondents had to answer a few questions (five in all). Results (Table 2) show that the effectiveness of the methodology. Participants in the safety talk conducted with multimedia tools have responded to the test with a 100% success rate.

Table 2. Case 1 - success rate

Group A	100%
Group B	84%

5.2   Case 2 – Serious Game for Skill Evaluation and Training

Twelve employees of Tower Automotive Italy Srl have been involved in the SG experimentation. They were divided into two groups (hereafter referred as group A and group B) of six people each.

The users of the group A are aged between 32 and 48, all male, including two workers, two lead managers, one team leader and one technologist. The users of the group B are aged between 39 and 48, all male, including three technologists, two workers and one team leader. All respondents have already been trained on this SOP.

In order to verify their skills, Group A was asked to use the SG, while Group B was asked to answer to a written questionnaire, structured in a similar way to the SG. In particular, participants of group A were selected among those who had the same frequency of use of video games (in fact very low) in order not to affect the results to the confidence of some workers with electronic entertainment devices.

As shown in Table 3, the results of group A are rather disappointing. In particular, the overall score was mostly influenced by the mistakes made in the sequence of commands to put the machine in a safe state before the operations.

Table 3. Case 2 – score and success rate

	PPEs selection	Crowbar selection	Buttons and sequence	Overall score
Group A	83%	100%	**33%**	**72%**
Group B	100%	83%	**83%**	**88%**

These results demonstrate that a theoretical knowledge (questionnaires) does not necessarily correspond to a real practical competence (serious game).

However, it is worth emphasizing that the workers had just one possibility to play with the videogame and, as mentioned, the experimentation involved just twelve workers.

6   Conclusions

This work focused on an innovative training methodology that uses Virtual Ergonomics and "serious games" in the field of occupational safety.

On the one hand, Virtual Ergonomics was chosen as an effective mean to convince the workers and disseminate the culture of safety among them. On the other hand, a so-called "serious game" was developed to train the operators on specific safety procedures and get a better feedback on the skills learned.

The results of the experimentation in a real industrial case study showed that, compared to the traditional training methodology, multimedia contents and quantitative ergonomic analyses improve the level of attention and the awareness of the operators about their safety.

With reference to the serious game, the score got by the testers showed that, despite the paper tests confirmed a good theoretical knowledge of the safety procedure considered, most users had difficulties in the practical application of the SOP (albeit in a simulated environment). Problems of usability of the instrument that could have significantly affected the results were excluded thanks to appropriate tests, which confirmed a good users experience. Likewise, disturbing factors, such as the degree of user familiarity with digital interactive tools, have been eliminated thanks to a careful choice of the sample of testers. However, it is worth emphasizing that just twelve workers were involved in the experimentation, future activities will thus involve more testers.

Future works will also concern the experimentation of a digital "hall of fame" to increase the competition among the workers and therefore their involvement.

Moreover, the next release of the serious game will be adapted to a commercial head mounted display for virtual reality. Then, comparative tests between the desktop and the VR-based application will be conducted.

Acknowledgements. The present work is part of the project *IDEE* (Interactive Design for Ergonomics) funded by *DII-Fraunhofer Joint Lab IDeas* and *INAIL - Direzione regionale della Campania*. The authors deeply thank *Tower Automotive Italy* for having provided the case study and all those involved in the experimentations, workers as well as trade union representatives, for their availability.

References

1. Aromaa S, Leino SP, Viitaniemi J (2014) Virtual prototyping in human-machine interaction design. VTT technology: 185. VTT, Espoo, Finland
2. EHS Today. http://www.ehstoday.com/safety-leadership-conference-2015/safety-productivity. Accessed 06 Apr 2018
3. Celentano MG (2014) Interfacce e Sistemi a Realtà Virtuale per un Apprendimento Esperienziale. Ital J Educ Res 4:21–33
4. Duffy VG (2009) Handbook of digital human modeling: research for applied ergonomics and human factors engineering. CRC Press, Boca Raton
5. Nérot A, Skalli W, Wang X (2015) An assessment of the realism of digital human manikins used for simulation in ergonomics. In: Ergonomics, vol. 58. Taylor and Francis, London
6. Pelliccia L, Klimant F, De Santis A, Di Gironimo G, Lanzotti A, Tarallo A, Putz M, Klimant P (2017) Task-based motion control of digital humans for industrial applications. In: 10th CIRP conference on intelligent computation in manufacturing engineering

7. Di Gironimo G, Lanzotti A, Melemez K, Renno F (2012) A top-down approach for virtual re-design and ergonomic optimization of an agricultural tractor's driver cab. In: ASME 11th biennial conference on engineering systems design and analysis, vol 3, pp 801–811

8. Di Gironimo G, Patalano S (2008) Re-design of a railway locomotive in virtual environment for ergonomic requirements. Int J Interact Des Manuf 2:47–57

9. Lanzotti A, Vanacore A, Percuoco C (2017) Robust ergonomic optimization of car packaging in virtual environment. Lecture Notes in Mechanical Engineering, pp 1177–1186

10. Chryssolouris G, Mavrikios D, Papakostas N, Mourtzis D, Michalos G, Georgoulias K (2009) Digital manufacturing: history, perspectives and outlook. Proc Inst Mech Eng Part B J Eng Manuf 223(5):451–462

11. Abt CC (1970) Serious games. Viking Press, New York City

12. Silva F, Almeida JE, Rossetti RJF, Coelho AL (2013) A serious game for evacuation training. In: IEEE 2nd international conference on serious games and applications for health, book of proceedings

13. Williams-Bell FM, Kapralos B, Hogue A, Murphy BA (2014) Using serious games and virtual simulation for training in the fire service: a review. Fire technology. Springer, New York

14. Chittaro L, Buttussi F (2015) Assessing knowledge retention of an immersive serious game vs. a traditional education method in aviation safety. IEEE Trans Vis Comput Graph 4 (4):529–538

15. Sacks R, Perlman A (2013) Construction safety training using immersive virtual reality. Constr Manag Econ 31(9):1005–1007

16. Argenton L, Schek E, Mantovani F (2014) Serious games as positive technologies. In: 6th international conference on virtual, augmented and mixed reality, VAMR. Applications of virtual and augmented reality, vol 8526, pp 169–177

17. Website. https://www.plm.automation.siemens.com/store/it-it/jack/index.html. Accessed 30 Mar 2018

18. Aiello A et al (2015) Finalization of the conceptual design of the auxiliary circuits for the European test blanket systems. Fusion Eng Des 96–97:56–63

19. Di Gironimo G, Mozzillo R, Tarallo A (2013) From virtual reality to web-based multimedia maintenance manuals. Int J Interact Des Manuf 7(3):183–190

20. ISO 9241-11 (2018) Ergonomics of human-system interaction—part 11: usability: definitions and concepts

21. Patalano S, Lanzotti A, Del Giudice DM, Vitolo F, Gerbino S (2017) On the usability assessment of the graphical user interface related to a digital pattern software tool. Int J Interact Des Manuf 11(3):457–469

22. Di Gironimo G, Matrone G, Tarallo A, Trotta M, Lanzotti A (2013) A virtual reality approach for usability assessment: case study on a wheelchair-mounted robot manipulator. Eng Comput 29(3):359–373

23. Website. http://www.unity3d.com. Accessed 21 Sept 2017

24. Website. http://www.towerinternational.com. Accessed 02 Sept 2017

25. Website. http://www.inail.it. Accessed 19 Apr 2018

26. Tarallo A et al (2015) Preliminary piping layout and integration of European test blanket modules subsystems in ITER CVCS area. Fusion Eng Des 93:24–29

27. Di Gironimo G, Marzullo D, Mozzillo R, Tarallo A, Villone F (2017) The DTT device: first wall, vessel and cryostat structures. Fusion Eng Des 122:333–340. https://doi.org/10.1016/j.fusengdes.2017.04.132

28. Labate C, Di Gironimo G, Renno F (2015) Plasma facing components: a conceptual design strategy for the first wall in FAST tokamak. Nucl Fusion 55:113013. https://doi.org/10.1088/0029-5515/55/11/113013

Healthcare in a Virtual Environment: Workload and Simulation Sickness in a 3D CAVE

Peter Hoonakker[✉], Gail Casper, Alex Peer, Catherine Arnott Smith, Ross Tredinnick, Nicole Werner, and Kevin Ponto

University of Wisconsin-Madison, Madison, WI, USA
Peter.Hoonakker@wisc.edu

Abstract. Much of patient care takes places in patients' homes, but we do know very little about how patients deal with their health and chronic illness condition(s) while at home and how the physical environment can have an impact on their care. In this study, we focus on patients' management of their personal health information management (PHIM) in the home.

To enable repeated assessment of a set of constant stimuli, we have scanned 20 different households that we subsequently rendered for viewing in a 3-D virtual cave (VR) CAVE. Study participants identified features in the virtual home models that they considered useful for PHIM.

Using the VR CAVE has many advantages. It enables all participants to experience the same stimulus in precisely the same condition, and it allows for standardization of the study procedures. However, we know relatively little about the impact the VR CAVE experience has on workload and simulation sickness, and if these interfere with task performance. In this study, we examine the relationship between time spent in the CAVE (duration), the number of frames rendered per second (framerate), the experienced workload and simulation sickness symptoms.

Results show that performing tasks in the CAVE required some effort, particularly mental workload. Only a few participants reported minor simulation sickness symptoms, such as dizziness, headache or eyestrain. Apart from a correlation between duration and workload, we did not find a significant relation between exposure, framerates, workload, and simulation sickness.

Keywords: Virtual reality · Personal health information management

1 Introduction

Virtual reality (VR), the computer-generated simulation of a three-dimensional (3D) image or environment that can be interacted with in a seemingly real or physical way, has made large progress in the past two decades. A person's experience in virtual reality is getting better and better, and participants are less distracted by technological shortcomings. Apart from entertainment (3D video games, 3D movies, etc.), virtual reality can for example be used for architecture and design, and education and training, including in healthcare [1, 2].

© Springer Nature Switzerland AG 2019
S. Bagnara et al. (Eds.): IEA 2018, AISC 822, pp. 281–289, 2019.
https://doi.org/10.1007/978-3-319-96077-7_29

However, VR technology still faces a number of challenges [3–7]. Most of these challenges are technical challenges that can have an impact on the end-user in the form of a type of simulation sickness, characterized by for example headache, eyestrain, blurred vision and nausea. When a large part of the area in our visual field is moving while the body is stationary, the illusion of self-motion, or vection, can appear [8]. The mind thinks that the body is moving while it is not. This is one of several causes of simulation sickness. Symptoms of simulation sickness are often the same as those of motion sickness but affect a smaller proportion of the population and the symptoms are in general less severe [9]. Users may become disoriented in a purely virtual environment, causing balance issues; and computer latency might affect the simulation, providing a less-than-satisfactory end-user experience. Several studies have demonstrated the relation between being immersed in virtual reality and simulation sickness [6, 10–13]. Several factors can have an impact on simulation sickness. Some studies have shown that both exposure duration (the length of the time spent in the 3-D cave) and the quality of the images, as for example measured as frame rates per second, are related to simulation sickness [12, 14, 15].

For example, Sharples et al. [12] compared the effects of 4 types of displays: desktop computer screen, head mounted display (HMD), projection screen, and reality theatre on simulation sickness. Results of the study showed that 60-70% of participants experienced an increase in symptoms pre-post exposure from HMD, projection screen and reality theatre as compared to the computer display. Overall, simulation sickness increases with exposure duration [11, 13]. Evidently, if symptoms are moderate or severe, they can have an important impact on performance of participants.

Kolanski et al. [15] examined the impact of image persistence and frame rates on simulation sickness. Persistence refers to the phenomenon that at low frame rates per second (fps) each frame is displayed for an extended period of time. As a result, movement may lose continuity, and the image jumps from one frame to another. Results of their study showed no statistically significant differences in simulation sickness between the three conditions that they tested: low frame rate (11 fps) and low persistence, low frame rates and high persistence, and high frame rates (55 fps).

Navigating through a virtual environment can sometimes be hard work. There is a risk that task load imposed on the user by the technology impedes actual task performance. There is relatively little literature on the workload created by doing tasks in virtual reality. In an early study, Riley and Kaber [16] examined the effect of display type (conventional monitor, large projection screen and virtual reality) on presence and workload. Results show that presence was highest with use of the conventional monitor, possibly because of the higher resolution of the monitor than either projection screen or head-mounted display. Display type did not have an effect on performance or workload as measured with the NASA Task Load Index (TLX).

To summarize, although much progress has been made in applications of virtual reality, virtual reality can have "side effects" that can negatively affect task performance. In this study, we examine the effects of performing tasks in the virtual reality environment. More specifically, we examine the relationships between of exposure time, frame rates, experienced workload and simulation sickness.

2   Methods

2.1   Study Design

In this study, 60 people with diabetes were asked to identify objects in virtual rooms (e.g. kitchen, living room, bedroom, etc.) that could be important for personal health information management (PHIM).

2.2   Setting

The study took place in a cave automatic virtual environment (CAVE, see Fig. 1). The 20 houses in this project were scanned using Light Detection and Ranging (LiDAR) to generate very accurate 3D spatial models of the interior spaces. To display the information, we used a VR CAVE. The VR CAVE visualization runs on a cluster of seven PCs using 12 projectors to project on six sides (four walls, floor and ceiling) at a resolution of 1920 × 1920 pixels per side. The VR CAVE is able to display spaces at a 1:1 scale. It automatically supports stereoscopic rendering to create depth perception and it supports natural exploration of the virtual space by walking within the CAVE confines. For a detailed description of the CAVE that was used in this study, see Brennan et al. [17].

Fig. 1. Participant in cave, selected boxed items (red lines) (Color figure online)

2.3 Sample

Sixty people participated in the study. They varied in age from 20 to 86 years (Mean = 58, SD = 15.7). Twenty-eight participants (47%) were female. Participants were relatively highly educated (1.7% of participants had some high school as highest level of education, 30.0% were high school graduates or equivalent, 38.3% had a Bachelor's degree, and 30% had a graduate degree. Fifteen percent of participants rated their own health as fair, 40% as good, 37% as very good, and 8% as excellent. Most of the participants use a glucometer (85%), 15% use an insulin pump, 68% take oral medications for their diabetes, 33% are on a special diet for managing diabetes and weight loss (55%) as part of their diabetes management. A quart of the participants had previous 3-D virtual reality experience.

2.4 Procedure

Sixty lay participants, who had been told that they had diabetes, were asked to explore five different rooms (e.g. kitchen, bedroom, and bathroom) from different virtual homes that were randomly assigned to them. In those spaces, a limited number of features (e.g. countertop, table) that had been selected in previous sub project of the study were segmented or enabled for selection "boxed".

Using a joystick, participants were asked to select tow features (tag them) and prioritize their usefulness (e.g. kitchen counter, chair in living room) for management of the following three PHIM tasks:

1. You have just developed a rash and are concerned it may be related to a medication you recently started taking. What boxed feature in this room would be most useful to you for *finding out if the rash could be a side effect of this medication*?
2. You are having an outpatient surgical procedure. In preparation for the surgery, you have been scheduled for 5 appointments for preoperative teaching and care at different locations. How would you *set up reminders to help yourself recall* the appointments? What boxed feature of this room would be most useful to you for setting up a reminder for yourself or someone else in your household to help you recall the appointment?
3. Your clinician told you that you should check your blood sugar four times per day and record the values until your next visit. What boxed feature of this room would be most useful to you for *checking your blood sugar and recording the value*?

No participant aborted a session for any reason, including inability to tolerate the simulation.

2.5 Data Collection Instruments

Participants were asked to fill out surveys after the activities. In the surveys, respondents were asked their age, gender, racial background and highest level of education, workload and simulation sickness symptoms.

The questions about personal health information management consisted of the following: Do you take any prescription or non-prescription medications on a regular basis (e.g. daily or weekly) (Yes/No)? Do you use any tools to help you take these medications? Do you monitor any information about your health (e.g. exercise, heart rate, weight, food intake)? Do you store records of your health information (discharge paperwork, prescription information, etc.)? Do you seek out information about your health outside of conversations or materials provided to you by your physician? For example, do you seek out materials about your health from books, pamphlets, Internet sources, etc.?

To measure workload we used the NASA Task Load Index (TLX). The NASA Task Load Index is a multi-dimensional rating procedure that provides an overall workload score based on a weighted average of ratings on six subscales: mental demands, physical demands, temporal demands, own performance, effort, and frustration [18–20]. The six items in the NASA TLX are scored on a scale from 0–100. NASA TLX has been used in many studies and has been proven to be reliable and valid [18–21].

We measured simulation sickness with items from the Simulator Sickness Questionnaire (SSQ) [9]. The full SSQ consists of 16 items that represent three underlying factors: (1) a general discomfort/nausea factor (e.g. general discomfort, nausea, stomach awareness); (2) a disorientation factor (dizziness, vertigo, etc.) and (3) an oculomotor factor (eyestrain, difficulty focusing, etc.) The oculomotor factors is sometimes split in two separate factors: disturbance of visual processing during a simulation (blurred vision, difficulty focusing, etc.) and symptoms caused by the disturbance (headache, eyestrain, fatigue, etc.). In this study, we used the four items that loaded highest on a factor analysis of SSQ data: nausea, dizziness, headache and eyestrain (for example "How much did you suffer from dizziness"?). The SSQ items have the following response categories: 0 (None), 1 (Slight), 2 (Moderate) and 3 (Severe). The SSQ scale was calculated by adding up all 4 items.

2.6 Data Analyses

Overall NASA TLX scores were obtained by calculating the average score of the six NASA TLX items. Descriptive statistics were used to examine the NASA TLX scores of the participants. An overall SSQ score was calculated by adding up the scores of the four SSQ items. The original SSQ manual suggested a weighting procedure for the SSQ items, but more recent publications suggested that using the unweighted items may be just as useful [22, 23]. We calculated descriptive statistics (mean, standard deviation, minimum and maximum) for the variables of interested using SPSS (v22). We also calculated the correlations between the different variables.

3 Results

Ninety-seven percent of the participants take prescription medications to help manage their diabetes. Fifty-three percent use tools (e.g. pill organizer or written schedule) to help them take these medications. Seventy-eight percent monitors information about their health (e.g. exercise, heartrate, weight, etc.). Seventy-seven percent of the participants store their health information (e.g. discharge paperwork, prescription information, etc.). More than two-thirds of the participants (68%), seek information about their health outside of the information provided by their doctor from books, Internet sources, etc. Tables 1 and 2 summarize the data on workload and simulation sickness.

Table 1. Mean score and [standard deviation] on items and scale of NASA TLX

	Mental workload	Physical workload	Temporal workload	Effort	Performance	Frustration	Scale score
NASA TLX	41.4 [27.0]	11.5 [10.6]	7.3 [9.1]	25.8 [23.2]	79.0 [20.2]	10.7 [17.4]	29.3 [10.9]

Table 2. Mean score and [standard deviation] on items and scale of SSQ

	Headache	Eyestrain	Nausea	Dizziness	Scale SSQ
SSQ	0.07 [0.31]	0.15 [0.36]	0.02 [0.13]	0.10 [0.35]	0.33 [0.86]

Results show that workload in the CAVE was relatively low. The overall NASA TLX workload score was 23.9 on a scale from 0–100 (minimum: 5.0, maximum: 55.0). Of the different aspects of workload, mental workload was relatively the highest (M = 41.4, minimum: 0, maximum: 90.0).

Results show that the scores on the SSQ were very low. Mean scale score was 0.33 on a scale from 0–12 (min: 0, max: 5.0). For example, out of 60 participants, only two reported a slight headache and one a moderate headache. The relatively highest score was for eyestrain: nine respondents reported slight eyestrain during the sessions (Table 3).

Table 3. Average time spent in cave (duration) and average framerates

	Mean	SD	Minimum	Maximum
Framerate	15.47	0.47	10.61	16.66
Duration (seconds)	594.2	203.87	278.06	1175.75
Duration (minutes)	9.9	3.4	4.63	19.6

Duration of the sessions in the CAVE varied from 594 s (4.6 min) to 1,175 s (19.6 min). On average, participants spent 594 s (9.9 min) per session in the cave. Framerates varied from 10.6 to 16.7 per second. Average framerate was 15.5 frames per second (Table 4).

Results show that overall workload is significantly related to exposure (time spent in the 3-D cave). Analyses of the relation between framerates, exposure and the individual workload items show that exposure is significantly associated with mental, physical and temporal workload as well as effort, but not with performance or

Table 4. The relationships (correlations) between age, exposure time, framerates, workload and simulation sickness

	Age	Exposure	Framerate	NASA TLX	SSQ
Age	1.00	0.19	0.25	−0.04	0.20
Exposure		1.00	−0.03	0.35**	0.07
Framerate			1.00	−0.01	0.20
NASA TLX				1.00	0.14
SSQ					1.00

**Correlation is significant at the 0.01 level

frustration level. Results do not show any significant relation between exposure, framerates and simulation sickness.

4 Discussion

While virtual reality (VR) applications are getting better and better, there are still many difficulties when using these types of environments. Moving around in a virtual environment may not be as effortless as moving around in the real world, and sometimes the experiences in VR can lead to simulation sickness. Further, because navigating in 3-D is still not effortless, there is a chance that workload interferes with task performance in the 3-D cave.

In this study, we examined the relationship between exposure duration, framerates, experienced workload and simulation sickness. Results show that experienced workload in the 3-D cave is not very high. Mental workload was rated highest by the participants, with an average of 41.4 on a scale from 0–100. Temporal workload (time pressure) received an average score of 7.3 and on average, participants rated their performance as 79.0.

Very few participants experienced symptoms of simulation sickness. Average score on the simulation sickness questionnaire was 0.33 on a scale from 0–12. Practically that means that out of 60 participants, only 4 participants reported a slight dizziness, 1 participant about slight feelings of nausea, nine participants about slight eye-strain, and 3 participants about headaches. Two participants reported a slight headache and one reported a moderate headache. In earlier studies [24–26], one as recent as 2003, about half of participants reported simulation sickness symptoms. In short, virtual reality experiences are getting better, but there are still (slight) side effects.

Average exposure duration in the cave was relatively low. On average, participants spent nearly 10 min per session in the cave. The literature shows that there is a relation between exposure duration and simulation sickness symptoms [26], with participants reporting more (serious) simulation symptoms after longer duration exposure in virtual environment, but in most studies, participants start reporting symptoms after a short (for example 15 min) exposure duration [27].

Framerates were on average 15.4 frames per second, which is not overly high, considering television uses 30 frames per second (FPS), console games 60 fps, and 3-D goggles such as the Oculus propose framerates of 90 frames per second. Studies have shown that low frame rates are associated with more and more severe symptoms of simulation sickness [28, 29]. In this regard, it is interesting to note that in a CAVE environment, the frame rate of 15 seemed to be acceptable. Furthermore, results of a review of several experiments suggested that changes in framerates (accelerations and decelerations) may be more important than the rate per second [27].

4.1 Study Limitations

The sample was relatively small (N = 60). We used mean framerates in the analysis. Framerates can vary substantially during a session. All measurements used in this paper were collected after the session. Exposure duration was relatively short.

4.2 Conclusion

Results of this study show that performing tasks in 3D requires some effort, particularly mental workload. Results of the study also show that a very small number of participants suffered from simulation sickness symptoms, such as dizziness, headache or eyestrain. However, apart from a correlation between duration and workload, we did not find a significant relation between exposure, framerates, workload, and simulation sickness.

References

1. Garrett B, Taverner T, McDade P (2017) Virtual reality as an adjunct home therapy in chronic pain management: an exploratory study. JMIR Med Inf 5:e11
2. Werner NE, Carayon P, Casper GR, Hoonakker PLT, Arnott SC, Brennan PF (2016) Affordances of household features important for personal health information management: designing consumer health information technology for the home. In: Healthcare systems ergonomics and patient safety conference. IEA Press, pp 390–394
3. Hansen MM (2008) Versatile, immersive, creative and dynamic virtual 3-D healthcare learning environments: a review of the literature. J Med Internet Res 10:e26
4. Stanney KM, Mourant RR, Kennedy RS (1998) Human factors issues in virtual environments: a review of the literature. Presence Teleoper Virtual Environ 7:327–351
5. Stanney KM, Cohn JV (2009) Virtual environments. In: Sears A, Jacko J (eds) Human computer interaction: design issues, solutions, and applications. CRC Press, Boca Raton
6. Naqvi SA, Badruddin N, Malik AS, Hazabbah W, Abdullah B (2013) Does 3D produce more symptoms of visually induced motion sickness? In: Proceedings of the 20th annual international conference of the IEEE engineering in medicine and biology society, pp 6405–6408
7. Nichols S, Patel H (2002) Health and safety implications of virtual reality: a review of empirical evidence. Appl Ergon 33:251–271
8. Arthur K (1996) Effects of field of view on task performance with head-mounted displays. In: Conference companion on human factors in computing systems, Vancouver, British Columbia. ACM, pp 29–30
9. Kennedy RS, Lane NE, Berbaum KS, Lilienthal MG (1993) Simulator sickness questionnaire: an enhanced method for quantifying simulator sickness. Int J Aviat Psychol 3:203–220

10. Bruck S, Watters PA Estimating cybersickness of simulated motion using the simulator sickness questionnaire (SSQ): a controlled study. In: 2009 sixth international conference on computer graphics, imaging and visualization, pp 486–488
11. Kennedy RS, Drexler J, Kennedy RC (2010) Research in visually induced motion sickness. Appl Ergon 41:494–503
12. Sharples S, Cobb S, Moody A, Wilson JR (2008) Virtual reality induced symptoms and effects (VRISE): comparison of head mounted display (HMD), desktop and projection display systems. Displays 29:58–69
13. Rebenitsch L, Owen C (2016) Review on cybersickness in applications and visual displays. Virtual Reality 20:101–125
14. Zielinski DJ, Rao HM, Sommer MA, Kopper R (2005) Exploring the effects of image persistence in low frame rate virtual environments. In: 2015 IEEE virtual reality (VR), pp 19–26
15. Kolasinski EM (1995) Simulator sickness in virtual environments. Defense Technical Information Center (DTIC)
16. Riley JM, Kaber DB (1999) The effects of visual display type and navigational aid on performance, presence, and workload in virtual reality training of telerover navigation. Proc Hum Factors Ergon Soc Ann Meet 43:1251–1255
17. Brennan PF, Ponto K, Casper G, Tredinnick R, Broecker M (2015) Virtualizing living and working spaces: proof of concept for a biomedical space-replication methodology. J Biomed Inform 57:53–61
18. Hart SG (2006) NASA-Task Load Index (NASA-TLX): 20 years later. In: 2006 human factors and ergonomics society (HFES) conference. HFES, pp 904–908
19. Hart SG, Staveland LE (1988) Development of NASA-TLX (Task Load Index): results of empirical and theoretical research. In: Hancock PA, Meshkati N (eds) Human mental workload. North Holland Press, Amsterdam, pp 239–250
20. Human Performance Research Group (1997) NASA Task Load Index (TLX). NASA Ames Research Center
21. Hoonakker PLT, Carayon P, Gurses A, Brown R, McGuire K, Khunlertkit A, Walker J (2011) Using the NASA Task Load Index (TLX) to measure workload of ICU nurses. J Healthc Eng 1:131–143
22. Bouchard S, Robillard RP (2007) Revising the factor structure of the simulator sickness questionnaire. Ann Rev CyberTherapy Telemed 5:117–122
23. Bouchard S, St Jacques J, Renaud P, Wiederhold BK (2009) Side effects of immersions in virtual reality for people suffering from anxiety disorders. J CyberTherapy Rehabil 2:127–137
24. Regan EC (1993) Side-effects of immersion virtual reality. In: International applied military psychology symposium, Cambridge, UK
25. Kennedy RS, Allgood GO, Van Hoy BW, Lilienthal MG (1987) Motion sickness symptoms and postural changes following flights in motion-based flight trainers. J Low Freq Noise Vib 6:147–154
26. Stanney KM, Kingdon KS, Nahmens I, Kennedy RS (2003) What to expect from immersive virtual environment exposure: influences of age, gender, body mass index, and past experience. Hum Factors Ergon Manuf Serv Ind 45:504–520
27. Cobb SVG, Nichols S, Ramsey A, Wilson JR (1999) Virtual reality-induced symptoms and effects (VRISE). Presence Teleoper Virtual Environ 8:169–186
28. Regan EC (1995) An investigation into nausea and other side-effects of head-coupled immersive virtual reality. Virtual Reality 1:17–32
29. Regan EC, Price K (1994) The frequency of occurence and severity of side-effects of immersion virtual reality. Aviat Psychol Environ Med 65:527–530

Does Preferred Seat Pan Inclination Minimize Shear Force?

Xuguang Wang$^{(\boxtimes)}$ (iD), Michelle Cardosso(iD), Ilias Theodorakos(iD),
and Georges Beurier

Univ Lyon, Université Claude Bernard Lyon 1, IFSTTAR, LBMC UMR_T9406,
69675 Lyon, France
xuguang.wang@ifsttar.fr

Abstract. Past biomechanical studies on seated postures showed that effects of seat parameters, such as seat pan angle, back angle and friction coefficient, on muscle activities, shear force between buttocks and seat and spinal loads are complex. Reducing all these biomechanical loads at the same time may not be possible. Lowered muscle activation may require higher frictional shear force. It is interesting to investigate how people behave compared to biomechanical simulations. In this paper, the question whether sitters prefer a seat pan angle for reducing shear force was investigated using the data collected from a multi-adjustable experimental seat. Two imposed seat pan angles (A_SP = 0°, 5°) and one self-selected were tested for two backrest angles (A_SB = 10°, 20°, from the vertical). A flat seat pan surface was used. Other seat parameters such as seat height, length and position of three back supports were defined with respect to each participant's anthropometry. As expected, results showed that shear force increased with backrest recline and decreased with seat pan recline. No significant difference in self-selected seat pan angle was found between two backrest angles. An average of 6.2° (±3°) was observed. The lowest shear was observed for the condition of self-selected seat pan angle, supporting the idea that seat pan should be oriented to minimize shear force. However, self-selected angle did not completely remove the shear. A zero shear would require a more reduced trunk-thigh angle, suggesting a minimum trunk-thigh angle should also be maintained.

Keywords: Seating · Biomechanics · Shear force · Discomfort
Airplane passenger

1 Introduction

Reed et al. suggested both pressure and shear force on the seat contact surface affect sitting discomfort [4]. It is generally recommended that peak pressure should be reduced and located at the area of the ischial tuberosities. Though large differences in pressure distribution and sensitivity among individuals make specifying a quantitative "optimal" pressure distribution difficult, Mergl et al. tended to determine ideal distribution. It is also recommended that surface shear on the seat cushion should be minimized by changing the cushion angle and/or contouring the cushion [3]. Goossens and Snijders theoretically and experimentally investigated the relationship between seat and

© Springer Nature Switzerland AG 2019
S. Bagnara et al. (Eds.): IEA 2018, AISC 822, pp. 290–295, 2019.
https://doi.org/10.1007/978-3-319-96077-7_30

backrest inclinations for removing shear force [2]. They found that a fixed inclination between seat and backrest should be chosen between 90° and 95°. Rasmussen and his colleagues developed a computational musculoskeletal seated human model and used it for investigating the effects of backrest inclination, seat pan inclination and seat surface friction coefficient on muscle activity, shear force in the buttocks and spinal loads by simulation [5–7]. It was shown that forward seat-pan inclination (up to about 15°) reduces the spinal-joint L4–L5 compression force for a fixed backrest inclination of 10° backward [7]. Muscle relaxation and shear force reduction may be conflicting [5]. A linear relationship between seat pan and backrest angles was found when both shear force and muscle activity were minimized [6]. An angle of less than 90° between backrest and seat pan was found without considering spinal load reduction. This seems quite unrealistic when comparing with existing aircraft or office seats which have a more opened angle between backrest and seat pan. By simulation, Rasmussen and his colleagues showed that the effects of seat parameters (seat pan and backrest angles, coefficient of friction) on muscle activities, shear force and spinal loads are complex. To date, few researchers have performed a parametric study of seat parameters on sitting discomfort. More specifically, few have verified experimentally whether people prefer a seat pan angle minimizing the shear force as it may be conflicting with other criterions such as reducing muscle activities and spinal joint load. This was one of the research questions we investigated experimentally using a newly built multi-adjustable experimental seat [1] in a research program for improving the comfort of economy class aircraft seats.

2 Materials and Methods

2.1 Participants

Thirty-six participants (18 males, 18 females), aged from to 19 to 56 years old, were recruited based on their body mass index (BMI) (healthy 18.5–25, obese > 30) and stature (small, medium and tall). Three stature groups were formed: 154–157 cm, 162–166 cm and 170–175 cm for females; 168–171 cm, 176–180 cm and 185–190 cm for males. A total of 12 groups were formed after considering sex, stature and BMI (3 individuals per group). Prior to the experiment, participants were screened using a health questionnaire. Participants who experienced any back injury or pain in the previous 3-months were excluded. The experimental protocol was approved by IFSTTAR (French Institute of Science and Technology for Transport, Development and Networks) ethics committee and informed consent was given prior to experiment.

2.2 Experimental Seat

The data was collected through use of the multi-adjustable experimental seat recently developed at IFSTTAR. The experimental seat had thirteen adjustable parameters directly controlled by a computer. Adjustable features included: fore-aft (x) and vertical position (z) of the foot support, seat pan and three back supports; rotational angle of the seat pan, backrest and global inclination of the whole experimental seat. Two armrests were also available and adjusted manually. Force sensors were mounted to measure contact forces in xz plane on the foot support, seat pan, three back supports and two armrests. The seat pan surface was composed of a matrix of 52 cylinders, each with a freely rotatable circular flat head of 60 mm in diameter. Each cylinder was equipped with a tri-axial force sensor, enabling the measurement of both normal and tangential forces. The height of each cylinder was adjustable with a maximum stroke length of 40 mm and pressure distribution could be controlled by changing seat surface. Pressure distribution on the seat pan surface was controlled using a uniform coupling law relating normal force and position for each cylinder. The coupling law enabled us to distribute normal contact force as uniformly as possible among the 52 cylinders (given the maximum displacement of the cylinders). A more detailed description of the experimental seat can be found in [1].

2.3   Experimental Conditions and Procedure

Participants were instructed to test a total of 40 seat configurations that simulated an economy class airplane seat. The H-point location of an existing airplane seat was used to define the x position of the middle back support. Two backrest angles (A_SB) from the vertical (10°, 20°) and three seat pan angles (A_SP = 0°, 5°, preferred) were used to define 6 different A_SP/A_SB combinations covering the range of variation of airplane eco-class passenger seats. For each combination, five conditions were tested successively. The first one, called 'reference position' with a flat seat pan surface, was used to determine seat pan length, foot support height and armrests position for each participant. The three backrest panels were positioned at specific anatomical points (occipital bone, T9 and L3). Their position in x was fixed at 135 mm in the seat back LCS. The seat pan length (X_SP_L, Fig. 1a) was fixed until there was approximately 70 mm (hand width) between the popliteal (behind the knee) and the front of the seat pan. Participants were asked to keep their back in contact with the lower and middle supports. The foot support was adjusted (Z_FS, Fig. 1a) until the knees were set at approximately 90 degrees. Participants were also asked to place rectangular foam of 100 mm (in thickness) between the knees to reduce postural variation (Fig. 1b). The armrests were self-positioned by subjects. Once participants were fitted to the seat, they were instructed to step off the experimental seat to zero all the force sensors. Then, they were asked to reposition themselves back on the experimental seat and look forward without use of the upper support. Measurements were recorded at a rate of 20 Hz for 1.25 s. The preferred seat pan angle was self-selected by participants for the reference position and kept unchanged for the four other test conditions with a fixed backrest angle. Four others conditions were aimed to study the effects of posture (relaxing with use of head support, looking forward without using head support) and pressure distribution on seat. In this study, only data from the reference position were used.

2.4   Data Processing and Analysis

The medians of the measurements of each trial were calculated at first. Inconsistent trials due to either measurement or manipulation errors were eliminated at first.

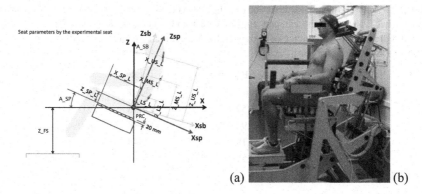

Fig. 1. Multi-adjustable seat. Definition of main adjustable seat parameters (a) and illustration of a participant sitting on the experimental seat (b)

Then, multifactor ANOVAs were performed using STATGRAPHICS Centurion XVII. Effects of independent variables were considered 'significant' when $p < 0.05$.

3   Results

To answer the research question whether participants preferred a seat pan angle which reduces shear force, the self-selected seat pan angle and global shear force applied on the seat pan surface were analyzed and results are shown in Tables 1 and 2. Preferred seat pan angles (A_SP) were 5.74° and 6.67° respectively for a seat back angle (A_SB) of 10° and 20°, showing that a slightly more reclined seat pan was selected for a more reclined backrest though no significant difference was observed between two backrest angles. The shear forces corresponding to the preferred seat pan angle are compared with two other imposed seat pan angles in Table 2 for two back angles. Both A_SP and A_SB significantly affected the shear force, which increased when increasing A_SB and lowering A_SP. As expected, the lowest shear force was observed for self-selected seat pan angle. The shear forces corresponding to preferred seat pan angle were respectively 4.61% and 6.19% of the body weight.

Table 1. Self-selected seat pan angle in degrees for seat back angle (A_SB) being fixed at 10° and 20° from the vertical backward

A_SB	Count	Average	SD	Minimum	Maximum	Range
10°	34	5.74	3.05	0.26	14.02	13.76
20°	36	6.67	3.01	0.45	13.84	13.39
Total	70	6.22	3.05	0.26	14.02	13.76

Table 2. Means and standard deviations of the normalized shear force applied at the seat pan surface when varying both seat pan angle (A_SP) and seat back angle (A_SB). PR is self-selected. Forces are in percentage of body weight. A negative force means the shear force applied on the seat is directed forward.

A_SP	A_SB = 10°		A_SB = 20°		Total	
	Mean	SD	Mean	SD	Mean	SD
0°	−8.56	1.78	−10.57	1.79	−9.58	2.05
5°	−4.73	2.17	−6.92	1.93	−6.38	2.20
PR	−4.61	2.97	−6.19	3.01	−5.42	3.07
Total	−5.97	2.98	−7.51	2.67	−6.94	2.88$^{A\_SB***,A\_SP***}$

*P < 0.05, **P < 0.01, ***P < 0.001

4 Discussion and Conclusion

In this paper, the question whether sitters prefer a seat pan angle for reducing shear force was investigated using the data collected from a multi-adjustable experimental seat. As expected, lower shear force at the buttock seat contact surface was observed when participants were allowed to self-select their preferred seat pan angle. Compared to a horizontal seat pan, shear force was reduced from 8.56% to 4.61% and from 10.57% to 6.19% of body weight on average respectively for a seat back angle of 10° and 20°. Our observations support the general idea that a reduced shear force is preferred.

Preferred seat pan angles were about 6° and 7° for a back angle of 10° and 20°. Compared to the recommendations by Rasmussen et al. [6] from simulations for minimizing both muscle activities and shear force, self-selected seat pan angles observed in the present study were much less reclined (backward), rather in agreement with the requirement of a minimum angle of 105° between seat pan and back (trunk-thigh angle) suggested by Keegan in 1953 [8]. Keegan considered that a good lumbar curvature could not be preserved with a trunk-thigh angle smaller than 105°. To fully remove shear force, a more reclined seat pan would be required with a more reduced trunk-thigh angle, as already predicted and observed by Goossens and Snijders [2]. This was not preferred by participants.

In the present study, a multi-adjustable experimental seat was used. Except for seat pan angle, other seat parameters such as seat height, seat pan length, position of the lumbar and thoracic supports were defined with respect to each participant's anthropometry. Thus their possible interactions with seat pan inclination were eliminated. For example, when changing seat pan angle, seat height was also adapted to avoid high compression of the posterior thigh. Results obtained in the present study can hardly be obtained without using a multi-adjustable experimental seat. The main limitation is also related to the use of an experimental seat. The preferred seat pan angle was obtained with a flat seat surface without any soft cushion. The backrest was simplified by three supports. A same initial seat pan angle was not imposed before asking participants to select their preferred seat pan angle. This may explain, at least partly, the large variation of self-selected angles. In a follow-up study [9], it was observed that preferred seat pan angle depended on initial seat pan inclination. An initially small seat pan angle would lead to a less reclined self-selected seat angle, suggesting that a range of seat pan orientation would be preferred.

In conclusion, sitters preferred a seat pan inclination with a reduced shear force, while a minimum angle of trunk thigh angle should be kept as suggested by Keagan [8].

Acknowledgement. The work is partly supported by Direction Générale de l'Aviation Civile (project n°2014 930818).

References

1. Beurier G, Cardoso M, Wang X (2017) A new multi-adjustable experimental seat for investigating biomechanical factors of sitting discomfort. SAE technical paper 2017-01-1393, https://doi.org/10.4271/2017-01-1393
2. Goossens RHM, Snijders CJ (1995) Design criteria for the reduction of shear forces in beds and seats. J Biomech 28(2):225–230
3. Mergl C, Klendauer M, Mangen C, Bubb H (2005) Predicting long term riding comfort in cars by contact forces between human and seat. SAE technical paper N° 2005-01-2690
4. Reed MP (2000) Survey of auto seat design recommendations for improved comfort. Michigan Transportation Research Institute (UMTRI), Michigan
5. Rasmussen J, de Zee M, Tørholm S (2007) Muscle relaxation and shear force reduction may be conflicting: a computational model of seating. SAE technical paper 2007-01-2456. https://doi.org/10.4271/2007-01-2456
6. Rasmussen J, de Zee M (2008) Design optimization of airline seats. SAE technical paper 2008-01-1863
7. Rasmussen J, Tørholm S, de Zee M (2009) Computational analysis of the influence of seat pan inclination and friction on muscle activity and spinal joint forces. Int J Ind Ergon 39:52–57
8. Keegan JJ (1953) Alterations of the lumbar curve related to posture and seating. J Bone Joint Surg 35A:589–603
9. Theodorakos I, Savonnet L, Beurier G, Wang X (2018) Can computationally predicted internal loads be used to assess sitting discomfort? Preliminary results. In: IEA 2018, Florence, Italy

Application of Virtual Reality to Improve Physical Ergonomics in a Control Room of a Chemical Industry

Mohsen Zare$^{(\boxtimes)}$ ⓘ, Maxime Larique, Sébastien Chevriau, and Jean-Claude Sagot

ERCOS Group (pôle), Laboratory of ELLIADD-EA4661,
UTBM-University of Bourgogne Franche-Comté, Montbéliard, France
Mohsen.zare@utbm.fr

Abstract. The control room is a critical workplace in a chemical industry or a nuclear power plant, where integrating human factors/ergonomics (HFE) could prevent accidents. Non-ergonomic situations in a control room have an adverse impact on operators, for example, cause anxiety, stress, and fatigue. Physical ergonomics such as computer workstations, sedentary postures, lighting, annoying noise, and communication difficulties are the common ergonomic factors in most of the control rooms. This study aims to use immersive virtual reality to improve physical ergonomics in a control room.

The study includes HFE evaluation, proposing a new design to improve work situations and modelization of the proposed concept in a virtual reality setting. We firstly performed various ergonomic analyses such as the interview with operators, activity analysis, lighting and noise measurement. Then, a meeting was organized to present the results of HFE evaluation for decision makers. A creative session was arranged with five experts to develop the idea for a new design in the control room by considering HFE principles. The control room workplace was visualized in a virtual reality setting to make a new concept more visible and tangible. Three operators of the control room were immersed in the virtual environment to verify the proposed configuration. The proposed design was modified based on the operator's feedback. This study showed that a comprehensive approach including ergonomic evaluation, virtual modelization, and stakeholder involvements provide a substantial improvement in a control room.

Keywords: Human factor · Ergonomics · Virtual reality · Control room

1 Introduction

The control room in a chemical industry or a nuclear power plant is a complex system, which an operator dynamically interface and interact with the machine to perform work routines [1–4]. The operators performed various critical tasks such as monitoring the process, controlling the system and identifying possible hazards or malfunctions that threaten the system safety. The purpose of their functions is to operate the system correctly and to prevent accidents [2, 4]. The control room design has to provide

© Springer Nature Switzerland AG 2019
S. Bagnara et al. (Eds.): IEA 2018, AISC 822, pp. 296–308, 2019.
https://doi.org/10.1007/978-3-319-96077-7_31

sufficient comfort and satisfaction for the operators to avoid human errors and accidents. The review of the past accidents in the chemical and nuclear industries such as Three Mile Island showed that ignoring the Human Factor/Ergonomics (HFE) principles were a reason of tension on operators, and accidents [2]. HFE in a control room aims to adapt system, layout, and equipment with the limitations and capabilities of human- designing a system compatible with the physical and cognitive characteristics of the operators [3]. The guidelines and standards (ISO 11064) for HFE issues in the control room can assist to identify the critical elements for safe operation and to design the system interface adapted to the operators [1, 2]. The HFE guidelines address the broad recommendations for design (such as human-system interface, workstation, and workplace design) [2, 5]. However, few published studies are available that evaluate the HFE principles and implement a new concept based on the HFE guidelines, mainly by using virtual reality (VR) environment. VR is a simulation method, which provides a virtual setting for a proposed design and illustrates a workplace, functions, and tasks. HFE engineers can visualize the workstations, for example, the control room workplace, in a VR and this tool allows verifying the HFE elements.

In a comparative study, Dos Santos et al. showed that the HFE evaluation of a nuclear control room had a good agreement in a real and virtual setting. This study concluded that visualization tool is reliable for verifying HFE elements in the design phase [2]. In another study, Aromaa et al. showed that VR is a useful tool for the design process [6]. Although several studies demonstrated the usefulness of VR for workplace design, few studies visualized a control room workplace and immersed the operators in this virtual setting to redesign the control room based on HFE elements. This study aimed to evaluate HFE factors in a control room and to develop a new design in a VR. We then immersed the operators in the virtual settings and proposed the final design based on the user feedback.

2 Materials and Methods

2.1 Context of the Study

This study was performed in a control room of chemical industry. Two teams shared this control room, and they processed the similar operation. Each side of the control room was dedicated to a team. Thirty operators rotated in three shift works (two shifts mornings, then two afternoons, and two nights) altering in different workstations inside and outside of the control room. This study focused on the inside of the control room. Eight similar workstations existed in the control room - four parallel workstations (one chief desk and three pilot operator's workstations) on each side. Each workstation was composed of several monitors arranged on two superposed rows, keyboards, mice, radio communications, and phones. Main tasks for the pilot operators were monitoring curves and values, managing their dedicated subsystem, communicating with external operators, dealing with alarms and incidents. Main duties for the chief were supervising the systems and the pilot operators, coordinating with the other team and supervising the works of subcontractors. Figure 1 shows the control room general layout.

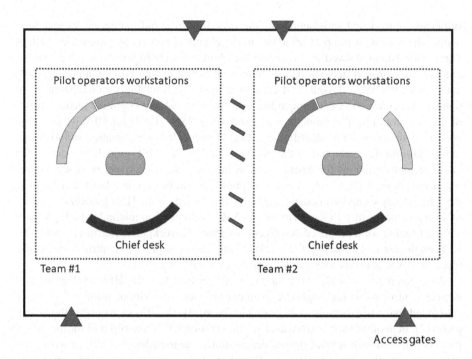

Fig. 1. The general layout of the workstations in the control room

Communication Means. A radio frequency was dedicated to each side of the control room, which was shared between all its operators. Two microphones, speakers, and a panel of buttons/lights allows communication with external stakeholders. Various alerts also existed that aimed to alert on the threshold limit values, incidents, and general evacuation alert.

Lighting System. The lighting system consisted of three types of luminaires: recessed round lighting (compact fluorescents bulbs), special ceiling lights (to provide a natural light), and spotlights (halogen bulb). The ceiling lights were supposed to produce a "natural" light, close to daylight (quality, intensity and color temperature).

2.2 Data Collection

Human Factor/Ergonomic Analysis. The study intended to establish an HFE evaluation to identify risk factors and to improve the future working conditions. We used the following tools to achieve to measure HFE elements:

- Preliminary ergonomic checklist or APACT (Agence Pour l'Amélioration des Conditions de Travail) checklist contains 22 criteria such as work posture, lighting environments, mental workload, and social environment. Handling criterion, for example, is evaluated by four parameters: awkward posture, frequency/weight, the quality of gripping and the distance. Each criterion gives a rating to represent the final score of evaluation.

- The dosimeters were placed on two pilot operators (one on the morning shiftwork and one on the afternoon shiftwork) to assess the level of noise exposure of each operator. A sonometer (model SL-451) was used to characterize the sound environment. Two sonometers were placed in the middle of each side of the control room near to the workstations.
- The pilot operators and the chief were interviewed about their activity, working conditions, the problems encountered and the possible solutions.
- Video recording was performed on two workstations (chief position and pilot operator position) to carry out the activity analysis by using the Kronos software.
- Illumination and luminance measurements were performed by lux meter and luminance meter to characterize the lighting environment.

2.3 Design

Based on the results of HFE assessment, we organized a creative session, which five experts (3 mechanical engineering, one ergonomist, and one designer) from outside the research group participated in this session. This session produced the principle solutions to improve HFE in the control room. We then proposed a new design including layout modification, modified lighting system, and noise reduction remedies.

The new design was simulated on a 3D modeling to validate the technical principles. The modeled design was then immersed in the virtual reality CAVE, and the control room staff (including three managers, two supervisors and two pilot operators) judged in an immersive experience how the proposed design could improve HFE condition. The final user feedbacks on the virtual reality CAVE were registered and used to modify the proposed design. Three mechanical engineers and one ergonomist followed and registered the feedback of the users.

3 Results

3.1 Ergonomic Analysis

Preliminary Ergonomic Analysis. This initial evaluation by APACT checklist identified seven principal risk factors (among the 22 criteria of the APACT checklist; Fig. 2). Two risk factors were inherent to the activity of control room, but we focused on the other five risk factors, in which improvements would be possible. Five HFE problems identified are as follows:

- Signals and information, because of the disturbing by the radio communications, alarms, and warning messages on the screens;
- Organization of the workstations and low autonomy of the operators because the workstation under study was dependent on the others;
- The mental requirement of the tasks was high;
- Sound environment reduced the level of concentration required, particularly in a crisis;

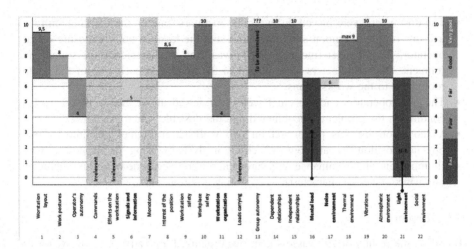

Fig. 2. Preliminary ergonomic evaluation by APACT checklist (Agence pour l'Amélioration des Conditions de Travail: APACT)

- Illumination level was weak, because of the low level of illumination and dazzling, poor contrasts on the screens, and the absence of natural light;

We only studied physical ergonomics elements founded by preliminary checklist as a critic and excluded the other aspects of HFE such as mental workload for this study.

Activity Analysis. Video recordings made over a half-hour on the workstations of the pilot operator and the chief to analyze their functions. The temporal distribution of different activities observed by a chronology software (Kronos) showed that more than 50% of looking direction of both pilot operator and the chief was at the screens of their workstations. The chief frequently moved to assist pilots and to organize outside work, and he monitored various screens of his workstation. However, he also spent 40% of the time looking at colleagues, twice as much as the pilot, which reflected the high need for oral communication- because of his supervisory task (Table 1).

The hands of pilot operator remained idle for 3/4 of the time analyzed (77%), which indicated the predominance of monitoring task. The chief used more frequently the keyboard because it proceeds various tasks on the computer (Table 1). The layout of the workstations must leave enough space for the use of documents by minimizing interference with keyboards and mouse.

The pilot operator communicated with his colleagues for more than half the time analyzed (38% oral and 17% radio communications). The chief communicated orally with his colleagues for 66.5% of the time, which showed the importance of his coordination task. No radio or telephone communication observed in the functions of the chief (Table 1).

The pilot operator and chief remained most of the time seated in front of various screens of the workstations. However, the chief moved more than pilot operator (moving on his chair through the workstation, communicating with his colleagues, and

Table 1. Activity analysis of the workstations of the pilot operator and the chief in the control room, 30 min video-recorded analyzed by chronology software (Kronos).

Activity	Subjects	Human factors Parameters analyzed				
		Monitor	Colleagues	Docu-ments	Others	
Looking direc-tion	Pilot oper-ator	55.3	19.6	11.8	13.4	
	Chief	52.2	39.7	2.5	5.6	
		Nothing	Mouse	Keyword	Pen	Others
Handling tools or objects	Pilot oper-ator	76.6	4.3	2.5	0.6	16
	Chief	57.3	6.2	32.8	2.5	1.2
		Nothing	Telephone	Oral	Radio	Others
Communication method	Pilot oper-ator	43.6	1.3	37.7	17.3	0.1
	Chief	33.5	0	66.5	0	0
		Sitting	Moving on the chair	Moving	Standing	Other
Main posture	Pilot oper-ator	99.5	0.3	0.2	0	0
	Chief	96.1	2.2	0.7	1	0

walking to meet outside staff; Table 1). Both operators spent significant time in a sitting position, which represented an increased risk of low back pain.

Lighting Measurement. The general level of illumination seems subjectively very low. A preliminary measurement without considering the possibilities of adjusting the level of lighting showed: the general lighting of the workstations was between 84 and 280 lx (the average of 158 lx), and the minimal illumination was 17 lx in the corner of the control room. This measurement highlighted the heterogeneity of the level of illumination on the workstations. We establish an illumination map by performing the second measurements, in various positions more precisely (Fig. 3). The analysis of the illumination map confirmed the heterogeneity of the illumination (between 68 and 344 lx measured on the workspaces; average = 146 lx). This measurement showed general low lighting on most of the workstations.

The luminance measurement highlighted the high luminance of the recessed lighting (13890–15320 cd/m^2). This high luminance was a source of glare on the screens. The glare of these lightings was even more annoying when the luminaires were directly located in the field of view of the pilot operators. On the other hand, the ceiling lights supposed to reproduce the natural light of the day (although some operators complained when they were in their field of view) did not present excessive luminance (less than 500 cd/m^2).

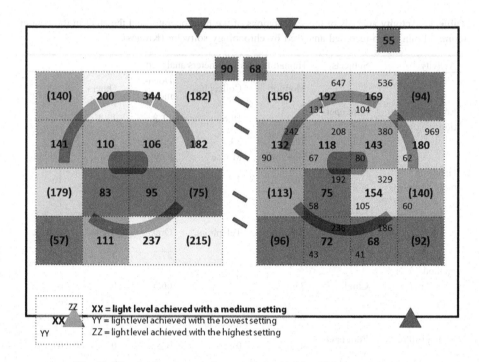

Fig. 3. Illumination map performed for both sides of the control room

Sound Measurement. The level of sound was very variable. The sound level quickly changed from quiet to very noisy due to many passages or simultaneous conversation in the control room. Radio communications with external operators, repeatedly, was another source of the noisy situation.

Dosimetric Measurements. Table 2 shows the results of dosimetry for two pilot operators during the morning and the afternoon. The results show the dose of noise exposure was higher in the morning (1.3%) than in the afternoon. The reason might be that the morning measurement included the time of changing shiftwork (at 12:00), which was noisy due to many passages and conversation.

Table 2. Dosimetric measurements performed for two pilot operators worked in two sides of the control room in the morning and the afternoon

Workstation	Period of the day	Measurement duration	LAeq	LEX, 8 h	Dose
Right side (HCN)	Afternoon	05:09:09	63.1 dB(A)	61.2 dB(A)	0.4%
Left side (AND)	Morning	00:52:46	76 dB(A)	66.4 dB(A)	1.3%

Sound Level Measurement. The logarithmic average of the equivalent continuous sound level (LAeq) was 64.65 dB(A) for the right side and 62.9 dB(A) for the left side. We measured both sides of the control room on the afternoon, and the duration of measurement was 455 min. Figure 3 shows the distribution of LAeq over the length of measurement. More than 5% of measurement was between 55 and 60 dB(A) at both sides of the control room. The sound level was below the limit of 85 dB(A), and it does not seem to endanger the hearing of the operators. Nevertheless, given the level of concentration required by the tasks, the recommendations of the French National Research and Safety Institute for the Prevention of Occupational Accidents and Diseases (INRS) is 55 dBA that was considered as the reference for comparison. Thus, the level of sound was higher than the standard for 5% of measurement. Frequency analysis also showed that the noise with the frequencies between 400 and 2000 Hz exceeded the INRS recommendation (55 dBA), and it would be annoying for performing the control room tasks, which require concentration (Fig. 4).

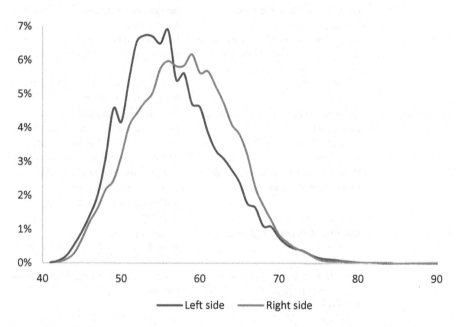

Fig. 4. The distribution of the equivalent continuous sound level (LAeq) measured over 455 min in both sides of the control room

3.2   Redesign of the Control Room

Creativity Session. After finding the main ergonomic problems, we organized a creative meeting to generate ideas for solutions. Five experts (Three mechanical engineering, one ergonomists, and one designer) participated in this session. After a presentation of the control room and the problems identified, each participant proposed solutions, display them on a board and discuss/challenge with the other participants. Table 3 shows the main proposed solutions, which are classified into six themes.

Table 3. The solutions generated by experts in the creativity sessions to improve human factors in the control room (* highlighted realistic solutions).

Ergonomic problems	Solutions proposed	
Noise solutions	**Noise reduction at the source:** Localize audible alarms Localize sound bells (for alarms and radio) Use vibrations, and silent keyboards Provide a sound meter showing the real-time sound level*	**Noise reduction in transmission:** Cover floor by absorbing materials* Reduce/remove the notice boards on the wall (reduce the absorption of perforated wall panels)* Provide moveable acoustic partition between the left and right sides of the room
Means of communication	**Substitution of radio:** Send a message to communicate with external operators Do text messages prewrote Do video chat Give video feedback to the control room through a wearable camera glasses carried by the external operators	**Modern technologies for communications:** Provide wireless headsets, personal headphones + speakers set at low-level volume* Give the ability to select the radio interlocutors (1, 2 or the whole team) Low-level communication except for selected operator
Management passages	**Alert the silence to the external personnel:** Provide signs on the floors at the entrances* Provide silent signs in access doors to the control room	**Modify organization:** Provide common zones near access doors, for control room operators and external staff to wait for the shuttle without disturbing their control room colleagues Reorganize the time of changing shiftwork
Lighting solutions	**Quality of general lighting:** Indirect lighting* Diffuse lighting* Provide wall lights, lighting simulating daylight Harmonize the different light sources	**Local lighting:** Provide local light with individual adjustment* Control and correct the brightness of the screens*
Workstation layout	**Screens:** Screens in table Bring the monitors closer to the operators* The top edge of the monitors must not exceed the height of the operator's eyes* Combine a large central screen and small supplementary screens Provide eyewear and screen-protector*	**Human Machin Interaction:** Wireless keyboard* Recess keyboard and trackball in the table Provide elbow and wrist rest* Allow sitting/standing position by adjustable workstation height Train operators (postures, adjusting)* Provide chairs with memory (memorize setting)

(continued)

Table 3. (*continued*)

Ergonomic problems	Solutions proposed	
The layout of the control room	**Re-arrange the control room:** Add a large shared wall screen for each side and arrange three pilot operators in front of the screen and the position of the chief behind them Flatten the cable tray to facilitate moving the chairs* Pass computer cabling in a network rather than on the ground	**Acoustic enclosure to reduce noise:** Separate both sides of the control room by acoustic enclosure or partition* Provide partial partition (up to the height of pilot operators)* Design moveable partition wall system*

Simulation on CAVE Virtual Reality. We presented on the virtual setting the following proposals that the users had to evaluate in the design process (Fig. 5):

- Four types of mobile acoustic partitions were presented so that users could perceive the real dimensions of the different solutions. The number of panels (3 or 4 panels) that affect the length of partition was also tested. Half height fixed partitions (1.4 m) and floor to ceiling sliding partitions were presented, and the users could verify the eye contacts with their colleagues from the different workstations.
- The users evaluated the readability and the position of sound meter displays and silent signs;
- The users assessed the field of view and postural comfort on the new workstations design implying lower displays positions.

Users' Feedback. The staff of the control room based on your experiences provided following feedbacks about the proposed design. The feedback about the mobile acoustic partitions was:

- Various meetings in one corner of the control room were a major source of noise for the other workstations,
- The chiefs have to discuss easily even when the partition is deployed,
- The pilot operators of the two sides of the control room need permanent eye contact, and barriers between two sides should be minimized,
- The users believed that the partitions would be folded most of the time,
- Three moving panels slide behind the fixed one when the partition is folded. This fixed part will not make a problem for operators' contact because two sides of the control room open sufficiently.
- The partition movement has to be as easy as possible (for example motorized) so that the partition won't stay folded all the time,
- Half height fixed partitions (1.4 m) would be inefficient against noisy conversations and meetings because many people talk together in standing position,
- The users suggested a vertical movement instead of a horizontal one

a.

b.

Fig. 5. Human factor element reviewed on the CAVE; a) sound meter display, b) acoustic partition and c) workstation design

c.

Fig. 5. (*continued*)

The feedback regarding sound level meter and silent signs were:

- A sound level meter and a display for each side would be sufficient,
- Users proposed an alternative position for the display (on the wall in front of the chief),

  Moreover, feedback about the position of the monitors was:

- Operators preferred the postural comfort provided by the lower position of the monitors.

Final Proposed Design. For the acoustic partition, the operators rejected the half-height fixed partitions (1.4 m), and we recommend a floor-ceiling partition composed of 4 sliding and motorized panels. This version has received the agreement of the operators. The operators proposed a sound level meter on each side of the control room, with a display located on the wall in front of the chief desk. Furthermore, the operators approved our recommendation to lower the two rows of monitors by 150 mm to improve postural and visual comfort.

4   Conclusion

This case study evaluated physical ergonomics such as lighting, noise, and workstation layout in a control room of chemical industry and used the VR CAVE to propose a new design. This study showed that the level of sound exceeded the threshold value of INRS for the control room (55 dBA). This level of sound disturb the operators who perform the tasks required concentration. The operators of the control complained of the excessive noise due to conversations/meetings, the passage of the external people, radio, and alarms. The users of the control room tested several solutions in the VR CAVE and, we finally proposed to install the motorized partitions (floor to ceiling). Furthermore, a sound meter and silent signs were installed in the control room to notice when the level of sound increase. Our finding showed the heterogeneity of the lighting and a high risk of dazzling in the control room. We, therefore, proposed new design of the lighting system that was adjustable between 300 and 600 lx. This system reduces the risk of dazzling. This study showed that VR CAVE is a useful tool that helps to visualize the new design and modify based on the final users. The next step of this study is to implement the new design validated on the VR CAVE and studying the real perceptions of final users.

References

1. ISO/TC 159/SC 4, Ergonomics of human-system interaction: ISO 11064-4:2013—Ergonomic design of control centres—part 4: layout and dimensions of workstations. https://www.iso.org/standard/54419.html
2. Dos Santos IJAL et al (2009) The use of questionnaire and virtual reality in the verification of the human factors issues in the design of nuclear control desk. Int J Ind Ergon 39(1):159–166
3. Simonsen E, Osvalder A-L (2018) Categories of measures to guide choice of human factors methods for nuclear power plant control room evaluation. Saf Sci 102:101–109
4. Crampin T (2017) Human factors in control room design: a practical guide for project managers and senior engineers. Wiley, Goodwood
5. U.S. Nuclear Regulatory Commission NUREG 700 (2002) Rev. 2. Human-system interface design review guideline
6. Aromaa S, Väänänen K (2016) Suitability of virtual prototypes to support human factors/ergonomics evaluation during the design. Appl Ergon 56:11–18

A Reach Motion Generation Algorithm Based on Posture Memory

Taekbeom Yoo[1] and Woojin Park[1,2]

[1] Department of Industrial Engineering,
Seoul National University, Seoul, South Korea
{tbyoo,woojinpark}@snu.ac.kr
[2] Institute for Industrial Systems Innovation,
Seoul National University, Seoul, South Korea

Abstract. Various models and algorithms (hereafter, simply algorithms) have been developed to simulate human motions. Most of these algorithms generate only a single "would-be-realistic" motion for a given scenario; thus, they cannot inform the designer of the full range of feasible human motions for the given scenario. In this paper, we present a novel reach motion generation algorithm based on the use of a posture memory, which aims to inform the range of feasible human reach motions for a given simulation scenario. In this algorithm, posture memory is constructed using a random posture generation and registration process. After memory construction, different paths connecting the starting and ending hand positions are created. Then, the human reach motion generation algorithm produces different "feasible" motions by selecting and connecting "connectable" postures found within the neighboring cells of the path. The algorithm proposed in this study generates feasible motions and an ability to generate and report the full range of feasible human motions for a given scenario allows a more complete understanding of the consequences of a design decision and also provides a basis for simulating human motions under different constraints.

Keywords: Motion generation algorithm · Posture memory
Virtual ergonomics

1 Introduction

Currently, various models and algorithms (hereafter, simply algorithms) have been developed for human motion simulation. Most of these algorithms generate, among many possibilities due to the kinematic redundancy of the human body, only a single "would-be-realistic" motion for a given scenario by optimizing a biomechanical or psychological objective function [1–7] or utilizing a certain learned statistical relationship [8].

These existing algorithms have been reported to be able to generate a realistic human motion for a given scenario, but have a limitation that they cannot inform the designer of the range of feasible (including realistic and unrealistic, and stressful and acceptable) human motions for the given scenario. An ability to generate and report a

© Springer Nature Switzerland AG 2019
S. Bagnara et al. (Eds.): IEA 2018, AISC 822, pp. 309–313, 2019.
https://doi.org/10.1007/978-3-319-96077-7_32

large range of realistic and unrealistic human motions for a given scenario has applications, including understanding the impacts of design decisions (e.g. the design of physical objects/obstructions in the workspace) and the worker's anthropometric characteristics on the range of feasible motions and thus work-related musculoskeletal stresses.

In this paper, we present a novel reach motion generation algorithm based on the use of a posture memory, which aims to inform the range of feasible human reach motions for a given simulation scenario.

2   Methods

The algorithm proposed in this study intended to generate two-dimensional, sagittally symmetric hand reach motions. To construct the human body model in the sagittal plane, body was divided into five segments (lower legs, upper legs, trunk-neck-head, lower arms and upper arms) and the joints between these segments (ankle, knee, hip, shoulder and elbow) are defined as shown in Fig. 1. The lengths of each body segments were calculated as ratios of body height based on Drillis and Contini [9].

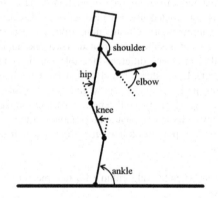

Fig. 1.   The human body in the sagittal plane

2.1   Posture Memory Construction

A posture memory is a collection of pre-generated feasible postures that satisfy basic physiological and biomechanical constraints, such as the constraints of joint range of motion and body balance maintenance. The current study adopted the posture memory construction process developed by Park et al. [10].

Posture memory is constructed using a random posture generation and registration process: the workspace is divided into small cells (width: 10 cm, height: 10 cm). Then, a physiologically and biomechanically feasible posture is randomly generated for a human figure and compared with the postures in the memory space corresponding to the hand position. If the memory space already contains a posture sufficiently similar to current generated posture, then the posture is discarded. To compare two postures,

Euclidean distance in the 5D joint space was used and the criterion was 20°. If there was no posture which is similar with new one, then the new posture is stored in the memory space. This generation and registration process is repeated for a predetermined number of times or until the posture memory is saturated.

2.2   Posture Memory-Based Motion Generation

For a given simulation scenario (the starting and ending hand positions, a human figure, and obstacle), the human reach motion generation algorithm produces different "feasible" motions using the following process: first, different paths connecting the starting and ending hand positions are created. In this step, a path is created by sequentially connecting neighboring cells from the starting point to the ending point. A cell occupied by the obstacle cannot be part of a valid path. Then, for each valid path, different reach motions are generated by selecting and connecting "connectable" postures found within the neighboring cells of the path. Two postures are considered connectable if their dissimilarity in the angle space is less than a certain predetermined threshold. In this paper, the threshold was set to 10°, 10°, 10°, 20°, and 30° for ankle, knee, hip, shoulder, and elbow joint, respectively.

3   Results

Some example motions generated through the proposed algorithm are shown in the figures below. The two red dots represent the starting and the end point, respectively, and the green lines show the trajectories of hand. The examples presented in Figs. 2 and 3 had the same starting and end points, and the workers were identical in the anthropometric dimensions. Figure 2 shows that obstacle with a width of 30 cm and a height of 100 cm is 30 cm ahead of the ankle point. Figures 2 and 3 show two motions, respectively. These figures demonstrate that different motions can be generated by the proposed algorithm. Also, the hand trajectories shown in Figs. 2 and 3 show the impacts of the presence of the obstacle.

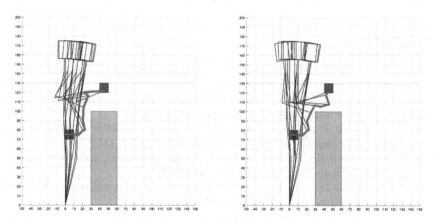

Fig. 2. Motion generation examples with obstacle

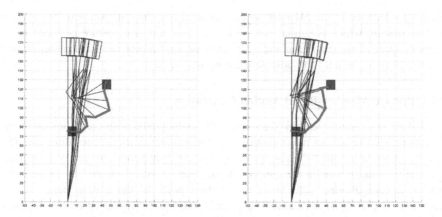

Fig. 3. Motion generation examples without obstacle

4 Discussion

A collection of motion simulation examples, including the ones subject to obstruction avoidance constraints, are provided to illustrate the working of the algorithm. The human reach simulation algorithm is expected to contribute to the virtual ergonomics analyses utilizing digital human models. In future works, the human reach simulation algorithm can be used in combination with different occupational biomechanics models to evaluate physical stresses of feasible motions in different workstation/workplace configurations and thereby improve their design.

The algorithm proposed in this study generates feasible motions and it helps to inform the range of feasible motions for a given scenario. However, some of the generated motions are not similar to motion of real humans. In addition, some of the motions generated by this method for a scenario were found to be too similar to one another. To overcome these limitations, methods for smoothing generated motions and merging similar motions are needed.

References

1. Nubar Y, Contini R (1961) A minimal principle in biomechanics. Bull Math Biophys 23(4):377–391
2. Röhrle H, Scholten R, Sigolotto C, Sollbach W, Kellner H (1984) Joint forces in the human pelvis-leg skeleton during walking. J Biomech 17(6):409–424
3. Seireg A, Arvikar R (1973) A mathematical model for evaluation of forces in lower extremeties of the musculo-skeletal system. J Biomech 6(3):313–322
4. Seireg A, Arvikar R (1975) The prediction of muscular load sharing and joint forces in the lower extremities during walking. J Biomech 8(2):89–102
5. Hsiang SM, Ayoub M (1994) Development of methodology in biomechanical simulation of manual lifting. Int J Ind Ergon 13(4):271–288

6. Chang C, Brown DR, Bloswick DS, Hsiang SM (2001) Biomechanical simulation of manual lifting using spacetime optimization. J Biomech 34(4):527–532
7. Lin C, Ayoub M, Bernard T (1999) Computer motion simulation for sagittal plane lifting activities. Int J Ind Ergon 24(2):141–155
8. Faraway JJ (1997) Regression analysis for a functional response. Technometrics 39(3):254–261
9. Drillis R, Contini R (1966) Body segment parameters. Office of Vocational Rehabilitation, New York, New York
10. Park W, Singh D, Martin B (2006) A memory-based model for planning target reach postures in the presence of obstructions. Ergonomics 49(15):1565–1580

DHM Based Test Procedure Concept
for Proactive Ergonomics Assessments
in the Vehicle Interior Design Process

Dan Högberg[1(✉)] ⓘ, Pamela Ruiz Castro[1], Peter Mårdberg[1,2],
Niclas Delfs[2], Pernilla Nurbo[3], Paulo Fragoso[4], Lina Andersson[5],
Erik Brolin[1] ⓘ, and Lars Hanson[1,4] ⓘ

[1] School of Engineering Science, University of Skövde, Skövde, Sweden
dan.hogberg@his.se
[2] Fraunhofer-Chalmers Centre, Gothenburg, Sweden
[3] Volvo Car Corporation, Gothenburg, Sweden
[4] Scania CV, Södertälje, Sweden
[5] Volvo Trucks, Gothenburg, Sweden

Abstract. The development of a digital human modelling (DHM) based test procedure concept for the assessment of physical ergonomics conditions in virtual phases of the vehicle interior design process is illustrated and discussed. The objective of the test procedure is to be a valuable tool for ergonomic evaluations and decision support along the design process, so that ergonomic issues can be dealt with in an efficient, objective and proactive manner. The test procedure is devised to support companies in having stable and objective processes, in accordance with lean product development (LPD) philosophies. The overall structure and fundamental functionality of the test procedure concept is explained by a simplified use case, utilizing the DHM tool IPS IMMA to: define manikin families and manikin tasks; predict manikin motions; and visualize simulations and ergonomics evaluation outcomes.

Keywords: Digital human modelling · Ergonomics · Design · Vehicle
Interior · Assessment · Test · Procedure · Lean product development

1 Introduction

Engineers typically utilize computer aided design and engineering (CAD/CAE) tools in contemporary product development (design) processes to reduce costs and time expenditures and to enhance product quality. Hence, objects like products and vehicles are to a large degree devised, detailed and evaluated within virtual worlds, where physical prototype solutions only are created and assessed when clearly gaining product quality, due to the associated monetary and time costs of making physical prototypes.

Central in product development processes is to create customer value [1]. The comprehension of what it is in a design that creates customer value is a central feature in product development processes. Supplementary is the skill to formulate such

© Springer Nature Switzerland AG 2019
S. Bagnara et al. (Eds.): IEA 2018, AISC 822, pp. 314–323, 2019.
https://doi.org/10.1007/978-3-319-96077-7_33

comprehension of customer value into constraints and target criteria, setting demands ('musts') and aims ('goals') for the object to be designed [2]. Some of those demands and aims are most likely related to the interaction between the user and the object being designed, i.e. having an ergonomics content. When treating ergonomics in design, it is beneficial from time, cost and quality perspectives that the consideration of ergonomics is done in a proactive manner, i.e. aiming to identify, and reduce or solve ergonomics related issues of a proposed design before user-product interaction related problems would occur in a real use situation. Hence, it is imperative for engineers to be able to consider the *users* and the *use* of the product within the development process. Consequently, and in accordance with Pheasant and Haslegrave [3], it is possible to identify three core components of a user-centred design framework: the *product* (the object being designed), the *user(s)*, and the *task(s)* that the user performs while interacting with the product.

Since products commonly are being designed in virtual worlds, digital human modelling (DHM) tools have been developed to assist engineers to consider ergonomics in virtual development processes [4–7]. The tools enable modelling and simulation of different *users*, where these virtual users, represented by digital human models, also known as computer manikins, perform different *tasks* while interacting with the *product* (the object being designed), within a virtual environment. Hence the tools need functionality to instruct the human models to interact with the objects in the virtual world, e.g. to perform tasks such as *sit in the seat* or *change gear*. The human model and the object being designed are the main components in the virtual system model. Commonly the virtual system model also contains objects that are not to be altered by the current design task, but act as control geometries. These control geometries act as a representation of the *environment*. The four components: *users; tasks; product; environment* is in accordance with an expanded view of the principal components of a human-machine system [8]. DHM tools typically also have a collection of ergonomics assessment tools in order to support the engineer to make decisions about the ergonomic quality of the object being designed. Hence, a DHM tool can be used for evaluative activities in the design process. A DHM tool can also be used to specify ergonomic requirements (constraints and target criteria), as input for generative design activities in the design process. This paper mainly covers the application of a DHM tool for evaluative activities in the design process. However, since iteration is a central feature in design processes, the outcome from such evaluations may give input to specifications or generative design activities [2].

An associated perspective on contemporary product development processes is that these processes often follow the philosophies and principles of lean product development (LPD) [1]. Central in this way to carry out product development is the focus to 'Do the right thing' and 'Do the thing right'. The 'Do the right thing' aspect has the concept of delivering value to all stakeholders as the central objective. The 'Do the thing right' aspect focuses on having efficient processes, with efficient tools and methods, which assist in creating value while reducing waste, i.e. reduce or remove activities that add no or little value. In addition to this, LPD highlights the importance of 'Continuous improvements', where the organization continuously strives to learn from and improve the product development process in respect to efficiency and improved understanding of how to deliver value [1].

This paper illustrates and discusses the development of a DHM based test procedure concept for proactive assessment of physical ergonomics assessments in the vehicle interior design process, expanding the work presented in [9]. Philosophies and principles of LPD are used in the paper to clarify and justify the DHM based test procedure concept.

2 Method

In line with the DRM (Design Research Methodology) framework [10], the work commenced with a descriptive study, aiming to enhance understanding about how DHM tools currently are used to support the assessment of ergonomics in the vehicle interior design process, and to identify opportunities and expectations for enhancing the functionality and usability of the tools. The method for this was discussions between researchers and engineers or ergonomists from collaborating industry partners, representing actual or potential end users of DHM tools for ergonomics in vehicle design. Based on the findings from the descriptive study, a DHM based test procedure concept for proactive ergonomics assessments in the vehicle interior design process is prescribed. The content of this paper represents the early stages of a prescriptive phase in the DRM process. The test procedure concept is illustrated in the context of using and developing the DHM tool IPS IMMA [11], even though the principles are meant to be general in terms of type of DHM tool used for ergonomic design.

3 Results

The prescriptive study led to a concept for a DHM based test procedure for proactive ergonomics assessments in the vehicle interior design process. It is inspired by how tests of vehicle ergonomics usually are performed in the real world, i.e. having real people doing real tasks in real vehicles. The rationale is that the DHM based test procedure enables proactive consideration of ergonomics, possible to perform in the virtual world even when no physical vehicle is produced or available, or where there may not be access to certain user representatives. Another benefit is to have full control of the tasks that are to be tested, which might be difficult when recruiting real people for user tests. And a clear benefit is the ease and cost-efficiency to iteratively change and test different design alternatives in the virtual world rather than in the real world. The overall aim with the test procedure is to enhance the quality, efficiency and objectivity of ergonomics assessments in comparison with current ways of using DHM in the vehicle interior design process. Linked to the quality objective is the ambition to do simulations with several well-conceived digital human models, leading to better consideration of user diversity. Another quality related objective is to simulate more use tasks, and to simulate these tasks in a manner that resembles how a real human would perform the tasks, i.e. containing predicted human motions required to perform the tasks. This to facilitate that all critical tasks, including associated motions and force exertions, are considered when assessing new design proposals, and also to get a richer comprehension of the aggregation and combination effects of these tasks, reducing the risks for sub-optimizations. To support both quality and efficiency, the DHM based conceptual test procedure also needs to have functionality to easily assess and compare several design proposals. In order to meet the aim associated with objectivity, the DHM based conceptual test procedure is to render the same, or very similar, outcomes regardless of who is performing the tests. Also, if tests are repeated, the test procedure is to render the same or very similar outcomes. The mentioned arguments for the DHM based test procedure concept go in line with the target in LPD to have stable processes [1]. Having stable processes for the use of DHM tools in design processes supports obtaining corresponding simulation results regardless of who is using the tool or when, and also obtaining comparable results when evaluating different design proposals. Figure 1 outlines the structure of the main components of the DHM based test procedure, where each component can be exchanged to represent different simulation and evaluation scenarios.

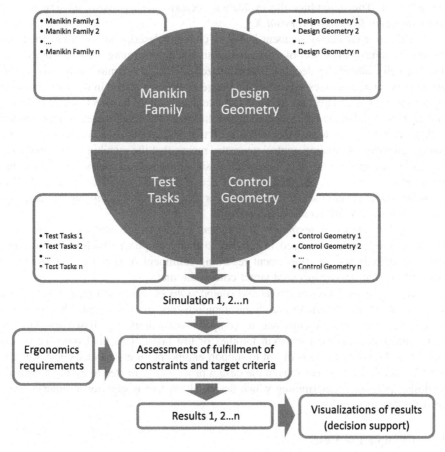

Fig. 1. Structure of the main components of the DHM based test procedure.

The *Manikin Family* segment in Fig. 1 contains descriptions of human models that represent certain user groups. Switching Manikin Family means that the digital human models used in the test procedure are altered, and that new simulation results are achieved. This resembles the evaluation scenario: *let us assess ergonomics of the proposed design when changing to manikin family X.*

The *Test Tasks* segment in Fig. 1 contains descriptions of tasks that are to be performed by the manikin family during a test. Switching Test Tasks means that the manikin family used in the test procedure follow the new tasks descriptions, and that new simulation results are achieved. This resembles the evaluation scenario: *let us assess ergonomics of the proposed design when changing to test tasks description X.*

The *Design Geometry* segment in Fig. 1 contains descriptions of alternative design proposals of an object, which has been generated in the development process. Switching Design Geometry means that the manikin family performs the test tasks, but now interacts with an alternative design of the object, and that new simulation results are achieved. This resembles the evaluation scenario: *let us assess the ergonomics when changing to design proposal X.*

The *Control Geometry* segment in Fig. 1 contains descriptions of geometries that the manikins have some sort of interaction with when performing the tasks, but where this geometry already is decided (settled/frozen). Thus, the control geometry is not meant to be altered or assessed per se, but the geometry is included in the virtual system model since it has an effect on the test. The interaction might be direct, where tasks are directly connected to the control geometry, or indirect, where tasks are connected to the design geometry, but where the location of the design geometry varies due to the control geometry. Changing control geometry means that the manikin family performs the test tasks, interacting with a specific design proposal geometry, but where the surrounding geometry is changed, and that new simulation results are achieved. This resembles the evaluation scenario: *let us assess the ergonomics of the suggested design when put in surrounding control geometry X.*

In the box named *Assessments of fulfilment of constraints and target criteria* in Fig. 1, simulations are assessed in relation to their compliance with specified ergonomics requirements, i.e. assessment regarding fulfilment of specified constraints or not, and to what degree specified target criteria are met.

In the box named *Visualizations and comparisons of results* in Fig. 1, results from the simulations in relation to ergonomic requirements are visualized. This to assess compliance with requirements, and to compare simulations, e.g. based on different design proposals, different manikin families or test tasks. This final outcome of the DHM based test procedure is meant to support engineers (or ergonomists or designers) to make well-informed decisions in the design process, e.g. related to giving approvals on design proposals, determining which design alternative is superior to others, or the need for design iterations.

3.1 Example Use Case

This simplified use case aims to exemplify the test procedure concept on principal level. The DHM tool IPS IMMA [11] is used as basis for the example. The description of the example is structured according to the structure in Fig. 1.

Manikin Family

Amanikin family, i.e. a collection of human models that represent a certain user group, is selected. One or several manikin families can be used in the test procedure, where a manikin family can aim to represent all targeted users, or aim to represent a sub-group of users desired to be treated separately, e.g. drivers from a certain nationality, or drivers of a certain anthropometric range. A collection of standardized manikin families can be established, e.g. within a company. This to have stable processes for DHM tool usage, in accordance with LPD. In IPS IMMA, manikin families are defined, stored and inserted through a dedicated anthropometric module, described in [12]. Figure 2 shows one member of the manikin family selected in this example use case.

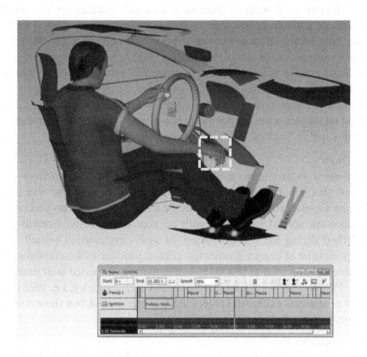

Fig. 2. Screen shot of a manikin family member performing part of a test procedure (CAD geometry courtesy of Volvo Car Corporation).

Test Tasks

Test tasks definitions can be seen as a standardized procedure for how to assess ergonomics. On a basic level, the definitions can consist of a number of primary tasks that are essential to consider from an ergonomics perspective, such as: sit in seat, reach steering wheel, put feet on pedals, adjust seat, adjust steering wheel, put on seat-belt, view mirrors, adjust mirrors, manoeuvre pedals, change gear, turn steering wheel, etc. On a more advanced level, a test tasks definition can form a longer virtual test route performed by the manikin family. Similarly to manikin families, a standard collection of test tasks can be established in a company to have stable processes, in accordance with LPD. In IPS IMMA, task definitions are made in a dedicated operation sequence editor, described in [13]. Only one member of the manikin family needs to be instructed, and then the other members of the family follow the same task instructions. The motions of the manikins when performing the defined tasks are automatically predicted by the IPS IMMA tool [14]. Figure 2 shows a manikin performing a certain test task, with the operation sequence editor window shown in the lower part.

Design Geometry
Design geometry is the object being designed. The core purpose of the DHM based test procedure is to provide decision support to engineers about how different design proposals fulfil specified requirements. A design geometry can be a set of items, like a new interior layout, or a separate item such as a new surface on a control device. Possibly there are a number of alternative design geometry proposals to assess, to see what would be a better solution from an ergonomics point of view, in balance with other design requirements. In the DHM tool IPS IMMA, a design geometry is imported from a dedicated CAD software. Ideally, properties relevant for the assessment of ergonomics are imported along with the geometry, such as the object's kinematic characteristics and forces or moments required for manipulation. In Fig. 2 (within the white dashed rectangle), a design geometry is exemplified with the gear shift design in terms of form, spatial location, motion and forces required.

Control Geometry
Control Geometry is geometry, including important characteristics from ergonomic point of view, that is not assessed by the test procedure as such, but that needs to be included in the simulation since it influences the test, i.e. the manikins and/or the design geometry has some sort of interaction with the control geometry. Control geometries can be switched during the test procedure, to assess ergonomics related to different surroundings. Similar to design geometries, control geometries are created in dedicated CAD software and imported in IPS IMMA, alternatively imported as scanned 3D point clouds [15]. Figure 2 shows a simplified control CAD geometry, i.e. here represented by all object geometries except the object in the white dashed box (hence representing the design geometry component).

Ergonomics requirements
The requirements stated in the product design specification document, that specifies constraints and target criteria for issues related to physical ergonomics, are used to assess the ergonomic quality of design proposals. In a real case, this list is likely to contain many items, but for the purpose of illustrating the principles, this use case only includes three hypothetical constraints and one target criteria.

> *Constraint C-1:* Gear shift in neutral position is within reach for all members in manikin family X when seated in preferred posture (without changing back posture).
> *Constraint C-2:* All gear shift manoeuvres can be performed by all members in manikin family X when seated in preferred posture (without changing back posture).

Constraint C-3: Summarized level of comfort for all members in manikin family X when reaching gear shift in neutral position using comfort assessment method Z shall be above 40.

Target criteria T-1: Summarized level of comfort for all members in manikin family X when reaching gear shift in neutral position using comfort assessment method Z. Target: as high summarized comfort level as possible.

Visualizations of results

Outcomes from different simulation setups are assessed in relation to fulfilment of constraints and target criteria. Table 1 illustrates assessment results from three different simulation setups, in this case by alternating the design geometry. The results are presented to the DHM tool user in order to support decisions for how to progress in the design process.

Table 1. Visualization of assessment results.

	Simulation setup			
	1	**2**	**3**	**n**
Manikin Family	MF-1	MF-1	MF-1	
Test Tasks	TT-1	TT-1	TT-1	
Design Geometry	DG-1	DG-2	DG-3	
Control Geometry	CG-1	CG-1	CG-1	

Constraint C-1				
Constraint C-2				
Constraint C-3				
Target Criteria T-1	46	25	58	

In this simple example the decision for how to progress in the design process could be to reject design proposal DG-2 since it does not fulfil constraint C-3, and to promote design proposal DG-3 over DG-1 since both proposals fulfil all constraints, but DG-3 performs better than DG-1 for target criteria T-1. Another decision can be to look closer at the simulations and investigate for which manikin in the manikin family ergonomic factors are reduced, and why. This simulation replay feature is provided by the DHM tool, enabling a better comprehension for the engineer, and can also be beneficial to show when explaining to others about issues identified, e.g. to other engineers, ergonomists or managers.

4 Discussion

Having a DHM based test procedure would integrate the use of DHM tools more tightly into the general product development process, supporting time, cost and quality objectives of product development. This paper presents a conceptual solution. A lot more research and development actions are needed to reach the state where the DHM based test procedure has the desired functionality and usability. But realized, it is argued that the solution will support more engineers, and not only DHM tool experts, to use DHM tools in the development process. This since the engineer will be guided through the test procedure, where pre-defined manikin families and test tasks can be inserted to represent *users* and *tasks*, and where the engineer can test outcomes from different design proposals of the *product,* within a certain control geometry, as a representation of the *environment.* This facilitates assessments of what-if scenarios for different design proposals, which supports the co-exploration of both the design problem space and design solution space, typical for design work [2], assisting the creative work of the engineer to find better design solutions. It also supports smaller iterative design loops, e.g. performed by engineers themselves, rather than being dependent on DHM tool experts to perform assessments. For more complicated design problems, a DHM tool expert is likely to be involved, but for basic design problems most engineers are assumed to be able to consider basic ergonomic factors. Such smaller iterations are believed to increase efficiency of the development process and gain product quality, e.g. by reducing risks that ergonomic design weaknesses are identified at major gates within the development process.

One challenge in realizing the test procedure is related to creating inter-connections between human models and the geometry when manikin family, user tasks, design geometry or control geometry are altered in the simulation model. In an ideal solution that would be automatic; but a more realistic solution, at least as a start, could be to provide some sort of wizard function that guides the DHM tool user to re-establish connections between human models and geometries. This is an area for deeper investigation, which may incorporate the use of metadata in CAD models. A complementary approach would be to make the digital human model understand its virtual environment to some degree, e.g. able to recognize CAD objects' perceived affordances [16]. Reaching a stage where re-establishment of inter-connections are automatic would support the introduction of simulation based multi-objective optimization techniques into the test procedure [17].

Acknowledgments. This work has been made possible with the support from The Knowledge Foundation and the associated INFINIT research environment at the University of Skövde in Sweden, in the project Virtual Driver Ergonomics (Dnr 20160296), and by the participating organizations. This support is gratefully acknowledged.

References

1. Pessôa MVP, Trabasso LG (2016) Lean product design and development journey: a practical view. Springer, Berlin
2. Cross N (2008) Engineering design methods: strategies for product design, 4th edn. Wiley, Chichester
3. Pheasant S, Haslegrave CM (2006) Bodyspace: anthropometry, ergonomics and the design of work, 3rd edn. Taylor & Francis, Boca Raton
4. Chaffin DB (2001) digital human modelling for vehicle and workplace design. SAE International, Warrendale
5. Bubb H, Engstler F, Fritzsche F, Mergl C, Sabbah O, Schaefer P, Zacher I (2006) The development of RAMSIS in past and future as an example for the cooperation between industry and university. Int J Hum Factors Modell Simul 1(1):140–157

6. Duffy VG (2009) Handbook of digital human modeling. Taylor & Francis Group, Boca Raton
7. Alexander T, Paul G (2004) Ergonomic DHM systems—limitations and trends—a systematic literature review focused on the 'future of ergonomics'. In: 3rd International digital human modeling symposium, Tokyo
8. Shackel B (1991) Usability—context, framework, definition, design and evaluation. In: Shackel B, Richardson S (eds) Human factors for informatics usability. Cambridge University Press, Cambridge, pp 21–37
9. Högberg D, Brolin E, Hanson L (2018) Concept of formalized test procedure for proactive assessment of ergonomic value by digital human modelling tools in lean product development. In: Cassenti DN (ed) Advances in human factors in simulation and modeling, AHFE Conference, pp 425–436
10. Blessing LTM, Chakrabarti A (2009) DRM, a design research methodology. Springer, London
11. Högberg D, Hanson L, Bohlin R, Carlson JS (2016) Creating and shaping the DHM tool IMMA for ergonomic product and production design. Int J Dig Hum 1(2):132–152
12. Brolin E (2016) Anthropometric diversity and consideration of human capabilities—methods for virtual product and production development. Doctoral thesis. Research series from Chalmers University of Technology, Series number 4035, Gothenburg
13. Mårdberg P, Carlson JS, Bohlin R, Delfs N, Gustafsson S, Högberg D, Hanson L (2014) Using a formal high-level language and automated manikin to automatically generate assembly instructions. Int J Hum Factors Modell Simul 4(3):233–249
14. Bohlin R, Delfs N, Hanson L, Högberg D, Carlson JS (2012) Automatic creation of virtual manikin motions maximizing comfort in manual assembly processes. In: Hu SJ (ed) Proceedings of the 4th CIRP conference on assembly technologies and systems, USA, pp 209–212
15. Mahdavian N, Ruiz Castro P, Högberg D, Brolin E, Hanson L (2017) Digital human modelling in a virtual environment of CAD parts and a point cloud. In: Proceedings of 5th international digital human modeling symposium, Bonn, Germany, June
16. Norman D (2013) The design of everyday things. Basic Books, New York
17. Wang L, Ng AHC, Deb K (eds) (2011) Multi-objective evolutionary optimisation for product design and manufacturing. Springer, London

Modeling the Range of Motion and the Degree of Posture Discomfort of the Thumb Joints

Natsuki Miyata[1]([✉]) [iD], Yuya Yoneoka[2], and Yusuke Maeda[2] [iD]

[1] National Institute of Advanced Industrial Science and Technology,
Tokyo 135-0064, Japan
n.miyata@aist.go.jp
[2] Yokohama National University, Yokohama, Kanagawa, Japan

Abstract. In this paper, we show a method to model coordinated joint range of motion (ROM) of the thumb with subjective discomfort by exhaustively testing various postures. Coordinated joint ROM was modeled by capturing twelve exercises for four adults without any hand disabilities. Posture candidates were then generated for the discomfort experiment by uniformly sampling each subject's ROM. Each subject was asked to answer if they felt it difficult to retain that posture for 10 s owing to pain or fatigue. The collected data from four subjects were converted into a discomfort possibility map.

Keywords: Joint range of motion · Reachability · Posture discomfort
Possibility map

1 Introduction

To assist ergonomic design of a product using a digital hand model, the possible natural postures required to use the product should be estimated and assessed. Especially when designing interface dispositions, a main concern might be regarding reachability, which requires appropriate joint range of motion (ROM). From the viewpoint of continuously using products for some period of time, avoiding discomfort postures is critical. In this paper, we concentrate on the thumb because it plays an important part to form various grasps by opposition and it is often used to operate control interfaces. Several studies have been conducted on modeling the thumb's carpometacarpal (CM) joint's ROM [1–3]; however, coordination with other joints such as interphalangeal (IP) or metacarpopha-langeal (MP) joint was not considered. A previous paper [4] collected postures that subjects felt comfort through their random posture trial, which unevenly sampled postures from its ROM in general. The purpose of this paper is therefore to exhaustively measure and analyze the thumb's posture discomfort with respect to the thumb's coordinated ROM.

2 Thumb ROM Modeling

Conventional ROM data was collected by measuring a joint angle in typically extended or bent posture independently for each joint [5]. To model ROM considering coordi-nation among joints, the authors have proposed to measure hand postures during a

© Springer Nature Switzerland AG 2019
S. Bagnara et al. (Eds.): IEA 2018, AISC 822, pp. 324–329, 2019.
https://doi.org/10.1007/978-3-319-96077-7_34

planned set of exercises by using an optical mocap system [6]. If the hand posture is expressed with n joint angle variables, the ROM of the hand is derived as an aggregation of $\binom{n}{2}$ boundaries of the projected postures on each plane of two of the joint angle variables. In this paper, to model the thumb ROM, we asked for four adults with no hand disabilities in their hands to execute twelve exercises.

On modeling CM joint ROM, we introduced new posture expression. The CM joint of the thumb is a saddle joint having two degrees of freedom but moves three-dimensionally. The direction of each anatomical joint axis is known to be different from person to person largely [7, 8], which leads to the difficulty in integrating different subjects' ROM in a joint angle space because the same joint angle could mean totally different postures. Therefore, a posture of the CM joint was expressed in terms of spherical coordinates of the MP joint with the center of the sphere located at the CM joint as shown in the right of Fig. 1. Then ROM boundary of the four subjects were derived as shown in Fig. 2, using the alpha-shape algorithm [9].

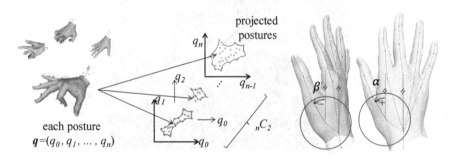

Fig. 1. Concept of coordinated ROM modeling (in the left) and CM joint posture expression (in the right).

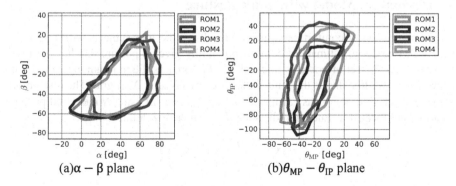

Fig. 2. Coordinated ROM of the thumb joints of four subjects.

3 Posture Discomfort Modeling

3.1 Discomfort Data Collection

The posture discomfort data was collected through experiments for the same four subjects in ROM measurement. To control the variety of the postures, a set of approximately 45 postures to be displayed was first generated for each subject by uniformly sampling on the CM joint ROM boundary as well as inside the ROM region of the subject. The subject was then asked to reproduce a displayed posture and to answer whether the subject felt any discomfort. Here, we defined "discomfort" as the state in which one feels it difficult to maintain a given posture for ten seconds owing to pain or fatigue. On reproduction, each subject was allowed to use a mouse to freely change the viewing direction of the displayed three-dimensional digital hand model (Fig. 3).

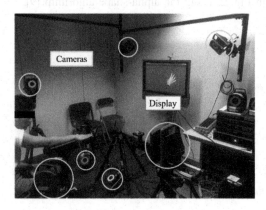

Fig. 3. Experimental setup.

4 Discomfort Model with Joint Posture

We combined the discomfort experimental results of the four subjects with respect to their ROM model to estimate discomfort possibility given a posture that has not been measured. As the set of postures to check discomfort was different among subjects in experiments, we first divide the joint angle space as a $N_{grid} \times N_{grid}$ grid. The discomfort value v_i at i-th grid point was calculated as Eqs. (1) and (2) using N measured posture data that are located at Euclidian distance d_{in} smaller than a threshold r.

$$v_i = \sum_{n=1}^{N} \left\{ e_{in} \left(1 - \frac{d_{in}}{r} \right) \right\} \tag{1}$$

$$d_{in} = \begin{cases} -1 & \text{(not discomfort)} \\ 1 & \text{(discomfort)} \end{cases} . \tag{2}$$

Then, the discomfort possibility at the i-th grid point was calculated as follows using the hyperbolic tangent function:

$$p_i = \frac{1}{2}\tanh \sigma v_i + \frac{1}{2}, \tag{3}$$

where $\sigma = 0.07$ in this paper. If no points were included within a circle with its radius r, $v_i = p_i = nan$ (not a number).

Figure 4 shows the discomfort possibility map built from experimental results for the four subjects. Many of the discomfort postures were observed near the ROM boundary and those were considered to be strongly affected by musculotendon or skin restraint. At around $(\alpha, \beta) = (30, -20)$, however, discomfort possibility was calculated to be low despite the region being close to the ROM boundary. This was considered to be natural because the ROM boundary around here was due to self-interference between the thumb and the palm. In contrast, low possibility ($p = 0.08$) at gridpoint $(\alpha, \beta) = (33, -45)$ despite several "discomfort" results were directly affected by subjects' preference and could be different with an increased number of experimental subjects.

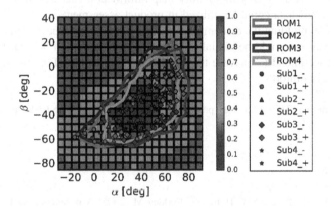

Fig. 4. Discomfort data of four subjects.

Figure 5 demonstrates the use of this possibility map when designing control interface disposition. The colored point cloud demonstrates a reach envelope of the thumb tip assuming a specific user's grip. Each color corresponds to the discomfort possibility as the same in Fig. 4. The average color of the points near the design target control interface can be used to check discomfort occurrence and to discover improved disposition to reduce discomfort occurrence possibility.

Fig. 5. A reach envelope colored according to discomfort possibility.

5   Conclusion

This paper demonstrated a method to model coordinated joint ROM of the thumb with subjective discomfort by exhaustively testing various postures. Modeled coordinated joint ROM from twelve exercises by four subjects were combined with discomfort experiment results that were then converted into a discomfort possibility map. A reach envelope with this discomfort possibility map information was shown to demonstrate how the collected data could be utilized in product assessment.

As stated in building the discomfort possibility map from experimental results, the possibility tendency could be partially different if more data will be added. Hence, in future work, we intend to increase the number of subjects both for coordinated ROM and posture discomfort.

Acknowledgments. This work was partially supported by JSPS KAKENHI (Grant Number JP17H05918).

References

1. Zhang X, Braido P, Lee SW, Hefner R, Redden M (2005) A normative database of thumb circumduction in vivo: center of rotation and range of motion. Hum Factors 47(3):550–561
2. Kuo LC, Cooney WP, Kaufman KR, Chen QS, Su FC, An KN (2004) A quantitative method to measure maximal workspace of the trapeziometacarpal joint—normal model development. J Orthop Res 22(3):600–606
3. Goubier JN, Devun L, Mitton D, Lavaste F, Papadogeorgou E (2009) Normal range-of-motion of trapeziometacarpal joint. Chirurgie de la Main 28(5):297–300
4. Tang J, Zhang X, Li ZM (2008) Operational and maximal workspace of the thumb. Ergonomics 51(7):1109–1118
5. American Academy of Orthopaedic Surgeons (1965) Joint motion: method of measuring and recording. American Academy of Orthopaedic Surgeons, Chicago
6. Miyata N, et al (2015) Modeling coordinated joint range of motion of the human hand–characteristics of coordination and grasping in relation to the ROM boundary. In: Proceedings of the 25th congress of international society of biomechanics, ISB, Glasgow, pp 469–470

7. Hollister A, Buford WL, Myers LM, Giurintano DJ, Novick A (1992) The axes of rotation of the thumb carpometacarpal joint. J Orthop Res 10(3):454–460
8. Crisco JJ, Halilaj E, Moore DC, Patel T, Weiss APC, Ladd AL (2015) In vivo kinematics of the trapeziometacarpal joint during thumb extension-flexion and abduction-adduction. J Hand Surg 40(2):289–296
9. Pateiro-Lopez B, Rodriguez-Casal A (2010) Generalizing the convex hull of a sample: The R package alphahull. J Stat Softw 34(5):1–28

Employing Game Engines for Ergonomics Analysis, Design and Education

Esdras Paravizo[(✉)] and Daniel Braatz

Department of Production Engineering,
Federal University of Sao Carlos, Sao Carlos 13 565 905, Brazil
esdras@dep.ufscar.br

Abstract. Game Engines (GEs) are digital platforms generally employed to develop computer games and 3D applications using pre-existing modules and functionality (and thus speeding up development process). Several studies already investigate GEs application for education and training, collaborative design, facilities simulation and design, operational simulations and for the development of virtual and augmented reality applications. In this paper we present three virtual environments developed in a free to use GE and analyze them in terms of their goals, intended audience, interaction possibilities and overall design. Results highlight GEs affordances such as the development of high-quality graphics 3D environments, their real-time system nature, the possibility for digital human manikin's customization, user agency, versatility in terms of possible interactions programming, among others, which make GEs powerful tools for ergonomics analysis, design and education.

Keywords: Game engines · Simulation · Virtual environments
Digital human modelling

1   Introduction

Game Engines (GEs) are software platforms originally created for the development of digital games which have a series of functionality and modules (e.g. game physics, input manipulation, 2D/3D rendering) ready for use, thus speeding up the design of new games and applications.

Developments in processing power, graphics rendering capabilities and increasing ubiquity of high-speed internet connections lead commercially available GEs to focus on delivering high-end visuals in immersive 3D environments. It also enabled the possibility of using GEs for developing Virtual Reality (VR), Augmented Reality (AR) applications and Virtual Environments (VEs) in general more easily, without having to build all the required functionality from scratch.

Researchers on several field have already employed GEs in a variety of contexts, from education and training (Borro-Escribano et al. 2014; Aziz et al. 2015; Koops et al. 2016) to collaborative design (Koutsabasis et al. 2012; Madni 2015) and facilities simulation and design (Braatz et al. 2011; Zamberlan et al. 2012; Gatto et al. 2013) to cite a few.

© Springer Nature Switzerland AG 2019
S. Bagnara et al. (Eds.): IEA 2018, AISC 822, pp. 330–338, 2019.
https://doi.org/10.1007/978-3-319-96077-7_35

GEs have already been employed in ergonomics and human factors research. Aromaa and Väänänen (2016) employed a commercially available game engine to develop VR and AR applications to evaluate reach, visibility and tools usage in a virtual prototyping process. Gatto et al. (2013) employed GE for developing an application for simulating control room operation and Zamberlan et al. (2012) employed GE for control room design. Researchers have also discussed GE's potential for professional training in assembly tasks, maintenance work and security agents training (Aziz et al. 2015; Mcnamara et al. 2016; Passos et al. 2016, 2017). Nonetheless, a more comprehensive understanding of GEs utility for ergonomics research and practice, in terms of analysis, design and education.

This paper aims to discuss the affordances GE present that can be gathered and applied to ergonomics research and practice whether on analyzing working conditions (or validating these analysis) or employing it as a simulation tool in the design stages of an ergonomics intervention or even in the education and training of students and future professionals in ergonomics. To achieve this goal, we analyze three virtual environments (VEs) developed using a game engine, highlighting the main characteristics of each of them and discussing how GE affordances can be fostered to ergonomics research and practice.

2 Methods

This paper takes a qualitative, exploratory approach to analyze three virtual environments created using GE. All three VEs were developed using the game engine Unreal Engine 4 (free to use) and were made available for users to explore them in their personal computers either using the keyboard and mouse or a joystick for navigate in the environment. Other software tools employed for the development of the VEs were Autodesk's AutoCAD and 3D Studio Max (for 3D modeling), Adobe Fuse CC (for designing digital human manikins) and Mixamo online database (for animating the manikins).

The first VE placed users in a multiplayer environment, in a fictitious industrial facility populated with machines. Users were free to explore the environment (either in first person or third-person views) and they could test different layout proposals by pressing a keyboard key. A more detailed account of the VE design is provided in Paravizo et al. (2016).

The second VE was developed based upon data and images from a real workplace in an oil refinery. Users were able to freely explore the environment in a first-person perspective and to interact with non-playable characters (NPCs), with the environment and machines (trough information spots) and with resources contextualizing the overall work condition represented in the VE. Users could also explore an alternative layout proposal for the workplace depicted, by pressing a keyboard key.

The third VE was developed in the context of a cooperative project among the authors, a team of ergonomists and a large steel industry. The goal of the project was the redesign of a specific sector of the company which was going through a technological change process. A VE was developed so users (in this case workers from the sector) could be able to explore and discuss the proposed layout and changes. This VE was relatively simple, the intention behind its development was to show the proposed changes in a platform which could be easier for worker to understand and visualize them.

The exploratory analysis of the VEs focus on the following aspects: their goals, intended users, interaction possibilities and overall VE design characteristics. The following section shows the results of the analysis of each o the VEs in terms of the aspects highlighted and Sect. 4 sums up the GE affordances that enabled these developments.

3   Results

3.1   Virtual Environment 1 – Fictitious Industrial Environment

The main goal of the first VE was to enable users to understand the concepts of different layouts (process layout vs. manufacturing cell layout). To that extent users could explore the environment and alternate between one or another configuration by pressing a button on their keyboard. Figure 1 shows the original configuration congregating the lathe machines in the same area (to the left) and, after pressing the button, the manufacturing cell (to the right) with a lathe machine, a saw and a bench press.

Fig. 1. Scenario before the layout change (left) and after the layout change (right).

The intended users of this scenario were originally undergraduate students in production engineering courses, hence the more didactic focus of this VE. Furthermore, regarding the overall VE design, it's worth noting that it supported up to six simultaneous users, sharing the same environment, over a local area network (LAN). Figure 2 shows four simultaneous users in the VE. In terms of the interactions users could have, the two main ones were with the environment (changing the layout) and among themselves (when sharing the environment exploration).

Fig. 2. Four users exploring concurrently exploring the same VE, concurrently.

3.2   Virtual Environment 2 – Local Control Room Scenario

The main goal of the second VE was to represent a real work situation existing in an oil refinery. The overall environment design was based on data and photos and videos collected by two ergonomists who participated in an ergonomics intervention at the workplace. The main goal of this VE was to accurately represent the current situation at the real workplace and to present a possible solution for discussion of its pros and cons. Intended users of this VE were both professional ergonomists, researchers and students, focusing the discussion of the tool as a possible medium for simulating workplaces and enabling users to understand some of the issues, constraints and challenges of the workplace. There several possible interactions possible in the VE, namely:

- Interacting with NPCs – interactive dialogues, inquiring about working conditions at the workplace depicted in the VE;
- Interacting with equipment and environment – information spots allowed users to gather more information about the physical aspect of the workplace and its equipment and machines;
- Interacting with resources available to the user – the user had a number of information sheets they could read to better understand aspects of work organization at the workplace and
- Layout change – by pressing a button to see the proposed change of the workplace layout and its implication to workers (NPCs).

Figure 3 shows the user interacting with a worker NPC (left) and with an infor-

Fig. 3. User interacting with NPC (left) and user interacting with information spot (right)

Fig. 4. Workplace layout before change (left) and after change (right)

mation spot (right) and Fig. 4 shows the sector layout before (left) and after (right) the user changed it.

3.3   Virtual Environment 3 – Sector in a Large Steel Industry

The main goal of the third VE was to represent a design proposal for workers and other stakeholders involved in the redesign of a sector in a large Brazilian steel industry. The overall VE design was simple, focusing on minimizing the time it would take to develop the VE. Thus, the users could only navigate around the scenario, in third person to investigate possible problems and positive aspects of the layout proposed. Figure 5 shows the VE developed.

Fig. 5. User exploring the VE depicting a new layout proposal an industrial sector.

4   Discussion

Based on the results shown in the previous sections it's possible to identify the main affordances GEs bring to the development of VEs which can be used for ergonomics analysis, design and education.

Firstly, GEs excel in delivering *high-quality graphics 3D environments*, being suitable for presenting and discussing design proposals and alternatives much more clearly than 2D technical drawings, for instance. Additionally, the level of abstraction required to participants involved in the evaluation of the design proposals using a VE built in a VE is much lower, since the realistic visuals of 3D models and overall environment are much closer to that of the reality. As such, GEs can be a useful tool for participatory design process by means of understanding the proposals and also for ergonomics analysis of workplaces, by exploring the environment and uncovering possible problems and issues workers may face.

The GEs are also *real-time systems,* where users interact with the environment in real time, receiving visual (and possibly other types of) feedback. This is the opposite of more traditional rendering techniques that generate still images or even videos of extreme high-quality visuals, but that are completely pre-defined and can only be passively analyzed. GEs on the other hand, allow for users to freely explore the environment, being able to spend as much time as wanted and the areas they judge most important in the depicted work-system.

Furthermore, GEs place the user as the protagonist of the VE exploration – the *agency* concept. Users are free to explore and interact with the models, characters and so on, but they have to choose to do so and they also can choose how they want to pursue that. Users are assigned avatars, either first-person or third-person ones (or both) which increase their awareness of their (virtual) surroundings and increase their sense of presence. In turn, these characteristics boost users' understanding of the scenarios and activities that must be performed in the represented workplace.

In terms of the avatars and other characters present in the VEs, GEs enable a great deal of *customization of digital human manikins*. In general, human manikins employed in GEs do not take into account aspects such as anthropometry and biomechanical functions of the human body, that is GEs are not substitutes of traditional Digital Modelling and Simulation techniques and software. Nonetheless, it's possible for developers to customize the manikins' appearance and dimensions if wanted to and the animation of the characters can even be performed through Movement Capture technologies (such as Microsoft's Kinect and Xsen's MVN, for instance). Creating a library of digital human manikins designed specifically for developing VEs in GEs is a possibility that in the long term could save time and improve the utility of the VEs.

GEs can also be seen as *simulation platforms* allowing the user to simulate operational procedures, specific tasks and emergency situations. Students can be instructed to explore a VE simulating a real workplace focusing to discover possible problems int the workplace, developing their understanding of the situation and their ability of what to do in a similar situation in real life.

Furthermore, GEs allow the development of *collaborative virtual environments* which can be can be employed in the design stages of ergonomics interventions. The LAN configuration for creating multiuser applications can be further improved since it's possible to develop and deploy systems for allowing users to connect over the internet. As such GEs can be beneficial for distributed located teams working on a design proposal or to validate the analysis and representation of a workplace with workers in a different location.

Another unique characteristic of GEs is the variety of possible *interactions programming*. The GE used for the development of the VEs presented in the previous section, has a visual programming tool which allows people that do not know a traditional programming language (such as C++ or C#) to build and deploy interactions. These interactions may come in a variety of forms such as the ones implemented in the VEs presented (layout change, interacting with equipment/environment, dialogues with NPCs, user resources/information) or even in different ways such as adding, deleting and moving objects around, picking up objects across the VE, positioning different manikins and different postures across the VE or even implementing traditional ergonomics evaluation tools and protocols (e.g. RULA, NIOSH lifting equation) in the VE.

Finally, GEs are becoming specialist tools for deploying *VR and AR applications* which can be created with ease by using the pre-existing modules and tools in the GEs. More broadly, GEs allow for a multitude of user interaction modes, from traditional keyboard and mouse, to head-mounted displays (either mobile, such as Samsung's Gear VR or not, such as Oculus Rift and HTC Vive), passing though joysticks, MoCap controls (Microsoft's Kinect, Leap Motion), and so on giving developers several possibilities which can be employed in accordance to the needs and degree of immersion intended with the VE.

5 Conclusion

GEs have particular affordances which make them suitable for being employed in ergonomics design, analysis and education. These affordances are general possibilities which developers of VEs must gather and apply to their designs, focusing on enabling ergonomics evaluation in the VEs.

Furthermore, the affordances highlighted are not exclusive to GEs as some of them appear in Computer Aided Design and Engineering software, in Digital Human Modelling and Simulation solutions and even Discrete Events Simulation software. Nonetheless, GEs are unique by having all of those affordances at the same time in a single platform which differentiates it from the other solutions mentioned.

Employing GEs however does not substitute the use of traditional applications generally used in the ergonomics field nor the need for going to the actual workplace and talking to the real workers following the various existing methods for ergonomics intervention. It can be seen as another tool, which is very powerful and versatile, that can be summoned upon need for complementing the traditional software and even physical prototypes and models for fostering stakeholder involvement into the situation analysis and redesign.

Specially regarding the educational aspects of GEs' application, the aforementioned affordances enable educators to develop simulation environments which can help students to understand how the ergonomics intervention may unfold, how the issues/constraints present in the workplace can be addressed through design and the importance of involving workers in the analysis and design of their workplace. Naturally, training in a simulation does not account for al the variability of the real situations students and professionals face in their daily routine, however it gets closer to it and complements the traditional, wide-spread, lecture-based style of instruction.

Nonetheless, challenges arise when in the development process of VEs in GEs. The learning curve for creating scenarios in these platforms is somewhat steep, although simpler environments with basic interactions are relatively easy to achieve. Building a 3D VE in a GE also requires (obviously), 3D models and thus researchers and practitioners should account for that when planning to implement a VE. Several free 3D models library currently exist (such as SketchUp's the 3D Warehouse) and also for characters and animations (such as Mixamo) and traditionally GEs have their own marketplaces where developers share assets (for free or at a cost) ready for use in the VE development.

Further studies can investigate how well GEs can depict a real work situation, develop and test open resources for ergonomics analysis, design and education, create the needed anthropometric libraries of manikins and animations/postures for utilization in ergonomics focused-VEs. Finally, it could be interesting to investigate the feasibility of developing an virtual-physical system, integrating the GEs affordances and other physical simulation tools such as 3D printed scale models.

References

Aromaa S, Väänänen K (2016) Suitability of virtual prototypes to support human factors/ergonomics evaluation during the design. Appl Ergon 56:11–18. https://doi.org/10.1016/j.apergo.2016.02.015

Aziz E-SS, Chang Y, Esche SK, Chassapis C (2015) Virtual mechanical assembly training based on a 3D game engine. Comput Aided Des Appl 12:119–134. https://doi.org/10.1080/16864360.2014.962424

Borro-Escribano B, Del Blanco Á, Torrente J et al (2014) Developing game-like simulations to formalize tacit procedural knowledge: The ONT experience. Educ Technol Res Dev 62:227–243. https://doi.org/10.1007/s11423-013-9321-6

Braatz D, Toledo FM, Tonin LA et al (2011) Conceptual and methodological issues for the application of game engines in designs of productive situations. In: 21st international conference on production research

Gatto LBS, Mól ACA, Luquetti dos Santos IJAA et al (2013) Virtual simulation of a nuclear power plant's control room as a tool for ergonomic evaluation. Prog Nucl Energy 64:8–15. https://doi.org/10.1016/j.pnucene.2012.11.006

Koops MC, Verheul I, Tiesma R et al (2016) Learning differences between 3D vs. 2D entertainment and educational games. Simul Gaming 47:159–178. https://doi.org/10.1177/1046878116632871

Koutsabasis P, Vosinakis S, Malisova K, Paparounas N (2012) On the value of virtual worlds for collaborative design. Des Stud 33:357–390. https://doi.org/10.1016/j.destud.2011.11.004

Madni AM (2015) Expanding stakeholder participation in upfront system engineering through storytelling in virtual worlds. Syst Eng 18:16–27. https://doi.org/10.1002/sys.21284

Mcnamara C, Proetsch M, Lerma N (2016) Investigating low-cost virtual reality technologies in the context of an immersive maintenance training application. In: VAMR 2016, pp 621–632

Paravizo E, de Sagae VS, Santoro FFD, de Moura DBAA (2016) Desenvolvimentos De Ambientes Virtuais Com Game Engine Para O Planejamento E Projeto De Situações Produtivas. In: Anais do XXXVI ENCONTRO NACIONAL DE ENGENHARIA DE PRODUÇÃO. ABEPRO, João Pessoa, pp 1–18

Passos C, da Silva MH, Mol ACAA, Carvalho PVRR (2017) Design of a collaborative virtual environment for training security agents in big events. Cogn Technol Work 19:315–328. https://doi.org/10.1007/s10111-017-0407-5

Passos C, Nazir S, Mol ACA, Carvalho PVR (2016) Collaborative virtual environment for training teams in emergency situations. Chem Eng Trans 53:217–222. https://doi.org/10.3303/CET1653037

Zamberlan M, Santos V, Streit P et al (2012) DHM simulation in virtual environments: a case-study on control room design. Work A J Prev Assess Rehabil 41:2243–2247. https://doi.org/10.3233/WOR-2012-0446-2243

The Evaluation of Existing Large-Scale Retailers' Furniture Using DHM

Carlo Emilio Standoli[1](✉) (iD), Stefano Elio Lenzi[2](✉) (iD),
Nicola Francesco Lopomo[2](✉) (iD), Paolo Perego[1](✉) (iD),
and Giuseppe Andreoni[1](✉) (iD)

[1] Dipartimento di Design, Politecnico di Milano,
Via Durando 38/A, 20158 Milan, Italy
{carloemilio.standoli, paolo.perego,
giuseppe.andreoni}@polimi.it
[2] Dipartimento di Ingegneria dell'Informazione,
Università degli Studi di Brescia, Via Branze, 38, 25123 Brescia, Italy
{s.lenzi002, nicola.lopomo}@unibs.it

Abstract. Digital Human Modeling (DHM) can support the design of the working environment, to avoid the risks of work-related musculoskeletal diseases and disorders (WRMSD). Large Scale Retail Trade involves a large amount of workers, performing a huge number of tasks and activities. The aim of this study was to analyze the working activities and tasks and assess existing furniture by the use of DHM, thus to evaluate the risk of developing WRMSD. Two case studies were analyzed: supermarket clerks and cashiers, involved in restocking and cash desk activities. By using DHM approach, this furniture was assessed in terms of reachability and adjustability. Outcomes of this evaluation were realized in terms of reaching maps, comparing different genders and height/weight percentiles. Preliminary findings suggested the use of dedicated guidelines to choose and set-up furniture in these specific applications, underling the variety of issues present in the large-scale retail trade.

Keywords: Digital human modeling · Human factors · Large scale retail trade
Work-related musculoskeletal disorders

1 Introduction

Large-Scale Retail Trade presents a great number of working tasks and furniture types, related to the specific department and to the assortment of different goods. In order to safeguard the workers' health and wellbeing and to prevent Work Related Musculoskeletal Disorders and Diseases (WRMSD), a rigorous check of all the elements related to the workplace and working conditions, such as environment, furniture, employees' activities and tasks, is needed.

According to International norms (ISO 11228-1, 2, 3) [1] and Italian National Standards (TUSL 81) [2] and through the evaluations carried out by the internal service, employers should avoid all work-related risks as better as possible: supplying

© Springer Nature Switzerland AG 2019
S. Bagnara et al. (Eds.): IEA 2018, AISC 822, pp. 339–350, 2019.
https://doi.org/10.1007/978-3-319-96077-7_36

aids, redesigning furniture, or changing the processes at the basis of workers' activities and tasks are the proposed solutions.

Digital Human Modeling (DHM) represents and effective tool to assess ergonomics of complex workplaces, such as those of Large-Scale Retail Trade. Our research adopted the proactive ergonomics approach through DHM simulation of standard working tasks in this field, to evaluate the environment and the type of activities daily performed by workers. Our activity supported the evaluation of existing furniture and the verification of the redesign of novel working environment and furniture. Moreover, DHM could be used in the work-related risks' definition and related analysis. The use of DHM was already applied in previous experiences related to different case studies [3–5], to analyze existing furniture and interfaces and the Human-Product Interaction in term of reachability and adjustability.

The methodology is generally composed of four different steps: (1) an initial observational phase with a detailed task analysis; (2) a digital and biomechanical modeling phase; (3) importing models into a DHM software suite; (4) the simulation of activities and the evaluation of postures and tasks. This paper describes the outcomes of these activities applied to the Large-Scale Retail Trade.

In this paper, we specifically present two working case studies: (a) the check-out assistant and (b) the restocking clerks. For each of these categories of workers, we analyzed their environment, existing furniture (cash recorder desk and shelves placed at different heights) and the supporting aids, through an ethnographic research phase. The DHM software solution (Santos, Santos Human Inc.) was applied to the ergonomic analysis. These ergonomic evaluations were performed considering 3 representative avatars: 5th percentile female, 50th percentile male and 95th percentile male (Caucasian population). The avatars were anthropometrically differentiated by gender, somatotype and percentile, according to the Standard ISO 3411 [6]. Assessment was evaluated in terms of reaching maps and comparing them with the different percentiles. The results of such analysis are presented in depth in the next paragraphs.

Obtained results offered the opportunity to refine the given requirements to redesign the existing furniture, and a forecast of the ergonomic situation of employees.

2   Materials and Methods

The methodological steps of this research are four, here below described in detail.

Firstly, an ethnographic research was conducted, to deeply understand working procedures, activities and tasks adopted by Large Scale Retail Trade companies. Numerous activities were mapped, each of them characteristic of a different department.

Secondly, the working environment and the analyzed furniture were digitally redesign and simplified, to be placed in the DHM system.

Thirdly, the simulation of working postures related to the two case studies were carried out, using the Santos system (Santos Human Inc.).

Finally, the ergonomic analysis and assessment for different tasks and furniture types by using different avatars was carried out, through a biomechanical software suite integrated in the DHM System. These ergonomic evaluations were performed considering 3 representative height/weight percentiles, according to the Standard ISO 3411 [3].

2.1   Observational Phase

In order to understand the complex working environment represented by supermarkets and, more in general, Large Scale Distribution retailers, an exploration phase was needed. This phase lasted more than 3 months, to map all the different activities and tasks that are performed during a working day in the supermarkets' departments. The analyzed retailer represents a complex environment in which coexists 8 departments, each of them with its own population of high skilled workers (at different career stage, e.g. foreperson, assistant, auxiliary.). These employees are exposed to a huge variety of work-related risks and, to prevent WRMSD, employers have to characterize in detail all the employees' activities, tasks, and working environment. In most of the observed supermarkets, there were all the 8 departments (i.e., fruit and vegetables, milk products, grocery and general merchandise, delicatessen, butchery, fishery, bakery and pastry). The restocking activity is the most common one, done by every workers in the different departments. Moreover, for the department requiring handmade activities (e.g., butchery, fishery and bakery), there were high skilled and qualified employees. During this phase, an ethnographic research was carried out. It involved 3 researchers, allowing them to visit and observe 6 stores of different sizes, placed in different city areas – from the city center to its suburbs. Small and medium size stores were located in the city center, whereas the large ones in the city suburbs. In each store, at least 1 employee of each department was recruited. In addition, 6 cashiers were recruited, to analyze in depth their working flow. Employees were observed during their daily working shift, recording their activities using a camera, to cover all main possible working situations. Researchers tried not to interfere in the employees' activities. Regarding the cashiers, they were recorded by using 2 fixed cameras, one placed on the frontal plane and one on the sagittal plane; their observation lasted a couple of hours, that means different "customer to customer" cycles (the frequency of these cycles depends on when the observation took place, e.g. during the early morning, there were few people; at 7 pm, the supermarket was full of people).

Figure 1 presents how the observation took place and the cameras' set-up for the cashier case study.

This phase was useful for investigating the employees' activities, tasks and attitudes. Cashiers usually spend most of their working shift waiting for customers at their cash desk. Sometimes, when customers are just a few, cashiers are asked to support clerks in restocking and sorting activities; when the customers' flow raises again, they have to return at their cash desk. Cashiers maintain a static seated posture at the cash desk, waiting goods carried by the conveyor belt. They have to pass goods from the right hand, through the scanner, to the left hand and release them to the customer. Nowadays, cash desks are equipped with tri-optics scanner, that allow cashiers to easily pass products under the scanner, avoiding wrists' twisting and bending. When they have passed all the products under the scanner, they have to wait for the customer's payment: when customers use cash payment, cashiers are asked to give them the change and the receipt; when customers use credit-card payment, cashiers just have to

Fig. 1. Samples of the observations that were carried out. On the left, a cashier waiting for a customer. Cameras' set up is highlighted with green circles: one focuses the coronal plane, the other the sagittal plane. On the right, an employee during a restocking activity of the lowest shelves.

wait for the money transaction and, at its end, give the customer the receipt. The described process is iterative; in this paper, we will analyze the reachability and adjustability of the cash desk for the three percentiles and we're not interested in tasks' repetitiveness.

Sorting and restocking activities take place everywhere in the supermarkets' area and involve most of the employees (depending on the daytime and the sale trade). In each department, there are different kind of furniture and displays, that influenced the restocking activities (e.g., displays with shelves at different heights, refrigerated displays, tables). Clerks have to manually load goods into the shelves. To ease this activity, a preparatory phase is needed, during which a dedicated employee places goods' packages in a cart. Preparing this cart, he takes into account the product group and the correct placement in the sales area, to avoid loss of time and chaos. Clerks take such cart to the proper supermarket area and start sorting and restocking goods. Sometimes they use the cart as a support, to avoid lifting heavy goods. In the store area, there are small ladders, that clerks use to restock items placed in the highest shelves.

2.2   DHM Environment's Set-up

For evaluating the supermarket environment, its furniture and the related tasks and postures, a virtual environment was settled up. First of all, three Avatars were selected, defining their gender and height/weight percentile, according to the Standards. Secondly, the simulation environment was prepared and the furniture and aids to be evaluated were redesigned in 3D CAD, simplifying their geometries, to ease and reduce the computational effort. After these steps, the parameters of interest were defined (i.e., furniture reachability and adjustability) and all the variables that could influence the simulation were excluded (e.g., items on the shelves could influence avatar simulation).

Regarding the Avatar definition, these evaluations were performed considering 3 representative avatars, belonging to the Caucasian population, covering 90% of anthropometric sizing: 5th percentile female, 50th percentile male and 95th percentile

Table 1. Details of the selected percentiles, according to ISO 3411

Percentile	Gender	Height (cm)	Weight (kg)
5th	Female	156.6	62.8
50th	Male	173	78.7
95th	Male	190.5	94.6

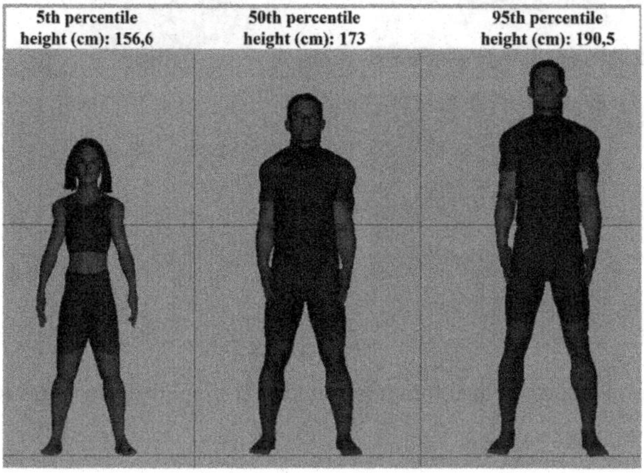

Fig. 2. Santos Avatars, corresponding to the selected percentiles. From left to. Right, the 5th, 50th and 95th percentiles, according to Standard ISO-3411

male. The avatars were anthropometrically differentiated by gender, somatotype and percentile, according to the Standard ISO 3411 (Table 1 and Fig. 2).

Regarding the implementation of the simulation environment, we simplified as much as possible the 3D models of the considered furniture. Simulation algorithm includes several functions related to the 3D geometry of the environment (e.g. collision avoidance, vision cone, reaching areas); a geometry simplification significantly reduces computational effort and time.

In addition, we avoided the use all the variables that could influence the simulation, such as the presence of displayed goods that could be considered additional constraints for the avatar movements. In real working conditions, the characteristics of goods (e.g., dimensions, weight, shape, where it is displayed, etc.) influence the working activities, workers' posture and strategies. For these simulations, we investigate the simple reachability tasks; further works will investigate more in details the working conditions. Furthermore, the main focus of this work was the general assessment of the workplace, considering posture as the most important parameter to be evaluated; for this reason, we did not evaluate grasping approaches (Fig. 3).

Cashiers. Each avatar was positioned in a sitting posture, in front of the cash desk, with one's arms on the cash-top. For each percentile, a stool and a footrest were provided. Both the stool and the footrest 3D geometries were simplified. The stool and

Fig. 3. Comparison among the real furniture and the 3D simplified versions, used for the DHM simulations.

footrest height were defined according to the percentile (the stool from 46 to 54 cm high; the footrest was from 7 to 17 cm high).

Avatars' hands had a related End-Effector; the final target to be reached was the area in front of the user, where the cash-drawer and the scanner are placed (Fig. 4).

Fig. 4. The green circles represent the End Effectors for both left and right hands.

Santos' Zone Differentiation tool was used to perform the evaluation. Using that tool, users can analyze information according to posture-based performance measures [7–9]. Zone Differentiation volume was created both for right and left hand; it was

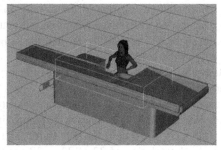

Fig. 5. On the left, the 5th percentile avatar positioned on the cash desk; on the right, the Zone Differentiation Volume used for DHM evaluation.

approximately $200 \times 170 \times 100$ cm, with a voxel resolution of $32 \times 32 \times 32$ cm (Fig. 5).

Clerks. Each avatar was placed in different postures and positions, according to the different shelves height. Three case studies were analyzed, grouping shelves at different heights: lower shelves, central shelves, higher shelves. Each shelf was 5 cm thick. The lowest shelf was at 20 cm off the ground; there were approximately 30 cm between the other shelves. The highest was at 165 cm off the ground. A ladder was provided, for the percentiles that cannot reach the higher shelves. As defined for the cashier's case study, 3D geometries were simplified, by using a cube 50 cm high.

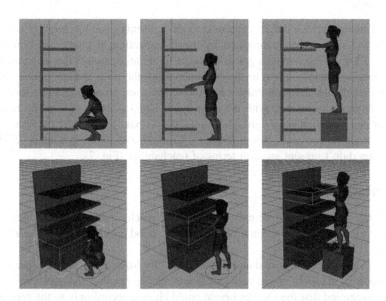

Fig. 6. In the first line of the figure above, the 5th percentile avatar positioned for the three case studies: lower shelves, central shelves and higher shelves. For the last case study, a ladder was provided. In the second line, highlighted in yellow, the Zone Differentiation Volume used for DHM evaluation.

The first avatar was in front of the display, in a crouch, with the arms close to the shelves. The second avatar was positioned in neutral standing posture, with the arms close to the shelves. The third avatar had the same posture of the previous one, but it was already on the ladder.

Each avatars' hand had a related End-Effector; the final target was the shelf, where goods are usually placed.

For each case study, Zone Differentiation volumes were created both for right and left hand. For the first case, they were approximately $150 \times 70 \times 80$ cm; for the second, $150 \times 70 \times 50$ cm; for the third, $150 \times 70 \times 30$ cm. For all them, we defined a voxel resolution of $32 \times 32 \times 32$ cm (Fig. 6).

3   Results and Discussion

A total of 24 simulations (6 for the cashiers and 18 for the restocking clerks) were conducted. Regarding the cash desk evaluation, in terms of reachability, the 5th percentile presents some issues compared to the other percentiles. Regarding the shelves, the 5th percentile could easily reach the lower shelves compared to the 95th percentile; for the higher shelves, the 5th percentile needs the use of the ladder. Results are described in detail below, using reaching maps for both right and left hand (for the cash desk). These maps define the comfort zone by means of a color gradient, that shifts from green (comfortable) to red (uncomfortable).

3.1   Cashier

Results demonstrate that the 5th percentile avatar presents good reachability values for both left and right sides, represented by the visualization of green and blue colour gradients. Cashiers have to pass goods from the right hand, through the scanner area, to the left hand and release them to the customer. Both the right and left conveyor belts are easily reachable; the cash-drawer and the scanner are properly placed. The receipt printer can be easily reached with the right hand. The 3 avatars could easily reach the conveyor belt placed at their right, pick up products, pass them under the scanner and leave them on the conveyor belt at their left. Compared to the 5th percentile, 50th and 95th avatars didn't present issue in terms of reachability (Fig. 7).

3.2   Clerks

Results underline a strong difference related to the reaching and comfort zones among the three groups of shelves. Central shelves present good results in terms of reachability (Fig. 8).

Lower shelves can be sorted and restocked hardly from all the different percentiles. Higher shelves, thanks to an aid such as a ladder, can be reached in a quite good way. It is useful to remind that the shelves height could change accordingly to the products to be sold and the display used for this evaluation represents just one of the possible combinations.

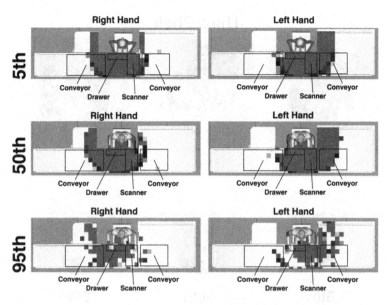

Fig. 7. Results of the comfort zone evaluation for the cash-desk. Results are expressed for all the analyzed percentiles, using reaching maps for both right and left hand.

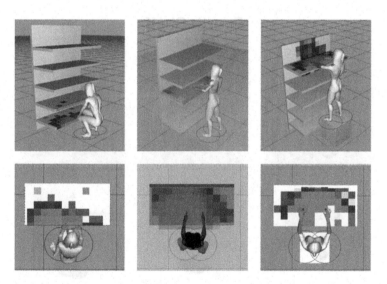

Fig. 8. Reaching maps of the analysis of the three case studies related to the 5th percentile. Results are shown using two different perspective.

50th and 95th avatars didn't present issue in terms of reachability of the central-higher shelves. As happened for the 5th percentile, they had problems both in postures and reachability for the lower shelves. The central shelves can be easily restocked from all the three percentiles, as compared in Figs. 9 and 10.

Third Shelf

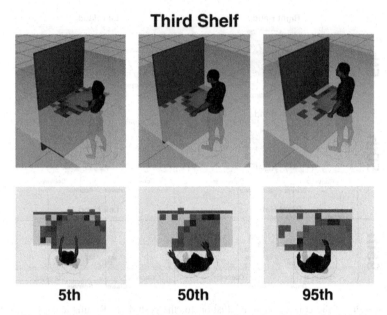

Fig. 9. Comparison among the three percentiles related to the third shelf.

Fourth Shelf

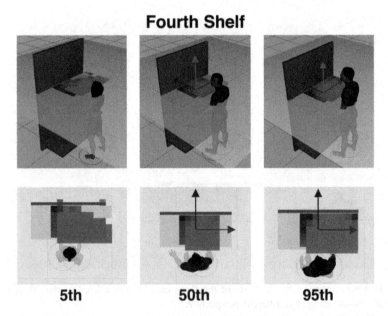

Fig. 10. Comparison among the three percentiles related to the fourth shelf.

Aids could ease the working activities related to the medium-high level of shelves; shelves placed at the lowest level represent a problem in terms of both posture and reachability.

Sometimes it is possible to use in the store furniture with more than 5 shelves, placed at different heights; it mainly depends on the goods typology. DHM can ease the evaluation of the WRMSD risks of these cases thanks to virtual simulation.

4 Conclusion

The methodological approach here described allows for the evaluation of working environments and furniture and the related WRMSD risks in a faster, accurate and detailed mode, compared to standard methods, such as recruiting a large population and making tests with real prototypes and furniture, collecting questionnaires and acquiring human kinematics.

The ethnographic research phase, done in readiness for the evaluation phase, revealed itself essential, in order to proper understand the working environments and furniture and related tasks and activities.

Performing tests at the CAD-level means a considerable cost advantage; if needed, it allows manufacturers to modify and re-design furniture. Especially in Large Scale Retail Trade, where the shelves' configuration need to be frequently changed according to the displayed goods, the use of a DHM evaluation represents an effective approach. Making the same tests in the real-world environment on real users (of the three percentiles), means high time-consumption, high costs and reduced possibilities to modify and implement both environments and furniture.

In conclusion, this paper demonstrates how DHM represents a promising and useful tool, to be used since the early design phases of working environment and furniture. Further developments and future works need to investigate how products and their features (i.e., typology, weight, dimension) can influence the working activities, postures and grasps.

References

1. International Standard ISO 11228—Parts 1, 2, 3 (2007)
2. D.Lgs. 81/08, Testo Unico in materia di salute e sicurezza nei luoghi di lavoro—TUSL8 (2008)
3. Mazzola M, Forzoni L, D'Onofrio S, Standoli CE, Andreoni G (2014) Evaluation of professional ultrasound probes with Santos DHM. Handling comfort map generation and ergonomics assessment of different grasps. In: Proceedings of the 5th international conference on applied human factors and ergonomics, AHFE 2014, Kraków, Poland (2014)
4. Mazzola M, Forzoni L, D'Onofrio S, Marler T, Beck S (2014) Using Santos DHM to design the working environment for sonographers in order to minimize the risks of musculoskeletal disorders and to satisfy the clinical recommendations. In: Proceedings of the 5th international conference on applied human factors and ergonomics, AHFE 2014, Kraków, Poland
5. Mazzola M, Forzoni L, D'Onofrio S, Andreoni G (2016) Use of digital human model for ultrasound system design: a case study to minimize the risks of musculoskeletal disorders. Int J Ind Ergon
6. International Standard ISO 3411 (2007) Earth moving machinery—Physical dimension of operators and minimum operator space envelope, 4th edn

7. Yang J, Marler T, Beck S, Abdel-Malek K, Kim H-J (2006) Real-time optimal reach-Posture prediction in a new interactive virtual environment. J Comput Sci Technol 21(2):189–198
8. Yang J, Marler T, Kim H-J, Arora JS, Abdel-Malek K (2004) Multi-objective optimization for upper body posture prediction. In: 10th AIAA/ISSMO multidisciplinary analysis and optimization conference, Albany, NY, USA
9. Yang J, Verna U, Penmatsa R, Marler T, Beck S, Rahmatalla S, Abdel-Malek K, Harrison C (2008) Development of a zone differentiation tool for visualization of postural comfort, SAE 2008 World Congress, Detroit, MI, USA

Ergonomic Evaluation of a Prototype Console for Robotic Surgeries via Simulations with Digital Human Manikins

Xuelong Fan[1]([✉]), Ida-Märta Rhén[1], Magnus Kjellman[2], and Mikael Forsman[1]

[1] Institute of Environmental Medicine, Karolinska Institutet, 171 77 Stockholm, Sweden
{xuelong.fan,ida-marta.rhen,mikael.forsman}@ki.se
[2] Department of Molecular Medicine and Surgery, Karolinska Institutet, Stockholm, Sweden
magnus.kjellman@ki.se

Abstract. Work-related musculoskeletal disorders impact surgical performance, which increase risks for patient safety. A new console has been designed to reduce workload for robotic surgery surgeons. Due to high costs and long waiting time of the production process, a pre-production ergonomic evaluation of the new design is preferable. In this paper, we evaluate if the new console at the pre-production stage by using an US checklist, and the Swedish standard for visual display unit work. A 3D model of the new designed console was introduced to the virtual environment of a digital manikin (Intelligently Moving Manikin, IMMA). The work-ranges of the console were calculated. Various individual work distances of 12 manikins (3 men and 3 women per each of the US and the Swedish population) were "measured". The data were integrated and used as an objective reference to compare with the Swedish standard, and the US checklist. The result shows that the criteria in the Swedish standard and the US checklist are fulfilled, except for those are related to the adjustable range of the screen view height, the height range of the armrest and the adjustable distance of the pedals. The new console fulfills most of the criteria in the checklist and the standard, but there is room for a few improvements. The DHM tool IMMA provides the possibility for a pre-production assessment. However, the limited virtual measurement tools of IMMA restrained the time efficiency of the ergonomic assessment.

Keywords: Ergonomic evaluation · Robotic surgery · Digital human manikins

1 Introduction

Work-related pain, fatigue, stiffness and numbness, especially related to the neck, back and shoulders, are frequently reported among surgeons in laparoscopic practice (74%) [1]. Such symptoms can affect surgical performance [2] and likely the patient safety.

Recent years' introduction of a robot-assisted laparoscopy approach provides ergonomics benefits for the surgeon [3, 4]. However, it still involves constrained and

© Springer Nature Switzerland AG 2019
S. Bagnara et al. (Eds.): IEA 2018, AISC 822, pp. 351–363, 2019.
https://doi.org/10.1007/978-3-319-96077-7_37

static working postures associated with risk factors for developing musculoskeletal disorders (MSDs) [5]. A new surgical system has been designed to allow the user more flexible working postures by e.g., offering an open display. The operation system that is so far not yet produced will consist of two main parts, a cart with interactive robotic arms which are in contact with the patient, and a console including hand controls, foot controls, and two screens (one 3D monitor in front of the surgeon, and one side screen for displaying medical images and to control the panel when necessary) which provides the surgeon to control the robotic arms and monitor the surgical site. Moreover, the console is equipped with an armrest frame. For flexibility between various users, the foot controls are adjustable in forward and backward direction and the armrest and the 3D monitor are adjustable in height. Both screens are also adjustable in angle position. The hand controls have seven degree of freedoms that allow for flexible control of the robotic arms.

However, the prototype with its promising improvements has not been ergonomically evaluated. The promising improvements have not been yet validated. Since the console is a complex and expensive system, manufacturing process can be both time- and finance-consuming. Any defects from the prototype can multiply the total cost. Therefore, a pre-production ergonomic evaluation is preferable to identify potential risks of the prototype.

Digital human modeling (DHM) tools have been used in human activity simulation within multiple scenarios in industry [6]. Most manikins require manually setting up every joint when using the model. To overcome the weakness of this method, a new IMMA (Intelligently moving manikins) was developed for simulation and visualization of human work and for ergonomic assessments [7]. The IMMA tool simulates body postures and motions required by assigning joint angels to the internal model of a manikin's skeleton [8, 9]. Within the IMMA tool, an anthropometric module allows the user to specify the requirements for anthropometric diversity in the assessment without repetitively setting up each individual manikin.

2 Aim

The aim of this project is to evaluate a prototype laparoscopy robotic console design at the pre-production stage, using a US checklist and the Swedish standard, that are relevant for visual display unit work, as references.

3 Method

A 3D model of a prototype console was introduced to the virtual environment of IMMA. Two manikin families were implemented, which represented 90 per cent of the US and Swedish population, respectively. Each family contained three digital manikins that referred to a low, medium and upper border of the targeted population. The manikin was controlled in a seating position in the measurement, where the angle between the lower leg and thigh was 90 degree, as recommended in both ergonomic standards.

3.1   Measurements in the Virtual Environment

Several landmarks were set up to conduct the measurements. Reference points and reference planes were set up to facilitate the comparison between the manikin's and the console's work range. The prototype was designed in a condition that the user's chair is equal or higher than 43 cm. Therefore, the manikin sitting height was regulated accordingly. All measurements were static. The possible work ranges of the manikins and of the console were calculated from the data and was normalized. The results were compared with each other.

3.2   Ergonomic Standards

Since ergonomic standards within surgery are lacking, and since the work posture in the console is similar to computer work, in this study ergonomic standards and checklists criteria for computer work were used for the ergonomic assessment as references. Swedish regulations for computer work places were used for comparisons with geometries from simulations using the manikins representing the Swedish population [10], which in some parts refer to the regulation for workload ergonomics [11]. The American checklist for computer workstations [12] was used in comparisons with geometries from simulations with US population manikins.

The Swedish standard is only available in the Swedish language and was therefore first translated into English. The text was then shortened and transferred into a table checklist to facilitate the comparison and the presentation of the outcomes. Parts and paragraphs of the checklist and standards not relevant for the present assessment, e.g., details related to the chair, glare from the monitor or windows as well as placement of accessories, were not considered in this study.

3.3   Comparison of Outcome Measures to Checklist and Standard

The ergonomic assessment was guided by the checklist and the standard, and was verified by comparison of the parameters of the optimized working postures of manikins and parameters that were allowed by the adjustment ranges of the design of the console.

4   Results

4.1   Work Ranges

The comparison between the work ranges of the console and the recommended work ranges of each population group was illustrated in Fig. 1 (Details of manikin measures and console measures were excluded for confidential issue).

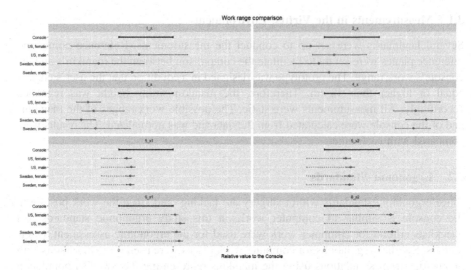

Fig. 1. Comparisons between the adjustable ranges of the console and the optimized work ranges of different manikins in the family. 1_x is between height of 3D monitor and height of manikin eyes; 2_x is between height of the top of the armrest and height of manikin elbow; 3_x is between height of the bottom of the armrest and height of manikin thigh; 4_x is between distance of the foot control and distance of manikin feet to the reference point; 5_x1 (5_x2) refers to the comparison between work depth of hand control and proximal work depth (intermediate work depth) of manikin (solid line represents the range of limits of the work depth of different manikin); 6_x1 (6_x2) refers to the comparison between lateral work range of hand control and proximal lateral work range (intermediate lateral work range) of manikin (solid line represents the range of limits of the work depth of different manikin).

4.2   Ergonomic Assessments

US Checklist. Figure 2 showed the results from US checklist. Criteria that were not fulfilled were listed here.

Original from evaluation were listed here.

Working Postures #1. Since the top height of the 3D monitor, when adjusted to lowest position, is in line with the eye height of the tallest female manikin, the shortest, and average female manikins, as well as the shortest male manikin, must bend their head backwards to be able to look at the top of the 3D monitor. For the average male manikin, the screen is within an adjustable range for the eye height, but for the tallest male manikin, the screen, when adjusted to the highest position, is too low, which means that the manikin always has to bend the neck forward when watching the screen.

Working Postures #6. None of the female manikins can achieve a relaxed sitting position for the forearm when using the armrest. When adjusted to the lowest height, the armrest is still too high (53 mm) for the tallest female manikin. For the shortest male manikin, the situation is the same, but both the average and tallest male manikins can achieve a relaxed sitting position.

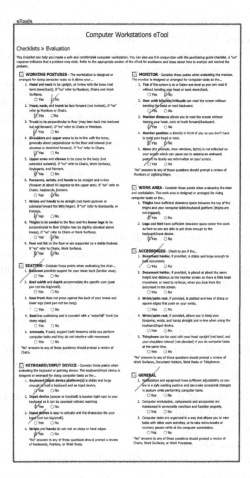

Fig. 2. Computer workstations eTool outcomes regarding to the US shortest, average and tallest male and female population. The crossed questions are considered irrelevant in this study.

Working Postures #8. While sitting with the thighs parallel to the floor, none of the manikins is able to reach the foot controls. The tallest manikin is within the closest range to reach, but still 28 mm in horizontal direction is missing. The missing distance is considerable longer for shortest female manikin, i.e., 141 mm. To be able to reach the foot controls, the knee angle between the thighs and the calves must increase. For interest, the knee angle was measured when the manikin was set to reach the foot control, and the knee angle, calculated as the angle between the calves and a line perpendicular to the floor, increased with 6° (tallest male manikin) to 30° (shortest female manikin) within the manikin family.

Working Postures #9. When placing the manikins in a seated position in front of the console, the feet can rest flat on the floor. But during surgery work, the user has to use the foot controls frequently, which means that only the heels may rest on the floor.

Monitor #1. See "Working postures #1".

Monitor #4. The 3D monitor is positioned directly in front of the user. The side screen, positioned at the side of the console armrest, is generally used only during a few minutes of the working time and therefore not considered appropriate to assess further.

Swedish standard. Table 1 showed the results from the Swedish standard. Original from evaluation were listed here.

Paragraphs within the standard not fulfilling the requirements in the Swedish standard

3. All screens, armrest and handles are adjustable in height, which means that the work posture, to a certain extent, can be changed while seating. However, it is not possible to raise the console to a level suitable for standing work.

4a. Only the tallest female manikin and the average and the tallest male manikin are able to use the armrest properly, i.e., while sitting with the forearm in a relaxed position. For the rest of the manikins, the armrest was too high, i.e., was not adjustable to a lower level.

4b. The armrest is narrow and tilted inward and cannot support the entire forearms.

5a. When adjusted to lowest position, the top height of the 3D monitor is above the eye level of the shortest and average female manikins as well as for the shortest male manikin. When adjusting the screen to the highest position, it is still too low for the tallest male manikin. For the tallest female manikin and for the shortest and average male manikins, the 3D monitor is within the adjustable range for the eye level.

5c. The screens are able to tilt but are not sufficient adjustable in height, see #5a.

5d. See question 5a.

6b and 11b. See question 4a.

12c and 13d. The console's limited adjustability in height, reduce the possibility to work in a standing posture.

13b and 13c. While having the manikins placed in front of the console, the feet can rest flat on the floor. However, during surgery work, the foot controls are used frequently.

Table 1. The Swedish standard for computer work summarized in a table together with the outcomes from the comparison with manikin-console measures

Question	Standard	Assessment	Requirements	The requirements are fulfilled YES or No
1	AFS 1998:5 Computer work §2	Screen and viewing conditions	— The screen is free from flicker, reflections and reflexes. — If characters on the screen, they are sharp, large and contrasted enough to be easily read. — There is enough space between characters and rows to enable good readability. — Brightness or contrast between characters and the background of the screen is easy to adjust. Luminance contrast between characters and background should not be less than 3:1.	— Not relevant — Not relevant — Not relevant — Not relevant
2	AFS 1998:5 Computer work §3	Lighting and viewing conditions	— Keyboard has matte finish to avoid glare. — The work surfaces are low reflective.	— Not relevant — Not relevant
3	AFS 1998:5 Computer work §4	Work postures and work movements	— The workplace is dimensioned, designed and equipped to enable comfortable and varying work postures and work movements. — It is important that the user can easily raise and lower the table to adapt the work posture if necessary. The display may need to be placed on an adjustable surface. Being able to switch between sitting, standing while working on a monitor contributes to more varied, flexible work postures, and work movements.	— Yes — No
4	AFS 1998:5, Computer work §4	Work postures and work movements	— The space at the keyboard is large enough to enable support of arms and hands on the table surface. — When using a control device, for example computer mouse, the work can be performed with the forearms relaxed on the table, so that the shoulder muscles are supported. — In "computer mouse work", the entire forearm is supported. — The control device can be placed and used in close connection to the keyboard to avoid movements with external rotation of the wrist. — Work with extended arm or external rotation in the shoulder joint can be avoided. — For more intensive use of the computer mouse, the keyboard may need to be slid or moved away easily so that the computer mouse can be placed directly in front of the worker. — An entire desk area in one plane allows keyboard and computer mouse to be easily placed after use. — Essentially, the computer mouse has a shape that prevents work with the wrist in bent position. — From a load point of view, it is also important	— No — No — Not relevant — Yes — Not relevant — Not relevant — Not relevant — Not relevant — Not relevant

Table 1. (*continued*)

			that the user can plan and put up his work and take breaks so that excessive use of computer mouse can be avoided.	
5	AFS 1998:5 Computer work §5	Work postures and work movements	— Screen and keyboard or equivalent are positioned so that they can be adapted to the individual worker's physical dimensions, i.e., working height and viewing angle of the screen becomes appropriate. — The distance between the user and the screen is app. 70 cm and the size of characters 3.5-4 mm.	— No — Yes
			— As far as possible, the keyboard and monitor are able to rotate, tilt and move according to the workers' needs. — The monitor is positioned so that the worker's neck is straight and slightly flexed when viewing the monitor. For normal office work or similar work situations, it is important that the entire display can be placed below eye level.	— No — No
6	AFS 2012:2 Workload ergonomics §5, Work postures and work movements	Work height	— A well-designed workplace is, among all, characterized by work in an upright position (most of the time) with lowered shoulders and upper arms close to the upper body. — The work height is approximately at an elbow height for the person performing the work, whether in the case of sitting or standing work.	— Yes — No
7	AFS 2012:2 Workload ergonomics §5, Work postures and work movements	Work area for the hands	The outer work area of the hands in the horizontal plane is limited by the length of the arm. Most of the hands' work is performed within the inner work area i.e., within a rectangle (measured from the nose of the user) of 30 cm (forward distance) and 60 cm (side width). See picture. (The more long-lasting and precise work tasks, the more important it is that the work is performed with entirely relaxed arms and shoulders close and in front of the body, i.e. central within the inner work area)	— Yes
8	AFS 2012:2 Workload ergonomics §5, Adjustment of the work equipment and work desk.		— It is easy and quick to change the setting on the desk and the work chair if several workers alternately use the same worktable more than temporarily.	— Not relevant
9	AFS 2012:2 Workload ergonomics §6, Manual	"Work with handheld controls"	In order to reduce the risk of stress disorder, the employer should provide the workers with handheld machines and hand tools that: … allows appropriate grips that are adapted to the requirements of power and precision, with good friction and where the grip is well spread	— Not relevant

(*continued*)

Table 1. (*continued*)

	handling and other loaded exertion.		over hand to avoid inappropriate point pressure, e.g., no sharp edges. ... fits different users hand sizes. ... are possible to use with both the right and left hand. ... allows a neutral position in wrist and arm (hands are relaxed, resting on a table). ... allows to see and reach the tool/work-piece. ... has pressure gears with reasonable operating resistance. ... vibrates as little as possible. ... is as lightweight as the function allows. ... are well balanced.	
10	AFS 2012:2 Workload ergonomics, Assessment model "sitting work"	Neck	— The neck is in the middle position — It is possible to freely move the neck	— Yes — Yes
11	AFS 2012:2 Workload ergonomics, Assessment model "sitting work"	Shoulder/Arm	— Working height and reaching area is fitted to the task and the individual — There is good arm support	— Yes — No
12	AFS 2012:2 Workload ergonomics, Assessment model "sitting work"	Back	— There are opportunities for free movements — The backrest is well-designed — Possibility to switch to standing	— Yes — Not relevant — No
13	AFS 2012:2 Workload ergonomics, Assessment model "sitting work"	Leg	— There is free space for the legs — Good footrest is available — Rarely leg or foot control work — Possibility to switch to standing	— Yes — Yes — No — No — No

5   Discussion

This study used the U.S. checklist and the Swedish standard requirements for computer work as references to evaluate the ergonomics of the prototype console from the outcome measures, i.e., distances and angles, from two manikin families (one based on the US population anthropometric data and the other on Swedish population data) placed in working position in the prototype console. The measures were obtained from the digital human modeling tool IMMA which provides objective and detailed information.

5.1   Comparison Between Console-Manikin Outcome Measures and Checklist/Standards Requirements

The outcome of the comparisons showed that several of the requirements (relevant for the study) were fulfilled, but also that a number of requirements were not satisfied.

Both manikin families were within the required viewing distance range, i.e., the preferred viewing distance, measured from the eyes to the front surface of the screen, which according to the OSHA should be within a range between 50 and 100 cm while the Swedish standard recommend approximately 70 cm. Moreover, all manikins could take advantage of the flexibility of the control handles, which allows the user to work with reduced load, with the arms close to the body (within the inner work area) and with the wrists in a neutral position.

The armrest allowed for a relaxed arm working posture, but it was not enough adjustable in height to suit all the manikins. Also, the adjustability of the screen height and tilt allowed for individually adapted working postures, however, the limitation of adjustability in height reduced the ability to achieve a good working position for some of the manikins.

Requirement regarding armrest height was not fulfilled while placing the manikin in an ergonomically correct sitting posture. Several of the manikins were not within the range of the adjustable height, and in about half of the cases, the height of the armrest was too high for the user to be able to sit in a relaxed position. However, if working in a position with arms stretched forwardly (i.e., upper arms not parallel to the back) the elbow angle increases, which means that the height of the forearm (measured from the floor) also will increase. In such a working position, the lowest height of the armrest may be suitable for those manikins where the armrests were too high while sitting with the upper arms ergonomically correct.

Neither the adjustability range of the 3D monitor height was suitable for all manikin members. However, we have not considered the preferred viewing angle to the center of the screen, which according to OSHA normally should be located 15°–20° below the horizontal eye level and should not to be greater than 60°. Considering this in the presented assessment, the user could possibly achieve a suitable working posture of the neck despite the limitation of screen adjustability in height.

The limitation of the adjustability of the 3D monitor and the armrest height also affects the possibility to work in a standing position. Long time sitting work is a risk factor associated with cardiovascular disorder, lower back pain and diabetes, which speaks for further investigation of the ability to switch between sitting and standing work. The use of foot controls while standing may be problematic, especially if both feet should be used simultaneously, but could be solved by also have some of the foot control functions combined with the hand control system.

Stretching the legs forwardly to reach the foot controls means that an ergonomically correct sitting posture may be difficult to maintain. Long time stretch of the legs in combination with a forward bent sitting posture may negatively affect the nervous system (e.g., the sciatic nerve) and the lack of support from the feet tend to affect the back posture causing long-term nerve irritation or/and back pain.

While sitting with the knees at a 90° angle, there is no strain on the low back. But when stretching for the foot controls, the thighs will be forced into a downward direction, which in turn will increase the knee angle and the back curvature, pulling on the low back and creating muscle strain.

Even though the armrest will support the forearm, support for the wrists is lacking. In computer mouse work, support for the wrist is essential, since long-term work without support is associated with overload of the wrist, potentially causing MSDs. In the present work with handheld controls, such support can both affect the work in a negative and positive direction. A wrist support may limit the flexibility of the control and the ability to move the wrist easily and freely. At the same time, such support can contribute to a more relaxed work posture for the wrist with reduced muscle activity, reducing the risk of wrist related MSDs.

5.2   The Use of a DHM Approach for Pre-production Assessment

The DHM approach allows for a pre-production ergonomic assessment of a product in the design phase. Here, the manikin families represented the stature characteristics of the populations required, and the size of the console prototype imported to the IMMA environment was consistent with the real measures of the product. Hence, the manikin postures in interaction with the console can represent the reality, which contributes to a good reliability of the results.

Accuracy and approximation. The population database that was applied was from 1989 (for US) and 2008 (for Sweden). It is known that anthropometry of a certain population evolves through year. Therefore, there could be a bias of the result due to the outdated data. Additionally, the operation in reality involves a series of complex movements and various possible postures. The assessment would be too complicated to perform if including all the factors and movements were taken into consideration. Therefore, certain approximations were applied before the measurement and assessment with two bottom lines. One is that a posture will be excluded if it is not maintained for significant long time during a typical operation. Since certain approximations have to be made due to the limitation of technical reasons, the quantity of the approximation was controlled to decrease effects on the outcome.

Limitation of the assessment. In the present study, we only assessed the manikin from an "upright sitting" posture. However, there are other ergonomically correct sitting alternatives, e.g., the declined sitting posture (OSHA) where the buttock is higher than the knee and the hip angle, i.e., the angle between the thigh and the spine is greater than 90°. In such a sitting position, the armrest, which in the upright sitting posture is slightly too high, may be in a suitable position. However, in order to objectively assess the working condition, clear definitions are required, and "declined sitting posture" is a far too wide term to set up the case.

Moreover, IMMA offers different view modes for better visualization of 3D objects. However, there is a lack of consistency among those modes. For example, the thigh of the manikin is modeled as a cylinder in the skeleton mode while in the rendering mode the thigh has an uneven thickness of the knee end and the hip end. It may confuse user when setting up the measurement markers for the estimation of sitting height when switching different view mode. It would be beneficial if an embedded measurement system is implemented to minimize the confusion. It would be even more useful if the embedded system is designed based on ergonomic checklists.

In general, IMMA was designed for better performance in dynamic ergonomic assessment. However, static postures were the "key frames" for motions. The difficulty of setting static postures may influence the effectiveness and efficiency of the total process. An improvement in functions regarding to manipulations of static postures of manikins can not only facilitate static ergonomic assessment such as this study, but also increase precision and accuracy of dynamic ergonomic assessment.

6   Conclusion

The console with its flexibility, i.e., the adjustable 3D monitor, foot controls, controlling handles and the armrest frame, fulfilled most of the requirements in the checklist and standards for all members of the two manikin families. There were, however, a few of the requirements that were not fulfilled for all the population representing manikin families, because of too limited adjustable ranges that restricted, especially the shortest manikins, to work in recommended postures. The limited adjustability also reduces the possibility of working in a standing posture. Hence, considering the design revisions, there is still room for a few improvements. The DHM tool IMMA provided the possibility to compare static surgery work in the digital prototype phase to ergonomic checklists and standards for visual display unit work, considering human diversity. The limited virtual measurement tools of IMMA restrained the time efficiency of ergonomic assessment. Improvements may also be made to the IMMA tool for these types of evaluations.

References

1. Alleblas CCJ, de Man AM, van den Haak L, Vierhout ME, Jansen FW, Nieboer TE (2017) Prevalence of musculoskeletal disorders among surgeons performing minimally invasive surgery. Ann Surg 266(6):905–920
2. Huysmans M, Hoozemans M, van der Beek A, de Looze M, van Dieën J (2010) Position sense acuity of the upper extremity and tracking performance in subjects with non-specific neck and upper extremity pain and healthy controls. J Rehabil Med 42(9):876–883
3. Zihni AM, Ohu I, Cavallo JA, Cho S, Awad MM (2014) Ergonomic analysis of robot-assisted and traditional laparoscopic procedures. Surg Endosc 28(12):3379–3384

4. Lawson EH, Curet MJ, Sanchez BR, Schuster R, Berguer R (2007) Postural ergonomics during robotic and laparoscopic gastric bypass surgery: a pilot project. J Robot Surg 1(1):61–67
5. Yu D, Dural C, Morrow MMB, Yang L, Collins JW, Hallbeck S, Kjellman M, Forsman M, Yu D (2016) Intraoperative workload in robotic surgery assessed by wearable motion tracking sensors and questionnaires. Surg Endosc 31(2):1–10
6. Demirel HO, Duffy VG (2007) Applications of digital human modeling in industry. In: Digital human modeling. Springer, Berlin, Heidelberg, pp 824–832
7. Hanson L, Högberg D, Carlson JS, Bohlin R, Brolin E, Delfs N, Mårdberg P, Stefan G, Keyvani A, Rhen IM (2014) Imma—intelligently moving manikins in automotive applications
8. Bohlin R, Delfs N, Hanson L, Högberg D (2012) Automatic creation of virtual manikin motions maximizing comfort in manual assembly processes. In: technologies and systems for assembly quality, productivity and customization: proceedings of the 4th CIRP conference on assembly technologies and systems, pp 209–212
9. Delfs N, Bohlin R, Hanson L, Högberg D, Carlson J (2013) Introducing stability of forces to the automatic creation of digital human postures. 2nd International Digital Human Model, 2013
10. Arbetsmiljöverket (1998) Arbete vid bildskärm, AFS 1998:5
11. Arbetsmiljöverket (2012) Belastningsergonomi, AFS 2012:2
12. U.S. Department of Labor (2018) "Computer Workstations eTool Checklists Evaluation," pp 3–5

Feasibility Evaluation for Immersive Virtual Reality Simulation of Human-Machine Collaboration: A Case Study of Hand-Over Tasks

Shang-Ying Hsieh and Jun-Ming Lu[✉]

National Tsing Hua University, Hsinchu, Taiwan
jmlu@ie.nthu.edu.tw

Abstract. The purpose of the study is to evaluate the feasibility of simulating human-machine collaboration via immersive virtual reality. six participants were recruited to collaborate on a hand-over task with a collaborative robot in the real world and Immersive Virtual Environments (IVE). There are five scenarios included in the experiment: real environment, IVE, IVE with auditory feedback for contacts, IVE with visual feedback for contacts, and IVE with both auditory and visual feedback for contacts. Although the results showed the difference of motion strategies between virtual and real environment, the task performance was no difference among scenarios.

Keywords: Human-machine collaboration · Immersive virtual reality
Motion analysis

1 Introduction

Nowadays, collaborative robots which work concurrently with human operators in a shared workspace have already become a feasible option for production lines. Compared with traditional industrial robots, collaborative robots have advantages of better flexibility, lower costs and lightweight. They are often used to help human operators carry out laborious and repeated jobs, such as pick-and-place, work pieces fixing and materials handling. Meanwhile, collaborative robots reserve some tasks relying on decision making for human to improve the flexibility. It is suitable for medium-sized companies to automate their production lines by introducing collaborative robots. According to ABI Research [1], collaborative robotics market will exceed 1 billion U.S. dollars in 2020, and the estimated units of collaborative robotics shipped will be 40,036 then. There is a huge market potential of collaborative robots.

Considering the direct interaction with operators, safety issues of collaborative robots should be concerned more. Also, how to make operators collaborate with machine fluently is important to production lines. Hence, prior evaluation should be set with the urgent priority to confirm the feasibility of introducing collaborative robots. Recently, several studies began to apply virtual reality to simulate human-machine collaboration. By using virtual reality, evaluation can be conducted in a safer and less expensive. Matsas et al. [2] developed a virtual training system for human-machine

S. Bagnara et al. (Eds.): IEA 2018, AISC 822, pp. 364–369, 2019.
https://doi.org/10.1007/978-3-319-96077-7_38

collaboration in manufacturing tasks. They integrated Kinect^{TM} motion sensing device and eMagin z8000 3DVisor^{TM} HMD to realize the interaction between human and virtual collaborative robots. Nevertheless, the feasibility depends on the similarity between real and virtual environment. When user wear a Head Mounted Display (HMD), their perceptual information is influenced by restricted field of view and the quality of interface. These technical issues cause users to generate different reaching behaviors [3]. Besides, Weistroff et al. [4] compared people's acceptability between real and virtual environment by investigating psychological and physiological measures. It was concluded that sense of presence in virtual reality influences the perception of the existence of robots.

To sum up, using virtual reality to simulate human-machine collaboration could be an alternative for evaluation or training. However, there is still a need to clarify the difference between real and virtual interactions when operators collaborate with robots. In this study, the feasibility of simulating human-machine collaboration via virtual reality will be investigated by analyzing motion strategies and subjective questionnaires.

2   Methods

Three males and three females (23.4 ± 1.3 years old) were recruited to perform a handover task with a collaborative robot in both virtual and real environments. None of them had experienced immersive virtual reality and industrial robots. Due to the restriction of experimental resources, the Wizard of Oz method was used in the real environment. A FARO Arm was operated manually by a test giver who hid behind a board, while participants were informed that they were working with an automated robot. In the virtual environment, the same experimental scene was reproduced by using an Unity3D. Participants wore a HTC Vive HMD and interacted with the virtual scene via gesture recognition by Leap Motion.

Participants were asked to pick 15 plastic balls delivered by the robotic arm and put them each of them into one of the three boxes with corresponding colors, including red, green and blue. There were five balls in each color. During the task, robotic arm consistently picked a plastic ball to a designated region which is called collaborative zone. When the robotic arm clamping a ball stopped at the collaborative space, participants should grab the ball off the arm (see Fig. 1) and put it into the adequate box (see Fig. 2).

Fig. 1. Participants grabbed the ball in the virtual and real environment.

Fig. 2. Participants put the ball into the box in the virtual and real environment.

In addition to Immersive Virtual Environment (IVE, scenario 1) and the real environment (scenario 5), there were three other scenarios in the experiment: IVE with visual feedback for contacts (scenario 2), IVE with auditory feedback for contacts (scenario 3) and IVE with both auditory and visual feedback for contacts (scenario 4). It aimed at investigating the difference between real and virtual environment during collaborative tasks. Meanwhile, which additional feedbacks made the interaction more realistic was discussed. Every participant finished the same task among five scenarios in random order.

During the experiment, the joint coordinates and angles of shoulders, elbows and wrists were captured by a motion capture system. Considering the large amount of data, data compression was performed in this study. Hence, only postures that represent consecutive motions were specified as key frames. In this experiment, the interval between grabbing two consecutive balls is defined as a cycle. In every cycle, the participant grabbed a ball and put it into the box, and then waited for the next ball. At first, the posture of grabbing the ball, putting the ball into the box and grabbing the next ball are chosen as the first, middle and last key frames of a cycle. Then, the period in between grabbing and putting the ball was segmented into four isometric time intervals (see Fig. 3). As a result, there are nine key frames chosen in each cycle (repeated frames of putting the ball were not chosen).

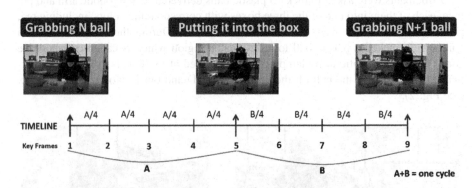

Fig. 3. The definition of keyframes

The kinematic data was then used to conduct Principal Component Analysis (PCA). Only two principal components, PC1 and PC2, would be the input of cluster analysis to identify different motion strategies performed by participants. Ward's method and K-means analysis were applied to determine the number of strategies and classify observations into different groups.

The task completion time and Simulator Sickness Questionnaire (SSQ) were measured in each scenario. The repeated measures ANOVA were applied to investigate the task performance and simulator sickness respectively. In addition, presence questionnaire was tested in every virtual scenario. The results were used to conduct chi-square analysis to compare the sense of presence among four virtual environments.

3   Results and Discussion

Since two participants' kinematic data were incomplete, the data of only four participants was analyzed (see Fig. 4). According to the results of PCA, PC1 is related to the hand participants used when grabbing or putting the ball. More specifically, higher PC1 score represents the use of right hand. Besides, PC2 is defined as the distance between the participant's reaching hand and body while grabbing and waiting for the ball. Higher PC2 score means reaching forward more at the moment of contacting the ball. Based on these results, the PC scores of all observations were treated as the input of Ward's method and K-means analysis for clustering. As a result, four motion strategies are determined. Strategy 1 is featured by reaching left hand more forward. Strategy 2

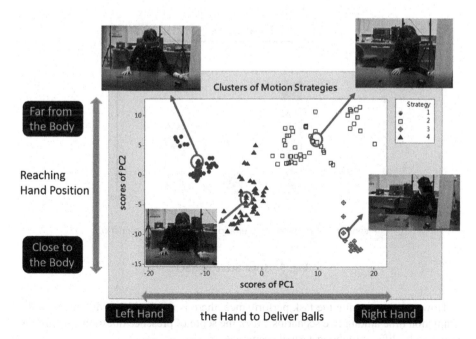

Fig. 4. The results of motion strategies

mainly is featured by the posture of reaching right hand more forward. Strategy 3 is characterized by the posture of reaching right hand to a shorter distance. Strategy 4 is featured by reaching left hand to a shorter distance.

The chi-square test was used in order to investigate the relationship between motion strategies and scenarios. Scenarios and motion strategies were taken as independent variable and dependent variable respectively. The results showed that there is a significant difference among scenarios (p value = 0.01). A bonferroni test was applied for post-hoc analysis and showed no significant difference among virtual scenarios (scenario 1, 2, 3 and 4). Real environment (scenario 5) is significantly different from other scenarios. The main reason is that all observations composing strategy 3 come from real environment. However, there is no significant difference in the composition of scenarios among strategy 1, 2 and 3. They are mainly influenced by the postures participants adopted no matter what the scenario is.

In SSQ, there are three subscores (Nausea, Disorientation, and Oculomotor) and a Total Score. Total Score means the overall severity of simulator sickness during the experience of virtual reality. The repeated measures ANOVA was used to clarify the relationship between scenarios and SSQ scores. Among four scores, there were significant differences in Total score and Disorientation (see Figs. 5 and 6) According to the results of LSD test, scenario 2 is significantly higher than scenario 5 in Total Score (p value = 0.023). In Disorientation, scenario 2 is significantly higher than scenario 1 (p value = 0.043). The visual feedback in scenario 2, the shade of colors of the ball changes along with the distance between reaching and virtual objects. It may influence participants' perception when they are delivering balls.

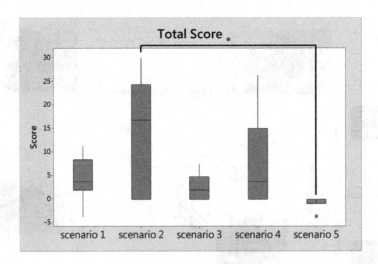

Fig. 5. Total Score of SSQ

However, concerning the task performance, there is no significant difference in the completion time among five scenarios. Also, the score of presence questionnaire are not significantly different among virtual scenarios.

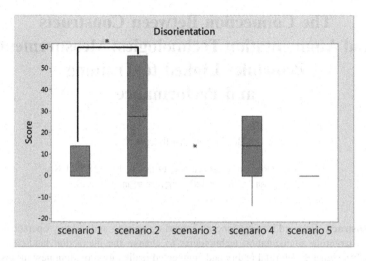

Fig. 6. Disorientation Score of SSQ

To sum up, significant differences of motions strategies between virtual and real environments were found in this study. However, the completion time keeps the same between two kinds of environment. It may be concerned as the similar performance between virtual and physical training. According to the results of cluster analysis, 3 out of four strategies included all kinds of scenarios respectively. The influence of individual difference may be higher than scenarios. Concerning the results of psychological measures, visual feedback is the least recommended alternative.

This study is an exploratory research for evaluating the feasibility of simulating human-machine collaboration by immersive virtual reality. It showed a potential of this method if an appropriate feedback is used in the virtual environment. By using virtual reality, prior evaluation or training of human- machine collaboration will be safer and more efficient.

References

1. ABI Research (2015) https://www.abiresearch.com/webinars/collaborative-robotics-market-opportunities/
2. Matsas E, Vosniakos GC (2017) Design of a virtual reality training system for human–robot collaboration in manufacturing tasks. Int J Interact Des Manuf (IJIDeM) 11(2):139–153
3. Ebrahimi E, Babu SV, Pagano CC, Jörg S (2016) An empirical evaluation of visuo-haptic feedback on physical reaching behaviors during 3D interaction in real and immersive virtual environments. ACM Trans Appl Percept (TAP) 13(4):19
4. Weistroffer V, Paljic A, Callebert L, Fuchs P (2013) A methodology to assess the acceptability of human-robot collaboration using virtual reality. In: Proceedings of the 19th ACM symposium on virtual reality software and technology. ACM, pp 39–48

The Connection Between Constructs and Augmentation Technologies: Measurement Principles Linked to Training and Performance

Benjamin Goldberg[✉]

U.S. Army Research Laboratory, Orlando, FL 32826, USA
benjamin.s.goldberg.civ@mail.mil

Abstract. Forward leaning concepts regarding training and military operations are targeting augmentation technologies to change the interaction landscape. Advancements in virtual reality and augmented reality are providing new means for immersing individuals in realistic experiences and providing access points to data and information previously not easily accessible. While these new inter-action modes evolve, measurement techniques are critical in understanding how best to apply these technologies. It is also important to investigate ways mea-surement techniques can influence the application of these technologies. In this paper, we present a high level overview of augmentation related research focused on human performance dimensions with a discussion on the role of construct measurement to inform and influence their utility. This is followed by a review of a three use cases that are applying the state of the art to advance the practice of training and on the job support tools.

Keywords: Augmentation · Training · Human performance · Adaptation Measurement

1 Introduction

As the U.S. Army engages in a new force modernization strategy, science and engi-neering will play a pivotal role. One area of modernization that is driving future research efforts is advancing simulation-based training practices to incorporate aug-mentation technologies and adaptation practices that maximize desired outcomes (i.e., performance, retention and transfer). This desire for a modernized program of record is evident in the U.S. Army's research and development investment in the Synthetic Training Environment (STE), which will serve as the next generation simulation-based training solution that supports collective scenario interactions across live, virtual and constructive platforms. Another high-demand focus area is linking these new training paradigms with human-agent teaming functions that account for new evolving vehicle interaction concepts, as well as teaming functions for dismounted units with integrated robotic assets. The goal is to leverage advancements across the learning sciences and augmentation research communities to provide a strategic advantage over our

© Springer Nature Switzerland AG 2019
S. Bagnara et al. (Eds.): IEA 2018, AISC 822, pp. 370–379, 2019.
https://doi.org/10.1007/978-3-319-96077-7_39

adversaries by extending operator functions to support seamless task execution in a complex multi-domain environment, and devising training programs that build relevant competencies for the future operational space.

In lieu of this, research is being performed across multiple nations to identify augmentation technologies that target human performance oriented constructs as they relate to enhanced perceptual cueing, information processing, knowledge/skill acquisition, and decision making. Enough international interest is garnered on the topic that a North Atlantic Treaty Organization (NATO) Research Task Group (RTG) was initiated in the fall of 2017 to report on the state of the art as it relates to augmentation technologies that impact human performance dimensions. A succinct focus of the RTG is to explore technologies that directly target training and job aid augmentation approaches that account for weaknesses associated with human cognition. A desired end-state is a system of systems approach that leverages multiple augmentation modalities for the purpose of optimizing mission and training outcomes on a task by task and individual by individual basis. This includes mechanisms to personalize training on individual strengths and weaknesses, monitoring cognitive and affective states for augmenting interaction components, and building tools that enhance the limitations of the human perceptual systems.

To level set some initial topic areas targeted in the RTG, a workshop was held at the United States Military Academy at West Point, NY in April 2018 to discuss current applications under development across various laboratories. In addition, current forward leaning futuristic concepts that are influencing research program strategies were discussed. Interestingly, regardless of the technology discussed, a common theme was the nature by which measurement techniques are applied to inform augmentation methodologies, as well as the role of measurement and assessment in gauging user reaction to various technological modalities as a form of an effectiveness measure.

In this paper, we review technologies presented at the workshop with a goal of using these use cases as a means to organize measurement principles commonly applied in augmentation research settings. In doing so, we establish a taxonomy that connects cognitive and performance oriented measurement constructs with established augmentation methodologies applied for training and job-aid contexts. This includes identifying empirically-derived methods for accurately measuring each construct, along with augmentation techniques recognized for manipulating and influencing those specified measures. Next, the relationship between the augmentation technology and its desired impact on training and mission-oriented objectives will be reviewed.

As a guiding function, we will present categorical use cases that highlight augmentation technology paradigms across both training & education and on-the-job performance aid factors. These two elements will serve as high-level variables by which to bin associated measurable constructs and augmentation tools and methods. The taxonomy will serve as a basis by which to explore the application of principles and heuristics for use by practitioners based on represented relationships between methods and desired outcomes.

2   Measurement and Training

Before discussing the use cases, it is important to review the basics of measurement and the various sources these measures originate from. When considering measurement techniques from an objective standpoint, it is important to understand the context by which that measurement is derived. We identify measurement in terms of data types and their relation to some formalized construct to orient what that measurement tells you. Historically, these constructs are ultimately used to inform decision points regarding training effectiveness and on-the-job performance. Now, these constructs are integral to research and development practices on technologies that advance human performance functioning on the cognitive level. They provide insightful data points that address constructs by which a technology aims to influence, or in some situations, compensate for. Therefore, in order to understand the objective target of an augmentation technology, it is important to establish a taxonomy of measurement types and the constructs they contextualize. Burford et al. [1] provide an excellent breakdown of measurement techniques and methods applied for performance indication research. Much of those findings are represented in the following sub-sections. In doing so, we review measurement types and those measurements are influencing current applications under development.

2.1   Construct Measurements and Their Impact on Outcome

From a training and job support stance, we distill measurement down into three categories: (1) performance measurement, (2) physiological measurement, and (3) behavioral measurement (see Table 1).

Table 1. Construct measurement taxonomy in relation to augmentation technology

	Data Source	Data Type Examples	Measurable Constructs
Performance	Training/Task Environment	Raw System Data, Transformed Variables, Outcomes, Percentages, etc.	Knowledge Acquisition, Skill Development, Competency, Trend
Physiological	Wearable Sensor Technologies	Heartrate Variability, Electroencephalogram (EEG), Galvanic Skin Response, etc.	Affect (mood/emotion), Workload, Attention, Fatigue
Behavioral	System Interaction, Wearable Sensors, Environmental Sensors	Motion tracking, interface tracking (i.e., embedded sensors), performance data types, physiological data types, etc.	Psychomotor Application, Help Seeking, Metacognition, Gaming Behaviors

These categories associate with data types that are produced through a direct source. In many instances, source data is further processed and transformed into a new measure for informing model development. As an example, heartrate data, collected via an Electrocardiogram (ECG) sensor, is transformed into heartrate variability to distill a sensors signal into a formalization of that signal over a designated time window. This can also be seen in performance related measures, where raw system data is used to compute some measure on pre-defined condition classes. An example is maintaining a distance measurement of an entity from a defined object (e.g., building entrance). The system collects location and movement data of interacting elements, with a requirement to dynamically compute distance measures used to assess performance. One effort looking at establishing standards around these inference procedures is the Human Performance Markup Language (HPML), where they provide methods for building measures out of data, assessments out of measures, and performance/competency ranks out of observed assessments [12].

From an application standpoint, these measures, either raw or transformed, are used to build modeled representations that are theoretically linked to performance and psychological constructs that apply meaning to measurement. These inferred meanings are then able to be used to inform adaptations and variations in augmentations. The goal is to monitor shifts in state on task relevant constructs and using that state shift data to guide system interventions that influence the task operator. With respect to training, a focus of this research area is to augment the interaction space through virtual and augmented reality technologies that replicate elements of a task environment in a safe and cost-effective manner. In the following subsections, we breakdown measurement types in the three categories listed above, and relate those measurements with common construct-derived meaning.

Performance Measurement Constructs. Performance measurement commonly associates with inference procedures that construct meaning out of interaction. In military context, performance is computed by accounting for the objectives of a task, the conditions by which the task is being executed within, and the designated standards that determine outcome and performance. In many instances, standards defined to inform performance can be represented in generic formats, while conditions determine the nature in which the standard is activated and scoring is applied. Performance measurement in the context of this research is an aggregate representation of an individual's interaction as it relates to some objective. As seen in Table 1, there are many representations of performance, with direct dependencies to the task being conducted and the type of processes and procedures applied to meet task objectives. While there are many forms of data that express performance, common aggregate measures include: percent correct across problem sets, time on task, number of errors, and knowledge and skill retention.

Tracking outcomes and shifts in performance over time are critical when comparing task execution procedures, especially when incorporating new technology that augments the experience. Regardless of training or job aids, performance constructs of knowledge acquisition, skill application, and competency are relevant to augmentation research and development. They are the primary variables of interest from a return on investment viewpoint, and serve as the basis by which the following measurement categories provide diagnostic information for what was observed across the performance related constructs. From a competency standpoint, knowledge and skill application observed across task scenarios provide evidence of designated competencies, with increases in confidence as more positive evidence events are observed. To support performance tracking across systems and applications, the eXperience Application Programming Interface (xAPI) was created, which provides a standardized format for producing experience related statements that are stored in some data record store. Systems that are xAPI compliant can produce and receive statements from data record stores, enabling easy tracking on a large temporal timescale.

Physiological Measurement Constructs. Monitoring physiological signals in human factors oriented research is not a new endeavor. There are multiple approaches and validated techniques that classify observable physiological signals into categorical states that apply meaning to a situation. While there are numerous physiological sensors applied in technology development, it is not the goal of this paper to review the variations in measurement and classification. The important take-away is to present the relevancy of physiologically informed state representations from two perspectives: (1) using physiological data to drive augmentation practices by building policies that act on shifts in user state, and (2) using physiological data as an additional data point in comparative evaluations of technology and using these inputs to contextualize observations in performance. The first applies physiologically informed constructs as active variables used for interaction with an augmentation technology based on observed shifts. The second utilizes physiology as a dormant variable applied for post-hoc analyses to determine the impact and effect on specified manipulations to have better granular information that may provide insight regarding observed outcomes and behavior patterns.

Much of the research surrounding physiological measurement centers on informing models that associate with cognitive and affective constructs. From a cognitive perspective, attention, workload and fatigue serve as the principal variables measures aim to inform [2]. These state measures can apply to any task situation and provide valuable insight into how an individual is performing from an effort standpoint. There are multiple methods for collecting variables that inform these states, with pros and cons across each (see Burford et al. [1] for an excellent review). Common approaches include Electroencephalogram (EEG), Electrocardiogram (ECG), Galvanic Skin Response (GSR), functional Near-Infrared Spectroscopy (fNIR), and eye tracking.

From the affective perspective, detecting emotional states and shifts in mood serve as the primary objectives of research surrounding physiological measurement [3]. Through observational shifts in physiological signal, scientists have developed reliable methods for classifying emotional state, with advancements in machine learning being critical to successful implementations. Research areas with significant contributions to the field include Intelligent Tutoring Systems and adaptive instructional systems [3, 4], autism research [5, 6], and market advertising research [7].

Behavioral Measurement Constructs. Behavioral measures differ from performance and physiological measures in the way the data is interpreted. Behavioral measurement approaches consume the same data types across the other two categories, but the data is assessed and refactored to establish behavioral patterns that can be used to further infer the context of a training or job task event. In this vein, behavioral measures can be applied for diagnostic purposes in assessing skill and procedural application [8], inferring strategy use and metacognitive understanding [9], and detecting gaming behaviors that associate with cheating a system [10]. Tracking behavior is also critical in user experience research, where studies can be applied to evaluate interface designs with tasks developed to observe behavior trends when meeting defined objectives. The objective is to establish data analysis methods that account for behavior patterns in context, where behavior can also be used to inform feedback and remediation practices.

3 Augmentation for Improving Human Performance: Use Cases

With an understanding of the data and measurement types made available for augmentation technologies, next we discuss how those measurement and assessment functions support the utility of augmentation development. The following applications are currently under development and were briefed at the NATO RTG workshop this past April. The use cases provide examples of state of the art applications that leverage advancements in virtual and augmented reality capabilities.

3.1 Use Case: Virtual Reality Pilot

With an identified need to support training in self-regulated settings, Biddle [11] and the Boeing Company are investigating the utility of a virtual reality agent that can support multiple interaction roles. The resulting virtual pilot is designed to support crew and flight management activities, with characteristics designed to invoke emotion and confrontation commonly occurring on the flight deck. Initial objectives of the project are to optimize training time on full flight simulators and building scenarios that target communication procedures through verbal speech and body language, while also learning coping mechanisms for varying personality types [11].

The virtual reality pilot's physical representation is currently rendered in the commercial off the shelf HTC Vive headset (see Fig. 1), with research examining what features of the pilot's appearance are most desirable to training pilots. Measurement plays a critical role in this form of end-user research. While common methods associate with self-report measures in a post-experience format, applying performance, physiological and behavioral data types in your methodology enables a richer data set for assessing pilot impact on training outcomes. This includes conversational and physical interaction components that can be used to better determine the reaction and impact of the training experience.

Future implementations look at rendering the character in augmented reality headsets (e.g., Microsoft HoloLens), where a pilot can train on a high fidelity simulator while directly interacting with an intelligent automated agent. In addition, there are opportunities to apply additional measurement techniques that can ultimately influence the agent's behavior during a training exercise. That is, measures of workload and emotional state can inform agent policies that adapt the experience on classified states represented in the measure.

Fig. 1. Virtual Reality Pilot seen through augmented reality headset

3.2   Use Case: Pilot Next Training

Another use case highlighting the incorporation of augmentation technologies into training paradigms is seen in the U.S. Air Force's Pilot Next Training program [13]. They are examining the use of low-cost virtual reality technologies to support initial exposure training on in-flight routines, maneuvers and communication protocols (see Fig. 2). This supports a new data driven training approach that is centered on adult learning methods. This involves establishing a new training culture with technology insertions where virtual reality and augmented reality can be applied for efficient and cost-effective skill development.

Fig. 2. Pilot training virtual reality headsets and low-cost interfacing

Initial implementation of the Pilot Next Training program has received positive results, both from a performance and trainee reaction standpoint. In terms of performance, instruction on maneuvers and procedural routines have shown positive transfer into a live cockpit environment. While the fidelity is not near the real thing, the virtual reality platforms have proven effective in collective training events with multiple individuals coordinating. Multiple measures are being collected to assess everything from performance to physiological response. In addition, the current system supports real-time feedback mechanisms based on the measurements being performed (e.g., correcting a barrel roll maneuver based on angular calculations of the aircraft's nose-tip). Other measures in play associate with physiological response to virtual reality headsets, with models in place to identify simulator sickness symptoms for early counter measures. The program continues to evaluate state of the art training technology to determine the best mix of simulation-based and live training exercises to build required competency sets.

3.3   Use Case: First Responder Virtual Command

The last use case presented associates with technology applied in developing a virtual command center for first-responder coordination and logistics operations. The concept is based on a 'Connected City' with a network of resources (see Fig. 3) and assets that can be used in planning action and movement involving multiple units (e.g., police, SWAT, medics, firemen, etc.). The idea is that an individual can tap into a network of resources to gather data, intelligence, and visual confirmations associated with specified areas of interest. This enables an individual to easily determine a plan of action by leveraging multiple points of view while reviewing critical data reports regarding number of resources that can be tasked and their locations.

Fig. 3. Emergency Operations Center virtual command map

This technology applies innovative virtual reality interfacing to enable a fully immersive experience with intuitive controls that allow easy shifts from resource to resource. Measurement approaches enable the system to adapt to both behavioral and physiological inputs and customizes visual fields based on relevant information channels commonly interacted with. This identifies an innovative application outside of the training space, where virtual reality technologies are applied to augment the work environment where by an individual can access an array of physical locations and resources from the comfort of an office. Identifying what measurement constructs can influence this operational space is a research question of interest, with the incorporation of automated agents acting on measurement inputs to drive decision support policies to optimize the operator's experience.

4 Conclusions and Future Work

Augmentation technologies continue to evolve, with applications being applied for training and technology assisted job aids. More and more applications of virtual reality and augmented reality are being utilized outside of labs, with new mechanisms in place to support efficient practices are being investigated. With such a vast research space, the NATO RTG aims to capture the current state of application to better influence future research and development interactions. While the projects reviewed show current efforts under development, a major thrust moving forward is the application of augmentation technologies to support human agent teaming and human robotic interactions.

References

1. Burford C, Reinerman L, Teo G, Matthews G, McDonnell J, Orvis K, Metevier C et al (2018) Unified multimodal measurement for performance indication research, evaluation, and effectiveness (UMMPIREE): Phase I Report. ARL-TR-8277
2. Carroll M, Kokini C, Champney R, Fuchs S, Sottilare R, Goldberg B (2011) Modeling trainee affective and cognitive state using low cost sensors. In: Proceedings of the interservice/industry training, simulation, and education conference (I/ITSEC), Orlando, FL, USA
3. Calvo RA, D'Mello S (2010) Affect detection: An interdisciplinary review of models, methods, and their applications. IEEE Trans Affect Comput 1(1):18–37
4. D'Mello SK, Kory J (2015) A review and meta-analysis of multimodal affect detection systems. ACM Comput Surv (CSUR) 47(3)
5. Picard RW (2009) Future affective technology for autism and emotion communication. Philos Trans R Soc B Biol Sci 364(1535):3575–3584
6. Picard RW (2016) Automating the recognition of stress and emotion: from lab to real-world impact. IEEE Multimed 23(3):3–7
7. McDuff D, El Kaliouby R, Cohn JF, Picard RW (2015) Predicting ad liking and purchase intent: large-scale analysis of facial responses to ads. IEEE Trans Affect Comput 6(3):223–235
8. Goldberg B, Amburn C, Ragusa C, Chen D-W (2018) Modeling expert behavior in support of an adaptive psychomotor training environment: A marksmanship use case. Int J Artif Intell Educ 28(2):194–224
9. Aleven V, Roll I, McLaren BM, Koedinger KR (2016) Help helps, but only so much: research on help seeking with intelligent tutoring systems. Int J Artif Intell Educ 26(1):205–223

10. Paquette L, Baker RS (2017) Variations of gaming behaviors across populations of students and across learning environments. In: Proceedings of international conference on artificial intelligence in education, China
11. Biddle B (2018) Building a virtual reality pilot: embedding the end user throughout the process. NATO RTG-297 workshop on augmentation technologies, West Pont, NY, USA
12. Human Performance Markup Language Webpage. https://www.sisostds.org/Standards Activities/DevelopmentGroups/HPMLPDG-HumanPerformanceMarkupLanguage.aspx. Accessed 20 May 2018
13. Pilot Next Training Program Webpage. http://www.aetc.af.mil/News/Article/1391431/aetc-explores-learning-possibilities-through-new-pilot-training-program/. Accessed 21 May 2018

A Comparative Study for Natural Reach-and-Grasp With/Without Finger Tracking

Huagen Wan$^{(\boxtimes)}$, Xiaoxia Han, Wenfeng Chen, Yangzi Ding, and Liezhong Ge

Zhejiang University, Hangzhou 310027, Zhejiang, China
hgwan@cad.zju.edu.cn

Abstract. Grasp interactions often involve both hand tracking and finger tracking to drive the virtual hand deformation and evaluate grasping conditions. With the more involvement of psychology into HCI technology, we are seeing more algorithms employing psychological finding. However, the performances and user experiences of these algorithms remain to be further explored. In this paper, a comparative study has been performed under the same grasping conditions between our formerly proposed method for reach-and-grasp tasks which needs only tracking the hand's 6-dof motions (Method A) and a typical forward-kinematics enabled virtual grasping method which needs both 6-dof hand tracking and a dataglove for finger tracking (Method B). Virtual spheres centered at the origin with different diameters (i.e., 6 cm, 8 cm and 10 cm) were used as the grasping targets. A panel of 12 participants were divided into two groups and employed in the comparative study on task completion time, accuracy and 3 subjective criteria. It is shown from the experimental results that Method A is better than Method B as far as the above 3 aspects were concerned for simple shapes such as spheres. A demo application was developed using both Method A and Method B, and users' preference evaluation was performed.

Keywords: Reach-and-Grasp · Grasp trajectory · Finger tracking

1 Introduction

Recent years have seen the rapid development of virtual hand interaction techniques and their applications in many fields such as evaluation of handheld information appliances, virtual assembly, virtual surgeries, and digital games. In these contexts, a dexterous virtual hand is often used as the avatar of the user's hand. Usually, both hand tracking and finger tracking are needed to drive the virtual hand. At present time, many kinds of techniques, including magnetic tracking, optical tracking, and inertia tracking, are capable of tracking the 6-dof human hand motion in real time with acceptable precision. They are mature enough to be applied in various applications. As far as finger tracking is concerned, although instrumented gloves are considered as popular finger tracking devices and have been employed in the community for more than twenty years [1], they are labeled as "expensive" and "cumbersome to wear." Recent development of Kinect sensors and LeapMotion provides promising ways for freehand

© Springer Nature Switzerland AG 2019
S. Bagnara et al. (Eds.): IEA 2018, AISC 822, pp. 380–388, 2019.
https://doi.org/10.1007/978-3-319-96077-7_40

finger tracking [2, 3], but they are limited by the tracking range and need more time to be matured enough. By utilizing the findings on the analyses of the kinematics of movement of all five digits during reach-and-grasp tasks for a variety of objects [4–6], a simple but effective approach was proposed for users by us to naturally reach and grasp a three-dimensional object via a dexterous virtual hand as an avatar within a virtual environment without finger tracking (Method A). However, its performance and user experience remain to be further explored. In this paper, Method A has been compared with a typical finger-tracking-based, forward-kinematics-enabled virtual grasping method (Method B), with a demo application being developed using both Method A and Method B.

2 Related Work

Real world grasp is dependent upon many complicated factors such as the shape, material and weight of the object, as well as the contact forces between the fingers and the object. However, in the virtual world, virtual grasping is mainly achieved through heuristics and/or physical simulations.

Heuristics-based virtual grasping depends on the assumption that a steady grasp can be performed by different finger combinations. For instances, Ullmann et al. established a grasping condition for one hand grasping so that a rigid virtual object can be grasped with one hand only if there exists an angle between the normals n_i of the contact faces f_i $(i = 1, 2)$ which is greater than a given, experimentally-determined angle β, and the thumb and at least one finger or the palm and at least one finger were involved [7]. Holz et al. presented a method for multi-contact grasp interaction, in which the concept of "grasp pairs" was proposed [8].

In comparison, simulation-based virtual grasping methods are more robust but more computational extensive [9, 10]. For instance, Jacobs et al. proposed a hand model based on soft bodies coupled to a rigid hand skeleton for physically-based manipulation of virtual objects. Though the virtual hand geometry was relatively simple and only the finger phalanxes were represented by a soft body mainly due to computational complexity, their method allowed for precise and robust finger-based grasping and manipulation of virtual objects [10]. Moehring et al. proposed to integrate pseudo-physical simulations into a heuristics-based approach to manipulate constrained objects [11].

Generally speaking, both heuristics-based methods and simulation-based methods need to track both the hand movements and the finger motions for online computations. These limit their applications since precise finger tracking remains a luxury either for economical or technical reasons [1, 2].

We formally proposed a hybrid method combining heuristics and physical simulation with the motivation to provide fast, stable and vision controllable multi-finger grasp interaction [6]. It firstly detected the collisions between objects and pre-defined finger trajectories, then deduced the grasp configuration using collision detection results, and finally computed feedback forces according to grasping conditions. As far as we know, it is the first and the only one method for natural reach-and-grasp tasks without finger tracking. However, its performances and user experiences need be further investigated. Interestingly, there have been many works on finger tracking in the past decade [12, 13].

3 Methods

3.1 Method A

Method A was formally proposed in [6]. For the completeness of this paper, we give a brief summary of the method.

The method focuses on the reach-and-grasp tasks to fully utilize neurophysiology findings of consistent and independent fingertip motion for reach-to-grasp movements which stated that fingertip motions for reach-to-grasp movements to a variety of objects tended to follow particular curved paths [4, 5]. It consists of the following three steps.

Constructing Grasp Trajectory. iFirstly, we construct each fingertip's grasp trajectory. A fingertip's grasp trajectory was defined as a series of line segments illustrating the trajectory of the fingertip when grasping an object. The construction of the fingertips' grasp trajectories is described as follows:

1. Ask a user to participate in a reach-and-grasp task and record his finger joint angles using a dataglove at approximately 50 Hz.
2. Manually adjust the virtual hand model to fit the shape of the user's hand and select a point for each fingertip surface as a seed point which is located in the center of the fingertip's surface.
3. Simulate the user's grasping process with those recorded joint angles and generate the trajectory of the seed point for each finger according to

$$P = l_{dp} * \sin(\theta_1 + \theta_2 + \theta_3) + l_{mc} * \sin(\theta_1 + \theta_2) + l_{pp} * \sin(\theta_1) \tag{1}$$

where l_{dp}, l_{mc}, and l_{pp} represent the lengths of the distal, middle, and proximal finger segments (Fig. 1(a)).

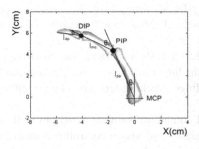
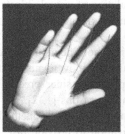

(a) Finger kinematics model   (b) Virtual hand with fingertips' grasp trajectories

Fig. 1. Virtual hand modeling.

4. Discretize the grasp trajectory of each finger into line segments by uniform sampling. Figure 1(b) shows the fingertips' grasp trajectories attached to the virtual hand from two different viewing angles.

Computing Intersection Points. Secondly, we compute intersection points of trajectories and an object when the virtual hand reaches the object. Since the grasp trajectory of each fingertip is represented by line segments, the problem of computing the intersecting points between the trajectories and the virtual object is converted to perform an intersection test between the line segments and the object.

To accelerate the intersection computation between the line segments and virtual objects, we modified the RAPID package [14] to make it suitable for the computation between line segments and triangles.

Grasp Condition Evaluation and Grasping Simulation. Once the intersecting points of the fingertips' grasp trajectories with the virtual object have been found during the reach-and-grasp, we evaluate whether the virtual object can be grasped according to the one-hand grasping condition of [7]. If the virtual object can be grasped, the grasping operation is driven by inverse kinematics till the fingertips move to the intersection points, and the virtual object is attached to the virtual hand for further manipulation.

3.2   Method B

A typical forward-kinematics enabled virtual grasping method (Method B) was employed in our comparative study. Both Method A and Method B used the same grasp conditions to evaluate whether a virtual object can be grasped except that Method B used a dataglove to online capture the user's finger motions and the intersecting points are computed between the virtual object and the virtual fingertips instead of the grasp trajectories. Specifically, Method B also consists of three steps [15].

1. Track the user's both hand and finger motions and use the tracking data to enable the virtual hand deformation.
2. Compute intersection points between virtual hand fingertips and an object when the virtual hand reaches the object.
3. Evaluate whether the object can be grasped in the same way as in Method A.

4   Hypotheses

As Method A employs only 6-dof hand tracking devices instead of both 6-dof hand tracking devices and finger tracking devices, we suppose that it would be better in performance and user experience in some applications (e.g. digital games) which need neither accurate nor direct mapping from user's finger motions to finger motions of the virtual hand. Specifically, we have following three hypotheses:

- H1: it would be faster to virtually grasp typical 3D objects using Method A than using Method B.

- H2: it would be more accurate to virtually grasp typical 3D objects using Method A than using Method B.
- H3: users would get better experience when using Method A than using Method B.

In this paper, comparative user studies were performed. Spheres with different diameters were employed as typical 3D objects.

5 Evaluation

5.1 Participants

The evaluation was conducted using a panel of 12 participants, 6 females and 6 males aged from 19 to 25. They are all right-hand dominant, and had no experience with either 6-dof hand tracking or finger tracking. The participants were divided into two groups: 3 females and 3 males in each group. One group was asked to first perform the reach-and-grasp operations using Method A and then Method B, while another group vice versa. Before the experiment, each participant was given 5 min to warm up.

5.2 Experimental Apparatus

The experimental apparatus consists of a 24-inch LED screen connected to a computer with an Intel(R) Core(TM) i7-4770 CPU @ 3.40 GHz and 8 GB RAM, an Optitrack motion capture system for capturing the participants' 6-dof hand motions and a dataglove for capturing the participants' finger motions. The user stands at about 110 cm in front of the screen. The experiment application written with OpenGL shows a virtual sphere centered at the origin of world coordinate system and a virtual hand whose position and orientation is determined by that of the user's hand. To make the user easily perform the reach-and-grasp operations, the coordinate system of the tracking system and the coordinate system of the virtual world were carefully calibrated. In the experiment, three virtual spheres of 6 cm, 8 cm, and 10 cm in diameters were used, but at each reach-and-grasp operation, only one sphere was shown. The presentation order of the spheres was determined by the 3-order Latin square. The size of the virtual hand is 10.64 cm $\times$ 8.23 cm $\times$ 20.46 cm. In Method B, the virtual hand is displayed as it is. But in Method A, it is displayed with the grasp trajectories attached to the fingertips. Figure 2 shows a snapshot of the experimental setting.

5.3 Tasks

The tasks of a participant were to reach and grasp a virtual sphere. At the beginning of each task, the instructor gave a command to inform the participant to begin the reach-and-grasp operation, in the meantime, she pressed a key on the keyboard to instruct the computer for timing. For both Method A and Method B, a participant was asked to successfully grasp each of the three virtual spheres for 5 times. As a result, each participant should successfully perform the tasks for $2 \times 3 \times 5 = 30$ trials. After the participant finished all the tasks, he/she was asked to evaluate the two methods by giving them scores.

Fig. 2. Experimental setting.

5.4   Data Collected and Results

The task completion time and whether a task is successful were recorded. Besides these data, a subjective questionnaire was proposed in which participants had to grade from 1 (lowest appreciation) to 5 (highest appreciation) the two methods according to 3 subjective criteria: (a) Ease of Use, (b) Interaction Naturalness, and (c) User Comfort. Figure 3 shows the results concerning the grades obtained by the two methods for each of the subjective criteria.

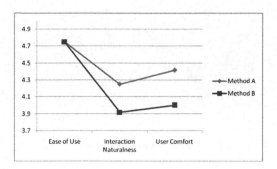

Fig. 3. Grades obtained by subjective questionnaire.

Table 1 shows the statistics of the task completion time for successful tasks, in which MVA and MVB represent in milliseconds the mean values using Method A and Method B respectively, and SDA and SDB represent the standard deviations using Method A and Method B respectively.

Table 1. Statistics of task completion time.

Sphere diameter	MVA	MVB	SDA	SDB
6 cm	3367.20	4867.42	1096.12	2117.16
8 cm	2546.87	4436.95	1382.45	1753.33
10 cm	2295.22	3038.38	847.57	1020.55
All	2736.43	4114.25	1213.84	1856.62

A one-way ANOVA was performed on the two different methods (Method A and Method B), and a significant effect was found ($F_{1,358} = 69.447$, p-value = .000). ANOVA performed on three different diameters (6.0 cm, 8.0 cm and 10.0 cm) of the virtual spheres also revealed significant effects for both Method A ($F_{2,177} = 14.764$, p-value = .000) and Method B ($F_{2,177} = 19.144$, p-value = .000). The rates of successful reach-and-grasp operations against all the operations in Method A and Method B are 82.95% (180/217) and 70.87% (180/254) respectively. In both methods, if a virtual sphere was successfully grasped within 15 s, then the reach-and-grasp operation was recorded as successful, otherwise it was recorded as failed.

6   Demo Application

To further evaluate both methods, we designed a demo application. In the demo, six red spheres were firstly distributed in the 3D space randomly, and a user was asked to arrange them into an array of spheres in whatever order through the reach-and-grasp operations. Figure 4 shows the initial frame and one of the end frames of the demo application respectively.

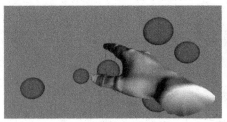
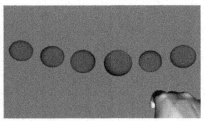

(a) The initial frame   (b) One of the end frames

Fig. 4. Demo application.

Like the experiments, the demo application was also implemented and evaluated using both Method A and Method B. The 12 participants were asked to practice the demo application as long as they wish and make their comments. If he/she thought Method A was better or worse than Method B, he/she gave an 'A' or 'B' respectively; if he/she thought the two methods were almost the same, he/she gave an 'C'. Finally, we got 7 'A's, 1 'B' and 4 'C's.

7   Discussions

From the above experimental data analysis and the demo application evaluation, we give a brief discussion on our comparative study.

Firstly, both statistical and ANOVA analyses on the task completion time show the advantage of Method A over Method B. Thus, hypothesis H1 was proved to be true.

Secondly, ANOVA analysis on the successful rates shows the advantage of Method A over Method B. Thus, hypothesis H2 was proved to be true.

Thirdly, subjective evaluations on "Interaction Naturalness" and "User Comfort" and demo application show the advantage of Method A over Method, while subjective evaluation on "Easy of Use" shows no significant difference between Method A and Method B. Thus, hypothesis H3 was partly proved to be true.

Finally but the most importantly, the advantages of Method A over Method B were achieved only on a shape as simple as a sphere, though different sphere diameters were investigated. For complex shapes with convexity, whether these advantages will be kept remains unclear. Nevertheless, we think that the convex hull can be used as an initiative for Method A to virtually grasp a complex shape with convexity.

8 Conclusion

In this paper, we have performed a comparative study under the same grasping conditions between a reach-and-grasp algorithm using only 6-dof hand tracking (Method A) and a typical forward kinematics-enabled virtual grasping algorithm using both finger tracking and 6-dof hand tracking (Method B). Experimental studies were conducted on the grasping of virtual spheres centered at the origin with three different diameters (i.e., 6 cm, 8 cm and 10 cm). It is shown from the experimental results and the demo application that Method A has the merits of simplicity, ease of use, naturalness, user comfort, and cost-effectiveness, compared with Method B. Although the advantages of Method A over Method B remains unclear for complex shapes with convexity, we think it is suitable for practical applications such as digital edutainment.

References

1. CyberGlove Systems. http://cyberglovesystems.com/. Accessed 26 May 2018
2. Qian C, Sun X, Wei Y, Tang X, Sun J (2014) Real-time and robust hand tracking from depth. In: Conferences on computer vision and recognition, IEEE, 2014, Columbus, Ohio, pp 1106–1113
3. Leap Motion. https://www.leapmotion.com. Accessed 26 May 2018
4. Kamper D, Cruz E, Siegel M (2003) Stereoscopical finger trajectories during grasp. J Neurophysiol 90(1):3702–3710
5. Maitland ME, Epstein MB (2009) Analysis of finger position during two- and three-fingered grasp: possible implications for terminal device design. J Prosthet Orthot 21(2):102–105
6. Zhu Z, Gao S, Wan H, Yang W (2006) Trajectory-based grasp interaction for a virtual environment. In: 24th international conference on computer graphics, pp 300–311. Springer, Berlin (2006)
7. Ullmann T, Sauer J (2000) Intuitive virtual grasping for non haptic environments. In: 8th Pacific conferences on computer graphics and applications, Hong Kong, pp 373–381
8. Holz D, Ullrich S, Wolter M, Kuhlen T (2008) Multi-contact grasp interaction for virtual environments. J Virtual Reality Broadcast 5(7):16–26
9. Borst C, Indugula A (2005) Realistic virtual grasping. In: Conferences on virtual reality, IEEE, 2005, Arles, pp 91–98

10. Jacobs J, Froehlich B (2011) A soft hand model for physically-based manipulation of virtual objects. In: Conferences on virtual reality, IEEE, 2011, Singapore, pp 11–18
11. Moehring M, Froehlich B (2005) Pseudo-physical interaction with a virtual car interior in immersive environments. In: 9th International workshop on immersive projection technology, 11th eurographics workshop on virtual environments, Aalborg, Denmark, pp 181–189
12. Vijitha T, Kumari JP (2014) Finger tracking in real time human computer interaction. Int J Comput Sci Netw Secur 14(1):83–93
13. Togootogtokh E, Shih TK, Kumara WGCW et al (2018) 3D finger tracking and recognition image processing for real-time music playing with depth sensors. Multimed Tools Appl 77 (8):9233–9248
14. RAPID - Robust and Accurate Polygon Interference Detection. http://gamma.cs.unc.edu/ OBB/. Accessed 26 May 2018
15. LaViola Jr JJ, Kruijff E, McMahan RP, Bowman D, Poupyrev IP (2017) 3D user interfaces: theory and practice, 2nd edn. ISBN-13: 978-0134034324 (2017)

Work With Computing Systems (WWCS)

Analysis and Design of a Cyber-Physical Production System (CPPS) in Sensor Manufacturing. A Case Study

Manfred Mühlfelder[(✉)]

SRH Fernhochschule – The Mobile University,
Lange Straße 19, 88499 Riedlingen, Germany
manfred.muehlfelder@mobile-university.de

Abstract. Cyber-Physical Production Systems (CPPS) combine methods of artificial intelligence with robotics, new ways of man-machine-man collaboration, and the "Internet of Things (IoT)" [1]. They create opportunities for individualized, one-piece-flow, and flexible manufacturing and assembly processes in tomorrow's manufacturing plants.

In this case study, a new production system for the assembly of sensors in a small/medium sized enterprise (SME) has been designed, implemented, and evaluated. As a method, the "complementary design of complex man-machine systems" [2, 3] has been used, adapted and expanded.

The results show that manufacturing processes in CPPS

- allow for more self-regulation and self-organization in semi-autonomous production teams,
- create higher complexity and dynamics in the work process,
- require more collaboration, communication and problem-solving skills of the operators,
- enable new ways of man-machine-man collaboration, and
- can significantly increase the productivity and quality of manufactured goods.

The case study has been carried out as part of the research project "PRADI-KATSARBEIT" ("Work of Excellence"), sponsored by the Federal Ministry of Research and Higher Education of Germany (Grant No. 02L14A093). The results shall be shared in the IEA community to promote new ways of man-machine-man collaboration in the "Industry 4.0". For more information about CPPS and manufacturing in digitally inter-connected production cells see [1, 4, 5].

Keywords: Cyber-Physical Production Systems (CPPS)
Man-Machine-Man-Interaction (MMI)
Complementary design of complex man-machine systems · Task analysis

© Springer Nature Switzerland AG 2019
S. Bagnara et al. (Eds.): IEA 2018, AISC 822, pp. 391–400, 2019.
https://doi.org/10.1007/978-3-319-96077-7_41

1 Essentials of Cyber-Physical Production Systems (CPPS)

1.1 Definition and Work System Characteristics

Cyber-Physical Production Systems (CPPS) are complex work systems, which can be characterized through the following features [6]:

- They are equipped with inter-connected sensors and actors.
- They encode, store, analyse and interpret production and product related data in real-time.
- The elements of a CPPS (e.g. machines, databases, transfer units, quality check units, packaging units) are connected with each other and can monitor and control each other.
- They are partially self-controlled and allow for semi-autonomous production planning based on current system parameters (e.g. cycle time, down time, production schedule, maintenance schedule).

CPPS enable a decentralized, contextual, flexible and adaptable control of production and logistics processes. Based on feedback mechanisms and intelligent learning algorithms the work system can optimise itself and improve the flow and the output of manufactured goods and products. For a more technical in-depth perspective on CPPS you can refer to [7] or [8].

1.2 Challenges for CPPS Design

The design of CPPS needs to address the complexity of the dynamics and states in the system. The overall model must cover and consider the consistency and integrity of the various parts, detect potential internal and external conflicts and bottlenecks and guide informed decision making. One key question in the design is the appropriate degree of computer/machine autonomy or, in other words, the synergy between artificial und human intelligence in production planning and monitoring. Man/operator and machine need to complement each other. They are constitutional elements of the total system and their smooth interaction is a pre-requisite for gains in productivity, economy, and quality.

2 Principles of Complementary Work System Design for CPPS

Complementary work system design is a framework for the balanced configuration of tasks, tools, and methods in complex work systems. The objective of this design approach is to ensure an optimal fit between human operators, supporting systems, and production control. It has been developed in the late 1990s by Grote and colleagues at the Swiss Federal Technical University (ETH) in Zürich [2, 3]. Fundamentally, there are three levels on analysis in this framework: (1) task analysis, (2) functional analysis of man-machine interaction (MMI), and (3) work system analysis (holistic view). For each level there is a specified set of criteria for identifying strengths, weaknesses, opportunities and threats in existing systems or new systems under design and construction.

2.1   Task Analysis

Human-centred task design aims to define "complete" tasks, which comprise planning, executing, and evaluating/feedback components. Such tasks require human cognitive skills such as planning, problem solving, monitoring, decision making, and communication and interpersonal skills. On this level, complementary work system design aims to define and design tasks which allow the human operators to use the full range of their abilities and skills. On the technical side, the system designers should investigate how they can support the task execution with assisting functions. For example, a machine operator in a CPPS environment might benefit from a planning function which machine can be used to process the next customer order, given the production schedule and the current conditions of the machinery, material flow and storage.

2.2   Functional Analysis of Man-Machine Interaction

Man-Machine-Interaction (MMI) is a crucial element in CPPS design. New technical opportunities for speech control, graphical data input and output, visualization of product and process parameters, and the connectivity of wearable control and monitoring devices empower the human operators to view and check the status of the work system, make informed decisions, and observe the effects of preventive, corrective and maintenance actions. The interaction between humans and machinery needs to be complementary in a sense that each technical function needs to be aligned with human action goals. For example, an input function for a machine needs to support the goal of the human operator to control the work process, start, stop or interrupt the machine. The work system designer needs to break down the work-related goals of the operators and design machine functions in such a way that the task executions is max. supportive and useful for the operators.

2.3   Work System Analysis

On this level, the whole work system should be analysed and designed in such a way that all material and information input and output are available in time, in good quality and with the required accuracy. Criteria for good work system design are checkability, transparency and predictability. The human operators should have full control of the work system and have the necessary tools and information available to make appropriate decisions. They should have easy access to relevant customer data, product data, machine data and material data.

In addition, the social relations both within the work system boundaries between the operators and outside the work system with other operators and supervisors need to be considered, e.g. task-related communication, collaboration and coordination/scheduling of tasks.

CPPS, like any other conventional production system, need to fulfil these psychological criteria for human centred work system design, respecting the cognitive, motivational and social relationships and requirements of human operators.

Figure 1 visualizes and summarizes these principles of complementary work system design for CPPS.

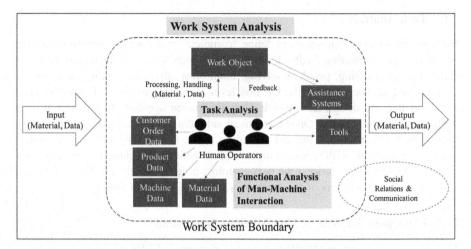

Fig. 1. Principles of complementary work system design for CPPS

In the following chapter, the application of these principles for the design of a CPPS in sensor manufacturing will be described in more detail.

3   Case Study: Design of a CPPS in Sensor Manufacturing

3.1   Case Description

A specialized manufacturing unit for electro-physical sensors which can be used for automation systems and control units wanted to improve the productivity in one of its production cells. The production cell has been designed for the small-scale batches of customer specific sensors in relatively small numbers (1 to 100 sensors per order). The human operators assemble and test the sensors on a work bench. They use various tools for the assembly and testing of electronic components and document their procedure by scanning and archiving electronic customer orders.

3.2   Objectives of the Work Systems Design

The overall purpose of the new work system design was to improve the productivity and work conditions of the human operators through the application of complementary work system design. At the same time, some new CPPS technologies like e.g. the equipment of the sensors with Active Digital Object Memories (ADOM) [7] should be tested. The new design should be evaluated to which extend the tasks, the human-machine interaction and the flow of material and data in the work system have been improved compared to the situation before.

3.3   Methods

Firstly, a task analysis was carried out to identify the critical process steps in the assembly process. Secondly, qualitative interviews with the human operators should indicate opportunities for improvement of the assembly process and potentials for CPS technologies in the production process. In addition, a functional analysis of man-machine interaction showed how the operators used the tools and assistance systems during the assembly process. The interviews followed a semi-structured interview guide with open and closed questions. Table 1 shows an extract of the relevant dimensions, categories and indicators of the interview guide.

Table 1. Structure of the interview guide for expert interviews with human operators in a sensor production cell (extract)

Dimensions	Categories	Indicators and Scaling
Completeness of work tasks	Scope of product handling (assembly, testing)	0: narrow scope… 3: broad scope
	Functions to be executed (programming, assembly, testing, documenting, quality checking)	0: limited functions… 3: complete functions
	Complexity of the product	0: simple part, no sub-components… 3: complex part, multiple sub-components
	Complexity of the assembly process	0: simple assembly, few connections between components… 3: complex assembly, multiple connections between components
	Learning opportunities for the human operators	0: little opportunities… 3: many opportunities
Autonomy of work system	Time autonomy	0: low autonomy… 3: high autonomy
	Coupling with other production cells	0: low coupling… 3: high coupling
Man-Machine Interaction	Transparency of production process	0: low transparency… 3: high transparency
	Process control	0: low control… 3: high control
	Authority of human operators	0: low authority of human operator, high authority of machine… 3: high authority of human operator, low authority of machine
Social Interaction inside/outside the work system	Cooperation between operators within the cell	0: no cooperation… 3: continuous cooperation
	Cooperation with operators of neighbour cells	0: no cooperation… 3: continuous cooperation
	Communication with supervisors	0: no communication… 3: continuous communication
	Communication with support functions (e.g. maintenance, quality assurance, shop floor logistics)	0: no communication… 3: continuous communication

The interviews were executed with ten interviewers to maximize the objectivity, reliability and the validity of the data. Comparisons of interview transcripts tested potential biases and errors in coding. The average inter-rater consistency of judgements was tested with spearman's rho = 0.73 [9].

3.4 Results

The task analysis followed the PDCA logic (Plan Do Check Act). Table 2 summarizes the results.

Table 2. Results of the task analysis

No. (Process Step)	Plan	Do	Check	Act
1	Plan assembly based on produc-tion plan		Check customer order specifica-tion	
2			Check sensor parts for assem-bly	
3			Check assembly tools	
4		Assemble parts (brazing and soldering of pins)		
5				
6			Mechanical check	
7			Functional check	
8				Scan order con-firmation
9				Scan quality assurance check list
10		Packaging of assembled sen-sors		

The results of the qualitative interviews indicated a high degree of completeness of the work tasks in the production cell. The operators need to plan and schedule the assembly process autonomously, check the availability of parts and tools, assemble the components and check the functionality and mechanical integrity of the assembled sensors. In addition, they confirm the correctness of the task execution and document the relevant check items, However, the coupling with other production cells and the communication with other operators both inside the cell and with other production cells and support functions could be improved. Figure 2 shows the profile of the analysed work system based on the interview data with ten experts in the sensor manufacturing unit.

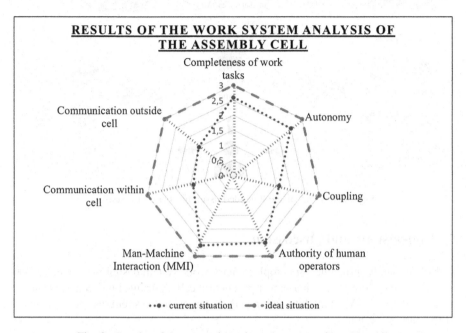

Fig. 2. Results of the expert interviews (average ratings; N = 10)

4 New Work System Design

Based on the results from the work system analysis, an improved design for the production cell was tested. A special focus by the designers was to enhance the communication both within and between cell operators, using audio-visual displays and a decentralized communication platform. In addition, the assembly parts were equipped with Active Digital Object Memories (ADOM) which allowed the human operators to identify which parts were needed for the fulfilment of a specific customer order. By scanning the parts with a 3D scanner connected with an automated testing system, functional and mechanical quality checks should be executed faster and smarter. Ergonomic measures, e.g. flexible seating and adjustable height of the assembly bench, guarantee a comfortable and safe work system (Fig. 3).

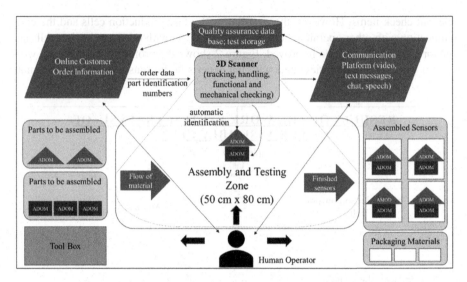

Fig. 3. Lay-out of the amended assembly bench for sensors

5   Conclusions and Discussion

In this case study, principles of complementary work system design have been applied to improve the productivity in a sensor production cell. A detailed task analysis and the analysis of Human-Machine Interaction (MMI) demonstrated strengths and weaknesses of the current design.

In the new design, the parts to be assembled in the production unit were equipped with Active Digital Object Memories (ADOM). These Memory Units can be used as parts identifiers for a 3D scanner. The human operators have access to online customer order information (e.g. order date, delivery date, quality requirements, special customer requirements) and can communicate with other operators, supervisors and support functions via a decentralized communication platform. All physical parts and components to be assembled are represented as 3D model in a central database. In addition, all functional and mechanical data, like e.g. electronic wiring, Input-Output (I/O) controls, pin layout, etc.) are stored and can be used for automatic quality checks during assembly.

The new work system design has been tested and evaluated in comparison with other, conventional production cells. The output increased by 20%. The number of defective sensors decreased by 50%. The complementary work system design allowed the human operators to focus on high quality "built-to-order" in small batches. Through direct access to the customer order data, the operator could schedule the assembly process based on order priority. The digital communication platform enabled direct/synchronous and indirect/asynchronous communication with other operators, supervisors and support functions. For example, at the end of shift an operator could leave a voice message for the operator of the next shift, explaining the status of an urgent customer order that needs to be fulfilled as soon as possible.

Overall, the operators were satisfied with the new work system design. The access to customer order data and the automatic testing of assembled sensors gave them more authority and process control. They could identify defects earlier and single out defective parts and components. Through the digital communication platform, they could communicate with colleagues and supervisors in a comfortable, fast and easy way.

As a conclusion, CPPS technologies like Active Digital Object Memories (ADOM) and automatic part identification can improve the productivity, quality and satisfaction in tomorrow´s industry. However, these technologies should not be implemented without a thorough work system analysis. The work system needs to be analysed on various levels (tasks, human-machine-interaction, storage and flow of data and materials). Digital tools and technologies should support and not replace specific human skills like work planning, problem solving and informed decision making.

6 Outlook

The future advancement of Cyber-Physical Production Systems in the industry will not only depend on technological progress, but also on the intelligent design of digitally enhanced work systems. The role of human operators in CPPS needs to be discussed and carved out with an emphasis on the principles of human-centred work design. The theoretical and empirical findings of industrial psychology related to the completeness of work tasks (Plan, Do, Check, Act), feedback, control, social relatedness and autonomy are still valid in CPPS. Hence, industrial engineers should always consider the principles of complementary work system design for the planning of new industry facilities and production systems to be successful [10].

References

1. Schwab K The Fourth Industrial Revolution: what it means and how to respond, World Economic Forum, 15 December 2015. https://agenda.weforum.org/2015/12/the-fourth-industrial-revolution-what-it-means-and-how-to-respond/. 16 Mar 2018
2. Grote G, Wäfler T, Ryser C, Weik S, Zölch M, Windischer A (1999) Wie sich Mensch und Technik sinnvoll ergänzen: Die Analyse automatisierter Produktionssysteme mit KOMPASS. vdf, Zürich
3. Wäfler T, Windischer A, Ryser C, Weik S, Grote G (1999) Wie sich Mensch und Technik sinnvoll ergänzen: Die Gestaltung automatisierter Produktionssysteme mit KOMPASS. vdf, Zürich
4. Gilchrist A (2016) Introducing Industry 4.0. In: Industry 4.0. Apress, pp 195–215
5. Porter ME, Heppelman JE (2015) How smart, connected products are transforming competition. Harvard Business Review. http://lp.servicemax.com/rs/020-PCR-876/images/HBR-Connected-Products-Summary.pdf. 16 Mar 2018
6. CyProS (Produktivitäts- und Flexibilitätssteigerung durch die Vernetzung intelligenter Systeme in der Fabrik) (2012) Research Project funded by the Federal Ministry of Education and Research, Germany. http://www.projekt-cypros.de. 16 Mar 2018

7. Bergweiler S (2015) Intelligent manufacturing based on self-monitoring cyber-physical systems. In: Proceedings of the ninth international conference on mobile ubiquitous computing, systems, services and technologies (UBICOMM 2015), pp 108–113
8. Hu F (2013) Cyber-physical systems: integrated computing and engineering design. Taylor & Francis, Abingdon
9. Gamsu CV (1986) Calculating reliability measures for ordinal data. Br J Clin Psychol 25(Pt 4):307–308
10. Rogelberg SG (ed) (2004) Encyclopedia of industrial and organizational psychology. Wiley-Blackwell, Oxford

Baiting the Hook: Exploring the Interaction of Personality and Persuasion Tactics in Email Phishing Attacks

Patrick A. Lawson, Aaron D. Crowson,
and Christopher B. Mayhorn[(✉)]

North Carolina State University, Raleigh, NC, USA
{Palawson, Adcrowso, Chris_mayhorn}@ncsu.edu

Abstract. Phishing is a social engineering tactic where a malicious actor impersonates a trustworthy third party with the intention of tricking the user into divulging sensitive information. Previous social engineering research has shown an interaction between personality and the persuasion principle used to generate non-electronic messages. This study investigates whether this interaction is present in the realm of email phishing. To investigate this, we used a personality inventory and an email identification task (phishing or legitimate). Our data confirms previous findings that high extroversion is predictive of increased susceptibility to phishing attacks. However, extraversion was also found to be associated with increased susceptibility to phishing emails that utilize specific persuasion principles such as liking. Findings are discussed in terms of potential approaches to anti-phishing interventions within organizations.

Keywords: Phishing · Email · Personality · Persuasion

1 Introduction

Phishing is a social engineering tactic designed to trick users into divulging sensitive personal information, such as one's social security or bank account numbers, through impersonation of a trustworthy third party (Jagatic et al. 2007). The primary phishing modes include mocking up a trusted website, emailing potential targets directly, or a combination of the two (Akbar 2014). This study focuses specifically on the emailing mode.

Past research suggests that there are various strategies that can be employed to persuade a target to divulge their sensitive information. Cialdini (1987) identified six broad persuasion principles. Four of these are far more common among phishing emails than the other two (Akbar 2014). Furthermore, those same four persuasion principles have been found to be increasing in volume over time in phishing emails (Zielinska et al. 2016); therefore, we will focus on them. The four principles and brief descriptions are provided below.

- *Commitment/consistency:* the concept of completing an action you previously initiated.
- *Liking*: trust due to a prior interaction or familiarity, such as for a largely recognizable brand.

© Springer Nature Switzerland AG 2019
S. Bagnara et al. (Eds.): IEA 2018, AISC 822, pp. 401–406, 2019.
https://doi.org/10.1007/978-3-319-96077-7_42

- *Authority*: an authority figure mandating an action, with consequences for failing to comply
- *Scarcity*: a short and specific time frame to complete an action. (Cialdini 1987).

Among these, authority is the most frequently utilized principle by volume in phishing emails (Akbar 2014; Zielinska et al. 2016). Many user characteristics also influence email phishing susceptibility. For instance, the personality profile of the victim plays a role in the likelihood of being phished. High distrust of others is significantly positively correlated with accuracy in identifying phishing emails (Welk et al. 2015). Generally, high extroversion is found to be the personality trait most predictive of increased phishing susceptibility (Halevi et al. 2013; Workman 2008). High agreeableness is also posited, but not proven, to be related to phishing susceptibility (Parrish et al. 2009).

The previously mentioned traits agreeableness and extroversion—in addition to neuroticism, openness, and conscientiousness—comprise the five personality constructs of the Five-Factor Model of personality, colloquially known as the 'Big Five' (Costa and McCrae 1992a). These traits have been shown to be stable over time, and universally identifiable regardless of language, race, culture, or gender (Costa and McCrae 1992b).

Notably, when looking at social engineering in the real world (i.e. not online), interaction effects between the persuasion principle used and the personality of the target have been demonstrated (Uebelacker and Quiel 2014). For example, extroverted individuals are especially susceptible to the liking and scarcity persuasion principles, while agreeable individuals are especially susceptible to the authority principle, among other such interactions (Uebelacker and Quiel 2014).

It has thus been demonstrated that (1) many different persuasion principles exist and are utilized in phishing emails, (2) potential victims' personality profiles are related to phishing susceptibility, and (3) that efficacy of real-world social engineering is modulated by an *interaction* between the persuasion principle used and the victim's personality profile. This paper investigates whether this interaction between the persuasion principle and the user's personality exists within the realm of email phishing attacks.

2   Method

This study involved completing a comprehensive personality inventory (NEO-FFI-3) derived from the "Big Five" personality traits (neuroticism, extroversion, openness, agreeableness, conscientiousness) framework championed by Costa and McCrae (1992a, b). The trust subsection of the IPIP NEO PI-R was used to assess trust and the impulse control subsection of the IPIP AB5C Facets Abbreviated Scale was used to assess impulse control.

2.1   Participants

One hundred and two participants (mean age 19.3 years old; SD = 2.8) were recruited from an undergraduate psychology course at a large Southeastern university in the United States, and given class credit for participation. Fifty-four participants were female.

2.2   Stimulus Materials

Forty-five legitimate emails were drawn from the experimenters' personal emails. Forty-five phishing emails were drawn from a corpus of confirmed phishing attacks compiled from three prominent universities (Zielinska et al. 2016). Only sensitive identifying information was removed, such as the recipient's name and email address; otherwise, the emails were unaltered. The emails were coded according to the personality principle (or principles) utilized. Three raters coded these emails, and there was an 87% agreement between raters (Zielinska et al. 2016). This analysis of phishing emails by division into categories according to all persuasion principles present was proposed and demonstrated to be of value by Ferreira et al. (2015). As mentioned above, four of Cialdini's persuasion principles were considered: commitment/ consistency (C), liking (L), authority (A), and scarcity (S). Emails were categorized according to all persuasion principles utilized.

After considering the prevalence of each principle and its likelihood of being combined with other principles, nine groups of persuasion principles (or combinations of principles) were derived: A, A/C, A/L, A/S, C, C/L, L, S, and Super (Su). The Super category was defined as using at least three of the four core Cialdini principles assessed in this study (A, C, L, S). The number of emails in each category can be seen in Table 1. Due to an inability to identify five legitimate emails exclusively using the Authority principle in a natural context, and a desire to maintain an equal number of legitimate and phishing emails, three legitimate groups comprised six emails rather than five. Thus, 90 emails were adopted as stimuli in this experiment.

Table 1. Number of emails (n = 90) by classification category.

Persuasion Principle(s)	A	A/C	A/L	A/S	C	C/L	L	S	Su
Legitimate Emails	2	5	5	5	5	6	6	5	6
Phishing Emails	5	5	5	5	5	5	5	5	5

2.3   Procedure

Participants completed a consent form before proceeding to the experiment, hosted online using the Qualtrics survey instrument. Participant demographics and personality measures were collected before they began the email classification task where they were asked to assess whether 90 emails were phishing attempts or legitimate emails. Upon completion of the experiment, participants were thanked for their participation and awarded class credit.

3 Results

An independent samples t-test was conducted to identify if there were differences between phishing and legitimate accuracies. Phishing accuracy (M = .66, SD = .13) was found to be significantly greater than legitimate accuracy (M = .62, SD = .13), t(202) = 2.21, p = .028. That is, participants were more likely to correctly label a phishing email as phishing than to correctly label a legitimate email as legitimate.

To assess the interaction of the various personality traits with the persuasion principle utilized, multiple linear regressions were conducted. In each of the following regressions the predictor variables entered into the model were impulse control, trust, neuroticism, extroversion, openness, agreeableness, and conscientiousness (i.e. all seven of the personality measures). All multiple regressions were conducted using the Enter method.

Inputting all of the listed predictor variables, a significant model was found for phishing accuracy F(7,101) = 3.37, p = .003, R^2 = .20. This model explained 20% of the observed variance. As can be seen in Table 2, it was found that high extroversion was a significant predictor of decreased phishing accuracy (β = −.33, p = .007). Notably, a significant model was not found with legitimate accuracy or overall accuracy as the outcome variable, and none of the predictor variables in these models reached significance.

Table 2. Beta coefficients for overall, phishing, and legitimate accuracies

Accuracy	Overall	Phishing	Legitimate
R^2	**0.16**	**0.20**	**0.08**
Impulse Control	0.15	0.11	0.07
Trust	0.09	0.05	0.07
Neuroticism	0.10	0.02	0.11
Extroversion	−0.15	**−0.33****	0.15
Openness	0.04	−0.10	0.15
Agreableness	0.22	0.19	−0.08
Conscientiousness	−0.12	−0.05	−0.09

**Indicates significance at the p = .01 threshold

Next, regressions were conducted with each of the nine individual persuasion principles (or combinations of principles) as the outcome variable, using the same predictor variables as above (i.e. all seven personality measures). Nine separate linear regressions were thus conducted, one for each persuasion principle (or combination of principles). We first looked at the phishing emails.

As can be seen in Table 3, extroversion was found to be associated with decreased detection of phishing attacks utilizing: authority & commitment/consistency persuasion (β = −.31, p = .015), authority & liking persuasion (β = −.29, p = .021), commitment/ consistency persuasion (β = −.30, p = .017), and liking persuasion (β = −.28, p = .024). In the five cases where extroversion was not *significantly* predictive of

Table 3. Beta coefficients for phishing accuracy by persuasion principle(s)

Persuasion Principle(s)	Phishing Emails								
	Authority	Authority & Commitment/ Consistency	Authority & Liking	Authority & Scarcity	Conscientiousness	Liking	Liking & Commitment/ Consistency	Scarcity	Super (3 +)
R²	**0.09**	**0.14**	**0.14**	**0.06**	**0.13**	**0.19**	**0.04**	**0.10**	**0.17**
Impulse Control	−0.03	−0.06	0.07	0.24	0.08	0.00	−0.03	0.06	0.27
Trust	−0.04	−0.05	0.05	−0.03	0.05	0.20	−0.11	0.02	0.16
Neuroticism	0.08	0.17	−0.12	0.05	0.09	0.03	−0.11	−0.15	0.10
Extroversion	−0.15	**−0.31***	**−0.29***	−0.011	**−0.30***	**−0.28***	−0.16	−0.19	−0.04
Openness	−0.14	−0.14	0.01	0.03	−0.13	−0.16	0.00	0.05	−0.07
Agreableness	0.25	0.08	0.15	0.06	0.01	0.20	0.03	0.14	0.12
Conscientiousness	0.09	0.10	0.10	−0.10	0.03	−0.15	0.09	−0.11	**−0.24***

*Indicates significance at the $p = .05$ threshold

Table 4. Beta coefficients for legitimate accuracy by persuasion principle(s)

Persuasion Principle(s)	Legitimate Emails								
	Authority	Authority & Commitment/ Consistency	Authority & Liking	Authority & Scarcity	Conscientiousness	Liking	Liking & Commitment/ Consistency	Scarcity	Super (3 +)
R²	**0.03**	**0.02**	**0.08**	**0.04**	**0.11**	**0.07**	**0.05**	**0.09**	**0.08**
Impulse Control	0.18	0.02	0.07	−0.06	−0.02	0.08	0.10	0.04	0.04
Trust	−0.03	0.14	−0.08	0.03	**0.25***	0.07	0.00	0.02	−0.04
Neuroticism	0.12	0.10	0.03	−0.12	0.02	0.18	−0.04	0.17	0.08
Extroversion	0.06	0.03	0.24	0.08	−0.15	0.20	0.03	0.16	0.09
Openness	−0.06	0.00	0.16	0.15	0.12	−0.11	0.20	0.04	**0.23***
Agreableness	−0.02	−0.03	0.11	−0.03	0.02	0.05	0.06	0.18	0.05
Conscientiousness	0.02	−0.02	−0.14	−0.10	0.13	0.03	−0.06	−0.15	−0.17

*Indicates significance at the $p = .05$ threshold

increased susceptibility to a persuasion principle, the results were trending in the direction of increased susceptibility. Conscientiousness was found to be associated with increased detection of phishing attacks utilizing super persuasion ($\beta = -.24, p = .031$).

The same steps were then used in the analysis of the legitimate emails. Here, trust was found to be associated with increased identification accuracy of legitimate emails utilizing commitment/consistency persuasion ($\beta = .25, p = .033$). Openness was found to be associated with increased identification accuracy of legitimate emails utilizing super persuasion ($\beta = .23, p = .035$). These results can be found in Table 4.

4   Discussion

Our findings indicate that extroversion was a strong predictor of overarching susceptibility to phishing attacks. It was the only factor significantly predictive of overall phishing identification ability, and achieved this significance at the $p = .01$ threshold. Extroversion was also significantly associated with increased susceptibility to four specific persuasion principles: authority & commitment/consistency, authority & liking, commitment/consistency, and liking. Also, conscientiousness was associated with an increased ability to detect phishing emails that utilized "super" persuasion using more than three principles. Thus, an interaction between the persuasion principle utilized and the target's personality characteristics is broadly supported.

Given the consistency with which extroversion predicted susceptibility to phishing emails, the potential for personality-based interventions should be explored further. The extroversion subsection of the personality assessment consisted of 12 items, taking only a minute or two to complete. It would not be too much of a burden, then, for organizations to screen employees for level of extroversion, and then direct phishing training to those scoring highly on this measure. Phishing training has been shown to be effective, but can be cumbersome to implement on an organization-wide scale; limiting training to highly extroverted individuals may be a cost-effective option.

References

Jagatic TN, Johnson NA, Jakobsson M, Menczer F (2007) Social phishing. Commun ACM 50 (10):94–100

Akbar N (2014) Unpublished master's thesis. Analysing persuasion principles in phishing emails

Cialdini RB (1987) Influence, vol 3. A. Michel

Zielinska OA, Welk AK, Mayhorn CB, Murphy-Hill E (2016) A Temporal analysis of persuasion principles in phishing emails. In: Proceedings of the human factors and ergonomics society annual meeting, vol 60, no 1. SAGE Publications, pp 765–769

Welk AK, Hong KW, Zielinska OA, Tembe R, Murphy-Hill E, Mayhorn CB (2015) Will the "Phisher-Men" reel you in?: Assessing individual differences in a phishing detection task. Int J Cyber Behav Psychol Learn (IJCBPL) 5(4):1–17

Halevi T, Lewis J, Memon N (2013) Phishing, personality traits and facebook. arXiv preprint arXiv:1301.7643

Workman M (2008) Wisecrackers: a theory-grounded investigation of phishing and pretext social engineering threats to information security. J Am Soc Inform Sci Technol 59(4):662–674

Parrish JL Jr, Bailey JL, Courtney JF (2009) A personality based model for determining susceptibility to phishing attacks. University of Arkansas, Little Rock

Costa PT Jr, McCrae RR (1992a) Normal personality assessment in clinical practice: the NEO personality inventory. Psychol Assess 4(1):5–13

Costa PT, McCrae RR (1992b) Four ways five factors are basic. Personality Individ Differ 13 (6):653–665

Uebelacker S, Quiel S (2014) The social engineering personality framework. In: 2014 workshop on socio-technical aspects in security and trust. IEEE, pp 24–30

Biomechanics of the Cervical Region During Use of a Tablet Computer

Grace Szeto[1(✉)], Pascal Madeleine[2], Kelvin Chi-Leung Kwok[1],
Jasmine Yan-Yin Choi[1], Joan Hiu-Tung Ip[1], Nok-Sze Cheung[1],
Jay Dai[1], and Sharon Tsang[1]

[1] Department of Rehabilitation Sciences,
The Hong Kong Polytechnic University, Hung Hom, Kowloon, Hong Kong
grace.szeto@polyu.edu.hk
[2] Physical Activity and Human Performance Group, SMI,
Department of Health Science and Technology,
Aalborg University, Aalborg, Denmark

Abstract. This study aimed to investigate the spinal muscle activity and postural variations across time in healthy young adults during a prolonged reading comprehension task using a tablet computer. Twenty healthy college students (10 males and 10 females; mean age = 21.5 ± 1.7 years) participated in this study. Subjects were seated on a standard office chair with adjustable height and allowed to move freely in their body postures during the 30-minute reading comprehension task. The subject was instructed to hold a Samsung Galaxy Tab S2 9.7 (SM-T810) with both hands. Surface electromyography and spinal kinematics were recorded simultaneously. The amplitude probability distribution function (APDF) was computed for each 10 min trial of the muscle activity data and postural angle data respectively. Amplitude measures of muscle activity using $50^{th}\%$APDF and APDF range (difference between 90th% and 10th% APDF) were examined in 3 time phases (T1, T2, T3) of 10 min each. Postural variation (using zero crossing algorithm) was also analyzed. There was a significant increase in median muscle activity ($50^{th}\%$APDF) of bilateral cervical erector spinae (CES) (left: $p = .002$; right: $p = .002$) and a decreasing trend in bilateral thoracic erector spinae (TES) (left: $p = .053$; right: $p = .068$) over time. Significant increase of cervical postural variation was also revealed across time ($p = 0.001$). Finally, sex differences regardless of time effect were shown, with females showing significantly higher left CES median muscle activity ($p = 0.044$) and muscle activity range ($p = 0.047$) when compared with males. Other muscles did not reveal such significant differences.

Keywords: Electromyography · Posture · Tablet computer

1 Introduction

1.1 Increasing Popularity of Tablet Computers

Tablet computer is a popular gadget in the modern technological era. Its global shipment reached almost 220 million units, ranking the highest among electronic products

© Springer Nature Switzerland AG 2019
S. Bagnara et al. (Eds.): IEA 2018, AISC 822, pp. 407–412, 2019.
https://doi.org/10.1007/978-3-319-96077-7_43

in 2013 [1]. This leading share in the market has been forecasted to remain in 2020 [1]. Among all age groups, young adults aged from 18 to 24 have been reported to be the most frequent tablet users with the longest average duration of use [2]. However, prolonged use of touchscreen devices has been associated with increased neck and shoulder muscle activity and neck-shoulder physical discomfort [3, 4]. Past research has reported that head and neck flexion angle was greater in tablet use compared to that in desktop computer operation [5]. Biomechanically, prolonged flexed neck posture is associated with increased gravitational demand, which may pose increased load in the cervical spinal muscles and eventually lead to increased risk of neck pain [6].

1.2 Association of Neck Pain with Spinal Kinematics and Muscle Activity During Computer Use

Recent research in the past few years have examined the cervical spine kinematics and cervical muscle activity during smartphone and tablet computer use. However, these investigations have mainly involved standardized postures and standardized tasks in using the electronic devices. Task durations usually involved 1 to 10 min at most. It is not known whether people have a natural tendency to vary their postures over longer periods of using tablet computers and whether this is related to their experience of discomfort or pain in the neck.

The present study examined the muscle activity and the spinal posture kinematics of the cervical spine during a 30-minute reading comprehension task using a handheld tablet computer. This task involves reading, tapping and scrolling components, which simulate the physical actions of the two most common tablet computer tasks such as email reading and internet browsing [2].

2 Methods

2.1 Study Design and Subjects

In this cross-sectional study, twenty healthy young adults (10 males and 10 females) were recruited from local universities by convenience sampling. Subjects had to use tablet computer for at least 30-minute continuously daily in recent 12 months. Subjects who had (1) history of neck-shoulder pain, (2) history of any neurological, orthopedic disorders or medical conditions which might affect the relevant regions, (3) chronic diseases affecting musculoskeletal system such as rheumatoid arthritis and other connective tissue disorders, were excluded. These exclusion criteria were screened according to subjects' past medical history before the study.

Demographic data and information about tablet computer usage pattern were collected. Ethical approval was sought from the Hong Kong Polytechnic University before launching of the study. Informed consent was obtained from each participant prior to their attendance in the laboratory session.

2.2   Experimental Procedures

Prior to the reading comprehension task, surface electromyography (EMG) electrodes and Inertial Measurement Unit (IMU) were applied to subjects' neck, shoulder and back region. A standard chair was provided with seat height adjusted to the level that subjects' knees were at approximately 90° flexion and feet were resting on the ground. During the task, free postural variation was allowed except rotating the chair and leaning onto the back support.

Noraxon Telemyo wireless EMG System (Noraxon USA Inc., USA) was used to record muscle activity. Three muscles, namely the cervical erector spinae (CES), upper trapezius (UT) and thoracic erector spinae (TES) were assessed bilaterally. Skin preparation procedure and electrodes placement were referenced to previous studies [7, 8]. Skin impedance below 10 kΩ was considered acceptable. Resting EMG was recorded in sitting position for 10 s before the task. The EMG sampling frequency was 1500 Hz. The EMG signals were digitized by a 16-bit analogue to digital (A/D) converter. ECG reduction was performed and all data were processed with a band pass filter of 20–300 Hz and a notch filter of 50 Hz to reduce signal noises. Full-wave rectification and signal smoothing with moving window of 50 ms were then applied. Two parameters of Amplitude Probability Distribution Function (APDF) 50th percentile (50th %tile APDF, or median muscle activity) and difference between the APDF 90th and 10th percentiles APDF, or amplitude range) were used [7, 8].

Noraxon wireless Myomotion System (Noraxon USA Inc., USA) with sampling frequency of 100 Hz was used to record spinal kinematics. The accuracy is $\pm\,2°$ of joint motion. The four inertial motion sensors measure the 3D movements of the cervical, thoracic and lumbar spine, and for the present study only the spinal movements in the sagittal plane will be reported. The sensors were calibrated to the standard sitting posture before the task. The mean angle data were computed for the three spinal regions in the three phases of 10 min each.

All subjects were instructed to perform the 30-minute reading comprehension task on the same tablet computer: Samsung Galaxy Tab S2 9.7 (SM-T810) (Wi-Fi version) (Width: 169.0 mm; Height: 237.3 mm; Depth: 5.6 mm; Weight: 389 g). Posture of holding the tablet and fingers for tapping were not restricted, however, zooming in or out on the screen was not allowed.

2.3   Data Analysis

The dependent variables were the mean amplitudes (50th %tile APDF) and the amplitude range (90th–10th%tile APDF) of the muscle activity in the three phases of 10 min each during the 30-min task. The mean flexion angle for each spinal region (cervical, thoracic and lumbar) were also examined in the same way using one-way repeated measure ANOVA. SPSS version 23.0 was used for all data analysis, and the significance criterion was $p < 0.05$ for all statistical analysis.

3 Results

The mean amplitudes (50th%APDF) of the three muscle pairs were examined in three phases of 10 min during the 30-min reading task (see Fig. 1). There was a significant increase in median muscle activity of bilateral cervical erector spinae (CES) (left: $p = 0.002$; right: $p = 0.002$) and a decreasing trend in bilateral thoracic erector spinae (TES) (left: $p = 0.053$; right: $p = 0.068$) over time. We also observed a significant increase in right CES muscle activity variation, or APDF amplitude range ($p = 0.042$) and an increasing trend in left CES amplitude range ($p = 0.056$) across time.

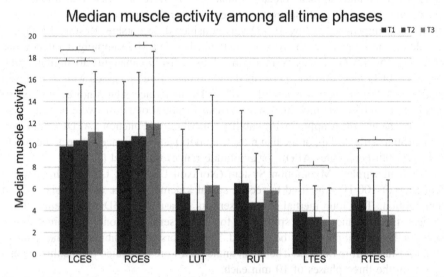

Fig. 1. Mean amplitudes of the three pairs of spinal muscles across three time phases (T1, 2, 3, 10 min each) in the 30-min reading task. *Denotes significant differences ($p < 0.05$)

When the EMG results were compared between the male and female subjects, the females showed significantly higher left CES median amplitude ($p = 0.044$) and APDF range ($p = 0.047$) when compared with males. Other muscles did not reveal such significant differences.

For the mean angle over the three 10-min phases, cervical flexion had an overall decreasing trend ($p = 0.087$) while thoracic and lumbar flexion had an overall increasing trend (thoracic: $p = 0.076$; lumbar: $p = 0.093$) across time. These results suggest that the participants gradually adopted less cervical flexion and more flexion in their thoracic and/or lumbar spine as they continued to interact with the tablet computer in the later part of the 30 min.

4 Discussion

The present results suggest that performing a prolonged reading task using a tablet computer is associated with increasing muscle activity in extensors of the cervical and thoracic region and increasing cervical postural variation. Past research has seldom involved electronic device use for such long durations. To our knowledge, this is the first study that examined the spinal biomechanics when individuals performed such a prolonged task in using an electronic device, and participants were allowed to vary their postures during the task.

However, the same model of tablet computer was used in the study, and it is possible that different screen sizes and weights of such handheld devices may affect the user's posture. Ning et al. [9] compared the kinematics and muscle activity in tablet computer users when they adopted a seated versus a standing posture. These variations in total body postures may affect the biomechanical loads. Our present study only examined the postural variations in a seated position and this may be a limitation for generalizing the results.

The participants in this study were mostly healthy painfree individuals. The present results suggest that they showed a tendency towards increasing their muscle activity variation as well as adopting greater thoracic/lumbar flexion instead of cervical flexion with the prolonged tablet task. It is plausible that such a motor control strategy is effective for distributing the mechanical loads and avoid excessive strain on the spinal structures. It would be interesting to compare the present results with a symptomatic group with neck or back pain while performing the same task. This future work will contribute towards establishing ergonomic recommendations for tablet computer users.

References

1. Statista Inc. (2016) Shipment forecast of laptops, desktop PCs and tablets worldwide from 2010 to 2020 (in million units). Statista Inc., Hamburg. http://www.statista.com/statistics/272595/global-shipments-forecast-for-tablets-laptops-and-desktop-pcs/. Accessed 20 Nov 2016
2. Salesforce (2014) 2014 Mobile behavior report. Salesforce, Singapore. http://www.exacttarget.com/system/files_force/deliverables/etmc-2014mobilebehaviorreport.pdf?download=1&download=1. Accessed 20 Nov 2016
3. Shin G, Zhu X (2011) User discomfort, work posture and muscle activity while using a touchscreen in a desktop PC setting. Ergonomics 54:733–744
4. Xie Y, Szeto GPY, Dai J, Madeleine P (2016) A comparison of muscle activity in using touchscreen smartphone among young people with and without chronic neck–shoulder pain. Ergonomics 59:61–72
5. Straker LM, Coleman J, Skoss R, Maslen BA, Burgess-Limerick R, Pollock CM (2008) A comparison of posture and muscle activity during tablet computer, desktop computer and paper use by young children. Ergonomics 51:540–555
6. Vasavada AN, Nevins DD, Monda SM, Hughes E, Lin DC (2015) Gravitational demand on the neck musculature during tablet computer use. Ergonomics 58:990–1004

7. Szeto GPY, Straker LM, O'Sullivan PB (2005) A comparison of symptomatic and asymptomatic office workers performing monotonous keyboard work-1: neck and shoulder muscle recruitment patterns. Man Ther 10:270–280
8. Szeto GPY, Straker LM, O'Sullivan PB (2009) Examining the low, high and range measures of muscle activity amplitudes in symptomatic and asymptomatic computer users performing typing and mousing tasks. Eur J Appl Physiol 106:243–251
9. Ning X, Huang Y, Hu B, Nimbarte AD (2015) Neck kinematics and muscle activity during mobile device operations. Int J Ind Ergon 48:10–15

Is CCTV Surveillance as Effective as Popular Television Crime Series Suggest? Cognitive Challenges

Fiona M. Donald[1](✉) [iD] and Craig Donald[2]

[1] University of the Witwatersrand, Johannesburg, South Africa
Fiona.Donald@wits.ac.za
[2] Leaderware, South Africa and Edith Cowan University, Perth, Australia

Abstract. The aim of this paper is to review the human and information processing factors that need to be addressed in order to improve closed circuit television (CCTV) surveillance effectiveness and to make recommendations regarding future research. This is done by contrasting the way in which CCTV is portrayed in popular television crime series with the challenges inherent in real world CCTV surveillance systems. Despite considerable amounts of money being spent on the equipment needed for CCTV systems, the work of operators is often poorly valued although the monitoring process is difficult and mentally demanding. There are many factors that affect the effectiveness of CCTV surveillance systems, and this paper focuses on the information processing demands on CCTV surveillance operators. Previous research on the human factors in CCTV were reviewed, and episodes in popular crime series which showed CCTV were observed and analysed. Key aspects that emerged were the cognitive demands made on CCTV surveillance operators by the design of the technical system and their job, the nature of scenes observed and the characteristics of significant events. These placed demands on the interaction of goal-directed and stimulus-driven attention, search strategies employed, distributed situation awareness, visual analysis, processes that impair detection, and the effects of certain job designs on monitoring. Recommendations for future research were made.

Keywords: CCTV · Monitoring · Information processing · Visual search
Visual analysis · Operator

1 Introduction

The aim of this paper is to review the human and information processing factors that need to be addressed in order to improve closed circuit television (CCTV) surveillance effectiveness and to make recommendations regarding future research. This is done by contrasting the way in which CCTV is portrayed in popular television crime series with the challenges inherent in real world CCTV surveillance systems. Studies of CCTV effectiveness have reported widely differing detection rates. Despite considerable amounts of money being spent on the equipment needed for CCTV systems, the work

© Springer Nature Switzerland AG 2019
S. Bagnara et al. (Eds.): IEA 2018, AISC 822, pp. 413–420, 2019.
https://doi.org/10.1007/978-3-319-96077-7_44

of operators is often poorly valued although the monitoring process is difficult and mentally demanding [19]. There are many factors that affect the effectiveness of CCTV surveillance systems [9], and this paper focuses on the information processing demands on CCTV surveillance operators.

2   Method

Searches of previous literature were conducted using the keyword 'CCTV' in conjunction with effectiveness, cognition, cognitive processes, visual search, information processing, attention, visual attention, detection, recognition, task engagement, boredom, vigilance and sustained attention. Search engines included Google Scholar, EbscoHost, Proquest, PsycInfo and ScienceDirect. Articles that did not refer to human factors or cognitive processes or which focused on digital recognition, algorithmic surveillance or biometrics were excluded. Regarding the comparison with television crime series, episodes involving CCTV were watched from a number of series, such as Castle, Chicago PD, Law and Order Special Victims Unit, Major Crimes, and Line of Duty. Observations regarding the purpose of the CCTV surveillance, the technical system and observers and operators were made and trends identified using thematic content analysis.

3   Results

Two key stereotypes regarding CCTV surveillance emerged from the analyses of the crime series—either CCTV is portrayed as being highly effective or highly ineffective. These are described and compared below.

3.1   CCTV System Design

When CCTV is portrayed as being effective in crime series, it is usually assumed that the CCTV system is well designed and aligned with human detection capabilities. The cameras are usually appropriately situated. Individuals on the screens can be seen clearly enough to be identified, and small details, can be seen. However, research has demonstrated that in real life, factors related to system design are at times problematic [29].

CCTV systems vary considerably regarding their design and the extent to which they facilitate detection. Features that influence image quality [21], and the number of cameras per operator impact on detection [29, 31, 34, 36, 39]. Information load has been identified as a factor that influences the difficulty of multiplex detection [34]. Additional factors that need to be considered in deciding on the number of cameras that operators should monitor concurrently include, amongst others, the importance of detecting significant events [7], and the nature of the scenes and incidents [10].

3.2   Nature of Scenes

The nature of scenes captured on camera is likely to influence the operators' ability to detect significant events [34]. This seldom comes across as a problem in crime series, where it is often assumed that if an event or person is captured on camera, the information can be used. However, it presents various detection challenges in real life. The scenes might be changing continuously and include multiple objects or people moving in different directions at different speeds [9]. When scenes change rapidly and unpredictably, even when an incident has not occurred, monitoring is likely to require more attentional resources [7]. Scenes are often cluttered, include objects that are occluded, and contain a large amount of visual information. Images with more relevant information are perceived as being more complex and difficult [4], and create a larger visual workload.

3.3   Nature of Significant Events

The nature of significant events contributes to detection difficulty. Significant events can consist of objects and people, events and behaviours, or patterns of behaviours and events. Consequently there is a wide range of visual stimuli that operators need to look for [8] and this is inefficient [26]. Significant behaviours are often camouflaged with targets trying to blend into the background [7]. Therefore a large amount of visual analysis and attention is required to detect certain events on CCTV.

The factors discussed above affect the effectiveness of the operator's ability to process information.

3.4   Operator Processes

The previous sections outlined the demands placed on operators by the nature of CCTV systems, scenes and incidents. These interact with the cognitive processes that influence where, when and how operators monitor and detect significant events.

In crime series, the observer often knows what to look for, and where and when the incident will or has occurred. This information is used to guide their visual search. This contrasts with proactive surveillance in real life, where there it is often not clear whether, where, when or how an incident will occur. This makes it more difficult for operators to know what search strategy to employ. The scanning strategies used by operators vary considerably, with some adopting a random approach, others relying on 'gut feel' and others using a more systematic approach [41]. Yet other operators are likely to use their attention sets, expectancies and situation awareness to guide their eyes and attention to areas that are likely to contain relevant information [35].

It is likely that goal-directed attention plays an important role in the successful detection of inconspicuous significant events. There is evidence that operators' goals of detecting incidents, their expectancies and understanding of incidents, influence what they look for, areas considered to be of interest, decisions regarding who or what to monitor [27], gestures and body position [38] on a goal-directed basis. The concept of 'suspiciousness' has been identified as a cognitive structure that influences surveillance operators' eye movements and search strategies, with observers directing their gaze at events they consider to be suspicious [21].

Observers in crime series usually have the advantage of only having to monitor one or at most a few cameras. This reduces the number of scenes to prioritise for visual search and alters their search strategy. However, when large numbers of cameras are shown in crime series, the operators are typically depicted as having instant recall of camera numbers, and accurate mental models of the locations of the cameras in relation to each other. They use the cameras to track the movements of suspects and guide response teams, moving seamlessly between cameras. They give the impression that they are aware of what is happening in many locations simultaneously. In real life, when operators observe large numbers of cameras, there is evidence that they select certain cameras to focus on via a spot monitor and pay less attention to the multiplexed cameras [35]. The decision regarding which camera to focus on is therefore very important and is typically based on the operators' expectancies regarding where incidents are likely to occur. When they do have to coordinate their search between different cameras, identifying camera locations on geographical maps can be problematic [24] and this reduces the effectiveness of CCTV surveillance.

In addition to strongly goal-directed attentional guidance, observers in crime series usually have the advantage of excellent distributed situation awareness. This is due their prior knowledge of the investigation, the high quality images provided by camera (or the capacity to enhance degraded images), alarms, excellent liaison with relevant personnel such as team members, dispatchers, response teams and instantaneous links with various databases. Their distributed situation awareness guides their search and assists with coordinating various stakeholders. These are aspects that do not always function as effectively in real life [23] as they do in crime series.

Many of the characters performing observation of CCTV in crime series have extraordinary abilities to recognise faces despite disguises and degraded images, as if they are super-recognisers. Research indicates that people are good at matching familiar faces on CCTV, even when the images are degraded [2]. In general, people are not good at recognising faces on CCTV [5] and this is likely to apply to many CCTV operators. The challenge of recognising people on CCTV is increased by the fact that targets have to be identified on different cameras, as they disappear from one camera's field and move to another camera. This creates different viewing conditions, such as different lighting, angles, and amounts of occlusion [25], creating difficulties in recognition. While being a super-recogniser could be advantageous when it is necessary to recognise individuals, people also use behavioural cues to recognise people and detect suspicious events [1, 6, 21].

Challenges regarding detection are at times aggravated by job design. Some operators are required to perform other activities in addition to monitoring cameras (e.g., writing reports, liaising with emergency services) and this removes their attention from the monitoring process. Alternatively, some operators deal with the monotony of their role by indulging in other activities (e.g., reading magazines, talking to colleagues, taking tea breaks) [33]. Yet others continue to look at the cameras, but are mentally disengaged from the detection task [11]. Not surprisingly, in crime series where CCTV is portrayed as being ineffective, the operators are often depicted as being incompetent and either not looking at the cameras, or gazing passively but not 'seeing.'

Change blindness [32] has been shown to occur in video footage, where people do not detect changes in unfamiliar scenes or in static inanimate objects [18].

The monitoring and detection tasks of CCTV surveillance operators require sustained attention. The fact that vigilance tasks place high demands on mental workload is well established [40]. In research on proactive surveillance and a single monitor, a quarter of the sample disengaged within the first 30 min of observation, but a third of the sample was able to maintain engagement for the duration of the 90-minute task [11]. Participants with higher disengagement had lower detection performance and came from generalist CCTV surveillance environments, while those who maintained task engagement had higher detection levels and worked in specialist, dedicated surveillance operations.

As with task engagement, there is evidence of detection performance decreasing over time (i.e., vigilance decrement) with CCTV footage, but this varies between individuals and contexts [12]. In crime series, the characters who detect significant events on CCTV are seldom operators who have been monitoring cameras throughout lengthy shifts. They are usually driven, highly goal oriented and extremely motivated to solve cases they are working on. They are relentless in their pursuit of solutions, have superhuman levels of energy, and can maintain their concentration and other cognitive abilities regardless of their personal problems, lack of sleep, injuries, or having to work multiple consecutive shifts. Real life operators cannot be expected to live up to these super-human qualities.

The reality is that operators have limited attentional resources, experience fatigue and are not always highly motivated. The monitoring task is arduous and there may be extensive periods when they do not detect significant events, giving them little feedback or reinforcement for their efforts. There is evidence that CCTV surveillance operators do not see their jobs as being highly valued [19], and that some become bored [33]. These circumstances are likely to affect their motivation levels and task engagement and need to be managed effectively.

4   Conclusion and Recommendations

CCTV surveillance is an arduous and difficult task, with demands arising from the numerous sources. Unfortunately these variables cannot be overlooked as they often are in crime series. Various guidelines include recommendations regarding CCTV system design and the working conditions of operators (e.g., [30]) and these are not repeated here. However, the following areas regarding the cognitive processes involved in surveillance are recommended for further research.

As the appropriate deployment of attention is a key process in surveillance, further research regarding search strategies and scanning patterns is recommended. The results of these studies could be incorporated into training programmes for operators. To prevent vigilance decrements, mechanisms for managing attention resources and re-engaging attention need to be identified. A greater understanding of the contribution of active, intentional visual analysis (as opposed to perception), the behavioural cues that predict aberrant behaviour, and the interaction between motivation, task engagement and the cognitive processes involved in CCTV surveillance would be useful.

Finally, as second generation or intelligent CCTV becomes more sophisticated, it is necessary to appreciate how these systems will interact with the cognitive and motivational processes of operators. These aspects would contribute to theory, as well as the selection and training of operators.

References

1. Blechko A, Darker I, Gale A (2008) Skills in detecting gun carrying from CCTV. In: Klima M (ed) IEEE international Carnahan conference on security technology. IEEE Press, New York
2. Bruce V, Henderson Z, Newman C, Burton AM (2001) Matching identities of familiar and unfamiliar faces caught on CCTV images. J Exp Psychol Appl 7(3):207–218
3. Castelhano MS, Heaven C (2010) The relative contribution of scene context and target features to visual search in scenes. Atten Percept Psychophys 72:1283–1297
4. Cole H, Wood J (2006) The relationship between image complexity and task difficulty as rated by experienced and novice viewers. In: Meeting diversity in ergonomics, international ergonomics association congress. Elsevier, Amsterdam
5. Davis J, Valentine T (2008) CCTV on trial: matching video images with the defendant in the dock. Appl Cogn Psychol 23(4):482–505
6. Donald C (2006) High performing CCTV operators. Hi Tech Security. www.securitysa.com/54993n
7. Donald C (2004) How many monitors should a CCTV operator view? Hi-tech Security Solutions. http://www.securitysa.com/article.aspx?pklarticleid=3313
8. Donald C (2001) Vigilance. In: Noyes J, Bransby M (eds) People in control: human factors in control room design. The Institute of Electrical Engineers, London
9. Donald F (2010) A model of CCTV surveillance operator performance. Ergon SA 22(2):2–13
10. Donald FM (2008) The classification of vigilance tasks in the real world. Ergonomics 51(11):1643–1655. https://doi.org/10.1080/00140130802327219
11. Donald FM, Donald CHM (2015) Task disengagement and implications for vigilance performance in CCTV surveillance. Cogn Technol Work 17:121–130. https://doi.org/10.1007/s10111-014-0309-8
12. Donald FM, Donald C, Thatcher A (2015) Work exposure and vigilance decrements in closed circuit television surveillance. Appl Ergon 47:220–228. https://doi.org/10.1016/j.apergo.2014.10.001
13. Eimer M, Nattkemper D, Schröger F, Prinz W (1996) Involuntary attention. In: Neumann O, Sanders AF (eds) Handbook of perception and action, vol. 3: attention. Academic Press, London, pp 155–184
14. Einhäuser W, Rutishauser U, Koch C (2008) Task demands can immediately reverse the effects of sensory-driven saliency in complex visual stimuli. J Vis 8(2):1–19
15. Evans KK, Treisman A (2005) Perception of objects in natural scenes: is it really attention free? J Exp Psychol Hum Percept Perform 31(6):1476–1492
16. Foulsham T, Underwood G (2008) What can saliency models predict about eye movements? Spatial and sequential aspects of fixations during encoding and recognition. J Vis 8(2):1–17
17. Gilchrist I, Harvey M (2000) Refixation frequency and memory mechanisms in visual search. Curr Biol 10(19):1209–1212

18. Gill M, Spriggs A, Allen J, Hemming M, Jessiman P, Kara D, Kilworth J, Little R, Swain D (2005) Control room operation: findings from control room observations. Home Office Research, Development and Statistics Directorate. Online Report 14/05
19. Helten F, Fischer B (2004) Reactive attention: video surveillance in Berlin shopping malls. Surveill Soc 2(2/3):323–345. http://www.surveillance-and-society.org/cctv.htm
20. Hillstrom AP, Hope L, Nee C (2008) Applying psychological science to the CCTV review process: a review of cognitive and ergonomic literature. Home Office Scientific Development Branch, St Albans
21. Howard CJ, Gilchrist ID, Troscianko T, Behera A, Hogg DC (2011) Task relevance predicts gaze in videos of real moving scenes. Exp Brain Res 214:131–137. https://doi.org/10.1007/s00221-011-2812-y
22. Itti L, Koch C (2000) A saliency-based search mechanism for overt and covert shifts of visual attention. Vis Res 40(10–12):1489–1506
23. Keval H (2008) CCTV control room collaboration and communication: does it work? In: 9th human centred technology group postgraduate workshop, Department of Informatics, University of Sussex, Falmer, Brighton, pp 45–48
24. Keval H, Sasse MA (2008) Man or a gorilla? Performance issues with CCTV technology in security control rooms. Perception 33:87–101. https://doi.org/10.1068/p3402
25. Layne R, Hospedales T, Gong S (2012) Person re-identification by attributes. Br Mach Vis Conf 2(3):1–8
26. Menneer T, Donnelly N, Godwin HJ, Cave KR (2010) High or low target prevalence increases the dual-target cost in visual search. J Exp Psychol Appl 16(2):133–144. https://doi.org/10.1037/a0019569
27. Norris C, Armstrong G (1999) The maximum surveillance society: the rise of CCTV. Berg, Oxford
28. Parasuraman R, Mouloua M (1996) Automation and human performance: theory and applications. Lawrence Erlbaum Associates, Inc., Mahwah
29. Pikaar R, Lenior D, Schreibers K, de Bruijn D (2015) Human factors guidelines for CCTV system design. In: 9th Triennial Congress of the IEA, Melbourne
30. Pikaar R (2015) Human factors guidelines for the design of CCTV-systems, revision 1.3. Published by HAN University of Applied Sciences, Arnhem; vhp human performance, The Hague; ErgoS Human Factors Engineering, Enschede; Intergo Human Factors & Ergonomics, Utrecht. https://www.sintef.no/globalassets/project/hfc/documents/cctv-hf-guidelines-revision-1-3-may2015.pdf
31. Schreibers KBJ, Landman RB, Pikaar RN (2012) Human Factors of CCTV: part 1 technology and literature review. ErgoS Engineering and Ergonomics, Enschede, Reference: S386-R1.CCT
32. Simons DJ, Rensink RA (2005) Change blindness: past, present, and future. Trends Cogn Sci 9(1):16–20
33. Smith GJD (2004) Behind the scenes: examining constructions of deviance and informal practices among CCTV control room operators in the UK. Surveill Soc 2(2/3):376–395
34. Stainer MJ, Scott-Brown KC, Tatler BW (2017) On the factors causing processing difficulty of multiple-scene displays. i-Perception. https://doi.org/10.1177/2041669516689572
35. Stainer MJ, Scott-Brown KC, Tatler BW (2013) Looking for trouble: a description of oculomotor search strategies during live CCTV operation. Front Hum Neurosci 7:615. https://doi.org/10.3389/fnhum.2013.00615
36. Tickner AH, Poulton EC (1973) Monitoring up to 16 synthetic television pictures showing a great deal of movement. Ergonomics 16(4):381–401

37. Torralba A, Castelhano MS, Henderson JM, Oliva A (2006) Contextual guidance of eye movements and attention in real-world scenes: the role of global features in object search. Psychol Rev 113(4):766–789
38. Troscianko T, Holmes A, Stillmanô J, Mirmehdi M, Wright D, Wilson A (2004) What happens next? The predictability of natural behavior viewed through CCTV cameras. Perception 33:87–101
39. Wallace E, Diffley C, Baines E Aldridge J (n.d.) Ergonomic design considerations for public area CCTV safety and security applications. Home Office, Police Scientific Development Branch, UK
40. Warm JS, Parasuraman R, Matthews G (2008) Vigilance requires hard mental work and is stressful. Hum Factors 50(3):433–441
41. Wells H, Allard T, Wilson P (2006) Crime and CCTV in Australia: understanding the relationship. Centre of Applied psychology and Criminology, Faculty of Humanities and Social Sciences, Bond University, Gold Coast
42. Yantis S (1996) Attentional capture in vision. In: Kramer AF, Coles GH, Logan GD (eds) Converging operations in the study of visual selective attention. American Psychological Association, Washington, DC, pp 45–76

Human Interaction Under Risk in Cyber-Physical Production Systems

Philipp Brauner[(✉)], Ralf Philipsen, André Calero Valdez,
and Martina Ziefle

Human-Computer Interaction Center, RWTH Aachen University,
Campus Boulevard 57, 52074 Aachen, Germany
{brauner,philipsen,calero-valdez,
ziefle}@comm.rwth-aachen.de

Abstract. The emergence of cyber-physical production systems poses new challenges for designing the interface between production systems and the human-in-the-loop. In this study, we investigate how human operators interact with risks in a supply chain scenario. We varied the financial magnitude and the expected value of the decisions, the combination of two types of risk (risk in delivery amount and risk in timeliness), as well as three different task displays as within-subject factors. As explanatory user factors we studied the influence of Need for Achievement and the Attitude towards Risk-taking on the dependent variables task speed and accuracy. Results of the user study with 33 participants show that each of the investigated factors either influences decision speed, decision accuracy, or both. Consequently, the human-in-the-loop profits from adequate decision support systems that help to increase decision efficiency and effectiveness and reduce uncertainty and workload. The article concludes with a research agenda to support the human-in-the-loop in production systems.

Keywords: Decision support systems · Cyber-physical production systems
Human factors · Decision under risk · Socio-technical system · Risk

1 Introduction

We are at the brink of the 4th industrial revolution and the convergence of information and communication technology with production systems along the value chain promises increased efficiency and effectiveness and overall competitiveness for individual companies, supply chain networks, and societies [1, 2].

Despite increased automation the human-in-the-loop remains a crucial component of cyber-physical production system (CPPS). Bainbridge's *Ironies of Automation* [3] postulate that automation is shifting the role of the human from manual activities to monitoring and controlling tasks. However, if automation fails, the capability to successfully intervene declines through lack of practice, lower knowledge of the underlying processes, or ill-aligned mental models.

In near future, a vast amount of fine-grained, heterogeneous data from different sources will be available in real for automatic machine control, for decision support systems, and the human-in-the-loop interacting with or supervising these systems [4].

© Springer Nature Switzerland AG 2019
S. Bagnara et al. (Eds.): IEA 2018, AISC 822, pp. 421–430, 2019.
https://doi.org/10.1007/978-3-319-96077-7_45

To bridge the gap between the human decision maker and the increasingly complex socio-technical cyber-physical production system (STCPPS) we need to understand their specific wants and needs, as well as their emotional and cognitive characteristics. While many generic norms (e.g., *EN ISO 9241*), heuristics [5], and guidelines [6] exist for designing user interfaces, the task-specific requirements for cyber-physical production systems are insufficiently understood. In this work, we focus on the specific application domain of inter-company flow of material and information. One key skill of the human-in-the-loop in this domain is decision making under risk and uncertainty, e.g., the ability to anticipate delivery variances and bottlenecks. To develop user-adaptive decision support systems that augment the decision makers abilities, we first need to identify their abilities, weaknesses, and potential biases by studying their decision-making processes under uncertain and risky conditions.

2 Interacting with Uncertainty and Risk

Knight distinguishes *Uncertainty and Risk* as two distinct types of uncertainty: When the potential outcomes of an event are not known in advance, this is referred to as Uncertainty. In contrast, the term risk is used when the potential outcomes of an event are known beforehand [7]. For example, the two possible outcomes of a coin-toss are known and the chances are quantifiable. Consequently, the outcomes can be maximized by maximizing the expected utility.

Tversky and Kahneman's *Prospect Theory* [8] showed that decisions under risk are prone to irrational behavior: For example, people prefer lower but likely profits over higher but unlikely profits. In contrast, higher but unlikely losses are preferred to lower but likely losses. Examples for decisions with uncertainty include decisions in complex, non-linear, dynamic systems, such as cyber-physical production networks with globally dispersed supply chain networks [9]. Managing uncertainty is more complex and—in the context of cyber-physical production systems—requires concepts such as resilient design of the CPPS, closer cooperation of the stakeholders, supply chain agility, and a culture for handling uncertainty [9].

In general, people are rarely objective in making decisions, but make use of a series of decision heuristics to compensate for abundant, incomplete, or inaccurate information [10]. Even though these heuristics often enable efficient and effective decision making, they still bear the risk of irrational or erroneous decisions that can negatively influence the outcome.

3 Method

With this study, we want to analyze the influence of risk on speed and accuracy of decision in the context of inter-company flow of materials and information. We embedded the study into a novel business simulation game, to both evaluate the factors and their interrelationships mentioned above and make it challenging and captivating for the participants. Section 3.1 presents the game and the decision tasks for the participants. Section 3.2 defines the within-subject variables considered in the study and

Sect. 3.4 presents the dependent variables from the experiment. Next, Sect. 3.3 elucidates the user factors studied in this experiment. Finally, Sect. 3.5 presents the hypotheses that guided the study.

3.1 Decision Game

Business simulation games facilitate studying human behavior in non-trivial, sufficiently complex, and experimentally controllable environments [11].

For this study, participants need to work on a series of decision tasks in the context of a purchasing department of manufacturing company. Each task resembles a business offer with a *required price* that must be payed and a *potential profit* that might be realized. The potential profit depends on two types of risk that must be taken into account: A risk in regard to the *quantity* of the delivered goods and a risk in regard to the *punctuality* of the delivery. If only a share of the delivery arrives (quantity risk), only an equal share of the revenues is generated. No profits are generated if the order does not arrive in time (punctuality risk).

For each decision task the participant needs to decide whether to *accept* or to *reject* the offers. Whether an offer has a positive, neutral, or negative expected value (*EV*) is not directly shown, but must be inferred from the presented risks, the price, and the potential profit. If an offer is rejected, nothing is won or lost. If an offer is accepted, its price is deducted, and a revenue is calculated depending on the quantity and punctuality of the delivery.

Finally, the overall *profit* of the participants accumulates through wins and losses across several decision tasks of the game. In the game the participants' task was to maximize their overall profit by *accepting* lucrative orders (*EV > 0*) and by *declining* unprofitable orders (*EV < 0*).

The business simulation game is designed for use in controlled laboratory environments as well as for widespread online studies. In this case, a laboratory experiment was used to obtain an initial evaluation of the framework.

3.2 Independent Variables

Within the experiment we varied the four within-factor variables time risk, quantity risk, magnitude, and expected value.

Time Risk: Based on chance, the delivery is either punctual or delayed. The chance for a punctual delivery is either 100%, 75%, or 50%. No revenue can be generated for a delayed delivery.

Quantity Risk: Based on chance, the delivery is made in full (100%) or partially (75% or 50%). For partial deliveries, profits can only be realized proportionately (75% for deliveries with 75% chance and 50% for deliveries with 50% chance).

Magnitude: To study the influence of the magnitude of the decision, the offers were scaled by 100 or 2000 for offers with a lower or higher financial volume.

Expected Value: The expected value of each offer is varied between −20%, −10%, 0%, + 10%, + 20%. Thus, gains or losses of ±20%, ±10%, or 0% can be realized based on chance and the offers' price.

Visualization: Three different risk visualizations were administered in distinct rounds of the business simulation game. First, a textual representation of the risk, a purely visual representation, and a combined textual and visual representation (Fig. 1).

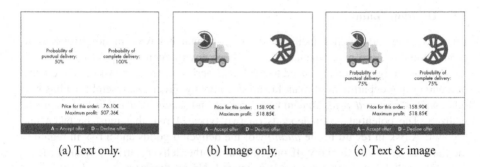

(a) Text only.   (b) Image only.   (c) Text & image

Fig. 1. Three task visualizations used in the experiment.

All factors are taken into account for calculating the required price and the potential profit of an offer. Each offer (i.e., decision task) then presents its price, potential profit, the time and quantity risk, as well as the magnitude of the decision. Apparently, the expected value is not presented and must be inferred from the other numbers. For each of the combinations decision tasks were constructed (uniformly distributed), shuffled, and presented to the participants.

3.3 Explanatory User Factors

We administered a survey before the experiment in the business simulation game to understand if individual user factors influence the interaction with risk in cyber-physical production systems. Besides the participant's age and gender the following latent constructs were queried:

Risk Attitude: An individuals' attitude towards taking risk may shape the performance in the decision tasks. In this explanatory study we used Beierlein et al.'s validated single item scale to capture this risk attitude [12].

Need for Achievement (NfA): Motivation plays a central role for the choice of actions and the performance in these actions. The Need for Achievement inventory by Schuler and Prochaska [13] is a psychometric instrument for assessing an individuals' need for achievement. We used a short scale with 6 items and an outstanding internal reliability of $\alpha = .977$.

3.4 Dependent Variables

As dependent variable we measured the accuracy, as well as the speed of the decision. The accuracy is based only on the expected value of the decision under risk and not the actual outcome: A decision is considered as accurate, if the $EV > 0$ and as inaccurate for $EV < 0$.

The participants were instructed to aim for highest profit. Speed was neither instructed, nor used for feedback. Consequently, the following evaluation focuses on the accuracy and speed is only reported for information.

3.5 Hypothesis

The experiment is guided by the following research questions, derived from literature and qualitative preliminary studies:

H1: Tasks with higher risks yield lower speed and lower accuracy. Tasks with combined risks are particular difficult for participants (lowest speed, lowest accuracy).
H2: Atask's expected value influences both speed and accuracy of the decision.
H3: Adecision's financial magnitude has a significant influence on accuracy.
H4: Double coding of risks through text *and* images increases speed compared to single coding.
H5: Higher need for achievement relates to higher accuracy and lower speed.
H6: Higher attitude towards risk relates to higher speed and lower accuracy.

The testing of these fundamental hypotheses is intended on the one hand to evaluate the relationships between risk and decision-making behavior in the context of inter-company flow and on the other hand to evaluate the research framework developed.

3.6 Statistical Procedures

We used parametrical and non-parametrical methods (Person's r, Spearman's ρ correlations, single and multivariate (repeated measures) analyses of variance (RM-M/ANOVA). The level of significance is set to $\alpha = .05$ and Pillai's V is used for multivariate tests. if sphericity is not met, Greenhouse-Geisser corrected values are used but uncorrected dfs are reported or legibility. Arithmetic means are reported with standard deviation denoted as $\pm$. Only trials with an expected value $\neq 0\%$ are considered, as otherwise accuracy is not well defined.

3.7 Description of the Sample

33 subjects in the age from 22 to 45 years ($M = 26.4$, $Md = 26$, $SD = 2.3$) have participated in the study, 28 were female and 5 were male. Within the sample age and gender is not correlated [$\rho(n = 33 - 2) = .069$, $p > .05$].

4   Results

This section presents the results for each of the investigated factors. First, the influence of the system factors are presented, followed by a presentation of the interface factors and concluded with an analysis of the individual user factors. Overall, the mean median reaction time is 2.851 ± 1.404 s and the average mean accuracy is $73.1 \pm 7.1\%$.

4.1 Effects of the System Factors

The **Magnitude** of the orders has a significant overall effect on the participant's decision as affirmed by a RM-MANOVA [$V = .416$, $F_{2,31} = 11.050$, $p < .001$, $\eta^2 = .416$]. Specifically, **Magnitude** influences the reaction time [$p < .001$, $\eta^2 = .388$], but not the accuracy of the decisions ($p = .272 > .05$). Decision tasks with a higher financial volume were carried out slower than tasks with lower volume, but with about the same accuracy (see Table 1a).

Table 1. Speed and accuracy of decisions depending on...

(a) Level (η^2=.416)

	Speed [ms]	Accuracy [%]
100€	2739 ± 1353	72.7 ± 6.9%
2000€	2972 ± 1473	73.6 ± 8.0%

(b) Expected Value (η^2=.780)

	Speed [ms]	Accuracy [%]
-20%	2793 ± 1376	70.4 ± 18.5 %
-10%	2881 ± 1405	59.1 ± 16.7 %
+10%	2888 ± 1428	64.7 ± 10.9 %
+20%	2920 ± 1592	71.5 ± 10.8 %

The **Expected Value** of the tasks has an overall effect on the decisions [$V = .780$, $p < .001$, $\eta^2 = .780$] and it influences the accuracy [$p = .014$, $\eta^2 = .159$], but not the speed of the decisions [$p = .295 > .05$]. Decisions with a lower absolute expected value (i.e., ±10%) had a lower accuracy than decisions with higher expected values (i.e., ±20%). Whether the expected value of the decision task had a positive or negative outcome had no significant influence on the accuracy (see Table 1b).

A multivariate repeated measures ANOVA with the two factors **risk time** and **risk quantity** yield a significant effect for both main factors [risk time $V = .772$, $F_{4,29} = 24.491$, $p < .001$, $\eta^2 = .772$; risk quantity $V = .708$, $F_{4,29} = 17.606$, $p < .001$, $\eta^2 = .708$] and the interaction of both within-subject factors [$V = .789$, $F_{8,25} = 11.696$, $p < .001$, $\eta^2 = .789$]. A further analysis shows that risk in time, quantity, and their combination influences decision accuracy, whereas decision speed is only affected by risk in quantity. As expected, the participants were fastest and achieved highest accuracy scores (91.1 ± 9.3%) for tasks with no risk (see Table 3). They were equally fast but achieved lowest accuracy scores (68.1 ± 5.8%) for tasks with the highest risk.

4.2 Effects of the User Interface

Presentation of the task (image, text, or combined) had a medium overall effect on the participants decision performance [$V = .397$, $F_{4,29} = 4.778$, $p = .004 < .05$, $\eta^2 = .397$]. The average **Speed** of the decisions was not significantly different for the three task visualizations [$F_{2,64} = 2.102$, $p = .139 > .05$, $\eta^2 = .062$], but task **Accuracy** was significantly higher for the image condition than for textual or combined task presentation [$F_{2,64} = 4.885$, $p = .012 < .05$, $\eta^2 = .132$] (see Table 2a).

Table 2. Speed and accuracy of decisions depending on...

(a) Task visualization (η^2=.397)

	Speed [ms]	Accuracy [%]
Text	3066 ± 1956	72.3 ± 7.9%
Image	2676 ± 1052	75.0 ± 8.1%
Text/Image	2879 ± 1531	72.2 ± 7.7%

(b) Repetition (η^2=.400)

	Speed [ms]	Accuracy [%]
1st round	3351 ± 1850	73.9 ± 7.2 %
2nd round	2785 ± 1552	72.9 ± 8.6 %
3rd round	2484 ± 1060	72.8 ± 8.2 %

Table 3. Effect of risk in delivery time (η^2 = .772) and quantity (η^2 = .708).

Risk time [%]	Risk quantity [%]	Speed [ms]	Accuracy [%]
50	50	2517 ± 1210	68.1 ± 5.8
	75	2814 ± 1459	70.0 ± 9.8
	100	3005 ± 1586	71.8 ± 10.7
75	50	3208 ± 1921	68.5 ± 10.6
	75	2936 ± 1444	68.8 ± 8.8
	100	3019 ± 1699	69.6 ± 9.1
100	50	3044 ± 1339	70.9 ± 11.2
	75	2848 ± 1489	79.7 ± 10.9
	100	2450 ± 1104	91.1 ± 9.3

4.3 Effects of Personality Factors

First, we checked for the influence of an individual's **Need for Achievement** on the decisions using an ANCOVA and **Need for Achievement** as covariate and speed and accuracy as dependent variables. Both variables were significantly influenced [$F = 5.213, p = .011 < .05, \eta^2 = .258$] and mean median correctness was significantly higher [$p = .009 < .05, \eta^2 = .202$] whereas decision speed [$p = .014 < .05, \eta^2 = .179$] were significantly lower for participants with a higher **Need for Achievement**.

A RM-MANOVA with **Round** as independent within-subject variable and **Speed** and **Accuracy** as dependent variables yield a significant model. Thus, practice has a significant effect on the overall decisions [$V = .400, F_{4,29} = 4.837, p < .001, \eta^2 = .400$]. speed of the decisions [$p < .001, \eta^2 = .314$], but not on their accuracy [$p > .564$]. Consequently, the decision speed increases with practice, while the decision accuracy remains constant.

5 Discussion

We developed a novel evaluation framework for studying the influence of personality (*user factors*), the interface between the human operator and (*interface factors*), and parameters of the underlying production system (*system factors*) on decisions in cyber-physical production systems.

Within this explanatory study, we could show that *all* investigated factor domains influence decision efficiency and/or effectiveness in our material disposition scenario.

Our first hypothesis (H1)—higher risk have a negative effect on decisions speed and accuracy—has been confirmed. Our second assumption—the combination of different risks has a negative impact—could not be shown in this study, as even single risks had a disastrous effect on performance. A more granular gradation of the risk levels in future experiments may still corroborate the presumed effect of the combined risks.

Our study partially falsifies the second hypothesis (H2) that a task's expected value has an influence on performance. Apparently, decision tasks with a lower expected value (i.e., ±10%) are more difficult and yield higher error rates in contrast to tasks with higher expected values (i.e., ±20%). Surprisingly, speed was not affected by the expected value.

In contrast, we found that the financial volume of a decision had a significant role on performance (speed, but not accuracy). Consequently, H3 is falsified. The study shows that the participants invested more times on decisions with a larger volume. It can be assumed that larger amounts of money and thus larger possible losses could lead to greater care in the sense of a higher time-investment in the decision. However, if the difficulty of the task remains the same, longer processing only does not automatically lead to higher accuracy in the present context.

Within this study, we focused on the visualization of the risk as one of the many imaginable interface factors. To our surprise, the presentation had no effect on the decision speed, but on the accuracy of the decisions. We have assumed that a double coding of the task (*text & image*) would yield highest performance (H4). However, pure image coding had the highest accuracy within this study. Why this is the case cannot yet be finally explained on the basis of the available data. Limitingly, it has to be considered that the pure image representation has led to significantly higher accuracy values, but the absolute accuracy values of all forms of representation differ only slightly.

Looking at user factors in terms of personality traits, it turned out that both H5 and H6 have been confirmed. As expected, a high need for achievement leads to longer processing times and more accurate results (H5). Conversely, a higher personal willingness to take risks leads to faster decisions, but also to less accurate results (H6).

In conclusion, the study shows that decision speed and most importantly decision quality is influenced by the personality of the participants as well as by the type of decision task. While the present study provides only a first glimpse at some of the numerous factors that influence human performance, it provides a viable experimental framework to study the interaction of personality, and interface design, and the various components of the underlying production system.

6   Limitations and Outlook

Obviously, the task in this baseline experiment is trivial to automate. But as Bainbridge's *ironies of automation* [3, 14] postulate, automation will not make the human actor in CPPS superfluous, but will rather shift its role to supervisory tasks. In case of failing automation or when the automation need to be supervised, the human agent will still need to process the available information, evaluate its meaning, and make a correct decision. In these cases, decision biases, the handling of incomplete information, personality, as well as interface influence the overall decision quality, the perceived workload, and also job satisfaction.

However, the number of investigated factors and the limited number of experimental trials forced us to analyze the data using a sequence of singular tests. Future work should integrate more system, interface, and human factors in a common statistical model that facilitates their prioritization in regard to their influence on efficiency and effectiveness. For example, the study was not controlled for perceptual speed, which was found to be important in similar studies on decision performance without risk. Likewise, the task complexity was rather low as only singular decisions had to be made at any given time. We expect that different and more pronounced effects will emerge in even more complex decision situations, when additional and more complex parameters have to be perceived, interpreted, integrated with prior knowledge and experience and correct decisions need to be inferred and communicated to the system. To fully understand how human operators interact with cyber-physical production systems this vast factor space must be described. Future research must therefore identify, quantify and weigh the components and their interactions.

The developed research framework can be the basis for this future research as it enables a thorough and systematic investigation of possible influencing factors and their interactions in the context of socio-technical cyber-physical production systems [15]. In addition to investigating the influence of user factors on decision-making, the research agenda focuses the development of suitable and user-adaptive decision support systems for both the inter-company flow of materials and information and the underlying production systems. In particular, trust in decision support systems, (blind) compliance and the restoration of trust in automation after a failure in different risk contexts are research topics to be addressed in order to bridge the gap between the human decision maker and the increasingly complex socio-technical cyber-physical production systems.

Acknowledgements. The work is funded by the German Research Foundation (DFG) as part of the German Excellence Initiative and the Cluster of Excellence "Integrative Production Technology for High Wage Countries" (EXC 128). We thank all participants for their willingness to support our study and Lara Mhetawi, Katharina Merkel, Fabian Comanns, and Wiktoria Wilkowska for research support.

References

1. Evans PC, Annunziata M (2012) Industrial internet: pushing the boundaries of minds and machines. Technical report, General Electric
2. Bruner J (2013) Industrial internet—the machines are talking. Reilly Media Inc., Newton
3. Bainbridge L (1983) Ironies of automation. Automatica 19(6):775–779

4. Schlick C, Stich V, Schmitt R, Schuh G, Ziefle M, Brecher C, Blum M, Mertens A, Faber M, Kuz S, Petruck H, Fuhrmann M, Luckert M, Brambring F, Reuter C, Hering N, Groten M, Korall S, Pause D, Brauner P, Herfs W, Odenbusch M, Wein S, Stiller S, Berthold M (2017) Cognition-enhanced, self-optimizing production networks. In: Brecher C, Özdemir D (eds) Integrative production technology—theory and applications. Springer, Berlin, pp 645–743
5. Nielsen J (1994) Usability engineering. Morgan Kaufmann Publishers, Burlington
6. Gould JD, Lewis C (1985) Designing for usability: key principles and what designers think. Commun ACM 28:300–311
7. Knight FH (1921) Risk, uncertainty and profit. Houghton Mifflin Co., Boston
8. Tversky A, Kahneman D (1973) Availability: a heuristic for judging frequency and probability. Cogn Psychol 5(2):207–232
9. Christopher M, Peck H (2004) Building the resilient supply chain. Int J Logist Manag 15 (2):1–14
10. Gilovich T, Griffi D, Kahneman D (2002) Heuristics and biases: the psychology of intuitive judgment, 8th edn. Cambridge University Press, Cambridge
11. Brauner P, Ziefle M (2016) Game-based learning in manufacturing and business. In: 6th international conference on competitive manufacturing (COMA 2016)
12. Beierlein C, Kovaleva A, Kemper CJ, Rammstedt B (2014) Eine Single-Item-Skala zur Erfassung von Risikobereitschaft: Die Kurzsskala Risikobereitschaft-1 (R-1) [A Single Item scale for measuring risk-taking attitude]
13. Schuler H, Prochaska M (2001) Leistungsmotivationsinventar [Need for achievement inventory]. Hogrefe, Göttingen
14. Baxter G, Rooksby J, Wang Y, Khajeh-Hosseini A (2012) The ironies of automation—still going strong at 30? In: Proceedings of ECCE 2012 conference, 29th–31st August, Edinburgh, pp 65–71
15. Brauner P, Valdez AC, Philipsen R, Ziefle M (2016) On studying human factors in complex cyber-physical systems. In: Workshop human factors in information visualization and decision support systems held as part of the mensch und computer 2016, Gesellschaft für Informatik

User Diversity in the Motivation for Wearable Activity Tracking: A Predictor for Usage Intensity?

Christiane Attig[1(✉)], Alexa Karp[1], and Thomas Franke[2]

[1] Department of Psychology, Cognitive and Engineering Psychology,
Chemnitz University of Technology, Chemnitz, Germany
`christiane.attig@psychologie.tu-chemnitz.de`
[2] Institute for Multimedia and Interactive Systems, Engineering Psychology
and Cognitive Ergonomics, University of Lübeck, Lübeck, Germany

Abstract. Wearable fitness devices (i.e., activity trackers) are increasingly popular for monitoring everyday activity. Research suggests that long-term adherence to activity trackers is relevant for their positive effects on health. Thus, it is essential to understand the factors that foster usage intensity and long-term adherence. Based on first research regarding users' motives for using activity trackers and self-determination theory, we examined usage motives as predictors for the current and estimated future usage intensity. In addition, we investigated the relation of usage motives and user diversity facets (affinity for technology interaction, geekism, and need for cognition). Results of an online study with $N = 58$ regular users of activity trackers indicated a substantial variation of users' intrinsic/extrinsic motivation for using an activity tracker. Further, positive relationships between intrinsic motivation, autonomous regulation and usage intensity were found. Regarding user diversity, affinity for technology interaction and geekism predicted the intrinsic motivation whereas need for cognition did not. Our results imply that, in order to obtain possible beneficial health effects of a more intensive activity tracker usage, users' intrinsic motivation and autonomy have to be supported.

Keywords: Activity tracking · Quantified self · Self-determination theory

1 Introduction

Activity trackers constitute a valuable tool for behavioral change and habit formation [24]. In spite of trackers' benefits for health and fitness (e.g., [7]), many users abandon their activity tracker after a few months [21]. However, continued use is of crucial importance for habit formation [26], therefore it is key to identify factors that are decisive for current and future activity tracker usage. From a perspective of self-determination theory (SDT; [9]) it can be argued that individually different usage motives likely influence activity tracker usage. Hence, the objective of the current research is to gain knowledge regarding the relationship between usage motivations and current as well as future usage intensity and the role of user diversity in activity

© Springer Nature Switzerland AG 2019
S. Bagnara et al. (Eds.): IEA 2018, AISC 822, pp. 431–440, 2019.
https://doi.org/10.1007/978-3-319-96077-7_46

tracking motivation. SDT [9] was applied as a basis for characterizing usage motives. Moreover, as it is essential to incorporate user personality into research in human-technology interaction [2, 3], we further examined whether usage motives can be explained by the personality constructs affinity for technology interaction (ATI; [13]), geekism [32], and need for cognition (NFC; [4]).

2 Background

2.1 Usage Motivation from a Perspective of Self-Determination Theory

According to SDT [9], goal-directed behavior can be performed out of intrinsic motivation (because it is intrinsically enjoyable) or extrinsic motivation (because it serves a superior goal). Moreover, there are four sub-facets of extrinsic motivation that vary from internal locus of causality and high level of autonomy (integrated and identified regulation) to external locus of causality and low level of autonomy (intro-jected regulation and external regulation [30]). First research indicates that activity tracker users differ in their usage motives. For instance, users stated to use their tracker to obtain beneficial effects for their fitness and health [8], to enjoy interaction with the device [21] or to gain new life experiences by interacting with a new technique and collecting personal data [6]. Moreover, it was possible to differentiate between more intrinsically and more extrinsically motivated users with regard to activity tracker usage [1]. Yet, how are these different motivations related to usage intensity?

2.2 Relation Between Motivation and Usage Intensity

Both intrinsic and extrinsic motivation are positively connected to performance mea-sures. However, intrinsic motivation has a stronger positive impact on behavioral persistence and long-term adherence than extrinsic motivation [31]. In addition, intrinsically motivated students have shown increased curiosity and exploratory behavior [23]. Regarding the sub-facets of extrinsic motivation, it has repeatedly been shown that autonomous motivations are more strongly positively related to behavioral intensity and persistence than controlled motivations [10, 15]. Positive relationships between motivation, usage intensity, and long-term adherence can also be assumed for the case of activity trackers. First research on activity tracking and motivational factors has shown that desired usage intensity is predicted by intrinsic motivation [28]. Moreover, motivation was positively related to intensity of tracking, and this effect was cumulative (i.e., the more motivations are fulfilled by the tracker, the higher the tracking intensity, [16]). Further research has found that high extrinsic motivation for tracker usage was associated with a stronger dependency effect (i.e., decreased activity behavior in situations when the tracker is not available) while no association was found for intrinsic motivation, indicating that extrinsic motivation might be disadvantageous for long-term behavioral change [1].

2.3 User Diversity as a Predictor for User Motivation

Personality-related individual differences have long been assumed to play a role regarding human-technology interaction [3]. We suggest that, for understanding motivations for using activity trackers comprehensively, it is essential to take user personality into account. While a focus on the broad Big Five personality dimensions is possible, it is usually more fruitful to focus on more specific personality facets [2]. ATI is defined as a persons' tendency to actively engage in intensive interaction with technical devices [13]. First research has found a positive relationship between ATI and the intrinsic motivation for activity tracker usage [1]. Moreover, affinity for technology measured with an attitude-focused scale [11] was positively associated with the acceptance of wearable devices [20]. A further construct related to ATI is geekism [32]. It is defined as computer enthusiasm and represents an intrinsic interest to explore and understand technical devices. There are yet no empirical findings regarding the relationship between geekism and intrinsic and extrinsic motivation. Based on the theoretical conception of the construct, we assume geekism to be positively related to intrinsic motivation for tracker usage. Finally, NFC [4] is a fundamental psychological construct that could play a particular relevant role in the context of activity tracker usage. NFC, the intrinsic motivation to engage in cognitively demanding tasks [4], has been found to be positively related to achievement motivation in technological-related situations [32], and computational, mechanic, and scientific curiosity [25].

3   Present Research and Hypotheses

With the present research, we aim at (1) characterizing intrinsic and extrinsic motivations for using activity trackers, (2) examining the relation of motivations to present and future tracking intensity, and (3) gaining insights into the extent to which user diversity variables can explain variance of usage motivation. The following research questions and hypotheses were addressed:

(1) Q1: To what extent are there individual differences in the motivation for using activity trackers?

(2) Q2: To what extent can the motivation variable explain variance in usage intensity? We hypothesize (H2a) intrinsic motivation to be positively related to current and future (i.e., estimated/intended) usage intensity. Moreover, we hypothesize (H2b) extrinsic motivation to be positively related to current and future usage intensity. In addition, we hypothesize (H2c) that the positive relationship is stronger for more autonomous sub-facets of extrinsic motivation than more controlled sub-facets.

(3) Q3: To what extent can personality traits explain a variance in the motivation for using activity trackers? We hypothesize ATI (H3a), geekism (H3b) and NFC (H3c) to be positively related to intrinsic motivation for tracker usage. However, as NFC is, in contrast to ATI and geekism, not a domain-specific trait, we expect the relationship to be somewhat weaker.

4 Method

4.1 Participants

Participants were recruited via Facebook (group 'Fitness Tracker') and the student e-mail distribution list of Chemnitz University of Technology. They were not compensated for their participation, but had the chance to win a voucher for a major online-retailer. The $N = 58$ participants had an average age of 30.2 years ($SD = 9.9$), 40 of them were female (69%).

Regarding usage duration, 38% had been using their current tracker for more than one year and 66% had been using activity trackers in general for more than one year. Regarding their typical daily tracker usage, 50% stated to wear it for 24 h a day. Most of the participants (74%) stated to wear the tracker 7 days a week in typical weeks. Moreover, participants stated that they track the following type of data: step count (100%), calorie consumption (93%), heart rate (60%), distance (19%), sleep activity (16), and stairs (16%).

4.2 Scales and Measures

Answers for all self-generated items were provided on 6-point Likert scales (from *completely disagree* to *completely agree*). Internal consistencies are depicted in Table 1.

Table 1. Internal consistencies and descriptive statistics of all variables.

Variable	Number of items	Scale range	Cronbach's alpha	M	SD
Intrinsic motivation	2	1–6	.66	4.67	1.00
Extrinsic motivation	2	1–6	.77	4.78	1.23
Autonomous regulation	6	1–7	.88	4.70	1.31
Introjected regulation	2	1–7	.84	2.01	1.30
External regulation	4	1–7	.72	1.97	1.10
Current usage intensity	18	1–6	.91	4.37	0.85
Future usage intensity	4	1–6	.72	5.07	0.79
ATI	9	1–6	.94	4.28	1.14
Geekism	15	1–7	.94	3.61	1.39
NFC	4	1–7	.71	4.71	1.11

Motivation for Activity Tracker Usage. The motivation for using an activity tracker was measured by two questionnaires. First, four self-generated items were used to assess intrinsic and extrinsic usage motivation. The two items assessing intrinsic motivation focused on pleasure regarding data exploration (e.g., "I use the activity tracker because it is fun to deal with my data."). The two items assessing extrinsic motivation focused on meeting external goals or avoiding negative consequences (e.g., "I use the activity tracker because I want to avoid taking too little exercise"). Reliability was questionable (intrinsic motivation), resp. acceptable (extrinsic motivation). Note that a low Cronbach's alpha is not uncommon for very short scales (see e.g. [14]).

To assess sub-facets of extrinsic motivation, we used 10 items from the Treatment Self-Regulation Questionnaire (TSRQ; [29]), which were translated into German and linguistically modified for activity tracker usage. A principal axis factor analysis with scree test suggested a 3-factor solution. The items for integrated and identified regulation loaded on one factor which we termed, in accordance to [22], autonomous regulation. Hence, in our analyses, we refer to the three dimensions of autonomous, introjected, and external regulation. Reliability was good (autonomous and introjected regulation), resp. acceptable (external regulation).

Intensity of Activity Tracker Usage. We used self-generated items to assess the current intensity of tracking as well as the estimated future tracker usage and usage intention. To meet our aim of assessing current tracking intensity on multiple facets, we generated 37 items and selected 18 for assessing the following aspects: depth of data exploration (e.g., "I occupy myself in great detail with the tracked data."), exploration of data accuracy (e.g., "I deal with the question how precise the data measurement of my activity tracker is."), data comparison with other users (e.g., "I use the opportunities of my activity tracker to compare my activity to other people."), frequency of wearing the tracker (e.g., "I wear my activity tracker as often as possible."), frequency of checking tracked data (e.g., "When I wear my tracker, I look at my data very frequently."), amount of tracked data types (e.g., "Usually, I look at as many of the available activity tracker data types/depictions as possible."), amount of tracked activities (e.g., "I use my activity tracker for almost all activities."), and exploration of tracking functions (e.g., "I have intensively occupied myself with the possibilities and functions of my activity tracker."). Reliability was excellent.

Four items were generated to measure the extent to which the participants intent to use their activity tracker in the future (e.g., "I plan to continue using my activity tracker at least as intensively as I currently do."). Reliability was acceptable.

User Diversity Variables. To assess ATI, the 9-item ATI scale [13] was used. Reliability was excellent. To assess geekism, the 15-item GEX [32] was applied. Reliability was excellent. To assess NFC, the 4-item NFC-K [4] was used. Reliability was acceptable.

5   Results

5.1   Q1: Individual Differences in Tracker Usage Motivations

To explore Q1, we present descriptive statistics and frequency distributions for the different usage motivations. Regarding intrinsic and extrinsic motivation, descriptive statistics (Table 1) and frequency distributions (Fig. 1) showed a substantial variance in both motivations. Given that all participants were current activity tracker users, the general left-skew of the distribution is rather unsurprising. When calculating difference values between intrinsic and extrinsic motivation, 46% of the participants were classified as more extrinsically and 35% as more intrinsically motivated. Nineteen percent could not be classified (i.e., identical scores for intrinsic and extrinsic motivation).

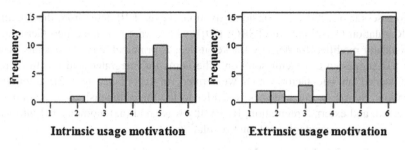

Fig. 1. Frequency distributions of intrinsic and extrinsic usage motivations.

5.2 Q2: Tracker Usage Motivations and Tracker Usage Intensity

To test H2a, H2b, and H2c, we conducted four multiple linear regression analyses (method: forced entry). In the first two analyses, current tracker usage intensity was criterion (see Table 2). In the second two analyses, future tracker usage was criterion (see Table 3). For both, we tested the superordinate motivations (intrinsic and extrinsic) and the sub-facets of extrinsic motivation (autonomous, introjected, external regulation) separately (therefore 2 × 2 regression analyses).

Table 2. Multiple regression models of predictors of current tracker usage intensity.

Predictor	$r_{\text{zero-order}}$	r_{partial}	B	β	p_β	R^2_{adj} (R^2)	p
Intrinsic motivation	.41**	.46	0.36	.43	<.001	.29	<.001
Extrinsic motivation	.36**	.35	0.22	.31	.007	(.31)	
Autonomous regulation	.49**	.47	0.31	.47	<.001	.22	.001
Introjected regulation	.22*	.14	0.10	.15	.294	(.26)	
External regulation	.14	−.04	−0.03	−.04	.795		

Note. (** = $p < .005$, * = $p < .05$)

Intrinsic motivation could positively predict the current intensity of tracking (medium-sized effect). In contrast, it could not predict the estimated future tracker usage, and only a weak correlation was found (H2a partially supported). Extrinsic motivation could positively predict the current and future tracker usage intensity (medium-sized effects; H2b supported). Of the three sub-facets of extrinsic motivation, only autonomous motivation could significantly predict the current and future tracker usage intensity (medium to large effect). Additionally, a weak significant correlation between introjected regulation and current usage intensity was found (H2c supported).

5.3 Q3: Personality Traits and Tracker Usage Motivations

To test H3a, H3b, and H3c, we conducted three simple linear regression analyses with ATI, geekism, and NFC as predictors and intrinsic motivation as criterion (see Table 4). Results showed that ATI was significantly positively related to intrinsic motivation for tracker usage (large effect, H3a supported). Geekism was significantly positively related

Table 3. Multiple regression models of predictors of estimated future tracker usage.

Predictor	$r_{\text{zero-order}}$	r_{partial}	B	β	p_β	R^2_{adj} (R^2)	p
Intrinsic motivation	.24*	.24	0.17	.22	.078	.17	.003
Extrinsic motivation	.38**	.37	0.23	.36	.004	(.20)	
Autonomous regulation	.47**	.48	0.29	.47	<.001	.23	.001
Introjected regulation	.19	.19	0.14	.23	.086	(.27)	
External regulation	−.02	−.20	−0.18	−.24	.103		

Note. (** = $p < .005$, * = $p < .05$)

Table 4. Simple regression models of predictors of intrinsic usage motivation.

Predictor	B	β	R^2_{adj} (R^2)	p
ATI	0.47	.54	.27 (.29)	<.001
Geekism	0.21	.30	.07 (.09)	.024
NFC	0.13	.15	.00 (.02)	.268

to intrinsic motivation for tracker usage (medium-sized effect, H3b supported). The relation between NFC and intrinsic motivation did not reach statistical significance. However, a small effect was found (H3c not supported).

6 Discussion

6.1 Summary of Results

The objective of the present research was to examine individual differences in the motivation for activity tracker usage, their relationships to current and future tracking intensity, and the impact of user diversity variables. Results revealed a substantial inter-individual variance regarding usage motivations, and more users stated to have a higher extrinsic than intrinsic motivation. Regression analyses revealed a higher intrinsic motivation to be associated with a higher current usage intensity. The same pattern was found for extrinsic motivation; in addition, extrinsic motivation could also predict estimated future usage intensity. Moreover, the more autonomous the extrinsic motivation is perceived, the higher the current and estimated future usage intensity. Regarding personality, higher ATI and higher geekism were related to higher intrinsic motivation for tracker usage, while NFC was not.

6.2 Implications

The present research provides insights into users' motivation for using activity trackers. Users indeed differed regarding their motivations and could be differentiated as having a higher extrinsic or intrinsic usage motivation. However, this differentiation was not possible for 19% of the participants, indicating that users can be simultaneously extrinsically and intrinsically motivated. This finding supports the notion that facets of motivation according to SDT do not lie on a one-dimensional continuum, but are rather separate dimensions [5].

Usage motivations were related to tracker usage intensity. The more self-determined and autonomous the motivation is perceived, the higher the current and estimated future tracking intensity is. This positive association between self-determined motivation and behavioral intensity has been found in prior research [10, 15, 29], also in the specific domain of activity tracking [28]. Hence, intrinsic motivation and/or more autonomous forms of extrinsic motivation seem to be crucial for intensive and persistent tracker usage and can therefore foster prolonged behavioral change. Moreover, a more differentiated view on the different sub-facets of extrinsic motivation rather than considering a definite dichotomy between intrinsic and extrinsic motivation might be decisive for investigating motivational strength and behavioral outcomes [27]. For manufacturers, this finding stresses the suggestion that design of activity trackers should strengthen intrinsic motivation for tracking and physical activity [1, 18], e.g. via gamification [19].

Users with a personal disposition to enjoy intensive interaction with technical devices (i.e., high ATI and geekism) more likely use their tracker out of intrinsic motives (i.e., because the interaction is inherently rewarding). Contrary to our hypothesis, no significant correlation between NFC and intrinsic motivation was found. Thus, when taking personality variables into account to explain technology-related behavior, we recommend to investigate domain-specific personality traits instead of broad personality traits (see also [13]).

6.3 Limitations and Future Research

All participants in our study were active tracker users. To better understand the relationship of usage motives and tracker usage, also former tracker users should be taken into account in future studies. Abandonment of activity trackers might not necessarily be caused by negative reasons such as measurement inaccuracy or privacy concerns [12]. Abandonment might also be a sign of successful behavioral change, that is, users might no longer feel the need to be assisted in meeting their activity goals [17]. Taking former users into account could deepen our understanding of usage motives that lead to successful behavioral change, thus, unfolding the various health benefits. A longitudinal study would be the ideal study design in this respect.

6.4 Conclusion

Tracker users differ regarding their usage motives. Intrinsic motivation, but also autonomous facets of extrinsic motivation were related to more intensive tracker usage. Further, the strength of intrinsic motivation was predicted by the domain-specific personality variables ATI and geekism. In sum, results highlight the crucial importance of self-determined motivation for activity tracker usage and long-term adherence.

References

1. Attig C, Franke T (2018) I track, therefore I walk–Exploring the motivational costs of wearing activity trackers in actual users. Int J Hum Comput Stud
2. Attig C, Wessel D, Franke T (2017) Assessing personality differences in human-technology interaction: an overview of key self-report scales to predict successful interaction. In: Stephanidis C (ed) HCI international 2017: posters' extended abstracts, part I, CCIS 713. Springer International Publishing AG, Cham, pp 19–29
3. Aykin NM, Aykin T (1991) Individual differences in human–computer interaction. Comput Ind Eng 20:373–379
4. Beißert H, Köhler M, Rempel M, Beierlein C (2015) Kurzskala need for cognition NFC-K. Zusammenstellung sozialwissenschaftlicher Items und Skalen
5. Chemolli E, Gagné M (2014) Evidence against the continuum structure underlying motivation measures derived from self-determination theory. Psychol Assess 26:575–585
6. Choe EK, Lee NB, Lee B, Pratt W, Kientz JA (2014) Understanding quantified-selfers' practices in collecting and exploring personal data. In: Proceedings of the 32nd annual ACM conference on human factors in computing systems. ACM, New York, pp 1143–1152
7. Coughlin SS, Stewart J (2016) Use of consumer wearable devices to promote physical activity: a review of health intervention studies. J Environ Health Sci 2(6):1–6
8. Day S (2016) Self-tracking over time: the FITBIT phenomenon. In: Verhaart M, Erturk E, Steele A, Morton S (eds) The 7th annual conference of computing and information technology research and education New Zealand (CITRENZ2016), pp 1–6
9. Deci EL, Ryan RM (1985) Intrinsic motivation and self-determination in human behavior. Plenum Publishing Co., New York
10. Duncan LR, Hall CR, Wilson PM, Jenny O (2010) Exercise motivation: a crosssectional analysis examining its relationships with frequency, intensity, and duration of exercise. Int J Behav Nutr Phys Act 7(7):1–9
11. Edison SW, Geissler GL (2003) Measuring attitudes towards general technology: antecedents, hypotheses and scale development. J Target Measur Anal Market 12:137–156
12. Epstein DA, Caraway M, Johnston C, Ping A, Fogarty J, Munson SA (2016) Beyond abandonment to next steps: understanding and designing for life after personal informatics tool use. In: Proceedings of the 2016 CHI conference on human factors in computing systems. ACM, New York, pp 1109–1113
13. Franke T, Attig C, Wessel D (2018) A personal resource for technology interaction: development and validation of the Affinity for Technology Interaction (ATI) scale. Int J Hum Comput Interact 1–12
14. Freudenthaler HH, Spinath B, Neubauer AC (2008) Predicting school achievement in boys and girls. Eur J Pers 22:231–245
15. Gardner B, Lally P (2013) Does intrinsic motivation strengthen physical activity habit? Modeling relationships between self-determination, past behavior, and habit strength. J Behav Med 35:488–497
16. Gimpel H, Nißen M, Görlitz RA (2013) Quantifying the quantified self: a study on the motivation of patients to track their own health. In: Proceedings of the 34th international conference on information systems (ICIS), pp 1–16
17. Gouveia R, Karapanos E, Hassenzahl M (2015) How do we engage with activity trackers? A longitudinal study of Habito. In: Proceedings of the 2015 ACM international joint conference on pervasive and ubiquitous computing. ACM, New York, pp 1305–1316
18. Hagger M, Chatzisarantis N (2008) Self-determination theory and the psychology of exercise. Int J Sport Exerc Psychol 1:79–103

19. Hamari J, Koivisto J, Sarsa H (2014) Does gamification work? A literature review of empirical studies on gamification. In: Proceedings of the 47th Hawaii international conference on system sciences. IEEE, pp 3025–3034
20. Kelly N (2016) The WEAR Scale: Development of a measure of the social acceptability of a wearable device. Graduate Theses and Dissertations 15230
21. Lazar A, Koehler C, Tanenbaum J, Nguyen DH (2015) Why we use and abandon smart devices. In: Proceedings of the 2015 ACM international joint conference on pervasive and ubiquitous computing. ACM, New York, pp 635–646
22. Levesque CS, Williams GC, Elliot D, Pickering MA, Bodenhamer B, Finley PJ (2007) Validating the theoretical structure of the Treatment Self-Regulation Questionnaire (TSRQ) across three different health behaviors. Health Educ Res 22:691–702
23. Martens R, Gulikers J, Bastiaens T (2004) The impact of intrinsic motivation on e-learning in authentic computer tasks. J Comput Assist Learn 20:368–376
24. Nelson EC, Verhagen T, Noordzij ML (2016) Health empowerment through activity trackers: an empirical smart wristband study. Comput Hum Behav 62:364–374
25. Olson K, Camp C, Fuller D (1984) Curiosity and need for cognition. Psychol Rep 54:71–74
26. Renfree I, Harrison D, Marshall P, Stawarz K, Cox A (2016) Don't kick the habit: the role of dependency in habit formation apps. In: Proceedings of the 2016 CHI conference on human factors in computing systems. ACM, New York, pp 2932–2939
27. Rigby CS, Deci EL, Patrick BC, Ryan RM (1992) Beyond the intrinsic–extrinsic dichotomy: self-determination in motivation and learning. Motiv Emot 16:165–185
28. Rupp MA, Michaelis JR, McConnell DS, Smither JA (2018) The role of individual differences on perceptions of wearable fitness device trust, usability, and motivational impact. Appl Ergon 70:77–87
29. Ryan RM, Connell JP (1989) Perceived locus of causality and internalization: examining reasons for acting in two domains. J Pers Soc Psychol 57:749–761
30. Ryan RM, Frederick CM, Lepes D, Rubio N, Sheldon KM (1997) Intrinsic motivation and exercise adherence. Int J Sport Psychol 28:335–354
31. Ryan RM, Deci EL (2000) Intrinsic and extrinsic motivations: classic definitions and new directions. Contemp Educ Psychol 25:54–67
32. Schmettow M, Drees M (2014) What drives the geeks? Linking computer enthusiasm to achievement goals. In: Proceedings of HCI 2014, Southport, UK, pp 234–239

Readability as a Component of the Usability of a Web Presentation - A Case Study from the Banking Sector

Aleksandar Zunjic[1(✉)] and Sylvain Leduc[2]

[1] Faculty of Mechanical Engineering, University of Belgrade,
11000 Belgrade, Serbia
azunjic@mas.bg.ac.rs
[2] LPS, Aix Marseille University, Aix-en-Provence, France
sylvain.leduc@univ-amu.fr

Abstract. In many comprehensive and overall considerations of usability of web presentations, readability as a factor that potentially influences usability is not represented at all, or it is represented unsystematically and superficially. The main goal of this paper is establishing a connection between usability and readability. There is no known research that previously dealt with this topic. In order to establish the above correlation, firstly the classification of usability features was performed. This original classification has had an important function in identifying the major website usability constructs. Based on the identified major usability and readability features (constructs), an original questionnaire for assessing the usability and quality of websites was formed. This tool, together with the interviewing method was used in the case study, which referred to the website of a company operating in the banking sector. As a result of the case study, a high correlation of 0.943 is established between the organization of information on the screen (a usability feature) and the structure of headings, sub-headings, and positioning of the text (a readability feature). The existence of a strong link between the possibilities for finding information (a usability feature) and the content (explanations, arguments, coherence) of the textual presentation (a readability feature) has been established. This and other results obtained indicate that complex considerations that refer to the usability of websites, as well as tools that are used for usability assessment should involve in a systematic way the readability as an important factor, which can have the influence on usability.

Keywords: Usability · Readability · Web design

1 Introduction

Usability is a concept that has wide application in the ergonomic designing and evaluation of different products and systems. Every company that wants to promote sales should have a website, through which it offers its products or services. However, it is not uncommon for web designers and developers who design websites for companies to have no knowledge in the area of ergonomics and usability, or have minimal experience in this field. For this reason, many companies' websites often have different

S. Bagnara et al. (Eds.): IEA 2018, AISC 822, pp. 441–456, 2019.
https://doi.org/10.1007/978-3-319-96077-7_47

shortcomings in the usability domain, which can be reflected in the success of the business. In addition, design and usability assessment methods also need to be improved to ensure that the websites are as customized as possible to the users and meet their requirements.

1.1 Website Usability and Website Usability Features

Website usability can be defined as the extent to which a web presentation can be used by a specific user in order to achieve a specific goal, while achieving appropriate effectiveness, efficiency, and satisfaction in use. Numerous methods can be used to evaluate usability. Some of the most commonly used are [1–4]: questionnaires, checklists, question-asking protocol, feature inspection, screen snapshots, interviews and focus groups, user and task observation, scenarios, heuristic evaluation, cognitive walkthrough, pluralistic walkthrough, consistency inspections, standards inspections, surveys, formal usability inspections, thinking aloud protocol, co-discovery method, performance measurement, eye tracking, prototyping, affinity diagrams, blind voting, card sorting, contextual inquiry, ethnographic study/field observation, journaled sessions, self-reporting logs. Most of the above methods can be used successfully to evaluate the usability of websites. The choice of the method that will be used in the specific case depends primarily on the objective of testing, but may also depend on other factors that should take into account in accordance with the circumstances. Often a combination of several methods is used to balance the shortcomings of one method with the benefits of another method.

Usability features can be identified for each type of product individually. Like any other product, websites also need to possess certain usability features, in order to be convenient for use. For the purposes of this, as well as other theoretical and practical research, it is necessary to make the difference between the individual usability features. In connection with this, when it comes to the websites, it can be distinguished:

- global usability features
- major usability features
- specific usability features
- optional usability features.

The global usability features are those usability features which are necessary that any product own, regardless of its type or purpose. These usability features stem from the definition of usability, that is, the basic attributes of usability:

- easy to use
- easy to learn
- easy to remember
- no errors (or as few errors as possible)
- subjectively pleasing.

Major usability features are those usability features that are essential for the certain type of product (software), such as for example websites. If the global usability features are excluded from consideration, for a particular product they are of the hierarchically highest importance. They can include one or more specific usability features. In the case of websites, for example, navigation can be considered one of the major usability features. If this feature is not met, it is understandable that the website will be unusable.

Specific usability features refer to some specific need or use of a product (software). They are hierarchically less important in relation to the major usability features. They are not, in their character, common to all products. They can also vary depending on the types of one and the same product. When it comes to websites, then specific usability characteristics may vary depending on the purpose of the websites. In addition, specific usability features may vary depending on who is the end user, terms of use, or the type of tasks the user should perform. For example, no broken links or no misleading links are the specific usability features of websites (while navigation, as it is previously mentioned is the major usability feature, which includes the two mentioned specific usability features).

Global, major and specific usability features according to the way they are presented have a general form. They do not contain a specification of how these features will be practically realized. Optional usability features are usability features that are usually given in the form of certain features (functions) that are desirable that a particular product (software) possesses. Optional usability features have a lower hierarchical character than specific usability features. If we consider the submission of payment in the case of commercial websites as a specific usability feature, then the availability of currency converter and multiple payment options are the optional usability features.

It should be mentioned that below the lowest hierarchical level there are design solutions that enable the realization of any usability feature at any of the above-mentioned hierarchical levels. These are in essence the practical interface solutions. They define how the user will use the interface. For example, for a particular website, it can be selected the use of text fields instead of checkboxes with the options offered, for the implementation of a predefined usability feature.

In the literature, no distinction has been made between the hierarchical levels identified here and which relate to usability characteristics. In addition, it should be noted that from the practical and terminological aspect, between the term usability features, and the terms such as usability characteristics, usability attributes, usability factors, usability determinants or usability dimensions there is essentially no difference. All of these terms that are mentioned in the literature and in tools for assessing usability relate to the same thing.

When it comes to websites, major website usability features are of particular importance because they directly determine the quality and use of websites (as opposed to global usability features that define the quality and use of any product, including websites). However, there is no unique point of view which are the major usability features. In some papers related to the usability features of websites, depending on the problems, major usability features are not considered at all. For example, in [5], primarily were used and analyzed the specific usability features, as well as optional usability features. On the other hand, there are papers, such as [6] where there was taken into account a mix of various usability features, whereby those usability features belong to different, previously defined hierarchical levels.

Except in papers, various website usability features can be noticed above all from the structure of web usability surveys, website usability questionnaires and checklists.

These tools often contain a clearly discernible hierarchical structure of usability features. The top level of usability features in such tools in essence is related to the major (or global) usability features. In the paper [7] a review of major (top level) usability features for a large number of tools that are used for the evaluation of the usability of websites has been given. A review of these usability features once again confirms the aforementioned statement that there is no unique point of view about usability features which should be adopted as major, that is, which usability features have essential significance for websites.

1.2   Readability and Readability Features

The readability of alphanumeric symbols is a feature that allows the recognition of the information content of the presented material composed of alphanumeric characters, whereby they are grouped into recognizable entities such as words, sentences, or whole text [8]. Readable material is one that is easy and pleasant to read, which may be read with satisfaction or interest, as well as that is attractive in style or treatment [9]. Readability can most easily be defined as ease of reading and understanding of a particular material.

The readability concept has a very long history. It is known that considerations of this phenomenon date back to the late nineteenth and early twentieth century. Readability depends on a very large number of factors. Gray and Leary identified 228 elements that affect readability [10]. Although readability and visibility are often treated separately, some authors view these two aspects in the context of readability, bearing in mind that they have a big impact on readability.

The readability can be evaluated in several ways. A very common approach to readability assessment is by using formula. Some of the frequently used readability formulas are the formulas defined by Lorge, Flesch, Dale-Chall, Fog, Fry, McLaughlin [11]. Since the formulas cannot cover all aspects of readability, other methods described in [10, 12] and other literature can also be used to estimate readability. One of the frequently used approaches for readability assessment is through questionnaires.

Readability can be tested on different media. Certain experimental readability studies on VDT screens were also performed, e.g. [13–18], and on small displays [19]. These studies have shown how individual readability features affect reading from the electronic displays. A number of studies have also been conducted with aspects of readability related to the visibility on VDT screens [20] and the legibility of alphanumeric symbols, for example [21, 22].

1.3   Readability as a Component of the Website Usability

Although based on the previous exposition it can be assumed that readability features can affect the usability of websites, there are many examples when, from the consideration of usability, completely were omitted features relating to readability. This applies even to the literature in the form of books in which different aspects of usability were the subject of comprehensive considerations [23–27], but in which the term readability has not been mentioned. In addition, there are books in the field of usability [28–31] where the notion of readability is barely mentioned, but where the importance of the concept of readability is not explained in more detail, as well as the function of readability in the context of the impact on usability.

As mentioned above, questionnaires and checklists are tools that are often used to estimate usability. Most questionnaires and checklists that are used to assess the usability, especially those more extensive (which contain a larger number of questions), in the frame of the titles of the main areas (constructs) of the questionnaire contain certain questions, relating to these main areas. In the case of usability assessment, the main areas (constructs) primarily correspond to the global or major usability features, according to the previously adopted classification of usability features. In [32], an original questionnaire for assessment of the usability of websites has been presented. From this questionnaire, as well as from the questionnaires cited in this paper, it can be noticed that in any of these questionnaires, readability is not used as a construct, in terms of assessing the usability of websites. Similarly, a usability assessment questionnaire containing 3 constructs was used in [33], whereby none of these three major usability features contain readability as a feature. Various tools are also considered in [34] for evaluating the usability of software and websites, primarily based on the use of the questionnaire. In the analysis of these tools, among other things, readability is also not mentioned as a factor that is present in the questionnaires. In [7] a review of constructs for 17 questionnaires that are used for assessment of the quality and usability of the software and websites has been given. None of these questionnaires contains readability as a construct. It follows that specific issues within these tools also do not relate directly to the assessment of readability as a construct (although there is a possibility that in some of the usability estimation tools a question from the domain of readability is within some another construct, due to the connection of that construct with readability).

1.4   The Aim of the Research

Based on the conducted analysis and insights into the literature, it can be concluded that readability is often not included and perceived as a factor that has an impact on usability, even when multi-component and complex considerations of website usability are taken into account. The impression is that some researchers who deal with website usability problems are not aware of the potential impact of readability on website usability. An additional problem is the fact that the vast majority of questionnaires and other tools of similar purpose do not contain readability as a component, i.e. feature of the usability.

In addition, a number of studies were carried out in the field of ergonomics and usability, for example [35–38], which dealt with the identification of problems of usability that are present on the websites of companies. These studies indicate that the presence of problems of usability on websites of companies can negatively affect the business of these firms. In view of this, improving readability of websites can also have a positive effect on the ability to use websites in the banking sector, as well as in other branches of industries.

Bearing in mind the foregoing, the main goal of this paper is to examine the possibility of the existence of the interconnection and the influence of readability features on usability features. In addition, as part of a case study from the domain of the banking sector, an additional goal is to identify shortcomings related to the usability of a website, which also include readability features. It is not known any research that has previously dealt with the establishment of a link between usability and readability features.

2 Method

From the previous consideration, it can be noticed that a large number of methods can be used for determination of usability, as well as readability. However, all these methods are not equally suitable for the established research goal, where it is necessary to establish a correlation between website usability and readability. In view of this, the most appropriate approach for establishing a connection between usability and readability of websites is based on usability and readability features. If it turns out that there is a correlation between the individual usability and readability features, then it will be possible to make certain conclusions regarding the relationship between these two entities.

The most straightforward way to establish a connection between usability and readability features is through their constructs. As already mentioned, in accordance with the established classification, usability constructs are in fact formed on the basis of global and major usability features. However, in the case of website questionnaires, global usability features are not suitable for use, because on the basis of them it is hard to identify directly, in which segments of the design and use of websites there is a concrete problem. For this reason, we decided to form usability constructs on the basis of the major usability features.

As mentioned above, there is no unified stand around the choice of items that will constitute the usability constructs for websites. When creating usability constructs (major usability features), we focused on several essential criteria that should be met when designing and using any type of website. In connection with this, in terms of usability, a website should meet the usability criteria that refer both to textual and graphic elements. In addition, a website should allow the user to easily use various navigation and control options in order to be able to quickly get the information requested, or to perform a specific task. Searching information and performing tasks should be based on user expectations. Intuitive use is also important because it speeds up work, reduces the potential for the appearance of errors and eliminates the need for additional help.

In order to fulfill the aforementioned criteria, 7 major usability features have been selected. These usability features have been formed in accordance with the literature previously mentioned, including additional literature, such as [39–41]. Selected major website usability features (constructs) are:

1. Easy navigation and control
2. Easy finding the information requested
3. Fast loading of web pages
4. Satisfaction with graphic elements
5. Intuitive use

6. Consistency with the expectations of the user
7. Organization of information on the screen (web page).

Each of the selected major usability features relates and determines some of the essential segments of the use of websites. Some of the identified major usability features are present in other questionnaires from the usability domain of the websites under different names. It is possible that some of the selected major usability features include several constructs from other questionnaires that have similar names, whereby they relate to the same or similar phenomenon. For example, the major usability feature under number two in the upper list includes constructs from other questionnaires that have names such as information quality, content quality and clarity of site organization. All three previously mentioned constructs are prerequisites for the easy finding of the required information.

It should be mentioned that there may be some degree of connectivity between some usability constructs, even within the same tool for usability assessment. The reason is that one or more specific usability features (or optional usability features) can positively or negatively affect the fulfillment of different conditions that are covered by different constructs. Such a situation can also be expected in the case of the constructs that are identified herein.

When it comes to readability, 4 constructs have been selected. These constructs are identical to the main constructs that have been selected in a comprehensive readability study carried out by previously mentioned Gray and Leary [10]. These readability constructs are:

1. Content
2. Style
3. Format
4. Features of organization.

In order for the selected constructs of usability and readability to get an operational dimension, based on the previously defined 11 major features of usability and readability, a questionnaire was created. In accordance with the basic aim of this research, the basic purpose of the questionnaire is to establish a correlation between usability features and readability features. For this reason, for the formation of the questionnaire, for each individual construct one question was selected, which represent as much as possible each selected major usability and readability feature. The first 7 questions from the questionnaire refer to usability features, while the last 4 questions from the questionnaire refer to readability features. The sequence of questions corresponds to the order of the selected constructs. The questionnaire consists of the following questions, which are presented in the form of statements:

5. Navigation and control of this website are simple and easy
6. It is easy to find information that I'm looking for on this website
7. All web pages on this website load fast
8. All graphic elements on this website are pleasing and support the content
9. This website is easy to use on an intuitive basis
10. In all aspects, this website is in accordance with my expectations

11. The organization of information on each screen (web page) of this website is appropriate
12. Explanations, arguments and coherence of the textual presentation is appropriate
13. Style of expression and presentation is appropriate
14. The general design of the content, including typography, layout, illustrations, tables are appropriate
15. Structure of headings, sub-headings and positioning of the text is appropriate and reflects the organization of ideas.

When formulating the question, the issues were simply formulated so that the subjects could easily understand them and that they can give a precise answer. It is envisaged giving the answers to the questions asked on the Likert scale with five levels. The offered answers to the questions were:

16. Strongly disagree
17. Disagree
18. Neutral
19. Agree
20. Strongly agree

In order to additionally check the validity of the designed questionnaire, it was distributed to evaluators, who use the internet on a daily basis for their work or studies. For the evaluation of the face validity, 23 evaluators were asked to state whether items appear relevant, important and interesting to them for evaluation of different aspects of the use of websites. All evaluators have given positive statements in relation to the questionnaire, emphasizing the importance of its consistency and compactness.

Additionally, evaluation of the content validity of the questionnaire was performed. Five evaluators of different specialties (three were experts in the field of ergonomics and two were specialists in the field of web design) participated in this assessment. The participants of this evaluation voted for each item of the questionnaire, assessing whether it is important and should remain as a part of the questionnaire, or should be omitted as less relevant. Lawshe method was applied for the purpose of quantifying of the assessment of the content validity. The mean CVR for entire questionnaire was 1. Although the primary goal was to establish a correlation between usability and readability features, the value obtained for CVR indicates a satisfactory level of content validity of this concise questionnaire for assessing the usability and quality of the design solution of websites. The information received regarding the validity of the questionnaire was important to us, in order that we could start collecting data related to the main goal of the research.

In order to test the assumptions about the interconnection of usability and readability features, a case study was formed. For this purpose, for the assessment of usability and readability features, the website of a multinational company operating in the field of banking was selected. This bank has in total nearly 50,000 employees, service 16,5 million customers through more than 2,400 business outlets.

Seven subjects were selected for testing. In the case of usability assessment, most authors agree that most of the usability problems can be detected even with five respondents. This number of respondents provides an excellent ratio between the achieved effect of testing on the one hand, and the time and cost involved on the other hand.

Test participants were required to fulfill several important conditions. It was necessary that they were not users of the services of the mentioned bank. In addition, the condition for participating in the test was that all subjects actively use the Internet for more than a year. The respondents were also required that they are employed, that they are adults, and to have any previous experience with the use of services (online or offline) in any of the banks (receiving salaries, taking loans, paying, etc.). The respondents also needed to have a completed English language course (i.e. that they are in the ability to use English). In accordance with these conditions, seven respondents were selected, whose average age was 42.7 years (SD = 18.4). The average number of years of internet use by the participants was 9.85 years (SD = 5.58). The test participants were of different professions. They all had a minimum BSc degree. All subjects had normal visual acuity (6/6) or better.

In connection with the primary goal of the research, before the start of the test, the participants received instructions. Also, the purpose of the test was explained to them. Starting from the homepage of the mentioned bank, the task of the respondents was to find all the information that are interesting to them, especially those related to the transactions they normally perform in any other bank. Then they needed to perform fictitiously an online transaction they normally perform in any other bank, if this option is available online. After that, the respondents were supposed to open the other parts of the web presentation of this bank and to get acquainted with the displayed content and options in the textual and graphical form. The respondents used the English version of the bank's web presentation, but they were also allowed to use the Serbian version as needed (both versions are almost identical).

In connection with the secondary objective of the research that consisted in the identification of specific problems related to the use of the internet presentation of the bank, as well as in the detection of shortcomings in the design of this web presentation, the task of respondents was to record the URL of the web page where they noticed some deficiencies and to write down concisely what was the problem. The respondents had available 45 min to perform the above tasks. After the completion of the test, respondents first filled out the questionnaire with 11 items. Then, interviews were conducted with respondents, which focused on the specific problems that they reported during the testing. During this interview, respondents needed to describe more precisely the problems which they encountered in using the internet presentation of the bank.

The test was performed on the LG E2241S 22 inch monitor. The monitor resolution was 1920 × 1080. The processor clock speed was 3.66 GHz (4 cores), while the internet connection speed was 50 Mbps. All respondents were allowed to use the preferred internet browser.

3   Results and Analysis of Results

Table 1 shows the mean values and standard deviations of the answers of 7 respondents to all 11 questions from the questionnaire. The ordinal numbers of questions are indicated in the table with Q. The standard deviation of the sample is marked with SD. On the Likert scale of five levels, respondents gave the highest marks for the question from the questionnaire under number 3, which refers to the speed of loading of the web pages of the website. Respondents gave the lowest ratings to the question under number 2, which refers to the ease of finding certain information on a bank's website, as well as to the question under number 7, relating to the organization of information on the screen.

Table 1. Means and standard deviations for each question from the questionnaire

Q	1	2	3	4	5	6	7	8	9	10	11
Mean	4.0	3.571	4.714	4.571	4.571	3.714	3.571	3.857	3.857	3.714	3.714
SD	0.816	1.272	0.487	0.534	0.534	0.487	0.975	0.377	0.690	1.253	1.112

To check the assumption about the mutual dependency between usability features and readability features, Pearson product moment correlation test was used. Table 2 shows the values of the correlation coefficient r, taking into account the answers of all 7 respondents to 11 questions from the questionnaire. From Table 2, a strong connection between some basic characteristics of readability and usability can be noticed. In connection with this, the highest correlation of 0.943 has been established between the organization of information on the screen (a usability feature) and the structure of the headings, the sub-headings, and positioning of the text (a readability feature). The existence of a strong link between the possibilities for finding information (a usability feature) and the content (explanations, arguments, coherence) of the textual presentation (a readability feature) has been established ($r = 0.891$). In both previous cases, the

Table 2. Correlation coefficients for questions from the questionnaire obtained using the Pearson product moment correlation test

Q	2	3	4	5	6	7	8	9	10	11
1	0.160	0.000	0.382	−0.382	0.418	0.418	0.000	0.000	−0.163	0.550
2		0.0383	0.175	0.420	0.844	0.633	0.891	0.488	0.328	0.605
3			−0.548	0.0913	0.300	0.0500	−0.258	0.354	−0.156	−0.175
4				−0.167	0.0913	0.548	0.471	0.258	0.533	0.600
5					0.0913	0.228	0.471	0.710	0.0355	0.040
6						0.750	0.645	0.354	0.389	0.745
7							0.710	0.636	0.701	0.943
8								0.548	0.603	0.679
9									0.330	0.372
10										0.649
11										

p value is below 0.05, which indicates the statistically significant relationship. A high correlation of 0.745 has also been registered between the compliance with the user's expectations (a usability feature) and the structure of the headings, the sub-headings, and positioning of the text. However, in this case, the p value is slightly above 0.05 (it equals 0.054).

In addition, a relatively strong relationship (r = 0.701) has been established between the organization of information on the screen (usability feature) and the format of the content (readability feature). However, obtained value for p, in this case, is above 0.05 (0.079). Besides, two additional relationships have been found. A correlation of 0.710 can be noticed between the use of the website on an intuitive basis (usability feature) and the style of the presentation (readability feature). Also, a correlation of 0.710 has been obtained between the organization of information on the screen (usability feature) and the style of the presentation (readability feature). However, in both last mentioned cases, the p value is above 0.05 (it equals 0.074 for both relationships).

Based on interviews with respondents and insights into their notes made during the testing, certain shortcomings from the readability domain were discovered. These shortcomings also affected the marks given by respondents, in their words, to certain issues that were primarily related to usability. A few such examples are given below (some images are blurry at places where the name of the bank was located).

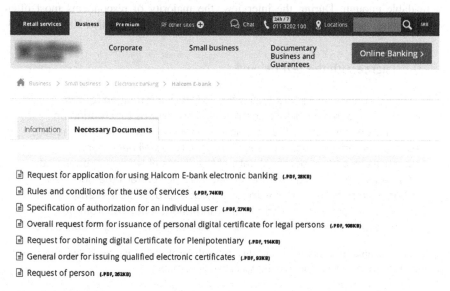

Fig. 1. Primer of a web page of the bank's website where the alphanumeric characters of inadequate contrast and size have been used.

Figure 1 shows an example of inadequate readability. The dark yellow color of the parts of the text that relates to the file type and its size is inadequate because it produces an inappropriate contrast in relation to the white color of the surrounding. In addition, the alphanumeric characters used are 6 pt. This character size is not recommended for

Corporate
Cash Management
Loans
Factoring
Payment cards
Electronic banking

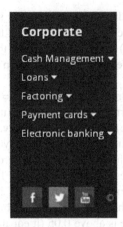

Fig. 2. An example of a graphic element of inadequate readability.

designing websites. During the interview, the respondents were most often connected this disadvantage with the answers to question number 10, as well as 6.

Some graphic elements also attracted the attention of the test participants. Figure 2 shows an icon colored in red, which according to the statements of most respondents is not readable enough. During the interview, the majority of respondents most often associated this disadvantage with the answers to questions under number 10, as well as 4.

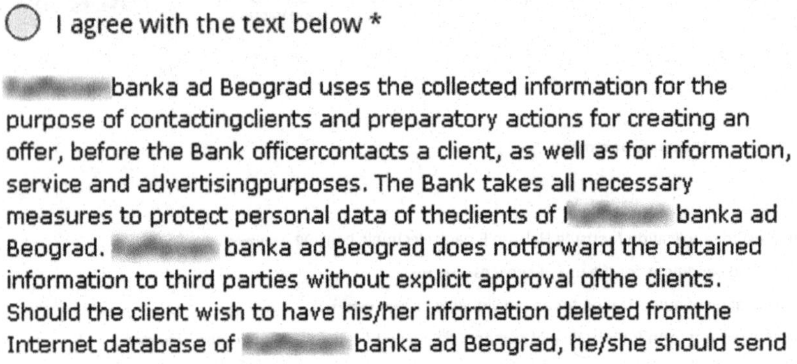

I agree with the text below *
banka ad Beograd uses the collected information for the purpose of contactingclients and preparatory actions for creating an offer, before the Bank officercontacts a client, as well as for information, service and advertisingpurposes. The Bank takes all necessary measures to protect personal data of theclients of
banka ad Beograd.
banka ad Beograd does notforward the obtained information to third parties without explicit approval ofthe clients. Should the client wish to have his/her information deleted fromthe Internet database of
banka ad Beograd, he/she should send

Fig. 3. An example of a lack of readability from the website of the bank that refers to the merging of individual words into entities that have no meaning.

Figure 3 shows a part of the web page from the bank's internet presentation that also possesses certain problem connected with readability. It can be noted that some words were merged, so it has appeared non-existent terms such as "contactingclients", "officercontacts", "advertisingpurposes", "theclients", "notforward", "ofthe", "fromthe" etc. This phenomenon extends reading time and can adversely affect the understanding of the text. During the interview, the majority of respondents most often associated this disadvantage with the answers to questions under number 9 and 10, as well as 2 and 6.

4 Conclusion

A high-quality website design implies not only the application of theory and practice in the programming of websites. The design and content of the entire displayed material on the website of the company affects the usability of the website. The overall features of the design solution affect the scope of the use of a website of a particular company, which ultimately influences the financial effect that a company can expect as a result of this type of interaction between users and the company.

However, when designing and evaluating the usability of websites, some aspects that have an impact on usability have been neglected completely or are superficially and unsystematically were taken into consideration. One of these aspects of usability is related to readability. This research has been primarily focused on establishing the connection between readability and usability.

In order to be able to establish a connection between usability and readability, it was necessary at the beginning to perform the classification of usability features. This original classification had an important function in identifying the major constructs of usability. Based on selected major constructs of usability and readability, an original questionnaire was created. The main purpose of this questionnaire was to establish a correlation between the usability and readability features. In addition, this questionnaire was intended to identify key flaws in the design of a company's website that operates in the banking sector.

The conducted case study has shown a clear correlation between certain factors of usability and readability. In connection with this, a high correlation of 0.943 is established between the organization of information on the screen (a usability feature) and the structure of the headings, the sub-headings, and positioning of the text (a readability feature). The existence of a strong link between the possibilities for finding information (a usability feature) and the content (explanations, arguments, coherence) of the textual presentation (a readability feature) has been established (r = 0.891). A high correlation of 0.745 has also been registered between the compliance with the user's expectations (a usability feature) and the structure of the headings, the sub-headings, and positioning of the text. The results of this research suggest that readability factors should be systematically included in considerations relating to design and assessment of the usability of websites. This means that tools for the design and assessment of the usability of websites, such as questionnaires, should contain readability features.

As stated above, the purpose of the formed questionnaire was primarily to establish a link between usability and readability features. However, the newly formed questionnaire was also useful for identifying main areas of usability (including readability) where improvements of the website are possible. Results of the research can serve as a guideline for web designers in which segments should be made improvements to the website.

Given its conciseness, this questionnaire is not recommended for independent use in order to assess in more detail the usability of websites. However, its use can be recommended in combination with an additional method for assessing usability, for example in collaboration with the interviewing method. This approach can provide revealing a greater number of details connected with the usability problems.

Companies that offer certain product or service over the internet should be aware of the importance of identifying and solving the problem of the usability of their websites. If a website of a company is not adequately designed, there is a probability that the user will quit searching and look for the website of other company to meet a particular goal. For this reason, in order to achieve positive financial effects, the company's management, and particularly people involved in designing websites should be aware of the importance of incorporating ergonomic factors that affect the quality of a website's design, such as readability.

References

1. Geisen E, Bergstrom JR (2017) Usability testing for survey research. Elsevier Inc., Cambridge
2. Rubin J, Chisnell D (2008) Handbook of usability testing: how to plan, design, and conduct effective tests, 2nd edn. Wiley, Indianapolis
3. Hom J (2018) The usability methods toolbox handbook. http://usability.jameshom.com/. Accessed 3 Mar 2018
4. Nielsen J (1993) Usability engineering. Morgan Kaufman, San Francisco
5. Wei J, Ozok A (2005) Development of a web-based mobile airline ticketing model with usability features. Ind Manag Data Syst 105(9):1261–1277
6. Raju NV, Harinarayana NS (2008) An analysis of usability features of library web sites. Ann Libr Inform Stud 55(2):111–122
7. Sauro J (2015) SUPR-Q: a comprehensive measure of the quality of the website user experience. J Usability Stud 10(2):68–86
8. Sanders MS, McCormick EJ (1992) Human factors in engineering and design, 7th edn. McGRAW-HILL, New York
9. Gray WS, Leary BE (1935) What makes a book readable. The University of Chicago Press, Chicago
10. DuBay WH (2007) Unlocking language: the classic readability studies. Impact Information, Costa Mesa
11. Bailin A, Grafstein A (2016) Readability: text and context. Palgrave Macmillan, Basingstoke
12. Rush RT (1985) Assessing readability: formulas and alternatives. Read Teach 39(3):274–283
13. Dyson MC, Haselgrove M (2001) The influence of reading speed and line length on the effectiveness of reading from screen. Int J Hum Comput Stud 54:585–612
14. Kolers PA, Duchnicky RL, Ferguson DC (1981) Eye movement measurement of readability of CRT displays. Hum Factors 23(5):517–527
15. Duchnicky JL, Kolers PA (1983) Readability of text scrolled on visual display terminals as a function of window size. Hum Factors 25:683–692
16. Mills CB, Weldon LJ (1987) Reading text from computer screens. ACM Comput Surv 4:329–358
17. Bernard M, Fernandez M, Hull S (2018) The effects of line length on children and adults' online reading performance. http://usabilitynews.org/the-effects-of-line-length-on-children-and-adults-online-reading-performance/. Accessed 4 Mar 2018

18. Dillon A, McKnight C, Richardson J (1988) Reading from paper versus reading from screens. Comput J 31(5):457–464
19. Tikka P (2013) Reading on small displays: reading performance and perceived ease of reading. PhD thesis. University of Northumbria, Newcastle
20. Zunjic A, Ristic L, Milanovic DD (2012) Effects of screen filter on visibility of alphanumeric presentation on CRT and LCD monitors. Work 41(S1):3553–3559
21. Zunjic A (2004) Software-generated factors of discrimination of alphanumeric symbols (in Serbian). In: Proceedings of the Yugoslav scientific symposium on Ergonomics. Ergonomics Society of F.R. Yugoslavia, Belgrade, pp 49–52
22. Erdogan Y (2008) Legibility of websites which are designed for instructional purposes. World Appl Sci J 3(1):73–78
23. Krug S (2006) Don't make me think: a common sense approach to Web usability, 2nd edn. New Riders Publishing, Berkeley
24. Bawa J, Dorazio P, Trenner L (2001) The usability business: making the Web work. Springer, London
25. Reiss E (2012) Usable usability. Wiley, Indianapolis
26. Liao H, Guo Y, Savoy A, Salvendy G (2010) Content preparation guidelines for the web and information appliances: cross-cultural comparisons. CRC Press, Boca Raton
27. Norlin E, Winters CM (2002) Usability testing for library Web sites: a hands-on guide. American Library Association, Chicago
28. Jarrett C, Gaffney G (2009) Forms that work: designing Web forms for usability. Morgan Kaufmann, Burlington
29. Chandler K, Hyatt K (2003) Customer-centred design: a new approach to Web usability. Prentice-Hall Inc., Upper Saddle River
30. Thurow S, Musica N (2009) When search meets Web usability. New Riders, Berkeley
31. Dillon A (2003) Designing usable electronic text: ergonomic aspects of human information usage. Taylor & Francis, London
32. Kuan HH, Vathanophas V, Bock GW (2003) The impact of usability on the intention of planned purchases in E-commerce service websites. In: Proceedings of the 7th Pacific Asia conference on information systems. Association for Information Systems, Adelaide, Australia, pp 369–392
33. Nahm ES, Barker B, Resnick B, Covington B, Magaziner J, Brennan PF (2010) Effects of a social cognitive theory based hip fracture prevention website for older adults. Comput Inform Nurs 28(6):371–379
34. Hayat H, Lock R, Murray I (2015) Measuring software usability. In: Proceedings of the BCS software quality management conference, BCS, Loughborough
35. Zunjic A, Spasojevic Brkic V (2011) Business improvement through application of ergonomic ISO guideance on World Wide Web user interface. In: Proceedings of the 6th international conference on total quality management: advanced and intelligent approaches. Faculty of Mechanical Engineering, Belgrade, Serbia, pp 282–287
36. Capell JJ, Huang Z (2007) A usability analysis of company websites. J Comput Inf Syst 1:117–123
37. Post GV, Kagan A, Sigman BP (2009) Usability investigation of E-business Web-based forms. J Int Technol Inf Manag 18(1):35–58
38. Downing CE, Liu C (2014) Assessing web site usability in retail electronic commerce. J Int Technol Inf Manag 23(1):27–40
39. Marsico MD, Levialdi S (2004) Evaluating web sites: exploiting user's expectations. Int J Hum Comput Stud 60:381–416

40. Naumann A, Pohlmeyer AE, Husslein S, Kindsmüller MC, Mohs C, Israel JH (2008) Design for intuitive use: beyond usability. In: Proceedings of the CHI 2008 conference, Florence, Italy. ACM, pp 2375–2378

41. Owoyele S (2016) Website as a marketing communication tool. Thesis. Centria University of Applied Science, Kokkola, Finland

Two User-Friendly Digital Tools for Multidimensional Risk Assessment Among Workers with Display Screen Equipment

Delaruelle Dirk, Pollentier Gerrit, Acke Sofie[(⊠)], De Leeuw Tine,
Goddet Cindy, Knops Kristel, Eerdekens Karine,
and Schmickler Marie-Noëlle

Mensura Occupational Health Services, Brussels, Belgium
sofie.acke@mensura.be

Abstract. Introduction: Workers using Visual Display Units (VDUs) face multiple risks, including upper limb discomfort and pain, mental stress, and visual fatigue. Since 2016, the Belgian Welfare Law has obliged companies to conduct risk analyses at least every five years to assess employees according to three dimensions: physical strain, mental stress, and visual load. To perform these assessments, researchers from Mensura Occupational Health Services developed two tools: (1) an online questionnaire (E-survey) and (2) a self-assessment tool (E-coach).

Methods: Both tools were developed by a multidisciplinary team of experts, including one European-registered ergonomist; one occupational physician; one occupational nurse; two prevention advisers, one of which specialised in ergonomics; and one epidemiologist.

Validation: After Dutch-to-French and Dutch-to-English translation and back-translation were performed, the questionnaire was validated in a two-stage Delphi process by a panel of 20 experts. Subsequently, a pilot test was conducted using all Mensura computer office workers. Written feedback was collected and used to develop the final version of the tool.

Results: From October 2016 to August 2017, 17 companies implemented the questionnaire and 203 companies implemented the self-assessment tool. This resulted in 2,088 completed questionnaires and 875 completed self-assessments.

Conclusion: Both tools were easy to use and effectively assessed risks across multidimensional levels, enabling organisations to map priorities and target actions to increase employees' well-being. Furthermore, they provided employers with an extensive database for self-monitoring and continuous benchmarking.

Keywords: Visual display units · Risk analysis · Digital tool

1 Introduction

1.1 Scope

Workers using Visual Display Units (VDUs) face multiple risks, including upper limb discomfort and pain, mental stress, and visual fatigue. Besides, total time spent per day

© Springer Nature Switzerland AG 2019
S. Bagnara et al. (Eds.): IEA 2018, AISC 822, pp. 457–460, 2019.
https://doi.org/10.1007/978-3-319-96077-7_48

in front of display screen equipment is increasing, and physical inactivity is an important cofactor in the pathway to possible musculoskeletal complaints and related sickness absence.

1.2 Aim

Since 2016, the Belgian Welfare Law has obliged companies to conduct risk analyses at least every five years to assess employees according to three dimensions: physical strain, mental stress, and visual load. To perform these assessments, researchers from Mensura Occupational Health Services developed two tools:

(1) an online questionnaire (E-survey);
(2) a self-assessment tool (E-coach).

2 Methods

2.1 Project Organisation

Both tools were developed by a multidisciplinary team of experts, including one European-registered ergonomist; one occupational physician; one occupational nurse; two prevention advisers, one of which specialised in ergonomics; and one epidemiologist.

2.2 Two Online Tools

E-survey: The close-ended questionnaire was partially based on two validated risk assessments: a standardised Nordic questionnaire for the analysis of musculoskeletal symptoms [1] and a Dutch checklist for work involving VDUs [2]. It consisted of the following five parts (10 scales, 105 items):

(a) Lifestyle
(b) Psychosocial well-being
(c) Occupational hygiene (indoor climate, noise, and illumination)
(d) Ergonomics (work tasks, working hours, work pressure, and workplace)
(e) Possible work-related complaints (neck, shoulders, upper and lower back, elbows, wrists/hands, hips/thighs, knees, and/or ankles/feet)

At the end of the survey, employees could download their corresponding infographics.

E-coach: The self-assessment tool (14 items) was developed in accordance with the concept of cognitive ergonomics (e.g., tested for usability; built according to user-centred design methods). By asking users to visualize their workspace when answering questions, the survey examined the ergonomics of individual chair and desk settings as well as employees' working methods. To aid employees in optimising their workspaces quickly and easily, advice was given after each question and customised infographics were provided, e.g. "areas for improvement", "give your eyes a break", "adjusting the chair and dynamic sitting", and "how to be more assertive in the workplace". At the end, a report with individual advice was provided.

3 Validation

After Dutch-to-French and Dutch-to-English translation and back-translation were performed, the questionnaire was validated in a two-stage Delphi process by a panel of 20 experts. This panel consisted of Dutch- and French-speaking prevention advisers, occupational physicians and nurses, ergonomists, marketing assistants, a consultant on strategic product management and a director of administration.

Written feedback was requested by the European ergonomist about the clarity and unambiguity of the instructions and the items, clarity of the objective, accessibility of the introduction and the instructions, overlap, structure, completeness and answer possibilities. After the first Delphi round, the checklist was adjusted based on the feedback from the experts. In the second Delphi round, the adjusted checklist was again submitted to the panel of experts and written feedback was again collected by the European Ergonomist.

Subsequently, a pilot test was conducted using all Mensura computer office workers, resulting in 425 completed questionnaires (response rate 74%). They were asked to fill in the questionnaire and to evaluate it on the above criteria. Based on their written feedback, the final version of the tool was developed.

4 Results

From October 2016 to August 2017, 17 companies implemented the questionnaire and 203 companies implemented the self-assessment tool.

This resulted in 2,088 completed questionnaires and 875 completed self-assessments. The mean age of the questionnaire participants was 41.4 years (Standard Deviation = 10.4); 48.5% were men and 51.5% were women. When asked whether they had experienced any pain or discomfort over the previous 12 months, 52.9% of respondents reported neck problems (95% CI 50.8–55.1) and 47.1% reported lower back problems (95% CI 44.9–49.2). A number of employees (13%) even reported being absent from work during the previous 12 months due to lower back problems (95% CI 11.7–14.5).

Most respondents (54%) spent a total time of at least 10 h per day in front of display screen equipment (95% CI 51.8–56.1), both during work and leisure. For this question, no difference was revealed between age groups younger than 45 years versus age groups 45 years and older. Up to 60.3% worked continuously in the same position during at least one hour (95% CI 37.9–42.1), while only 64.8% declared to have healthy eating habits (95% CI 62.9–67). Employees who didn't work at least one hour in the same position reported significantly less lower back pain (adjusted odds ratio for age and gender = 0.57; 95% CI 0.47–0.68; p ≤ 0.000).

5 Discussion

The prevalence of neck pain (52.9%) was within its reference value for the European population (33–54.5%), while the prevalence of lower back pain (47.1%) was a little higher (compared to its European reference value of 46–47%). The total time spent in front of visual display units during work and leisure was, indeed, very high and equal for younger versus older age categories. Employees who didn't work for at least one hour in the same position were significantly less likely to report lower back pain.

Both the E-survey and the E-coach tools were easy to use and effectively assessed risks across multidimensional levels, enabling organisations to map priorities and target actions to increase employees' vitality and well-being. Furthermore, they provided employers with an extensive database for self-monitoring and continuous benchmarking.

6 Links to the Videos

E-survey: https://www.mensura.be/fr/e-enquete-travail-sur-ecran
E-coach: https://www.mensura.be/fr/e-coach-travail-sur-ecran

References

1. Kuorinka I, Jonsson B, Kilbom A, Vinterberg H, Biering-Sørensen F, Andersson G, Jørgensen K (1987) Standardised Nordic questionnaires for the analysis of musculoskeletal symptoms. Appl Ergon 18(3):233–237
2. Peerenboom KJ, Huysmans MA (2002) Handboek RSI. Risico's, oplossingen, behandelingen, 3rd edn. Sdu Uitgevers, Den Haag

Acceptance for Technological Solutions Targeted at the Mastery of Terrorist Attacks: Experience and Innovation

Sonja Th. Kwee-Meier[1]([⊠]) [iD], Jochen Nelles[1] [iD], Axel Knödler[2],
Joachim Schulz[3], and Alexander Mertens[1] [iD]

[1] Institute of Industrial Engineering and Ergonomics,
RWTH Aachen University, Bergdriesch 27, 52062 Aachen, Germany
s.meier@iaw.rwth-aachen.de
[2] Forensics Department, State Office of Criminal Investigation
Baden-Wuerttemberg, LKA BW, Taubenheimstr. 85, 70372 Stuttgart, Germany
[3] Federal Criminal Investigation Department, BKA,
KT13 – SG Render Safe Procedure, 65173 Wiesbaden, Germany

Abstract. The authorities and emergency forces are challenged with increasing terrorism threats. The present paper presents the results from a workshop and interviews with experts for the neutralization of improvised explosive devices (IEDs) regarding the introduction of a robotic sensor system, which is developed within the project. On the one hand, we found great openness and enthusiasm for innovations and new technologies. On the other hand, experience plays a key role in the job of IED neutralization experts. This implies the need for training phases in case of the introduction of new systems in this domain.

Keywords: Terrorism · Technology · Innovation · Acceptance
Experience

1 Introduction

Terrorist attacks have increased during the past years, also in Europe. Thus, the work of authorities and organisations with security tasks such as the police has become much more demanding. The support of their tasks by technological innovations is highly desirable. In our project, we aim at developing a robotic sensor system to detect improvised explosive devices (IEDs) in suspicious objects such as abandoned luggage. Although the new system offers many advantages from an objective point of view, the subjective point of view of the different stakeholders is the determining criterion for the success of introducing the system in a later developmental stage. The foreseeable acceptance issues were evaluated together with application partners in an early stage of our project in order to avoid failure. The design freedom in the early stages of a project offer greater flexibility to exert influence on design issues to improve technology acceptance according to the human-centered design for interactive systems postulated in ISO 9241-210 [1].

© Springer Nature Switzerland AG 2019
S. Bagnara et al. (Eds.): IEA 2018, AISC 822, pp. 461–468, 2019.
https://doi.org/10.1007/978-3-319-96077-7_49

The paper comprises two data collections. Firstly, results of a workshop with two application experts with management functions are described. Secondly, results from an interview study with eleven experts from the emergency forces are presented.

2   Workshop

2.1   Workshop Methodology

For the workshop, insights into praxis were of very high importance. Therefore, two IED neutralization experts with management functions were acquired for the workshop with other project partners from sensory development, human–machine interaction and legal sciences, further described in Kwee-Meier et al. [2]. We were interested in identifying the various internal and potential external stakeholders and their immanent supporting and impeding factors of influence for technology acceptance. Therefore, we applied an adapted form of Lewin's [3] force field analysis (see Fig. 1). The key idea of this model are the competing factors. The goal was to reveal as many supporting factors of the future robotic sensor system as possible to outbalance unavoidable inhibiting factors.

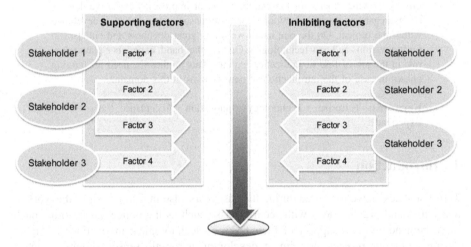

Fig. 1. Force-field analysis based on Lewin [3] and adapted for identification of stakeholders and their immanent supporting and inhibiting factors regarding technology acceptance of new innovations.

The image of the competing factors motivated an intensive discourse with everyone participating in the brainstorming process. The process was supposed to be a two-step process, first, identifying the stakeholders, and, second, the supporting and inhibiting factors. As it turned out, iterations supported creativity and discourse. Thus, stakeholders and factors were iteratively identified and cards on moderation boards were restructured several times to create a comprehensive overview with rows for the stakeholders and two columns for supporting and inhibiting factors. The restructuring led to better refinement of the factors together with all workshop participants.

2.2   Workshop Results and Discussion

Various stakeholders as well as supporting and inhibiting factors were revealed. For further analysis, stakeholders are differentiated in stakeholders within the emergency forces, i.e. internal stakeholders, and others, i.e. external stakeholders (see Fig. 2).

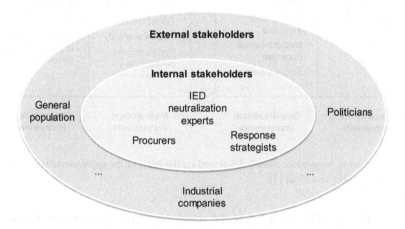

Fig. 2. Model of the identified most important internal and external stakeholders for the context of neutralization of improvised explosive devices (IEDs).

Workshop Results Regarding Internal Stakeholders. The most prominent group of internal stakeholders are the *experts for IED neutralization*. Another internal stakeholder is, for instance, the general population, discussed below.

IED Neutralization Experts In case of an emergency, time is a critical resource to them. Thus, all possibilities to enhance efficiency during the emergency response actions are of high relevance for these experts. Of course, the *perceived usefulness* is a prerequisite, as already established in the Technology Acceptance Model (TAM, [4]). The *perceived usefulness* was further found to be influenced by the perceived effectiveness during the IED detection process and the perceived safety, which is partly a result of the before described dimensions. All of these dimensions refer to the *intrinsic motivation* as aspect of *innovation readiness*, which refers to *implementation readiness* framed by Bernecker and Reiß [5] (see Fig. 3). Their work was conducted within the field of change management. However, later on it was transferred to concepts such as *implementation acceptance* by Schneiders et al. [6]. Change management and innovation or acceptance of the implementation of new technology are highly linked to one another. For instance, the phases of the AKDAR model by Hiatt [7], reaching from awareness and desire over knowledge and ability to reinforcement, reveal great similarity with factors for innovation acceptance (see Fig. 3).

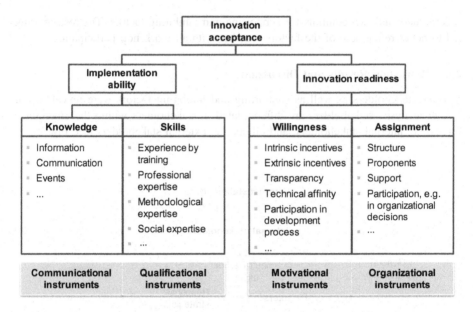

Fig. 3. *Innovation acceptance model*, developed on the basis of the *implementation acceptance model* by Bernecker and Reiß [5].

In contrast, the time effort before the actual emergency response, for instance for qualification measures to acquire knowledge and skills, relates to the *implementation ability* [5]. However, in the context at hand, these aspects are complemented by the factor of experience, which was revealed to be of very high importance in this domain and, therefore, integrated into the implementation acceptance model in Fig. 3. The experience of IED neutralization experts will be discussed in more depths also in the interview Sect. 3.

Regarding the aforementioned TAM [4], it should be noted that *perceived usefulness* is only one of two factors of influence that are widely acknowledged to determine the main share of technology acceptance. The other factor is *perceived ease of use*. The workshop with the IED neutralization experts showed the relevance also of this aspect. However, *perceived ease of use* is much more difficult to determine, before the system is developed, than *perceived usefulness*. Nevertheless, our projects aims at developing a demonstrator with high perceived ease of use, which is ensured by applying a highly participatory design process according to ISO 9241-210 [1].

Furthermore, a factor was identified for implementation acceptance, which cannot be influence by the design of the system nor of the interface. This factor is the technical affinity, which was defined by Karrer et al. [8] as personality trait as attitudes towards, enthusiasm for and trust in technology. This factor is regarded as factor for *willingness* to implement a new system and leave status quo, thus, as factor of *innovation readiness*. The resistance against the implementation of new systems "due to the bias or preference to stay with the current situation" [9] was described in detail in Kim and

Kankanhalli [9] with regard to information systems. Mahmud et al. [10] provided one of the rare papers on resistance, or *end user grumbling,* in the pre-implementation phase. This scarcity of research during the pre-implementation phase complicates model development unlike in traditional technology acceptance models (e.g. Venkatesh and Davis [11], Venkatesh and Bala [12]).

In the groundwork by Bernecker and Reiß [5] for Fig. 3, participation was considered as factor for assignment. We acknowledge that *participation,* for instance in organizational decisions, is a factor of assignment. However, based on the human-centered design process in ISO 9241-210 [1], *participation in the development process* can be assumed to likely contribute to the willingness to introduce a technical or even organizational innovation. Participation was therefore described and reassigned to both dimensions.

Other internal stakeholders. Two main groups of other internal stakeholders were identified within the workshop, i.e. *procurers* and *response strategists.* The key factor regarding procurers was seen in the costs related to innovation procurement. Response strategists, who make decisions about technology usage during emergencies responses, are interested in similar factors like operative IED neutralization experts, such as time and efficiency issues. However, they were perceived to be more engaged in external communication, rendering also the factor *image,* framed by Venkatesh and Davis [11], interesting for them.

As already mentioned above, experience was viewed as highly relevant for the work of IED neutralization experts. Therefore, it was assumed that the meaning of experienced IED neutralization experts about innovations would be of high relevance for procurers and response strategists, allowing them to act as proponents due to their credibility within these organizations. This further emphasizes the importance for participatory design processes in this domain.

Workshop Results Regarding External Stakeholders. The general population and industrial companies were found to be the most relevant stakeholder during the workshop.

General Population. The general population was considered the most important external stakeholder. The key aspect for innovation acceptance was seen in the safety factor. This incorporates also the reduction of material damage. It should be noted that the perception of innovations by the general population is widely mediated by media reports and the way it is reporting about innovations. Hence, the publication of research results from safety-critical domains for content and societal discourses should be considerate and society sensitive.

Industrial Companies. The obvious factor for innovation for industrial companies is financial interest. However, cooperations between the represented authorities and industrial companies have shown mutually interesting knowledge transfer in the past.

3   Interviews

The interviews with eleven IED neutralization experts were conducted on the basis of the afore described workshop and another workshop to investigate the factors for innovation acceptance, system and interface design. For the present paper, the interviews were analyzed with regard only to attitudes towards innovation and the relevance experience in the job of IED neutralization experts. This approach evolved from the workshop results for internal stakeholders (see Sec. 2.2), as the relevance of experience is seen as a critical factor to be considered for the introduction of new systems, for instance by prolonged hands-on training phases instead or in addition to instruction trainings.

3.1   Interview Methodology

Interviews were conducted with eleven IED neutralization experts in Germany. Ten out of the eleven experts allowed audio recording. Detailed transcripts of these ten interviews are the data basis for the following analysis. The interviews started with information of the interviewees and signing an informed consent form. They took about one to one and a half hours and covered socio-demographic, professional and content aspects regarding innovation acceptance, system and interface design. For further analysis, the experts were grouped by age in three categories: young ($N = 3, M = 37.33$ years), middle ($N = 4, M = 50.25$ years) and old ($N = 3, M = 57.00$ years). Term counts were collected for the terms new (relating to system, technology or similar), experience, training/train, knowledge, future, (not) used to, skills, creativity/creative, innovation and improvisation (German: neu*, Erfahrung, Übung/üben, Wissen, Zukunft, (un)gewohnt, fähig*, kreativ*, innovati*, improvis*). The counts within groups were averaged and summed up for different groups. Admittedly, the analysis is based on a heuristic approach based on perceived important terms during the workshop, the interviews and while reviewing the transcripts. Each term occurrence was manually checked to ensure that the term was not part of a different word or idiomatic term without reference to the actual topic.

3.2   Interview Results and Discussion

Figure 4 shows the final term counts of the ten interview transcripts. The high interest of IED neutralization experts in *new* systems and technologies is striking, especially in the young and old group. In line with this, *future* was an important issue for quite some experts. However, the words *innovation* or *innovative* rarely occured, which is probably due to their rare usage in spoken language in German.

The words *experience* and *training* or *train* had very high occurrence frequencies. Their usage was similarly pronounced in experts of young, middle and old age. *Knowledge* and *skills* seem far less important to the experts than long-term experience. Many experts also referred to the fact that they would *not be used* to a new system or would have to get *used to* the system first.

Although talent for improvisation was told to be an important factor for IED neutralization, *creativity* and *creative* as well as *improvisation* and *improvise* had very

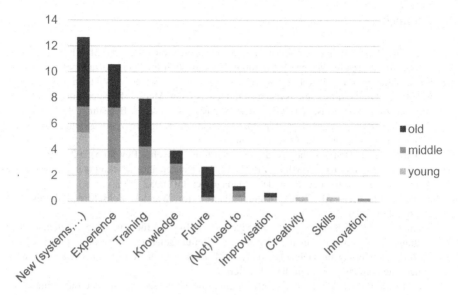

Fig. 4. Term counts of the interview transcript. The term counts were averaged within age groups and summed up for different age groups.

low word counts. This might have been caused by the rather technology-than process-oriented questions.

4 General Discussion and Conclusion

An intensive workshop with two IED neutralization experts with management functions and ten interviews with audio recording with IED neutralization experts were conducted. The insights from the workshop were generally confirmed by the interviews. Experience and hands-on training is of very high relevance for these experts. This implies the need for training phases if new systems are introduced. Experience further renders them proponents for new systems within their organizations if they are convinced. Otherwise, a status quo bias [9] in these experts might hinder innovation. However, the interest in new systems and technologies also in old experts shown by the interview results, was foreseeably high based on the workshop. It should be noted that this only refers to attitudes towards and enthusiasm for new technologies but not to abilities to deal with these as this aspect was not part of the interviews and prototype explorations are not possible at this stage in the pre-implementation phase. Explorations, structured usability tests and evaluation studies with emphasis on the visual interface for sensor data analysis are planned for the next project phases.

Acknowledgements. The workshop, the interviews and the their analysis were performed within the project DURCHBLICK (FKZ: 13N14329), funded by the German Federal Ministry of Education and Research (BMBF) in the context of the national program "Research for Civil Security" and the call "Civil Security – Aspects and Measures of Coping with Terrorism".

References

1. ISO 9241-210 (2010) Ergonomics of human-system interaction-part 210: human-centred design for interactive systems. International Organization for Standardization
2. Kwee-Meier S, Nelles J, Knödler A, Schulz J, Victoria H, Stefan M, Viktoria R, Mertens A (2018) Integrierte Informationsvisualisierungen bei der Mensch-Roboter-Kooperation zur Unterstützung von Einsatzkräften unter erhöhter psychischer Belastung. In: Arbeit(S). Wissen.Schaf(f)t – Grundlage für Management & Kompetenzentwicklung. 64. GfA-Frühjahrskongress, pp. 1–5. GfA, Dortmund
3. Lewin K (1943) Defining the "field at a given time". Psychol Rev 50(3):292–310
4. Davis FD, Bagozzi RP, Warshaw PR (1989) User acceptance of computer technology: a comparison of two theoretical models. Manage Sci 35(8):982–1003
5. Bernecker T, Reiß M (2002) Kommunikation im Wandel. Zeitschrift Führung und Organisation 71(6):352–359
6. Schneiders M-T, Schilberg D, Jeschke S (2014) Einführung eines Telematischen Rettungsassistenzsystems in die Organisation Rettungsdienst – zur Rolle der Akzeptanz im Implementierungsmanagement. In: Jenk M, Ellebrecht N, Kaufmann S (eds) Organisationen und Experten des Notfalls. LIT, Berlin
7. Hiatt J (2006) ADKAR: a model for change in business, government, and our community. Prosci, Loveland
8. Karrer K, Glaser C, Clemens C, Bruder C (2009) Technikaffinität erfassen – der Fragebogen TA-EG. In: Lichtenstein A, Stößel C, Clemens C (Hrsg.) Der Mensch im Mittelpunkt technischer Systeme, 8. Berliner Werkstatt Mensch-Maschine-Systeme. VDI Verlag GmbH, Düsseldorf, pp 196–201
9. Kim HW, Kankanhalli A (2009) Investigating user resistance to information systems implementation: a status quo bias perspective. MIS Q 33(3):567–582
10. Mahmud I, Ramayah T, Kurnia S (2017) To use or not to use: modelling end user grumbling as user resistance in pre-implementation stage of enterprise resource planning system. Inform Syst 69:164–179
11. Venkatesh V, Davis FD (2000) A theoretical extension of the technology acceptance model: four longitudinal field studies. Manage Sci 46(2):186–204
12. Venkatesh V, Bala H (2008) Technology acceptance model 3 and a research agenda on interventions. Decis Sci 39(2):273–315

Believability of News

Understanding Users Perceptions of Fake News and the Effectiveness of Fact-Checking Badges

André Calero Valdez$^{(\boxtimes)}$ (ID) and Martina Ziefle (ID)

Human-Computer Interaction Center, RWTH Aachen University,
Campus Boulevard 57, 52074 Aachen, Germany
calero-valdez@comm.rwth-aachen.de

Abstract. When social media becomes a dominant channel for the distribution of news, manipulation of the news agenda and news content can be achieved by anyone who is hosting a website with access to social media APIs. Falsehoods masked as legitimate news with the intent to manipulate the public are called Fake News. This type of propaganda is disseminated by sharing of individual social media users. Fake news pose a threat to democracies as they influence the public agenda and contribute to polarization of opinions. To limit the dissemination of fake news, social media websites utilize fact-checking badges to flag possibly fabricated content. It has however not been investigated whether these badges are effective and who responds to them. In a survey study with 120 participants we found little evidence for the effectiveness of such badges. However, believability of news in a social media sites were generally rated rather low.

Keywords: Fake news · User study · Fact checking · Personality
Social media · Opinion formation

1 Introduction

In June 2016, the population of Great Britain decided on whether to remain a part of the European Union or not. About 51.9% no longer felt to be a part of EU and voted with leave. Shortly before, all opinion polls suggested the opposite outcome. A similar mistake was repeated on November 8th, 2016 during the presidential election of the United States of America. Donald Trump was elected president unexpected to polls and media alike. Opinion polls were reported on heavily in the last few hours before the vote ended, possibly causing more voters to go to the booth. One explanation is looked for in the different uses of social media by Trump and Clinton during the campaign. Trump's campaign relied heavily on social media and more postings associated with Trump's hashtags were found online than for Clinton [1].

With the increasing importance of social media, two phenomena have become increasingly important as well. The use of social bots to promote posts beyond humanly possible activity levels is one way of shifting the perception of majority and minority in social media. Another way of affecting the public opinion is the utilization

© Springer Nature Switzerland AG 2019
S. Bagnara et al. (Eds.): IEA 2018, AISC 822, pp. 469–477, 2019.
https://doi.org/10.1007/978-3-319-96077-7_50

of Fake news [2]. Fake news are news posts that seem like authentic news posts, but have been fabricated to affect the public opinion. Such posts often address controversial topics and intentionally report falsehoods to manipulate opinions. The bigger problem with these posts is that not only social bots share them, but also users. Users seem to share fake news to a large extent—up to 20% more than regular news posts [3]. But, who are those users that repost fake news? Do users verify content before sharing? Do services such as "fact-checker" from snopes or the associated press help users in determining what is true and what is fabricated?

In this paper, we investigate how fake and authentic news regarding refugees in Germany affect opinion forming by measuring the impact of different fabricated news articles on different topics in comparison to authentic articles. In an experimental study with 142 participants we measure the believability of fake news and how the knowledge of the news being fake influences the believability afterwards.

2 Related Work

In the time of digitization, media, and mass communication have such short-circuit feedback loops, where people respond to content of news quickly, thus affecting the content being reported on. When Social Media is used, these feedback loops become ever shorter and feedback becomes almost immediate. The public's reaction to news itself becomes important on its own. In the case of fake news, this can cause problems. We define fake news in accordance with Allcot et al. [1], namely as deliberate falsehoods being reported seemingly objectively to manipulate the publics opinion with the purpose of disinformation by misinterpreting, manipulating or fabricating content. But who reads these "obvious" or not so obvious fabrication? And who interacts with it? Does it affect opinion?

How opinions are formed has been studied since the 1960s extensively in subjects such as political opinion formation and the formation of opinions on novel products. The theory of opinion leadership by Rogers and Cartano [4] postulates that a certain subgroup of the population characterized by high levels of domain knowledge and social status determine opinion formation in the population. By addressing opinion leaders first, a shift in the public opinion can be achieved most easily. Resulting from this research Childers [5] investigated the psychometric properties of opinion leaders to allow the identification of opinion leaders more easily.

The theories of opinion formation have become increasingly important with the rise of mass media and the public news broadcast and their perception in the population [6]. The media has the potential to shift opinion not only by setting the agenda, but also by creating a spiral of silence by not reporting minority opinions. In the digital age, digital media has similar influences. Yet, content creators are no longer people with broadcasting licenses, but everyone.

Dimitrova [7] found that an increase in usage of digital media is not necessarily linked to an increase in political knowledge, while the contrary was true for classical media. Social media comes with the challenges and opportunities that a decentralized, global system brings along [8]. Small minorities find likeminded others even when they are scattered across the globe. While from the perspective of, e.g., a self-help group this is beneficial, it is problematic when enemies of democracy organize themselves on social media. One of the dangers stems from the so-called filter bubble proposed by Pariser [9]. The filter bubble is the phenomenon that algorithms affect the information that is being presented to users in social media in such a way that new content matches the preferences of the users. Right-wing radicals are exposed to right-wing fake news by liking right-wing content. Bozdag and Hoven found [10] that the filter bubble poses several threats to democracy and that users should aim at breaking their filter bubble.

How can these challenges be addressed? Kosinski et al. [11] found that algorithms are able to determine private traits such as personality from user behavior online and might consequently adapt their own behavior to provide content that matches the receptive patterns of users. If social media providers understood how these patterns can trigger, e.g., fact checking behavior, social media providers could steer users to refrain from reposting fake news. The open questions are: What determines the believability of news posts and who are the users that repost such content. In this paper we investigate the effect of fact-checking badges on believability and how their effectiveness is affected by user factors.

3   Method

In order to understand who is affected by fact-checking badges such as the ones provided by snopes or the associated press, we conducted an online survey that was distributed using surveymonkey.com. We recruited 142 participants using convenience sampling and asked them to rate the believability of 13 news items embedded in a social media context (see Fig. 1). Participants were asked to rate the believability using a slider on a 1–100 scale. If the news items was considered fake news by snopes we would then show a banner (see Fig. 2) indicating this. We would then ask the participant to reconsider their judgment. In order to not make the participants assume, all news items are fake, half of the items were real news items.

The set of news messages were all selected from current political topics, addressing the refugee "crisis" in Germany during the time of conducting the survey in early 2017. The different news messages that were used contained the following headlines (F marks faked articles):

1. Lesbos: Refugees set refugee camp on fire (F)
2. The amount of refugees applying for asylum has been stagnating for 2 years despite the refugee crises. (F)
3. Jamaica-Coalition agrees on an upper limit of 1 Mio. refugees. (F)
4. Merkel: 100,000 refugees will leave Germany voluntarily. (F)
5. Integration commissioners demand integration tax (F)
6. 700 Euro Christmas bonus for refugees (F)
7. France wants to take in 10,000 refugees in the next 2 years under the UN programme.
8. Refugees increase economic growth (F)
9. The number of asylum proceedings in the administrative courts has increased fivefold in recent years.

Fig. 1. Example news post: The headline reads "Lesbos: Refugees set refugee camp on fire" (translated from German).

Fig. 2. Banner indicating the previous entry was considered to be fake.

10. The Aachen police donated lost bicycles to refugees.
11. Security risk: 20 disguised jihadists came to Germany as refugees
12. Refugees visit their home countries despite their asylum status.
13. Refugees thank Aacheners, give roses, say thank you.

Special effort was taken to not pick fake and real news from a certain political angle. The political "tone" of the news does not help in estimating, whether the news are correct or fake. The order was randomized for all participants.

To investigate who believes what type of news and who changes their opinion about the believability of a given news post, we assessed demographic variables such as age, gender, income and education and we asked participants how often they used different types of media (social media, online news, online yellow press, message boards, radio, political blogs, local newspaper, newspaper, TV, discussion groups). We further asked them to complete a short personality test called BFI-10, measured their political interest using 4 items, and measured two of the BIDR social desirability scales (self-deceptive enhancement and impression management) as well as political judgements regarding refugees (nationalistic judgments, consequential judgments). As a last step we asked users, whether they had fact-checked some of the news while answering the survey.

4 Results

Results were generated using R, the tidyverse, and the psych package. The data is analyzed using descriptive statistics with 95% confidence intervals and correlation analysis. Factor structures are tested using minimum residuals using ordinal least squares. Factor counts are decided using the Bayesian information criterion (BIC). Reliability of resulting scales was measured using Cronbach's alpha. We report t-tests and correlation tests as null-hypothesis significance tests and set the level of significance to $\alpha = .05$.

4.1 Description of the Sample

From the 142 participants, 120 participants completed the survey with sensible results. Twenty-two were removed because their believability ratings were always 1. Of the remaining 120, 60 were male, and 60 were female. The average age was $M = 29.6$ years ($SD = 11.4$). Thirty-seven reported to have 700€ or less, 23 between 700€ and 1500€, 20 between 1500€ and 2500€, 19 between 2500€ and 3500€, 12 between 3500€ and 5000€ and 9 over 5000€ of monthly gross salary.

Men showed higher conscientiousness ($t(116.02) = -3.8$, $p < .01$, $M_{male} = 3.3$; $M_{female} = 2.65$), higher neuroticism ($t(116.86) = -4.05$, $p < .01$, $M_{male} = 4$; $M_{female} = 3.24$), and a lower political interest ($t(112.98) = 4.79$, $p < .001$, $M_{male} = 2.33$; $M_{female} = 3.3$) than women in our sample. Men further used social media to a larger extent than women ($t(117.76) = 2.3481$, $p < .05$, $M_{male} = 3.63$; $M_{female} = 2.86$).

We found that age negatively correlated with conscientiousness ($r = -.22$, $p < .05$), but also negatively with nationalistic judgments ($r = -.26$, $p < .01$), and positively with consequentialist judgements ($r = .27$, $p < .01$). Older users used classical media more frequently ($r = -.27$, $p < .01$) and had larger incomes ($r = .31$, $p < .001$).

Classical media consumption is associated with lower nationalistic judgments ($r = -.2$, $p < .01$), with lower levels of education ($r = -.29$, $p < .01$), with higher levels of income ($r = .2$, $p < .05$), and social media consumption ($r = .19$, $p < .05$). Social media consumption on the other hand is highly negatively correlated with political interest ($r = -.46$, $p < .001$). Political interest is negatively correlated with nationalistic judgments ($r = -.33$, $p < .001$).

From the social desirability scales impression management correlates negatively with conscientiousness ($r = -.44$, $p < .001$) and agreeableness ($r = -.2$, $p < .05$). Self-deceptive enhancement correlates with nationalistic judgments ($r = .23$, $p < .05$), consequentialist judgments ($r = .24$, $p < .05$), extraversion ($r = .35$, $p < .001$), conscientiousness ($r = .32$, $p < .001$), and negatively with neuroticism ($r = -.29$, $p < .01$).

4.2 Believability of News Posts

We first look at how the believability of our 13 news posts was seen by our participants. For this purpose we plot the 95% confidence intervals on the half scale range of the slider in the survey (see Fig. 3). Interestingly, most believability ratings are rather low—lower than a third of the scale. The believability of fake news was, generally

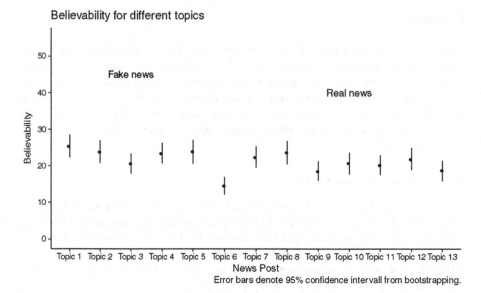

Believability for different topics

Fig. 3. Believability of our thirteen news posts and their confidence intervals. Topic 1 has the highest believability and topic 6 the lowest. The y-axis scale extends up to 100.

speaking, no higher or lower than authentic news posts. The most dramatic difference in believability can be seen between post 1 ("Lesbos: Refugees set refugee camp on fire") and post 6 ("700 Euro Christmas bonus for refugees").

When looking at how the fact-checking badge affects the believability we look at fake news only. Here we see that only a few news posts are affected by the badge at all (see Fig. 4). In particular post 1 sees a shift of believability that is downwards, as does post 8 ("Refugees increase economic growth").

Lastly, only eight participants reported to have checked the facts during filling out the survey. Thus, we can assume that a large proportion of the outcome is due to the spontaneous reaction of the participants filling out the questionnaire.

4.3   Effects of User Diversity on Believability

We found only little evidence of impact of user diversity factors on the believability ratings of our fake news topics. Of all our independent variables only nationalistic judgments, political interest, education, and self-deceptive enhancement affect the rating of the believability of our topics. Nationalistic judgments increases the believability of topics 1 and 5 ("refugees set camp on fire": $r = .23, p < .05$, and "integration tax": $r = .20, p < .05$). Political interest increases the believability of topic 2 ("stagnating refugee counts": $r = .25, p < .01$), and education of topic 1 ($r = .2, p < .05$). Self-deceptive enhancement increases the believability of topic 5 ($r = .25, p < .01$). Overall, there is only little evidence for a systematic influence of believability of fake news.

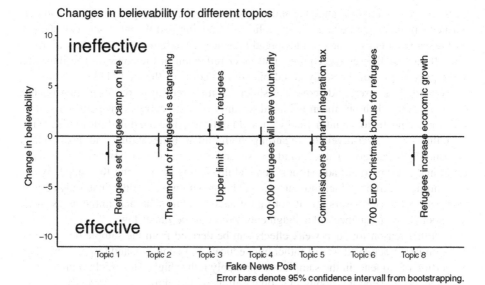

Fig. 4. Changes in believability depending on topic after a fact-checking badge was shown. Only two topics are positively affected by the badge (topic 1 and 8).

When looking at the *changes of believability* individually, we found that the strongest correlations was between self-deceptive enhancement and the changes regarding topic 5 ($r = -.27$, $p < .01$), indicating that people who tend deceive themselves about their own behavior change their believability of a news post in accordance with the badge (i.e., they decrease the believability)—yet, only if the topic is the introduction of an integration tax.

Similarly, we find that education affects believability of topic 1 ($r = -.18$, $p < .05$, higher education, stronger decrease in believability), classic media usage affects topic 4 ($r = -.21$, $p < .05$), conscientiousness affects topic 2 and 3 ($r = -.18$, $p < .05$ and $r = -.19$, $p < .05$), agreeableness affects topic 3 ($r = -.21$, $p < .05$), and nationalistic judgments affect topic 6 ($r = -.19$, $p < .05$). These findings, however, provide only little evidence for the influence of user-diversity on shifts in opinions with a fact-checking badge, when looking at the set of all possible interactions.

5 Discussion

We have seen that the participants in our study rated most of our news posts—both fake and authentic—relatively low in terms of believability. The changes in believability after adding a fact-checking badge were also very small and in some cases even increased believability on our scales. What does that mean?

First of all, we should look at the way our measurements were conducted. We used a 100-point slider, which defaulted to 1. This could make it very easy for users, who distrusted the news to leave the slider where it was. Believability in our experiment was a conscious decision to change the slider. In other words, we nudged participants to rate our news posts as not believable. In reality the nudging is different. Users can share a post using two clicks (share, ok) to spread fake news. Moving a slider all the way from 1 to 100 across the screen requires a stronger reflection of the content. The interface and representation of the study could bias the outcome of the study [13].

Nevertheless, users did move the sliders, some even up to 65 depending on the topic. So even when our mean is biased towards zero the correlations between scales, which are agnostic to the numerical values, do indicate some but only little evidence for an influence of user-diversity regarding both initial believability and post-badge-exposure believability. There could be several explanations. Cognitive dissonance could nudge users to not adjust their believability rating after seeing the badge. When committing to certain value, showing a small badge will not affect the "naturally certain judgment" of a user. Users might not find it easy to admit that their initial assessment was poor. In consequence, later judgments would all be rather low.

Another reason for such weak effects can be derived from the sample description. The sample we achieved was of rather high education, young age, and low income. The proportion of students in the sample was probably rather high. This might dramatically shift the preconceptions about the topic refugees and thus could have caused the influence of a confirmation bias. Also most participants did have very low nationalistic judgments, so even if the variance in this variable did influence some of the believability ratings, the variance was in the low end of the scale. To improve the data here, extending the sample to a larger and more diverse group of people would be necessary.

Methodologically speaking, we must consider that the procedure (i.e., within-subject, pre- and post-badge judgment) could not be ideal for such an experiment. The error rates are lower in within-subject designs, however, effects of order and across trials are also expected. An improved study would rather rely on fewer judgments, with a larger sample size. Still, a shift in believe change can then not be traced back to individual differences or user diversity.

The results do have an interesting implication. If setting a slider across the screen, causes users to reflect on the content they are about to "believe" or "share", such input metaphors could be used to cause reflection in a social media website. Similar to the "slide to unlock" metaphor a "slide to share" metaphor could very implicitly reduce sharing of fake news. To investigate, whether this effect is present, one could compare a slider-based version of this research with a single-click-based version and compare the tendency to rate believability between these versions.

The larger question, how does a decision-support system ("is this fake news?") help users in making better judgments and how responds to such systems still requires further research efforts that link the findings to user-factors and factors of interface design [14].

5.1 Conclusion

In this article we have shown that fact-checking badges have little effect on the believability of news posts in social media sites. Generally, news that are posted on social media are distrusted by a young, educated, and generally left-oriented group of users. The little effects that we found were related to agreeableness and conscientiousness.

Putting a fact-checking badge on social media could be seen as preaching to the choir for those who already rate fake news as non-believable, and be seen as paternalism by the left-wing media by those who don't. Testing such badges in real-life conditions should shed light onto this question.

Acknowledgements. We would like to thank the anonymous participants for sharing this intimate data with us and Isabell Busch, Claudia Gerke, Vanessa Götzl, Nicole Kuska and Vivian Lotz for designing the news items and helping with sampling. This research was supported by the Digital Society research program funded by the Ministry of Culture and Science of the German State of North Rhine-Westphalia. The authors thank the German Research Council DFG for the friendly support of the research in the excellence cluster "Integrative Production Technology in High Wage Countries".

References

1. Allcott Hunt, Gentzkow Matthew (2017) Social media and fake news in the 2016 election. J Econ Perspect 31(2):211–236
2. Lazer DM, Baum MA, Benkler Y, Berinsky AJ, Greenhill KM, Menczer F, Schudson M (2018) The science of fake news. Science 359(6380):1094–1096
3. Berghel H (2017) Lies, damn lies, and fake news. Computer 50(2):80–85
4. Rogers EM, Cartano DG (1962) Methods of measuring opinion leadership. Public Opin Q 435–441
5. Childers TL (1986) Assessment of the psychometric properties of an opinion leadership scale. J Mark Res 184–188
6. Zaller J (1990) Political awareness, elite opinion leadership, and the mass survey response. Soc Cogn 8(1):125–153
7. Dimitrova DV et al (2014) The effects of digital media on political knowledge and participation in election campaigns: evidence from panel data. Commun Res 41(1):95–118
8. Kaplan AM, Haenlein M (2010) Users of the world, unite! The challenges and opportunities of Social Media. Bus Horiz 53(1):59–68
9. Pariser E (2011) The filter bubble: what the Internet is hiding from you. Penguin, London
10. Bozdag E, van den Hoven J (2015) Breaking the filter bubble: democracy and design. Ethics Inf Technol 17(4):249–265
11. Kosinski M, Stillwell D, Graepel T (2013) Private traits and attributes are predictable from digital records of human behavior. PNAS 110(15):5802–5805
12. Hart CM, Ritchie TD, Hepper EG, Gebauer JE (2015) The balanced inventory of desirable responding short form (BIDR-16). Sage Open 5(4):2158244015621113
13. Calero Valdez A, Ziefle M, Sedlmair M (2018) Priming and anchoring effects in visualization. IEEE Trans Visual Comput Graph 24(1):584–594
14. Calero Valdez A, Brauner P, Ziefle M, Kuhlen TW, Sedlmair M (2016) Human factors in information visualization and decision support systems. In: Mensch and computer 2016–workshop band

The Evolution of Computed Tomography (CT) and Its User Interface: A Contextual and Comparative Analysis of Some of the Most Used Solutions

Oronzo Parlangeli[1]([⊠]), Alessandra Giani[2], Federico Baccetti[1],
Ilaria Bonanno[1], Ylenia Iervolino[1], Marino Todisco[1], Renzo Ricci[3],
and Stefano Guidi[4]

[1] Università degli Studi di Siena, 53100 Siena, SI, Italy
oronzo.parlangeli@unisi.it, {federico.baccetti,
ilaria.bonanno,ylenia.iervolino}@student.unisi.it,
todisco.marino@gmail.com
[2] Azienda Ospedaliera Universitaria Senese, Siena, Italy
a.giani@ao-siena.toscana.it
[3] Azienda USL Centro toscana, Florence, Italy
renzo.ricci@uslcentro.toscana.it
[4] Centro Gestione Rischio Clinico e Sicurezza del Paziente, Florence, Italy
stefano.g73@gmail.com

Abstract. The study presented in this paper was aimed at analyzing the user interfaces of Computed Tomography (CT) scanners and comparing three different systems used in two hospitals in Tuscany, Italy. Heuristic evaluations were conducted to measure the level of usability of these systems. The results show that the three systems have similar, but not equal, usability, that they rely heavily on operators' competences, and are not designed and implemented to facilitate the execution of complex tasks.

Keywords: Computed Tomography · CT · Healthcare technology
Graphical user interface · Heuristic evaluation · HMI · Design thinking

1 Introduction

Rapid and complex developments in digital tools have a profound impact in healthcare. Most of the improvements, however, concerns the hardware components of the Human Healthcare systems, more than the software components which define the User Interface (UI) of these systems. In the design of many technologies now in use in the medical domain, in facts, it seems that Human Machine Interaction aspects are little considered. It would seem that not much has changed since the early nineties, when Vicente and Rasmussen [24] noticed that while new technologies offer possibilities that just a little before were unthinkable, they also open whole new sets of problems. All in all, the main reason for this seem to be the fact that technologies that have a mass diffusion evolve in a different way and at a different pace than technologies used by a

© Springer Nature Switzerland AG 2019
S. Bagnara et al. (Eds.): IEA 2018, AISC 822, pp. 478–487, 2019.
https://doi.org/10.1007/978-3-319-96077-7_51

restricted user base. While the former ones tend to be improved continuously in all their aspects, in the latter ones improvements tend to pertain mainly to the technological aspects more than the interactive ones. In a review of the evolution and the state-of-the-art of Computed Tomography (CT) systems, Kohl [13] summarized the main evolutionary trends in terms of increase in slice size and scanning rate, and not a word was spent on interactivity and user-system interaction. The same could be said about several other studies on the evolution of medical technologies similar to CT [6, 10].

It is clear from these studies that CT technologies are part of socio-technical systems in continuous evolution; nonetheless, in these systems, the interaction between the human and the technological components have been so far partly, if not completely, disregarded.

The origin of Computed Tomography dates back to the studies of John Radon, which in 1917 devised efficient mathematical solutions for image reconstruction. Starting from these basis, new algorithms were then invented to improve image quality and efficiency. But it was only in the 70s, and precisely in 1972, that CT systems really emerged. In that period, in fact, the first patent for an X ray CT scanner was filed by Godfrey Cormack, later invested with Allan Cormack of the Nobel Prize in Medicine for this invention. Since then continuous and important innovations emerged one after the other allowing progressively to reduce the time needed for image acquisition – nowadays a CT scanner can gather images in less than 1-2 s – and to increase the image quality.

As Shital [20] noticed Computed Tomography (CT) has evolved: today non-invasive medical imaging of internal body organs are more accurate and easier to perform. However, this latter expression "easier to perform" it would seem to require still more attention. The ease of execution of a scanning task, in facts, is directly related to the interaction between the user and the system, and to the opportunities provided by the user interface. And, so far, there haven't been any studies devoted to describe or push forward the development of the user interfaces of modern CT scanner systems.

User Interface design, as a discipline, is located within the wider frameworks of User Experience design [9] and Design Thinking [18] which study the needs, abilities and requirements of the end users of a product. In the case of CT scanners these users are technicians that interact with the patient and manage the system to acquire images. Within this very specific usage domain, a UX/UI design perspective must take into account both the different use cases for the scanners (e.g. first aid/ER vs routing examinations) and the different requirements for each type of examination (e.g. speed vs precision). The wide range of potential patients itself opens up different use cases that require great familiarity and promptness on the operating technicians side. A technology designed by taking all these variables into account should offer a better support for this important and delicate activity, which is not always free of adverse events [4].

A possible way forward for improvement could thus be that of adopting a different approach to the problem, and embracing User-Centered design [16, 17] so that software development could be informed by established UI design standards [1, 19, 21].

To verify how current interactive aspects of CT scanners are adequate to operational user requirements in terms of efficiency and safety we have conducted an evaluation of the user interfaces of three CT systems by different manufacturers.

The assessment was conducted by experts and was based on established guidelines for the measurement of the usability of complex systems [3].

2 The Study

Biomedical engineering is developing diagnostic and therapeutic tools that are every day more efficient and less invasive. In the case of CT scanners, the main design goals are those of increasing the scanning rate, reducing the emission of radiations, and developing software solutions allowing the highest possible spatial resolution in the lowest possible time [12]. Great leaps forward have been made in recent years, so that current generation systems are emitting up to 50% less radiations than previous generation ones.

The cutting-edge technology in CT scanners is known as "Dual Source", which uses two radiation sources and two detectors simultaneously. This allows to overcome issues related to patient movements within the scanner, making possible "free-breathing" scan [22]. The scanning rate has increased up to 737 mm/s, and this allows to see up to 50 cm inside the human body, a threshold which in many cases, and particularly in pediatrics, make it possible to have a complete visibility of the regions of interest.

Notable improvements have also been achieved in "Single source" CT scanners, that is scanners that have a single detector, which are now capable of scanning 192 slices in 0.25 s, generating 3D diagnostic images with resolution below a millimetre.

CT systems manufacturers have then patented several solutions for specific needs, from technologies to minimize electronic noise (*True signal*), to focal point improvements, 0.5 mm slices (*Edge Technology*) and details acquisition form low kV data sets (*HiDynamics*).

Concerning the most delicate aspect of CT technology, that of radiations emission, research follows the ALARA principle (As-Low-As-Reasonable-Achievable). The so-called *filtered back-projection* (FBP) model, considered the best for fast image reconstruction but still too prone to noise issues and unreliable, has been surpassed and research is now refining "fourth generation" systems. Currently, the most widely adopted solution is ASIR (Adaptive Statistical Iterative Reconstruction) which implements iterative and retro-projective algorithms for image reconstruction, thus allowing radiation dose reduction. Images obtained through this technique are just a little less appreciable from a qualitative point of view, however the noise is notably reduced, so they preserve their diagnostic value [23]. ASIR is today used in many imaging centers and inserted in an increasing number of diagnostic protocols. As of 2014 ASIR was installed in more than 4,200 CT systems, being in this way the world most used IR method [7].

All these developments in CT technology have allowed big steps forward in Human Healthcare, but are still often challenging for hospitals. And it must be noticed, again, that the technical improvements that we have reviewed were mostly centered on the "engineering" side, while the human factor was basically disregarded, and implicitly considered as perfectly reliable once equipped with efficient tools.

Over this background, we have analyzed and compared three different CT scanners in use in Tuscany hospitals, with the aim of assessing whether these machines are aligned with the current market trends and, more importantly, with the operational user requirements.

2.1   Procedure

Four observers visited two different hospitals that have CT scanner for diagnostics. The assessment method that was used is akin to heuristic evaluation based on guidelines [11, 15, 21]. More specifically, each observer analyzed the scanner based on the guidelines with the support of a scanner operator. Data were collected in a heuristics assessment table for each observer, and then averaged across observers to derive an overall usability assessment for each system analyzed.

Three different tasks were considered for the assessment:

- The analysis of Gantry (i.e. the ring that holds the radioactive source within which the bed moves) and of the buttons to control patient positioning
- The scan set up
- The execution of the scan

For each of these phases we have used the following heuristic assessment criteria, as proposed by Bastien and Scapin [3]: (i) Guidance; (ii) Workload; (iii) Explicit Control; (iv) Adaptability; (v) User Control; (vi) Error management; (vii) Consistency; (viii) Code meaning; (ix) Compatibility.

Guidance comprises four different aspects: *prompting, items grouping, immediate feedback*, and *legibility*. *Prompting* concerns everything that indicates the status/state of the system, windows titles and data input. *Items grouping* refers to the hierarchical or logical organization of items, and the way these are separated by area and function. Lastly, *immediate feedback* and *legibility* concern the speed at which the systems provide a visible feedback in response to the user's actions, as well as the font brightness, the color content homogeneity and the dimension of the elements on the screen.

The second criterion is that of *workload*, and it comprises two aspects: brevity, which refers to the conciseness of the information, and *information density* which concerns the ratio between the amount of information displayed to the available screes size.

The third criterion, *explicit control*, is also divided into two sub-criteria: *explicit user action* (on which we have focused in the analysis, particularly the direct ENTER mode), and *user control*. In the latter case the assessment concerns the degree to which the user can control pages and processes, and the availability of clearly visible controls to "undo" previous actions. These criteria are used to understand how much effort is required to the user to have the total control of the system, and how responsive the system is to the user actions.

The next criterion concerns *error management*, and it comprises three sub-criteria: *error protection, quality of error messages* and *error correction*. The level of protection is assessed based on the ability of the system to predict and prevent errors, either by notifying to the user possible malfunctioning/anomalies, or by prompting the user to explicitly confirm their actions. In case or errors, the system should provide minimal yet direct error messages that offer directions for recovery.

Another criterion is that of *consistency*, and it refers to the presence of similar parameters, graphic design elements and procedures. A correct degree of consistency tends to make the systems functions easier to understand and to identify.

Lastly, *code meaning* concerns to the choice of symbols, acronyms/abbreviations, and expressions. The evaluator assesses *direct signification*, the *explicit presence of abbreviation rules*, and whether the code used are standard or specific to the system under investigation.

3   Results

The first part of the assessment concerned the Gantry of three different CT systems, hereafter referred to as system A, B and C. In Table 1 are reported the number of Usability problems (Ups) identified in each area for each system and their average severity.

Table 1. Number of usability problems (Ups) found in the gantry interfaces of systems A, B and C and their average severity level.

CT	A		B		C	
	N°	Severity	N°	Severity	N°	Severity
Guidance	3	3			4	3
Workload					1	3
Explicit control	1	2	1	1		
Adaptability	1	2	1	2		
Error management						
Consistency	1	2			1	3
Code meaning						
Compatibility						

As show in the table, for system A the analysis of the Gantry interface did not highlight any particularly severe problem. The most significant problem, perhaps, was related to guidance, and consisted in the lack of directions for hearing impaired patients about when to hold their breath. In addition, the pedals to control the manual positioning of the stretcher were also almost completely hidden, not very useful and the icons on the buttons difficult to understand. It was also found a low degree of control on the speed of the moving bed (it is not possible to adjust the speed of the bed, which is quite high, but only to stop it from moving, resulting in an abrupt stop). Lastly, it was found that there are three different labels for the same "stop" control, and thus a problem of consistency.

Even fewer Ups, and less severe ones, were found for the Gantry interface for system B. Concerning explicit control, the only problem identified was minor, and concerned pedal controls for manual position, which allow only to move the patient forward or backward but not to adjust the vertical position. To move the patient up/down the operator must use the buttons on the Gantry. Beside this, the most relevant problem found in system B concerned once again the lack of icons for hearing-impaired patients.

The analysis of system C revealed several Ups related to *guidance*, and particularly concerning items grouping. It was found, in facts, that the hierarchical and logical organization of the elements was not very clear, and that labels and data were not optimally distinguished. We also found issues in the legibility dimension, as some of the icons were not easy to understand and some indicators of the system status and of measurement taken were not clear. Moreover, the interface presented information elements that were of little value for the user, buttons that were never used, numbers without labels or measurement units and graphic elements that were too similar among them. All these Ups, however, should not be considered as serious, also given the high level of expertise of the operators.

Overall, the analysis of the Gantry interface shows that system B is the easiest to use, even for operators with less expertise. Despite not being completely free of Ups it was in fact the system with less issues among the three under comparison.

The results of the analysis of the interfaces for the scan set-up, in general, high-lighted a higher number of Ups than those found for the Gantry interfaces (Table 2).

Table 2. Number of usability problems (Ups) found in the scan set-up interfaces of systems A, B and C and their average severity level.

CT	A		B		C	
	N°	Severity	N°	Severity	N°	Severity
Guidance	2	2	6	2	7	3
Workload	1	2			1	3
Explicit control					1	3
Adaptability	1	1	2	2	3	2
Error management						
Consistency	1	2			1	2
Code meaning					1	2
Compatibility						

Concerning system A it was found, first of all, that the spatial layout and placement of the elements on the screen was sub-optimal. The interface was also rated low in legibility, due to the low luminance contrast between the text and the background, the information density was too high, and the interface presented never-used controls. A last further problem was then found concerning input forms, which only had English labels, without the possibility for the user to change the language.

In the interface of system B the main issues identified were the low prominence of dialog windows titles, a sub-optimal placement of the items, the presence of unused options in a prominent position. Legibility was also a problem, due to the color and size of the fonts, and like in system A all the labels were in English. An interesting finding was the presence of shortcuts, although these are completely unused as they are unclear and not presented in a sequence corresponding to the flow of the activity.

Several Ups of medium severity were finally found for the interface of system C. First of all, the items are not grouped hierarchically or logically. Labels are not clearly distinguishable from data points, and spatial placement of information is confusing and do not support the user in the task. There are many clickable buttons on the screen but the system does not display any feedback. The size of the font is small and the text/background contrast is too low; information that is logically linked is not enclosed by any frame or border. Like the other systems, the interface presented Ups of information overload. Moreover, the analysis highlighted control problems: to go back the user need to close the interface and start again the task, and there are no shortcuts available.

Overall, the interface of system A seemed more advanced in terms of items legibility and order. The interface of system C, instead, had many Ups that, although not particularly serious, make using the system less than optimal.

The analysis of the interface for executing the scan in system A revealed a substantial lack of logical and hierarchical organization in items grouping (see Table 3). Luminance contrast was again sub-optimal and icons were all too similar among them, opening the possibility of perceptual confusion errors. The interface of system B, beside the previously highlighted issues of legibility concerning colors, contrast, font and icons, presented also code meaning issues in the controls for shortcuts, since it used a particular code that is not immediately clear to novice users. The interface also presents consistency problems since, in some cases, the same objects or the same actions are represented in different dimensions and with different colors.

Table 3. Number of Usability problems (Ups) found in the scan execution interfaces of systems A, B and C and their average severity level.

CT	A		B		C	
	N°	Severity	N°	Severity	N°	Severity
Guidance	2	2	5	2	6	2-3
Workload					1	2
Explicit control						
Adaptability						
Error management						
Consistency	1	1	1	1	1	2
Code meaning						
Compatibility						

Lastly, system C interface presents, as the other interfaces, Ups with logical and hierarchical organization, but also the lack of visual feedback on mouse-over the buttons. Luminance contrasts, moreover, were sub-optimal and in some cases the text overlapped the images acquired during the scan. Finally, one of the most important problem was a very high information density in the interface.

Overall, despite the Ups identified in the analysis, none of the three systems is really characterized by a very low usability level. System C is the one that presents the biggest information organization Ups, which could slow down learning for trainees. But none of the systems is completely free from item grouping problems, although sometimes, such as in the Gantry interface of system B or in the scan set-up interface of system A, the analysis at least revealed design efforts aimed at a good organization of the information. The fact that users are still able to operate the systems quickly, despite the high workload, however, is most likely due to the high level of expertise acquired throughout the years.

The HMI of system C also presented the highest number of Ups, resulting not intuitive to use, was overcrowded with more or less useful information, had sub-optimal choices of colors, contrasts, icons design and layout.

Systems A and B have certainly less problems, and their HMIs were clearly designed taking the organization of information into considerations, and thus, despite being far from perfect, results more intuitive and less complicated than system C, and more accessible to less expert users.

4 Discussion

The results of the analyses conducted during this study suggest that today in manufacturing and design of complex systems such as CT scanners engineering requirements overcome the attention to use practices. Generally speaking, it would seem that engineers and system designers take for granted and assume that the users of their systems are sufficiently trained and able to use complex interfaces. Like in other domains, designers leverage on the high competence levels of the users to design systems that have low usability. While there might be good reason for this, both theoretical and practical [5], systems that rely so much on users' competences, and that do not consider the way an artifact itself can facilitate the execution of complex tasks, cannot be considered truly balanced systems [14]. We could summarize the current state of affairs in the evolutionary trends of CT scanners by saying that, on the one hand, the danger for the patients are being reduced more and more, as effective barriers are placed in the systems to prevent risks; on the other hand, the demands for the CT operators are increasing, as the amount of theoretical and practical training required to operate the system are increasing. Almost paradoxically, however, the interface between the user and the system (i.e. the mean by which users and systems are related) is being substantially neglected, possibly due to the high levels of safety and efficiency that both the human and the technological components can provide on their own. And this could negatively affect the efficiency and operativeness of the CT operators, which might in turn propagate to the whole hospital system [2].

The root of this phenomenon could be traced back to the fact that CT systems are not consumer products, conceived for a very large user base, but they instead target very specific customers and are used by highly skilled and trained users. The user interface, therefore, is not considered a discriminative factor in the decisions that hospitals or research centers make about the type of system to buy.

Nonetheless, a good usability, and the possibility offered by a system to effectively perform examinations, and to discriminate quickly among different possible complications, should be important factors in the choice of a diagnostic imaging system, if only because this choice might directly impact the level of safety for the patients and the operators [4].

Moreover, one should not overlook the impact that usability can have on operators training efficiency. As repeatedly stated by the CT operators during our field observations, learning to use all the functionalities of the CT scanners is long and difficult. It is clear that more usable systems could cut the time needed to train operators and transfer them the competences required to operate the systems effectively and efficiently [8].

Acknowledgements. The authors would like to thank all the staff of the two hospitals were the study took place, for the kindness shown and for welcoming and guiding us in this research.

References

1. Abujarad F, Breslin M, Guo G, Hess EP, Lopez K, Melnick E, Pavlo A, Powsner S, Post LA (2017) Patient-centered decision support: formative usability evaluation of integrated clinical decision support with a patient decision aid for minor head injury in the emergency department. J Med Internet Res 19(5):e174

2. Bagnara S, Parlangeli O, Tartaglia R (2010) Are hospitals becoming high reliability organizations? Appl Ergon 41(5):713–718

3. Bastien JMC, Scapin DL (1993) Ergonomic criteria for the evaluation of human computer interfaces, INRIA, Technical report

4. Bogner MS (2017) Error as a behavior and Murphy's law: implications for human factors and ergonomics. In: Carayon P (ed) Handbook of human factors and ergonomics in health care and patient safety

5. Card SK, Moran TP, Newell A (1983) The psychology of human-computer interaction. Lawrence Erlbaum Associates, Hillsdale

6. Carney JPJ, Hall NC, Townsend DW, Yap JT (2004) PET/CT today and tomorrow. J Nucl Med 45(1 suppl):4S–14S

7. Fan J, Melnyk R, Yue M (2014) Benefits of ASiR-V* reconstruction for reducing patient radiation dose and preserving diagnostic quality in CT exams. www3.gehealthcare.co.uk

8. Guidi S, Mengoni G, Parlangeli O (2011) The effect of system usability and multitasking activities in distance learning. In: Proceedings of the CHItaly conference 2011, 13–16 September, Alghero, ACM Library, pp 59–64

9. Hassenzahl M (2013) The encyclopedia of human–computer interaction, 2nd edn. Interaction Design Foundation, Aarhus, Denmark

10. Hell E, Knüpfer W, Mattern D (2000) The evolution of scintillating medical detectors, nuclear instruments and methods in physics research section A: accelerators, spectrometers, detectors and associated equipment. Elsevier, New York

11. Hollingsed T, Novick DG (2007) Usability inspection methods after 15 years of research and practice. In: Proceedings of the 25th ACM international conference on design of communication 2007, El Paso, TX

12. International Society for Computed Tomograpy's site, half a century in CT: how computed tomography has evolved. https://www.isct.org/computed-tomography-blog/2017/2/10/half-a-century-in-ct-how-computed-tomography-has-evolved. Accessed 28 Apr 2018
13. Kohl G (2005) The evolution and state-of-the-art principles of multislice computed tomography. In: Proceedings of the American Thoracic Society
14. Linblom J, Susi T, Ziemke T (2003) Beyond the bounds of cognition. In: Proceedings of the 25th annual conference of the cognitive science society 2003, Erlbaum, Mahwah
15. Mack RL, Nielsen J (1994) Usability inspection methods. Wiley, New York
16. Norman D (2017) Design of everyday things, 2nd edn. Time Warner, United States
17. Norman D (2010) Living with complexity, 1st edn. The MIT Press, Cambridge
18. Norman DA, Verganti R (2014) Incremental and radical innovation: design research vs. technology and meaning change. Des Issues 1(30):78–96
19. Olsen DR Jr (2007) Evaluating user interface systems research. In: UIST '07 proceedings of the 20th annual ACM symposium on user interface software and technology
20. Shital A (2009) Graphical user interface (GUI) to study different reconstruction algorithms in computed tomography, Wright State University
21. Shneiderman B, Plaisant C (2010) Designing the user interface: strategies for effective human–computer interaction. Addison Wesley, Boston
22. Siemens AG/Divisione Healthcare, Gherardelli A (2009) La tecnologia dual-source si fa largo nella tomografia, Milano. www.stampa.siemens.biz
23. Singh S et al (2010) Abdominal CT: comparison of adaptive statistical iterative and filtered back projection reconstruction techniques. Radiology 257(2):373–383
24. Vicente KJ, Rasmussen J (1990) The ecology of human-machine systems II: mediating "direct perception" in complex work domains. Ecol Psychol 2:207–250

Developments and Problems in the Man-Machine Relationship in Computed Tomography (CT)

Claudia Cassano[1]([⊠]), Antonella Colantuono[1], Guido De Simone[1],
Alessandra Giani[2], Paul M. Liston[3], Enrica Marchigiani[1],
Gislain Talla[1], and Oronzo Parlangeli[1]

[1] Dipartimento Di Scienze Sociali, Politiche E Cognitive,
Università Degli Studi Di Siena, Siena, Italy
{claudia.cassano, antonella.colantuono, guido.desimone,
gislain.talla}@student.unisi.it,
{enrica.marchigiani, oronzo.parlangeli}@unisi.it
[2] Azienda Ospedaliera Universitaria Senese, Siena, Italy
a.giani@ao-siena.toscana.it
[3] Trinity College Dublin, The University of Dublin, Dublin, Ireland
pliston@tcd.ie

Abstract. Over the last few years the use of Computed tomography (CT) has become increasingly widespread. Despite the common usage of CTs, their usability and their suitability as safe diagnostic tools have not been subjected to scrutiny. The purpose of this study is to understand the evolution of the relationship between users (i.e. technical operators) and the complexity of a technological system such as that of the CT in a normal operational context. To this end, two studies were conducted employing ethnographic observations seven years apart. This paper compares these two studies from 2011 and 2018. The results show a significant difference in the timing of the activity relative to the interaction with the technology, while there are no significant differences in relation to the interactions with the patient and the problem situations that may occur during the CT examination.

Keywords: Computed Tomography (CT) · Usability · Execution time
Interruption · Human Machine Interaction (HMI)
Healthcare technology · Technicians

1 Introduction

Over the last two decades, medicine has undergone a significant transformation due to scientific and technological advancements, and in this context, radiology has experienced progress due to changes in hardware and software used for image acquisition and reproduction [1]. The use of Computed Tomography (CT) for diagnostic purposes has certainly contributed in a disruptive way to these advances.

CT is a technique introduced in the late 70 s, which has revolutionised the field of diagnostic imaging and which facilitates the reproduction of specific areas of the body,

© Springer Nature Switzerland AG 2019
S. Bagnara et al. (Eds.): IEA 2018, AISC 822, pp. 488–496, 2019.
https://doi.org/10.1007/978-3-319-96077-7_52

such as the skull, chest, abdomen, skeleton, using three dimensions to reproduce the area of interest. Through the use of a computer equipped with specific software, it is possible to view the virtual image and proceed with any subsequent reconstruction [2].

The use of CT has increased considerably in recent years, but the same cannot be said of the studies performed to analyse operator performance and the adequacy of the systems and processes that can support an operator in performing safely. Currently this diagnostic test is usually requested by the clinical physician, and then the radiologist will decide what type of diagnostic technique to use (radiography, ultrasound, CT, etc.). The complexity and the number of intervening variables in the choice of the right protocol to use demands for a deep competence on the CT operator side about physics principles and image reconstruction algorithms. The role of the radiologist technician has thus evolved in recent years, in much the same way as the technology and the whole system that revolves around it has evolved. The digitalisation of radiology systems, the ease of exporting images and the speed of communication have catapulted the radiology technician into a deeply complex interactive system [3].

Prior to this technological revolution, the interpretation of images fell within the exclusive competence of the operator [4], who also took on a more decentralized role. Today the role of the technician no longer has exclusivity over the immediate interpretation and management of the image, but they still have new responsibilities, and above all their role is an integral part of a complex system which extends from patient management to the management of IT systems; from the assurance of adequate levels of patient safety and quality control to maintaining relationships with medical and nursing staff; and from maintaining relationships with the organisation to, in some cases, the management of supplies.

Since it is also the starting point compared to a possible series of interventions and diagnostic tests, the technician assumes a role of great responsibility in the patient's care path. Indeed, it is often the results of a CT that determine the requests for therapies or subsequent diagnostic tests [5]. Performing the most appropriate examination, whilst trying to achieve a good cost-benefit ratio, can save the patient's life, or at least safeguard their health, avoiding subjecting them to unnecessary tests that could mean greater radiation exposure [3]. Therefore, even if technological advances have altered and improved the operator's task, the interaction with the system as a whole, the decision-making process and the problem-solving aspects of the task have become wider and more complex [6–10].

Given the complexity of the system in which the radiologist is working and in which the CT belongs, the planned man-machine interaction must take into account many factors which however are situated along a continuum that sees at its extremes a minimum or a maximum level of automation. In fact, it is possible to distinguish between situations in which a totally manual control over certain parts of the system may be preferred, as in the case of communication with the patient at the time of the examination (e.g. "do not move, breathe, do not breathe" etc.); or other situations in which these parts of the interaction are automated, (as in the case where the "do not move" is managed by the machine in interactions with hearing impaired patients). In different circumstances, with the most advanced systems, the technician has the discretion to decide how to operate "automatically", i.e. the extent to which they should rely on technological skills and how the extent to which they should proceed "manually", i.e. assuming responsibility for the management of the exam. With the current technology, some radiographers, thanks to their skills and experience, create their own protocols to optimise the relationship between image quality and radiation exposure. In daily practice, in short, that which the designer or the trainer cannot predict, that is to foresee all the possible situations which can occur, is solved by the radiologist owing to their experience [11].

In the case of a first aid CT, the system is even more complex because the operator in many cases has no control: the patient may be alert, semi-conscious or completely unconscious; they may be male, female, elderly, adult or child; they may present different degrees of physical and cognitive ability. The potential damage that can result from X-ray exposure therefore depends on many factors and the data available to the operator often leads him/her to solve complex problems in order to obtain the best images and to ensure a correct diagnosis, all the while adopting strategies to minimise radiation exposure [2, 12, 13].

In all this, the flow of communication between the members of the medical, technical and nursing staff is of crucial importance and even more so in an emergency department where the need to operate quickly can contribute to operational and communication-related cognitive overload [14]. In the case of CT it is the task of the radiologist to ensure the flow of activity and communications is in the interests of diagnostic effectiveness and efficiency regardless of the complexity of the emergency and the host of doctors, nurses, health workers and specialists who at times populate the department.

Despite the fundamental role that radiological technicians have within the socio-technical healthcare system, there is a lack of research studies which seek to better understand the operational realities and the interactions with the complex technological system which they manage (CT). Additionally, this lack of knowledge seems even more incomprehensible given the current situation of the health systems in which these emergency situations are being managed. It is for these precise reasons that this study seeks to investigate the activity of the radiology technician in interaction with a complex system such as Computed Tomography in a hospital emergency department.

2 The Study

This study was conducted using an observational study methodology focusing on the complex activity of technicians working in the Emergency Department of the hospital "Santa Maria alle Scotte" in Siena, Italia. The choice of an ethnographic study was mainly due to the context. Observational research within healthcare environments is a method which facilitates the elicitation of a better understanding of the context in which work takes place (in the case of complex work environments), focusing attention on the activity, the tasks and the problems that may arise in the interactions inside the normal, everyday setting [15, 16].

The research presented herein can also be described as a longitudinal study as it reports an observational study conducted on the 13th, 16th and 19th March 2018. The data thus obtained will then be compared with another observational activity that has been conducted in the same context and in the same manner on November 16th and 17th of 2011. This allowed us to highlight possible differences from 2011 to 2018, and to then identify some of the probable changes which have emerged in recent years.

3   Method

3.1   Study Setting - Computed Tomography in the Emergency Department

The environment in which the CT examination takes place is divided into four rooms (see Fig. 1). A first room with the emergency trolley. The second room (the controlled area) is where the patient is positioned for the examination, and where the CT machine and the injector (for the possible use of contrast and the need for resuscitation) are located. The third room is the operating room, connected to the second by a door. A large glass wall allows you to see everything that happens inside the second room. In the operating room the technician starts the scan, manages it and displays the exam in real time. A fourth room, which can only be accessed from the third room, and where a 'work station' is located, is where the rough images, obtained from the CT examination, are reconstructed by the radiologist.

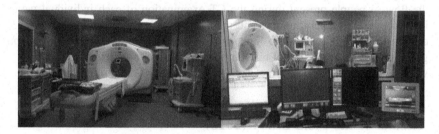

Fig. 1. CT setting in "Santa Maria alle Scotte" hospital in Siena

The technology for conducting the exam consists of a scanning unit called a Gantry, a generator, the table on which the patient is placed, a computer for the acquisition of patient data, a command console where the images are displayed and finally, a recording system for the data.

The interaction with the different parts of the technology for the performance of the examination takes place primarily in the second and third rooms and can be divided into four phases: entry of patient data, positioning of the patient on the table, execution of the examination (image acquisition), and the discharge of the patient.

There were no substantial differences in the layout of the work environment between 2011 and 2018.

3.2   Implementation of Ethnographic Observation

Prior to beginning the study, both in 2011 and in 2018, the research team participated in a meeting with the Chief of Emergency CT, which provided researchers the opportunity to clarify any queries about the diagnostic process and to plan the study. All technicians were informed about the research prior to their participation.

Similar to the study performed in 2011, in 2018 the ethnographic observation was performed by six observers divided into three groups of two people each. The observations were carried out in three days during the three shifts, morning, afternoon, night, covering the hours from 10:00 to 22:00, for a total of 26 h of observation. The research shifts included four hours of observation for each group (10:00–14:00, 14:00–18:00, 18:00–22:00).

The work of three technicians was observed (one female and two males out of the six who are on shift for first aid CT); their on-the-job experience ranged from 12 to 20 years. During the observations, there were sometimes also trainee students, health workers, nurses and doctors present. The presence of the trainees was constant while the presence of medical doctors occurred only when the examination required the use of the contrast agent or in those cases in which the patient was a 'code red' or in an emergency situation.

For ease of data collection a coding scheme was structured according to the following parameters:

- the duration of the examination divided in time by (i) data entry, (ii) patient positioning, (iii) the execution of the examination and (iv) patient discharge;
- the flow of people present during the CT examination of each individual patient, indicating the minimum number and maximum number of persons;
- conversations, subdivided into functional (i.e. relative to the examination in progress) and distractive (i.e. not relevant to the context of the examination);
- work interruptions; critical issues and the resolution of the problems identified.

4   Results of the Field Study (2018)

Duration of the Exam. In 2018, 43 examinations were observed, 20 (46.5%) in the 10–14 h shift, 21 (48.8%) in the 14–20 h shift, and 2 (4.6%) in the 20–24 h shift.

A total of 482 min of examinations were analysed. Of these, 51 min (10.5%) were dedicated to data entry, 142 min (29.4%) to the placement of the patient, 206 min (42.7%) constituted the execution of the exam, and 83 min (17.2%) concerned the discharge of the patient.

There are no significant differences with respect to the timing of the examination in consideration of the different shifts of the operators - morning, afternoon or night.

The duration of a single exam varies from a minimum of 6 min to a maximum of 28 min, with an average duration of 11.2 min.

As evidenced by the results, the activity of the technician is dedicated to a greater extent to the examination - choosing the protocol for the acquisition of images and scanning them in digital format (4.7 min). This task may involve, depending on the case, a consultation with the team of the department, for example in the case of the use of contrast dye, but also involving external doctors, as in the case of cerebral stroke in which the intervention of a neurologist was sought.

This is followed, in terms of duration, by the positioning of the patient (3.3 min) – a task that is done with extreme care by the technician him/herself and in some cases with the help of health workers, for example when the patient is not able to position him/herself unaided. In some cases times are lengthened, for example when the technician becomes aware of a problem situation, for example the presence of a metal body that could interfere with the result of the examination, in this case the patient must be positioned again, after solving the problem.

Data input (1.8 min), and patient discharge (1.9 min) take up less of the technician's, but they are no less important. During the data entry, as attested to by the operators themselves, great care is needed to avoid mistakes that would be irrecoverable and that could lead to errors of pairing the examination to the patient. Data entry, over the years, has been completely digitized using software that transfers the patient's data from the first aid department (or from the department of origin) directly to the CT department. Nonetheless, the technician must check the data entered to make sure that they match.

Flow of People Regarding the flows of people the observation recorded an average of 4 people located within the CT department in the course of exam execution. The maximum number of people within the room was 17, while the minimum number was only 1 person, the technician. During the morning shift, the maximum flow of people was recorded, with a significant difference (t-test, p < .001) between the mean minimum flow average of the morning (average 3.9), and the afternoon (average 1.7) shifts.

Conversations Observation results indicate an average of 5 conversational exchanges per exam, with a peak of 25 for a 'code red' exam. Overall, 62.12% of the conversations were functional and 37.87% distractive. No significant differences emerge between functional and distractive conversations in the three shifts.

The functional conversations concern in particular the communication with the patient for confirmation of personal data, for communicating the patient's state of health at the time of the examination and the request for clinical information necessary for the exam execution. In addition, the patient is given instructions for correct positioning on the couch. The conversations then concern consultation with the radiologist, in particular when the case requires the use of contrast agent or necessitates the advice of other specialist doctors, for code red situations or if the situation is considered critical for cardiac or neurological problems.

Critical Issues Out of 43 exams, 24 cases of critical issues were identified (55.81%), in particular due to patient positioning (14 cases, 58.3%), the temporary absence of the doctor who has to attend to other emergency cases (3 cases, 12.5%), the presence of objects on the patient's person that prevented the examination (3 cases, 12.5%), the patient's state (2 cases, 8.3%), data entry (1 case, 4.1%), and the presence of trainees which slowed down the work (1 case, 4.1%).

A correlational analysis, expected, highlighted a significant positive correlation between the maximum flow of people and the number of critical issues (r = .233; p < .05). The critical issue often requires a consultation with the departmental team and in some cases also with specialists from other departments.

A positive correlation was also found between the maximum flow of people and distractive conversations ($r = .546$; $p < .001$), the greater the number of people in the room the higher the rate of communicative flow, including distractive communication. In this case it is up to the operator to filter the stimuli and process only those functional to the activity.

4.1   Results of the Field Studies Comparison (2011–2018)

The results of the longitudinal analysis, which compares the two observational activities carried out at a distance of almost seven years from each other (November 2011– February 2018), highlight interesting and not entirely predictable results.

The three operators observed in 2018 had also participated in the 2011 observational study.

It should be underlined that during the intervening years the software for the acquisition of images had been updated. In fact, in September 2017 the Imago package (Ris) - CT Software was replaced by ASIR (Adaptive Statistical Iterative Reconstruction), a system designed to reconstruct the rough image data using iterative algorithms based on statistical modelling. ASIR allows for a more accurate reconstruction of the raw data acquired by the scanner, facilitating an important reduction of the radiation dose delivered to the patient and maintaining the diagnostic value of the images.

The main results of the first observational study as compared to the second can be found below. This allows us to understand if and how the activity of the CT department has evolved.

In 2011 the ethnographic observation was conducted over two days in which 30 examinations were observed for a total of 643 min.

On average, an examination lasted 17.5 min. That was made up of 3.3 min of data entry, 4 min for positioning the patient, 8.1 min for the examination, and 2.1 min for patient discharge (Fig. 2).

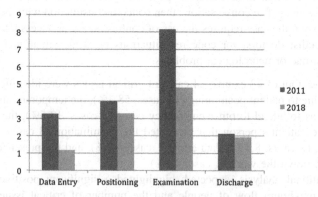

Fig. 2. Comparison between observations from 2011 and 2018 for mean times (in minutes) of technician's activity divided into data entry, positioning, examination and discharge phases.

Comparing this older observation with that performed in 2018, likely due to the software upgrade and the experience of technicians, the working times for data input (t-test; $p < .001$) and execution of the examination (t-test, $p < .005$) have accelerated significantly – these aspects of the task deal exclusively with the machine. On the other hand, the times relative to the patient's positioning and to their discharge have remained virtually the same – these aspects of the task are more related to interpersonal interactions between the technician and the patient.

Also in the comparison between 2011 and 2018 there are no significant differences in relation to the flows of people present in the room during the exams, to both functional and distracting conversations, and to critical issues. Regarding critical issues however, in the first observation (2011) 28.57% of the critical issues were due to technical problems, while no such problem was observed in 2018.

5   Conclusion and Discussion

The work of an emergency department CT technician is complex, characterised by a not insignificant workload subjected to time pressure and aggravated by a dynamic and unpredictable work environment. There are several factors that can create problematic contexts for this work. For example, time pressure is often exacerbated by the presence of many people who can interfere with the performance of the activity.

The study carried out here shows that increased work experience and the technological advancement of recent years, have brought obvious improvements in the activity of the operator. The results show that in the period from 2011 to 2018 the execution times of the examination have been reduced, especially in relation to interactions with the technology. The times dedicated to patient care have remained substantially the same. This is to be considered as a positive: we observed rigorous attention to detail in relation to interactions with the patient, not only for their correct positioning, but also in eliciting information on their state of health and for any reassurances needed prior to the exam, and in checking and confirming personal data.

In the 2011–2018 comparison, no change was observed in relation to the flows of people present at the time of the examination and the type of communication with the medical, nursing and trainee personnel. This paints a picture of a mature and resilient context, capable of managing emergencies and at the same time providing a learning environment for trainees.

The longitudinal analysis also shows that the rate of critical issues has not increased or decreased over the years. This is particularly interesting because it suggests that steps can be taken to make it an increasingly smooth operation with less bottlenecks. In this regard, it is illuminating to consider the qualitative aspects of the critical issues. In 2011 these were largely due to technology, which was the second critical factor. In 2018, on the other hand, thanks to the increase in operators' experience, the technology no longer creates problematic situations. The critical issues that are still encountered are those due to the intrinsic variability of the context, to the dynamism of the systems with which the operator interfaces, to the healthcare staff and to the patients.

Overall, it can be said that in this context the technology and the human element appear to be well aligned within a complex, unpredictable and changeable environment such as that of a CT in an emergency department. And it is precisely in reducing the level of unpredictability of events that there is still room to improve levels of safety and patient care. And this can only happen, in recognition of the centrality that the technician has within the complexity of the socio-technical system of a CT in an emergency department.

Acknowledgments. Te Authors would like to thank all the staff of the "Santa Maria alle Scotte" hospital who greatly helped us in conducting this study and those who, in 2011, collaborated with us in collecting the observational data.

References

1. Bellolio MF, Bellew SD, Sangaralingham LR, Campbell RL, Cabrera D, Jeffery MM, Shah ND, Hess EP (2018) Access to primary care and computed tomography use in the emergency department. BMC Health Serv Res 18(154):1–10
2. Ning P, Zhu S, Shi D, Guo Y, Sun M (2014) X-ray dose reduction in abdominal computed tomography using advanced iterative reconstruction algorithms. PLoS ONE 9(3):e92568, 1–5
3. Knechtges PM, Carlos RC (2007) The evolving role of radiologists within the health care system. J Am College Radiol 4(9):626–635
4. Rana AQ, Zumo LA, Sim V (2013) Neuroradiology in clinical practice, 1st edn. Springer International Publishing, eBook ISBN: 978-3-319-01002-1
5. Forrest CB, Shipman SA, Dougherty D, Miller MR (2003) Outcomes research in pediatric settings: recent trends and future directions. Pediatrics 111(1):171–178
6. Vicente KJ, Rasmussen J (1990) The ecology of human-machine systems II: Mediating "direct perception" in complex work domains. Ecol Psychol 2(3):207–250
7. Parasuraman R, Barnes M, Cosenzo K (2007) Adaptive automation for human-robot teaming in future command and control systems. Int. J. Command Control 1(2):43–68
8. Cummings ML, Bruni S, Mercier S, Mitchell PJ (2007) Automation architecture for single operator-multiple UAV command and control. Int Command Control J 1(2):1–24
9. Cummings ML, Mitchell JP (2008) Predicting controlling capacity in supervisory control of miltiple UAVs. IEEE Trans Syst Man Cybern Part A Syst Hum 38(2):451–460
10. Bagnara S, Parlangeli O, Tartaglia R (2010) Are hospitals becoming high reliability organizations? Appl Ergonom 41(5):713–718
11. Ramli MN, Faridah Y (2010) The boiling frog syndrome: a radiologist's perspective. Biomed Imag Interv J 6(4):e36
12. McCollough CH, Primak AN, Braun N, Kofler J, Yu L, Christner J (2009) Strategies for reducing radiation dose in CT. Radiol Clin N Am 47(1):27–40
13. Tipton K, Launders JH, Inamdar R, Miyamoto C, Schoelles K (2011) Stereotactic body radiation therapy: scope of the literature. Ann Intern Med 154(11):737–745
14. Woloshynowych M, Davis R, Brown R, Vincent C (2007) Communication patterns in a UK emergency department. Ann Emerg Med 50(4):407–413
15. Goodson L, Vassar M (2011) An overview of ethnography in healthcare and medical education research. J Educ Eval Health Prof 8(4):1–4
16. Soukup T, Lamb WB, Sevdalis N, Green JSA (2017) Undertaking field research. J Clin Urol 10(1):58–61

Overload of Technological Connections for Communicating at Work

Ophélie Morand[1,2,3(✉)], Béatrice Cahour[2],
Marc-Eric Bobillier-Chaumon[3], and Vincent Grosjean[1]

[1] INRS, Rue Du Morvan, 54500 Vandœuvre-Lès-Nancy, France
ophelie.morand@inrs.fr
[2] CNRS I3 Nat.Sup.School Télécom ParisTech, Paris, France
[3] University of Lyon2, GRePS, Lyon, France

Abstract. Connection has become an essential component of work. It appears to be an essential resource for efficiency and rapidity, as well as for accessibility and transmission of information. However, it may have harmful effects on individuals, teams or organizations because of its continuous nature. Our study aims to characterize factors that may contribute to a positive or a negative experience related to this connection at work by managers, and ways of having such factors taken into account by the organization. Our long term aim is to help the organization to find solutions to improve the situation. We put together a questionnaire about practices, uses and feelings related to connection and about potential effects on health in a large French company, and we also conducted in-depth interviews. The analyses focus on the individuals' connection practices and on their feelings about them in specific situations. We also analyzed the existing individual, collective, and institutional strategies used for regulating the connection process. The analyses of the questionnaires indicate the proportion of managers who experience difficulties related to their connections, the nature of the troubles evoked and their impacts on work activities and on life. The interviews allow us to go deeper into contextual elements and their feelings about these practices. The connection experience depends on ICT characteristics, experience of connection, team dynamics and psychological factors. Our analyses have shown, among others things, the role of working together as a team and how the team could regulate connection overload where neither the individuals nor the institution can succeed.

Keywords: Connection · ICT · Activity · Feelings · Regulation

1 Introduction

Connection has become an essential component of work, in a changing world of work, where work is no longer limited to one place but could now also be done at home or while travelling, and where connection is what keeps us "in touch" with colleagues, clients and the firm. Nevertheless, working with a constant connection has impacts on work activity, personal life and even on health, in particular for managers and executives who use a lot of ICT (Information and Communications Technology). Disconnection appears to be one way of regulating this issue, as it is illustrated by the

© Springer Nature Switzerland AG 2019
S. Bagnara et al. (Eds.): IEA 2018, AISC 822, pp. 497–506, 2019.
https://doi.org/10.1007/978-3-319-96077-7_53

"right to disconnect" ("*droit à la déconnexion*") brought into force by the French legislation (law known as the "*loi travail*" or "*loi El Khomri*", passed in August 2016), and how this "right to disconnect" has been taken up in the company-level agreements of many companies in France. However, what happens to this right when the employees are in a firm who provides them with business Smartphones? Moreover, how can such employees be expected to cope well with such a level of connection? The aims of this study is to characterize over-connection, to identify the factors that affect people's experience of connection, and to discuss it in a group including employees who might feel difficult to cope with the experience of connection. The initial results we are presenting are the findings of an exploratory study aimed at identifying other avenues for thought and analysis on over-connection and on how it can be managed.

2   Theoretical Framework

Firstly, we tried to understand what could explain and motivate a high degree of connection. The literature provides some possible answers: first of all, connection seems to have emerged as an implicit rule of work (Boudokhane-Lima and Felio 2015). Indeed, several studies show that any disconnection leads at least to the employee having to give a justification, and possibly even to disciplinary action being taken against them (Bobillier-Chaumon 2012), and if disconnection can entail such consequences, it is because connection has become an (implicit) social norm (Jauréguiberry 2014). However, it is also because connection is perceived as a great resource.

Connection allows employees to be responsive, efficient and fast, like the technologies they use, and also highlights the work done (Andonova and Vacher 2013; Domenget 2014; Bobillier-Chaumon et al. 2015). Indeed, whenever work activity is no longer carried out in one place only, connection enables the worker "to stay in touch" with the company via social networks, emails, phone calls, or shared calendars and diaries. However, while being able to work anywhere opens up the possibility of working at any time, many studies also show that connection generates a certain temporality of urgency (Laïdi 1999) because it allows non-desired solicitations during and outside conventional working hours (Haroche 2004; Jauréguiberry 2006; Aubert 2008). The constant presence of notifications via, in particular, the smartphone, generates what is sometimes called "obstinate presence" solicitations for the person (Datchary and Licoppe 2007; Day et al. 2010; Brooks 2015).

It is then a question of employees finding their own ways of regulation in order to establish borders between their work lives and their personal lives, even if ICT and especially the smartphone (considered as a boundary object between the two spheres of lives) is conducive to blurring that boundary. Several strategies are described in the literature to control this border: segmentation, integration, hybridization and, when it is not possible to regulate, overflow (El Wafi and Brangier 2013, 2016). Overflow corresponds to an unwanted invasion of work life into personal life and this causes high stress and a feeling of inability to manage the flow of information and solicitations conveyed by ICT (Lahlou 2000; Jauréguiberry 2014; Créno and Cahour 2016). Solicitations are also a stress factor in the workplace, and the vast majority of interruptions are linked to notifications (emails or instant messaging). In fact, a study shows that 70% of emails are taken into account 6 s after their arrival and this fact should be weighed against the losses in efficiency resulting from the fact that it takes from 64 s (Jackson et al. 2003) to 25 min (Brooks 2015) to resume an interrupted task. To do their work under such conditions, individuals must acquire dispersal skills, which are "the ability to distribute one's attention" (Datchary and Licoppe 2007). They suggest that in a complex work environment such as with ICT, multi-activity has often become a normal way of operating that is required by the diversity of the demands. However, the same author (Datchary 2007) also quotes a psychiatrist, Hallowell, that compares hyperactivity in children with dispersion capacities in adults at work. It is therefore surprising that, in order to cope with a stimulating work environment, the employee must put in place a strategy that is described as pathological for a child.

In attempting to manage such connection and the influx of information related to it, we can describe three levels of regulation:

(1) the individual level with technical filtering (activate "silent" mode, disable push mode, and turn the smartphone over) (Datchary 2007; Felio 2013; Boudokhane-Lima and Felio 2015) or professional filtering (employees, often the oldest ones, select the mails to open according to the potential interest they might have - according to the sender, the title, or the attachment - or else they manage the mails by answering only if chased up (do not answer immediately and wait for a chase-up email or phone call) (Bobillier-Chaumon and Eyme 2011) or they preserve themselves by determining a time without solicitations, either virtually (absent mode of instant messenger), or physically (room without a phone) (Felio 2013; Boudokhane-Lima and Felio 2015), which corresponds to partial disconnections).

(2) Secondly, there is regulation at institutional level (training, agreements, good practice guidelines), but that is not unanimously accepted by employees because rules are sometimes too vague and established by the organization without prior consultation with their employees (Boudokhane-Lima and Felio 2015; Créno and Cahour 2016).

(3) Finally, there is collective regulation, such as working groups that establish rules of use in matters of (dis)connection.

Our objective is to define what over-connection is and what its determinants are. To our knowledge, there is no clear definition of what over-connection is, and we want to define the factors that cause over-connection to greater or lesser extents, and how to conduct prevention on this issue. Addressing this issue involves answering the following questions: what exactly is over-connection for employees, when does it start and what does it trigger? How does it materialize and manifest itself? How can it be regulated?

3 Methodology

The study takes place in a large French company where we intervened at the request of occupational physicians and psychologists of this company who are directly concerned with these issues. In order to study the question of connection, we implemented a methodology combining both a quantitative and a qualitative approach.

Our first step was to develop a *questionnaire* to which 436 executives responded (during periodic medical check-ups conducted by occupational physicians). The questionnaire had 3 main themes and 29 questions, organized into the following three theme sections:

- a part for characterizing ICT practices and uses: questions on the use of ICT outside working hours (evenings, and weekends), and on whether or not disconnection practices were implemented.
- a part about the feelings related to such use: effect on privacy, quality of work, etc.
- a health part that investigated firstly perceived effects of connection on employee health (stress, fatigue, vision problems) and secondly physiological indicators (blood pressure, Body Mass Index).

The objective of this questionnaire was firstly to obtain a sort of "snapshot" of the current situation relating to use of ICT in this company and the impacts of that use, and secondly to "recruit" participants to probe deeper into the topics covered through individual interviews. The sample consisted of 139 women and 296 men, the average age was 48.26 years (SD 8.48), the minimum age was 24 years and the maximum was 62 years. As regards employment positions, 48% technical functions, 29% transversal functions, 15% sales, and 8% "others". The sample broke down as follows: 58% qualified professionals, 19% first-line managers, 18% middle managers and 5% senior managers. We chose to work with executives only, because they are the largest users of ICT.

In a second phase, we conducted and analyzed ten *semi-structured interviews* with volunteer employees, lasting from one hour to one and a half hours, in isolated offices, that we recorded and transcribed, using:

- firstly general questions about employee practices in terms of tool management, multi-activity, logging in and logging out feelings related to connection and activity (impact on privacy, stress/anxiety related to connection); and the individual, collective and organizational regulations put in place.
- and then the technique of Explicitation Interview (Vermersch 1994; Cahour et al. 2016), applied to a specific an representative day of the employee. They could use at first their agenda to remember their planned activities during this day. This method enabled us to gain a better understanding not only of the work activity and of how the various ICTs were used for it, but also of the subjective experience, i.e. the flow of thoughts, sensorial perceptions, actions and emotions related to those uses during moments of a specific day.

The interview sample consisted of 7 people working in transversal function (2 men aged 40 and 50 and 5 women aged 50 to 58), 6 men working in technical jobs (aged from 40 to 55) and 7 in commercial function (2 women, 30 and 56 years old, and 5 men aged from 45 to 57).

The questionnaire gave us an overview of off-hours work and of the impacts it has on health for example, while the interviews informed us about how the work activity proceeds with ICT and gave us more in-depth information about feelings and perceptions.

4 Results

Our first analyses showed that several factors influence the feeling of connection, at work level, at individual level, and at regulation level. We will look at the following in more depth below: (1) tool management and how this implies multi-activity, (2) the experience of connection during and out of work, (3) how employees define over-connection, (4) managing boundaries between work and private life, and (5) how individual and collective regulations play a role in the experience of this connection. We will use in parallel some of the results of the questionnaire and the 10 interviews analyzed at this stage of the study.

In order to refine our comprehension of the answers to the *questionnaire*, we created three groups by clustering the individuals who gave the closest answers to certain items. This allowed us to build the following 3 groups: people with a high level of connection (122 people), people with a low level of connection (143) and people with an intermediate level of connection (171). The questionnaire indicates that 52.30% of the salespeople are in the hyper-connected class compared to 23.80% for the other jobs.

For the *interviews*, in order to determine connection levels, we referred to four general questions directly related to connection. Thus, we have a group of people with low connectivity, a group of people with intermediate-to-high connection practices with predominantly *positive* feelings, and a group of high connection practices people with predominantly *negative* feelings (i.e. feeling over-connected).

4.1 Tools Use and Perception

In the company studied, *Skype* is mainly used for its instant messaging, and it is perceived quite positively as allowing short and precise questions, in a relatively informal mode, and as being a less intrusive way of interaction than the phone call for 5 of the 10 executives interviewed. It is the tool most used internally.

Conversely, *emails* are perceived rather negatively, as a "*scourge*", and a "*drag when on vacation*", but they remain indispensable for all the interviewed executives because, among other things, they procure traceability of information, and also visibility of work, especially in the run-up to holidays.

While emails are used to keep track of interactions, *phone calls* are used when topics are sensitive or to avoid misinterpretation related to an email. The tools are seen generally as resources for efficiency and speed, yet all seem to generate constraints when they are described one by one. The relationship with the *smartphone* plays a special role in connection: the stronger the emotional ties to that object, the more the person will find it difficult to disconnect, since the smartphone is perceived as "*an extension of the hand*" (3 executives out of 10 uttered that phrase). And this is what we find particularly in the definition of over-connection, with the object being considered as "*a baby*" or "*[a] little creature*", i.e. with the smartphone sometimes being considered as being alive, or even as being a part of the person.

4.2 Multi-activity

The multi-activity induced by joint activation of several tools (i.e. Skype, email, videoconferencing) is very common, especially for the sales representatives, one of whom (57 years old, male) told us "*Sometimes even during a presentation, I answer a Skype call while talking to the customer*". This multi-activity stems not only from many internal requests but also from the workload or the number of visits. It may create collective tensions : sales people reported not having a lot of time to process email and therefore doing such processing regularly during meetings, while a non-sales executive (working in real-estate support and less connected) told us it was "*particularly annoying to see people doing something else during meetings.*" Several executives (4/10) told us about the effects on concentration, ranging from "*we can lose the thread*" to "*it's physically impossible*".

4.3 Connection Dynamics

To address the connection issue, we separated working-time connection and outside of working time connection. Regarding experience of connection at work, the less-connected executives admitted a propensity to "disconnect" or "log off" to protect themselves. For the people (3 technical executives) who were highly connected but who had positive feelings (theyr explicitly answered that connection did not pose a problem), connection was described as voluntary and not imposed on them. For people with more negative feelings (who explicitly said that connection was currently making them suffer, 3 commercials and 1 transversal function), connection was difficult for several reasons: (1) pressure related to customers, when the interactions are negative "*when it's okay, it's okay, but when things go wrong, we take it very badly*". (2) Peer pressure ("*our internal demands are extremely strong*"). The feeling of guilt is described in the questionnaire where 50.8% of the highly-connected people compared to 9.8% of the less-connected ones talk of having this guilt feeling when they are not connected in non-working hours. In addition, some are blamed for non-connection: an executive (trainer, 50 years old) talked, for example of disconnection when leading a session of training and reported that when she is called, her colleagues complain that she "*never answers*". The results of the questionnaire point to a paradox, since 47.5% of the highly-connected people say they have already observed negative consequences of not having logged in (remarks, etc.) as against 17.5% for the less-connected people.

At non-work connection level, the questionnaire indicates that the 122 highly-connected people work on average "frequently" in the evening (6 p.m. to 9 p.m.) and on the move, and "sometimes" at weekends, and during their holidays. On average, the 143 least-connected "never" work in the evening, and "rarely" work in the evening, at weekends or during holidays. Thus, a less-connected woman (50 years old, property billing manager) stated that "*there is no need for a service that requires me to look at my emails during holidays, weekends and evenings*". Conversely, two highly-connected people with positive feelings even reported that connection at breakfast "*improves the quality of life*" in the sense that they save time on the processing of their emails since they can do it without being engaged elsewhere (by appointments, or meetings).

For highly-connected people with negative feelings, connection is almost constant. They rarely disconnect and when they do, the 122 hyper-connected people on average "often" find themselves consulting their emails reflexively (while non-connected people do so rarely). In an interview, a manager reported he never disconnected because "*I always need to reassure myself that I did not miss something, and that no one called me*". Similarly, the highly-connected people said they "sometimes" felt stress when they do not have the opportunity to connect, while that feeling was shared, on average, only "rarely" by the less connected people.

4.4 Over-Connection

We define here over-connection as the subjective evaluation of being too much connected and with problematic consequences. The way people consider over-connection based on usage experience varies greatly depending on connection level. Thus, for a person who is not very connected, over-connection begins when "*we are in a meeting and we answer an email*". Conversely, a highly-connected person described herself as "*naked*" without her computer and her laptop. Over-connection is mainly described by the feeling of being reachable all the time and being always tied to your phone "*it's when you let go that you think you can live without your mobile, but you have to accept letting go of it. And that's the hardest thing.*" (F, 50 years, transversal function). Over-connection is illustrated by the development of a certain degree of fear when people do not have their phones, as applied to three of the executives interviewed (3 women) ("*the total freakout*" "*I get withdrawal symptoms*"), by health impacts, as applied to 2 commercials (a 57-year-old man and a 30-year-old woman) like fatigue, weight gain or concentration problems "*sometimes it happens at 6 p.m. [...] you understand nothing that you are told actually*". According to those accounts, over-connection would appear to be based both on reflexive, or indeed compulsive, consultation of ICT tools -which must be constantly accessible-, and also on constant availability and consent of the person via such tools, with deleterious effects on health and well-being of users.

4.5 The Boundary Between Private Life and Professional Life

The boundary between work life and personal life appears porous to varying degrees. Some people do not separate these two spheres at all for the reason that they feel no need to separate them: "*I'm not just helping my colleague, I'm helping my friend*" (F, 50, transversal function), whereas a person who segments private and professional life (low level of connection, without Internet at home) said "*in 35 years, I have never had personal bound with people from the company outside work time*". Close family and friends also play a role in establishing a border, and several executives (4 out of 10) indicated they were "*told off*" by their spouses: "*I get shouted at by my wife*", "*[my husband] hides everything*" (phone, PC). This is consistent with the data from the questionnaire, which shows that 66.4% of highly-connected people have already received negative comments from family or friends. These results are linked to the fact that all of the participants (436) in the questionnaire consider that the possibility of connecting out of hours is rather positive, except for the family and personal life aspects. The less people are connected, the more negative their feelings are about this issue of the influence of work life on privacy through digital tools.

4.6 Regulation

The perception of connection is also influenced by how individuals regulate. We find partial disconnections (put the mobile down, put it into airplane mode in the evenings or at night) for 5 out of 10 executives interviewed, and forced disconnections, not necessarily taken well by two executives (technical and sales) such as *"I'm on vacation in a place that is not well covered by the mobile phone network, there is no Internet connection, and there's nothing I can do about it!"*. Different strategies appear in the interviews for handling the flow of emails: handling by thresholds (*"when I see more than 50 emails I become more and more anxious"*), a commercial (57 years old) told us he teleworks in order to deal with emails without being *"pestered by his colleagues"* via his instant messages. A manager deployed another strategy by abandoning emails or by answering them only if chased up: *"There are 1212 that I have not read... and that I will never read"*. Because of her status and (senior) experience, she can use that regulatory strategy.

It should be noted that one of the possible ways of coping better with connection would appear to be distancing oneself from work. Indeed, it appears that highly-connected executives who cope with it quite well establish a boundary with and take a step back from work situations. In the interviews, such distancing was cited by half of the respondents and seems to be related to older age and experience, but further studies are needed to prove it. Two people explained to us that they had distanced themselves from situations of over-connection or sometimes after a burn-out.

At collective regulation level, some teams (2 out of 10 respondents) set up weekly meetings to discuss and control the flow of exchanges related to communication tools (e-mails and Skype). Other teams also highlighted how sharing can help in difficult situations (*"we share tips"*). For two commercials, in the same type of department (commerce with corporations), their experiences of connection were contrasted: one of them is in a supportive team with a manager who facilitates things, i.e. *"tips for difficult situations"* are exchanged between colleagues and the manager takes over when the situation becomes too uncomfortable; in the other case, the team is perceived as fragmented and exhausted, and the manager as out of his depth and unable to cope (at the time of the interview, he had just gone off on sick leave). These experiences can be expressed for the first rep as *"we do not feel alone in the world"* and for the second rep as *"we are all in the shit here so we cannot talk with colleagues"*. This example of more or less successful consultation shows the utility of the team and the importance of working on this regulation resource in particular.

5 Conclusion

The initial findings of this study therefore indicate that the experience of connection depends firstly on the job itself and on the need for connection (salespeople seem to be more affected), secondly on the potential modes of use of the tool (communication mode enabling the life systems to be hybridized), thirdly on the characteristics of ICT, and fourthly on the level of teamwork and of the feeling of belonging. Indeed, a supportive team works to reduce over-solicitations, and managerial support make it possible to cope with difficult interactions marked by a quasi-permanent connection. People who distance themselves from their work, having established a border between their personal lives and their work lives, through disconnecting at the weekend, being involved in extra-professional activities or through a vigilant family environment, would also appear to be less affected by an exhausting and demanding level of connection. While one purpose of this exploratory study was to define this over-connection, i.e. to determine the tipping point at which we go into over-connection, it also aimed to determine the factors enabling us to cope well with such over-connection, so that, in a later phase of the study, such healthy and not-so-healthy practices can be discussed within groups of employees to find ways of ameliorating this problem of over-connection at work.

References

Andonova Y, Vacher B (2013) Nouvelles formes de visibilité des individus en entreprise: technologie et temporalité. Communication et Organisation 44:5–14

Aubert N (2006) Hyperformance et combustion de soi. Études 405(10):339–351

Aubert N (2008) Violence du temps et pathologies hypermodernes. Cliniques méditerranéennes 78(2):23

Bobillier-Chaumon ME, Eyme J (2011) Le cadre décadré: quand les TIC désarticulent le travail. In: Jeffroy F, Garrigou A (eds) L'ergonomie à la croisée des risques. SELF'2011, Congrès International d'Ergonomie, Paris

Bobillier-Chaumon M-E (2012) Nomadisme et dépendance. Revue CFDT Cadres, Des pratiques professionnelles reconfigurées, p 448

Bobillier-Chaumon M-E, Cuvillier B, Sarnin P, Bekkadja S (2015) Pour un usage responsable des TIC. Technologies de l'Information et de la Communication (TIC) et conditions de travail des cadres. (PhD Thesis). Guide de réflexion. Eurocadres. CEE

Boudokhane-Lima F, Felio C (2015) Les usages professionnels des TIC: des régulations à construire. Commun Organ (2):139–150

Brooks S (2015) Does personal social media usage affect efficiency and well-being? Comput Hum Behav 46:26–37

Cahour B, Salembier P, Zouinar M (2016) Analyzing lived experience of activity. Le travail humain 79(3):259–284

Créno L, Cahour B (2016) Les cadres surchargés par leurs emails : déploiement de l'activité et expérience vécue. Activites 13(1)

Datchary C (2007) Dispersion au travail, entre pathologie et compétence professionnelle. L'harmattan

Datchary C, Licoppe, C (2007) La multi-activité et ses appuis: l'exemple de la « présence obstinée » des messages dans l'environnement de travail. Activités 4(4–1)

Domenget J-C (2014) Formes de déconnexion volontaire et temporalités de Twitter. Réseaux 186 (4):77

El Wafi W, Brangier E, Zaddem F (2016) Usage des technologies numériques et modèles de la perméabilité des frontières entre la vie personnelle et la vie professionnelle. Psychologie du Travail et des Organisations 22(1):74–87

El Wafi W, Brangier E (2013) How ICT change borders between personal life and professional life? Understanding four models of boundaries permeability. In: Proceedings of the international conference e-Society 2013

Felio C (2013) Visibilité numérique des cadres d'entreprise. Un choix stratégique au risque de l'isolement. Communication et Organisation 44:123–132

Felio C (2015) Les stratégies de déconnexion des cadres équipés en TIC mobiles. Nouvelle revue de psychosociologie 19(1):241

Haroche C (2004) Manières de regarder dans les sociétés démocratiques contemporaines. Communications 75(1):147–169

Jackson T, Dawson R, Wilson D (2003) Reducing the effect of email interruptions on employees. Int J Inf Manag 23(1):55–65

Jauréguiberry F (2014) La déconnexion aux technologies de communication. Réseaux 186(4):15

Laïdi, Z. (1999). La tyrannie de l'urgence. Les Editions Fides

Lahlou S (2000) Attracteurs cognitifs et travail de bureau. Intellectica 30(1):75–113

Sarnin P, Bobillier-Chaumon M-É, Cuvillier B, Grosjean M (2012) Intervenir sur les souffrances au travail: acteurs et enjeux dans la durée. Bulletin de psychologie, Numéro 519(3):251

Vermersch P (1994) L'entretien d'explicitation, vol 2003. Esf, Paris

Playing for Real: An Exploratory Analysis of Professional Esports Athletes' Work

Esdras Paravizo[(✉)] and Renato Rodrigues Luvizoto de Souza

Department of Production Engineering,
Federal University of Sao Carlos, Sao Carlos 13565 905, Brazil
esdras@dep.ufscar.br

Abstract. In recent years competitive computer gaming—esports—is becoming increasingly mainstream as audiences, general interest and acceptance of its status as actual sports grows. Its representativeness is backed by numbers: more than 33 million unique viewers on average watched the *2017 World Championship* (a *League of Legends* championship) and the prize pool for *Dota 2, The International* championship earlier this year topped the 20 million dollars mark. As eSports consolidate itself as a possible career path, it is a natural step to start investigating its particularities regarding athletes' work, from an ergonomics perspective. This exploratory study aims to achieve a better understanding of the constraints and issues faced by esports athletes in their work. To that extent, we analyze 36 interviews conducted with professional players of *Counter Strike: Global Offensive* (CS:GO) in 2017 by the HLTV.org website, a news organization focused in covering CS:GO competitions. The interviews' analysis followed the general inductive approach for qualitative data analysis and focused on gaining insight on players' work, uncovering possible issues related to ergonomics and human factors aspects on the physical, cognitive or organizational domains. Although the interviews were not designed by the researchers and hadn't ergonomics as their main subject, they still provided useful information about players' work, the dynamics in their teams and their overall work strategies. This exploratory study can be seen as a first step for an ergonomics and human factors understanding of this emerging work situation. Further studies on this theme are necessary to understand the work of professional eSports' athletes and, possibly, improve its conditions.

Keywords: Esports · Professional game playing · Counter strike

1 Introduction

The game industry is developing at an accelerated pace. In 2012 this industry had a revenue of approximately US$ 70.6 bn, jumping to an expected US$ 137.9 bn by the end of 2018 and it is forecasted to surpass the US$180 bn mark by 2021, a 255% increase over 9 years (Wijman 2018). Several factors influence the game industry development such as: the popularization and increased access to gaming platforms, technological advances in hardware and software, globalization and the competitive environment associated to the games. Using gaming platforms for organizing competitions (both professional and amateur ones) converged to the emergence of a new type

© Springer Nature Switzerland AG 2019
S. Bagnara et al. (Eds.): IEA 2018, AISC 822, pp. 507–515, 2019.
https://doi.org/10.1007/978-3-319-96077-7_54

of sports, the esports. Esports are already popular; more than 33 million unique viewers on average watched the *2017 World Championship* (a *League of Legends* championship) and the prize pool for *Dota 2, The International* championship earlier this year topped the 20 million dollars mark.

Hamari and Sjöblom (2017) focusing on highlighting the digital nature of the esports define them as *"a form of sports where the primary aspects of the sport are facilitated by electronic systems; the input of players and teams as well as the output of the eSports system are mediated by human-computer interfaces"*. This definition renders esports as a subset of general sports, differentiating them through their inherent digital dimension.

Recognizing esports as a legitimate sport is a natural step as highlighted by Kane and Spradley (2017) although it's still a topic of debate (Hallmann and Giel 2017). The esports' sector comprises game developers, competitive leagues, competition organizers, teams organizations, brands (usually of electronics consumer goods) that also act as sponsors of championships and events, broadcasting services via streaming and the viewers who consume products and content from this sector. Naturally, legislators in different countries have started to discuss and propose bills regulating, for instance, the "electronic sports practice" in Brazil (Brasil 2017) and the "organization of videogame competitions" in France (France 2017).

Researchers from different areas have started to show interest in esports as a possible research field. Wagner (2006) for instance discuss that esports research could lead to better understanding or novel approaches for managing high-performance teams in the management field or even that a game theory approach could be applied to esports. Other studies on esports particularities and opportunities are found in the field of sport management (Cunningham et al. 2017; Funk et al. 2017; Hallmann and Giel 2017), and also in the fields of marketing (Seo and Jung 2016) and human behavior (Martončik 2015).

In the field of ergonomics and human factors there are a number of studies dealing with aspects related to traditional sports practice (Macquet et al. 2015; Olaso Melis et al. 2016; Hulme et al. 2017). However few studies have looked into esports specifically, with the exception of the ones reported by Freeman and Wohn (2017) and Lipovaya et al. (2018). In parallel, the fields of Human–Computer Interaction and Computer Supported Cooperative Work deal with questions regarding the design and use of devices and systems that incorporate computation (Carroll 2006) and understanding cooperative working practices to develop technologies and facilitate, work, communication and regulation (Schmidt and Bannon 2013), respectively. These fields are especially well-positioned to enable researchers to achieve an understanding of esports, creating possible approaches for ergonomics research to take on the esports field.

This paper aims to gain initial insight on esports athletes' work and the main issues they face from the analysis of interviews conducted with professional players of the game *Counter-Strike: Global Offensive* (CS:GO). Players in this game are in constant need of communication and information sharing among teammates and heavily relying on team strategy, thus being representative of (collective) esports in general. Furthermore, for this exploratory study, the interviews analyzed were all conducted by the same news organization which focus on covering CS:GO competitions.

The next section details the data collection and analysis process. Section 3 presents the main results from the interviews analysis and Sect. 4 discusses its implications and particularities. Section 5 presents the main conclusions, study limitations and future questions that could be addressed regarding this topic.

2 Methods

In this exploratory, qualitative study, a total of 36 interviews were retrieved from the HLTV.org website,[1] a Danish news organization focused in covering CS:GO competitions. The interviews selected were conducted during international competitions carried out in the first semester of 2017, with professional players from different nationalities. Only text interviews were included into the sample and only competitions which had a prize pool of US$ 150.000 or more were considered. Two interviews were excluded from the sample, one for being a collective interview and the second for not having the direct responses of the interviewee.

The analysis of the interviews followed the general inductive approach for qualitative data analysis devised by Thomas (2006). Three main categories were created based on the broad ergonomics' domains (physical, cognitive, organizational) as defined by the International Ergonomics Association (IEA 2018). The first author closely read all the interviews and coded relevant segments to emerging subcategories.

The coding process was conducted in the QDA Miner version 5. Furthermore, in order to assess the reliability of the coding process, the second author of the paper read a subset of the interviews, coding the segments to the categories and subcategories created and a Scott's pi statistic (Scott 1955) was calculated. The following section shows the results of the interviews retrieval and analysis process.

3 Results

The 34 interviews retrieved from the HLTV.org website (following the criteria presented on the previous section) covered the period from 12/01/2017 to 25/06/2017 encompassing a total of 8 international events. In total, thirty different players from twenty different teams were interviewed (four players were interviewed twice in the period).

The coding process highlighted 303 relevant segments that were coded to one of the available 35 subcategories (which were grouped under the three main categories related to the ergonomics domains). Most of the segments coded discussed organizational aspects (76%, n = 231), followed by the cognitive aspects with 23% (n = 70) and physical aspects only appeared in two segments (1%, n = 2). Table 1 summarizes the

---

[1] The interviews were retrieved from https://www.hltv.org/events. The researchers chose not to provide in the paper direct links to each individual interview in order to not provide identification of either players or organizations. A full list of all the interviews considered can be obtained from the authors upon request. We stress that all interviews were conducted by HLTV.org journalists and made available on their website.

subcategories identified during the coding process. The seventeen subcategories shown on Table 1 cover 83.2% of the total of segments coded. The remaining 18 subcategories represent less than 17% of the total of coded segments and thus were grouped together under the "others" label on Table 1.

Table 1. Summary of the number of segments coded to each subcategory and category.

Category	Subcategory	# Segments coded (%)
Organizational	Team training	40 (13.2%)
Organizational	Preparation for opponents	34 (11.2%)
Organizational	Roster changes reasons/impact	23 (7.6%)
Organizational	Roles of team members/team strategies	22 (7.3%)
Cognitive	Confidence on individual skill	21 (6.9%)
Organizational	Map pool aspects (changes, veto)	18 (5.9%)
Organizational	Goal setting for the team	13 (4.3%)
Organizational	On the fly adaption	13 (4.3%)
Organizational	Aspects related to coaches	12 (4.0%)
Organizational	Team communication	9 (3.0%)
Cognitive	Earning confidence from victories	8 (2.6%)
Organizational	In game leader (IGL) style/role	8 (2.6%)
Cognitive	Pressure affecting performance	7 (2.3%)
Cognitive	Importance of 'mentality'	6 (2.0%)
Cognitive	Poor decision making affecting performance	6 (2.0%)
Organizational	Support staff impact	6 (2.0%)
Cognitive	Poor morale due to losses	6 (2.0%)
–	Others	51 (16.8%)
	Total	303 (100.0%)

To assess the reliability of the coding process, the Scott's pi statistic was calculated to verify intercoder agreement. This statistic was calculated in the QDA Miner software. The value of 0.811 obtained for the Scott's pi was deemed to be satisfactory to the present study.

4   Discussion

The results from the interviews analysis uncovered a broad range of issues faced by professional esports players. The aspects reported on the interviews mostly focused on the organizational domain, comprising the training of the team, how does the preparation for opponents unfold, issues related to team management such as roster changes and the roles each player has in the game, the responsibilities of the in-game leader (IGL), coaches and support staff. Cognitive aspects highlighted the importance of players' "confidence" in their skillset and as a team which is achieved by winning matches, however it's not clear on the interviews what gives them these "boosts" in confidence. Furthermore, the pressure on players, either from their organization, themselves or the community is also a factor that influences their performance.

Training for esports usually consists of regular training sessions throughout the week or intensive training periods ("bootcamp") before an important tournament. Furthermore, a usual arrangement between players and their organizations is setting up a "gaming house", where the players and support staff will live and train blurring the boundaries between work and personal live. The following excerpts highlight these aspects, the third one specially highlights the importance of the gaming house

> We'll now finish our mini bootcamp in Dallas, then we'll go back home and get a couple of days off again. We didn't really talk about specifics yet, but we won't underestimate anything, we won't take too much time off, we'll work hard and play a little bit more relaxed at [the next tournament]. (Player A, 25 years old).
>
> After France, we took two days off and then played every day including weekends. We practiced mostly T sides and we prepared for some opponents, because we already knew the groups and who we were going to play against. [...] But the preparation was exhausting, we played at least eight hours every day. (Player B, 25 years old).
>
> Having a gaming house would be a great step, but that's really hard and I understand that. (Player C, 20 years old).

Another aspect that deeply affects teams' dynamics and performance are the changes to the active roster. In esports it's relatively common for organizations to change players and coaches. A rough estimate based on the roster transfers tracked by the Score eSports website[2] for the Summer of 2017 onwards, show that from the roughly 186 players and coaches considered (from 31 teams) there were about 78 players moves between teams, amounting to more than 40% of the total. Team communication which is an essential aspect of esports is also hindered due to frequent player changes and team's "atmosphere" (as the third following excerpt highlights). These frequent changes ultimately pose yet another challenge to teams to incorporate the new player into the existing system and regain coordination as highlighted in the following excerpts.

> Well, mainly we will have to remake our entire game, because our game was based on [Player Z, who left]. Him leaving will hurt us, but maybe not as much as people think. We will try to prepare as much as we can, change everything and we'll see how it goes. (Player D, 25 years old).
>
> Another thing is that [the new player] is not used to the role he's playing with us, it's a different thing, in [his previous team] and other teams he was playing with, he was always going in the front, making his own plays, trying to do some impact in the game. There are a lot of maps where he's still doing that, but on other maps he's [...] doing something different than he's used to. So I think it's more about him adjusting to us as well. But so far it has been perfect, although we can't really say from only playing against North American teams, I think we have to put him to a test on LANs. (Player E, 25 years old).
>
> Also, just the communication level between the two teams is like night and day, these guys almost sometimes talk too much, I think that's great, you kind of wish that happens in a team. Unfortunately with [previous team], I think our comms took a big blow towards the end when people were unhappy with the environment. So, I think in general the environment here is a lot better than it was in [previous team]. (Player F, 27 years old).

Regarding teams' internal organization during the matches, due to the nature of the game, each player has a specific role to play. Teams strategies can be built around a collective strategy for gaining control over strategic areas for instance or around players in specific roles. It's usual that teams have an in-game leader (IGL) who can be seen as the captain of the team. In general, the IGL is responsible for deciding which approach the team will take in a specific round, what play they'll try to employ, to keep track of the economic aspects of the game and other on the fly adaptations. Being the IGL takes an extra toll on players, since they have to be aware of the overall situation (based on info passed by team members) as well as perform the general player role. The excerpts highlight the different roles in a team and the demanding characteristics of the IGL position.

Like you said, AWPing [sniping] in [the team] hasn't been good. I don't think he has been comfortable, he has been sacrificing too much, he has been more of a support AWPer [sniper] rather than us supporting him, which is how it should be. (Player G, 23 years old).

Unfortunately for me, and for my career, I took a lot of heat and flak for being the in-game leader of [a past team], my individual level dropped significantly, even the way I thought about the game changed. I actually think that helped me develop more as a player, but like the raw stats, playing the support role, having the flashes out every round, not lurking, not having any aggression in my style of play, no matter what happens, it's gonna hurt your individual level. Stats-wise, teamplay-wise, it definitely hurt me a lot, I'm very happy that on [current team] I don't have to lead, [Player W] is here [...]. I'm like the right-hand man to [Player W], so I still give my ideas, but I'm able to focus more on my style of play, on my lurking, I can free up myself that I don't have to think about everything that's going on in the middle of the round. (Player I, 27 years old).

Besides players, the teams usually have coaches and additional support staff who may assist both during the training and preparation for tournaments but also during the competition in itself. The roles of coaches in the team is particularly interesting to analyze since it differs from traditional sports coaches in the sense that they have little to no power on changing the roster of the team (usually, there is no bench) as pointed out by one of the interviewees. Nonetheless, the "outside" perspective of the coach helps in the identification of improvement areas and issues that can be revisited later on during training. Furthermore, support staff such as analysts who look for information on how opponents (teams or players) usually play a particular map is seen as a valuable source of information for players. There also appeared in the interviews verbalization regarding the need of a sports psychologist to assist players' mental preparation for the games. The following excerpts are representative of the issues highlighted.

It definitely hurts having no coach now, because it's very hard for me to see the mistakes when I'm having to play my own game. I can't see what they are doing at all times. Now it's very hard for me to actually fix problems in the team. [...] There are some teams that do have coaching power, this team is not one. I think the biggest problem is that we don't have anything to threaten players with. Traditional coaches have a whole bench. If you mess up, you can get replaced at any moment. And here, you can't replace anyone, that's the biggest problem. And then, what I was mentioning earlier, when I felt kinda useless was... I have six instances in a match where I can give input, four timeouts, before the first [round] and before the second half. So you don't really feel too involved in it, to be honest, it's kinda boring. You do a lot of prep work, and even sometimes your prep work is ignored by the players. They don't use information that you give them so a lot of times you feel kinda useless. (Player J, 27 years old).

We've been working with our sports psychologist about that as well, how people affect us before and after the game. (Player K, 21 years old).

He [the analyst] is doing amazing work. We all have a lot of respect for him, he is working really hard, putting a lot of hours to prepare for tournaments. He is basically analyzing all of the opponents and even our game. [...] For example watching demos for the players, like players ask him questions how for example a certain player plays a position or if they have a strategy on A and what they are doing. [...] But yeah, I think he is really important, I think in other teams a lot of coaches do that, I couldn't do that though, it's way too much work and I think it's really good to separate two things, to have a coach and an analyst. (Player L, 28 years old).

Yet another aspect that emerged from the interviews analysis is related to the tournaments management in terms of which maps (sort of scenarios for the matches) are going to be considered for selection (usually tournaments have a standardized group of 7 available maps – the map pool). From time to time, changes are made in the map pool, removing one map in favor of another that wasn't being played competitively, thus changing teams' strategies to select which maps they'll play (veto process). Maps-related aspects are also linked to the game developers which are a crucial stakeholder of the eSports environment. Furthermore, tournaments' organization sometimes fail to provide a satisfactory physical environment for players to compete, affecting players health and performance. The excerpts below are representative of these aspects.

By the end of the last tournament, we started seeing a couple of different things on [Map X], so we started to see people adapting a little more, but I agree that it needed some changes. For us, it's sad, because we've been working hard to get [Map X] into our map pool, and now that we think it's decent and we're competitive on it, it's just not here anymore. [...] No, you still have the veto, in group stage you have three vetoes, so if you didn't have time to practice on [New Map], you can veto it in the beginning. But I think as unified and as fast we are with the change, the better, if one tournament had [Map X] and another had [New Map], then suddenly we need to practice eight maps, so the best thing to do is to change it as soon as you can so everyone changes together. (Player M, 25 years old).

I guess everyone has run out of energy by now, it's so hot in those booths, it's so hard to concentrate, everyone is having headaches. Hopefully, tomorrow they fix the booths so it's not so warm in there because now you can't play in there and focus at the same time. (Player N, 24 years old).

Moving on to cognitive aspects that appeared in the interviews, players reported the "confidence" factor as being an important one for their individual performances. Pressure either self-imposed or from community or organizations is a factor that affects players. Morale is also affected when players lose "badly" (from a great point difference, for instance). The following excerpts highlight these aspects.

Because when he has confidence it shows, he can hit any shot and can make any play he wants. What I've been trying to do with him and he's been trying to really improve on is to get his confidence back and also just make a lot more plays for himself. When he gets in the rhythm he is a lot better, we just have to support him more instead of him supporting us. (Player O, 23 years old).

Yeah, coming into the tournament, we had a lot of pressure. Of course, we won the Major, we're top one in the world and now we're looking to establish ourselves there for a long period of time. Coming in, we tried to not put so much pressure on ourselves and just try and take it like a normal tournament. (Player P, 21 years old).

Personally, I felt bad after that map because my performance was very bad. I had much higher expectations for myself? If I would've played Mirage a little better, I think we would've won that map. (Player Q, 20 years old).

You know, it's unfortunate that none of the games were close because that really did hurt our morale. That's the one thing we did not want to happen, to get blown out like we did. The next few weeks is really going to show whether or not we have the mental fortitude to continue basically. (Player R, 27 years old).

It was tilting, it was hard to come back from that, because we talk about these mistakes after just about every match, we've been dealing with this after every match we have played since Tours. In every match, there are situations like this from which we need to learn and avoid making the same mistakes in the future. But since Tours to now it went the same way and we were rarely able to avoid them. (Player S, 25 years old).

5 Conclusion

This exploratory study on the work conditions of professional esports players albeit somewhat limited on the data collection method provides an overview of the main aspects affecting players' work from an ergonomics standpoint. The recurrence of topics related to organizational aspects (team coordination and communication, roles of players and staff members) and cognitive aspects (e.g. pressure, mentality) can be particularly interesting for various ergonomics subfields.

The lack of comments on physical issues (pain, injuries and other sorts of work related musculoskeletal disorders) may indicate that other data collection methods (surveys, epidemiological studies) should be employed to further investigate this aspect on professional esports.

Nonetheless, it's possible to stablish that professional competitive gaming is a form of computer mediated work, carried out collaboratively among team members that play specific roles in the game and that players are exposed to a range of factors (organizational, cognitive or physical) that may affect their performance. Intersections with the fields of Human–Computer Interaction and Computer Supported Cooperative Work should also be further explored.

Going forward, a series of questions persist. How does different games affect players work, team dynamics and overall work organization? How the different emerging formats for competitive professional play (for instance the *Overwatch League*) affect players in comparison with the endemic tournaments and leagues of others esports? What are the long-term prospects of an eSports career? How does game development affect players and how it can be improved? How one can assess the mental workload of a professional esports player? How the gaming houses set-up affect players' work? Many other questions may be raised upon a closer examination of the subject which presents itself as a prolific field of research and practice for ergonomics and human factors.

References

Brasil (2017) Projeto de Lei do Senado No 383 de 2017. Senado, Brasília
Carroll JM (2006) Human–computer interaction. Encycl Cogn Sci 1–4
Cunningham GB, Fairley S, Ferkins L et al (2017) ESport: construct specifications and implications for sport management. Sport Manag Rev 21:1–6. https://doi.org/10.1016/j.smr.2017.11.002

France (2017) Décret no 2017-871 du 9 mai 2017 relatif à l'organisation des compétitions de jeux vidéo

Freeman G, Wohn DY (2017) Understanding eSports Team Formation and Coordination. Computer Supported Cooperative Work (CSCW)

Funk DC, Pizzo AD, Baker BJ (2017) ESport management: embracing eSport education and research opportunities. Sport Manag Rev 21:7–13. https://doi.org/10.1016/j.smr.2017.07.008

Hallmann K, Giel T (2017) ESports-competitive sports or recreational activity? Sport Manag Rev 21:14–20. https://doi.org/10.1016/j.smr.2017.07.011

Hamari J, Sjöblom M (2017) What is eSports and why do people watch it? Internet Res 27:211–232. https://doi.org/10.1108/IntR-04-2016-0085

Hulme A, Salmon PM, Nielsen RO et al (2017) From control to causation: validating a 'complex systems model' of running-related injury development and prevention. Appl Ergon 65:345–354. https://doi.org/10.1016/j.apergo.2017.07.005

IEA (2018) Definition and domains of ergonomics. http://iea.cc/whats/index.html. Accessed 31 Mar 2018

Kane D, Spradley BD (2017) Recognizing ESports as a Sport. Sport J 1

Lipovaya V, Lima Y, Grillo P, et al (2018) Coordination, communication, and competition in eSports: a comparative analysis of teams in two action games. In: Proceedings of the 16th European conference on computer-supported cooperative work—exploratory papers, reports of the European Society for Socially Embedded Technologies, Nancy, pp 1–26

Macquet AC, Ferrand C, Stanton NA (2015) Divide and rule: a qualitative analysis of the debriefing process in elite team sports. Appl Ergon 51:30–38. https://doi.org/10.1016/j.apergo.2015.04.005

Martončik M (2015) E-Sports: playing just for fun or playing to satisfy life goals? Comput Hum Behav 48:208–211. https://doi.org/10.1016/j.chb.2015.01.056

Olaso Melis JC, Priego Quesada JI, Lucas-Cuevas AG et al (2016) Soccer players' fitting perception of different upper boot materials. Appl Ergon 55:27–32. https://doi.org/10.1016/j.apergo.2016.01.005

Schmidt K, Bannon L (2013) Constructing CSCW: the first quarter century. Comput Support Coop Work 22:345–372. https://doi.org/10.1007/s10606-013-9193-7

Scott WA (1955) Reliability of content analysis: the case of nominal scale coding. Public Opin Q 19:321. https://doi.org/10.1086/266577

Seo Y, Jung S-U (2016) Beyond solitary play in computer games: the social practices of eSports. J Consum Cult 16:635–655. https://doi.org/10.1177/1469540514553711

Thomas DR (2006) A general inductive approach for analyzing qualitative evaluation data. Am J Eval 27:237–246. https://doi.org/10.1177/1098214005283748

Wagner MG (2006) On the scientific relevance of eSports. In: Internet computing and conference on computer games development, ICOMP 2006, Las Vegas, pp 437–442

Wijman T (2018) Global Games market revenues 2018 | Per Region and Segment | Newzoo. https://newzoo.com/insights/articles/global-games-market-reaches-137-9-billion-in-2018-mobile-games-take-half/

What Do You Need to Know to Stay Healthy? – Health Information Needs and Seeking Behaviour of Older Adults in Germany

Sabine Theis[1(✉)] (iD), Dajana Schäfer[1], Katharina Schäfer[1],
Peter Rasche[1], Matthias Wille[1], Nicole Jochems[2],
and Alexander Mertens[1]

[1] Chair and Institute of Industrial Engineering and Ergonomics,
RWTH Aachen University, Bergdriesch 27, 52062 Aachen, Germany
s.theis@iaw.rwth-aachen.de
[2] Institute for Multimedia and Interactive Systems (IMIS),
University of Lübeck, Ratzeburger Allee 160, 23562 Lübeck, Germany

Abstract. Deep understanding of users' needs is crucial for developing successful digital health technology. At the beginning of system development, it is thus important to analyze and specify the context of use within the users-centered development process. Knowing what patients need to know about their health and which information sources they apply to find those, bears implications for personal health ICT conveying health information to the elderly patient. Present results from a survey on health information need and seeking behaviour of N = 551 older adults in Germany suggest that older adults are fairly satisfied with the information they get, indicating a low need to acquire health information. Higher health information need corresponds with a larger amount of health apps installed on older adults tablet PC and with the usage of smart-watches and apps in general. Finally, results support the theoretical influence of demographic variables. Here, educational attainment significantly revealed to be a main influence on information need.

Keywords: Health information need · Information seeking behavior
Gerontology · eHealth · Context analysis · Older adults · Ageing

1 Introduction

In times of demographic change, the need for innovative and comprehensive care concepts for older adults has increased. This is because people are not only getting progressively older, but also because the incidence and prevalence of illnesses and the resulting limitations can greatly reduce the quality of life for the people affected [1]. The application of digital information and communication technology (ICT) and related services in existing and future health services represents an extremely promising opportunity. Digital health is thus expected to meet the challenges posed by the societal age shift. This does not only apply to Germany but to most western countries [2]. The decoupling of medical care from the local and temporal availability of medical personnel, which is associated with digital health systems, makes it possible to provide

© Springer Nature Switzerland AG 2019
S. Bagnara et al. (Eds.): IEA 2018, AISC 822, pp. 516–525, 2019.
https://doi.org/10.1007/978-3-319-96077-7_55

patient-specific and cost-efficient services and therapies. Unfortunately, the nationwide distribution of so-called eHealth systems is the exception in Germany and other European countries. Not only legal and organizational obstacles but above all a lack of knowledge of the requirements of older adults regarding the design of eHealth and eHealth services could be identified in post-hoc analysis of facilitators and enablers of digital health systems [3]. Therefore, existing systems and services often lack acceptance and adherence. A means to support the acceptance and permanent use of digital health information systems is the user-centered development process [4, 5]. It begins with an understanding of the user, his tasks and the environment. Information demand and information search behaviour describes the context of digital information systems. Accordingly, identifying the users' goals, their information needs and the different ways via the user, obtains the information needed, would help to better understand the context of ICT systems and thus support its design. In this paper, we would like to examine in more detail the potential relationships between ICT use and information requirements as well as information search behaviour using the example of the application context of health. The following section, in this regard, deals with a theoretical model illustrating the concepts of health information need and seeking behaviour and the variables that are expected to influence them.

2 Related Work

Various theoretical models exist to describe information need and information seeking behaviour. One incorporating health variables is Wilsons model of information seeking behaviour. The goal of Wilsons model of information seeking behavior [6–8] as depicted in Fig. 1, is to provide a general overview of information behaviour and the relationship between the user and the main concepts of information need and information seeking behavior. Information seeking behaviour describes human behaviour as part of the search for information, which usually represents the consequence of a previously existing need for information. To give an example, imagine the case of a patient receiving a medical diagnosis. Consequently, this patient may now develop a need to seek additional information to his or her diagnosis. Accordingly, to satisfy this need, the patient will want to employ information sources or information retrieval systems (IR-systems) such as the internet, using a computer, a smartphone or other compatible devices. This can lead to a success or a failure of this source. In the case of success, the patient will use the gathered information and will either have satisfied the initial need for more information or not. Success or failure of one information source also influences the use of other sources. Additionally, the model shows that the process of seeking information can involve the exchange of information with other people. In addition, the use and subsequent processing or transfer of information can be part of information behaviour that again can be accomplished by incorporating the use of technical devices. The concept of information seeking—as mentioned previously—is based on the concept of information need. Need is a subjective experience which can have underlying physiological, cognitive and affective motives that on part of the information user can influence the seeking process. Next to those three kinds of need, also personal circumstances such as the social role and environmental factors can exert influence on the seeking behaviour. Other intervening factors that were later included in the revised model can be personal characteristics of psychological, cognitive or emotional kind. In addition, demographic variables, educational variables, social/interpersonal variables, economic variables and source characteristics should be considered as being possibly intervening. Moreover, the revised model now also takes into account in what way information-seeking behavior can occur, ranging from passive attention to more active and ongoing search. Furthermore, it incorporates different stress- coping mechanisms that can arise when e.g. getting an unpleasant medical diagnosis. Some people may show an avoidant behaviour while others orientate towards the possible threat. Additionally, the construct of self-efficacy that can affect how successful information sources are used, is also considered.

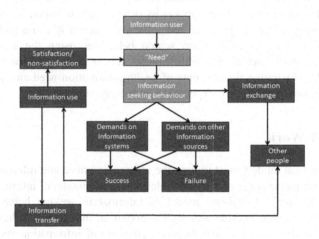

Fig. 1. The general model of information behaviour [6]

Information need and seeking behaviour was investigated for numerous application contexts. Greysen et al. [13] have investigated how functional impairment affects internet use among older adults and found that this had negative implications for meaningful use of patient portals. Yagil et al. [14] also found that using everyday technologies generated stress for older adults, which also suggested a potential barrier to patient portal adoption and use. Ancker et al. have investigated how disadvantaged populations use patient portals, based on system access logs of more than 74,000 patients at a Federally Qualified Health Center in New York City [15]. They found good adoption of the system by low-income participants, but small persistent racial disparities in both the likelihood to be offered an access code and the likelihood to actually use the system. Thus, income itself did not appear to be a limiting factor in adoption of patient portals, but may have a modifying influence in combination with other demographic factors. However, these results may not generalize to rural populations. In a qualitative study dealing with eHealth users in rural Nebraska, Fruehling reports that usability and clarity were important in the development of such systems, and also noted that security and privacy issues were a major concern with this population [16]. It can be concluded, that most studies show that information need and technology usage correspond in some way. Additionally, while the investigations of health-related information need primarily focus on specific diseases and contexts [17–21], the information need and information seeking behaviour of older adults has so far not been described in general. This holds especially for a German population of older adults. Therefore, in order to close this knowledge gap, we want to find out (RQ1) Where do older adults seek information on their personal health and on health topics in general? (RQ2) How do older adults use information on their personal health and on health topics in general? (RQ3) What demographic factors influence health information need and seeking behaviour? (RQ4) Finally, we are interested in statistically quantifying the relationship between technology usage and information need by asking if there is a relationship between health information need/seeking behaviour and technology usage.

3 Method

Aim was, among others, to describe the health information need of older adults as starting point and motivation for investigating ergonomic aspects of personal health data access [9–11]. A detailed description of the study method and descriptive results can be found in [12].

3.1 Sample

The mean age of the participants (N = 551) was 69.17 (SD = 5.787) and ranged from 60 to 90 years. The gender ratio of the sample was balanced including 51.3% males and 48.7% females. 441 participants (80%) were already retired, while 109 (19.8%) were still working.

3.2 Procedure

A 15-page paper-based questionnaire was sent to N = 5000 randomly selected German adults aged 60 years and older in June 2016. 586 participants responded to our questionnaire. Thirty-five participants had to be excluded because they were younger than 60 years, resulting in a final sample size of N = 551 and a response rate of 11%.

Investigating elderly's health information need and seeking behaviour started with an adapted version of the EU-HLS-47, consisting of 15 instead of 47 items to assess participants' health competence selected in an expert workshop. Following this, general health information need and trust in information sources (TV, magazines, internet, doctor and pharmacist, family and friends or other) were queried. Subsequently, questions were asked on topic-related information need (diagnosis, therapy, interdependency of medicine). This section closed with a question on information sources participants use to access (personal) health information and with a question on characteristics of health information search behaviour considered as relevant in Wilsons model on information seeking behaviour.

Returned surveys were digitalized through Remark Office software. SPSS 24 served to conduct the statistical analysis of the data resulting from survey digitization with the help of Remark software.

4 Results

4.1 Health Information Need

Older adults in Germany are satisfied with the information they get about (personal) health. Sixty-four percent claim to be 'very satisfied' and 'satisfied' while 27.6% consider their satisfaction with the health information they get as 'neutral'. Only 33 participants are not satisfied. The most important health topics are information on health billing from doctors and health insurances (n = 516, M = 3.05, SD = 0.93). Besides billing transparency, older adults are interested the most in medical diagnosis (n = 540, M = 2.27, SD = 0.75), the meaning of examination results (n = 535, M = 2.37, SD = 0.72) as well as therapy and treatment options (n = 530, M = 2.66, SD = 0.83).

4.2 Health Information Seeking Behaviour

Descriptive results suggest that older adults in Germany get information about health primarily from their doctor and pharmacist or from TV. The internet in contrast is the least used information source when it comes to searching and finding health-related information. In total 45.6% of N = 551 valid answers find health information on TV, 41.6% find health information in magazines, while 37.4% look on the internet. Doctor and pharmacist are the primary source for 47.9%, while friends and family are only consulted by 38.8%. Participants trust their doctor and pharmacist the most (M = 1.96, SD = 0.67, n = 539), while their trust in magazines and newspapers is the lowest (M = 3.30, SD = 0.80, n = 492).

Older adults primarily share their information with family and friends (89.7%) and with their doctor or their pharmacist (73.5%). Just over a quarter (25.2%) of older adults in Germany share, health-related information with acquaintances and 11.8% share it with health insurance companies. A negligible number of participants share health-related information on social networks, with self-help groups or with people, they do not know at all. From N = 551 valid answers only twenty people do not share health-related information at all.

Older adults in Germany describe their search for information about health in general or their personal health as active and incidental. On a scale where 1 = 'fully agree' and 5 = 'fully disagree' the average rating of the statement that their information search is active is M = 2.7 (SD 1.3, n = 387) while it is M = 2.7 (SD 1.22, n = 424) for the statement that their information search is causal. The least consent is received by the statement 'I search regularly for health information' with a rating of M = 3.3 (SD = 1.9, n = 361).

4.3 Relation Between Demographic Variables, Information Need and Information Seeking Behaviour

Health Information Need. There was a significant association between the information need and the educational attainment: People with lower educational attainment are less satisfied with the information they have regarding their personal health.
Older adults without education were satisfied the least with the information they receive about medical billing. Differences between all other levels of educational attainments were non-significant. Another significant association could be identified regarding the satisfaction with information about the experience that others have with their health and their education. People with a lower level of education were much more satisfied with the information available to them than people that are more educated. As there was also significant association between the satisfaction concerning information about (interaction of) medications and the number of chronic diseases our results suggest that people with a higher amount of chronic diseases are much more satisfied with this kind of information they get than people with no or less chronic diseases. The same applies to satisfaction concerning information about the course of diseases. Here, too, the group of participants without qualifications is satisfied the least. Finally, we found women are more satisfied with the information they get than men are. This is the opposite for the understanding elderly's have about their medical diagnosis. Women indicated that understanding the meaning of a diagnosis is very easy while more men than women indicated a neutral position to this ease of understanding.

Trust in Health Information Sources. Surprisingly, people in the eastern part of Germany rate the internet as slightly less trustworthy than people in the western part. Similarly, we found a significant association between the trustworthiness in doctor/chemist and their education. Older adults with university degree rate this information source as less trustworthy while people with lower educational level trust this source more.

Health Information Sharing. The older people are the less they share health related information with family and friends. In nearly all age groups the amount of people who share their information was higher than the amount of people who do not so. Furthermore, results reveal that people who live in a supervised community for older adults, share health information with acquaintances. There was also a significant association between sharing health information with health fund and age group. The amount of older adults who do not share health information is in all cases higher than the amount who do not share their health information with their health fund. The only exception here is the age group of the 85 to 90 year old adults. Here more people share information about their health with their health fund. Older adults in the east of Germany share health information less with their doctor or chemist than in the western part.

Health Information Search Characteristics. Significantly, more women than men describe their search for health information as active. Finally yet importantly, results indicate that the major part of older adults without a chronic diseases report that they do not regularly search health information while the people with two or more chronical diseases report the opposite.

4.4   Relation Between Health Information Need and Technology Usage

There was a significant negative relationship between information need and the possession of a computer or laptop $r_s = -.106$ T 95%, BCa CI $[-.193, -.011]$, $p = 0.017$. Hence, based on the point biserial correlation we conclude that the possession of a laptop or computer accounts to 0.11% to the information need represented by the satisfaction with available health information. We additionally found a significant negative relation between information need and frequency of using a smartwatch. $r_s = -.301$ T 95%, BCa CI $[-.505, -.101]$, $p = 0.037$. That means the lower the satisfaction with available health information the more frequent older adults use smartwatches. Furthermore, a significant negative relationship could be identified between information need and older adults frequency of app usage on tablet PCs. $r_s = -.146$ T 95%, BCa CI $[-.288, .001]$, $p = 0.046$ as well as between information need and amount of health apps older adults have installed on their tablet PC $r_s = -.170$, T 95%, BCa CI $[-.335, .003]$, $p = 0.041$. This indicates that lower satisfaction with health information available corresponds with a higher amount of health apps installed on older adults tablet PC. This corresponds with the relation between information need and frequency of app usage (Tables 1 and 2).

Table 1. Health information need of older adults (1 = very satisfied, 5 = very unsatisfied)

Education	Information need (N = 520)					
	1	2	3	4	5	Total
None	0	0	0	0	1	1
Certificate of secondary education	4	21	18	9	0	52
General certificate of secondary education	2	42	36	5	1	86
Professional education	12	109	34	8	1	164
A-Levels	0	17	12	1	0	30
University degree	9	109	49	8	0	175
Other	1	8	2	1	0	12

$\chi^2 (24) = 209,385$, $p < .000$; Phi $= .635$; Cramer-V $= .317$; Contingency coefficient $= .536$

Table 2. Chi-square test statistics of associated variables (crosstabs).

Associated variables: information needs	χ^2	DF	p
General health information need × education	209.38	24	<.001
Billing information × education	60.48	24	<.001
Experiences of others × education	52.46	20	<.001
Medication interactions × number chronic diseases	47.93	24	<.001
Course of disease × education	36.44	24	<.05
Therapeutic options × gender	12.05	4	<.05
Meaning of examination results × gender	11.08	4	<.05

(*continued*)

Table 2. (*continued*)

Associated variables: information needs	χ^2	DF	p
Associated variables: trust in sources			
Internet trustworthiness × postal codes west/east	12.08	4	<.05
Trustworthiness doctor/chemist × education	59.25	18	<.001
Associated variables: information sharing			
With family/friends × age group	15.2	6	.05
With acquaintance × living situation	9.76	4	<.05
With health fund × age group	23.97	6	.001
With doctor/chemist × postal code (west/east)	5.519	1	<.05
Associated variables: search characteristics			
Activity of search × gender	11.56	4	<.05
Longevity of search × number chronic diseases	47.93	24	<.01

5 Summary

Results of a large-scale survey with adults older than 60 years in Germany suggest that older adults are mainly satisfied with the information they get about health in general as well as about their personal health. As a result, in addition to usability and legal hurdles, a need for health information that is sufficiently addressed by traditional information sources could also be attributed for the comparably slow spread of digital health technology into the German market. Besides doctors or pharmacists, the television is the most important source for health information. As older adults consider professional medical sources as most trustworthy, involving these actors in the introduction and application of digital health technology might increase the acceptance and adherence of digital health systems. The results further support the theoretical influence of demographic variables on health information need and seeking behavior. It is not surprising that both objectives strongly relate to educational attainment. Furthermore, the understanding of health information and level of activity in information seeking revealed to be different depending on gender. In addition, the occurrence of chronic diseases plays a major role for the information search continuity. Particularly unexpected were the results, which suggest that the trust in information sources as well as the information sharing with the doctor seem to be influenced by the political history. In the areas of the former DDR, the mistrust for the internet revealed to be higher. In addition, older adults coming from those areas share personal health information to a smaller extent with the doctor or pharmacist. However, it has to be pointed out that a relation between variables may not be equated with causality. Further investigations on this aspect are necessary. Sharing information about one's own health relates demonstrably to age. The older people get, the less they share information with family and friends. However, it seems reasonable to assume that this is not only due to the willingness and motivation, but also to mortality. Finally, our results empirically demonstrated a relationship between health information need and seeking behaviour and older adult's technology usage. Only in cases where existing sources did not deliver the desired information, health applications and devices were applied. Here it also has to be considered that the significant relation of mentioned variables does not necessarily implicate causation. However, it indicates the theoretical relationship describing that a sources failure to deliver desired information influences the usage of other sources. Further investigation is needed in order to describe the nature of this relation in depth, as the amount of people who used health apps was low in our sample. This might be attributable to the acquisition method, meaning that an online sample might have exhibited a more frequent use of health technology due to a higher technical affinity of participants. Presented results thus not only show that information need and information seeking behaviour relate to technology usage, they also proofed a feasible mean to describe the context of digital health systems beyond disciplinary boundaries.

Acknowledgements. This publication is part of the research project "TECH4AGE," financed by the Federal Ministry of Education and Research (BMBF, under Grant No: 16SV7111) and promoted by VDI/VDE Innovation + Technik GmbH.

References

1. Akner G (2009) Analysis of multimorbidity in individual elderly nursing home residents: development of a multimorbidity matrix. Arch Gerontol Geriatr 49(3):413–419
2. Pack Jochen et al (1999) Future report demographic change: Innovation ability in an ageing society. DLR-PT, Bonn
3. Ziefle M, Röcker C (2010) Acceptance of pervasive healthcare systems: a comparison of different implementation concepts. In: Pervasive computing technologies for healthcare (PervasiveHealth), 2010 4th international conference on-no permissions, pp 1–6
4. Nielsen J (2010) Usability engineering, 18th edn. Morgan Kaufmann, Amsterdam u.a.
5. Norman DA Psychology of everyday things: the design of veryday things—revised and expanded edition. 1. Industrial design-Psychological aspects. 2. Human engineering
6. Wilson TD (2000) Human information behavior. Informing Sci 3(2):49–56
7. Wilson TD (2008) Activity theory and information seeking. Ann Rev Inf Sci Technol 42(1):119–161
8. Wilson TD, Walsh C (1996) Information behaviour: an inter-disciplinary perspective: a review of the literature. British Library Research and Innovation Centre, London
9. Theis S, Rasche P, Mertens A, Schlick CM (2017) An age-differentiated perspective on visualizations of personal health data. In: Advances in ergonomic design of systems, products and processes. Springer, Berlin, pp 289–308
10. Theis S et al (2016) Ergonomic considerations for the design and the evaluation of uncertain data visualizations. In: Human interface and the management of information: information, design and interaction. Springer, Berlin, pp 191–203
11. Theis S, Wille M, Mertens A (2016) Age-dependent health data visualizations: a research agenda. In: Mensch und computer 2016-workshopband
12. Mertens A, Rasche P, Theis S, Bröhl C, Wille M (2017) Use of information and communication technology in healthcare context by older adults in Germany: initial results of the Tech4Age long-term study. i-com 16(2):165–180
13. Greysen S, Chin GC, Sudore R, Cenzer I, Covinsky K (2014) Functional impairment and internet use among older adults: Implications for meaningful use of patient portals. JAMA Internal Med

14. Yagil D, Cohen M, Beer JD (2013) Older adults coping with the stress involved in the use of everyday technologies. J Appl Gerontol. https://doi.org/10.1177/0733464813515089
15. Ancker JS, Barron Y, Rockoff ML, Hauser D, Pichardo M, Szerencsy A, Calman N (2011) Use of an electronic patient portal among disadvantaged populations. J Gen Internal Med 26 (10):1117–1123
16. Fruhling, A. Perceptions of e-health in rural communities. Patient-Centered E-Health (2008), pp 157–167
17. Azadeh F, Ghasemi S (2016) Investigating information-seeking behavior of faculty members based on wilson's model: case study of PNU University, Mazandaran. Iran. Global Journal of Health Science 8:9. https://doi.org/10.5539/gjhs.v8n9p26
18. Harland JA, Bath PA (2008) Understanding the information behaviours of carers of people with dementia: a critical review of models from information science. Aging Mental Health 12(4):467–477. https://doi.org/10.1080/13607860802224300
19. Kostagiolas PA, Lavranos C, Korfiatis N, Papadatos J, Papavlasopoulos S (2015) Music, musicians and information seeking behaviour: a case study on a community concert band. J Doc 71(1):3–24. https://doi.org/10.1108/JD-07-2013-0083
20. Tury S, Robinson L, Bawden D (2015) The information seeking behaviour of distance learners: a case study of the University of London International Programmes. J Acad Librariansh 41:312–321. https://doi.org/10.1016/j.acalib.2015.03.008
21. Cao W, Zhang X, Kaibin X, Wang Y (2016) Modeling online health information-seeking behavior in China: the roles of source characteristics, reward assessment, and internet self-efficacy. Health Commun 31(9):1105–1114. https://doi.org/10.1080/10410236.2015.1045236

An Ad-Hoc 'Technology-Driven' Creativity Method

Cécile Boulard-Masson[1]([⊠]), Sophie Zijp-Rouzier[2],
and Olivier Beorchia[3]

[1] Naver LABS Europe, Meylan, France
cecile.boulard@naverlabs.com
[2] Orange Labs, Meylan, France
sophie.zijprouzier@orange.com
[3] Altran ID, Lyon, France
olivier.beorchia@altran.com

Abstract. In innovation, many actors look for ways to generate new ideas i.e. to support creativity. In this paper we present a work conducted in a project that is "technology-driven". A new haptic technology is in search of a context of use. As ergonomists and designers we tried two methodologies to support ideation. The first one is a rather classical creativity session in a 'focus group'. The second one is 'technology-driven' as it is first based on the experience with the prototype. The technology-driven methodology is divided in two steps. In the first one the participants experience the new haptic technology and try to imagine relevant use cases. The second step is another interview occurring a couple of days after the first one, where the participants report if they found any new use case between the two interviews. The conclusions show that both methodologies are very complementary. In the creativity session, the quantity of produced use cases is higher and more 'out of the box'. In the technology-driven, the ideas of use-case are more aligned with the haptic technology.

Keywords: Creativity methodology · Haptic · Technology-driven

In many project, ergonomists and designers may find themselves in a situation where engineering teams create a breakthrough technology, without any well-defined use case. This situation is not the one preferred by the ergonomists where prospective ergonomics would rather aim at identifying needs at earlier stage in design projects before finding appropriate technologies [2]. Nevertheless, this situation may occur quite naturally in any innovation Center. The work described in this paper relates to a new haptic technology and our goal as ergonomists and designers in the project is to identify relevant use cases for this technology. It occurs in a wider project with many actors working as engineers, researchers in various context of work (start-up, industry, research environment) addressing hardware, software, integration, tests, usability. In this project, we are typically in a technology-driven context. The technology targeted involves the sense of touch and is about reproducing textures.

© Springer Nature Switzerland AG 2019
S. Bagnara et al. (Eds.): IEA 2018, AISC 822, pp. 526–532, 2019.
https://doi.org/10.1007/978-3-319-96077-7_56

1 Related Work

1.1 The Modality of Touch

The modality of touch encompasses three distinct sensory systems:

- The cutaneous system that receives sensory inputs from mechanoreceptors that are embedded in the skin,
- The kinesthetic system that receives sensory inputs from mechanoreceptors located within the body's muscles, tendons and joints,
- The haptic system that combines sensory inputs from the cutaneous and kinesthetic systems [10]

The haptic perception allows to discover objects and perceive texture (with lateral motion), hardness (with pressure), temperature (with static contact), weight (with unsupported holding), global shape or volume (with enclosure) and global shape (with contour following) [10].

In literature, we find a distinction between active and passive tactile perception [1, 10]. The passive tactile perception can be defined as a stimulation applied on an immobile part of the body. In that case, the perception is limited and requires a contact between the stimulus and the body. Passive tactile perception is reduced to cutaneous touch, without any motion. On the other side, active tactile perception is based on active movements in order to explore and grasp objects integrally [1]. In the context of texture perception, it can only be active tactile perception.

When we consider sense of touch, it is also necessary to deal with the socio-cultural context of touch [7]. Indeed, the use of the sense of touch as a mean of interpersonal communication is not developed in the same way in the various regions of the world.

1.2 Haptic Technologies and Use Cases

For the interaction with electronic devices the main channels used are vision and audio. For the sense of touch, besides vibration of devices, very little information is conveyed by this channel. It is surely related to the limitation in hardware, the costs and development and the level of reality such technology can provide [6]. In any case, haptic modality allows to convey information discreetly avoiding hearing or visual perception [3, 16]. To go further, haptic technologies seems to be too inexact to convey explicit messages or complex information. It seems rather appropriate for conveying ambient information, communicating urgency or an emotional reaction [5].

Many applications based on haptic modality have been designed in a variety of fields.

First, haptic modality presents a great interest in specific tasks such as navigation tasks. Information for guidance and navigation can be conveyed via a wearable device (bracelet, walking-stick) or via a car seat [3, 11, 16]. Wearable devices can also notify specific information such as a rendezvous in the calendar [8].

Many applications dedicated to couples are also based on the sense of touch. For 'long-distance' couples, the use of haptic feedbacks has been tested to provide information on the current context of the partner such as availability or location [9].

For expectant parents, a device has been designed where foetal movements are translated into a light message for the mother-to-be and into a haptic one on a bracelet for her partner [15].

The future applications based on haptic modality targets e-commerce, games, teaching, re-education and data representation [6].

2 Methodology

Given the 'technology-driven' context of the project, the attempt is to engage in an ideation and creativity methodology. Therefore, the plan is to organize a creativity session (as a focus group) with a group of participants on haptic technologies in order to generate ideas[1]. This kind of creativity session in a group is based on Osborn's rules of idea generation: as many ideas as possible, no criticism or judgement, free-wheel and wild ideas, develop and elaborate on other's ideas [13]. The creativity session is organized over a day with two main phases, a divergence phase where idea generation in quantity occurs followed by a convergence phase where selection and fine-tuning of ideas occurs.

The preparation of the creativity session and the first results of the perception tests of the haptic technology brought to us two main difficulties for the creativity methodology on a haptic technology.

- The first one relates to the specificities of the haptic technologies and our ability to discuss on sense of touch. It is quite obvious that we can share with others what we can see, hear, smell or taste. For the sense of touch, it is not that easy or obvious and we tend to not describe much or share what is experienced when we touch something. In a similar way, literature shows that we have difficulties to discuss and that we miss vocabulary to describe our feelings with the sense of touch [4, 12, 14]. The first results of preliminary perception tests on the technology trying to identify what is perceived by users demonstrate that people may experience very differently one texture. It appears that there is little "shared" representation on feelings with the sense of touch. It also recalls that the sense of touch is more intimate.
- The other difficulty is the interaction situation that is imposed by the technology. If one wants to feel a texture, he or she needs to touch the surface while having a lateral movement. The specificities of this interaction situation make it difficult to bring people discuss potential use cases while the context of interaction is so specific and limited.

In order to overcome those identified difficulties, we put in place in the project, in addition to the creativity session, an 'ad-hoc' technology-driven (TD) creativity methodology that starts from the technology. The aim is to ask participants experiencing a prototype of the technology to think about possible use cases. The TD creativity methodology is in two stages:

---

[1] Example of creativity methodologies: Synectics Creative-Problem-Solving Methodology, TRIZ & ASIT methodology, Method HMW (How Might We?).

- After having experienced the prototype, the participant is interviewed. He or she is invited to describe the possible use cases they can identify or imagine. At the end of the interview, the participant is asked to think about the technology and possible use cases for a couple of days.
- Two days later, the participant is contacted for a second interview to provide possible new use cases.

There are three hypothesis supporting this TD creativity method. The first one relates to the specificity of interaction of experiencing texture (contact + lateral motion) which represents a constrained context to consider use cases. Therefore, as the technology is presented first and the participants could experience it in first place, they are well aware of the potentiality of the technology. The second hypothesis is that as it is difficult to discuss sense of touch, maybe the best option is to experience first. In that way, there is no need to describe and define what is going to be felt. It is just experienced. The third hypothesis is that by going back to their everyday life for two days, participants may behave as researchers from the project. With some time, they may face or recall specific situations where a technology of texture rendering may fit to answer specific needs.

3   Creativity Session in a Focus Group

3.1   Participants

The participants recruited for the creativity session come from private industry, with a variety of profiles: engineers, designers and researchers. The creativity session is led by a senior designer and a total of 12 participants are involved.

3.2   Preparation

In order to overcome the limitations identified to discuss the sense of touch, an important preparation work is done. In anticipation of the session, some knowledge on the sense of touch is gathered:

- Touch Diaries: In a day identify where/when sense of touch is solicited: to feel if something is wet or dry, to find something in a bag without looking in it.
- State of the art regarding haptic technologies
- Interview of professionals working with the sense of touch: a worker from textile industry, a physiotherapist and a woodworker.

During the focus group, some material is shared with the participants to facilitate discussion on touch and also to create a common referential:

- Touch stands, where participants can touch and feel various textures in order to ease verbalization on the topic
- Recall of some of the verbatim gathered during perception tests of the haptic solution and link them to everyday objects

All the work helped to provide references and ideas on why, when and where the sense of touch is used, and what kind of information it provides.

3.3 Results

During the creativity session, the group has generated 193 ideas. All the work of selection and structuration of ideas leads to the identification of 4 main themes of applications: E-commerce, Game, Sex and Emotion, Education and re-education. Interestingly the identified themes are relevant with the one identified in the literature [6].

In all the ideas that have been developed and fine-tuned by the group only part of it targeted a type of interaction involving texture feeling. Many of them were more aligned with other types of haptic interaction including resistance to motion for instance for a computer mouse or vibration.

Some difficulties were experienced by designers and ergonomists during the creativity session. First there was a confusion for the participants between the words tactile, touch and texture and second, there was lack of a shared semantic field to discuss on the sense of touch.

4 The Technology-Driven Creativity Method

4.1 Participants

20 participants, involved in a perception test, have been recruited to follow our TD creativity methodology. The first interview occurred just after the perception test and is led by one researcher. First the participants are invited to imagine different use cases that could be relevant with the technology. Then there is a recall of the specificities the interaction to experience texture rendering (contact + lateral motion). Two days later, the participants are solicited for another interview, either face-to-face or on the phone. They are invited to share again with the researcher if they have found other use cases.

4.2 Results

On the ideas of use cases generated by the participants following the TD creativity methodology, the scope of ideas and domain is less broad than for the creativity session. Nevertheless, all the ideas are relevant for the technology. For the 20 participants, over the 2 interviews, we were able to collect: 19 mediums where to display the technology, 34 context of use and 40 usages. Participants were able to identify the limits of the technology.

Concerning the mediums, participants identified the cars, wearable objects, objects in house (walls, surface), objects at the desktop and all kind of tactile screens (smartphones, tablets). The context of use targeted by several participants: shadow environment, game, e-commerce or household appliances. There were also very specific domains identified only by some participants, such as, art, telemedicine or 3D printing. The main use cases were: improve tactile interaction on smartphones, ease the functionality of remote control on tactile display or support dermatology training.

5 Conclusion and Discussion

The two methodologies presented, creativity session and TD creativity method, allowed us to collect a great number of relevant use cases. We have done the choice to have different participants for each creativity methodology. Our goal was to limit any interference of one methodology on the other.

5.1 Discussion

When trying to compare both methods, it is interesting to notice that the context of use of household appliances has not been discussed at all during the creativity session whereas some participants highlighted it during the TD creativity methodology. This may be explained by the fact that with the TD creativity methodology, participants are closer to their everyday life. We can assume that this specific methodology is strongly impacted by the everyday life of participants. Even though it might also be true for creativity sessions, it is very possible that the impact is higher for TD creativity methodology. It can be perceived as a limit in terms of creativity, this outcome can also be interpreted as a way to gather real needs that are closely related to one's life. The other point is that the TD creativity is appropriate to discover use cases that are closer to what can be done with a technology compare to creativity session.

The TD creativity methodology can be attractive as it is quite easy to put in place and eventually not really expensive. If some user tests are performed with a given technology, there is no much added cost to perform this TD creativity methodology and gather ideas of use cases.

To go further, it could be interesting to combine both methodologies with a same group of participants in a sequential mode. For instance, the participants first participate to the TD creativity methodology and then to a focus group, so that we benefit from the group discussion that improves creativity and idea generation. The other way around could also be interesting, having first a focus group and then involve the participants in the TD creativity methodology.

5.2 General Conclusion

As a conclusion we can say that both methodologies can be seen as complementary. The ad hoc methodology permitted to have more targeted use cases and very specific niches were identified. The creativity session in focus group allowed a wide range of ideas with more "out of the box" use cases.

On the side of haptic technology, the work presented in the paper brings information on the difficulties associated to haptic technology regarding creativity and idea generation. The preparation of the creativity session in focus group, as well as, the TD creativity methodology helped to partially overcome these difficulties.

References

1. Benali Khoudja M (2004) VITAL: un nouveau systeme de communication tactile. Diss. Evry-Val d'Essonne
2. Brangier E, Robert JM (2012) L'innovation par l'ergonomie: éléments d'ergonomie prospective. In: Llerena D, Rieu D (eds) Innovation, connaissances et société: vers une société de l'innovation. L'Harmattan, Paris, pp 59–82
3. Brunet L et al (2013) "Invitation to the voyage": the design of tactile metaphors to fulfill occasional travelers' needs in transportation networks. In: World haptics conference (WHC). IEEE, 2013
4. Dagman J, Karlsson M, Wikström L (2010) Investigating the haptic aspects of verbalised product experiences. Int J Des 4:15–27
5. Dobson K et al (2001) Creating visceral personal and social interactions in mediated spaces. In: CHI'01 extended abstracts on human factors in computing systems. ACM
6. El Saddik A (2007) The potential of haptics technologies. IEEE Instrum Meas Mag 10(1):10–17
7. Gumtau S (2005) Communication and interaction using touch-examine the user before you reproduce his hand! Cogn Sci Res Pap-Univ Sussex CSRP 576:18
8. Hansson R, Ljungstrand P (2000) The reminder bracelet: subtle notification cues for mobile devices. In: CHI'00 extended abstracts on human factors in computing systems. ACM
9. Hinds P et al (2011) Proceedings of the ACM 2011 conference on computer supported cooperative work
10. Klatzky RL, Lederman SJ (2003) Touch. In: Handbook of psychology, vol 4: experimental psychology, chap. 4. Wiley, London
11. Lin M-W et al (2008) Investigation into the feasibility of using tactons to provide navigation cues in pedestrian situations. In: Proceedings of the 20th Australasian conference on computer–human interaction: designing for habitus and habitat. ACM
12. Obrist M, Seah SA, Subramanian S (2013) Talking about tactile experiences. In: Proceedings of the ACM annual conference on Human factors in computing systems, pp 1659–1668
13. Osborn AF (1957) Applied imagination. Scribner's Sons, New York
14. O'Sullivan C, Chang A (2006) An activity classification for vibrotactile phenomena. In: Haptic and audio interaction design. Springer, Berlin, pp 145–156
15. Righetto M, Crampton Smith G, Tabor P (2012) Aura: wearable devices for non-verbal communication between expectant parents. Studies in Material Thinking, vol 7
16. Williamson J et al (2010) Social gravity: a virtual elastic tether for casual, privacy-preserving pedestrian rendezvous. In: Proceedings of the SIGCHI conference on human factors in computing systems. ACM

Conditions of Use and Adoption of Digital Tools: Results from a Field Study on Airport Reception Agents

Nina Barera[1(✉)], Leduc Sylvain[2], Gérard Valléry[1], and Sonia Sutter[3]

[1] EA7273 CRP-CPO, Université de Picardie Jules Verne, 80025 Amiens, France
nina.barera@etud.u-picardie.fr
[2] Université d'Aix Marseille, 13007 Marseille, France
[3] Air France-KLM, Roissy-Charles de Gaulle, Tremblay-en-France, France

Abstract. The use of information and communication technologies (ICT) has spread widely in our personal life as well as in the workplace. The design of these tools can contribute to increased business productivity. In a service relationship context, the challenge is not only to deliver a service but also to improve customer satisfaction. Becoming a relational brand implies an exemplary level of listening, service and responsiveness across all channels of interaction. In order to deal with these expectations, some airline companies have developed software on digital tablets. The aim of this tool is to allow sedentary workers, in charge of meeting customers at airports, to access all the data relating to them in real time, such as their journeys and their luggage for example. This situation raises several issues related to models of task definition and working organization. A study was carried out in order to explore these "digitized" environments and particularly to understand how this technology has been adopted by airport reception agents. The goal is to identify the factors that encourage, or hinder, their use. The method is based on a field observation of real work activity in order to analysis the user's tasks by using this tool in a real life situation. The data collected has been combined with individual and collective feedback during interviews that were held. The key factors of use were also collected via an on-line survey. Results show how use of the tablet, in particular centralizing information contributed to helping airport agents to increase their responsiveness and performance for passengers. In addition, the agents claimed that the tool facilitates the accomplishment of their tasks and improves their ability to react. The development of digital tools also brings a positive image: one of a modern and reactive company. For the agents, it is essential, they felt that they gain in credibility and that they are valued. Nevertheless, although a majority of agents integrated the tablet into their daily practices, misappropriations of use were created. These adjustments led to a profound change in the work, which led to a re-consideration of the work of ground staff. The development of digital tools implies a real challenge to the job, knowledge and know-how. Regarding the speed at which technological developments are growing and the major changes that they bring, the question of the connections between work, health and use seems important.

Keywords: Work analysis · Acceptance · Technology

© Springer Nature Switzerland AG 2019
S. Bagnara et al. (Eds.): IEA 2018, AISC 822, pp. 533–538, 2019.
https://doi.org/10.1007/978-3-319-96077-7_57

1 Introduction

In recent years, a digitization movement is underway in most large companies. Digital tools are everywhere, they reinvent daily life by creating new uses and offering new possibilities for experimenting. The workplace is also changing, many tools, initially intended for personal use, have changed the professional world. Companies must keep their eye on new developments in technology to see what emerges in terms of use, in order to adapt them to new modes of communication and work, to increase productivity and to protect themselves from competitors. Today, the digital revolution is an essential tool for development. This is a key performance and profitability vector: it is a challenge for businesses and involves a real change in the way they operate. It's not a question of whether companies need digital transformation, but how to implement and encourage it [1].

As part of its digital transformation, the airline company in which our study took place decide to take a chance on new developments in tools to support its work. They chose a digital tablet (iPad-®). It was proposed to airport agents. Airport agents are in charge of welcoming passengers from their arrival at the airport until the time when the plane takes off. The new tool gives agents real-time access to information about passengers, their profiles, their luggage, and their flights. The main aim is to develop customer relationship through a tool which allows agents to help passengers and reply to their questions. It is also better to know your customers well in order to anticipate their needs with one-to-one attention. The developmental aim of the digital tablet project was to improve day-to-day work with the use of practical tools such as the being able to see timetables, mails and access internal contact lists.

In this context of change, an ergonomics study was requested. The aim of the ergonomics study was to understand the evolution of work and practices created by these new tools to determine the conditions, the criteria and the factors promoting the acceptance of technologies. This is firstly to assess the functional quality of new tools in a work situation, i.e., confirm that the tool is adequate for the task at hand but also to take into account how users will understand the new tools and their motivation for improving these devices.

2 Adoption Process of Technologies

The adoption process of technologies can be seen as a continuum between *acceptability* and *acceptance* [2]. Acceptability sends back to an individual's representation of a future technology before its development. While acceptance concerns people's real-life experiences once the technological system has been implemented.

2.1 From Acceptability

Technological acceptability offers the possibility of assessing and/or predicting the conditions and factors that could make technology acceptable or not for its future users. Researches on the study of acceptability identify two theoretical orientations [3]. **Social acceptability** would be the initial step in the process of adopting technology. Models of social acceptability refer to "*the examination of the conditions that make a product (or a service) acceptable or not for the user before its actual and effective use*" [4]. These models are based on subjective perceptions towards technology to identify the factors determining intentions of its use. The aim is to predict the acceptability *a priori*. One of the most famous model is the TAM, standing for Technology Acceptance Model [5]. TAM explains people's behaviors by perceived utility and perceived ease-of-use before putting the technological solution into service.

In order to complete this approach based on the perception evaluation, the **operative or practical acceptability** seeks to apprehend first uses of technology. Work on operative acceptability seeks to explain difficulties in use considering the characteristics of the technology. Contrarily to the first orientation, it is not only a question of predicting the intentions of use, but also to evaluate it after a first experiment [6]. The models provided by operative acceptability seek to design devices that are more accessible and usable from criteria, requirement and principle of valuation. The aim of theoretical orientation is to improve human-machine relations working under constraints of use to ensure functional quality of the device.

2.2   To Acceptance

The models' acceptability allows us to apprehend the future users' perceptions, if we look at their activity « as it is ». It seems necessary to widen understanding of acceptability to real situations of use and to understand the processes involved [3]. There is always a difference between what I say and what I really do. In fact, the real situation controls and makes the behavior possible or not. Running experiments with the article in question in a real environment could influence and determine a favorable attitude or not regarding a tool [3]. On-site acceptance is centered on the users' activity and real-life experience with technology. It requires the real conditions of use. On-site acceptance is evaluated by a describable and comprehensive approach of the work activity. Four dimensions have been identified for on-site acceptance [3]. **The individual or personal aspect** involves studying the constraints' appreciation and the obligations generated by the new uses: how it helps agents to improve performance, efficiency and effectiveness. **The organizational aspect** concerns the system of control exerted by the technology on the employees' autonomy and their room for manoeuver to carry out their tasks. **The relational aspect** focuses on collective and collaborative activities and the tools' effects on interpersonal relationships at work. **The professional aspect** refers to the estimation of what the technology will recognize, or not, in terms of knowledge, skills and experiences. The acceptance is located at the level of practical use. Therefore, it is complementary to acceptability.

3   Method

Our study takes place at an intersection of multiple possibilities in terms of appropriation of new tools, prevention of work hazards and more widely technological acceptance. The intervention carried out is based on ergonomic work analysis [7]. It consists of analysis of the current work situation of the stabilized digital tablet implementation through the data collection on technical, human and organizational aspects. Our approach consisted in combining the various methods to allow us to study situated acceptance. Each method brings elements of knowledge at the same time about the feelings, the experience and the ambitions of new uses. Adding all this together makes it possible to understand what the agents experience is, in the use of the digital tablet and what they really do with them.

3.1   Observations

Observations in work situations are an essential source of information in this study. They make it possible to appreciate issues concerning agents' real activity, especially in connection with the use of technological tools. The observation days took place in two Parisian airports and two different French provincial stopovers. To obtain as many different situations, the analyses were carried out during peak times (school holidays, weekends) and in nominal mode in quiet periods (outside school holidays, mid-week). These observations were accompanied by interviews in situ with the agents that is to say while they were working.

3.2   Online Survey

The main aim of the online questionnaire was to have a better knowledge of the tablet's use. The mains themes concerned the biographical profiles of the population, general knowledge level and experience of technologies, conditions of use of the digital tablet and their opinion on its uses. When finished, from the 946 people who answered, 938 replies were analyzed. The typical respondent profiles are women (75%) between 41 to 50 years old who have held their current position for more than 11 years (53%). "Customer pole" (check-in, boarding the plane...) is the most representative role. After, we find complaints on baggage position (11%), lounge reception (10%) and tickets sales (9%). The majority of those who replied work at Roissy-Charles de Gaulle airport.

3.3   Working Groups

Performing the ergonomics intervention requires an investigation on the ground. It also puts into perspective the conclusions of the study. We animated debates during a working group, which bringing together project stakeholders (design, change management, deployment, users, etc.). This group's mission is to discuss the conclusions of the first phase in order to identify areas for improvement and measures to be implemented consecutively. From a presentation of the field study findings, a discussion was started on the elements to be capitalized on and those for which could be improved. It was a privileged moment of exchange, dialogue and collective appropriation.

4 Main Results

About two years after the beginning of deployment, the majority of reception agents use a digital tablet for their daily work (60%). It is considered as indispensable for 57% of users. In this section, we present the results of our analysis. Data comes from a broader study about the process of adopting technologies, only the element influencing the situated acceptance will be exposed.

4.1 Performance vs Reliability

Using tablet computers, agents can personalize exchanges with customers through instant access to their travel history, the continual updating of their passenger profiles, the possibility of sending them a message if they are late to the boarding gate, and access information about loyalty card members. All this information associated with the use of a mobile device allows to agents to respond more effectively and quickly to their customers' demands: "it is time saving". Performance gain and customer satisfaction are acceptance factors. The unreliability of data is irritating: 91% of agents are aware of it. Acceptance is based on the trust that user has in the technology [2].

4.2 From Valorization to Worry About Loss

For 62% of agents, having a digital tablet is rewarding. First of all, from the customer's point of view, it makes them more visible. On the other hand, to the company, individual ownership is taken as a sign of trust: "it is very gratifying", "now, we have a tool of our time". But this valuable tool becomes an object of desire for others. Fear of a loss or theft becomes a worry for 76% of the agents. At the time of this study, there is no support and means to keep the tool safe leading to some agent's reluctance to use a digital tablet.

4.3 The Prescribed Autonomy

The agents judge the tablet as an advantage that allows them to gain autonomy (90%) in their activity. Company people are able to instantly solve their passengers' problems before and after their flight by offering them alternative solutions or additional services. The advantage of this information is to be able to anticipate needs and go ahead to meet them. It is a major asset in customer relationship. The treatment of passengers becomes very personalized and interactive. It is the responsibility of the officer to take initiatives in his actions and work. For some taking initiatives is very rewarding. For others, too much flexibility leads them to lose their bearings.

4.4 Between Professional Abilities and Technical Skills

The use of a new tool radically changes positions and workplaces, but also the attitude of the agents and their role. New tools requires new business skills. For example, agents must know how to sell even if they are on the reception desk. The digital tablet is an effective tool for agents, but their personal professional expertise and experience are no longer valued. In parallel, the use of technology also requires new knowledge in terms of technology control. The reception agents are concerned: "how we are going to adapt to the digital evolution". The development of technology creates a gap between the most hardened and the least comfortable with technology that could lead to the tool being rejected.

5   Conclusion

New technologies bring changes to professional activity, they can improve it or counteract it. The aim of using new technology is really to make work easier. However, when new technology is opposed to the essentials of the quality and the meaning of work, it can undermine effectiveness, well-being of users and their health. Insufficient support during the introduction of technologies can complicate and intensify work, resulting in the emergence of new psychological and physical constraints. Consequently, it is necessary to focus on understanding user's work and the use they make of working tools, in order to provide effective technological solutions.

References

1. Champeaux J, Bret C (2000) La cyberentreprise. Dunod, Paris
2. Bobillier-Chaumon ME, Dubois M (2009) L'adoption des technologies en situation professionnelle: quelles articulations possibles entre acceptabilité et acceptation? Le Travail Humain 72(4):355–382
3. Bobillier-Chaumon ME (2013) Conditions d'usage et facteurs d'acceptation des technologies de l'activité: Questions et perspectives pour la psychologie du travail (Doctoral dissertation, Ecole doctorale Sciences de l'Homme, du Politique, et du Territoire)
4. Terrade F, Pasquier H, Boulanger J, Guingouain G, Somat A (2009) L'acceptabilité sociale: la prise en compte des déterminants sociaux dans l'analyse de l'acceptabilité des systèmes technologiques. Le Travail Humain 72(4):383–395. https://doi.org/10.3917/th.724.0383
5. Davis FD (1989) Perceived usefulness, perceived ease of use, and user acceptance of information technology. Inf Technol MIS Q 13(3):319–340
6. He W, Qiao Q, Wei KK (2009) Social relationship and its role in knowledge management systems usage. Inf Manag 46(3):175–180
7. Daniellou F, Rabardel P (2005) Activity-oriented approaches to ergonomics: some traditions and communities. Theor Issues Ergon Sci 6(5):353–357

Digital Sketching for Distant Collaborative Diagnosis in Neurosurgery: An Experimentation

Stéphane Safin[1(✉)], Felix Scholtes[2], Pierre Bonnet[2],
and Pierre Leclercq[3]

[1] I3-SES, CNRS, Télécom ParisTech, 75013 Paris, France
stephane.safin@telecom-paristech.fr
[2] CHU, University of Liège, Liège, Belgium
[3] LUCID, Faculty of Applied Sciences, University of Liège, Liège, Belgium

Abstract. This study relates an experiment about multimodal distant collaboration in neurosurgery. We observed a two-hours distant meeting between neurosurgeons, supported by an original platform allowing to convey remotely speech and gesture (videoconferencing) but also images and hand drawn sketches. We use a bottom-up approach, analyzing and understanding the spontaneous activity set up by the professionals. We show that conversations about diagnosis and operational strategies are intimately linked, that different communication modalities are preferred to support different communication topics, and that digital Sketch is a flexible tool in remote collaboration in medicine.

Keywords: Remote collaboration · Digital sketch · Neurosurgery

1 Introduction

Nowadays, in a wide range of activity sectors, collaboration between actors has intensified. In particular, in the medical field, the over-specialization of professionals, the merging of hospital institutions, and the need to cross skills to refine diagnoses and define optimal therapeutic strategies, necessarily raise the issue of effective remote collaboration. In this context, and with a solid experience of distant collaboration in the field of architecture, we are specifically interested in understanding and supporting remote synchronous collaboration, particularly in the field of neurosurgery.

In this study, we address the following questions: how can digital sketch effectively support distant collaboration in medical domain? How do expert practitioners rely on several communication modalities (verbal, gestural, visual) to build, remotely, a joint understanding and a common approach to medical challenges? What is the relation between communication modalities and contents?

The study consists in the observation of a remote collaborative session between two neurosurgeons based in Liège, and one in Montreal, all three expert practitioners. The discussion is free, addresses three clinical cases and is supported by SketSha, a digital freehand sketching software. Originally developed for architecture, combined with a conventional video conferencing system, it constitutes a multimodal communication device conveying speech, gestures, drawing and professional content (medical imaging). Neurosurgery, as a discipline that relies heavily on medical imaging, is an appropriate field for testing new forms of graphical interaction.

The objective of this two-hours meeting was multiple: to experiment the device in the medical field, to mobilize it in a real medical collaboration activity and to document this activity, to identify the potentialities of the system in the field and to imagine uses for longer term.

2 Collaborative Practices in Medicine

Diagnostic and therapeutic medical decisions are based on a bundle of diverse information most often analyzed collegially, in real time, only if the stakeholders are together. A legal framework now imposes these collegial practices in certain situations such as the management of oncological pathologies (Multidisciplinary Oncology Consultation). Thus, recommendations for good medical practice go through a Peer-review system that sometimes takes on an international dimension. These collegial analyzes therefore imply that physicians (most of whom are specialists with high added value) would have either to travel over sometimes enormous distances, which implies an important cost, or to make these meeting in degraded conditions, by phone for example. In view of the difficulties of time and resources, these collegial meetings are limited in number and/or participants, to the potential detriment of the medical quality of patients' care.

In addition, the hospital landscape has evolved. The institutions have merged into hospital groups located in several geographically distant locations. Each location often has a specific, oriented activity.

Medical practice, especially that of adequate diagnosis, leading to the possibility to devise treatment strategies, relies on medical data of various kinds: biological data, medical history and medical imaging. Medical imaging has evolved considerably. This includes the way high resolution tomographic images are delivered to the practitioner. Today, diagnosis is based on the viewing of series of images, merged images and reconstructions. Analyzing a complex imaging case may require navigating through hundreds of images, using multi-planar reconstructions, and measurement tools in some cases. More and more often medical experts analyze these images collegially. The written analysis from the medical imaging specialist is not sufficient, the decisions being often the result of a collective intelligence. In addition, the images are often manipulated and annotated to integrate therapeutic decisions. In the field of surgery, this makes it possible, for example, to delineate the area of tumor resection or sketch the approach for vascular lesions.

This evolving landscape and potential requirements of new practices require the use of adapted exchange tools. Videoconferences which allow synchronous collegial work between geographically distant teams partially respond to these requirements, but lack flexibility and functionality: the need for dedicated adapted premises, the impossibility of producing graphic annotations collaboratively and remotely. In order to fulfil existing requirements and potentially drive development of multi-modal long-distance collaboration between experts, It is therefore crucial to have effective remote collaboration tools, and in particular synchronous collaboration tools, to equip these growing collaborative practices and support the transitions in the hospital field.

3 Digital Sketching and SketSha Software

Sketching has been recognized as a powerful tool for expressing ideas in the design domain: by its ambiguous character, it supports creativity, and by its easiness of expression, it allows problem decomposition (Goldschmidt 1991). Therefore, several software and environments tend to combine these intrinsic properties of hand-drawn sketch with digital support. This is known as the paradigm of Digital Sketch.

In order to support remote synchronous collaboration, the LUCID-ULg lab has developed a software called SketSha (for sketch sharing), as a shared drawing environment allowing several users to be connected to the same virtual drawing space. Various functionalities, such as a panel of colored pens (and an eraser) and a navigation tool (to zoom, translate, rotate), are available through intuitive graphical widgets and a digital pen. Some layout facilities have also been included in the prototype, such as the possibility of drawing and managing different sheets of virtual paper, of deleting or duplicating them, and of managing their transparency. The software is installed on a graphic tablet and completed with a video-conferencing system, for remote collaboration (see Fig. 1)

Fig. 1. SketSha environment (used in design domain)

The system has been intensively used in architectural professional and educational settings (Dondero and Shirkhodaei 2014; Safin et al. 2011). These uses have proven its efficiency to support distant flexible and rich collaboration in design domain (Safin 2011; Safin et al. 2012), but have also opened the doors to the exploration of digital sketch as a support to remote collaboration in other domains, such as medicine. This is the aim of this study.

4 Research Issue

In this study, we experiment the use of SketSha environment in a medical context. We use a bottom-up approach: we propose the system, support the technical setting and let the users build their own collaborative activity around the system. The research questions are twofold

- On a pragmatic/applied point of view, we aim to identify the potentialities of digital sketch for collaboration in the medical domain and to define requirements for effective sketch-based remote medical collaboration, by documenting spontaneous collaborative activities in the field.
- On a fundamental point of view, we aim to analyze an original collaborative practice (remote sketch-based collaboration in medicine) to address the following questions
  - How do expert practitioners rely on several communication modalities (verbal, gestural, visual) to create a common diagnosis remotely?
  - What is the relation between communication modalities and contents?
  - How is the remote collaboration activity spontaneously structured?

For this purpose, following Rabardel (1995) notion of *instrument*, we provided users with an *artifact* (the SketSha environment, initially developed for architecture and design, and probably poorly fitted to medical practices) and document the *instrumental genesis* processes, i.e. the way this artifact is modified, diverted, appropriated, and the way uses and activities are re-built around this artifact.

5 Setting

The study consists in the observation of a remote collaborative session between two neurosurgeons based in Liège, and one in Montreal, all three confirmed practitioners (Fig. 2). The discussion is free and addresses three patients' cases. The goal for practitioners are to confirm the diagnosis, discuss operative indications and strategies and share best practices.

Fig. 2. Remote collaboration between Montreal (left) and Liège (right)

During the meeting, which lasted about 2 h, the observed professionals discuss cases, manipulate videos (collection of images from medical imaging devices[1], extract still images on the system, draw on these images, point to elements with the pen (the cursor is reproduced in the remote environment) and make gestures through video-conferencing (Fig. 3). The three patient cases discussed at this meeting allowed practitioners to discuss about care strategies (embolization, surgery, etc.) and about best practices in surgery (position of the surgeon for the operation, type of craniotomy, instruments used, etc.)

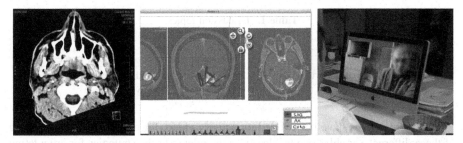

Fig. 3. Information channels: videos composed of medical imagery (left), digital sketches on imagery (center), videoconference (right)

Beyond case discussions, we observe reflexive moments where the participants explain of the use of the device, they express comments about it, and where they consider intensive use of the device for their professional practice.

6   Methodology

The video of the collaborative activity was analyzed in depth, based on ad-hoc coding grids, covering the following aspects

At first, we identify the **type of activities** which take place, with the following categories: Case discussions (the core activity); Organization of the collaboration (explicitation of collaboration strategies); Preparation of discussion material (choice of videos, selection, import of images into SketSha); Logistical feedbacks (explanation on the device, spontaneous comments from users on the device, technical problems); Projection in the future (moments of activity recomposition, feedback on the activity progress and projection in probable future activity); and Informal discussions (humor, etc.).

Secondly, we note **multimodal actions**: Manipulation of videos; pointing gestures (pen passing over the image); annotations (drawings); figurative gestures, i.e. which participate in an expression of a content, and not only in the punctuation of the

---

[1] Each image is a cross-section of the considered anatomic element, and the viewing of the video can simulate the navigation on the set of images, which is a usual way for neurosurgeons to apprehend the three-dimensional aspects of a case in a large anatomic imaging set.

communication. It should be noted that, for technical reasons, these actions have only been identified for the Montreal partner.

Third, for the sequences bearing only case discussions (the majority of the activity), we operate a very simple classification of **topics of discussion** according to two crossed dimensions (4 categories) (1) Epistemic vs. pragmatic (Folcher and Rabardel 2004): on the one hand the elaboration or the evocation of knowledge (diagnosis, anatomy, etc.), on the other the discussions on action modalities (surgical procedures, materials to use, etc.) and (2) Particular vs. General depending on one refers to the current case or to general questions. By crossing these two dimensions we obtained in Table 1.

Table 1. Discussion topics.

	Epistemic	Pragmatic
General	Anatomy	General requirements on surgery processes
Particular	Current case diagnosis	Actions to be planned for the current case. Preparation of surgery

All these actions have been coded on activity video, for each second. By crossing of these different variables, we tackle questions related to the strategies for articulating different contents in medical collaboration, to the contribution of the different communication modalities in distant communication, to the evolution of communication and system appropriation, to the issues and opportunities related to the mobilization of the digital sketch for remote medical collaboration. When appropriate, we calculate Relative Deviations (RDs)[2] that measure local associations between specific modalities of each of the two variables

7 Results

7.1 Global Activity Structure and Description

Figure 4 shows the sequence of the types of activities. Globally, task-oriented activities represent 46% of the time, process-oriented activities (operating synchronization: preparation and organization) represent 23% of the time, communication management

Fig. 4. Timeline of actions: Discussion (green), organization (red), preparation (black), logistics (blue), projection (purple) and informal (orange)

---

[2] RDs measure the association between two nominal variables. They are calculated on the basis of a comparison between observed and expected frequencies (i.e. those that would have been obtained if there was no association between the two variables). There is attraction when the RD is positive, and repulsion – when it is negative. We considered a threshold of 0,5 (in absolute value) to consider it as a strong association.

activities and feedbacks represent 26% (logistic and projection) and social-emotional processes represent 5% of the discussion time. By comparing the proportion of different activities in the two halves of the session, we obtain the following observations (Table 2). The proportion of time devoted to the core of the activity remains stable, but the logistics and preparation activities are drastically reduced in the second half of the activity, indicating a habituation to the system. Informal activities (humor), on the other hand, increase as does the organization (which is mainly a long discussion on the actions to be put in place in the short and medium term, closing the meeting). So there is a fairly clear learning curve of the use of the system.

Table 2. Types of activities and evolutions.

	First half		Second half
Discussion	0.47	=	0.45
Organization	0.03	+	0.14
Preparation	0.20	–	0.10
Logistic	0.20	–	0.12
Projection	0.09	=	0.11
Informal	0.02	+	0.09
Total	1		1

The activity can be divided into five phases (Fig. 5).

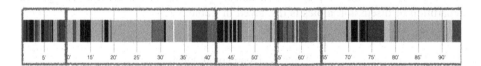

Fig. 5. Five phases of the activity

A first phase, at the beginning of the meeting, is logically dominated by logistical issues and meeting preparation. This phase consists in the training of the users, as well as the preparation of the first images. This training session lasts about 10 min, which seems sufficient to apprehend the system.

There follows a first round of discussion on the first case: subarachnoid hemorrhage with large aneurysm. This first cycle starts with preparation, continues with the discussion of the case (on the therapeutic strategy, the operative strategy, the operating modes, and the instruments and tools to be used for this type of case), interrupted by a moment when the users spontaneously propose uses of the communication device and ways of using it, especially for teaching. The cycle closes with an evaluation of the system by users and suggestions for improvement (logistics), followed by proposals for uses (projection).

The second round of discussion is about another case of aneurysm (reflection on the strategy to use, surgical or endovascular, treatment delays, recovery between two surgeries) contains a preparation step followed by a long discussion. We also see in this second cycle "informal" moments, humor and off-topic discussions, reflecting a relaxed atmosphere. This social element is a crucial issue of remote collaboration, and seems appropriately supported by the system.

In the next stage, there is an explicit feedback on the collaborative experimentation, with a mixture of projections into the future, logistical feedbacks or clarification, and organization of the continuation of the session.

The third round of discussion, preceded by a very short preparation, is essentially productive. It focuses on an intra-axial tumor with edema, and professionals discuss the therapeutic strategy and the operative strategy. It is interrupted by a logistic cut to select other images. There are only a few moments of projection in the possible future activity and informal communications continue to punctuate the sequence.

7.2   Topics

In the first round of case discussions, the collaborators make a global movement from the particular/heuristic to the general/pragmatic through episodes of particular/pragmatic. In other words, they rely on information about the current patient's case (part/heur), then discuss the modes of intervention to be mobilized for this case (part/prag) to finish with more general considerations on operations of this type (gen/prag).

The second cycle, shorter, also includes the same movement, but the points of view discussed are a little more mixed up with each other. In the third round of discussion, most of the time is spent on the diagnosis on the specific case (part/heur) before concluding very briefly by discussions of type part/prag then gen/heur. Finally, the fourth cycle is essentially pragmatic, from the particular to the general. The general repartition of discussion topics is displayed in Table 3.

Table 3. Proportion of topics

	Epistemic	Pragmatic
General	0.04	0.38
Particular	0.30	0.28

7.3   Actions and Communication Modalities

The work of the neurosurgeons is essentially verbal: 78% of the working time is done without other actions. The handling of videos is 9% of working time, drawings 7%, gestures 4% and pointing 2%. The videos are composed of a succession of cross-section. The manipulation of a set of images is a usual way for neurosurgeons to understand the three-dimensional aspects of the cases. The drawing proportion concerns only the moments where sketches are traced and do not include moments where sketches are viewed. The gestures refer to figurative gestures (to mimic a surgical gesture for example). Finally, these actions only concern the Montréal partner. Although being quite relative, these numbers clearly show a predominance of verbal interactions punctuated by complementary actions of different types.

These actions differ depending on the type of discussions. The intensity of the link is strong (link rate, V2 = 0.16). The relative deviations are indicated in colors (0.25 threshold) in red for repulsions and in green for attractions in Table 4.

Table 4. Links between actions and topics

	Drawings	Gestures	Pointing	Video
General/epistemic	0.01	0.04	0.36	0.10
General/pragmatic	0.32	0.85	0.25	0.09
Particular/epistémic	0.43	0.00	0.24	0.59
Particular/pragmatic	0.23	0.11	0.15	0.23
Total	1	1	1	1

Drawing thus has a polyvalent function, and serves to support all types of speech, with the exception of general considerations on anatomy. Gestures are clearly preferentially mobilized to support the discourses concerning general considerations on the neurosurgical practice. It is essentially used to figure and mimic operative gestures. The pointing gestures are quite infrequent, and support essentially general topics, to support discourses by highlighting some elements of the image. It seems that sketching and pointing are complementary: when a user needs to highlight information related to the current case, he draws, leaving a trace, but when he uses current image to support general considerations, he points with the stylus, leaving no trace on the drawing area. Finally, videos are mostly used to talk about the current case and support very few general pragmatic considerations.

On the use of drawing, users make spontaneous comments quite positive: "*For the past few months, videoconferences have been used, a monthly meeting with our colleagues [...] And that's pretty much what we do, we present each other patients and then we discuss. And I see the difference. I am very happy to use this system today because I see the difference and the potential that we can have in drawing at a distance, because we have one more element that I think could be very important, especially if we had an image database*".

In addition, the users look a lot through videoconferencing, unlike what has already been observed in collaborative activities in the field of architecture (Mayeur et al. 2010). They explain this because the images used and manipulated "do not represent exactly what one is talking about" they are a support for the discussion, but the information is primarily verbal.

8 Conclusions

The SketSha system, originally designed for architecture, has a number of domain-specific features: the system is designed to import a small number of large images (plans) on which professionals must work for relatively long durations. However, in the field of Neurosurgery, we see that it is necessary to manipulate many images (scanner cross-sections) and work on it for shorter periods. This requires, for an adaptation to the field, important modifications of the device, such as a specific management of medical images. At this stage, and unsurprisingly, the relative importance of logistics and preparation activities is detrimental to the smooth use of the communication device. On the other hand, the relatively short appropriation time of the device and the spontaneous construction of new communication practices based on the sketch testify to a promising nature for this type of device. In addition, the social-affective processes of collaboration can be supported with this type of multimodal environment.

From a more fundamental point of view, this study is a unique experiment providing some insights on the collective practice of medical diagnosis and treatment strategies elaboration, when it is held by remote practitioners, and aims to understand how professionals spontaneously construct uses of a multimodal device and structure their collaborative activity. We identify different movements in the discussion of a case in neurosurgery. Although each case is approached differently, we see movements starting with the diagnosis of the current case, supported by the images, continuing towards the determination of the operating strategies for this case, supported by the freehand drawing, and ending with the sharing of general experiences on the modes of operation, supported by the gestures. Our study thus shows the potential of the sketch for distant communication in neurosurgery, supporting a versatile role in communication, complementary to pointing gestures. In addition, the sketch appears the privileged means, in combination with the dynamic medical images, to work on the particular/pragmatic dimensions, i.e. on the definition of strategies and gestures to carry out during surgery. This is precisely the general objective pursued by meetings of specialists to prepare surgical procedures.

Finally, the users propose also to use the device for distance training in anatomy. It would involve setting up discussions with less experienced professionals around an images database. As we have seen in this study, pragmatic reflections are largely supported by the system: the gesture and the drawing allow to support long discussions useful for the learning of the profession of neurosurgeon.

Obviously, the study has several limitations: only one session was observed, with few training from the participants. Results may be different with "expert" use of the system, with more drawing and pointing, less logistical issues, etc. We need to engage in a more longitudinal study of the system use.

References

Dondero MG, Shirkhodaei S (2014) Appropriation et gestion d'une activité collaborative distante. Le cas d'un atelier pédagogique de la conception architecturale. In: Proceedings of COMMON 2014, Liège

Folcher V, Rabardel P (2004) Hommes, artefacts, activités: perspective instrumentale. In: Falzon P (ed) Ergonomie. PUF, Paris, pp 251–268

Goldschmidt G (1991) The dialectics of sketching. Creat Res J 4(2):123–143

Mayeur A, Ben Rajeb S, Darses F, Lecourtois C, Caillou S, Guéna F, Honigman A, Leclercq P, Safin S (2010) Concevoir à plusieurs et à distance en architecture : vers de nouvelles pratiques professionnelles ? Présentation au séminaire Globalisation et Territorialisation: questions de travail, Université Paris 1 Sorbonne, 7–9 juin

Rabardel P (1995) Les Hommes et les Technologies. Approche Cognitive des Instruments Contemporains. Armand Colin, Paris

Safin S (2011) Processus d'externalisation graphique dans les activités cognitives complexes: le cas de l'esquisse numérique en conception architecturale individuelle et collective. PhD Thesis, University of Liège

Safin S, Kubicki S, Bignon J-C, Leclercq P (2011). Digital Collaborative Studio: 4 years of practice. Poster presented at the CAAD Futures 2011 Conference, Liège, Belgium, July 2011

Safin S, Juchmes R, Leclercq P (2012) Use of graphical modality in a collaborative design distant setting. In: Proceedings of COOP 2012, Marseille, France, June 2012

Security and Usability: A Naturalistic Experimental Evaluation of a Graphical Authentication System

Moustapha Zouinar[1(✉)], Pascal Salembier[2], Robin Héron[1],
Christophe Mathias[1], Guirec Lorant[1], and Jean-Philippe Wary[1]

[1] Orange Labs, Paris, France
moustafa.zouinar@orange.com
[2] Institut Charles Delaunay (UMR 6281 CNRS),
Université de Technologie de Troyes, Troyes, France

Abstract. This article sums up results from an experimental study conducted in the context of a research project on the design and assessment of a graphical authentication system (HSA® for Human Semantic Authentication). The experiment addresses different issues related to the usability of the system (performance), and its cognitive demands (memorisation and categorisation). The results put into evidence that, despite the fact that this type of system presents some advantages in terms of use (providing some requirements are observed), it raises nonetheless different problems which put uncertainty on its potential acceptability and use in real world situations. The results illustrate the traditional tension between security and usability.

Keywords: Graphical authentication · Usability · Security

1 Security and Usability: A Challenge for Ergonomics

The security problems raised by the use of authentication systems have led to a significant development of work combining computer security and human-machine interaction [1]. The most commonly used authentication scheme (alphanumeric password) proves to be more and more vulnerable to sophisticated attacks, whether automatic or human. This vulnerability is largely related to the uses of this type of scheme (e.g. the tendency of users to choose easily identifiable passwords). The multiplication and disparity of protection measures sometimes lead users to adopt strategies to circumvent them, which make their daily activities more difficult, to the detriment of security [2]. Thus, the issue of the security of computer systems can not be reduced to a problem of a purely technical nature and that it must integrate the human component [3]. The search for solutions that optimize both security and usability has led researchers in the field to explore new authentication systems. Graphical authentication for example has been proposed as an alternative which offers a reinforcement of the level of resistance to attacks (traditional passwords are predictable when the user chooses them) and, most importantly, increased ease of use, especially in terms of memory retention.

© Springer Nature Switzerland AG 2019
S. Bagnara et al. (Eds.): IEA 2018, AISC 822, pp. 550–558, 2019.
https://doi.org/10.1007/978-3-319-96077-7_59

This article sums up results from a study conducted in the context of a research project on the design and assessment of a graphical authentication system (HSA® for Human Semantic Authentication). This system was designed to make it more difficult for automatic spy attacks (via malicious tools installed on the client terminal, such as spyware or key loggers) that operate by data capture or by random authentication attempts. The authentication task with HSA® can be described as a perceptual and cognitive task of associating concepts (i.e. meaningful words) – which constitutes the password - with pictorial elements.

From a more general perspective, the results illustrate the traditional tension between security and usability since they put into evidence that, despite the fact that this type of authentication system may present some advantages in terms of usability it raises nonetheless different problems which put uncertainty on its potential acceptability and use in real world situations.

2   Graphical Authentication

In the literature, three main categories of graphic authentication systems are distinguished according to their characteristics in terms of memorization [1]: (1) in systems based on recognition, called cognometrics, the user must memorize a set of images, for example faces as in the Passfaces system [4]; the authentication procedure then consists in recognizing them among another wider set of decoys. (2) Systems based on recall require the memorization of a graph or drawing that the user must reproduce using a mouse or stylus for example, *Draw-A-Secret* [5] is a canonical example of this type of system called drawmetrics. (3) In systems based on the indexed reminder, the user has clues that facilitate the task of retrieving the password from memory. This is for example the case of *PassPoints* [3] or *Passhints* [6].

The HSA® System. This system was designed to make automatic spy attacks more difficult (via malicious tools installed on the client terminal, such as spyware or key loggers) that operate by data capture or random authentication attempts (for more details on security, see [7]). The operating principle of this system from the point of view of the authentication procedure is as follows: a password consisting of four concepts (for example, Yellow, Tool, Animal, Food) is assigned to a user who must remember them. These concepts are "coded" in images in which instances of these concepts are present. Coding consists of defining the zones that correspond to these concepts and thus making them active. This coding is invisible to the end user. The authentication phase consists of clicking on the areas that correspond to the password (i.e. to each of the four concepts), knowing that the image changes with each new authentication for security reasons. The authentication task is successful when the user clicks or points to areas that have been pre-coded as instances of concepts, in the order defined at the time the password was created. The task of authentication can thus be described as a task of associating concepts with pictorial elements, which involves the implementation of visual search and categorization processes based on concepts). Compared to other systems, with HSA® authentication is based on both recall (there are no explicit clues in the image) and recognition (the user must recognize visual elements, such as objects or colors, from words). However, a specific feature of this system is that it is not entirely graphical since it includes words; it is the authentication that is done graphically. From the "user" point of view, this system raises a series of questions related to: (1) usability of the authentication device: does the system allow the user to authenticate effectively and efficiently? Is the interaction mode by pointing adapted to different types of terminals: tablet, smartphone? What are the memorization requirements? (2) intelligibility of the logic of concept-image associations: what types of concepts/words should be provided to users? How to choose them? (3) compatibility between usability and safety criteria.

Previous Results. Following two previous experiments [7, 8] several critical elements can be highlighted:

- The mode of authentication of the HSA® system does not present difficulties as for its understanding; the participants very quickly understood the task.
- The overall results of the concept/body association task can be seen as generally satisfactory (82% of correct associations).
- The correct association of 4 concepts to an image, which approximates an authentication procedure based on the graphic password principle as instantiated in the HSA® demonstrator, is insufficient (40% of authentications passed the first time).
- The time taken by participants to identify four instances (28 s on average) also seems unsatisfactory in relation to the performance thresholds considered acceptable in the literature (generally less than 20 s [2]). More satisfactory authentication times were recorded (13 s), but this result must be qualified insofar as the participants were not subjected to the memorization constraint (the concepts were presented in clear text), which can be assumed to influence performance.
- Finding visual instances of concepts in a rich image can be especially difficult under certain conditions (for example, when concepts are too abstract or require significant inferential work, when instances are not prominent enough, or when the subject's categorical system and that of the coder are not congruent).
- The second experiment made significant progress in terms of performance compared with the first experiment. In particular, there was an improvement in the correct association rate (88 vs. 82%) and the average completion time (5.8 s. vs. 7.2 s). But many questions remain, in particular the consequences on memorization and usage in a realistic authentication framework. To address these points, a third experiment was carried out.

3   Testing Graphical Authentication in Real-World Setting

3.1   Method

Subjects. The population selected for this experiment consisted of 22 subjects familiar with the regular use of alphanumeric authentication systems. The sample consisted of engineering, humanities and social sciences students, teacher-researchers, engineers and administrative staff. The population consisted of 12 women and 10 men; the average age was 33 years (minimum: 17; maximum: 61).

Materials. The concepts were selected according to the results of the first two experiments, in particular in relation to the observed correct association rates and the association difficulties pointed out by the subjects.

Three fictitious websites were created for the experiment: social network (AMICO), online banking (BANCO), online shopping (SHOPPING). Each site was assigned a password consisting of 4 concepts (Table 1). The choice to test three sites was mainly guided by the desire to bring us closer to real situations in which users often use several passwords. This point has to be considered when analysing the effects on memory.

Table 1. List of concepts used in the experiment 3.

Sites	Concept 1	Concept 2	Concept 3	Concept 4
AMICO	Transport	Lighting	Vegetal	Writing
BANCO	Food	Hooked	Glass	Device
SHOPPING	Liquid	Animal	Metal	Wood

The concepts were selected on the basis of previous experiments. Each password concept had to be associated with areas of an image from a list of 48 images designed and selected for the experiment (24 images for AMICO, 16 for BANCO and 8 for SHOPPING). These images were of 3 types: photographs (33), comic strips (14) or drawings (1). The variation in the number of images is related to the fact that we wanted to study the effects of the connection frequency and therefore the use of passwords on memorization. Usability criteria were also taken into account following the first two experiments: avoid image overlay; optimize instance visibility (contrast, sharpness, contour); minimize instance ambiguity.

The images were coded by the three experimenters in charge of the construction and handing over of the experiment, according to the following method: a first phase of individual coding followed by a collective confrontation to check the inter-coder agreement, and/or to carry out a collective coding.

Procedure. The experiment was conducted over a period of approximately 3.5 months. The instructions were initially sent by e-mail to the participants who became familiar with the task (with the help of one of the experimenters). They would then receive the passwords by e-mail.

After this phase, the participants had to connect and authenticate 48 times on the 3 sites designed for the experiment (24 times for AMICO, 16 times for BANCO and 8 times for SHOPPING i.e. for each site respectively: 6 times, 4 times and 2 times per week). They received an electronic message at a fixed deadline indicating the connections to be made during the current day and the number of connections made/remaining for each of the 3 sites. During a connection and after having indicated their e-mail address, the subjects had to enter their password by associating the 4 concepts composing this password to an image (each time new), and by respecting the predefined order; they had 3 attempts. They were then asked to complete a short online questionnaire that aimed to collect information on difficulties encountered (forget password, association problem, etc.). If they failed, the site would remind them of their password. Participants could use a PC screen or tablet but not a smartphone to perform the test as the system was not yet suitable for this type of terminal. At the end of the 4 weeks of experimentation, an interview was conducted with the subjects on the basis of the data collected and the experimental material used.

Data Collection and Analysis. The data collected were the following: time-stamped logs, responses to online questionnaires, and verbal reports collected during semi-directive post-test interviews.

3.2   Results

Association Task. The total number of associations carried out by the subjects during the experiment was 4134. As in the first experiment, the task was correctly understood and performed by all participants. The success rate (correct associations) is 91%. There is some variability depending on the concept: from just over 80% for Metal and Wood to over 95% for Transport on average; Fig. 1).

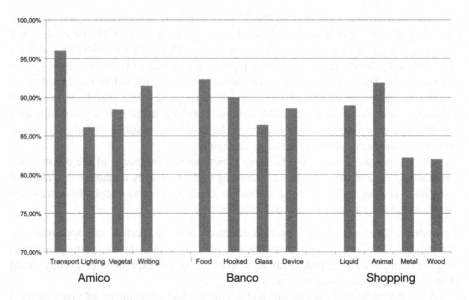

Fig. 1. First try match rate by concept for the three websites.

When we examine the time taken by subjects to find instances by concept (Fig. 2), we notice that the first concepts of passwords take significantly longer than the following concepts (on average 14 s for the 1st; 4 s for the 2nd; 4.5 s for the 3rd and 4.2 s for the 4th). Several explanations are possible: the subject looks at the image as a whole, he looks for all the concepts before clicking, the search for the password in memory or in external supports takes time.

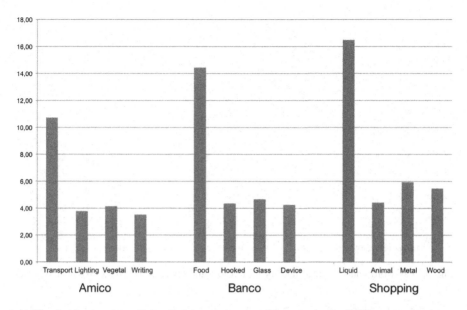

Fig. 2. Average association time per concept to first test per site (1005 connections).

Here too, the subjects encountered difficulties in associating concepts and instances, but much less so than in previous experiments.

Finally, it should be noted that we observed far fewer phenomena of interaction between concepts and images than in the first two experiments from the point of view of the association task. This is probably due to the collective coding procedure that avoided the coding and instance selection errors observed in the first two experiments.

Authentication. The results show an average correct authentication rate of 94% for all attempts combined. This result breaks down as follows: 81.5% passed the first test; 90.5% passed the second test; 94% passed the third test. It can be seen that the success of authentication varies between sites (and therefore the number of connections): as might be expected, the more connections there are, the higher the correct authentication rate increases. This phenomenon is particularly noticeable when only the first tests are taken into account.

Concerning the authentication time per connection, the average observed is 31.25 s. for all attempts combined (m: 23, sd: 18.4, min 4, max 220). The average time obtained on the first test is 24.75 s. We can notice that the minimum value obtained seems to show that authentication can be very fast. We also observe that, as for the correct authentication rate, there is a significant difference according to the sites, here inversely proportional to the number of connections: the average authentication time goes from 26 s. for Amico, to 37 s. for Banco and 40 s. for Shopping. A link between the number of connections and the authentication performance therefore seems to be confirmed.

Memorization. The recall rate of concepts at the end of the test during the interviews was 73% for the AMICO site, 55% for BANCO and 50% for SHOPPING (Fig. 3). This result, which must be taken with caution (experimental situation, recall performed

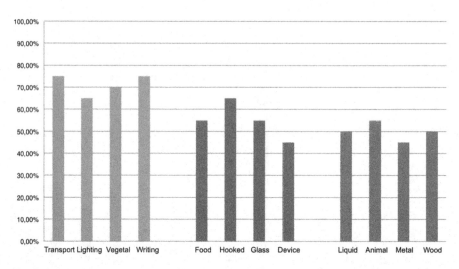

Fig. 3. Password concept recall rate. From left to right AMICO, BANCO, SHOPPING sites.

more or less long after the end of the test, passwords imposed), indicates the existence of a relationship between recall rate and connection frequency, the regular use of passwords seems to favour their retention.

The difficulties of memorization resulted in omissions and confusion (passwords or order of concepts). This problem arose especially for subjects who made the effort to memorize passwords (a significant proportion of subjects used external media on an ad hoc or regular basis). Moreover, the interviews show that the memorization aspect was experienced as the most problematic by the subjects.

4 Discussion and Conclusion

Compared to the first two experiments, the third experiment shows a clear improvement in usability results, in particular the successful association rate for four concepts in the first trial: 40% for the first, 60% for the second, 81.5% for the third (Fig. 4). This result is largely related to the three-step approach that refined the coding procedure and the progressive selection of concepts more suited to the task.

Compared to those of previous studies on graphic authentication systems, the authentication rates obtained in the 3rd study are rather encouraging (in the literature these rates vary between 57 and 100% [1]). However, the average authentication time (24.7 s.) remains relatively high compared to what is generally considered acceptable (less than 20 s.) even if it has improved compared to the first study. It should be noted that in the 3rd experiment, only 5 subjects felt that authentication took too long.

While the overall results obtained are rather encouraging, the fact remains that the system raises a number of difficulties that question its viability from the point of view of usability. As it stands, it does not seem suitable for the use of several passwords, especially from the point of view of storage when passwords are imposed. Besides the

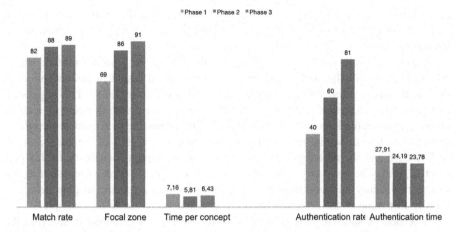

Fig. 4. Summary of results for the 3 experiments from left to right: overall correct association rate; correct association rate in focal zone; association time per concept; authentication rate; authentication time.

fact that it remains to be verified that allowing the choice of password would have a positive effect on memorization, the associated risk in terms of weakening security cannot be ruled out.

Another difficulty, related to this tension between security and usability, concerns concepts and images: if we follow security criteria, instances must not be easily identifiable (avoid well delimited objects, limit empty surfaces), and concepts must not be too specific (for example "Dog"). However, in terms of usability, the study shows that the best scores are obtained when the instances are well delimited and/or when the concepts do not require significant inferential work or do not generate uncertainty (contrary, for example, to the "Heavy" concept which can involve complex operations of comparison between the various objects present in the image). This raises the question of the ideal compromise to be found between safety and usability, and of the nature of this compromise in the particular case of the HSA® system. Another notable result of these experiments is that the interpretation by users of the concepts composing the password may be different from that of the designers, hence the importance of the systematic-matization of the coding procedure that was implemented in the 3rd experiment.

Although it has provided results that provide food for thought on the design of the system and its potential interest, this study does have some limitations. First, the analysis of the consequences in terms of memory was made difficult because of the methodological choice made. In the 3rd experiment, we favoured the ecological aspect, which resulted in the occasional or regular use by some of the participants of memory aids when they were asked by instructions to avoid using them. It was therefore difficult to obtain precise quantitative information on the effects on memorization. We could have retained the traditional method used in this field of study (by inserting "distracting" moments between authentication sessions), which would have required remaining in the context of a laboratory experiment. But such a method seemed too far away from the real use of passwords. Despite this limitation, the results, particularly in terms of the participants' lived experience, provide us with significant elements on memorization.

Second, we were not able to explore the usability of the system on other smartphone devices while the implementation of the principle of graphical authentication on these terminals is giving rise to an increasing number of studies [3, 9]. This suggests a first perspective for continuing the study that would incorporate the current state of knowledge on tactile interaction. It would indeed be interesting to document the effects of a more or less important reduction of the screen size, and thus of the image, on the performance and the user experience, compared to PC screens or tablets. In this perspective, it may be necessary to zoom in on the image. In the same vein, adapting the system to visually impaired users will require the use of a specific display device to identify lost areas of the screen (translation in a non-visual mode for example).

A second perspective, linked to the security aspect, would consist in studying the vulnerability of the system to "shoulder surfing attacks" (observation over the shoulder) and automatic attacks based on the use of image recognition algorithms.

Finally, general questions relating to the industrialization of the image generation and coding, and to the concepts research process remain open.

Acknowledgements. This study was carried out within the framework of a collaborative research contract between Orange Labs and the Troyes University of Technology.

References

1. Biddle R, Chiasson S, Oorschot PC (2011) Graphical passwords: learning from the first twelve years. Technical Report TR-11-01, School of Computer Science, Carleton University
2. Norman DA (2009) The way I see it when security gets in the way. Interactions 16(6):60–63
3. Wiedenbeck S, Waters J, Birget J-C, Brodskiy A, Memon N (2005) PassPoints: design and longitudinal evaluation of a graphical password system. Int J Hum Comput Stud 63(1):102–127
4. Davis D, Monrose F, Reiter MK (2004) On user choice in graphical password schemes. In: 13th USENIX security symposium (2004)
5. Jermyn I, Mayer AJ, Monrose F, Reiter MK, Rubin AD (1999) The design and analysis of graphical passwords. In: USENIX Security Symposium
6. Chowdhury S, Poet R, Mackenzie L (2014) Passhint: memorable and secure authentication. In: Proceedings of the SIGCHI conference on human factors in computing systems (CHI'14), pp 2917–2926
7. Salembier P, Zouinar M, Mathias C, Lorant G, Wary J-P (2015) Evaluation ergonomique d'un système d'authentification graphique Actes de la conférence EPIQUE 2015, Aix-en-Provence, 8–10 juillet: Arpege Publishing
8. Salembier P, Zouinar M, Mathias C, Lorant G, Wary J-P (2016) Etudes expérimentales d'un système d'authentification graphique basée sur la catégorisation sémantique Actes de la conférence IHM'2016, 26–28 octobre, Fribourg, Suisse. ACM
9. Schaub F, Walch M, Könings B, Weber M (2013) Exploring the design space of graphical passwords on smartphones. In: Proceedings of the ninth symposium on usable privacy and security, SOUPS'13, pp 1–15

Interaction Options for Wearables and Smart-Devices While Walking

Jessica Conradi[1(✉)], Martin Westhoven[2], and Thomas Alexander[1]

[1] Fraunhofer FKIE, Fraunhoferstr. 20, 53343 Wachtberg, Germany
jessica.conradi@fkie.fraunhofer.de
[2] German Federal Institute for Occupational Safety and Health,
Friedrich-Henkel-Weg 1-25, 44149 Dortmund, Germany

Abstract. Using smartphones and wearables in parallel activities, e.g. while walking, is a widespread phenomenon. In different individuals we furthermore find different interaction styles which involve one or both hands. To study the effects of different interaction styles, we carried out a study involving three different interaction styles for touch-sensitive devices compared to a HMD operated by an additional controller providing passive haptic feedback. The experimental task was carried out while walking on a treadmill. In addition to the primary task of using the interaction device, a secondary task was administered which competed for the participants' visual attention.

We found an impact of the different interaction styles on the input performance: Time on task proved to be faster with the HMD combined with a haptic input device, but error count increased significantly as well. Using the thumb for input resulted in a longer overall time during which the visual attention was focused on the interaction device.

Keywords: Mobile interaction · Walking · Interaction styles

1 Introduction

Ubiquitous usage of mobile devices, like smartphones or wearables, calls for fast and dependable means of interaction. This is especially important in case of secondary interaction, e.g. to facilitate interaction while navigating in unfamiliar or dynamic environments. This includes frequent information exchange with the device, e.g. by extracting information about waypoints or by inserting data concerning the navigation, like destination changes. In this kind of interaction, touch input is most common. It can be operated with one or both hands. Two-handed interaction often constrains one hand to a purely supportive task, i.e. holding the device. To avoid this, a common solution is to attach the device to e.g. the forearm or wrist. Alternatively, smart glasses can be used as an entirely hands-free display. These goggles help to further perceive the surroundings visually while interacting with the system. However, in contrast to smartphones which combine in- and output functionality smart-glasses only provide output functionality and require an additional interaction metaphor. Feasible options are gestures captured by an additional camera-system, voice input or devices operated by hand, like touchpads or haptic controllers. Pros and cons of this kind of controllers are

© Springer Nature Switzerland AG 2019
S. Bagnara et al. (Eds.): IEA 2018, AISC 822, pp. 559–568, 2019.
https://doi.org/10.1007/978-3-319-96077-7_60

scarcely known. This is especially true for the use while walking. Although interacting with mobile devices while walking is widespread, there are few studies focusing on optimizing the devices for this scenario.

Due to their versatility, there are various options to interact with smartphones. The touch-sensitive surface allows for interaction with any finger or other body part. The primary input procedures rely on the thumb or index finger. Especially in applications, which demand many input actions, e.g. texting, thumb-driven interaction is frequently used. Gold et al. [1] observed college-aged persons (n = 859) typing on a mobile device. 32.1% of them used a touchscreen, the others used keypads. They found 51.1% to use both hands to hold the device and to type with one or both thumbs, 44.4% to hold the device with one hand and to type with the thumb of this hand, only 2.1% carried the device with the left hand and typed with the right index finger. Altogether, 53.2% of the observed users engaged both hands in the interaction, while 44.4% engaged only the left or the right hand. However, this study focused on typing on mobile devices while standing (64.4%) or sitting (35.4%).

Other research is performed regarding the interaction performance of one- and two-handed interaction, focusing on thumb inputs. E.g. Trudeau et al. [2] compared one-handed and two-handed thumb-interaction in a tapping task with fast reciprocal tapping. They found higher performance and less movement of the device in two-handed interaction tasks than in one-handed. For Thumb interaction, the size and position of the button is of high relevance. E.g. Park et al. [3] determined the area on the surface of mobile devices, which supports correct and fast thumb interaction. The minimal button size has been determined by Parhi et al. [4]. They recommend a size of at least 9.2 mm for discrete tasks and 9.6 mm for serial tasks on one-handed thumb use on touchscreen-based handhelds. However, these findings only consider applications used while standing or sitting.

Ng et al. [5] looked into the effects of encumbrance on one- and two-handed interactions with mobile devices while walking. They found the targeting performance to be affected by carrying a bag in each hand and additionally walking while interacting with a smartphone. Unencumbered, target accuracy was evenly matched between the input postures one-handed (thumb), two-handed (thumb) and two-handed (index finger). Encumbered, accuracy decreased by 16.7% for two-handed (index finger), 9.9% (one-handed thumb, and 8.7 for two-handed (thumb). Error increased by 40% for one-handed thumb, 32.3% for index finger and 22.6% for both thumbs.

There is a significant influence of walking speed on index finger interaction. A comparison of interaction performance of standing and walking (5 km/h) showed a significant influence of the independent variables walking speed and button size on the dependent variables error count and time on task. Consequently, button sizes of 11 mm for standing and 14 mm for walking are recommended, with at least 8 mm as the very minimum [6].

Interaction with mobile devices is often neither the only nor the main task of the user. While walking he or she has to be attentive to the surroundings to be able to navigate and to avoid obstacles and collisions. Therefore, the distraction caused by interacting with the mobile device should induce minimum distraction from the environment. This distraction may also be influenced by the interface design. To address this problem, in another study a pen was used as input-device and sitting, standing and walking on a treadmill and walking on an obstacle course were considered. Walking was found to have no effect on time on task, in contrast to button size, where time on task decreased with increasing button size. The error rate was highest on the obstacle course and lowest in seated position [7].

The biomechanics of walking affect visual acuity [8]. However, because of multiple changing translational and angular speeds of the hand-arm-shoulder system, the amount of influence on visual acuity is uncertain and varies [9]. Therefore, it is hard to calculate the effect precisely. Conradi and Alexander [10] quantified the impact of walking and found visual acuity to decrease by about 20% compared to standing.

All the above mentioned factors impact interaction during walking. To contribute to the understanding of this impact and to help finding solutions to enhance interaction in this widespread scenario, we carried out a study on different interaction styles for the same task while walking. We examined single-handed interaction, a style involving one hand and the wrist of the other arm as well as an interaction style which engaged both hands. Additionally, we explored an interaction metaphor involving an Augmented Reality display which was operated by means of a controller providing passive haptic feedback. While walking, the attention of the user usually should be on the environment to avoid collisions and to perceive and be able to react on any occurrence. To answer to this practical implication, the participants' attention was drawn to a secondary task in the environment. For this scenario, we hypothesized a difference in terms of interaction performance for these interaction metaphors.

2 Method

2.1 Participants

16 volunteer participants (7 female, 9 male; aged 27–33 years) were recruited. Prior to participation, color blindness was excluded by means of a Color Vision Test [11]. Only right-handed volunteers were included in the study, since the developed interfaces were optimized for right-handed interaction. Each participant was tested in each condition, resulting in a repeated measurements design. The conditions were administered on subsequent days. The order of the conditions was varied according to a Latin square.

2.2 Variables

The independent variable *interaction style* was varied on four levels. For each style an adapted interface was designed and implemented, taking ergonomic principles into account. Three different styles were developed for a touch-sensitive hand-carried device. Additionally, a head-mounted device was used in combination with input-device (Wii-Controller) providing passive haptic feedback.

The first interaction style uses a smartphone carried in *portrait* mode in the left hand. Interaction was carried out with the right index finger. In portrait mode, the interaction area was in the lower part of the display, it consisted of four equally-sized rectangular buttons. The visual representation of the interaction was placed in the top section (see Fig. 1). In this interaction mode, both hands are engaged. The left hand

Fig. 1. Interaction (left) and interface (right) in portrait mode.

holds the smartphone all the time, while the right hand is intermittently used for interaction.

The second interaction style is the thumb mode. In this mode, the smartphone is carried in the left hand, like in portrait mode, but interaction is done by the thumb of the carrying hand. So this hand has a double function in holding the device and inserting data as well. In this style, the other hand is completely free.

As the used smartphones has a big display, the thumb cannot reach the whole surface. Therefore, an adapted interface was designed. The interaction buttons were placed on a circular segment to be accessible for the thumb while remaining visible. The round buttons were big enough to be used easily. The visual feedback was placed above the interaction zone (see Fig. 2).

Fig. 2. Interaction (left) and interface (right) of thumb mode.

The third style had an interface in landscape mode, whereas the smartphone was attached to the left *wrist* and lower arm. Interaction was carried out with the right index finger. This relieved both hands from a permanent carrying task. Outside interaction, both hands can be engaged in other activities. However, during interaction both extremities are engaged in the interaction, but the left hand can still be used in other non-interfering tasks like carrying smaller items. In this mode, the visual representation is depicted in the left part of the display and the interaction buttons in the right part. So, the visual part is not obscured by the interaction activities of the right hand (see Fig. 3).

Fig. 3. Interaction (left) and interface (right) of wrist mode.

The last interaction style was included to provide insight into another interaction and display style and to explore differences and pros and cons of the different styles. The visual feedback was provided by a Head Mounted Display (*HMD*), a controller was used for input which includes passive haptic feedback. In this case, the visual feedback is always visible, regardless where the participant's visual attention lies. In contrast to the other interaction styles, the input device provides passive haptic feedback, which allows interaction without visual control. The devices as well as the visual feedback on the HMD are depicted in Fig. 4.

Fig. 4. HMD (left) haptic input controller (middle) and interface (right) of *HMD* mode.

Dependent variables were used to gain insight into the participants' performance. Therefore, mean interaction *time on task* for an interaction cycle was used. An interaction consisted of about 12 interaction steps. However, is some cases, less steps were needed and in other cases, errors occurred, which led to additional interaction steps to correct the errors. We were interested in these errors and correcting actions, therefore we analyzed the *number of interaction steps* per interaction cycle as second dependent variable.

We also were interested in the distraction induced by the interface. Therefore, we used an eye-tracking device (Dikablis, Ergoneers, Germany). This allowed us to calculate the total time during which the visual attention was focused on the interaction device (*overall glance time*). However, the eye-tracking system was not compatible with the HMD, so this data was collected only for the three touch-sensitive conditions.

2.3   Task

In the primary task, the participants had to fulfill interaction sequences involving a hierarchical menu with $3 \times 3 \times 3$ interaction options or 27 different combinations. To trigger the interaction, the participants were presented with different human characters in a virtual environment. In randomized intervals, additional information was presented. The participants had to input this information as fast as possible. In a secondary task, visual distractors were used, to keep the participants' attention on the environment. They had to be detected and reacted to in a limited time span.

The whole task was performed by all participants in all conditions; order and timing of the subtasks were varied randomly. Each task took about 12 min in all conditions.

2.4   Apparatus

The experimental task was administered through a large-scale display (ca. 2.30 m 3.6 m). The virtual reality environment consisted of a small village with about 150 characters populating the scene. The scene was built by means of the CryEngine 3. During the task, the participants walked on a treadmill (H/P/Cosmos pulsar, h/p/cosmos sports & medical GmbH, Nussdorf, Germany) at a speed of 5 km/h.

The participants used a Samsung S3 smartphone for the touch sensitive conditions and a Lite-Eye LE-750 A in combination with a Wii-Controller for the HMD condition.

Eye motion was recorded by a head-based eye tracking system (Dikablis, Ergonoeers, Germany).

2.5   Analysis

All data was tested for normal distribution by means of the Kolmogorov-Smirnov-test. In case of normality, the data was tested for sphericity (Mauchly-test). In the next step, an Analysis of variance (ANOVA) with repeated measures was administered. In case of significant results, pair-wise comparisons with Bonferroni-correction were performed.

3   Result

Performance in the main task was measured by the dependent variables *time on task* and *number of interaction steps*. In the following Table 1 the numbers of mean and standard deviation of both variables are given.

Table 1. Means and standard deviation of performance measures.

		Portrait	Thumb	Wrist	HMD
Time on task [s]	Mean	10.61	11.61	11.06	9.87
	SD	2.37	3.37	2.86	2.09
Number of interaction steps	Mean	12.37	12.49	12.39	13.81
	SD	0.28	0.72	0.35	1.17

For *time on task* we found significant influences of the *interaction style* ($F_{(3,45)}$ = 5.591; $p = 0.011$; $\eta^2 = 0.272$). The post-test revealed no significant differences but some statistical trends for wrist and HMD ($p = 0.057$) and thumb and wrist ($p = 0.081$). Figure 5 (left) depicts these results.

Number of interaction steps was influenced as well ($F_{(3,45)}$ = 15.85; $p < 0.01$; $\eta^2 = 0.514$). Pairwise comparisons showed highly significant differences between HMD and portrait ($p < 0.01$), HMD and thumb ($p < 0.01$), and HMD and wrist ($p < 0.01$). See Fig. 5 (right) for a depiction of the results.

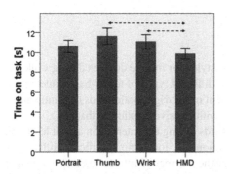 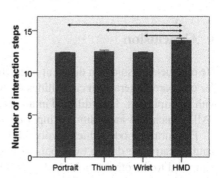

Fig. 5. Mean and standard error and statistical trends (dotted arrows) of the variables *time on task* (left) and mean and standard error and significant differences (arrows) of *number of interaction steps* (right).

The variable *overall glance time* yields insights into how long the visual attention was needed on the device to fulfill the tasks. In Table 2 means and standard deviations are given. It was not possible to combine HMD and the eye tracking system, therefore no data is given for HMD.

Table 2. Means and standard deviations of *overall glance time*.

		Portrait	Thumb	Wrist
Glance time [s]	Mean	151.3	172.0	152.3
	SD	34.5	44.4	29.2

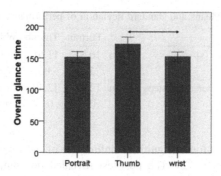

Fig. 6. Means and standard errors of the variable *overall glance time* and pairs showing significant differences (arrows).

In the analysis of variance we found significant differences $(F_{(3,30)} = 3.44;$ $p = 0.045;$ $\eta^2 = 0.187)$. Pairwise comparisons found the difference to be between thumb and wrist $(p = 0.033,$ see Fig. 6).

4 Conclusion

We conducted a study on different interaction styles for mobile devices during walking. We compared interaction capabilities of three different types of touch-based interaction with smartphones. They differed in involvement of left/right/both hands or thumb and a HMD which had an additional input-device with passive haptic feedback.

For portrait mode, which involves both hands during the interaction and at least one hand to carry the device even outside interaction, we found no significant differences compared to the other touch conditions. Only the number of interaction steps appeared to be lower than for the HMD. However, this lower number of interaction steps found no equivalent in the other performance measure. This may be due to the usage of the index finger in his usual role as index instrument.

The one-handed interaction with the thumb of the device-carrying hand showed no significant differences in the performance measures compared to the other conditions with touch-sensitive devices, but overall glance time was significantly higher than at the wrist condition. So, participants needed more visual control to operate with the thumb. Additionally, time on task was longer and number of interaction steps was lower than for the HMD condition. The participants used more time to operate more accurately.

Attaching the smartphone to the wrist offers the benefit that both hands are free outside interactions and even during interaction, the left hand is free to carry smaller objects. For the performance we found significant differences compared to HMD, i.e. longer time on task but a lower number of interaction steps. In this respect, wrist is similar to the thumb condition. However, overall glance time was lower than in the thumb condition; the participants needed less visual feedback for equally fast and correct interaction.

The HMD condition featured an interaction style which is different from the other conditions in two respects: Firstly, visual feedback is given in a HMD and not via the interaction device itself. Secondly, the controller is a separate device, which has to be carried in one hand and offers passive haptic feedback. Compared to the other interaction styles, HMD showed a lower time on task but at the same time a higher number on interaction steps. Participants interacted faster and risked more errors which triggered correction actions. But still the overall interaction time was lowest.

To summarize, we found an impact of the different interaction styles on the input performance. Time on task proved to be lower with the HMD combined with an input device featuring haptic feedback, but error count increased significantly as well. Using the thumb for input triggered a longer overall time focusing on the interaction device.

All in all, the tested interaction styles are feasible for interaction with wearables while walking, but provide different benefits and drawbacks to consider for the concrete application.

References

1. Gold JE, Driban JB, Thomas N, Chakravarty T, Channell V, Komaroff E (2012) Postures, typing strategies, and gender differences in mobile device usage: an observational study. Appl Ergon. https://doi.org/10.1016/j.apergo.2011.06.015
2. Trudeau MB, Asakawa DS, Jindrich DL, Dennerlein JT (2016) Two-handed grip on a mobile phone affords greater thumb motor performance, decreased variability, and a more extended thumb posture than a one-handed grip. Appl Ergon. https://doi.org/10.1016/j.apergo.2015.06.025
3. Park YS, Han SH, Park J, Cho Y (2008) Touch key design for target selection on a mobile phone. In: ter Hofte H, Mulder I (eds) Proceedings of the 10th conference on human–computer interaction with mobile devices and services, Mobile HCI 2008, Amsterdam, The Netherlands, p 423. https://doi.org/10.1145/1409240.1409304
4. Parhi P, Karlson AK, Bederson BB (2006) Target size study for one-handed thumb use on small touchscreen devices. In: Nieminen M, Röykkee M (eds) Proceedings of the 8th conference on human–computer interaction with mobile devices and services. MobileHCI '06, Helsinki, Finland, 12–15 September 2006, p 203. https://doi.org/10.1145/1152215.1152260
5. Ng A, Brewster SA, Williamson JH (2014) Investigating the effects of encumbrance on one- and two-handed interactions with mobile devices. In: Jones M, Palanque P, Schmidt A, Grossman T (eds) Proceedings of the SIGCHI conference on human factors in computing systems, CHI'14, Toronto, Ontario, Canada, pp 1981–1990. https://doi.org/10.1145/2556288.2557312
6. Conradi J, Busch O, Alexander T (2015) Optimal touch buttons size for the use of mobile devices while walking. Procedia Manuf. https://doi.org/10.1016/j.promfg.2015.07.182
7. Lin M, Goldman R, Price KJ, Sears A, Jacko J (2007) How do people tap when walking? An empirical investigation of nomadic data entry. Int J Hum Comput Stud. https://doi.org/10.1016/j.ijhcs.2007.04.001
8. Hillman EJ, Bloomberg JJ, McDonald PV, Cohen HS (1999) Dynamic visual acuity while walking in normals and labyrinthine-deficient patients. J Vestib Res 9(1):49–57
9. Pozzo T, Berthoz A, Lefort L (1990) Head stabilization during various locomotor tasks in humans. Exp Brain Res. https://doi.org/10.1007/BF00230842

10. Conradi J, Alexander T (2014) Analysis of visual performance during the use of mobile devices while walking. In: Hutchison D, Kanade T, Kittler J, Kleinberg JM, Kobsa A, Mattern F, Mitchell JC, Naor M, Nierstrasz O, Pandu Rangan C, Steffen B, Terzopoulos D, Tygar D, Weikum G, Harris D (eds) Engineering psychology and cognitive ergonomics, vol 8532. Lecture notes in computer science. Springer, Cham, pp 133–142
11. Kuchenbecker J, Stilling J, Broschmann D, Hertel E, Velhagen K, Kuchenbecker J (2014) Plates for color vision testing. Georg Thieme Verlag, Stuttgart

Invisible Touch! – Design and Communication Guidelines for Interactive Digital Textiles Based on Empirical User Acceptance Modeling

Philipp Brauner[✉], Julia Offermann-van Heek, and Martina Ziefle

Human-Computer Interaction Center, RWTH Aachen University,
Campus Boulevard 57, 52074 Aachen, Germany
{brauner,vanheek,ziefle}@comm.rwth-aachen.de

Abstract. Interactive digital textiles are both light and shadow in regard to users' perception and technology acceptance. To understand people's perceived barriers and benefits in regard to interactive digital textiles and to derive empirically founded design and communication guidelines, we conducted an empirical study based on a synthesis of the Unified Theory and Acceptance and Use of Technology 2 (UTAUT2) and the Smart Textile Technology Acceptance Model (STTAM). In a scenario-based quantitative user study with N = 324 participants, we evaluated the projected acceptance of two exemplary textile products in a between-subject design. The first product addressed wearables in form of a smart jacket and the second product referred to technology augmented living environments in form of a smart armchair. Regression analyses revealed that the combined model (UTAUT2 and STTAM) explained over 80% of the variance of the participants intention to use the smart armchair (80.3%) as well as the smart jacket (84.7%): For the smart armchair, the model dimensions performance expectancy, hedonic motivation, washability, and input modality were decisive, while performance expectancy, hedonic motivation, and connectivity were relevant for the smart jacket. We conclude with empirically based communication and design guidelines to increase the acceptance of interactive digital textiles.

Keywords: Digital textiles · Smart textiles · Technology acceptance
User modeling · User diversity · Wearables

1 Introduction

Textiles have been an integral part of our lives for thousands of years [1]. With the advent of the integrated circuit smaller – but more powerful – information and communication technology (ICT) became available [2]. Smart objects are on the rise and the once keen visions of Ubiquitous computing [3] and Ambient Intelligence [4] are becoming increasingly real. While it is feasible to integrate sensors, actuators, and networking into fabrics and textile-based everyday objects [5–8], the question of the acceptance of such a system is not sufficiently elaborated.

In particular, neither the requirements from the perspective of potential users are described, nor are any barriers to acceptance identified. This work closes this gap and

© Springer Nature Switzerland AG 2019
S. Bagnara et al. (Eds.): IEA 2018, AISC 822, pp. 569–578, 2019.
https://doi.org/10.1007/978-3-319-96077-7_61

models the acceptance of two textile devices in the domestic environment, describes commonalities and differences, and provides guidelines for a user-centered design of novel textile-based human-computer interfaces.

2 Related Work

This section presents the theoretical background of this study, starting with theories and models to assess acceptance of a technology. Afterwards, we introduce current technical developments in the area of interactive textile user interfaces as well as research results concerning their acceptance and perception.

2.1 Measuring Technology Acceptance

One prominent model to assess the overall acceptance of technical systems is Davis' Technology Acceptance Model [9]. It postulates that the actual use of software applications is predictable in advance by surveying the intention to use from potential users. Moreover, the model has shown that the intention to use is determined by the usefulness and the ease of use as perceived by the users. The model was validated for different types of software and extended and refined to capture their acceptance. A prominent TAM successor is Venkatesh el al.'s Unified Theory of Acceptance and Use Of Technology 2 (UTAUT 2) [10]. It predicts the voluntary adoption of consumer technology and incorporates several new concepts into the evaluation framework, such as the price-value trade-off and the hedonic evaluation of the technology. Despite the astonishing predictive power of TAM and its successors, these models must be used carefully: The evolution of these models show that each novel technology requires a custom adaption, validation of existing and integration of novel evaluation criteria.

2.2 Digital Textile Interface

The integration of electronics into fabrics was first investigated by Post et al. [11]. A variety of sensors, actuators, and power distribution technologies [8] facilitate the development of novel interactive artifacts. In contrast to conventional human-technology interfaces made of metal, plastic, or glass, these can be warm and soft, and they can even offer new interaction paradigms.

For example, Rekimoto et al. suggested an interactive textile surface that facilitates touch interactions [7]. These textile touch-pads can be integrated in, for example, furniture: Rus et al. [12] augmented textile interaction surfaces on a couch than can be integrated in smart home environments. Heller et al. embroidered conductive yarn into curtains and used conductive sensing to open or close the curtain using swipe gestures [13]. Textile interaction paradigms were explored, for example, in Lee et al.'s deformable interfaces [14]. These interfaces based their interaction on textile specific affordances such as folding, bending, or stretching. Karrer et al.'s [6] Pinstripe explored folds in textiles. Rolling a textile fold can be used as a one-dimensional input device. Brauner et al. [15] created tangible folds in the textile surface of a recliner armchair.

The folds resemble the form of the armchair and facilitated effective and intuitive adjustments of the chair's back and foot rest.

While the aforementioned projects refer to singular prototypes, Poupyrev et al. showed recently that conductive yarns can be integrated into fabrics with industrial looms, which facilitates manufacturing smart textiles at scale [5]. Consequently, we can expect that smart digital textiles will soon appear in a variety of consumer products, such as in furniture, clothing, or gadgets.

2.3 Acceptance of Digital Textiles

Although the technical development of interactive digital textiles gains traction, the research on acceptance of these novel interfaces has many blind spots. This ignores the perspective of potential users and represents a considerable risk for the success of the products and future textile human-technology interfaces.

Hildebrandt et al. elaborated parts of the design space of textile input devices in home environments [16]. The conjoint based study found that the technical realization of the interface is the most decisive decision criterion for potential users. The integrated technology must be invisible and unnoticeable to the users, in a way that the textile character of the object is retained. The second most important criterion was the planned location of the textile interaction device. Here, the living room as an apparently public space was clearly preferred over other rooms, such as the bedroom. The least important criterion was whether the interface was integrated into a household item, such as a pillow or a chair, or into clothing and the participants of the study had no preference for either.

Brauner et al. suggested a Smart Textile Technology Acceptance Model (STTAM) based on a scenario-based survey [17]. It combines textile-specific evaluation dimensions with dimensions from UTAUT2 [10] (see above) and showed that over 86% of the variance in intention to use could be predicted using the model. The influence of user diversity was also investigated and a positive attitude towards technology as well as a positive attitude towards automation had a strong positive impact on projected acceptance of textile interfaces.

Brauner et al. studied the acceptance of an armchair with a textile fold as input systems in a user test [18]. The findings show that acceptance and the desire for using the textile interactions were generally high. Moreover, acceptance was more strongly influenced by hedonic aspects than by pragmatic ones.

3 Method

This section presents the empirical scenario-based approach of the study focusing on two textile devices – the smart jacket and the smart armchair. Afterwards, the structure of the online questionnaire and the study's sample are described.

3.1   Empirical Design

An online questionnaire was conducted to examine relevant parameters for the evaluation and acceptance of different smart interactive textiles. Subsequent to a short introduction of the topic smart textiles, its possibilities, and options, the participants were asked for demographic information (i.e., Gender, Age, and level of education) in the first part of the questionnaire. Further, the participants evaluated several items concerning individual attitudes: they assessed their self-efficacy in interacting with technology (SET) using 4 items (α = .883) based on Beier [19], their affinity towards textiles using 5 items (α = .708), and their potential previous experience with smart textiles using 5 items (α = .883).

Next, the two textile products were introduced and then evaluated in random order using the same evaluation parameters and items described in the following. To evaluate the textile devices, a combination of the UTAUT2 and the STTAM model was applied. Overall, the adapted model contained 10 technology-related dimensions as well as the behavioral intention to use the textile devices (measured by using 3 items). Five of the technology-related model-dimensions were taken from the original UTAUT2-model (each measured by 2 items), as a previous study confirmed those as relevant factors for the acceptance of smart textiles [17]: Performance Expectancy, Effort Expectancy, Hedonic Motivation, Social Influence, and Facilitating Conditions.

Based on STTAM [17] and numerous other studies on the acceptance of smart interactive textiles (e.g., [16, 18, 20]), the Washability of textile devices was proven to be an influencing factor for smart textiles' acceptance and was therefore integrated in the present study (measured by 3 items).

Further, an expert workshop on interactive textiles (n = 3) revealed additional parameters which were relevant for the evaluation of different smart textile devices (jacket vs. armchair). Within the expert workshop, the aspects Innovation (2 items), Functionality (2 items), Connectivity (3 items), and Modality (2 items) were intensively discussed and thus integrated into our research design.

At the end of the questionnaire the participants had the opportunity to give feedback or express their opinion on smart interactive textiles. All items were assessed using 6-point Likert-scales (min = 1; max = 6).

3.2   Sample Description

The respondents were invited to participate in the study by a link distributed via email and in technology-mediated social networks. Overall, 584 participants took part in the study. As only complete data sets could be used for further analyses, a sample of n = 324 data sets remained. Gender was equally distributed with 49.4% males and 50.6% females and the mean age of the participants was 26.9 years ($\pm$ 9.1; min = 15; max = 60). As 33.3% of the participants hold a university degree and 42.6% a university entrance certificate, the educational level of the participants was high with regard to attitudinal characteristics, the participants reported on average a high SET (M = 3.76 $\pm$ 1.05), a rather neutral affinity to textiles (M = 2.61 $\pm$ 1.04), and low experiences with smart textiles (M = 2.17 $\pm$ 0.97).

4 Results

This section is structured as follows: First, we present the participants' absolute evaluation of the two smart textile products on the model dimensions. Subsequently, data is analyzed by correlation and linear regression analyses to identify key predictors of acceptance. Finally, we investigated the influence of user diversity on overall product acceptance by means of a cluster analyses that identified likely users and non-users.

We analyzed the data with parametric (RM-ANOVA, step-wise multiple linear regressions) and non-parametric methods (Spearmen's ρ, Cramer's V). The level of significance was set at $\alpha = 5\%$. Prior to descriptive, correlation, and regression analyses, item analyses were calculated to ensure measurement quality.

4.1 Absolute Evaluation Oft He Jacket and Armchair

The absolute evaluations of the model dimensions for both textile devices are shown in Table 1. The participants assessed their overall intention to use the smart jacket (M = 2.49, SD = 1.39) and the smart armchair (M = 2.48, SD = 1.30) rather neutrally (with 2.5 being the center of the scale). Considering the model dimensions, effort expectancy was rated significantly highest for the armchair, but received also high evaluations for the jacket. Facilitating conditions and innovation were also evaluated rather high, each having a significantly higher evaluation for the smart jacket. The textiles' input modality was rated on a rather positive and similar level for both textile devices, while Hedonic motivation was rated slightly more positive for the smart armchair than the smart jacket. Further, functionality had a significantly higher meaning for the smart jacket than the smart armchair. The dimension performance expectancy was rated rather neutrally for both textile devices, while social influence was evaluated slightly negatively for both devices, and the textiles' connectivity was

Table 1. Evaluation of the smart jacket and the smart armchair.

	Smart jacket		Smart Armchair		p
	M	SD	M	SD	
Effort expectancy	3.40	1.10	3.21	1.15	0.003 < .05
Facilitating conditions	3.09	1.51	3.30	1.47	0.002 < .05
Innovation	3.02	1.33	3.26	1.25	<.001
Modality	2.99	1.07	3.00	1.08	0.979, *n.s.*
Hedonic motivation	2.98	1.33	2.87	1.40	0.085, *n.s.*
Functionality	2.61	1.01	2.77	1.03	0.006 < .05
Performance expectancy	2.55	1.35	2.47	1.42	0.25, *n.s.*
Social influence	2.44	1.36	2.36	1.31	0.18, *n.s.*
Connectivity	2.26	1.25	2.38	1.25	0.01 < .05
Washability	2.25	1.23	1.62	1.12	<.001
Intention to use	2.48	1.30	2.49	1.39	0.891, *n.s.*

rejected even more clearly for the armchair than for the jacket. Finally, the washability of the textile devices was rated significantly more negatively for the smart jacket than for the smart armchair.

4.2   Acceptance Model Jacket vs. Armchair

Figure 1 illustrates the influence of the evaluation dimensions on the intention to use the smart armchair (left) or the smart jacket (right). All model dimensions but washability correlated with the intention to use the smart jacket and the smart armchair. Washability plays no significant role for the acceptance of the armchair but has a strong influence on the acceptance of the jacket. Confirming previous studies, the model dimensions Performance Expectancy, Hedonic Motivation, and Social Influence showed the highest correlations with the intention to use both textile devices. As a quite new result, also the added model dimensions were in parts strongly related with the intention to use: e.g., Innovation or Connectivity. In contrast, the added model dimensions Washability (armchair and jacket), Functionality (jacket), and Modality (armchair) showed the lowest correlations with the intention to use the textile devices.

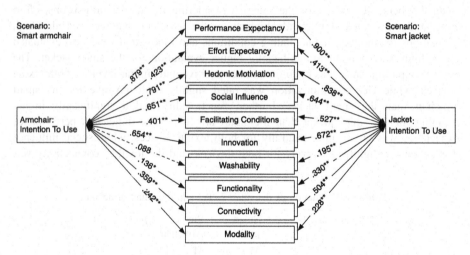

Fig. 1. Relationship of the evaluation criteria to the intention to use both products.

4.3   Identifying Key Predictors for Acceptance

In a next step, we identify the factors with the strongest influence on the intention to use. Hereto, we calculated two stepwise multiple linear regression analyses for the jacket and the armchair, each with the respective evaluation criteria as independent variables and the intention to use as dependent variable.

For the armchair, the regression analysis yields a model with four significant predictors that explain a total of 80% of the variance of the intention to use the armchair (see Table 2a). The key predictor is performance expectancy, followed by hedonic motivation, perceived washability, and input modality. Note that the influence of perceived washability is only discovered through the regression analysis that controls for the effect of the prior identified factors.

A similar pattern emerges for the smart jacket. Here, the regression yields a model with the three factors performance expectancy, hedonic motivation, and connectivity that explains 85% of the intention to use the jacket (see Table 2b).

Table 2. Regression tables for *intention to use* the jacket or the armchair.

	B	SE	β	p
(const)	-.352	.176	—	> .05
PE	.650	.039	.675	< .001
HM	.236	.040	.241	< .001
WASH	.098	.098	.093	< .001
MOD	.066	.032	.054	.037 < .05

	B	SE	β	p
(const)	-.069	.102	—	> .05
PE	.608	.038	.623	< .001
HM	.286	.038	.288	< .001
CONN	.102	.027	.092	< .001

(a) Model for the armchair (r^2=.803)   **(b)** Model for the smart jacket (r^2=.847)

4.4 Influence of User Diversity on the Evaluations

Now we want to analyze if the intention to use the smart textile products is rather homogeneous or if distinct user segments can be identified. Hereto, we calculate two hierarchical cluster analyses on the items from the intention to use scale. Both analyses for the armchair and the jacket identified a set of two clusters with a good silhouette coefficient of >.7, indicating both cohesion and separation of the clusters.

By construction, the clusters correlate well with the respective intention to use. Therefore, we call the first group *"rejectors"* and the second *"adopters"*. Astonishingly, cluster membership is strongly related (Cramer's V = .411, p < .001). Hence, if the smart jacket is accepted, it is likely that the smart armchair is also accepted.

Besides the cluster variable intention to use, both clusters differed significantly (p < .01) with regard to the evaluation of all model dimensions for both textile devices. For both textile products, the cluster with a higher intention to use the textile device showed significantly higher evaluations of all model dimensions compared to the group with a low intention to use.

Besides technology-related criteria, the clusters were analyzed for potential user diversity influences. With regard to demographic characteristics, the two clusters (for the jacket as well as the armchair) did not differ in age, gender, or education level. Furthermore, the participants' SET did also not influence the cluster member-ship. Instead, the results showed significant influences of the participants' smart textile expertise on the cluster membership referring to the smart armchair (F(1, 323) = 5.517; p < .01). The participants' affinity towards textiles impacted the armchair's (F(1,322) = 24.409; p < .01) as well as the jacket's cluster membership (F(1,323) = 15.004, p < .01). Here, people with a higher intention to use the jacket and the armchair were characterized by a higher affinity towards textiles than the participants belonging to the group with a low intention to use both textile devices.

5 Discussion

In this study, we assessed peoples' opinions on smart textile interfaces for two different potential products using a scenario-based approach.

Contrary to other studies, the overall intention to use the jacket respectively the armchair was relatively low. We expect that this is caused by the unfamiliarity with these novel interaction devices and the scenario-based approach. This can be seen, for example, in a subsequent study with "hands-on" experiences of a smart armchair that yielded significantly higher acceptance of the users and the novel textile interaction surface was preferred over a conventional control currently available at the market [15]. The greatest absolute differences between the two investigated products were found for the dimensions washability, facilitating conditions, and innovativeness and were presumably affected by the individual characteristics of the two products. For instance, jackets are usually cleaned more often than furniture and smart wearables have a higher media coverage than smart furniture which might thus be perceived as more innovative.

Despite the mediocre overall desire to use the smart textile interfaces, this study still sheds light on the participant's evaluation criteria that are crucial for acceptance. Influencing those acceptance and perception parameters has the strongest impact on the overall intention to use:

1. For both investigated technologies, performance expectancy is the strongest predictor for acceptance (i.e., changes in performance expectancy has a stronger influence on the intention to use than changes in all other evaluation criteria). Hence, designers must focus on providing actual benefits of using the product and must convey the merits and benefits of using the smart textile interfaces.
2. Furthermore, the hedonic evaluation of the interfaces was identified as the second-most important acceptance criterion. This finding is in line with previous research, which sees beauty, fun, and elegance of products and services as a significant counterweight to purely functional and pragmatic aspects [21]. In this respect, smart textile products should be developed with an early focus on aesthetics and hedonic aspects, for example by involving designers in the development process at an early stage.
3. For the armchair, the aspects washability and input modality were also relevant for acceptance. Thus, product development and design should focus on communication with regard to the cleaning of the armchair. Further, the armchair should be operable by gestures on the textile surface, but should also allow other input modalities, e.g., voice control.
4. In contrast, connectivity was an important aspect for the smart jacket's acceptance. Hence, the product development should enable a connectivity of the jacket at least with smart phones. Then, the connectivity of the jacket should also be focused during communication and marketing.

6 Limitations and Outlook

Of course, this study is not without limitations. First, as digital textiles are not yet available in everyday products and most of our participants haven't had hands-on experiences with these, we had to use a scenario-based approach for assessing the overall acceptance. Hence, future work should evaluate tangible products with digital textile interface in more detail. However, while the scenario-based approach obviously influences the absolute evaluations of the textile devices, current studies indicate that the identified relationships within the acceptance model and the key predictors of intention to use remain the same.

Second, our sample is not representative of the general population. While this certainly limits easy transferability of the findings, it still offers valuable insights into the relevant evaluation criteria for assessing digital textiles and the influence of user diversity on these evaluations. Consequently, the study should be replicated with a larger sample and include people from all ages and more diverse backgrounds.

Acknowledgments. The participants' willingness to share their thoughts and the outstanding research support of David Peters are highly acknowledged. The German Ministry of Education and Research (BMBF) funded this work as part of the project Intuitex (16SV6270) [18].

References

1. Kvavadze E, Bar-Yosef O, Belfer-Cohen A, Boaretto E, Jakeli N, Matskevich Z, Meshveliani T (2009) 30,000-year-old Wild Flax Fibers. Science (New York, NY) 325 (5946):1359
2. Moore GE (1965) Cramming more components onto integrated circuits. Electronics 86 (1):114–117
3. Weiser M (1991) The computer for the 21st century. Sci Am 265:94–104
4. Aarts EH, Encarnacão JL (eds) (2006) True visions: the emergence of ambient intelligence. Springer, Berlin
5. Poupyrev I, Gong NW, Fukuhara S, Karagozler ME, Schwesig C, Robinson KE (2016) Project jacquard: interactive digital textiles at scale. In: Proceedings of the 2016 CHI conference on human factors in computing systems, pp 4216–4227
6. Karrer K, Glaser C, Clemens C (2009) Technikaffinität erfassen Der Fragebogen TA-EG [Measuring Affinity to Technology]. ZMMS Spektrum - Der Mensch im Mittelpunkt technischer Systeme. 8. Berliner Werkstatt Mensch-Maschine-Systeme 22(29):196–201
7. Rekimoto J (2001) Gesturewrist and gesturepad: unobtrusive wearable interaction devices. In: Proceedings of 5th international symposium on wearable computers. IEEE, pp 21–27
8. Cherenack K, van Pieterson L (2012) Smart textiles: challenges and opportunities. J Appl Phys 112(9):091301
9. Davis FD (1989) Perceived usefulness, perceived ease of use, and user acceptance of information technology. MIS Q 13(3):319–340
10. Venkatesh V, Thong JYL, Xu X (2012) Consumer acceptance and use of information technology: extending the unified theory of acceptance and use of technology. MIS Q 36 (1):157–178

11. Post ER, Orth M (1997) Smart Fabric or "Wearable Clothing". In: Proceedings of the 1st IEEE international symposium on wearable computers, ISWC '97, Washington, DC, USA. IEEE Computer Society, pp 167–168

12. Rus S, Braun A, Kuijper A (2017) In: E-textile couch: towards smart garments integrated furniture. Springer, Cham, pp 214–224

13. Heller F, Oßmann L, Al-huda Hamdan N, Brauner P, Van Heek J, Scheulen K, Möllering C, Großen L, Witsch R, Ziefle M, Gries T, Borchers J (2016) Gardeene! Textile controls for the home environment. In: Short paper at Mensch und Computer 2016, Gesellschaft für Informatik (2016)

14. Lee SS, Kim S, Jin B, Choi E, Kim B, Jia X, Kim D, Lee K (2010) How users manipulate deformable displays as input devices. In: Proceedings of the SIGCHI conference on human factors in computing systems. ACM, London, pp 1647–1656

15. Brauner P, van Heek J, Ziefle M, Al-huda Hamdan N, Borchers J (2017) Interactive FUrniTURE evaluation of smart interactive textile interfaces for home environments. In: Proceedings of the 2017 ACM international conference on interactive surfaces and spaces, Brighton, England. ACM Press, New York, pp 151–160

16. Hildebrandt J, Brauner P, Ziefle M (2015) Smart textiles as intuitive and ubiquitous user interfaces for smart homes. In: Zhou J, Salvendy G (eds) Human computer interaction international - human aspects of IT for the aged population. Springer, Cham, pp 423–434

17. Brauner P, van Heek J, Martina Z (2017) Age, gender, and technology attitude as factors for acceptance of smart interactive textiles in home environments – towards a smart textile technology acceptance model. In: Proceedings of the international conference on ICT for aging well (ICT4AWE 2017). SCITEPRESS Science and Technology Publications, pp 53–56

18. Brauner P, van Heek J, Schaar AK, Ziefle M, Al-huda Hamdan N, Ossmann L, Heller F, Borchers J, Scheulen K, Gries T, Kraft H, Fromm H, Franke M, Wentz C, Wagner M, Dicke M, Möllering C, Adenau F (2017) Towards accepted smart interactive textiles - the interdisciplinary project INTUITEX. In: HCI in business, government, and organizations (HCIGO), held as part of HCI International 2017. Springer, Berlin, pp 279–298

19. Beier G (1999) Kontrollüberzeugungen im Umgang mit Technik [Locus of control when interacting with technology]. Rep Psychol 24(9):684–693

20. Van Heek J, Schaar AK, Trevisan B, Bosowski P, Ziefle M (2014) User requirements for wearable smart textiles. Does the usage context matter (medical vs. sports)? In: Proceedings of the 8th international conference on pervasive computing technologies for healthcare (2014)

21. Hassenzahl M (2006) Hedonic, emotional, and experiential perspectives on product quality. In Ghaoui C (ed) Encyclopedia of human computer interaction. Number 2000. Idea Group, pp 266–272

Analysis of 'Quantified-Self Technologies': An Explanation of Failure

Cécile Boulard-Masson[✉], Tommaso Colombino,
and Antonietta Grasso

Naver LABS Europe, Meylan, France
{Cecile.boulard, Tommaso.colombino,
Antonietta.grasso}@naverlabs.com

Abstract. This paper presents the results of a long-term study of the use of fitness trackers, with particular focus on the factors that influence sustained use or abandonment. The analysis focuses on the consequences of adopting a reductionist quantification of general fitness, and how it fails to provide a coherent and understandable measure from the users' point of view.

Keywords: Activity trackers · Quantified self · Measurement
Ethnomethodology

1 Introduction

One of the driving principles of the quantified-self (QS) movement is that knowledge is power. To have fine-grained and objective measurements of our body and its functions, and of our routine activities should, in theory, give us better control over them. For instance, *"The Withings Pulse O_x can help you be more active and improve your health"* [12]. But what happens when quantification gives us a representation of ourselves that we don't understand? Do we question the quantified model of self, its objectivity and accuracy, or do we question ourselves? Drawing on an ethnographic study of QS inspired technologies, we want to provide some reflection on why there may be a mismatch or misunderstanding between measurement and self-representation.

2 Related Work

In response to the development of sensing technology, research into wearable devices and their uses has become a significant area of interest within the HCI and ubiquitous computing communities [1, 10]. All those studies focus on the use of these devices for the practice of collecting a great amount of data on oneself, typically as part of a self-improvement regime. However, many studies on activity trackers show a short rather than long-term use, and a large number of reasons for early abandonment have been identified [8].

Studies pinpoint the difficulties encountered in the quantified-self practices or in the use of smart devices [1, 6]. For instance, having objective knowledge on oneself is not

© Springer Nature Switzerland AG 2019
S. Bagnara et al. (Eds.): IEA 2018, AISC 822, pp. 579–583, 2019.
https://doi.org/10.1007/978-3-319-96077-7_62

enough to motivate change in the "right" direction. While it possible to find people that use smart devices over a long period of time [1, 3], initial adoption is strongly dependent on the novelty and the curiosity of the technology [7]. Others identify some reasons why users abandon the use of smart devices. Some feel that the device does not fit their conception of themselves, others find the data collection not useful, or the extra work associated too large [2].

The question we target in this paper relates to the reasons of abandonment. Our understanding is that the use of activity trackers may lead to feelings of frustration and unfairness because it essentially reduces a complex and contextual concept such as fitness to a step-count. Therefore, when users explain that it requires too much extra-work to get accurate data, we would rather revert the analysis and suggest that trackers are far too simplistic to represent a complex concept of 'physical activity'.

3 Methodology

The technology considered in this paper is an "all-in-one" step, distance, calorie and sleep tracker. The interview-based study was done with 10 volunteers participants, and involved minimal direction, simply explaining what the device was for and how to use it, how to register on-line to look at their data, and then basically asking the participants to explore usage. After leaving the participants with the devices for a number of months we arranged debrief interviews to try to understand their motivations in signing up for the study and using the activity tracker, how they understood, interpreted and reacted to the device, the data and the feedback and whether and in what way they had achieved the effective use of the device. The interviews were semi-structured with a series of topics but the participants were encouraged to share whatever opinions and insights they had. Interviews are a good means to gather insights on how the device was used and understood [5].

The interviews lasted between 20 and 50 min and were audio-recorded. All the interviews were entirely transcribed and analysed from an ethnomethodological per-spective [4] – we had no theoretical orientation and were interested in how participants expressed their usage, activities, understandings and opinions about the devices in their own words.

4 Findings

The use of an activity tracker provides the user with a rough understanding and awareness of their daily activity in terms of their movements (particularly their steps and exertion) during the day and duration and quality of sleep during the night.

The results of our study depict that the trackers and the information they provide are not fully accepted by the users. The main raison seems to relate to a mismatch between the reductionist way the technologies present potentially complex issues and the users' understanding and self-perception in isolation. Partially as a consequence of this mismatch, all the 10 participants reported that they stopped using the tracker within three months.

4.1   "In Fact It's Only a Pedometer"[1]

As Participant 1 says, the activity tracker is essentially a connected pedometer with a gamification layer. Which means that it does not consider all kind of physical activity. Some participants who do a lot of physical activity reported that the device can't measure it. *"I was doing a lot of workout that the device doesn't take into account. In the end it demotivated me because I couldn't reach the goal proposed even though I was doing far enough physical activity"* (Participant 6). The device sends back to the participant a "bad" image of her/his physical activity that is further more incoherent with what the participant feels.

The distorted picture creates disappointment because it is very far from the marketing presentation of activity trackers. *"When I go to the swimming-pool with the kids, the device doesn't consider the activity of swimming. So in the end the device doesn't reflect reality"* (Participant 9), or we could emphasize that it doesn't reflect "my" reality. In any case, the tracker cannot track well anything else besides steps.

In order to get more accurate data, one participant explained the tip to put the device in her socks when she was biking so that some steps were tracked for her bike activity. In that case, it means that if a user wants to have a fair representation of her physical activity, at some point, she has to "trick" the tracker.

4.2   When Fitness Improvements Become Arbitrary Goals

The tracker used in the study suggests default milestones of 10000 steps per day and 8 h sleep per night. These milestones are like goals and are bound to a gamification layer which aims at supporting change behavior toward a "healthier" direction. The initial impact of the set-up is that participants feel really motivated, at least at the beginning. What quickly becomes less obvious for users is where to go from there. For example, two participants claimed to be bored by the mechanistic gamification. *"Rapidly I got fed up that the device always relaunches me like you did 10000 [steps] now your goal is 12000 and then you did 12000 now your goal is 14000"* (Participant 5). Real-life fitness milestones cannot be infinitely incremental, but quickly become repetitive if they do not evolve. The gamified layer of virtual rewards, and the social layer of sharing (or competing on) achievements within a social network do not provide users with clear, long-term end-game. This type of mechanistic gamification setup seems to never consider a limit in the activity users can achieve in a day and in the end participants lost the motivation.

4.3   Sleep Analysis

Several participants expressed that they were also disappointed by the simplistic models of the device's sleep analysis algorithm. For example, Participant 4 had to wake up at night several times to take care of his children and the device considered that the night ended as the participant stood up, so it didn't count the sleeping hours after the brief wake-up at night. Participant 2 was falling asleep on the sofa before going to bed,

---

[1] Participant 1.

"my sleeping hours in the sofa are not recorded by the pulse, I slept but it's not recorded". Before using the device Participant 3 had strong expectations toward the sleep analysis, *"But the device doesn't analyze the sleep, only your movements during the sleep".* That is, you need to tell the device when you are going to sleep and it measures sleep through a crude indication of deep sleep – the lack of significant bodily movement. The measure can easily fail to differentiate between sleep and non-sleep and any duration is very approximate. It cannot tell you in any meaningful way whether you got a good or bad nights' sleep, and it certainly can't help you sleep better.

4.4   Lay Measurement and (Pseudo)-scientific Measurement

As Sacks pointed out in his analysis of therapist-patient interactions [11], people routinely monitor their own state with respect to broad range of items (for example appetite and sleeping), and the key property of this self-monitoring is that its qualifiers (good, bad, etc.) tend to be directional with respect to a notion of normalcy which both personal and contextual to a conversation or interaction. Self-measurement of this sort essentially operates as an interactional resource that allows us to monitor and communicate progress (feeling better) or deviations from an expected state of affairs. While these lay measurement categories do not operate as a lay approximations of scientific measurements of fitness or health, they nevertheless provide a "baseline" awareness one's own condition and progress (or lack thereof), and are routinely used by doctors, for example, to inform clinical diagnosis [9].

While in doctor patient interactions patients usually acknowledge scientific evidential sources (such as x-rays or other clinical tests) that may contradict their own perceptions of themselves, the fitness measurements provided by a tracker, as we discussed earlier in the paper, lie on much shakier foundations. And the veil of objectivity provided by the reductionist quantification of step counts into a general fitness measure does not hold up to sustained use, especially where highly decontextualized and de-personalized measures of progress come into conflict with users' own self-monitoring. Consequently, these types of fitness trackers fail on what should be their main selling point: providing measurements that are more structured, objective and systematic and that allow users to take their own understanding of their fitness levels and their personal fitness regimes to the next level.

5   Conclusion

From this analysis, we understand a need for the users to get accurate and personalized data in a more holistic way that can provide a full account and assessment of a user's life or life-style. This need can't be provided by the device in its current version.

We do not deny the usefulness of such trackers, we rather think that it may be useful within the context of a health intervention targeted to a user. On its own, however, the activity tracker tends to give users the perception that it provides an "unfair" characterization of their efforts and progress.

Overall we found that with QS technologies, there is a risk of decontextualizing and reducing complex activities to simple calculations which encourages binary true-false thinking on the part of users. This leaves little room for a nuanced understanding of the underlying problem and of the specific circumstances and requirements of any individual user. We would like to propose that quantified-self technologies may benefit from less simplified models, even at the expense of more complexity, but be able to provide more contextual and ultimately understandable quantifications for the users.

References

1. Choe EK, Lee NB, Lee B, Pratt W, Kientz JA (2014) Understanding quantified-selfers' practices in collecting and exploring personal data. In: CHI, pp 1143–1152
2. Epstein DA, Caraway M, Johnston C, Ping A, Fogarty J, Munson SA (2016) Beyond abandonment to next steps: understanding and designing for life after personal informatics tool use. In: Proceedings of the 2016 CHI conference on human factors in computing systems. ACM, pp 1109–1113
3. Fritz T, Huang EM, Murphy GC, Zimmermann T (2014) Persuasive technology in the real world: a study of long-term use of activity sensing devices for fitness. In: Proceedings of the SIGCHI conference on human factors in computing systems. ACM Press, New York, pp 487–496
4. Garfinkel H (2002) Ethnomethodology's program: working out Durkheim's aphorism. Rowman & Littlefield, Lanham
5. Klasnja P, Consolvo S, Pratt W (2011) How to evaluate technologies for health behavior change in HCI research. In: Proceedings of CHI'11
6. Lazar A, Koehler C, Tanenbaum J, Nguyen DH (2015) Why we use and abandon smart devices. In: UbiComp'15 Adjunct, Osaka, Japan, 07–11 September 2015
7. Macvean A, Robertson J (2013) Understanding exergame users' physical activity, motivation and behavior over time. In: Proceedings of the SIGCHI conference on human factors in computing systems. ACM, New York, pp 1251–1260
8. Meyer J, Beck E, Wasmann M, Boll S (2017) Making sense in the long run: long-term health monitoring in real lives. In: ICHI 2017 - IEEE international conference on healthcare informatics (to appear)
9. Peräkylä A (1997) Conversation analysis: a new model of research in doctor–patient communication. J R Soc Med 90(4):205–208
10. Rooksby J, Rost M, Morrison A, Chalmers M (2014) Personal tracking as lived informatics. In: Proceedings of the SIGCHI conference on human factors in computing systems, Toronto, Ontario, Canada, April 26–May 01 2014
11. Sacks H (1992) Lectures in conversation. Blackwell, Oxford
12. Withings. http://www2.withings.com/us/en/products/pulse/. Accessed 2 Jan 2016

User-Interface and Operators: Evolution in the Perception of Computed Tomography (CT)

Oronzo Parlangeli[1]([⊠]), Alessandra Giani[2], Margherita Bracci[1],
Ilaria Bonanno[1], Antonio Conte[1], Veronica Del Priore[1],
Anna Di Genova[1], and Angela Lucia[1]

[1] Università degli Studi di Siena, Siena, Italy
{oronzo.parlangeli,margherita.bracci}@unisi.it,
{ilaria.bonanno,antonio.conte,veronica.delpriore,
anna.digenova,angela.lucia}@student.unisi.it
[2] Azienda Ospedaliera Universitaria Senese, Siena, Italy
a.giani@ao-siena.toscana.it

Abstract. Computed tomography (CT) is the most used clinical examination to produce cross-sectional images of the body. The main objective of the study reported here is to analyze the CT operators' perceptions in relation to the evolution of the technology they use, with a specific focus on the usability of the user interface. The study is based on the consideration that the performance of the whole man–machine system is strictly dependent on the perception/experience that the operator has of the machine. To this aim, current opinions expressed by 6 operators have been compared to those collected in a study conducted in 2011. Results reveal a high level of satisfaction with the technology in use, but the hope for a even better operability in the future, which was expressed in 2011, seems to be no longer valid today.

Keywords: Computed tomography · CT · Perceptions · Situation awareness
User experience · HMI · Healthcare · CT operators

1 Diagnostic Imaging

1.1 Evolution of the Technology and Increase of the Use of Computed Technology

The computed tomography (CT) is a diagnostic procedure in which a narrow beam of X-rays is directed and quickly rotated around the body of a patient, thus producing signals that are processed to generate cross-sectional images or "slices" of the body. The introduction of the CT scanners in the early 70s has been a revolution for medical radiology. Although at the beginning it was mostly used to validate already established diagnosis, the CT has become today the leading technology for the evaluation of patients in emergency departments. The exam rapidity and his affordable costs, together with the increase in the availability of equipments, have witnessed a corresponding increase in the utilization of CT scanners in the developed world [1–4]. In Italy, around

S. Bagnara et al. (Eds.): IEA 2018, AISC 822, pp. 584–594, 2019.
https://doi.org/10.1007/978-3-319-96077-7_63

24 millions of adults and 1.6 million of children have access to the emergency department every year. From these figures it could be said that, in principle, one out of three persons in Italy accesses the emergency department every year and that each patient who comes to the need 1,1 CT imaging on average. In fact, the data show a gradual increase in the global number of the emergency department services (+31%) and an impressive growth of the Computed Tomography (+107%) in the last seven years [5].

In this scenario, the role of the radiology operator, the technician responsible for the acquisition and processing of imaging services in the emergency department become extremely important and critical for an appropriate choice of the diagnostic intervention with respect for the patient's need and medical context [6]. This means, primarily, shorter exposure times to the radiations and, at the same time, an image quality sufficiently good to produce an accurate diagnosis. This can be easily described as a compromise between dose and image quality [7, 8].

Currently, CT operators are entrusted with the highest responsibility, despite the high level of automation, the human factor remains decisive [9, 10] for the patients' safety and health. It has been clearly shown that the implementation of information systems for Healthcare (Clinical Information Systems) is not enough to produce an effective improvement on quality or safety if they are not designed to meet the user's needs [10].

1.2   The CT System

A CT system consists of a machine body including the *Gantry* – a circular rotating frame with an X-ray tube mounted on one side, and a detector on the opposite side –, a table for the patient; and a console with an integrated operating system.

The console generally allows the operators to:

- track the entire workflow, from the management of patient's personal data to the execution and reporting of the examination (Radiology Information System, RIS system);
- manage the system settings to perform the scan by choosing the most appropriate protocol, the acquisition and reconstruction of the images (CT software);
- archive and manage radiological images (PACS - Picture Archiving and Communication System).

Interactive systems, as the one described above, in which the operators are continuously required to make decisions can be particularly complex for non-experts users [11]. In the emergency department the complexity is increased by the frequent need to operate under time pressure, in high-risk situations, where errors can have serious consequences. In this context it becomes necessary to operate as quick as possible and to coordinate the decisions of the working team. The interaction between the operator and the technological system is influenced by the operator's experience and by the "Situation Awareness", that is the set of cognitive processes aimed at the perception of the elements of a situation, their comprehension and the anticipation of their evolution in the near future [12]. The way in which the software is designed and implemented, its user interface, can thus directly and indirectly influence the Situation Awareness and the performance of the operator [13, 14].

2 The Study

The main objective of the study reported here is to analyze the CT operators' perceptions in relation to the evolution of the technology they use, with a specific focus on the evaluation of the level of usability of the user interface. It is assumed that the performance of the entire man–machine system depends on the perceptions and on the cognitive representations, these being mediated by the ongoing experience, the operator has of the machine [15, 16].

2.1 Subjects, Tools and Procedure

In March 2018 a study was conducted to collect the opinions of CT technicians in regard to characteristics of the system they were using. In order to get a view on the evolution of opinions, the study has been compared to a previous one, conducted in the same way, and asking the same questions, in 2011. The study presented here can therefore be considered a longitudinal study.

Participants in the study were 6 CT scanner operators of the Emergency department of the Hospital *S. Maria alle Scotte* in Siena. 5 of them had participated in the previous study in 2011.

Their average age was 45 years, and they had, on average, 20.8 years of working experience.

A structured interview was administered to collect data and information. The same interview had already been used in 2011.

2.2 The Interview

Interviews were conducted while operators were on shift and no patients were present in the CT rooms.

The first part gathered demographic information (age, experience, training course). The second one collected subjective opinions on the CT system they were using on nine factors:

- complexity;
- automation level;
- precision;
- level of maneuverability;
- level of attention required to properly manage the system;
- mental workload;
- level of multitasking;
- intuitiveness of the system;
- amount of errors committed.

CT operators were asked to express their opinions on these factors focusing their answers in relation to the past (2011), the actual system, and a possible future scenario. (E.g.: How complex do you think the CT were seven years ago? How much complex is the CT you are currently using? Do you think that in the future the CT you will be using will be equally complex?). Answers were collected and interpreted according to a 5-point Likert scale. CT operators were also invited to freely express their opinions, which were then took into account to better qualify the information gathered.

3 Results and Discussion

In the following, the graphical representations of the mean evaluations obtained both in 2011 and in 2018 for each of the system characteristics are reported. For all of them a comparison between the data obtained in the current study and those obtained in 2011 is shown.

Complexity
In 2011 it was generally believed that the complexity of CT technology was affected by two main factors:

(1) The level of complexity of the hardware;
(2) The level of operational complexity caused by the user interface.

The former was expected to increase with the creation of more complex technologies, while the latter was expected to decrease (Fig. 1). The common thought was that machines were going to become more and more simple to be used thanks to the introduction of better user-friendly interfaces. In 2018 these expectations appear to have been quite disappointed: complexity, globally understood, is evaluated equivalent to 2011. The relative weight between hardware and software is reversed and operators see scant possibilities in lowering the complexity level.

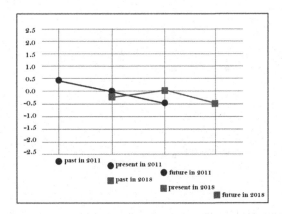

Fig. 1. Perceptions related to the level of hardware complexity and operations made possible by the user interface.

Automatism

Judgments are nearly the same and trends do not seem to change substantially. There is a lasting thought that thing were worse and will not improve much in future (Fig. 2). The past is seen as worst than today. However, the comparison between 2011 and 2018 reveals that the automatism was judged better than it is remembered today.

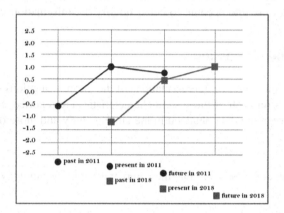

Fig. 2. Perceptions of the automatism level, considered as the ability of the machine to work without the intervention of the operator.

Accuracy/Precision

All the opinions about accuracy are positive (Fig. 3). The judgments in 2011 were pessimistic in relation to possible future improvements of the CT technology. It should be noticed that the variable "accuracy" has always been interpreted as a synonym for "capacity of the images to represent realistically what they have to represent".

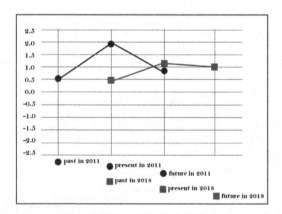

Fig. 3. Perceptions related to the precision level, considered as accuracy and ability to realistically represent the images.

There is, also for this factor, a gap between the memory of the past and what had been said in 2011. It seems that the operators do not see as possible any future improvement.

Level of Maneuverability
In 2011 did not emerge any kind of physical discomfort that could be ascribed to operational efforts (Fig. 4). The machine was perceived nearly as an extension of the body and it seemed that there were no possibilities for any positive change in the future.

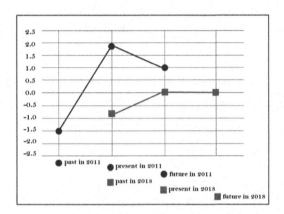

Fig. 4. Perceptions related to the level of maneuverability, i.e. the ability of the system to limit physical efforts, as in the case of patient positioning.

The opinions collected in 2018 confirm the positive evaluation of the level of maneuverability, though it now seems quite worse than in the past. CT operators do also underline that the physical effort of positioning the patient on the table is likely to be the same in the future, despite possible technology developments. Once again the past is remembered as worse than it had been actually evaluated in 2011.

Attention
The results of 2011 must be considered not sufficiently informative because CT operators did not expressed a common opinion on the level of attention required both in the past and in the future (Fig. 5). However, there was a general agreement that the system of the time then required high, or even very high, levels of attention.

Today, technicians do agree on the level of user-friendliness and intuitiveness of the machine, and the level of attention required to efficiently operate the system does not seem to be that high. On the other hand they believe that the most important factor to avoid problems during the execution of an examination is the operator's competence. In this case the remembering of the difficulties 2011 seem to be attenuated.

Mental and Memory Workload
The study in 2011 highlighted, in some cases, an excessive memory load both due to the amount of information and the complexity of procedures (Fig. 6). However, the

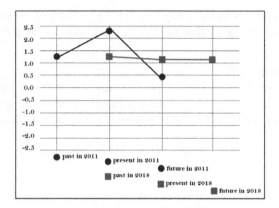

Fig. 5. Perceptions of the attention level considered as the need for cognitive effort and concentration.

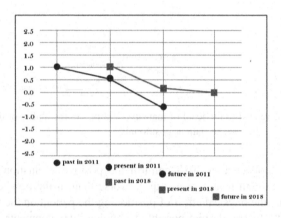

Fig. 6. Perceptions related to the mental and memory workload considered as cognitive commitment related to the amount of things to keep in mind during an examination.

levels did not seem to express any very problematic situation and revealed a certain confidence that things would improve in the future.

In 2018 the quantity of thing to keep in mind is considered adequate. When compared to the past, the current system is seen as more efficient and allows to keep visible all the information about the patient in the screen were the chosen protocol is displayed. This makes the first part of the job easier.

In regards to the positive expectations expressed in 2011 about a lower mental workload in the future, it is possible to say that they have been partially disappointed. The mental workload is perceived today nearly equal as in 2011.

Multitasking

The results in 2011 were generally positive. The system was not considered as excessively demanding from an operational point of view: it did not request to perform multiple things at the same time (Fig. 7.) The future was foreseen in a positive trend.

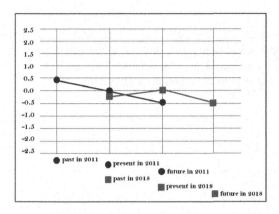

Fig. 7. Perceptions related to the level of multitasking considered as the request to perform multiple things at the same time.

In 2018, the memory of the past seems to be correct but the present is not as positive as expected. There is still a confidence in future improvements.

Intuitiveness of the System

The opinions expressed in 2011 were based on the relationship between the complexity of the technology and its usability (Fig. 8). The system was considered more intuitive than in the past, but there were not positive expectations for future improvements. Today there is a positive evaluation of the evolution from the past to the future. Once again, the past is remembered as a little more negative than it actually was.

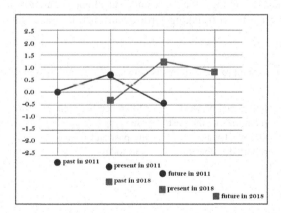

Fig. 8. Perceptions related to the level of intuitiveness considered as the ability of the system to be understood.

All the operators agree that the system in 2018 in more intuitive. They do not think possible a further increase in the intuitiveness of the software, but are convinced that even the actual level is suitable for a good CT exam.

Errors

In 2011 the technology was considered an actual error free system by the operators (Fig. 9). Operators' inattention was indicated as the primary source of error and, in addition, this was considered as unavoidable. The only chance to mitigate errors occurrence was attributed to the technological system.

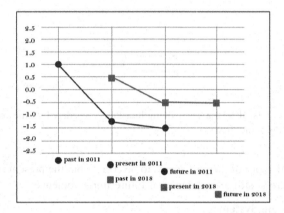

Fig. 9. Perceptions relating to the number of errors that may occur during the execution of an examination.

Today CT operators still think that the main cause of errors is the human fallibility, but they do not consider any longer the machine as a perfect technology. Some errors can come from the technology they use and can affect negatively the activity of the technicians, who have to work taking error occurrence into account.

Table 1. Differences found in the judgments between 2011 and 2018 and between the present and the future.

	Difference between memories of the past and what was reported at the time	Difference between what was foreseen in the past and what is said today	What is said today	Expectations
Complexity	=	–	+	=/+
Automatism	–	=	+	=/+
Accuracy	–	=/+	+	=
Maneuverability	–	–	=	=
Attention	+	–	+	=
Mental workload	–	–	=	=
Multitasking	=	–	=	+
Intuitiveness	–	+	+	=
Errors	–	–	=/+	=

The most common mistake is about the scout. Other errors can be facilitated by patients' homonymy, the wrong patient positioning, the choice of the wrong protocol, the administration of the contrast liquid when not required or, conversely, an examination in which the contrast liquid should be provided but it is not.

Again, the confidence for a better future does not seem to find any reason, and the memory of the past is, as for other factors, worse than it was expressed at the time.

4 Conclusion

Table 1 summarizes some of the differences found in the judgments between 2011 and 2018 and between the present and the future.

The first column reports the differences between the judgments from 2011, as remembered today, and what really was said at the time. It is interesting to see how memories for almost factors are worse than what was then said: today things are remember worse than how they really were. This could be a general distortion of long term memory; but, since these are all comparative evaluations, it seem more plausible to explain this as a perception of improvements over the years.

However, expectations about the future seem to be generally disappointed (Table 1, second column). This probably happened because some of the idealizations were excessive and unrealistic for today's technology. Though it could also be that CT producers have not dedicated the necessary efforts to improve these systems (see, in this volume, the paper by Parlangeli et al. about the evaluation of the user interface of Computed Tomography).

The current situation (Table 1, third column) is always positively or neutrally evaluated, but never negatively.

As in 2011 there is a positive general perception of the technology, in particular concerning *precision, level of attention required, intuitiveness, number of errors produced.*

Overall, the expectations for a future in which the factors considered have a positive trend has partially fade away, and it has become difficult for the operators to predict further positive evolution for the next years.

Acknowledgements. The Authors would like to thank the operators who with competence, courtesy, sensitivity, and availability, took part in this study. Our admiration goes to them, as they perform every day a very precious work.

References

1. Raja AS, Ip IK, Sodickson AD, Walls RM, Seltzer SE, Kosowsky JM, Khorasani R (2014) Radiology utilization in the emergency department: trends of the past 2 decades. Am J Roentgenol 203(2):355–360
2. Berdahl CT, Vermeulen MJ, Larson DB, Schull MJ (2013) Emergency department computed tomography utilization in the United States and Canada. Ann Emerg Med 62(5):486–494

3. Bellolio MF, Heien HC, Sangaralingham LR, Jeffery MM, Campbell RL, Cabrera D, Shah ND, Hess EP (2017) Increased computed tomography utilization in the emergency department and its association with hospital admission. West J Emerg Med 18(5):835–845
4. Kocher KE, Meurer WJ, Fazel R, Scott PA, Krumholz HM, Nallamothu BK (2011) National trends in use of computed tomography in the emergency department. Ann Emerg Med 58 (5):452–462
5. http://www.quotidianosanita.it/lavoro-e-professioni/articolo.php?articolo_id=55650.   Last Accessed 15 May 2018
6. Salvadori P. L'appropriatezza prescrittiva nella specialistica salverà le liste di attesa. Toscana Medica. http://toscanamedica.org/edizione-mensile/category/4-2017
7. Yu L, Liu X, Leng S, Kofler JM, Ramirez-Giraldo JC, Qu M, McCollough CH (2009) Radiation dose reduction in computed tomography: techniques and future perspective. Imaging Med 1(1):65–84
8. Parakh A, Kortesniemi M, Schindera ST (2016) CT radiation dose management: a comprehensive optimization process for improving patient safety. Radiology 280(3):663–673
9. Saleem JJ, Russ AL, Sanderson P, Johnson TR, Zhang J, Sitting DF (2009) Current challenges and opportunities for better integration of human factor research with development of clinical information systems. Yearb Med Inf 18:48–58
10. Patel VL, Kannampallil TG (2014) Human factors and health information technology: current challenges and future directions. Yearb Med Inf 9(1):58–66
11. Vicente KJ, Rasmussen J (1990) The ecology of human–machine systems II: mediating 'direct perception' in complex work domains. Ecol Psychol 2(3):207–249
12. Endsley MR (1995) Toward a theory of situation awareness in dynamic systems. Hum Factors 37(1):65–84
13. Ziemke T, Schaefer KE, Endsley M (2017) Situation awareness in human–machine interactive systems. Cogn Syst Res 46:1–2
14. Stanton NA, Stewart R, Harris D, Houghton RJ, Baber C, McMaster R, Salmon P, Hoyle G, Walker G, Young MS, Linsell M, Dymott R, Green D (2006) Distributed situation awareness in dynamic systems: theoretical development and application of an ergonomics methodology. Ergonomics 49(12–13):1288–1311
15. Bate P, Robert GR, Maher L (2007) Bringing user experience to healthcare improvement: the concepts, methods and practices of experience-based design. Radcliffe Oxford Publisher
16. Garrett JJ (2011) The elements of user experience, 2nd edn. New Riders, Berkeley

A Toolkit for Studying Attention and Reaction Times to Smartglass Messages in Conditions of Different Perceptual Load

Tilo Mentler[(⊠)] and Daniel Wessel

Institute for Multimedia and Interactive Systems,
University of Lübeck, Lübeck, Germany
mentler@imis.uni-luebeck.de

Abstract. Smartglasses can provide safety-relevant information during tasks, e.g., by displaying warnings or important updates during search operations in crisis management, or about patients while treating them in healthcare. It is both necessary for the desired outcomes and frequently taken for granted that users perceive these messages reliably and in a timely manner. However, research on inattentional and change blindness has shown that visual stimuli—even "obvious" stimuli directly "in front of one's eyes"—can be overlooked. Thus, in safety-critical situations, in which the user is focused on a task in the environment, instructions or warnings can be overlooked, despite being displayed in the user's field of view—with potentially serious consequences. In this paper, we address the problem of taking perception of messages displayed on smartglasses for granted with respect to perceptual load theory. We present the results of a study comparing reaction times to smartglasses messages in conditions of low and high perceptual load with 24 participants in order to analyze how well users can notice messages in different load conditions. We also describe the implementation of an application to conduct these studies and possible designs for future studies.

Keywords: Smartglasses · Visual attention · Perceptual load
Safety-critical systems

1 Introduction

Smartglasses (optical head-mounted displays) are increasingly introduced in various mobile contexts, especially safety-critical ones like aviation, crisis management or healthcare [6]. Such wearable devices can provide safety-relevant information during task performance, e.g., by displaying warnings or important updates during search operations in crisis management, or about patients while treating them in healthcare [1]. It is crucial and usually taken for granted that messages are perceived reliably and in a timely manner.

However, research on inattentional and change blindness has shown that visual stimuli can be overlooked, esp. if a primary task requires high visual attention (high perceptual load, e.g., due to a complex visual search task). Even "obvious" stimuli directly "in front of one's eyes" can be overlooked if the person's attention is focused

S. Bagnara et al. (Eds.): IEA 2018, AISC 822, pp. 595–604, 2019.
https://doi.org/10.1007/978-3-319-96077-7_64

on another task. Thus, in safety-critical situations, in which the user is focused on the environment, instructions or warnings could be overlooked despite being displayed in the user's field of view—with potentially serious consequences.

In this paper, we examine whether the quick and reliable perception of messages displayed on smartglasses can really be taken for granted. We look at the role of visual information in safety-critical contexts and the role of smartglasses, before referring to perceptual load theory. We then introduce the overall research strategy and describe the toolkit developed to enable further studies. Subsequently, we report our study, provide suggestions for further research, and draw first conclusions regarding the role of attention and smartglasses.

2   Background and Related Work

In the following subsections, background and related work with respect to role of visual information and smartglasses in safety-critical contexts as well as research on visual perception are summarized.

2.1   Role of Visual Information in Safety-Critical Contexts

In safety-critical contexts, "*knowing what is going around*" [4] often is a major challenge. "*Usually, circumstances are too complex and too dynamic. Many important elements and processes might be out of sight, range or mind*" [8]. However, situation awareness (see Fig. 1) is a premise for proper decision making.

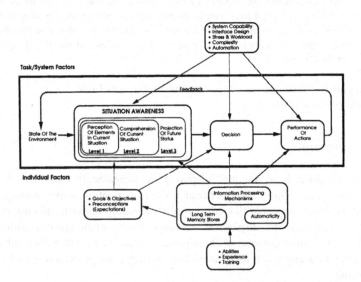

Fig. 1. Model of situation awareness in dynamic decision making [3]

In this regard, visual information, both in the way of personal perception of the environment and provided by checklists, forms, maps, tables or warning signs, is of utmost importance. While paper-based documents allow for flexible ways of working (e.g. tearing one paper in two, writing something in the margin) to a certain degree, their overall utility for time-critical decision making has to be assessed critically for at least two reasons:

- Some of them are stationary or have to be delivered by messengers.
- Searching and filtering of information is hampered.

Computer-based mobile devices and applications, e.g. smartglasses, could enable task-specific information access and exchange near real-time and from remote locations.

2.2 Role of Smartglasses in Safety-Critical Contexts

Smartglasses, and Google Glass as a representative product, have become increasingly important in various safety-critical contexts. They allow for

- using computer-based applications hands-free,
- having information available ubiquitously, and
- augmenting reality with context-related information.

However, actors in safety-critical contexts (e.g. crisis managers, pilots, surgeons) are often subject to high cognitive and physical workload, because they have to deal with extraordinary and sometimes unique challenges. They have to decide under uncertainty and solve tasks in parallel. Working environments vary from well-defined ergonomic workstations to remote outdoor locations. In any case, they are characterized by various, frequently overwhelming and new stimuli, which pose high demands on the actors. Therefore, visual information displayed in the user's field of view by smartglasses (see Fig. 2) could be overlooked, if the person's attention is focused on another task or the environment.

Fig. 2. Information about hazardous material in emergency medical care [1]

Before continuing we need to deal with a common objection: the use of audio to increase salience of message. While the use of audio might be obvious to increase salience of messages, it should be treated with care, as there is ample research on how auditory stimuli can also be missed. While the use of the participant's name might increase attention to auditory stimuli, in safety-critical contexts the use of auditory stimuli for messages might be impractical. Users might need to listen closely to the environment—or alarms and other noises might make noticing the alarms difficult. Additionally, simply escalating the salience of alarms (e.g., via increased volume) might prove detrimental if the current environment requires the full attention of the user and highly distracting alarms might interfere with actually dealing with the object of the (warning) message. Similarly, the usefulness of tactile stimuli (as employed by some smartwatches) should be examined empirically.

2.3   Research on Attention in Visual Perception

There has been research suggesting that noticing visual stimuli requires attention, or at least, some attentional resources. If our attention is focused on one object or event, we may overlook unexpected stimuli. This holds true even when the stimuli are in close spatial proximity to or occupy the same attentional focus of the attended object or event [9]. Research also suggests that this inattentional blindness might be a quite frequent phenomenon. While the first studies go back to the 1970s, the impressive example used in the study of Simons and Chabris (1999, "The Invisible Gorilla") has made the phenomenon more known to the general public (to—not—see it to believe it, see http://www.theinvisiblegorilla.com/videos.html). They also concluded that stimuli are more likely to be noticed if they are visually similar to events the users pay attention to (e.g., a black gorilla is more easily noticed when attention is focused on black players than white players), which has implications for message design and their integration into the environment.

Research on focused attention (vigilance, avoiding distraction) by Lavie (e.g. [7]) provides a possible explanation when and how stimuli are overlooked (perceptual load theory). Conditions of low perceptual load leaves attentional resources to detect new/distracting stimuli, while in high perceptual load conditions (e.g., complex visual task), all resources of perceptual attention are spend on the task, thus leaving no resources to notice other stimuli.

While using different visual stimuli (e.g., object displayed for 200 ms while doing a perceptually demanding or easy main task, [2]; prolonged dynamic events taking around five seconds, [9]), the research on inattentional blindness suggests that similar effects could hold true while using smartglasses. A study examining visual search tasks with optical-see-through displays [5] additionally found switching costs when attention was switched between presentation devices.

Thus, especially safety-critical contexts with high perceptual load might make it likely to overlook stimuli displayed in the field of view of the participants.

3   Research Questions and Research Strategy

Perceptual load theory indicates that in conditions of high perceptual load, new and unexpected stimuli can be overlooked. Additionally, new stimuli are more likely noticed if they are perceptually similar to the target stimuli. However, it is unclear to which extend this holds true for messages displayed on smart glasses while the user is focused on a primary task in the environment. Thus, we ask the following research questions:

Q1:

Influence of perceptual load on visual attention with smartglasses

> **Q1-1:** Are reaction times to messages displayed on smartglasses longer under conditions of high perceptual load compared to low perceptual load?
>
> **Q1-2:** Are messages more likely to be overlooked under conditions of high perceptual load compared to low perceptual load?

Q2:

Influence of message design on visual attention with smartglasses

> **Q2-1:** Are perceptually similar messages (compared to the environment) more quickly noticed than perceptually dissimilar messages (i.e., high-contrast) messages?
>
> **Q2-2:** Are perceptually similar messages (compared to the environment) less likely to be overlooked than perceptually dissimilar messages (i.e., high-contrast) messages?

To answer these questions we need to examine conditions with a high vs. low perceptual load environmental search task on an external display (factor 1) and a high vs. low contrast smartglass message reaction task (factor 2). In order to perform such an experiment an experimental software system was needed that allows for:

- designing search task images and their frequency,
- adapting the allowed time to solve the search task to the particular participants perceptual speed,
- configuration of the design of the messages and both their frequency and timing,
- accurate measurement of reaction times to both search tasks and smartglass message, and
- a close integration of the search task and the smartglass messages.

As such a system was not available, it was developed first.

4 Toolkit Development

The (high-level) system architecture of the experimental toolkit is shown in Fig. 3. Its basic components are an application server with a display and an optical head-mounted display (in this case: Google Glass). They are connected to a wireless network.

Application server and image search task application have been realized with the programming language C# and the web application framework ASP.net. Therefore, they could be used with different operating systems. However, the Google Glass application can't be used with other smartglasses in a straightforward manner because it is based on device-specific development environments (Android Studio) and frameworks (Android SDK 4.4). While this lack of device compatibility is unfortunate, it relates to a more general problem: Smartglasses of different manufactures vary with respect to input modalities (gestures, speech, and touch), screen sizes and display

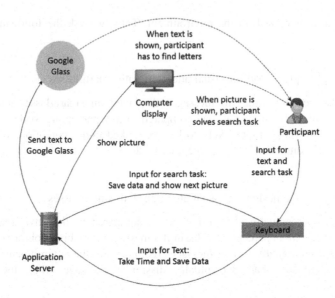

Fig. 3. System architecture of the experimental toolkit

characteristics (binocular, monocular). Therefore, development of a device-independent experimental toolkit would require several configuration and calibration features.

5   Experimental Study

5.1   Method

Design: A 2 × 2 within-subjects design was used. We examine conditions with a high vs. low perceptual load environmental search task on an external display (factor 1) and a high vs. low contrast smartglass message reaction task (factor 2). Participants performed 100 search tasks, 50 high vs. low load, with 25 each with the search stimuli present vs. absent. Messages were displayed during 10% of the search tasks.

Sample: 25 participants were recruited for the experiment. One participant was more than twice the age of the next younger participant and was excluded. Thus, 24 participants were analyzed. The age varied between 21 and 33 years ($M = 25.25$, $SD = 2.72$), 19 male and 5 female, and 8 participants needed glasses. 14 participants had no experience with HMDs.

Instruments—Search Task: To ensure accurate measurement and high experimental control, the search task was done on a computer screen. Similar to [5], a matrix of Os was used. Participants were tasked with quickly finding out whether a 0 (a zero, which is perceptually similar to an O) was present in the matrix (hit f key) or not (hit d key). For high perceptual load a 10 × 10 matrix was used, for low perceptual load a 6 × 6 matrix.

No 0 s were present in the outer rows or columns. The matrix was moved after each answer to avoid making the change of the search stimuli instantly salient. Thus, participants answered the search task continuously without the need of a break (visual masking) between the stimuli, which participants could have used for checking for messages. Time for the search task was adapted to users' perceptual speed and an auditory warning was given when three or less seconds were left. The next search stimuli was automatically presented when the time was up.

Instruments—Message Reaction Task: Messages were configured to be either perceptually similar to the search task (white text on black) or dissimilar (white text on red, high contrast to search task display). As participants should actually read the message, not react to a flicker in their peripheral vision, they had to check whether the message contained a specific letter and press j if so, and k if not. Thus, they actually needed to focus their attention on this task. Message frequency was set to 10% of the search tasks to make them rare and discourage deliberate checking after each search task was finished. A delay was introduced to ensure that a message never appears directly with a new search task. The message stayed until reacted to or overwritten by the next message. Participants were instructed that answering the messages had priority over the search task—they should interrupt the search task immediately when they noticed the message and answer it as quickly as possible.

Setting: To avoid unrelated distractions, the experiment was carried out in a usability lab with constant lighting conditions and no distracting stimuli.

Procedure: Participants filled in a questionnaire (sociodemographic variables, questions about their visual perception incl. color-blindness, typing skills, HMD and computer game experience), received instructions about the task, tried out the search task and message reaction task (as demonstration of the task, to get used to the smartglasses, and to calibrate the allowed search time to their perceptual speed), and performed the main task. Afterwards they filled in an evaluation questionnaire (among others, difficulties, rating of the message saliency).

5.2 Results

Due to the limited number of participants (n = 24), the research questions were assessed separately via direct comparisons (t-tests for paired samples). Given the random distribution of messages, trials were excluded if the displayed message was overwritten by another message in less than the shortest response time of the participant. In four cases the main task had to be restarted due to a connection loss, likely due to Google Glass overheating.

Manipulation Checks: To ensure that the different search task conditions and message designs actually had an effect, we looked at reaction times to high vs. low perceptual load search tasks and participant rating of the messages. Reaction times for high perceptual load search tasks were significantly longer ($M = 2870$ ms, $SD = 1144$) than low perceptual load search tasks ($M = 1681$ ms, $SD = 437$, $t_{(23)} = 7.684$, $p < .001$). For the messages, participants rated how fast they thought they noticed the messages (1–5, with 5 being very quickly). Ratings differed for the message designs with red (M = 4.12, SD = 0.9) being regarded as more quickly to notice than black (M = 3.5, SD = 0.98, $t_{(23)}$ = 2.333, p = .029). Message designs did not differ in ratings regarding readability, difficulty, or overall grade. However, participants rated the black design as more pleasant (M = 3.71, SD = 0.75) than the red design (M = 3.25, SD = 0.74, $t_{(23)}$ = 3.412, p = 0.002).

Q1:
Influence of perceptual load on visual attention with smartglasses

Regarding **Q1-1**, the reaction times to messages (not to the search task) under conditions of high (M = 2039 ms, SD = 648) vs low perceptual load (M = 2093 ms, SD = 956), no significant difference was found ($t_{(23)}$ = 0.269, p = .79).

For **Q1-2**, regarding the average amount of not responded to messages (overwritten by the next message), no significant difference was found under high (M = 0.18, SD = 0.16) vs low perceptual load conditions (M = 0.25, SD = 0.18; $t_{(23)}$ = 1.593, p = .125).

Q2:
Influence of message design on visual attention with smartglasses

Regarding **Q2-1**, comparing the reaction times to red (M = 1785 ms, SD = 403) vs. black (M = 2364 ms, SD = 1202) messages, a significant difference could be found ($t_{(23)}$ = −2.357, p = 0.027). Participants answered Google Glass messages faster if they were displayed with white on red text (high contrast to main task display) than with white on black text.

For **Q2-2**, no significant difference ($t_{(23)}$ = −1.326, p = 0.198) was found regarding the average amount of messages that were not responded to when red (M = 0.38, SD = 0.28) vs. black (M = 0.51, SD = 0.49) messages were used.

6   Discussion

Results indicate that the experimental manipulation worked—the search tasks took significantly longer under conditions of high than low perceptual load and the perceptually dissimilar red message design was rated as quicker to notice than the black one.

As for the influence of perceptual load on visual attention for messages displayed on the smartglasses, no statistically significant differences could be found, nor were messages more likely overlooked under high load conditions. Regarding the message design, reaction times to the message did differ depending on the design (red being faster to notice than black), but they were not overlooked more or less frequently.

There are several possible explanations for these results which warrant further research.

The search task might not have taxed the user to their perceptual capacity. While the high perceptual load task was more difficult (longer reaction times to the search task), participants might still have had enough perceptual resources left to notice the messages. There are a few possible ways to examine this hypothesis. For example, using prolonged dynamic events like finding objects in a video would increase perceptual load. Especially compared to a static search task in which users could check the smartglasses and immediately return to the same, unchanged place in the static matrix. With moving stimuli (e.g., quickly identifying objects in a video), these quick checks—even without a message having been displayed—could be discouraged. Also, the search task consisted of white text on black background. A more visually rich search task (e.g., a series of "Where is Waldo?" images) or increased time pressure might make the task more challenging.

Another explanation is the prior knowledge of the participants. In contrast to, e.g., [2] or [9], participants knew that they would see stimuli in a particular place and were specifically instructed to give these stimuli precedence over their search task. In this sense, the stimuli were rare but not unexpected. A way to prevent this knowledge is to ostensibly interrupt the experiment while asking the participants not to remove the smartglasses (e.g., ostensibly due to calibration issues). This break could be used to get the participants engaged in a highly demanding perceptual task (e.g., helping the experimenter to quickly pick up dropped objects that will quickly deteriorate otherwise), during which a message is displayed on the smartglasses for a few seconds. In these conditions, participants would not expect a message.

Also, while allowing for high experimental control, further studies should use larger focus distances. Participants sat close to the computer screen. To increase external validity, larger focus distances like in natural contexts should also be used.

An open question is why the red message design was rated as quicker to notice by the participants and led to faster reaction times to these messages. It is likely that it is not the red color per se (even if it is widely regarded as a warning color), but any contrasting colors to the black and white search task would have gotten similar ratings. Looking at safety critical contexts, choosing a contrasting color might prove difficult due to the visually rich environment. A possible design could use the camera of the smart glass and select a contrasting design to the currently viewed environment. The use of contrasting messages might contradict [9] finding that stimuli are more likely to noticed if they are visually similar to events the users pay attention to. However, the stimuli in [9] was part of the same situation, while the stimuli on the smartglasses is separate from it. However, in augmented reality with tight integration of messages into the perceived situation, implications might be different.

7 Conclusions

The developed toolkit is well-suited to examine the role of perceptual load on the salience of smartglasses messages. A first experiment comparing two factors with two conditions each showed the feasibility of this line of research. The manipulation checks indicated that perceptual load and message salience could be influenced. As for the lack of differences under conditions of high vs. low perceptual load on message reaction times, possible explanations were discussed. The effect of message salience shows potential ways to increase visual attention to messages. Further research questions were also discussed that should be pursued, especially when smartglasses become frequently used in safety-critical contexts and messages salience becomes crucial.

Acknowledgements. The present paper is based on an unpublished Bachelor thesis by Leif Jonas von Koschitzky. We thank him and all participants of the study for their contribution.

References

1. Berndt H, Mentler T, Herczeg M (2016) Optical head-mounted displays in mass casualty incidents: keeping an eye on patients and hazardous materials. Int J Inf Syst Crisis Response Manag (IJISCRAM) 7(3):1–15
2. Cartwright-Finch U, Lavie N (2007) The role of perceptual load in inattentional blindness. Cognition 102(3):321–340
3. Endsley MR (1995) Toward a theory of situation awareness in dynamic systems. Hum Factors 37(1):32–64
4. Endsley MR, Garland DG (2000) Situation awareness analysis and measurement. Lawrence Erlbaum, Mahwah
5. Huckauf A, Urbina MH, Grubert J, Böckelmann I, Doil F, Schega L, Tümler J, Mecke R (2010) Perceptual issues in optical-see-through displays. In: Proceedings of the 7th symposium on applied perception in graphics and visualization, Los Angeles. ACM, pp 41–48
6. Kim S, Nussbaum MA, Gabbard JL (2016) Augmented reality "smart glasses" in the workplace: industry perspectives and challenges for worker safety and health. IIE Trans Occup Ergon Hum Factors 4(4):253–258
7. Lavie N (2010) Attention, distraction, and cognitive control under load. Curr Dir Psychol Sci 19(3):143–148
8. Mentler T, Herczeg M (2015) Interactive cognitive artifacts for enhancing situation awareness of incident commanders in mass casualty incidents. J Interact Sci 3(7):1–9
9. Simons DJ, Chabris CF (1999) Gorillas in our midst: sustained inattentional blindness for dynamic events. Perception 28(9):1059–1074

The Effect of Virtual Environment and User/Designer Collaboration on the Creative Co-design Process

Peter Richard[1]([✉]), Jean-Marie Burkhardt[2], Todd Lubart[1],
Samira Bourgeois-Bougrine[1], and Jessy Barré[3]

[1] Université Paris Descartes, LATI, Boulogne-Billancourt, France
peter.richard90@yahoo.com
[2] IFSTTAR, Laboratoire Psychologie des Comportements et des Mobilités,
Versailles, France
[3] ENSAM, Laboratoire Conception de Produits et Innovation, Paris, France

Abstract. Users have been for a few decades recognized as precious contribution to the creative design process. However, the gap of knowledge and motivation between users and designers may weaken the collaboration between participants of co-design meetings. With the emergence of new technologies, design meetings inside virtual environments are developing. This study aims to determine whether virtual environments enhance the creative performance and the quality of collaboration inside co-design workshops. Besides, we tried to determine whether user-designer collaboration inside a virtual environment was more or less effective than "only-user" collaboration. Thirty teams of three participants each took part to creative design workshops whose purpose was to create a new solution to improve mobility in Paris. The teams were distributed into 3 experimental conditions: mixed teams in virtual environment, user teams in virtual environment and user teams in real environment. The workshops consisted in a 10-min step of idea-generation followed by two 10-min steps of solution-selection (one short-term and one long-term solution). In the idea-generation step, user teams produced more ideas in the virtual environment than in the real one. In the solution-selection step, teams in real environment performed a better time management than teams in virtual environment. Finally, mixed teams produced more useful solutions than user teams, and teams in real environment produced more useful solutions than teams in virtual environment, while no effect of the conditions has been evidenced on the level of originality of the produced solutions.

Keywords: Co-design · User participation · Virtual environment

1 Introduction

Different kinds of user involvement can be considered in design process [1, 2]: informative (users give and/or receive information), consultative (users rate or judge products) or participative (users influence design decisions). For a few decades, users have been recognized as potential contributors to the creative design process, especially

© Springer Nature Switzerland AG 2019
S. Bagnara et al. (Eds.): IEA 2018, AISC 822, pp. 605–614, 2019.
https://doi.org/10.1007/978-3-319-96077-7_65

due to their "user-experience", i.e. their knowledge of products which are similar to the product to be designed, and their motivation to create something that fits their needs [3, 4]. For example, Von Hippel's [5] "Lead-users" approach specifies that experienced users (close enough to innovators, or early adopters from Rogers' definition [6]) should be involved in the design process in order to benefit from their expertise and creativity. Future users can also participate in Role Playing method to create requirements and needs or solutions during the design process [7, 8]. However, the gap of knowledge, power and motivation between users and designers may weaken the collaboration between participants during co-design meetings [9, 10].

With the emergence of new technologies, design meetings in virtual environments are developing and users are more and more involved in these design meetings [11]. These technologies can support positively collaboration and creativity [12, 13]. In addition to the reduction of costs related to the organization of physical meetings, these virtual environments can provide participants an anonymity which has a positive effect on the creative process [14].

2   Research Question

This study aims to explore two main research questions. First, do virtual environments enhance the creative performance and the quality of collaboration during co-design workshops? Second, is user/design collaboration in virtual environments more or less effective compared to collaboration involving only users?

3   Procedure

3.1   Participants

A total of 90 participants (62 women and 28 men), aged from 19 to 39 (median age: 22, quartiles: 21 and 23) took part to the experiment. Eighty of these participants were psychology students at the Université Paris Descartes, and 10 were engineering students at the Ecole Nationale Supérieure d'Arts et Métiers (ENSAM) in Paris. Sixty-seven students participated in exchange for credits in a course unit, the 23 other students participated on a voluntary basis.

3.2   Design

The participants were distributed into 30 teams of three participants each, and the 30 teams were distributed into three experimental conditions: mixed teams in a virtual environment, user teams in a virtual environment and user teams in a real environment. The mixed teams were composed of two students in psychology ("users") and one student in engineering ("designers"). The user teams were composed of three students in psychology. The real environment was a meeting room inside the university at the psychology faculty. The virtual environment was a reproduction of the meeting room in the Second Life virtual world (see Fig. 1).

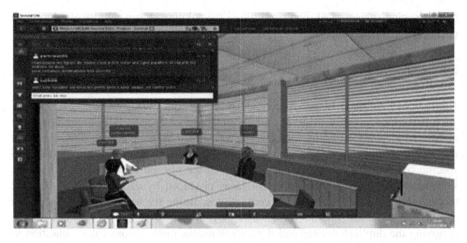

Fig. 1. A "mixed" team (and the experimenter) in the virtual environment

3.3   Procedure

Introduction. In real condition, participants signed the consent form to be video recorded. In virtual condition, participants were encouraged to become familiar with the Second Life (SL) environment by moving their avatar from a starting point to the virtual meeting room. Then, all the participants read the instruction sheet which introduced the mobility problems in Île-de-France area, the goal and the general conduct of the workshop, and the four brainstorming rules [15]. The design problem was introduced in the following way: *"Together, you will have to imagine new solutions to improve the daily mobility of the transportation users in Île-de-France"*. The teams were informed that they had to produced one solution feasible in a one-year horizon and one solution feasible in a ten-year horizon.

Idea-Generation Step. Participants were given 10 min to generate a maximum of ideas in response to the design problem. In real condition, each time they had an idea, they were writing it down on a Post-It (one idea for one Post-It), read it aloud and gave the Post-It to the experimenter who placed it at the center of the table. In virtual condition, they wrote their ideas through the messaging system of SL (one idea per one line).

Solution-Selection Step. Firstly, teams were given 10 min to pick one solution to be implemented in a one-year horizon, based on the ideas generated in the first step. Then they were asked to fill out the corresponding solution sheet, which was composed of a title, a short description of the solution, the list of features of the selected solution and the motivations for its development. Secondly, teams were given 10 more minutes to pick one solution to be implemented in a ten-year horizon and fill out the second solution sheet.

In real condition, participants were interacting orally while one of them was in charge of filling the solution sheets. In virtual condition, participants were interacting through the messaging system of SL and were filling the solution sheets collectively through a collaborative writing software.

3.4   Collected Data

Number of Produced Ideas. According to Osborn rules [15], we measured the creative performance firstly using the number of produced ideas. In real condition, the number of ideas produced by each team was recorded by counting the filled Post-It at the end of the idea-generation step. Each participant was given a set of Post-It of a specific color so that the identification of the author of each idea was made easy.

In virtual environment, the number of ideas produced by each team was recorded through the file which recorded the conversation during the idea-generation step. The number of lines written between the starting signal and the ending signal was counted, removing the duplicated ideas and the miscellaneous interventions (like a question to the experimenter of the judgement of an idea).

Quality of Collaboration During the Solution-Selection Step. The quality of collaboration (QC) in the solution-selection steps was assessed with a method adapted from the QC method [16] that assesses four collaboration categories (see Table 1): information exchange, argumentation and decision making, work process and time management, cooperative orientation.

Table 1. Collaboration assessment grid.

Category	Question
Information exchange	Have participants suggested at least one alternative to the selected solution?
	Had the selected solution been detailed (specifications on its processing, etc.)?
	Had all the participants only discussed about the ideas, without any digression?
Argumentation and decision making	Were there at least one argument presented to justify the choice of the solution?
	Had one participant asked or given information to determine whether the solution was feasible or not?
	Had all the participants explicitly agreed on the selected solution?
Work process and time management	Had one participant suggested a method to select the solution (like a vote)?
	Had the team selected a solution before the ending signal of the experimenter?
Cooperative orientation	Had all the participants took part to the discussion?
	Were there at least two participants who contributed to the selection of the solution?
	Had one participant emphasized a positive contribution of another participant?

Two excerpts were selected for each team: the first excerpt started when the instruction to choose the one-year solution was given and ended when participants agreed on the selected solution. The second excerpt started when the instruction to choose the ten-year solution was given and ended when participants agreed on the selected solution. The collaboration inside all the excerpts was assessed by the first author through 11 questions distributed in four categories.

Each positive answer to these questions was rated 1, and each negative answer was rated 0, without any intermediary score. Thus, in each excerpt, the dimensions "Information exchange", "Argumentation and decision making" and "Cooperative orientation" were rated between 0 and 3, the dimension "Work process and time management" was rated between 0 and 2, and the overall collaboration was rated between 0 and 11 (adding up the scores of each dimension).

Level of Creativity of Solutions. Given that the 30 teams produced 2 solutions (a one-year horizon and a ten-year horizon), we collected 60 solutions. These solutions were rated by 11 researchers from the Institut Vedecom, a laboratory specialized in transportation and new systems of mobility, who participated on a voluntary basis. These raters had the status of users/experts, both because they were doing research in the field of transportation and mobility and because they were using transportation systems in Ile-de-France.

The solutions were presented to the raters in a short form, containing the title and the description of the solution, as written by the teams. As suggested by [17], the raters had to evaluate each solution, according to three criteria: the level of originality, the level of usefulness, and the level of feasibility in the given horizon (1 year or 10 years). The used scales ranged from 0 to 5.

4 Results

4.1 Number of Produced Ideas in the Idea-Generation Step

The three experimental conditions were compared according to the average number of ideas produced by the teams in the idea-generation step (Table 2).

Table 2. Average number of ideas produced by the teams according to their condition

Condition	Mean (SD)
Real/users (n = 10)	20.2 (4.54)
Virtual/users (n = 10)	24.8 (3.52)
Virtual/mixed (n = 10)	27.0 (7.82)

Because of the low number of teams in each condition, we used the Wilcoxon test as a nonparametric test of mean comparisons for more than 2 groups. Results indicated no significant difference between Virtual/Mixed and Virtual/Users conditions in terms of number or produced ideas. In contrast, teams in the Virtual/Users condition produced more ideas than teams in the Real/Users condition ($Q = 78, p = 0,035$), and teams in the

Virtual/Mixed condition produced more ideas than teams in Real/Users condition ($Q = 24$, $p = 0053$). In short, teams working in a virtual environment produced more ideas than teams in real environment, confirming previous results [14, 18, 19].

4.2   Number of Produced Ideas in the Idea-Generation Step

The three conditions were compared on the four dimensions of the collaboration and on the overall collaboration. Each team obtained two scores: one for each selected solution (1-year horizon and 10-year horizon) (Table 3).

Table 3. Average scores (and standard deviation) of collaboration inside the teams according to the condition

Dimension	Real/users	Virtual/users	Virtual mixed	F	df	p	η^2
Information exchange	2.15 (0.59)	2.25 (0.79)	1.85 (0.59)	1.988	29	0.146	6.5%
Argumentation & decision making	2.05 (0.69)	2.20 (0.77)	2.25 (0.64)	0.443	29	0.645	1.5%
Work process & time management	1.25 (0.55)	0.90 (0.55)	0.65 (0.49)	6.432	29	0.003**	18.4%
Cooperative orientation	1.35 (0.88)	1.25 (0.72)	1.25 (0.85)	0.100	29	0.905	0.3%
Overall collaboration	6.80 (1.58)	6.60 (1.54)	6.00 (1.08)	1.733	29	0.186	5.7%

The results of the Analysis of Variance (ANOVA) indicated that the only significant difference between the three conditions was in the "Work process and time management" category. Post-hoc comparisons indicated that users in real environment managed time better than mixed groups in virtual environment (1.25 vs 0.65, $p = 0.002$). The other comparisons are not significant.

4.3   Level of Creativity of the Selected Solutions

Inter-Rater Reliability. The reliability between the 11 raters was computed with the two-way mixed Intra-Class Coefficient (ICC) for single measurements (Table 4).

Table 4. Inter-rater reliability for the three dimensions of creativity

Dimension	ICC	F	p
Originality	0.319	6.149	0.000***
Usefulness	0.206	3.854	0.000***
Feasibility	0.350	6.919	0.000***

The ICC evidences a weak inter-rater reliability, which testifies a lack of a common representation of creativity inside the mobility area. It is worth noting that raters had different fields of knowledge (automobile industry, public transportation, etc.) and different profiles of transportation users: car drivers, public transport users, riders, or multimodal transport users. These results are proof that the representation of creativity is very subjective and depends on raters' knowledge and own needs in the mobility area. That emphasizes the importance of having as many raters as possible for evaluating creativity.

Originality, Usefulness and Feasibility of the Solutions. The 60 collected solutions were firstly analyzed in terms of correlations between the three dimensions. The results indicated that usefulness and feasibility were positively and moderately correlated (r = 0.40). Originality was negatively and moderately correlated with usefulness (r = −0.35) and feasibility (r = −0.51). Thus, there is a tendency of raters to associate usefulness and feasibility, and to oppose both from originality.

The 60 solutions were also compared according to the conditions of their production. The results showed no significant difference between the conditions for originality. However significant differences are observed for usefulness and feasibility (Table 5).

Table 5. Average level of originality, usefulness and feasibility of produced solutions according to the condition

Dimension	Real/users	Virtual/users	Virtual/mixed	F	df	p	η^2
Originality	2.82 (0.42)	2.60 (0.53)	2.84 (0.49)	1.449	59	0.243	4.8%
Usefulness	2.77 (0.38)	2.46 (0.51)	2.98 (0.48)	6.485	59	0.003**	18.5%
Feasibility	3.24 (0.46)	2.86 (0.51)	3.36 (0.46)	5.884	59	0.005**	17.1%

Concerning usefulness of solutions, post-hoc comparisons indicated no significant difference between the Real/Users and Virtual/Mixed conditions. However, users team in virtual environment (Virtual User) produced solutions judged less useful than users' teams in real environment (Real/Users; 2.46 vs. 2.77, p = 0,04) and mixed teams in virtual environment (Virtual/Mixed; 2.46 vs. 2,98, p = 0.002).

In a similar fashion, post-hoc comparisons indicated no significant difference regarding the dimension of feasibility of solutions between the Real/Users and Virtual/Mixed conditions. The solutions produced by users' teams in virtual environment (Virtual/Users) were judged as less easily feasible than the one produced by user teams in real environment (Real/Users; 2.86 vs 3.24, p = 0,02) and the one produced by mixed teams in virtual environment (Virtual/Mixed; 2.86 vs 3.36, p = 0.003).

5 Discussion

The main results indicated that, in the idea-generation step, user teams produced more ideas in the virtual environment than in the real one, which can be a positive effect of anonymity at this step. This result corroborated previous results on the positive effect of electronic brainstorming on the number of produced ideas [14, 19, 20]. It can be hypothesized, like in these studies, that the anonymity allowed by the virtual environment lifted the apprehension of judgement and facilitated the production of additional ideas that they might not have produced in a real environment. Indeed, we observed in the workshops in real environment an apprehension of the judgement: on several occasions, when participants were about to emit an idea they consider as "unusual", they started by justifying it: "The instructions indicate we can let go, so I propose...".

Second, in the solution-selection step, time management was better in real environment than in virtual one. Communication in the virtual environment seems to be more time consuming. This result could be explained by the fact that the teams in real environment communicated verbally, while they communicated in writing in a virtual environment. This methodological choice was made for ecological reasons: the fact of collaborating face-to-face generally implies a modality of oral communication, whereas the participants who collaborate remotely do it more often in writing, even in the tools of videoconference tend to increase the proportion of oral exchanges. In addition, the setting up of workshops in virtual environment with oral exchanges would have lifted, in part, the anonymity of the participants. Anonymity is one of the main factors in increasing the creative performance of remote collaboration. Another explanation could be the loss of time track in virtual environment because of feeling of freedom from constraints in isolated experimental box compared to a real meeting room with other participants and the experimenter. Future research in needed to verify this hypothesis.

Finally, virtual environment mixed teams and real environment user teams may have decided to develop more useful and feasible solutions than virtual environment user teams. The latter group proposed slightly more original ideas, but the difference failed to reach significance. To explain the positive effect of group heterogeneity on feasibility, we can hypothesize that the presence of experts (engineering students in our study) provided technical knowledge that allowed teams to better determine what was achievable on the horizon of given time and what was not. In mixed groups, the knowledge and the pragmatism of engineering students might have influenced the team members to choose feasible ideas in the convergence step. A study [4] demonstrated a positive effect of technological knowledge on the feasibility of the proposed ideas. However, this does not explain the difference between user teams in real and virtual environments in terms of feasibility and usefulness of the selected ideas. One hypothesis is that users in virtual environment might produce more original ideas during the idea-generation step than those in real environment and therefore selected original and less feasible ideas to be developed in the solution-selection step. This result suggests that the positive effect of virtual environment on creativity is more important for fluency and originality rather than for feasibility.

6 Conclusion and Perspectives

In this study, we observed that the collaborative virtual environment had an ambivalent effect on the creative co-design process. In the divergence step, it allowed the participants to produce more ideas than the participants in the real world. This can be explained by a lifting of self-censorship in virtual environment, which is allowed by the anonymity of the participants. In the convergence step, however, the written communication modality, although it allows anonymity, makes probably more difficult time management. A limitation of this study is the absence of mixed teams in real environment and the absence of real experts in mobility domain. Indeed, the presence of expert participants enables the production of solutions more useful and more easily achievable than the absence of experts, and the real environment allows the productions of solutions more useful and more easily achievable than the virtual environment. However, radical innovation implies a break with the past, either through the introduction of a new technology, a paradigm shift or through a new vision that comes from a deep reinterpretation of the meaning of a product or service in line with sociocultural changes [20].

In addition, our 2nd Life based virtual environment was reproducing as closely as possible the meeting room of the real environment, for the sake of comparison. However, the 2nd Life software makes it possible to develop other virtual environments that stimulate the creativity of co-design teams, such as a futuristic city where many modes of transport coexist. Appropriate use of co-creativity in virtual environment would probably encourage radical innovation by promoting the emergence of original ideas rather than feasible ones in the short term.

References

1. Damodaran L (1996) User involvement in the systems design process: a practical guide for users. Behav Inf Technol 15(6):363–377
2. Olsson E (2004) What active users and designers contribute in the design process. Interact Comput 16(2):377–401
3. Sanders EBN, Stappers PJ (2008) Co-creation and the new landscapes of design. Co-design 4(1):5–18
4. Kristensson P, Magnusson PR (2010) Tuning users' innovativeness during ideation. Creat Innov Manag 19(2):147–159
5. Von Hippel E (1986) Lead users: a source of novel product concepts. Manag Sci 32(7):791–805
6. Rogers Everett M (1962) Diffusion of innovations. Free Press of Glencoe, New York
7. Newell AF, Carmichael A, Morgan M, Dickinson A (2006) The use of theatre in requirements gathering and usability studies. Interact Comput 18(5):996–1011
8. Norman DA, Verganti R (2012) Incremental and radical innovation: design research versus technology and meaning change. In: Designing pleasurable products and interfaces conference. Milan
9. Buur J, Matthews B (2008) Participatory innovation. Int J Innov Manag 12:255–273

10. Richard P, Burkhardt JM, Lubart T (2015) Exploring the impact of users' participation to creative workshops for new mobility solutions. In: Proceedings of the 2015 European Academy of Design Conference

11. Kohler T, Fueller J, Stieger D, Matzler K (2011) Avatar-based innovation: consequences of the virtual co-creation experience. Comput Hum Behav 27(1):160–168

12. Maher ML, Rosenman M, Merrick K, Macindoe O, Marchant D (2006) Designworld: an augmented 3D virtual world for multidisciplinary, collaborative design. In: Proceedings of CAADRIA 2006, pp 133–142

13. Uribe Larach D, Cabra JF (2010) Creative problem solving in second life: an action research study. Creat Innov Manag 19(2):167–179

14. Pissarra J, Jesuino JC (2005) Idea generation through computer-mediated communication: the effects of anonymity. J Manag Psychol 20(3–4):275–291

15. Osborn AF (1957) Applied imagination: principles and procedures of creative thinking. Scribner, New York

16. Burkhardt JM, Détienne F, Hébert AM, Perron L, Leclercq P (2009) An approach to assess the quality of collaboration in technology-mediated design situations. In: European conference on cognitive ergonomics: designing beyond the product—understanding activity and user experience in ubiquitous environments. VTT Technical Research Centre of Finland, p 30

17. Casakin H, Kreitler S (2005) The nature of creativity in design. In: Gero JS, Bonnardel N (eds) Studying designers'05. University of Sydney, Sydney, pp 87–100

18. Connolly T, Jessup LM, Valacich JS (1990) Effects of anonymity and evaluative tone on idea generation in computer-mediated groups. Manag Sci 36(6):689–703

19. DeRosa DM, Smith CL, Hantula DA (2007) The medium matters: mining the long-promised merit of group interaction in creative idea generation tasks in a meta-analysis of the electronic group brainstorming literature. Comput Hum Behav 23(3):1549–1581

20. Norman DA (1988) The psychology of everyday things. Basic Books, New York

Text Input in Hospital Settings Using IoT Device Ensembles

Jasmin Wollgast, Andreas Schrader[(✉)], and Tilo Mentler

University of Luebeck, Ratzeburger Allee 160, 23562 Lübeck, Germany
schrader@itm.uni-luebeck.de

Abstract. Care processes in hospitals require intensive communication between stakeholders, interaction with technology and documentation of processes. In any of these cases, text plays a central role. Variable and adaptive text input methods can enhance workflows approaching quality, time, hygiene and resource management. Touchless technologies and speech input offer promising new input methods. Especially, the Internet of Things (IoT) offers plenty of opportunities for the separation of input and output interaction devices, and the adaptive and context-aware provision of wirelessly connected compositions of smart devices (IoT-ensembles) fulfilling the specific needs of all relevant stakeholders of the care process.

This paper analyzes empirically the requirements and possible integrations for text input device ensembles within hospitals using the process of Human-Centered Design (ISO 9241-210). The results were used as the methodical base to design a concept, to build an appropriate prototype, and to evaluate usability according to ISO 9241-11. The participants of the evaluation confirmed an improvement of efficiency or effectiveness for dedicated situations by using the provided devices.

Keywords: Internet of Things (IoT) · Text input devices · Clinical care

1 Introduction

Information, communication and documentation are of fundamental importance for stationary and ambulant care processes. Especially care documentation is one of the most important everyday tasks in hospitals. In all phases of the workflow of patient care, documentation is required for the accounting of services, to ensure reliability and to fulfill legal requirements.

There has been a number of work dedicated to developing automatic systems for documentation. The main idea is to incorporate various sensors at devices, patients and care givers in order to automatically detect actions and trigger respective entries into documentation systems. Usually referred to as hands-free clinical documentation systems, various types of sensors and approaches have been tested, including wrist-worn sensors [1] or camera-based systems [6].

However, fully automated systems have limitations and drawbacks. Sensors will probably not be able to detect every possible aspect of care processes, systems require ubiquitous instrumentation throughout the complete premise, and staff acceptance

© Springer Nature Switzerland AG 2019
S. Bagnara et al. (Eds.): IEA 2018, AISC 822, pp. 615–625, 2019.
https://doi.org/10.1007/978-3-319-96077-7_66

might be low due to suspected surveillance. Care process detection is a very complex task, where probabilistic classification and temporal modelling and petri nets might be used [2].

Alternatively, semi-automatic systems would enable to manually triggering documentation events, from simple hands free text entry to insertion of automatically generated documentation elements with pre-defined text. Altakrouri et al. [3] positively evaluated a combination of automatic detection using on-body sensors and manual trigger using smartphones. In addition, for communication purposes with patients, and for controlling medical devices by setting machine parameters, manual selection of text is crucial for the provision of qualified healthcare.

In previous work, we have proposed to develop a semi-automatic system. Instead of providing a static system with dedicated technology of choice, our paradigm supports the "come-as-your-are" principle and fosters an adaptable system automatically creating an appropriate context-aware IoT ensemble of interaction devices in-situ based on user preferences and abilities, available infrastructure and currently performed task [4]. We believe that the state-of-the-art method of designing specific devices (micro-adaption) to fit specific needs does not scale in those complex scenarios and propose a novel form of (semi-)automatic ad-hoc device ensemble construction on the fly (macro-adaptation). Important research questions are the cognitive load of users to deal with dynamically varying input methods, as well as the automatic creation of usage explanations ("manual on-the-fly").

In this paper, we limit our scope to text entry as a special and important subtask of clinical care documentation. Choosing the right method for a task or a situation requires to select appropriate means (e.g., pushing buttons for text elements or drawing strokes for written language elements) and to combine them with the goal of maximized usability. However, the working conditions for healthcare professionals impose numerous restrictions (hands-free operation, hygienic environments, noisy environments, multi-user settings, etc.) for the usage of conventional text-related technologies (e.g., keyboard, mouse, touchscreen, or telephone). Currently used input methods tend to interrupt workflows, decrease efficiency, increase the need for disinfection, and produce inaccuracy [5].

Therefore, we have performed an intensive investigation of existing and upcoming interaction techniques and input devices for text entry using literature review and market surveys. A variety of options is available today, ranging from hardware-based keyboards in various forms, joysticks, pens, virtual keyboards on touch-enabled displays, speech recognition, handwriting detection, gesture interfaces, wearables, and brain interfaces. A complete overview is beyond the scope of this paper, and the reader is referred to [5].

The remainder of this paper is organized as follows. In Sect. 2 we present results from our in-depth analysis of users, environments and tasks in hospitals. In Sect. 3 we outline our first prototype intended to support workshops with relevant stakeholders. Section 4 reports about the evaluation results. Finally, Sect. 5 summarizes our results and outlines future work.

2 Analysis

In order to identify the context requirements, situations, working environments and nurses' needs, we selected a qualitative type of empirical analysis. We performed a case study in one specific hospital to avoid heterogeneous technical standards of equipment, without losing generality. The hospital still uses a mix of paper-based and digital means and therefore represents a typical type of hospital in the transition phase to full digital information processing.

2.1 Method

We used a two phase-approach to explore the current daily routine of text input in hospital settings. First, we conducted several half-structured qualitative observations at one peripheral (mouth, tooth and jaw diseases) and one central clinic (combustion ICU). Second, three interview rounds with groups of nursing students were performed.

The observation focused on all kinds of communication and documentation as well as text or digit inputs on medical or other technical devices during nursing tasks. At the same time, we observed restrictions and problems.

We assumed that domain experts and researchers are more likely to imagine realistic technical solutions. Hence, we decided to discuss experienced problems and possible solving with nursing students who have work experience and a scientific background. Additionally, the students are still new in the hospital environment and generally less biased by the hospital system and negative or positive working experience.

2.2 Results

The observations and interviews delivered a number of valuable insights. First, users could be classified according to expertise and according to role. Differences in experiences played a major role for the amount of attention needed for text input. Role in the hierarchy decides about distributed care tasks and text input requirements. Additional subcategories include skills, gender and even religion. The users emphasized the support of interpersonal communication and the avoidance of disturbance of this communication as the most important aspects for IoT-technology in hospitals. There are two different types of communication to be considered: the communication amongst the hospital staff and the communication between staff and patients.

Beside of these social aspects we structured and classified the technical problems and requirements for text input in hospitals. The main problems observed were the asynchronous communication between nurses and doctors, the ubiquitous and instant need for documentation, interruptions during text inputs, the requirements for hygiene and the situations with stress, lack of time and limited cognitive or physical resources (e.g., hands in use for cleaning a wound). Additional influence factors were identified as high fluctuation of staff members, additional effort to retype handwritten notes, privacy issues in multi-room settings, information availability, and many more [5].

We also performed a task analysis, investigating standardized processes for care documentation, an organizational analysis identifying challenges of communication flows within the complex team hierarchy, and a context analysis considering the social, health and financial aspects as well as quality and authorization issues [5].

2.3 Concluding Requirements for Text Input Technologies

Based on the empirical analysis results, requirements for text input devices can be specified in five categories: IT-security (failure-proof, tracking, authorization, data security, etc.), transparency (manuals, feedback, support, etc.), functionality (error tracing, hands-free, consistency, easy installation, adaptation, etc.), usability and ergonomics, and context factors (hygiene, social acceptance, etc.).

To consider these categories and the users' needs we designed a three-step classification scheme (Fig. 1). This scheme supports the consideration of changeable contexts and the finding of adequate interaction techniques for a certain user and a certain context. If the text input problem concerns the user, the selection of interaction methods starts with step one. Dedicated to the requirements for each category a mapping and an adequate complexity of software and hardware is chosen. In step two the developers pick possible interaction techniques which suits the assumptions of the first step. Iterative adjustments should be minded. With the last step each of the possible text input mappings of step one and two has to be discussed with the context requirements.

If the classification is used the other way round, the context of the text input problem is emphasized. The intersection of possible text input methods of both directions is the solution which could corresponds to the design principles of ISO 9421-11. However, the purpose of this classification is to find a weighted trade-off if there is no intersection of context and user requirements.

Concerning this classification we also deduced innovative or existing technologies which could be used as context adaptable IoT-technologies in hospitals. HMD glasses, speech input, brain interfaces, adaptable wearables and mobile devices, keyboard for the feet, symbol input, gesture input (e.g., AirWriting), eye gaze input, input on projections, touchless nano displays [8], digitalized handwritten input, keyboards with adaptable keys, and key-symbols are promising candidates for ensemble inclusion.

3 Prototype

For evaluation purposes, we implemented a "Wizard of Oz" prototype in a fictive hospital patient room. The room contains a smart hospital bed, a set of typical furniture and a number of different input and output devices to demonstrate principle options for building interaction ensembles and to provide a realistic test environment inside the usability health lab[1] of the Center for Open Innovation in Connected Health (COPI-COH) at the University of Lübeck, Germany.

The setting supports the selection of different text input devices and to make a situation or context dependent choice. It also supports different kinds of display

---

[1] https://www.copicoh.uni-luebeck.de/forschung/health-lab.html.

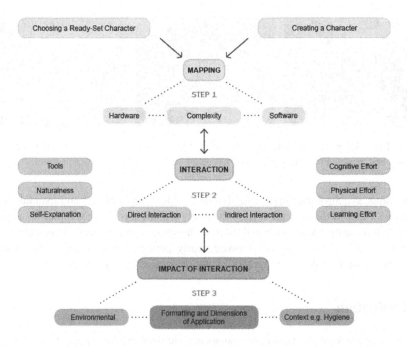

Fig. 1. Classification of text input techniques with two different directions depending of the dedicated problem.

techniques to imitate the seamless interconnection between input devices and simulated medical devices or screens.

The setup (see Fig. 2) included the following ten input devices: (1) speech input using a personal microphone (smart pen or wearable); (2) speech input at microphones placed in the environment; (3) wireless hardware-keyboard on a tripod; (4) virtual keyboard projected under the bed and operated by foot; (5) mouse on a moveable tripod; (6) handwriting using smart pen; (7) handwriting using fingers; (8) Leap Motion[2]; (9) virtual keyboard on a mobile touchscreen; and (10) virtual keyboard projected on a wall.

The following seven receiver and output devices have been integrated as well: (1) wall projection as display; (2) stationary touch screen; (3) smartphone; (4) laptop; (5) perfusor as an example of a medical device; (6) tablet on a moveable tripod; and (7) projected floor display. Touch interfaces acted as input and output devices at the same time.

The prototype uses a restricted version of IoT-ensembles by always connecting one type of input device to one specific receiver device at a time, e.g., a wireless keyboard for selecting parameters of a perfusor. All inputs from all input devices are always shown on all output devices, simulating ad-hoc ensemble building just be selecting a preferred input device. The setup uses three types of connections: direct HDMI cable

---

[2] https://www.myo.com/.

Fig. 2. Left: Setup of simulated patient room, Middle: Projected keyboard underneath the patients' bed. Right: Projected screen and keyboard on the wall of the patient room.

connections, synchronization over the Internet using TeamViewer,[3] and manual wizard insertion of text in the background by the study technicians. For example, we avoided any speech recognition software and inserted the spoken text by hand. This method mix ensures that all selected devices showed nearly perfect behavior and avoids participants' bias based on detection quality of the respective devices of choice.

4   Evaluation

The first objective of the prototype was to put the user in a situation of use were text input devices and output devices have no dedicated connection but an adaptable, selectable connection. The second objective was to evaluate a set of innovative text input technologies (e.g., feet keyboard, wall as a touchscreen) and their acceptance and approval by nurses.

4.1   Setup

Eleven subjects (students from nursing, medicine and media informatics between 18 and 25 years, 9 female) were introduced to the test setup and the functioning of all devices in the simulated patient room. After the introduction, they had to type a sentence (e.g., "stop medication") with each input device in any arbitrary order. To get information about their preferred text input devices in different situations, the subjects had to complete a standardized questionnaire of 29 items. For this, we specified eight typical situations in clinical care: (1) changing bandage; (2) delivering water to patient; (3) colleagues in the same room; (4) family members of the patient in the same room; (5) needing to communicate with colleague in remote room; (6) controlling of medical devices in same or remote room; (7) making notes; and (8) documenting performed care task.

Text fields were provided for the subjects' reasons and thoughts for a decision. Additional questions have been asked to explore, whether adaptable, ad-hoc connected IoT-technology would support the documentation task.

---

[3] www.teamviewer.com/de.

4.2 Results

The relatively small number of subjects in the evaluation does of course not support statistically significant evidences. Nevertheless, the results deliver a number of interesting qualitative trends and support for the design of the planned optimized clinical trial. A comprehensive overview of all evaluation results is given in [5]. Here, we can only report some of the major observations. The participants of the evaluation mainly confirmed the importance of adaptable IoT-devices and that they would use IoT-ensembles and innovative text input technologies in hospital contexts (see Figs. 3 and 5).

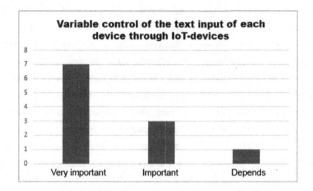

Fig. 3. Importance of adaptable text input using IoT-ensembles

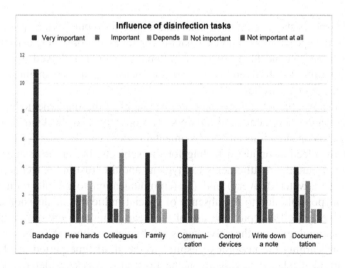

Fig. 4. Importance of disinfection for selection of devices

Figure 4 shows the influence of disinfection requirements for the selection of specific input devices for the eights care situation. Especially with bandage exchange, all subjects declared it as the only selection factor. Touch-free input devices have been preferred in this context. In all other situations except, when colleagues are in the same room, disinfection still is the most important influence factor. This reflects expectations, since the daily routines for disinfection are annoying, time consuming and potentially harmful for the environment.

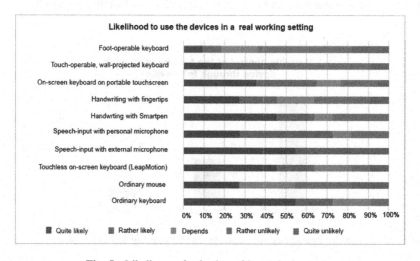

Fig. 5. Likeliness of selection of input device types

Figure 5 shows the likeliness of selecting a specific input device type. Somehow surprisingly, the speech input with external microphones is more likely selected than an own microphone. Second likely is a conventional keyboard, followed by virtual keyboards on a mobile touchscreen and handwriting using a pen. The smallest preference is given to the foot controller.

Figure 6 shows the advantages and disadvantages of the implemented devices due to the participants' comments and answers. The quantity and the kind of use of the devices depends on situations and environmental issues. The decision for one device is not a random selection or just a feeling for the subjects, but an assessment of complexity of the input, the requirements for hygiene and the trust in the technology. Only in situation (2) (water delivery) and (5) (communication to remote room), random selection was possible due to the absence of special influences as decision enablers.

Most subjects well accepted the use of the Leap Motion as the only implemented AirWriting-technology for the advantages of hygiene, although subjects admitted challenges in learning the necessary gestures in the short time period available. Foot input on virtual displayed keyboard on the floor also was considered positively in relation to hygiene, privacy and hands-free operation. On the other hand, it was the most challenging form of text input from motoric point of view. Subjects also did not

Text Input Method	Advantages	Disadvantages
Ordinary wireless keyboard	• no time for learning needed • suitable for extensive writing task at the desk	• limited ergonomics • inefficient for disinfection • additional device
Ordinary wireless mouse	• no time for learning needed • sequential selection of characters • lack of efficiency	• inefficient for disinfection • additional device
Touchless keyboard (LeapMotion)	• no disinfection task needed • natural interaction	• unusual interaction • time to learn
Speech input with personal microphone	• no disinfection task needed • natural interaction • hands-free • multitasking • place independent • fast and exact	• environmental impact • confidential data
Speech input with external room integrated microphone	• no disinfection task needed • natural interaction • hands-free • multitasking • fast and exact	• environmental impact • confidential data • requires exact speech recognition
Handwriting with smartpen	• no disinfection task needed • natural interaction • unprompted notes • availability of notes • pen useable as input tool for screens • hygienically	• easy to lose • additional device • requires exact handwriting recognition
Handwriting with fingertip	• no disinfection task needed • natural interaction • unprompted notes • availability of notes	• problems with hygiene • imprecisely
On-screen keyboard on touch-operable screen	• confidential • different input modalities • accustomed through private smartphones/tablet	• problems with hygiene • requires shelf
Wall-projected, touch-operable screen	• large surfaces • ubiquitious • different input modalities • hands-free	• problems with hygiene • problems with confidential data • requires appropriate walls/surfaces
Foot-operable keyboard	• multiasking • hygienically • challenge for physical skills	• very unfamiliar interaction • head is down to the ground • problems with hygiene • unintentional uses

Fig. 6. Importance of adaptable text input using IoT-ensembles

have any problems using well-known input devices as part of an IoT-ensemble. Display sizes were decisive for the effectiveness of handwriting and virtual keyboards. Although all subjects owned and used smartphones on a regular basis, the observed text input seems to be more challenging then at larger displays on a tablet or bigger touch screen.

Another important aspect of the evaluation was the question, how subjects would like to specify the device of choice and trigger the automatic creation of an ensemble connecting it to an output device. Here, the subjects very likely or likely preferred to use a dedicated (graphical) selection interface (91%), directly touch the respective devices of choice (82%), or use speech commands (73%). Using gestures or eye gazing (27%), automatic detection (36%) or proximity (45%) receive lower preferences. 82% of the subjects would also like to use pre-defined ensembles for certain tasks and situations.

5 Conclusion and Future Work

The users who participated in the evaluation of the prototype confirmed an improvement of efficiency and/or effectiveness for dedicated situations by using the provided devices. They confirmed the benefit of variable connections between input devices and operated devices just as the possibility to choose straightforward between input devices with different features. In contrast to conventional criteria for the definition of ergonomics of general human computer interfaces (ISO 9241-11) it turns out that in this special environment of hospitals, new and even imperfect device types would be accepted and appreciated as long as users can avoid disinfection routines and improve their arsenal of text input methods.

This study also delivered very insightful information for the design of appropriate IoT-ensembles for our upcoming clinical field trial. It is advisable to only installing those devices that subjects will use with a high probability. This will reduce mental and physical complexity and potentially increase acceptance amongst professionals.

We are currently working on the evaluation of additional device types and text input technologies in clinical settings, including writing in the air using finger scanner and thinking text input using brain interfaces. We are also developing improved evaluation setups in order to improve detailed investigation of learning rate, input speed, ergonomics and acceptance for additional health care tasks.

Further, we explore methods for authorization and security of confined mapping between authorized users, adapted device ensembles and respective input devices. Access to information systems and medical devices should be restricted to authorized staff members. Another research activity is currently ongoing, establishing improved self-description of input devices and automatic creation of explanation manuals of in situ ad-hoc ensembles in our Ambient Reflection framework [7].

References

1. Nguyen HP, Ayachi F, Lavigne-Pelletier C, Blamoutier M, Rahimi F, Boissy P, Jog M, Duval C (2015) Auto detection and segmentation of physical activities during a Timed-Up-and-Go (TUG) task in healthy older adults using multiple inertial sensors. J NeuroEng Rehabil 12(1):36
2. Pinske D, Weyers B, Luther W, Stevens T (2013) Personalassistenz im Krankenhaus durch intelligente und ambiente Pflegedokumentation. Kognitive Systeme (1), University of Duisburg-Esse
3. Altakouri B, Kortuem G, Grünerbl A, Lukowicz P (2010) The benefit of activity recognition for mobile phone based nursing documentation: wizard-of-oz study. In: 14th annual IEEE international symposium on wearable computers (ISWC), Seoul, pp 1–4
4. Altakrouri B, Schrader A (2012) Towards dynamic natural interaction ensembles. In: Ramduny-Ellis D, Dix A, Gill S (eds) Proceedings of the fourth workshop on physicality. Birmingham University, Birmingham
5. Wollgast J (2017) IoT-Ensembles für Texteingabe im klinischen Kontext. Bachelor Thesis, University of Lübeck (in German)
6. Fabbri D, Ehrenfeld JM (2016) Hands free automatic clinical care documentation: opportunities for motion sensors and cameras. J Med Syst 40(10):213

7. Burmeister D, Schrader A (2018) Runtime generation and delivery of guidance for smart object ensembles. In: 9th international conference on applied human factors and ergonomics (AHFE), Orlando

8. Max Plank Gesellschaft Homepage. https://www.mpg.de/9664205/beruehrungslos-touchless-bildschirm. Last Accessed 28 May 2018

Ideologically-Embedded Design

Françoise Détienne$^{(\boxtimes)}$, Michael Baker, and Chloé Le Bail

UMR 9217, CNRS - Telecom ParisTech, 46 rue Barrault, 75013 Paris, France
{francoise.detienne,michael.baker,
chloe.lebail}@telecom-paristech.fr

Abstract. In this paper we defend the position that design projects are not only socially embedded, but also ideologically-embedded, as far as the designers or other stakeholders consider necessary to take into account the insertion and future position of the artefact in its societal and cultural environment. This is illustrated by examples from studies of design of artefacts in three cases: (i) Wikipedia articles, (ii) production and use of local governments' open data, and (iii) co-design of social artefacts such as participatory housing. We conclude with some implications on the design process.

Keywords: Design · Values · Collaboration

1 Introduction

In this paper we propose to take into account the ideological dimension in the design process. First we will develop our theoretical background then we will exemplify the ideological-embeddedness of design on the basis of examples from studies of design of knowledge artefacts in three cases: (i) Wikipedia articles, (ii) production and use of local governments' open data, and (iii) co-design of social artefacts such as participatory housing.

2 From Social to Ideological Embeddedness of Design

It is now well established that design is a process of negotiating among disciplines. Solutions are not only based on purely technical problem-solving criteria. They also result from compromises between designers: solutions are negotiated [1]. Since the 1990s at least, the fact that design of artefacts (knowledge objects, tangible artefacts, digital artefacts) does not only involve fitting with psychological-physiological, æsthetic and economic characteristics of individuals, but must also be socially embedded, has become generally accepted [2]. Part of this concerns the role of values in design [3]: "[t]he designer or design team makes choices at every point in the design process and most of these are value laden. Every decision at each choice point will give priority to certain values over others".

In the design-thinking literature, priority values for decision-making in designing have been proposed to explain variant and invariant characteristics of designers' thinking and acting across disciplines, for example in graphic design, architecture,

© Springer Nature Switzerland AG 2019
S. Bagnara et al. (Eds.): IEA 2018, AISC 822, pp. 626–630, 2019.
https://doi.org/10.1007/978-3-319-96077-7_67

interaction design and engineering [4]. Le Dantec and Yi-Luen Do [5] and Llyod [6] applied this notion, extended to social and ethical considerations, to understanding interactions between architects and clients, highlighting thereby the characteristics of social embeddedness of design.

Our position is that any design project will be not only socially embedded, but also ideologically-embedded, as far as the designers or other stakeholders consider necessary to take into account the insertion and future position of the artefact in its societal environment, characterized not only by its users but also by historical, cultural aspects of society. Note that by "ideology" here we do not mean irrationally held or dogmatic beliefs, but more literally a system of ideas and ideals or values ("idea"-"logos") that are often held by and are the 'ciment' of particular social groups. It is a system of thoughts, beliefs and values, a set of social and moral ideas and opinions, specific to a group or an individual, and which guide action and/or appear in discourse [7]. Thus, designing artefacts involves systems of values or ideological systems developed at organizational, community, institutional or societal levels. This position implies two specific characteristics in the design process.

First, we should consider that the negotiation process evolves in two spaces of debate, not only on criteria and their hierarchization but also on a more 'organic' level of systems of values. The 'organic' or systematic character of systems of ideas and values means that "values" is not simply another class of criteria to be added at the end of a list of design 'constraints' or 'criteria'. Rather, values permeate, underlie and cohere with the ensemble of other criteria for preferring one design over another. Spaces of debate, referred to in the field of argumentation studies [8, 9] as relating to factual and axiological types of referents, entail a particular type of negotiation leading either to minor revisions of one's own attitudes -and system of beliefs- or to deepening of one's own ideology.

Second, as collective design involves design-oriented and group-oriented processes, this ideological dimension is mobilized at the articulation of these processes, and made visible through design and group co-evolution phenomena. In other words, it acts as a ciment of group(s) evolution (either ways, reinforcement of groups or scissions) in close relation to the negotiation process.

3   Three Examples of Ideological Embeddedness

3.1   Controversy in Wikipedia

In Wikipedia, ideology enters into design of both the communities themselves—for example, via the ideas/values of "free participation" and "neutrality of point of view", itself hotly debated—and within the production of particular articles on controversial subjects [10]. In this case, ideology shapes both the structure of the co-working communities and their particular products.

In a previous research, we analysed the evolution of the article "The Shroud of Turin" in the French language Wikipedia (Suaire de Turin). Two types of contributors, with strong and opposed ideological values, are particularly active in this article elaboration: "scientists" and "religious people". They have opposed points of views on the carbon 14 analysis that were done on the shroud, that revealed that it dated from the 14th century. Although one might expect that appeal to 'sources', 'established facts' or 'evidence' outside the debate itself could help to resolve it, this was not the case here, since the "scientists" disputed the validity of the sources evoked by the "religious people", and the other way around. In effect, the scientists criticized not only the content of sources, but also the credentials of their authors, their motivations (were not these 'sources' created simply to support the religious point of view?) and the sincerity of the participants in Wikipedia who evoked the so-called sources.

Rather than leading either to even minor revisions of one's own attitudes -and system of beliefs- the debate entailed deepening of each party's ideology. This 'religion versus science' conflict was never really resolved here. Rather, it was 'managed by the strategy of separating out the conflicting views, each of which was accorded its (undisputed) separate section in the article (e.g. "Viewpoint of the Catholic church", "Scientific viewpoint"). Thus, Wikipedia is not only a "socio-cognitive" online community [11] but can also be seen, in the case of articles on controversial issues, as a 'terrain' on which ideological 'battles' are fought, in the attempt to 'gain ground' on a global scale.

3.2 Open Data and Participatory Democracy

Over the past ten years, open data initiatives [12] have become decisive factors in the policy of transparency in governments, in order to promote open and even participatory democracy. Our work [13] revealed the mismatch between the ideology of political-institutional discourses, which aim to promote the values of "openness" and "transparency", and the actual activity of both open data producers (who in fact function according to "strategically opaque transparency") and open data users, whose interests (e.g. commercial) may be quite different.

As producers of open data, public administrations ("PAs") are supposed to take into account citizens' needs in order to produce open data understandable and useful for them. We can consider that process as design of artifacts by PA for citizens, in which the ideological dimension can influence both the shape and the content of the produced open data. We found out that PAs had tensions between their own ideological values, those of "serving the public and involving citizens in political decision-making" and "preserving economical and political interests". These internal tensions lead to processes of selecting which data should be published or not (e.g. PAs are reticent to publish data—such as on air pollution—that might have a negative economic effect on the attractiveness of the city).

So, on one hand PAs intend to involve citizens in political decision-making to restore public trust and to improve accountability of policy makers (European E-Government Action Plan 2011–2015), whilst, on the other hand, they do not provide full access to public data. In sum, they adopt what we termed "strategically opaque transparency", restricting the available data, or else spreading information across disparate data-sets, which renders understanding more arduous. This calls into question the nature of their collaboration with citizens, and thus disrupts the elaboration of collective solutions to societal problems.

3.3   Collaborative Design of Participatory Housing Projects

In the analysis of collaborative design workshops [14], whose aim was to produce blueprints for participatory housing projects (what is to be shared, what rules for the community?), we showed that for participants, the dimension of sharing an ideology (values of sharing, preserving the environment, etc.) was just as important as practical considerations (e.g. economizing money by sharing resources).

We proposed the use of a Role-Playing Simulation [15] for supporting assessment of a sustainable sociotechnical system, a neighborhood that encourages collaborative consumption practices. Results show that the Role-Playing Simulation stimulates debate about sociotechnical solutions, rules and rational criteria, in addition to sustainability values and ideological values. The RPS could thus be a means for making sociotechnical solutions evolve and for identifying what is "technically", "socially" and "ideologically" necessary for acceptance and success of the future system.

4   Discussion

To conclude, we have shown that ideology, in sense of values (e.g. ecology), can be a motivation for design of an artefact, but it also can function as a source of strong dissent and verbal conflicts where it is difficult, and sometime impossible, to find a compromise of agreed solution. Thus ideology can strongly influence the very composition and functioning of the collective itself.

However, it remains important, as shown in our last example on participatory housing projects, to organize the design process in such a way that stimulate debates, not only on criteria but also on values underlying the organization of criteria, so that the ideological dimension can be made explicit in the co-evolution of design and group-oriented processes.

References

1. Bucciarelli LL (1988) Engineering design process. In: Dubinskas F (ed) Making time: culture, time and organization in high technology. Temple University Press, Philadelphia, pp 92–122
2. Schmidt K, Bannon L (1992) Taking CSCW seriously: supporting articulation work. Comput Support Coop Work 1(1–2):7–40
3. Marshall T, Erlhoff M (2008) Design dictionary, perspectives on design terminology. Birkhäuser, Basel
4. Da Silva Vieira S, Badke-Schaub P, Fernandes A, Fonseca T (2010) Understanding how designers' thinking and acting enhance the value of the design process. In: Proceedings of the 8th design thinking research symposium (DTRS8), pp 107–120
5. Le Dantec C, Yi-Luen Do E (2009) The mechanisms of value transfer in design meetings. In: McDonnell J, Lloyd P (eds) About: designing, analysing design meetings. Taylor & Francis, London, pp 101–118
6. Lloyd P (2009) Ethical imagination and design. In: McDonnell J, Lloyd P (eds) About: designing, analysing design meetings. Taylor & Francis, London, pp 85–100

7. Gerring J (1997) Ideology: a definitional analysis. Polit Res Q 50(4):957–994
8. Golder C, Coirier P (1996) The production and recognition of typological argumentative text markers. Argumentation 10(2):271–282
9. Schwarz B, Baker M (2016) Dialogue, argumentation and education: history, theory and practice. Cambridge University Press, Cambridge
10. Baker MJ, Détienne F, Barcellini F (2017) Argumentation and conflict management in online epistemic communities: a narrative approach to Wikipedia debates. In: Arcidiacono F, Bova A (eds) Interpersonal argumentation in educational and professional contexts. Springer, Cham, pp 141–158
11. Détienne F, Baker M, Fréard D, Barcellini F, Denis A, Quignard M (2016) The descent of Pluto: interactive dynamics, specialisation and reciprocity of roles in a Wikipedia debate. Int J Hum Comput Stud 86:11–31
12. Denis J, Goëta S (2017) Rawification and the careful generation of open government data. Soc Stud Sci 47(5):604–629
13. Groff J, Baker M, Détienne F (2016) Aligning public administrators and citizens on and around open data: an activity theory approach. In: Nah FH, Tan CH (eds) HCI in business, government, and organizations: information systems. HCIBGO 2016. Lecture notes in computer science, vol 9752. Springer, Cham
14. Le Bail C, Détienne F, Baker M (2016) A methodological approach to the conceptualisation of a socio-technical system: a smart and collaborative neighbourhood. In: Proceedings of the European conference on cognitive ergonomics (ECCE'16). ACM, New York
15. Le Bail C, Détienne F, Baker M (2018) A role-playing simulation to support assessment of sustainable sociotechnical systems for and by citizens. In: Proceedings of the European conference on cognitive ergonomics (ECCE'18). ACM, New York

A Comparative Force Assessment of 4 Methods to Move a Patient Up a Bed

Mike Fray$^{(\boxtimes)}$ and George Holgate

Loughborough Design School, Loughborough University, Loughborough, UK
M.J.Fray@lboro.ac.uk

Abstract. This study compared four different postures and positions regularly suggested for moving a patient up towards the head of the bed, using both novice and expert users. The trial was carried out in a laboratory using 21 participants (10 novices and 11 experts). All participants completed all conditions (n = 4) three times each (n = 3 repetitions). The physical force at each hand was recorded using electronic four compression/tension meters, recorded on DasyLab software. After each condition a subjective review questionnaire was completed. The data was processed with excel and SPSS to evaluate the differences between the conditions. A significant statistical reduction was found when comparing combined force for all carers ($F(3,27) = 24.63$, $p < .05$) and the load per individual ($F(2.21,44.21) = 27.26$, $p < .05$). However there was found to be no statistical difference between left and right hand or upper or lower hand. Transfers carried out with the carer pulling the patient towards them corresponded with a lower force to complete the transfer. This study suggests that a position with an oblique offset base and an action of pull and push in line with the carer could be the preferred position for a wide range of patient transfers.

Keywords: Biomechanics · Patient transfers · Perceived effort

1 Introduction

Warming et al. (2009) defined patient handling as consisting of two tasks '(1) transfer tasks, which is when the nurse assists the patient moving from one position to another and (2) care tasks, which is when the nurse assists the patient in doing daily activities or necessary professional tasks for the well-being of the patient'. The focus of this report will be on patient transfer tasks, in particularly moving patients horizontally towards the head of the bed.

Care staff are frequently exposed to significant load during their daily work (Fragala et al. 2005). The handling and moving of patients in bed are part of these frequently carried out tasks, with Smith (2005) stating the 'act of pulling a patient up to the head of the bed, is a frequently performed patient handling task' and 'nurses are exposed to high risk patient handling tasks at a high frequency'. It is considered that nursing tasks can be high risk (Fragala et al. 2005; Owen and Staehler 2003).

Various studies have highlighted that carers are at risk of potential injuries when carrying out transfer tasks (Garg et al. 1991; Schibye et al. 2003; Waters 2007).

© Springer Nature Switzerland AG 2019
S. Bagnara et al. (Eds.): IEA 2018, AISC 822, pp. 631–640, 2019.
https://doi.org/10.1007/978-3-319-96077-7_68

Other supporting studies (Alamgir et al. 2007; Jäger et al. 2012) confirm the fact that carers are vulnerable to sustaining MSI due to 'transferring and repositioning tasks during patient/resident/client care'. The potential risk of injury is accentuated through not using slide sheets; with Jordan et al. (2011) stating transfer tasks not using slide sheets are responsible for the 'highest lumbar load of various patient-handling tasks'.

In order to reduce the impact of poor/awkward patient handling and in particularly patient transfer, training and certain patient handling techniques/aides have been put in place to help aid and inform carers (Smith 2005; Cohen et al. 2010; Vic 2009). Pellino et al. (2006) states using transfer aides reduce physical stress on the part of nurses (Nelson et al. 2003). Information, education and training in proper use of aids/equipment are essential to promote behavioral and attitude changes among staff (ISO 2012; ANA 2013). Several studies reported the combination of both using equipment and education is more effective safer patient handling (Black et al. 2011; Lim et al. 2011; Garg and Kapellusch 2012).

2   Methods

2.1   Aims

The overall aim of this study is to quantify and compare four different conditions in terms of the force required to transfer a patient up towards the head of the bed, using both novice and expert users.

2.2   Objectives

To quantify the amount of force required in each condition of transfer for both novice and expert users.

To compare and rank the different conditions from best to worst in terms of force needed.

2.3   Conditions

The postures/positions to move someone towards the head of the bed were selected having reviewed current best practices (Brooks and Orchard 2011; NHS 2010; Smith 2011). A focus group of professionals within the subject area was used to provide evaluation on the methods. The force required to move the patient up the bed was evaluated in pilot studies. Through testing a final suitable weight ($\approx$68 kg) was decided. On evaluation a weighted mannequin was used to standardise the trial. Market standard slide sheets were placed under the mannequin and the force devices were attached to standard positions on a non-slip under sheet. The conditions selected for the trial were:

1. Parallel Stepping - Two participants, on either side of the bed with hands at the shoulder and hip of the patient the carer's step up the bed sliding the patient along.
2. Rotation - Two participants, one on either side of the bed with hands at the shoulder and hip of the patient participants turn towards the head of the bed, without moving the feet.

3. 2 Person Oblique Pull Up Bed - Two participants at each top corner with an oblique offset base. To move the carers take a step backwards keeping their arms straight.
4. Single Person Pull Up Bed - This transfer uses one participant, at the head of the bed with hands shoulder width apart. When ready the participant steps backwards, keeping their arms straight.

2.4   Data Collection

In line electronic force meters allowed force data to be collected from each hand for the four conditions. Ethical approval was achieved through the University systems. The force devices were calibrated daily and between participants. Each condition was completed three times, with an average taken, any large variations in the data caused a repetition of the action. A convenience, non-probability sampling strategy was used, by sending a number of emails to potential participants and using those who replied in the trial.

During the trial participants were given time to familiarise themselves with the transfer. Each transfer would then be carried out three. Once completed the participants being asked to fill out either a short novice or expert review questionnaires. The transfer was repeated if any errors or significant differences from normal movement or participant body position were observed, to give consistency in the analysis.

2.5   Subjective Data

After the participants carried out each condition they were asked to complete a subjective review depending on their level of experience (novice & expert). These included a Borg rating scale and likert scale questions reporting the effort, security and safety of each condition, with the expert review also including how likely the particular condition was to be used in their regular practice and workplace.

2.6   Forces Assessment

Forces in Newtons were recorded at 20 reading per second in the software. Each transfer was examined with the length of transfer, average force and peak force being calculated and recorded. These data were used to compare the conditions:

• Total load per transfer - all hands all carers
• Individual load per transfer - both hands per carer
• Individual hand load for conditions 1, 2 and 4

2.7   Statistical Analysis

All Shapiro-Wilks test for normality was conducted on all data to evaluate for normal distribution. This implied the use of a repeated measures ANOVA test or a Friedman test. For the total load per transfer and also individual load per transfer, a repeated measures ANOVA was used to test for significance between the average force of each condition. Independent Samples T-tests were conducted to compare the average force exerted between the novice and expert participants. For non-normal distributions

Wilcoxon Signed Rank test or Mann-Whitney U test were used. Post hoc analysis used the Bonferroni corrections.

3   Results

A total of 21 participants were used. These were split into novices (n = 10) and experts (n = 11), with each completing each transfer three times.

3.1   Subjective Review - Qualitative Data

A repeated measures ANOVA for novice carers showed Borg scores significantly differed between the conditions (F(1.94,17.43) = 10.47, p < .05) (Table 1). Post hoc tests using the Bonferroni correction revealed significant difference between conditions 2 and 3 (p = .002) and also conditions 3 and 4 (p = .004). Therefore, perceived force for condition 3 was easier than conditions 2 and 4, but not condition 1. For the experts, the ANOVA determined the scores again significantly differed statistically between conditions (F(3,30) = 14.65, p < .05). Post hoc tests using the Bonferroni correction revealed significant difference between conditions 2 and all of the other conditions (1:p = .016, 3:p = .013, 4:p = .003). However, there was no significant difference between the other conditions. The perception of the rotation task showed it as much worse. Looking at the both groups together it can be seen that condition 2 scored the highest for both (4.0 and 6.1 respectively). This would suggest all participants felt they put more effort into this transfer. Alternatively, it could suggest novices felt condition 3 was the easiest whereas experts preferred condition 4 (Table 2).

Table 1. Mean response of the borg rating scale for all participants

Condition	Novice		Expert	
	Mean	SD	Mean	SD
1. Parallel stepping	2.8	0.75	3.7	0.96
2. Rotation	4.0	0.90	6.1	2.02
3. 2-person oblique pull up bed	1.9	0.75	3.0	1.57
4. Single person pull up bed	2.8	1.16	2.7	0.68

Table 2. Subjective response for all four conditions for novice participants

Condition	Comfort		Safety	
	Mean	SD	Mean	SD
1. Parallel stepping	3.6	1.08	3.7	0.95
2. Rotation	2.1	0.57	2.5	0.71
3. 2-person oblique pull up bed	4.3	0.95	4.1	0.99
4. Single person pull up bed	3.5	0.97	3.8	0.79

Table 3. Subjective response of the four conditions for expert participants

Condition	Comfort		Safety		Individual acceptance		Organisational acceptance	
	Mean	SD	Mean	SD	Mean	SD	Mean	SD
1. Parallel stepping	2.8	0.98	3.1	0.83	2.5	1.21	3.2	1.34
2. Rotation	1.7	0.91	1.7	0.91	1.5	0.69	2.9	1.70
3. 2-person oblique pull up bed	3.6	0.93	3.6	0.92	3.2	1.08	2.9	0.94
4. Single person pull up bed	3.8	0.54	4.1	0.54	3.4	0.92	3.4	1.03

Table 3 shows that novice participants felt condition 3 was the most comfortable (4.3) and safest (4.1), whereas condition 2 was the least comfortable (2.1) and least safe (2.5).

For the expert participants it was found that condition 4 not only scored the highest in terms of comfort (3.8) and safety (4.1), but also when considering individual (3.4) and organisational acceptance (3.4). However, resembling the novice participants, condition 2 also scored the worst across the different aspects. The Spearman's Rho test (Table 4) was repeated for the expert participants, for all conditions, to find the relationship between comfort, safety, personal and organisational acceptance. Similar to the novice participant, there was very strong positive correlation between comfort and safety. Personal acceptance was also found to have strong positive correlation with comfort, safety and organisational acceptance. With moderate positive correlation being found between organisational acceptance and comfort and safety.

Table 4. Result of the Spearman's Rho test for expert participants

		Comfort	Safety	Individual	Organisational
Comfort	Correlation coefficient	1.000	.950**	.755**	.499**
	Sig. (2-tailed)	.	.000	.000	.001
Safety	Correlation coefficient	.950**	1.000	.761**	.520**
	Sig. (2-tailed)	.000	.	.000	.000
Individual	Correlation coefficient	.755**	.761**	1.000	.637**
	Sig. (2-tailed)	.000	.000	.	.000
Organisational	Correlation coefficient	.499**	.520**	.637**	1.000
	Sig. (2-tailed)	.001	.000	.000	.

3.2 Force Data

The force data was investigated as; total load per transfer, individual load and individual hands. Although the peak force would be the major cause of injury, it was not used in the analysis as, for all transfers, this force was far away from the 1.8 kN limit value set out in ISO 11228-2 (2007) and reported in Jäger et al.'s (2001) study.

3.2.1   Total Load Per Transfer

Shapiro-Wilks test found normality in all conditions (p > .05). The mean average total load per transfer differed significantly between the four conditions (F(3,27) = 24.63, p < .05). Post hoc tests using the Bonferroni correction revealed significant differences between condition 1 and 3 (p = .001) and 4 (p = .001). Also between condition 2 and 3 (p = .002) and 4 (p = .003). However there was no significant difference between condition 1 and 2 (p = .487) and also condition 3 and 4 (p = 1.000). Therefore conditions 3 and 4 were similar, but significantly lower in the total amount of force exerted per transfer than conditions 1 and 2, which were also similar. Thus suggesting it is easier to carry out transfers from the top of the bed as opposed to at the side of the bed (Table 5).

Table 5. Mean average total load per transfer

Condition	Novice		Expert		Combined	
	Mean	SD	Mean	SD	Mean	SD
1. Parallel stepping	184.5	19.2	127.1	26.4	155.8	37.3
2. Rotation	219.0	70.5	164.4	52.4	191.7	65.2
3. 2-person oblique pull up bed	84.5	18.6	79.3	5.7	81.9	13.2
4. Single person pull up bed	78.6	21.9	68.0	13.0	73.0	20.9

An Independent Samples T-test (Table 6) was conducted to compare novice and expert participants in terms of total load per transfer for each condition. It was found that the total load per transfer for condition 1 was significantly different for novice and expert participants (p < .05). For the other conditions there proved to be no difference between novice and experts (p > .05). This suggests that novice participants exert more force per transfer than experts for all of the conditions.

Table 6. Independent samples T-test for novice vs. experts for total load per transfer

Condition	t	df	Sig. (2-tailed)
1. Parallel stepping	3.929	8	.004
2. Rotation	1.388	8	.202
3. 2-person oblique pull up bed	.594	8	.569
4. Single person pull up bed	1.362	19	.189

3.2.2   Individual Load Per Transfer

Shapiro-Wilks test showed normal distribution (n = 21, p > .05). A repeated measures ANOVA showed the mean average individual force differed significantly between the conditions (F(2.21,44.21) = 27.26, p < .05). Post hoc tests revealed significant differences between condition 3 and the other conditions (1:p = .000, 2:p = .000, 4:p = .000) and condition 2 and 4 (p = .047). The two person oblique was significantly lower in force exerted, with the same being said for condition 4 and 2. An Independent-

Samples T-test compared the individual forces between novice and expert participants. For condition 1, though Levene's test was violated (F = 5.49, p = .03), there proved to be a significant difference in force for novice and expert participants (p = .001). Although all of these calculations suggest novice participants exert more force during each condition than their expert counterparts, only conditions 1 and 2 alter significantly (Tables 7 and 8).

Table 7. Mean average individual load

Condition	Novice		Expert		Combined	
	Mean	SD	Mean	SD	Mean	SD
1. Parallel stepping	92.3	12.3	61.8	21.7	76.3	23.4
2. Rotation	109.5	35.9	80.5	27.5	94.3	34.3
3. 2-person oblique pull up bed	42.3	10.8	38.3	6.1	40.2	8.7
4. Single person pull up bed	78.6	21.9	68.0	13.0	73.0	18.2

Table 8. Results from independent samples T-test for novice vs. experts for individual force per transfer

Condition	t	df	Sig. (2-tailed)
1. Parallel stepping	4.011	16.135	.001
2. Rotation	2.091	19	.050
3. 2-person oblique pull up bed	1.054	19	.305
4. Single person pull up bed	1.362	19	.189

3.2.3 Individual Hand Load

For the individual hand load data, a Wilcoxon Signed-Rank test was used to compare the hand force data. The results indicated that for conditions 1 (Z = −2.03, p = .042) and 2 (Z = −2.2, p = .028) the left hand scores were, statistically higher than the right hand scores. However for condition 3 (Z = −2.0, p = .046), right hand scores were, statistically higher than the left hand scores.

To compare novice and expert participants a Mann-Whitney U test showed that for the left hand in condition 1 the force exerted by the novice participants was significantly greater than that of the expert participants (U = 8, p = .001). However for all of the other sets of data, there was no statistically significant difference between novice and expert participants (p > .05). This suggests the level of experience does not have an effect on reducing the force exerted when considering each hand individually.

The same process was completed to compare the hand up vs. down the bed for conditions 1 and 2. The ANOVA determined that the force exerted significantly differed statistically between conditions (F(1,20) = 6.97, p < .05). However there was no statistically significant difference between the force exerted and hand positions (F(1,20) = .128, p = .724), as well as no significant interaction between the condition and hand position (F(1,20) = .186, p = .671).

A Paired Samples T-test was conducted to compare hand positions the force exerted for up vs. down the bed. This analysis suggested that the location of the hand in relation to the bed does not have an impact on the amount of force exerted during the transfer, even though in these conditions the force exerted by the bottom hand was slightly greater than that exerted by the hand up the bed.

4   Discussion

The rotation movement was clearly perceived to be the transfer that required the greatest amount of force to be exerted. It was also understood to be the least comfortable and safe, which was due to the fact that it caused the participant to twist their body. These findings are backed up by various regulations and guidelines which state the dangers of twisting the body when handling a load/patient. There was a very strong correlation between comfort and safety. Meaning the more comfortable the participant felt completing the transfer, the safer they felt the transfer was. With patients feeling safer and more comfortable if the transfer used a safe technique according to a work technique score. The expert participants recorded a positive correlation between individual and organisational acceptance, however further research would have to be conducted to explore the cause of this correlation.

There was a large variation in the recorded data, which was due to individual differences between participants. Other similar studies quantifying the force of patient handling tasks have similarly reported variations and inconsistencies due to the natural form of humans. Even with these differences there was differences between each condition and the novice and expert participants. Physically there were interesting findings between the conditions: when comparing the total load per transfer the transfers with the participants at the head of the bed showed lower forces; similarly for individual load per transfer it was found that condition 3 was lower than the other conditions. This suggests that less force is needed in transfers where the patient is pulled towards the carer. This statement is backed up by the findings of McGill and Kavcic (2005); unsurprisingly, the use of two carers reduced the force needed to complete the transfer per individual.

The results from individual hands were less convincing. When looking at individual hand loads and comparing the force exerted by the left and right hands, there were some differences between the different positions. The data suggested that when participants are at the side of the bed they exerted more force using the left hand/arm when compared to the right hand/arm. However, when at the top of the bed the right hand/arm, was more dominant. This could suggest that for 'pulling' transfers the dominant hand contributes more, however more research would have to be conducted to confirm this.

4.1   Concluding Remarks

It was found that novice participants exerted more force during the transfer than their expert counterparts, however only some of these factors proved to be significant. This suggests that the level of experience and training could be related to the amount of muscle activity used in the transfer (Keir and MacDonell 2004). Nevertheless, as only some were significant it suggests training does not influence force exertion resulting in musculoskeletal problems (Johnsson et al. 2002). From the results it was found that, although none of the conditions posed an immediate threat to the carers wellbeing, there was a statistically significant reduction in the amount of force exerted for the two conditions at the head of the bed when compared to the two at the side. In particular the rotation task was significantly harder and perceived as such by the participants and should be avoided in transfers tasks. When taking these results into the healthcare environment, it would suggest that, where possible, carers partake in transfer from the head of the bed, using their own weight as an aid to pull the patient up the bed. Doing so would help reduce the risk of developing musculoskeletal injuries.

References

Alamgir H, Cvitkovich Y, Yu S, Yassi A (2007) Work-related injury among direct care occupations in British Columbia, Canada. Occup Environ Med 64(11):769–775

American Nurses Association (ANA) (2013) Safe patient handling and mobility, 1st edn. American Nurses Association, Silver Spring. http://www.nursesbooks.org/ebooks/download/SPHM-Standards.pdf

Black T, Shah S, Busch A et al (2011) Effect of transfer, lifting and repositioning (TLR) injury prevention program on musculoskeletal injury among direct care workers. J Occup Environ Hyg 8(4):226–235

Brooks A, Orchard S (2011) Core person handling skills. In: Smith J (ed) The guide to the handling of people: a systems approach, 6th edn (see moving up the bed). BackCare, Middlesex, pp 139–142

Cohen MH, Nelson GG, Green DA et al (2010) Patient handling and movement assessments: a white paper. The Facility Guidelines Institute, Dallas

Fray MJ, Hignett S (2015) An evaluation of the biomechanical risks for a range of methods to raise a patient from supine lying to sitting in a hospital bed. In: 19th triennial congress of the IEA, Australia, 9–14 August 2015

Garg A, Kapellusch JM (2012) Long-term efficacy of an ergonomics program that includes patient-handling devices on reducing musculoskeletal injuries to nursing personnel. Hum Factors 54(4):608–625

Garg A, Owen B, Beller D et al (1991) A biomechanical and ergonomic evaluation of patient transferring tasks: bed to wheelchair and wheelchair to bed. Ergonomics 34:289–312

ISO 11228-2 (2007) Ergonomics-manual handling–part 2: pushing and pulling. International Organization for Standardization, Geneva

ISO (2012) Ergonomics-manual handling of people in the healthcare sector: an edited summary of ISO Technical Report 12296

Jäger M, Jordan C, Theilmeirer A, Wortmann N, Kuhn S, Nienhaus A, Luttmann A (2012) Lumbar-load analysis of manual patient-handling activities for biomechanical overload prevention among healthcare workers. Ann Occup Hyg 57(5):528–544

Johnsson C, Carlsson R, Lagerström M (2002) Evaluation of training in patient handling and moving skills among hospital and home care personnel. Ergonomics 45(12):850–865

Jordan C, Luttmann A, Theilmeier A, Kuhn S, Wortmann N, Jäger M (2011) Characteristic values of the lumbar load of manual patient handling for the application in workers' compensation procedures. J Occup Med Toxicol 6(1):17. https://occup-med.biomedcentral.com/articles/10.1186/1745-6673-6-17

Keir P, MacDonell C (2004) Muscle activity during patient transfers: a preliminary study on the influence of lift assists and experience. Ergonomics 47(3):296–306

Lim H, Black T, Shah S et al (2011) Evaluating repeated patient handling injuries following the implementation of a multi-factor ergonomic intervention program among health care workers. J Saf Res 42:185–191

McGill S, Kavcic N (2005) Transfer of the horizontal patient: The effect of a friction reducing assistive device on low back mechanics. Ergonomics 48(8):915–929

Nelson A, Lloyd JD, Menzel N, Gross C (2003) Preventing nursing back injuries: redesigning patient handling task. AACHN J 51(3):126–134

NHS (2010) Guidance for the moving and handling of patients and inanimate loads (version 3). http://www.leicestercity.nhs.uk/library/hs010guidanceformovingandhandlingv3d2final.pdf

Owen BD, Staehler KS (2003) Decreasing back stress in home care. Home Healthc Nurse 21(3):180–186

Pellino TA, Owen B, Knapp L, Noack J (2006) The evaluation of mechanical devices for lateral transfer on perceived exertion and patient comfort. Orthop Nurs 25(1):4–10

Schibye B, Hansen AF, Hye-Knudsen CT, Essendrop M, Bocher M, Skotte J (2003) Biomechanical analysis of the effect of changing patient-handling technique. Appl Ergon 34(2):115–123

Smith J (2005) The guide to the handling of people, 5th edn. BackCare, Teddington

Smith J (ed) (2011) The guide to the handling of people: a systems approach, 6th edn. BackCare, Middlesex

Warming S, Precht D, Suadicani P, Ebbehoj N (2009) Musculoskeletal complaints among nurses related to patient handling tasks and psychosocial factors—based on logbook registrations. Appl Ergon 40:569–576

Waters TR (2007) When is it safe to manually lift a patient? Am J Nurs 107:53–58. (quiz 59)

Interpretation of Eye Tracking Findings in Usability Evaluation

Lin Wang$^{(\boxtimes)}$

U.S. Census Bureau, Washington, DC 20233, USA
lin.wang@census.gov

Abstract. The eye tracking technique has been widely used in usability evaluation of screen-based user interfaces. Eye tracking collects user's visual scanning behaviors during task performance (e.g., shopping holiday gifts on an online store). While eye tracking data contain rich information about user's performance, meaningful use of eye tracking requires accurate recording of eye movements which has been a challenge to usability practitioners. In the present paper, we will (1) analyze the structure of a saccadic eye movement and its association with cognitive processes, (2) describe today's most commonly used eye tracking technique, and (3) propose a systematic approach to improving spatial accuracy. The approach includes 6 steps: (1) test design, (2) participant's seating, (3) test instructions, (4) manufacturer-provided calibration, (5) customer-tailored calibration, and (6) statistical adjustment to calibration.

Keywords: Eye tracking · Usability testing · Spatial accuracy

1 Introduction

The eye tracking technique has been widely used in usability evaluation of screen-based user interfaces. Eye tracking collects user's visual scanning behaviors during task performance (e.g., shopping holiday gifts on an online store). While eye tracking data contain rich information about user's performance, meaningful use of eye tracking requires accurate recording of eye movements which has been a challenge to usability practitioners. In the present paper, we will (1) analyze the structure of a saccadic eye movement (saccade) and its association with cognitive processes, (2) describe today's most commonly used eye tracking technique, and (3) propose a systematic approach to improving spatial accuracy.

2 Role of Saccades in Visual Scanning

When analyzing eye tracking data in usability studies, we are mostly dealing with saccades and fixations. Saccades are quick and jerky eye movements that are one of the five types of eye movements (smooth pursuit, saccadic, vestibular, optokinetic, and vergence) in the human oculomotor system which is part of the central nervous system. A fixation (aka, fixation point or a gaze) is a point on the visual field at which the fovea vision is directed. One of the primary functions of the oculomotor system is to direct

© Springer Nature Switzerland AG 2019
S. Bagnara et al. (Eds.): IEA 2018, AISC 822, pp. 641–647, 2019.
https://doi.org/10.1007/978-3-319-96077-7_69

the gaze to a target of interest. When a user interacts with a screen interface, visual scanning is primarily accomplished by saccadic eye movements in conjunction with head movements (depending on the span of a gaze shift). Saccades thus play a crucial role in visual information acquisition and processing.

There are two types of saccadic eye movements, reflexive and voluntary. A reflexive saccade is elicited by a sudden occurrence of a visual object in the peripheral visual field; while a voluntary saccade is made endogenously without the elicitation of a peripheral stimulus. Like a ballistic missile, the production of a saccade undergoes three phases: launching, flying, and landing, as depicted in Fig. 1. Each phase can be characterized by one or several parameters. Launching is characterized with latency (the time taken to prepare a saccade) which can be measured, in a reflexive saccade paradigm, as the duration between the onset of the peripheral stimulus and onset of a saccadic eye movement. Flying is characterized with amplitude (the degrees of visual angle through which the eye ball rotates) and peak velocity (the highest speed at which the eye ball rotates). Landing is characterized with fixation accuracy (the visual angle between the center of the visual target and the fixation point). In task performance with a screen-based interface, usually, the operator is continuously making voluntary saccades to search and process information. Fixation accuracy is particularly important.

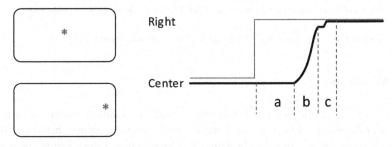

Fig. 1. A graphical sketch of a reflexive saccadic eye movement. The two panels to the left represent a paradigm of a visual target (asterisk) displacement from the center of the visual display (upper panel) to the right side of the display (lower panel). The graphic to the right shows the position change of the visual target from center to right (thin red line) and gaze position of the tracking eye (thick black line). The dotted lines delineates the time course of a saccade's launching phase (a), flying phase (b), and landing phase (c). In a large saccade, a second small corrective saccade may be made to adjust fixation point after initial landing. (Color figure online)

The interest in applying eye tracking to usability assessment stems from the evidence that voluntary saccades are influenced by high-level cognitive processes. In his seminal work, Yarbus [1] showed that individual's visual scan pattern is modulated by viewer's attention allocation. Yarbus' findings laid the foundation for using eye tracking to investigate high-level cognitive processes. There is a large body of theoretical and applied research on visual scanning and usability [e.g., 2–10].

3   Pupil-Center/Corneal-Reflection Recording Technique

With the advance in technology, eye tracking equipment is becoming better, cheaper, and easier to use. The pupil-center/corneal-reflection technique is at present the most popular method of recording eye movements. In this method, the operator, who is interacting with a screen interface, faces a video camera and an infrared light source. The infrared light impinges onto the operator's eye while the camera is capturing the pupil image and a glint (highlight) on the corneal. When the eyeball rotates, the glint stays relatively stationary while the pupil center location changes with the eyeball rotation. A vector between the glint and the center of the pupil can be calculated, and the vector changes as a function of eyeball rotation. The change of the vector is thus translated to the change of the fixation point on the screen (the visual field), and consequently, the gaze is tracked. The diagram in Fig. 2 demonstrates this translation.

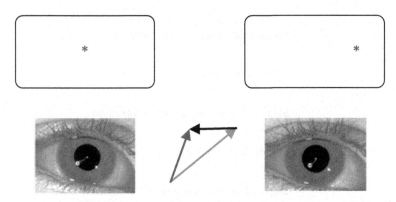

Fig. 2. The top two panels show a visual target (asterisks) moving from the center of the display (left panel) to the right (right panel). Each eye image shows the pupil center (small whit dot at the center of the black disk) and corneal glint (larger white dot near the edge of the black disk) when the eye is looking at the visual target in the above panel. The black disk is the reflection from the pupil. The red line and blue line represent the vectors of fixation at the center (red line) and to the right (blue line), respectively. The graphic between the two eye images depicts the result (black line) of the addition of the center fixation vector (red line) and right fixation vector (blue line). The arrow represents the direction of the vector. The black arrow points to the opposite direction of the target movement because the eye images are horizontally rotated 180 degree. The magnitude of the vectors are not proportional for the ease of demonstration. The eye images were captured from Google Images search [11]. (Color figure online)

A sound use of eye tracking technology in usability assessment is based on accurate recording of eye movements during task performance. One must notice in Fig. 2 that, accurate tracking of fixation point depends on accurate measurement of the vector between the corneal glint and pupil center across the entire screen display.

4 Systematic Approach to Improving Spatial Accuracy

A major challenge in the application of eye tracking to usability testing is to ensure spatial accuracy of fixation point. This challenge is particularly pronounced in relatively long sessions of usability testing, e.g., a 20-min session. In a long testing session, the test participant is likely to move their head and/or body, which may compromise the established spatial relationship between the eye locations and the screen interface. To mitigate this problem, we propose a systematic and practical approach. The approach includes (1) test design, (2) participant's seating, (3) test instructions, (4) manufacturer-provided calibration, (5) customer-tailored calibration, and (6) statistical adjustment to calibration. Figure 3 shows the workflow of this approach.

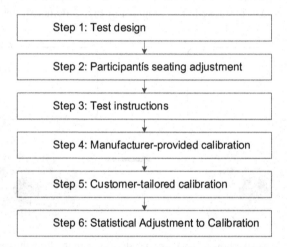

Fig. 3. The implementation workflow of the systematic approach to improving spatial accuracy.

4.1 Test Design

In designing a usability test using eye tracking, one should take into account of the impact of test duration on eye tracking accuracy. Prolonged testing increases the likelihood of a change in the subject's sitting posture, resulting in inaccurate measurement of glint-pupil-center vector. We would suggest a 5-min recording period as a rule of thumb for test design. If the test requirements call for a long test session, one shall make effort to divide a large task into a series of small tasks and calibration trials between the tasks.

4.2 Participant's Seating

This is a seemingly trivial, but critically important matter. A subject in an uncomfortable sitting posture will likely adjust his/her posture during the test, resulting in inaccurate recording of eye positions and data loss. Among various posture changes, head movement is a serious threat to spatial accuracy and data lost though the current eye trackers are able to tolerate relatively large head movement. Our method to reduce subject's tendency of moving the head and body around during the test is, before the commencement of the test, to ask the participant to sit comfortably at the beginning of a session. Subjects usually will make more than one adjustment after being instructed until they find a comfortable position.

4.3   Test Instructions

To reinforce subject's awareness of the importance of keeping the head stable during testing, the test administrator asks subjects to avoid moving their head as long as they feel comfortable. In the meantime, the instructions should be clear that subjects are not prohibited from moving their head. Otherwise, the subjects' visual scanning behaviors will not reflect their true performance.

4.4   Manufacturer-Provided Calibration

This is a critical step for spatial accuracy. The calibration establishes a one-to-one mapping from the glint-pupil-center vector to a fixation point in the visual field. The general procedure goes through these steps: (1) The eye tracker camera first captures the subject's eye(s); (2) the eye tracker software presents a dot on the interface screen, and the subject is instructed to look (fixate) at the dot; (3) the dot (seemingly) randomly moves around on screen, and the subject's eyes follow the dot until it disappears. In order to get a good mapping between eye position and fixation point on the screen, the dot must travel through the center region of the screen as well as along the border. The calibration procedure usually samples a few data points at each dot location, and takes the centroid of the gaze cluster as the fixation point. The calibration quality is determined by the spread of the cluster. The smaller the spread, the better quality of calibration. Calibration quality reflects the subject's eye tracking behavior. One shortcoming with manufacturer-provided calibration is that it is difficult, if not impossible, to re-assess calibration during or after test session.

4.5   Customer-Tailored Calibration

To address manufacturer-provided calibration's inability to reassess eye-scene mapping, we developed a customer-tailored calibration paradigm. In this method, the researcher designs a display of words placed at different places on the screen. Preceding the formal usability test session, the subject is instructed to read aloud the words on the screen, and subject's gaze is recorded. This process is repeated again following the usability test session. The two recordings are compared. If there are significant discrepancies between the two recordings, a calibration factor will be calculated and applied to the eye tracking data set of the usability testing session (will be further discussed in the following section). This method provides an opportunity for post-hoc statistical adjustment to the eye tracking data. Figure 4 shows a sketch of calibration display superimposed with fixations and scan paths.

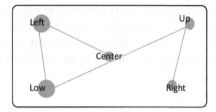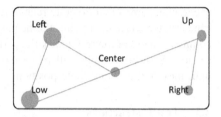

Fig. 4. The customer-tailored calibration paradigm. The left panel shows the gaze plot over the words display during pre-session calibration. The right panel shows the gaze plot over the words display during post-session calibration. A red dot represents the fixation point when the subject is reading aloud the word. The size of a red dot depicts relative (comparing other dots) fixation duration. The right panel shows a shift of gaze pattern from the left panel, indicating that the subject's head position has been changed during the usability test session. (Color figure online)

4.6   Statistical Adjustment to Calibration

After the usability test session is over, we use the customer-tailored calibration data to calculate the estimate of spatial discrepancy between pre- and post-test calibrations, and then apply the results to the eye tracking data. The following are the procedure of this statistical adjustment to calibration.

1. Calculate the differences between pre- and post-test calibrations for each fixation point in horizontal (x) and vertical (y) dimensions, respectively.
2. Eliminate the two extreme fixation points which have smallest and largest distance from the corresponding word on the screen.
3. Calculate the mean of the differences between pre- and post-test calibrations for each remaining fixation point in horizontal (x) and vertical (y) dimensions, respectively.
4. Apply the mean differences between pre- and post-test calibrations in x and y dimensions, respectively, to the entire eye tracking dataset for the test session.

5   Conclusion

The present paper discussed an important and practical matter in usability application of eye tracking technique – spatial accuracy. To address this issue, one needs to take into account of three factors: Equipment, subject, and testing design. A systematic approach is proposed.

References

1. Yarbus A (1967) Eye movements and vision. Plenum, New York
2. Li CL, Aivar MP, Tong MH, Hayhoe MM (2018) Memory shapes visual search strategies in large-scale environments. Sci Rep 8:4324
3. Rogers SL, Speelman CP, Guidetti O, Longmuir M (2018) Using dual eye tracking to uncover personal gaze patterns during social interaction. Sci Rep 8:4271

4. Ranzini M, Lisi M, Zorzi M (2016) Voluntary eye movements direct attention on the mental number space. Psychol Res 80(3):389–398
5. Hafed ZM, Krauzlis RJ (2006) Ongoing eye movements constrain visual perception. Nat Neurosci 9:1449–1457
6. Martinez-Conde S, Macknik SL, Hubel DH (2004) The role of fixational eye movements in visual perception. Nat Rev Neurosci 5:229–240
7. Schiessl M, Duda S, Thölke A, Fischer R (2003) Eye tracking and its application in usability and media research. MMI Interaktiv
8. Duchowski AT (2003) A breadth-first survey of eye-tracking applications. Behav Res Methods Instrum Comput 34(4):455–470
9. Wang L, Stern JA (2001) Saccade initiation and accuracy in gaze shifts are affected by visual stimulus significance. Psychophysiology 38:64–75
10. Rayner K (1998) Eye movements in reading and information processing: 20 years of research. Behav Res Methods Instrum Comput Psychol Bullet 124:372–422
11. Google Images Homepage. https://www.google.com/search?tbm=isch&sa=1&ei=xsj4 WsKDPK-Rgg-fi5rX4Dg&q=pupil+center+corneal+reflection&oq=pupil+center+&gs_l=img. 1.0.0j0i8i30k1l4j0i24k1l2.633037.637451.0.640232.13.13.0.0.0.0.239.1378.10j2j1.13.0....0... 1c.1.64.img..0.13.1334...0i67k1j0i10k1j0i5i30k1.0.zw5hz2lesvw#imgrc=JVIY_FUbnBf1oM: &spf=1526254408070. Accessed 15 May 2018

Ring the Alarm - Identification and Evaluation of Users' Requirements for a V2X-Smartphone App

Teresa Brell$^{(\boxtimes)}$, Ralf Philipsen, and Martina Ziefle

Human-Computer Interaction Center,
RWTH Aachen University, Aachen, Germany
brell@comm.rwth-aachen.de
http://www.comm.rwth-aachen.de

Abstract. The fast developing and intertwining fields of mobility and technology focus human-centered designs more than ever. V2X-technology is one of the major players on that research field, which will influence the people's behavior in traffic in terms of safety and efficiency. To include all participating members of traffic, four important message features for a V2X-smartphone application are analyzed. Using a two-tiered research approach, we focused on possible information scenarios in traffic, the most important features (type of data, correctness of warning, battery consumption and data recipient) and a practical recommendation for interaction from a users' perspective.

Keywords: V2X-technology · Smartphone application · Mobility
Human-centered design

1 Introduction

Today's vast technological developments in the mobility and communication sector are widely observed and focused in media. A way to travel autonomously will be reality by 2050 and all communicational devices will enhance a new level of connection. With these opportunities, future challenges of transport due to the increasing world population can be addressed today. Alongside the urbanization comes an increasing proportion of the elderly living in cities [1], which will necessitate novel mobility infrastructure solutions, in which a diverse population can travel safe, clean and efficient to ensure a sustainable care-taking. Safety and safety in traffic are still two of the most important and discussed issues [2], which need to be addressed in forthcoming transport technologies. Indisputably, today's numbers of traffic accident fatalities are decreasing, but safety in traffic overall is still one of the biggest challenges to address [3] - also from a user's perspective. The technological development of driver assistance systems is focused worldwide and connecting infrastructure and vehicles opens new possibilities to observe and intervene in critical situations. With these technological possibilities of connecting not only cars and infrastructure, but also (smart) watches and smartphones, vulnerable road user (VRU) can be integrated into the technological

© Springer Nature Switzerland AG 2019
S. Bagnara et al. (Eds.): IEA 2018, AISC 822, pp. 648–657, 2019.
https://doi.org/10.1007/978-3-319-96077-7_70

development. A promising way to include road users like pedestrians, bicyclists and wheel-chair user is via smartphone app [4]. With a two-tiered research approach, we focused on users' requirements for and modalities of a V2X-smartphone App in order to understand their interaction behavior and needs.

2 Using Information and Communication Technology to Protect VRUs

The causes of accidents of VRU are largely linked to human error. Essentially, the underlying erroneous human decisions are based on distraction, a lack of information about the traffic situation or a misinterpretation of the available information [5]. In addition to conventional measures, such as safety education, speed limits or mitigation of accident focal points through construction measures [6], enlarging the information base available to VRU in a suitable format with as little additional distraction as possible could therefore reduce the accident rate.

Modern network and sensor technology offers the possibility to share the knowledge of different road users in order to involve all of them in accident prevention in a cooperative approach [7, 8]. While the networking between vehicles and infrastructure is progressing, pedestrians and cyclists in particular have so far been largely excluded from this development due to the lack of sensors and digital communication channels. The latter has changed with the widespread use of smartphones to the extent that in principle the position of each smartphone user could be determined and communicated. In such a car2pedestrian system, both a car and its driver as well as the VRU could be warned of an imminent collision by exchanging information [9]. Liebner, Klanner and Stiller were able to show as early as 2012 that the GPS information from smartphones can basically be used for security applications for VRU, even if the precision of location recognition was still relatively inaccurate at the time [10]. A concrete implementation of the technology is e.g. *WiFiHonk*, which allows the exchange of information about position, speed and direction of movement of different road users and can theoretically reduce the probability for different collision scenarios with VRU even at high vehicle speeds [11]. Further, the driver can also be informed about a critical, conflicting movement of a VRU [12, 13]. However, an evaluation of the visualization, the reaction times of car drivers or the VRU and the potential distraction by the system is still pending. First steps towards this can be found at Schmidt et al., who developed and evaluated an icon and map design for traffic and warning displays in VRU smartphone safety applications [4]. The above-mentioned approaches concentrate primarily on the technical aspects and prove a general feasibility. However, it is currently still unclear, which requirements users have regarding this type of smartphone-based safety systems, e.g. regarding data transfer and data use by third parties, and how these affect the willingness to use the technology.

3   Methodology

The following research objectives were addressed due the current state of research:

- Identification of realistic traffic scenarios for information distribution
- Identification of most important V2X-App features
- Assessment of the importance of App features for the decision to use it

As can be seen from Fig. 1, the methodological approach was two-tiered: Possible realistic traffic scenarios for a V2X-smartphone App were identified. Second, the most important features for the application (from a human-centered perspective) were inquired. Further, the derived application features were quantified to derive practical recommendations for possible interaction modalities.

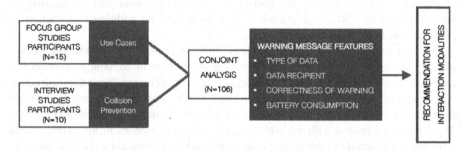

Fig. 1. Methodological concept of research model.

3.1   Qualitative Preliminary Studies

To address the research objectives, we conducted focus group studies ($N = 15$) for the mentioned situation-based distribution of information in traffic as well as task-based interview studies ($N = 10$) to evaluate which possible App-features are preferred for collision prevention on a smartphone. Both preliminary studies lasted between 30 and 50 min. The audio recordings were fully transcribed and analyzed, using qualitative content analysis.

3.2   Quantitative Study

Taking all results into account, a quantitative online study ($N = 106$) focused on the interaction frequency and identified important features of the V2X-App. Through a conjoint analysis, it was possible to weigh in on which type of data would be shared (position, demographical data) with whom (data receiver), how correct messages need to be (correctness of warning) and how strong the influence on the used smart device may be (battery consumption).

Demographics, Digital Profile and Traffic Scenarios. The quantitative study consisted of five parts. First, demographic data was assessed, e.g., age, gender, technical self-efficacy [14]. Second, respondents had to evaluate their mobility and digital profile.

In a third step, the identified attributes of the smartphone App and their levels were introduced. Afterwards, the scenarios were presented and participants were to imagine the following: They were walking along a road and want to cross it. There's no traffic light nearby. The only way to cross the road is to walk between some parked cars on the other side. This makes it difficult to see for passing vehicles and also limits the visibility. A truck approaches their position from the left.

- The **first** scenario addressed the participant as primary recipient of the information (warning of others/hazardous situation). An App on the participants smartphone receives a warning of the approaching truck and the participant can react accordingly.
- The **second** scenario addressed all other road users as primary recipients of the information (warning for others/hazardous situation). An App on the participants smartphone sends out a warning to the approaching truck and the truck can react accordingly.

Then, participants should select the scenario that meets their needs for safety and privacy best and most. In the fourth part, the choice-based-conjoint (CBC) tasks with four attributes and four levels each (see next section) were presented. The CA simulates various decision processes as well as fragmentations of scenario preferences into separate part-worth utilities of attributes and their levels are enabled. Aiming the respondent evaluation of specific scenario-based features or configurations which consist of multiple attributes and differ from each other in the attribute levels. Each respondent rated 12 random tasks. A test of design efficiency confirmed that the reduced test design was comparable to the hypothetical orthogonal design (median efficiency of 99%).

Attributes and Levels. Out of former qualitative studies, we identified the following relevant App-features, which may influence the willingness to use a V2X-smartphone App:

Battery Consumption: The attribute battery consumption describes how much higher the phones energy (battery) consumption (in %) will be when the application is used. Here, the attribute contains the levels 5%, 25%, 50% and 100% (see Fig. 2).

Data Recipient: The attribute recipient describes different variations of who may collect the data. It contains the levels (1) other road user (involved in the situation), (2) infrastructure (e.g. traffic lights), (3) police and rescue service, (4) service provider (commercial). The previous level is included in the next level.

Correctness of Warning: The attribute correctness of warning describes the error messages in the application, i.e., how often incorrect data and warnings are issued. Based on the desired error-free operation of the app, the attribute was contrasted from 5% to more than 50% erroneous information. It contains four levels altogether.

Type of Data: The attribute type of data describes different ranges of data packages delivered to the receiver. It contains the levels (1) position/velocity/movement direction, (2) planned route, (3) personal data (demographic), (4) physiological data (reaction time, mood) of driver. Here again, the previous level is included in the next level.

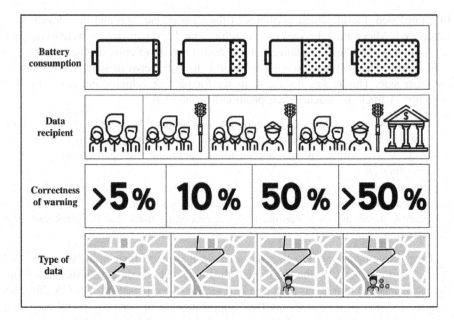

Fig. 2. Visual display of attribution levels

The last section of the questionnaire addressed the perceived usefulness and the willingness to pay for the application (both scales: 6-point Likert scale, 6 = full agreement).

3.3   Data Acquisition and Analysis

The study was designed as an online questionnaire. The participants were acquired in the university environment and in thematic forums. The quality of the responses was checked with regard to processing time and contradictory response behavior. The resulting data were analyzed by descriptive analysis and, with respect to the effects of user diversity, by uni- and multivariate analyses of variance ((M)ANOVA). The relative importance of attributes and part-worth utilities were computed on the basis of Hierarchical Bayes estimation. The level of significance was set to $\alpha = .05$.

4   Sample

In total 106 participants took part with an age range of 17 to 78 years (*Mean (M)* = 35.2; *Standard Deviation (SD)* = 16.0). The gender distribution is symmetrical with 58 men (54,7%) and 45 women (42,5%). Most participants hold a driving license (94,3%) and possess a smartphone (89,6%). The sample contains 32,1% with a university degree or higher (n = 34), followed by 41,5% with a high school diploma (n = 44) and 25,5% have a secondary school certificate (n = 27). All participants reported a rather high technical self-confidence with 4.58/6 (*SD* = 1.1). Here, men are

Table 1. Overview of the user characteristics of the groups identified within the two scenarios by latent class analysis.

Group		Scenario 1		Scenario 2	
		1	2	1	2
N		49	57	49	57
Gender	Male	59.6% (28)	53.6% (30)	57.4% (27)	55.4% (31)
	Female	40.4% (19)	46.4% (26)	42.6% (20)	44.6% (25)
Age		35.6	34.9	34.1	36.2
Technical self-efficacy		4.5	4.5	4.5	4.5
Education	University	20.4% (10)	42.1% (24)	22.4% (10)	40.3% (23)
	High school	26.5% (13)	19.3% (11)	26.5% (13)	19.3% (11)
	Sec. school	28.6% (14)	22.8% (13)	26.6% (13)	24.6% (14)
Drivers License		93.9%	94.7%	93.9%	94.7%

significantly more technical affine ($M = 5.0$; $SD = 1.0$) than women ($M = 4.1$; $SD = 1.1$) ($t(101)= -4.569$, $p < .001$). For further research, users were classified into subgroups with differing preferences and estimated part worths for each segment, called latent class estimation (Table 1).

5   Results

The following section displays the perceived usefulness and willingness to pay an extra for the application in both introduced information distribution traffic scenarios. Afterwards, the importance of the individual attributes and levels for decision making are addressed. Finally, the results of the latent class estimation groups are compared.

5.1   General Evaluation and Relative Importances of Attributes

The general evaluation focuses on the perceived usefulness and actual willingness to use the App. Most of the participants (44.3%, n = 47) stated, that they would use the App to be warned and give a warning to other traffic participants (in both information directions). On the contrary, 40.6% (n = 43) would refuse to use an App of this kind.

Looking closer at both traffic scenarios, the overall picture shows that the participants would mostly agree on using a smartphone App to receive a warning of others (1.: $M = 3.51/SD = 1.53$) or as a warning for others (2.: $M = 3.64/SD = 1.59$). 55.7% (n = 59) of the participants agreed, that the app would be useful in both traffic situations, followed by 24.5% stating, that the app is not useful at all. Hence, the willingness to pay for the app was questioned and slightly rejected (1. warning of others: $M = 2.45/SD = 1.31$, 2. warning for others: $M = 2.54/SD = 1.36$).

Average importance scores were calculated, to quantify how all attributes contribute to the participants' decision making. The battery consumption was actually the most important factor in the decision of the participants, as can be seen in Fig. 3. It scores a relative importance of 27.76%. The correctness of warning was the second

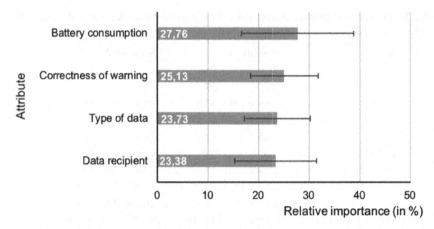

Fig. 3. Relative importances and standard deviations of attributes for decision making.

most important criterion in the decision process. Here, the relative importance score was 25.13%. The type of data and the data recipient are the two least important factors, although the importances show no significant distance between one another.

5.2   Latent Class Estimation

This section displays the comparison between the two beforehand introduced scenarios with focus on both estimation groups.

Scenario 1. As can be seen in Fig. 4 the estimation group scores for the first traffic scenario (warning of others) differ between the two groups. The average importance scores for Group 1 (black bars in Fig. 4 show that battery consumption is the most important factor (with 33.02%), followed by correctness of warning. The type of data is

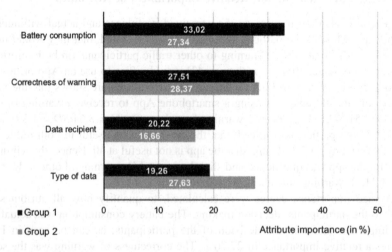

Fig. 4. Relative importances clustered with latent class estimation for scenario 1.

here the least important factor. As for Group 2 (grey bars), the correctness of warning is also the most important factor (with 28.37%), followed by the type of data (27.63%). Here, the least important factor is data recipient. Both groups differ significantly with regard to relative importances ($F(3,102) = 7.757, p < .001$). This main effect is mainly due to the different perceived importance of the data type for the two groups ($p < .001$).

Scenario 2. The second traffic scenario showed slightly different estimation group scores for Group 1 as in the first scenario. Again, battery consumption is the most important factor (see Fig. 5) with 33.52%, but data recipient was the second most important criterion in this decision process, with 22.68%, followed by correctness of warning. The results are again different for Group 2. As can be seen in Fig. 5 correctness of warning scores 29.84% and is the most important factor for this group in this scenario.

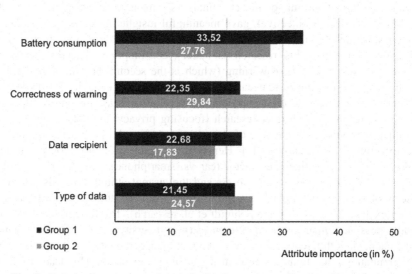

Fig. 5. Relative importances clustered with latent class estimation for scenario 2.

Further, battery consumption was the second most important factor (27.76%) in the decision of the participants, followed by type of data. Similar to scenario 1, there was a main effect of group affiliation on the relative importances (F(3,102) = 14.935, p < .001). However, this time there were significant differences between the groups regarding the relative importance of the data type (p = .003), the data users (p = .005) and the error messages (p = .031).

6   Discussion and Conclusion

The two-tiered research approach was driven from the fact, that lack of information about the traffic situation and distraction are two of the underlying erroneous human decisions, which lead to accidents with VRUs [5]. With the idea of a V2X-Smartphone App, which provides the necessary information to reduce road accidents and to integrate VRUs, we have set ourselves the goal of increasing road safety.

On basis of our former work, in which we evaluated the warning message type and needed information of and in the App [4], we set direction to the identification and evaluation of a realistic traffic scenario to distribute needed information. This first research objective was addressed through qualitative preliminary studies (focus groups discussions). The result (i.e., traffic scenario) was further used to evaluate the parallel identified App features (second research objective). The four features (attributes) were battery consumption, correctness of warning, data recipient and type of data. The need and appropriateness of the provision of information was generally recognized - as can be seen in the results. Focusing on different possible scenarios to distribute the information (i.e., warning for/of others) no significant differences in the participants answers could be identified. This hints a clear signal: whether I am the only one receiving or also sending information about the traffic situation is not - at least for this sample - a crucial turning point. The final assessment of the importance of these features (last research objective), gave meaningful insights into the users wishes and needs. Nevertheless, addressing the App features, it is surprising, that the most anticipated feature in this research is battery consumption. Thereby being more important than correctness of warning (which is the second most important feature), which seem to mirror the current state of distraction in traffic behavior [5]. Type of data and data recipient are the least important features, also an astonishing result. It contradicts the results from former research (focusing privacy and data security), which indicated that most of all the type of shared information plays a crucial role for using and accepting V2X-technology [2]. Hence, the difference of the information device could hereby be an influencing factor (car vs. smartphone). The participants of this study were almost all smartphone users and the current attitude towards information policies of smartphone applications might differ from the attitude towards car information exchange (as used in the Schmidt et al. research). The further analyzes aim to identify possible human-centered differences (user diversity) for a better understanding of the latent class differences. However, at a first glance, the latent class results point at a differentiating educational level as distinguishing user-factor. This leads to the presumption, that other user characteristics are not yet identified as explanatory influences. Future research will therefore look at data security profiles, risk-taking traffic behavior and cultural differences in information and communication. To conclude, the most wanted modality and therefore final recommendation for the App is a combined and customizable feature for battery consumption (possibility of accu-friendly setting). The differentiating importance of all identified features (and their specification) showed a not yet identified user diversity, which needs to be focused in future research.

Acknowledgements. We would like to thank the research group on mobility at RWTH Aachen University. This research was supported partly by the Excellence Initiative of German State and Federal Government (Project CERM - Center for European Research on Mobility) and partly by the German Ministry of Research and Education (Project I2EASE: reference number 16EMO0142 K). Our gratitude goes also to Manuel Scherberich, Chen Tao, Susanne Gohr, and Imke Haverkämper for their valuable research input.

References

1. United Nations (2017) Department of Economic and Social Affairs: Population division: world urbanization prospects, the 2017 revision: highlights
2. Schmidt T, Philipsen R, Themann P, Ziefle M (2016) Public perception of v2x-technology-evaluation of general advantages, disadvantages and reasons for data sharing with connected vehicles. In: 2016 IEEE intelligent vehicles symposium (IV). IEEE, pp 1344–1349
3. Statistisches B (2017) Mehr Unfälle, aber weniger Verkehrstote als jemals zuvor (more accidents, but less fatalities). Pressemitteilung 6, Juli 2017, p 230/17
4. Schmidt T, Philipsen R, Dzafic D, Ziefle M (2017) Watch out! In: International conference on digital human modeling and applications in health, safety, ergonomics and risk management. Springer, pp 365–383
5. Otte D, Jänsch M, Haasper C (2012) Injury protection and accident causation parameters for vulnerable road users based on german in-depth accident study gidas. Accid Anal Prev 44(1):149–153
6. Constant A, Lagarde E (2010) Protecting vulnerable road users from injury. PLoS Med 7(3): e1000228
7. Andreone L, Visintainer F, Wanielik G (2007) Vulnerable road users thoroughly addressed in accident prevention: the watch-over european project. In: 14th world congress on intelligent transport systems
8. Scholliers J, Bell D, Morris A, García AB (2014) Improving safety and mobility of vulnerable road users through its applications. Transp Res Arena 14–17
9. Engel S, Katzsch C, David K (2013) Car2pedestrian-communication: protection of vulnerable road users using smartphones. In: Advanced microsystems for automotive applications 2013. Springer, pp 31–41
10. Liebner M, Klanner F, Stiller C (2013) Active safety for vulnerable road users based on smartphone position data. In: 2013 IEEE intelligent vehicles symposium (IV). IEEE, pp 256–261
11. Dhondge K, Song S, Choi BY, Park H (2014) Wifihonk: smartphone-based beacon stuffed Wifi car2x-communication system for vulnerable road user safety. In: 2014 IEEE 79th vehicular technology conference (VTC Spring). IEEE, pp 1–5
12. Wu X, Miucic R, Yang S, Al-Stouhi S, Misener J, Bai S, Chan W (2014) Cars talk to phones: a dsrc based vehicle-pedestrian safety system. In: 2014 IEEE 80th vehicular technology conference (VTC Fall). IEEE, pp 1–7
13. Anaya JJ, Talavera E, Giménez D, Goméz N, Felipe J, Naranjo JE (2015) Vulnerable road users detection using v2x communications. In: 2015 IEEE 18th international conference on intelligent transportation systems (ITSC). IEEE, pp 107–112
14. Beier G (2004) Kontrollüberzeugung im Umgang mit Technik: Ein Persönlichkeitsmerkmal mit Relevanz für die Gestaltung technischer Systeme [locus control in a technological context]. Ph.D. thesis, dissertation premium. dissertation.de, Berlin

A Conceptual Framework of Collective Activity in Constructive Ergonomics

Sandrine Caroly[1(✉)] and Flore Barcellini[2]

[1] PACTE Laboratory, Grenoble Alps University, IEP,
BP 48, 38040 Grenoble Cedex 09, France
Sandrine.caroly@univ-grenoble-alpes.fr
[2] Le Cnam CRTD, 41 rue Gay Lussac, 75005 Paris, France
flore.barcellini@lecnam.net

Abstract. The aim of this presentation is to present a constructive view of the development of collective activity. The concept of collective activity is a proposal of Caroly [1] aiming to connect cooperative work of operators with work collective – collectif de travail in French - that they belong to. These tow concepts are described to understand the organizational conditions to support the development of collective activity: recognition of skills, arenas for debate of criteria for quality of work, reelaboration of rules, intermediaries objects, technical systems, projects design, reorganization with managers.

Keywords: Collective activity · Collective work · Constructive ergonomics

1 Objectives

The aim of this presentation is to present a constructive view of the development of *collective activity*. The concept of *collective activity* is a proposal of Caroly [1] aiming to connect cooperative work of operators with *work collective – collectif de travail* in French - that they belong to. CSCW Research dealing with cooperative work are interested in the way operators cooperatively reach their goals. They reveal that coordination and cooperation is based on co-elaboration of knowledge such as construction of a common frame of reference, operative synchronisation and implies the development of situation awareness. However, other research, mainly grounded in Activity Theory [2–4] proposed the concept of work *collective* to understand the health protecting function of groups at work. The goal of this communication is to articulate these two concepts to understand the proposed concept of collective activity.

In a first part of this paper, we will present the concept of collective work- seen as a resource for the development of performance and work collective - seen as a resource for the development of health and skills, the concept of collective activity is a proposal to link together collective work and work collective [1], and finally the approach of constructive ergonomics to understand the component of effective and efficient collective activities.

In a second part, we will describe the organizational and material conditions that are crucial to develop collective activity. Finally, we will highlight the need for Ergonomics to focus on the work that consists in organizing environments so that they may

S. Bagnara et al. (Eds.): IEA 2018, AISC 822, pp. 658–664, 2019.
https://doi.org/10.1007/978-3-319-96077-7_71

enable the development of this collective activity – and, as a consequence, the need to focus on the activity of the people who organize these environments.

2 Construction of Performance and Health in Cooperative Work Situations

Cooperative work, is here seen as a resource for performance, where as *work collective/Collectif de travail*- is seen as a resource for the development of health and skills among operators involved in a cooperative work. However, all cooperative work does not necessarily involve the existence of *work collective/collectif de travail* that may be construct thanks to development of *collective activity* [1].

Collective work refers to the ways in which operators may cooperate, in a more or less effective and efficient manner, in a work situation [5]. It is therefore defined in relation to the task which the partners of collective work are involved in, and relates to their performance with respect to achieving the goals of this task. Collective work implies processes of task allocation and knowledge sharing. These processes are related to the implementation of adjustments within the activity.

Many kinds of sociocognitive resources can foster the production of effective collective work [6–9]: opportunities for operative synchronization – i.e. coordination – between participants, the construction of a Common Frame of Reference (COFOR), reciprocal knowledge of the work of all the persons involved, and a shared reference concerning the state of progression of the process, which implies the development of situation awareness.

Operative synchronization defines the possibilities for coordination between participants involved in collective work [7]. This coordination is never completely pre-specified (e.g. through prescribed procedures). It is constructed by the partners and involves communication – both verbal and nonverbal – between them [10]. The *cognitively synchronize* with one another [7, 8] is constructed to maintain, and to update a set of "shared knowledge" that allows the partners of collective work to manage the dependencies connecting their individual activities. This knowledge is based on a set of situations experienced together, and on trade-specific knowledge and beliefs that are historically and culturally constructed [8].

Two types of knowledge appear to be essential for effective collective work:

– On the one hand, participants must be able to construct shared knowledge regarding their field of activity (technical rules, objects of the field and their properties, problem-solving procedures, etc.). This shared knowledge is also termed "Common Frame of Reference", or COFOR. This framework comprises the *"functional representations shared by operators, that guide and control the activity which they carry out as a collective"* [11, 12].

– On the other hand, in the "here and now" of a particular task, participants must be able to construct a representation of the current state of the situation which they are involved in (knowledge of facts related to the state of the situation, of the contributions of partners involved in the task...), also known as "awareness" [8, 13] The construction of "awareness" is not just an opportunistic process resulting from the affordances of a situation. It also relies on the ability of the partners of collective work to recognize, interpret, and understand each other's conducts and the resources that are available to them [8, 10].

3 The Work Collective/Collectif de Travail a Resource for the Development of Health and Skills

The existence of a *work collective/collectif de travail* protects the individual's subjective relationship with action. The realization of «quality work» depend on the existence of such *work collective* that supports: debates among operators about the meaning and criteria of their action, sharing of resources that help them to cope with situation of conflict of goals within their activities, or creation of various ways of «doing work» (reelaboration of rules).

Collective work must be distinguished from the concept of work collective. Indeed, all cooperative work does not necessarily involve the existence of *work collective/ collectif de travail* that may be construct thanks to development of *collective activity* [1], Yet, many studies tend not to make a clear distinction between what is related to the work collective versus what is related to collective work. For Activity-Centered Ergonomics, a work collective is constructed between operators who share goals related to the realization of activity with the criteria that *they themselves* ascribe to "effective work", and with the meaning that they give to this work. The collective's ability permits to construct – or reconstruct – standards and rules to frame action, in compliance with the criteria of quality of work, to manage conflicting relationships at work, and finally, to give meaning to the work. These rules are grounded in a history which structures exchanges between people at work, and fosters the mobilization of the subject in his/her own activity. In this sense, the concept of "work collective" refers to something more than just a collection of individuals. The collective is a part of the activity, and not just a determining factor of the work situation. Following a constructive view, the work collective emerges as a resource for the development of health: contributes to individual health and fosters learning and the development of skills.

The work collective allows the preservation of the health of its members, in the sense that it ensures that the debate about work does not focus directly on issues related to personality, but on issues related to activity and work organization. These may help operators find, in their activity, ways and means of doing things that are suited to the situation, such that they might foster the preservation of health and the construction of a meaning of work. More specifically, the work collective plays a part in preserving psychosocial resources [14] and preventing Musculo-Skeletal Disorders or MSDs [15]. The work collective permits to "take care" of their own work, and whose subjectivity is constantly put at odds by the the contraints or conflicts of goals. The work collective provides a support for innovating regarding the various ways of "doing work". It allows innovative forms of learning because it supports inquiry, confrontation, and debate between the members of the work collective.

4   The Concept of Collective Activity

In order to account for this connection between the work collective and collective work, we have proposed the concept of *collective activity* [1, 16, 17]. The implementation of a collective activity pursues the goals of health, effectiveness, and the development of values that are specific to that activity (i.e. the meaning of work for operators who are involved in exchanges with their colleagues about the quality of work in the trade). This collective activity allows the development of individual skills, allows these skills to complement each other in work, and enriches the liveliness of the work collective [16]. Thus, collective activity cannot be built solely on the basis of a sum of different individual activities, but through constant toing and froing between the activity of a subject, the implementation of collective work, and the operation of the work collective.

The concept of *collective activity* outlines that the construction of a *work collective/collectif de travail* implies the performance of a specific activity – the *collective activity* – referring to the co-elaboration of shared criteria by operators to assess a high-quality work and to re-elaborate rules framing their work [1]. Performance of *collective activity* requires that work organization provide specific arenas supporting debates among operators, arenas that may be equipped by intermediaries' objects to improve co-elaboration among operators [17].

Collective work and the work collective are the two linchpins in the production of a high-quality collective activity. The work collective supports the development of skills, learning, and the preservation of health; effective collective work supports the achievement of task goals. But this is only possible under specific conditions. Following a constructive view, ergonomics must therefore foster the development of a collective activity by acting on the organizational and material conditions of work that foster the construction of collective work and of the work collective.

5   Supporting the Conditions to Develop Collective Activity

Following a developmental perspective, the goal of ergonomics is to support collective activities by creating the tools and the means that are required for its development [17]:

- to support the development of representations of competence and quality in the work of others: Recognizing competence in another person, such as a colleague, is a necessity for collective work, and it enriches the work collective. Thus, this recognition can foster collective activity because it gives rise to cooperation in action and implies effectiveness in the work collective. If collective work is acknowledged and supported, this can contribute to the implementation of meta-functional activities regarding specific work situations. This can help operators to become aware of their own experience and to formalize their own skills – possibly in order to pass those skills on;
- to construct spaces in which to share the criteria of the quality of work: In a constructive approach, ergonomics must take responsibility for supporting debates about the work activity, so that members of the work collective might enter into a

dialogue, regarding both the difficulties encountered at work and the internal and external resources of activity. This implies providing the collective with specific methodological tools, particularly with discussion spaces for operators from a same craft, that allow the criteria of work effectiveness and the values mobilized in a work activity to be discussed;

- to develop organizations supporting the processes of "reconstruction of rules": It is not just to alleviate the constraints of work that derive from prescriptions of the hierarchy, but also to manage conflicts within the goals of the activity by finding ways of circumventing them to complete "a job well done". The operational leeway provided by the organization of work must help the implementation of operative adjustments, and the construction of meta-rules that define the collective rules for using the prescribed rules;
- to design tools to support the development of resources for collective activities, via intermediary objects and technological devices: In a constructive approach, these intermediary objects constitute instruments of the collective activity that ergonomists can contribute to develop. Ergonomics must make these instruments visible to agents who are not always conscious of their existence, in order to support a confrontation between them regarding the goals of production and quality. Furthermore, to support the development of collective activity, ergonomics must also take part in designing intermediary objects that will make it possible to debate various points of view and to foster controversies;
- to design projects with methods of simulation of future activity integrated the work collective: proposing an approach to accompany socio-technical systems projects, so that they might aim to jointly develop the technologies and the organizations in which they are to be integrated, and finally, the future collective [18]. For e.g. to construct representations focusing on the skills, roles, and expertise of the other protagonists – to facilitates in system of interactions to have a "social conscience" [19] for helping collective work;
- to design enabling organization supported by activity of proximity manager: The implementation of organization could take better integration of collective activity in the reorganization of work into account [20, 21]. One of these means is to help managers to be able to recognize in what ways operators reorganize their own work, and how this allows them to fuel a debate about the meaning of work and of quality work; to take this into account in the process of redesigning an organization, and to design organizations that leave room for these debates and allow for the reconstruction of these rules.

This also means that the equipment used to support a collective activity (e.g. technical devices, actor networks, discussion spaces, means to pass on experience) should be investigated directly by ergonomists in the course of their interventions. In this presentation, we will give several illustrations regarding operationally of the concept of *collective activity*: among same trade professionals (e.g. police officers); heterogeneous trade professionals (e.g. designers and engineers in project); various stakeholders (e.g. citizen, enterprises, student or researcher, etc.) such as in fab lab or online epistemic communities (Open Source, Wikipedia).

References

1. Caroly S (2010) Activité collective et réélaboration des règles: des enjeux pour la santé au travail. Document d'habilitation à diriger des recherches en ergonomie, Université de Bordeaux 2. http://tel.archives-ouvertes.fr/tel-00464801/fr/
2. Engeström Y, Miettinen R, Punamäki RL (1999) Perspectives on activity theory. Cambridge University Press, Cambridge
3. Daniellou F, Rabardel P (2006) Activity-oriented approaches to ergonomics: some traditions and communauties. Theor Issues Ergon Sci 6(5):355–357
4. Clot Y (2008) Travail et pouvoir d'agir. PUF, Paris
5. de la Garza C, Weill-Fassina A (2000) Régulations horizontales et verticales du risque. In: Weill-Fassina A, Hakim Benchekroun T (eds) Le travail collectif: perspectives actuelles en ergonomie. Octarès Editions, Toulouse, pp 217–234
6. Caroly S, Weill-Fassina A (2007) How do different approaches to collective activity in service relations call into question the plurality of ergonomic activity models? Activités 4(1):99–111. http://www.activites.org/v4n1/v4n1.pdf
7. Darses F, Détienne F, Falzon P, Visser W (2001) COMET: a method for analysing collective design process. INRIA Publications, Rocquencourt. http://hal.inria.fr/inria-00072330/
8. Salembier P, Zouinar M (2004) Mutual intelligibility and shared context. Conceptual inspirations and technological reductions. Activités 1(2):64–85
9. Schmidt K (2002) The problem with 'awareness': introductory remarks on 'awareness in CSCW'. J Comput Support Coop Work 11(3–4):285–298
10. Heath C, Svensson MS, Hindmarsh J, Luff P, Von Lehn D (2002) Configuring awareness. JCSCW 11(1–2):317–347
11. Leplat J (1991) Organization of activity in collective tasks. In: Rasmussen J, Brehmer B, Leplat J (eds) Distributed decision making. Wiley, Chischester, pp 61–74
12. Hoc JM, Carlier X (2002) Role of common frame of reference in cognitive cooperation: sharing tasks between agents in Air Traffic Control. Cogn Technol Work 4(1):37–47
13. Carroll JM, Rosson MB, Convertino G, Ganoe CH (2006) Awareness and teamwork in computer-supported collaborations. Interact Comput 18:21–46
14. Caroly S (2011) Collective activity and reelaboration of rules as resources for mental health: the case of police officers. Le Travail Humain 74(4):365–389
15. Lemonie Y, Chassaing K (2015) From the adaptation of movement to the development of gesture. In: Falzon P (ed) Constructive ergonomics. CRC Press, Boca Raton, pp 49–64
16. Caroly S (2009) Designing collective activity: a way to workers' health. In: Proceedings of the IEA 2009. Congress IEA, Beijing, 10–14 August 2009
17. Caroly S, Flore F (2015) The development of collective activity. In: Falzon P (ed) Constructive ergonomics. CRC Press, Boca Raton, pp 19–32
18. Barcellini F, Van Belleghem L, Daniellou F (2015) Design projects as opportunities for the development of activities. In: Falzon P (ed) Constructive ergonomics. CRC Press, Boca Raton, pp 171–204
19. Barcellini F, Détienne F, Burkhardt JM (2010) Distributed design and distributed social awareness: exploring inter-subjective dimensions of roles. In: Lewkowicz M, Hassanaly P, Rodhe M, Wulf V (eds) Proceedings of the COOP 2010 conference. Springer, Berlin
20. Petit J, Coutarel F (2015) Intervention as dynamic processes for the joint development of

agents and organization. In: Falzon P (ed) Constructive ergonomics. CRC Press, Boca Raton, pp 171–204

21. Falzon P (2005) Ergonomics, knowledge development and the design of enabling environments. In: Presented at humanizing work and work environment conference (HWWE 2005), Guwahati, Inde

Requirements for a Sociotechnical Support System for the Critically Ill – A Qualitative Study on the Needs and Expectations of Patients, Relatives and Health Professionals

Susanne Krotsetis[1], Adrienne Henkel[2], Björn Hussels[1],
and Katrin Balzer[2(✉)]

[1] University Hospital Schleswig-Holstein,
Ratzeburger Allee 160, 23538 Lübeck, Germany
{susanne.krotsetis,bjoern.hussels}@uksh.de
[2] Institute for Social Medicine and Epidemiology,
University of Lübeck, Ratzeburger Allee 160, 23562 Lübeck, Germany
{adrienne.henkel,katrin.balzer}@uksh.de

Abstract. Mechanical ventilation and subsequent weaning from the respirator are linked to high levels of physical and emotional stress for critically ill patients. Due to the severity of illness, impact of sedatives and the endotracheal tube, these patients cannot orally express their feelings and needs. In clinical practice, effective tools to facilitate early communication and re-orientation in these patients are lacking. To address this lack, a multidisciplinary project was set up to develop an "Ambient System for Communication, Information and Control in Intensive Care" (ACTIVATE). The present study was the first step and aimed to identify needs and expectations of patients, relatives and health professionals regarding ACTIVATE. A qualitative study involving 16 patients recently weaned from a ventilator, 16 relatives and 34 health professionals was conducted. The results show that for patients the weaning period is dominated by unanswered needs for effective communication with health professionals and relatives. These needs include the desire to directly express physical symptoms, receive re-orienting information and get into touch with relatives. For large parts, these needs were confirmed by the relatives and the health professionals, although some were associated with a lesser relevance. In addition, all interviews revealed user- and context-specific requirements for the design and functionalities of ACTIVATE. The results highlight the strong need for an innovative sociotechnical system to facilitate early and effective communication with patients undergoing weaning. We derived typical communication needs to be supported by ACTIVATE as well as user- and context-specific design requirements and potential ethical, legal and social implications.

Keywords: Sociotechnical support systems · Mechanical ventilation Communication

© Springer Nature Switzerland AG 2019
S. Bagnara et al. (Eds.): IEA 2018, AISC 822, pp. 665–671, 2019.
https://doi.org/10.1007/978-3-319-96077-7_72

1 Introduction

1.1 Background

Mechanical ventilation represents a central treatment condition in the care for critically ill patients. However, patients receiving invasive mechanical ventilation also need to receive high levels of sedative medication, inducing (almost) a loss of consciousness to ensure patients' tolerance of this treatment. Termination of mechanical ventilation, called weaning, usually happens in a stepwise manner. During this process, the sedative medication and the amount of mechanical support are gradually reduced so as to the patients become increasingly awake and able to ventilate by themselves. Although these weaning procedures aim to promote patients' health and well-being, the procedures themselves may cause high levels of physical and emotional stress to patients in intensive care units (ICU). Due to the severity of illness, impact of sedatives and the endotracheal tube, patients undergoing invasive mechanical ventilation and successive weaning cannot orally express their feelings and needs. Inadequately treated pain, physical burden, fear and feelings of strangeness, coupled with a lack of communication, are the most often reported strains [1] and are associated with delays in physical and cognitive recovery [2]. Recent research emphasizes the importance of positive communication activities between critically ill patients and the care team [3]. However, health professionals often experience feelings of failure and frustration when they try to communicate with ventilated patients, since easily to use communication tools are lacking [4].

1.2 Objectives

To address this lack, a multidisciplinary project was set up to develop an "Ambient System for Communication, Information and Control in Intensive Care" (ACTIVATE). The present study was the first project step, targeting following research questions for the initial requirement analysis: Which patient needs and symptoms or health care situations are experienced as being most burdensome by patients, relatives and health professionals during the waning period? Which purposes, functions and design features do patients, relatives and professionals expect from a socio-technical system intended to support nursing care for weaning patients.

2 Methods

2.1 Study Design

We conducted a qualitative study consisting of semi-structured topic-guided individual or group interviews with patients successfully weaned from the mechanical respirator, relatives of such patients and health professionals involved in the ICU care for patients undergoing weaning.

2.2 Study Setting and Population

The study was conducted in a university hospital in the North of Germany. It involved two surgical and one medical ICU as well as three related intermediate care (IMC) units and one medical ward where ICU patients are used to be transferred to after successful weaning from the mechanical ventilation.

The study targeted three sub-populations: (i) patients recently weaned from the mechanical respirator, (ii) relatives of such patients and (iii) ICU staff members involved in the care for patients undergoing weaning. Table 1 summarizes the eligibility criteria for these sub-populations.

Table 1. Eligibility criteria of the target sub-populations.

Sub-population	Inclusion criteria	Exclusion criteria
ICU patients	• Age ≥ 18 years • Admission to one of the three participating ICU for ≥ 48 h during the current hospital stay • Successful weaning from ≥ 24 h mechanical ventilation • Sufficient vigilance • No symptoms of delirium	• Inability to communicate in oral German language • Inability to participate in 20-min interview due to clinical reasons (judgement by treating physicians or nurses)
Relatives	• Primary significant others irrespective of the degree of kinship • Related patients: ≥ 24 h mechanical ventilation in one of the three participating ICU during the current hospital stay	Inability to communicate in oral German language
Health professionals	• Professionals currently involved in direct ICU patient care or having been involved for ≥ 8 weeks in recent 6 months • Qualifications: registered nurses, 3rd year nursing students, medical doctors, 5th or 6th year medical students, completed degrees in other health professions	Inability to communicate in oral German language

For all sub-populations, written informed consent was required for study participation. To recruit patients and relatives, members of the study team visited participating wards each second day. At these visits, they screened all patients currently treated in these wards for study eligibility and informed potentially eligible patients orally and in written about the study. Invited patients could take as much time as they required for decision-making on study participation, with the restriction that the interview should take place before discharge. During the ward visits, the study team members also provided oral and written information about the study to relatives regularly visiting potentially eligible patients.

For the recruitment of the health professionals, we used various channels of information, among them posters and leaflets, electronic mailing to staff members and students as well as face-to-face talks with single staff members.

2.3   Data Collection

We conducted face-to-face interviews with patients and relatives, and focus groups with health professionals, except medical doctors who were also interviewed face to face due to organizational reasons. All interviews, be it individual or focus groups, were guided by a semi-structured questionnaire which consisted of open-ended questions reflecting the research questions underlying this study.

All interviews with patients and relatives were conducted before the respective patient's hospital discharge. The estimated length of individual interviews was 20 to 30 min, the focus groups were scheduled for 60 min. All interviews were audio recorded and transcribed verbatim.

2.4   Data Analysis

We conducted a content analysis using the thematic framework analysis methods [5]. Following this framework, we iteratively sifted through the transcripts for meaningful statements, summarized the meanings to themes and sub-themes at more abstract levels and compared the derived themes and sub-themes both across and within the sub-populations.

While our analysis was primarily driven by the research questions, it was also open to any potentially relevant information concerning the needs of patients, relatives and health professionals and the potential use of sociotechnical supports systems.

2.5   Ethical Considerations

The study protocol was approved by the ethical research committee of the University of Lübeck, Germany (reference number 17-098). All data were collected in a pseudonymized manner and anonymized after completion of data collection. ICU patients recently weaned from mechanical ventilation represent a highly vulnerable population. The interviews were conducted by nursing researchers (SK, AH, BH) all being experienced in critical care. No adverse events did occur during this study.

3   Results

3.1   Sample Characteristics

The data collection period went from May to June 2018. In this period, 16 out of 26 patients and 16 out of 28 relatives, respectively, assessed as being eligible consented to participate. Furthermore, three focus groups involving 28 health professionals and 6

Table 2. Sample characteristics.

Subpopulation	Sample size	Sample characteristics
Patients	N = 16	• Mean age 58 years, Min–Max 21–82 • 4 females, 12 males • 12 emergency admissions, 4 elective surgery • 13 surgical treatment, 3 medical treatment • 11 mechanical respiration via tubus, 5 via tracheal cannula
Relatives	N = 16	Degree of kinship: • 7 acquaintances • 5 spouses • 3 children • 1 mother
Health professionals	N = 17 registered nurses N = 9 nursing students (3rd year) N = 4 medical doctors N = 2 medical students (6th year) N = 1 chaplain N = 1 physiotherapist	Professional experiences of registered nurses: • Mean length of experience 9 years, Min–Max 21–82 Degree of qualification of medical doctors: • 2 residents • 2 senior physicians

individual interviews with medical doctors and advanced medical students were conducted. Table 2 displays basic characteristics of these samples.

3.2   Themes and Subthemes

The analysis revealed following themes and sub-themes: (i) support needs of patients, (ii) desired purposes, functions and designs features of the ACTIVATE system, and (iii) potential barriers and facilitators for the implementation, including ethical, legal and social requirements (ELSI).

The support needs are dominated by the patients' unmet needs to express oneself and to receive re-orienting information, e.g. concerning place and time, the current health status or treatment conditions, or domastic or family issues. The patients' desire to express oneself comprises both the need to communicate with others and the desire for experiencing feelings of being heard and recognized and able to report burdening symptoms more precisely to health professionals, especially nurses, and relatives. Although most of these needs were described by all sub-populations, slight differences in the frequency of mentions were noted for certain needs and symptoms (see Fig. 1). Whereas the patients' and relatives' views hardly differed, the health professionals less frequently mentioned basic needs and symptoms like thirst or patients' desire for information.

The interviewees linked several purposes, functions and design requirements to the ACTIVATE system. They unanimously stressed the importance of providing information (e.g. on place and time) for patients' re-orientation and facilitating communication with others, especially for the early and valid assessment of patient needs.

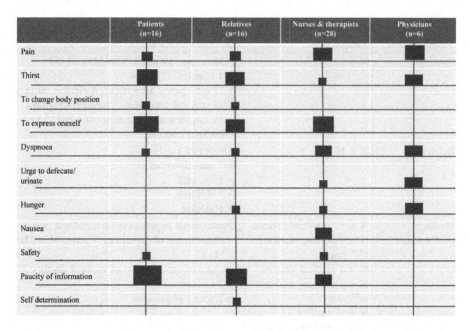

Fig. 1. Frequencies of mentions of patients' needs and symptoms across sub-samples. Largest boxes represent mentions by 11 participants, smallest by 1 participant.

To realize such purposes, following functions were suggested: standardized easy-to-recognize input icons for patients, automatized, individually adjusted output information stimulating the wakening and re-orientation, e.g. by biographically preferred music or videos. Furthermore, the interviewees demanded that the ACTIVATE system should be easily and intuitively to use, both for patients and professionals. They also regarded "press" modes (patients pressing the input device by the hand) as the most suitable input option for patients at early weaning stages. With respect to potential output modes no preferences could be noted.

However, also a number of context factors facilitating or hampering the use of a sociotechnical support system like ACTIVATE was mentioned. Patients and relatives expressed concerns that such a system may cause an overload of signals and increase the risk of false positive or negative signals. Professionals were concerned about noise and working load added to their existing burdens. They also stressed the need for compatibility with other medical technologies in use in ICU patients as well as requirements due to infection protection, occupational safety and data protection and security. In the professionals' view, there may also be patient-related factors that hamper the use of ACTIVATE, e.g. impaired cognition due to delirium or limited fine motor skills. All sub-populations furthermore raised concerns that the sociotechnical support system may replace personal contact between patients and health care staff.

4 Discussion

The interviews revealed important requirements for the development of the intended support system ACTIVATE. While the findings confirmed the need for an effective sociotechnical support of early communication and re-orientation in ICU patients undergoing weaning, they also provided some unexpected insights. In particular, the patients' needs to express more accurately very basic physical symptoms like thirst or body position changes have not yet been stressed in the existing literature where much research is reported on weaning patients' pain and related burden instead. This divergence is also reflected by our findings suggesting that professionals' are well aware of patients' burden due to pain, lack of information and communication and physical needs related to excretions but not so much of other needs important to patients as well, e.g. like thirst. This divergence weakens the validity of professionals' communication with the patients and the accuracy of their symptom assessment [6]. Altogether, existing empirical evidence suggests that the ACTIVATE system may have large potentials to improve the quality and outcome of professionals' communication with patients undergoing weaning.

However, aside a number of potential barriers at patient and ward environment levels for ACTIVATE use, all sub-samples also articulated concerns that the sociotechnical system may reduce the quantity and quality of nurse-patient contacts. We will take these concerns (and all other mentioned potential barriers) very seriously throughout the whole project up to the final clinical evaluation of the first prototypes. The findings of our qualitative study will also inform the ongoing systematic analysis and reflection of potential ethical, legal and social implications.

As confirmed by the findings of our study, our project strives to improve communication and re-orientation in non-vocal ICU patients, and we will early and robustly evaluate and report the ACTIVATE system's effects with regard to these impacts.

References

1. Tsay SF, Mu PF, Lin S, Wang KW, Chen YC (2013) The experiences of adult ventilator-dependent patients: a meta-synthesis review. Nurs Health Sci 15(4):525–533
2. Rose L, Dainty KN, Jordan J, Blackwood B (2014) Weaning from mechanical ventilation: a scoping review of qualitative studies. Am J Crit Care 23(5):e54–e70
3. Happ MB, Sereika SM, Houze MP, Seaman JB, Tate JA, Nilsen ML et al (2015) Quality of care and resource use among mechanically ventilated patients before and after an intervention to assist nurse-nonvocal patient communication. Heart Lung 44(5):408.e2–415.e2
4. Abuatiq A (2015) Patients' and health care providers' perception of stressors in the intensive care units. Dimens Crit Care Nurs 34(4):205–214
5. Spencer L, Ritchie J, O'Connor W (2003) Analysis: practices, principles and processes. In: Ritchie J, Lewis J (eds) Qualitative research practice: a guide for social science students and researchers. Sage, London, pp 199–218
6. Schindler AW, Schindler N, Enz F, Lueck A, Olderog T, Vagts DA (2013) ICU personnel have inaccurate perceptions of their patients' experiences. Acta Anaesthesiol Scand 57(8):1032–1040

Age-Appropriate Design of an Input Component for the Historytelling Project

Torben Volkmann[✉], Friedemann Dohse, Michael Sengpiel,
and Nicole Jochems

University of Lübeck, Ratzeburger Allee 160, 23562 Lübeck, Germany
volkmann@imis.uni-luebeck.de

Abstract. As a result of the demographic change, the population of the older adults within our society increases steadily. If done right, information and communication technology offer great potential benefit for older users. Thus, we focus on a human-centered design approach for aging (HCD+). This Paper describes the development of an input component for the Historytelling project, a cooperative interactive platform for older adults to connect life stories with historic events and appreciate the potential for older adult's life experience. We conducted a task analysis based on requirements for the Historytelling project and developed a high-fidelity prototype which was tested with eleven older adults. The prototype was valued highly overall but still had minor usability issues. Thus, we could show that a rather complex task such as telling a life story with various multimedia objects can be executed by older adults if they are integrated into the design process from early on.

Keywords: Aging users · Human-centered design · Demographic change
Intergenerational communication · Social isolation

1 Introduction

Due to the demographic change, the percentage of older adults within society increases continuously [1]. The Historytelling (HT) project addresses these demographic changes and in particular older adults' social integration by fostering communication across generations and among older adults. It aims to mend the often-lamented deficit orientation in design for older adults by minding age-related gains and losses. At the core of HT, life stories connected to historic events and other stories appreciate the vast potential of older adults' life experience, (re-) creating history as a network of authentic subjective stories, told from multiple perspectives in a way that children, grandchildren, and generations to come will find fascinating, worth sharing and asking their parents about - and eventually contributing to themselves. At the same time, the design process for the HT follows state of the art and science in designing for older adults to account for age-related user characteristics.

HT is developed step by step within a component-based human-centered design approach and plugged into the different platforms with little changes to apply to various contexts and user groups [2]. This paper focuses on the challenges in HT design and the development of an age-appropriate input component for multimedia content following

© Springer Nature Switzerland AG 2019
S. Bagnara et al. (Eds.): IEA 2018, AISC 822, pp. 672–680, 2019.
https://doi.org/10.1007/978-3-319-96077-7_73

a human-centered design approach (HCD+ [3]) analogous to typical procedures for age-appropriate design [4]. We used well-known design guidelines for older adults alongside results of a prior task analysis to design the component and validated our design decisions with a usability evaluation with eleven older adults.

2   Literature Review

Much progress has been made analyzing age-related sensory, cognitive and motoric decline, yet there is great need of research regarding technology usage and the interplay with personal and social development in the later stages of life [5]. As Brewer and Piper [5] stated, blogging, for instance, can lead to the fostering of self-expression, provide meaningful engagement and increase social interaction.

To ensure the accessibility for a broad user range, there are fundamental guidelines for public website design (i.e. WCAG) [6]. Evidence shows that many users benefit from these guidelines, but that they are often not implemented, especially on corporate websites [7]. Evaluations show that in particular, the usage of larger texts, clean user interfaces, adapted use of words and consistency are important usability aspects [8]. Additionally, according to the universal usability approach, the thus improved usability should lead to better usability for all users [9].

For participation in digital communities, it is especially important to provide usable ways to publish content. Several high-level guidelines exist to help design and develop reminiscence systems for older adults [5, 10, 11]. In general, most of these guidelines share aspects which the design described in this paper should address: (1) Support as many different media types as possible, (2) make the reminiscence process unique and personal, (3) encourage sharing and linking between content and people and give space to comment the content. These requirements were heavily emphasized during the development of the component.

3   Methodology

Following the human-centered design approach (Fig. 1), in a first step interviews were conducted with the potential target group of six older adults aged between 61 and 72 years (M = 67, SD = 3.4) to pinpoint the context of use and first design implications for the input form. This analysis is later used for evaluation purposes to see if the design decisions and tasks match. Design development was conducted in two steps. First, a low-fidelity paper prototype was created and tested by six usability experts and second, a high-fidelity clickable prototype was developed and tested with eleven older adults.

3.1   Requirements and Tasks

Literature states several general requirements for an multimedia input component, among them: Selection of existing digital media, such as photos, audio, and videos [12, 13], capturing of photos, audio and videos within the input component [12], tell stories with

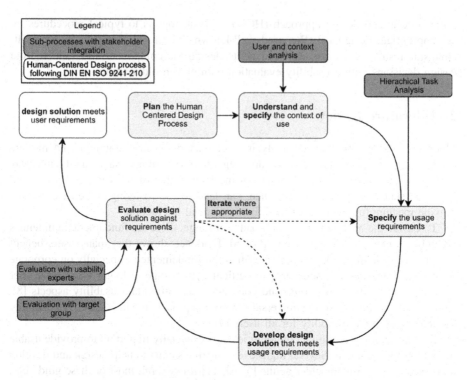

Fig. 1. The human-centered design process following DIN EN ISO 9241-210

text elements [12], selection of individual fonts and font sizes and text styling options [12], integration of an undo and redo function [12, 14], possibility to discard the story at any time [14], possibility to store a story temporarily [14]. With these requirements in mind, we conducted a hierarchical task analysis (HTA) for further development and evaluation. Figure 2 shows a small part of this analysis. The former identified high-level guidelines (see Literature Review) are implemented throughout the task analysis but can especially be found in the media-based specification of the story (1 & 2) and the linking of the story to a historical event (3).

3.2 Design Development

For design development, we used a two-step approach. First, we designed a low-fidelity prototype based on the tasks and literature guidelines and evaluated it with usability experts (N = 6). Second, we developed a high-fidelity prototype based on findings of this formative evaluation. Before each evaluation, we verified design decisions with personas based on former interviews.

The different design aspects derived from literature and former interviews were divided into three different sections: (1) Navigational pattern: Use a home site and place on it all main topics [15], Do not use a panorama or pivot control [15], (2) Interaction pattern: Try to avoid drag and drop, instead use drop down menus [16, 17], The needed

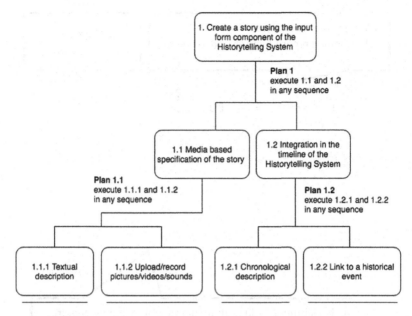

Fig. 2. HTA showing the underlying tasks of the input component

usage of the keyboard should be minimized, instead use radio buttons, check boxes and pickers [15], Use explanations suitable for the target group [15], (3) Design pattern: Use at least 12 pt. font size [12], Use enough white space to avoid wrong clicks and declutter the UI [15], Use a combination of text and images for UI elements [12, 15], Do not place elements at the edges of the screen especially for mobile devices [15]. These were used in combination with common design guidelines [18] to develop both prototypes.

Low-Fidelity Prototype. The first prototype was developed digitally with the Samsung application "S-Note" and was printed out for evaluation purposes. Figure 3 shows one page of the finished low-fidelity prototype with text, button and icon design referencing popular word processors.

Usability problems in effectiveness and efficiency, as well as missing functionality, were qualitatively assessed, leading to an improved clickable high-fidelity prototype. The most frequent responses regarding usability were about the labeling of functions such as "taking a picture" and inconsistency of highlighting and layout. Also, the design elements derived from popular word processors were criticized because of small button size and incomprehensible labeling. Missing functionality was assessed especially regarding editing functionality of the multimedia content. Participants stated that cropping, editing, and labeling of video and audio footage should be possible within the input component.

High-Fidelity Prototype. Thus, we integrated the evaluation learnings into our clickable high-fidelity prototype designed with Axure RP. We added new screens to edit video and audio material as shown in Fig. 4. With this prototype, users have the

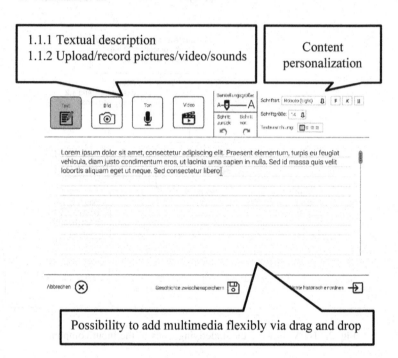

Fig. 3. Low-fidelity prototype of the input component.

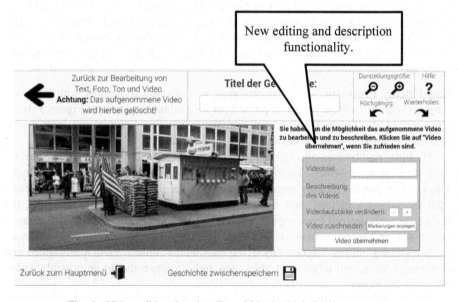

Fig. 4. Video editing functionality within the high-fidelity prototype

option to upload a video, give it a title and a description, crop the video and change the overall audio volume. Although not recommended in age-related design guidelines, we used drag and drop gestures for cropping media [9, 19] because it is commonly used for this kind of interaction.

Another view was added to see the progress of the ongoing story project and all added elements of the story. Each element can be placed individually by the user on a grid to make every story a personal memory. Figure 5 shows a nearly finished story with an added picture, audio, and video.

Fig. 5. Full view of a nearly finished story within the high-fidelity prototype

4   Evaluation

A summative evaluation was conducted to test the usability of the high-fidelity prototype of the input component. Eleven older adults between the age of 61 and 77 years (M = 69, SD = 4.6) with an average computer literacy (Mean score = 17, SD = 4.5) [20] participated in a summative evaluation. Seven participants carried out the study within a workshop for older adults in a media and computer science institute of a German university. Four participants participated at home at their own computer.

4.1   Methods and Procedure

First, the participants completed a questionnaire about their computer usage and demographic information. Then participants were asked to complete several tasks with the high-fidelity prototype based on the HTA (Fig. 2). Tasks were performed with a think-aloud protocol to get more precise information about usability problems.

The following tasks had to be executed: Name the story, Change the size of the input mask, write a story, capture a photo with a webcam and modify it, describe the story with recorded audio, link the story with a historical frame, publish the story.

4.2   Results

In the summative evaluation, older adults generally valued the prototype highly. Judged very positively were especially the usage of text on every control element (stated by 5 participants), the consistency and scale of the interface (3 participants) and the step-by-step guidance through the interface (3 participants). However, some elements were judged negatively: Of 11 participants, 4 were not able to find a symbol to play a sound, and 7 took very long to find it. Also, explanations hidden in tooltips resulted in ineffective and inefficient task completion. In another task, drag and drop gestures were difficult to execute since the participants were not aware of this interaction gesture and some of them had problems with fine motor skills, resulting in 4 out of 10 participants being unable to solve the task and 6 taking very long (inefficient). Finally, three participants had difficulties differentiating between similar functions, such as increasing the size of the interface and the size of the text in the textbox.

5   Discussion and Further Research

In this paper, we described the development of an input component for the History-telling (HT) project, which aims to provide digital, interactive platforms to connect life stories to historic events and other stories to appreciate the potential of older adult's life experience and connect different generations through these stories.

We were able to show that even rather complex tasks as telling a life story with various multimedia objects can be executed by older adults if they are involved through the whole design process and if guidelines for older adults are considered.

However, it needs to be pointed out that participants of the summative evaluation had a positive affinity towards ICT and thus are not representative of the target group which could bias these findings.

We were also able to show the demand for the design of an age-related input component for the HT. For further research, we want to confirm our initial findings through comparing the component with various other input components. Although similar in several aspects, there are some essential differences regarding the tasks of other systems: (1) HT needs to link (historic) events with user-generated content within the input component. (2) HT sees every story as an unfinished ongoing project so that every story must have the possibility to be changed and extended. (3) HT should provide the functionality to create and share multimedia content on as many devices as possible.

The developed and here presented input component is one small step in an ongoing project. Next steps for research are the expansion of the context and thus deployment in museums and conducting field studies with different developed components to gain insight into daily usage of HT.

References

1. Preißing D (2014) Erfolgreiches Personalmanagement im demografischen Wandel. Walter de Gruyter GmbH & Co KG, Berlin
2. Volkmann T, Sengpiel M, Jochems N (2016) Historytelling: a website for the elderly a human-centered design approach. In: Proceedings of the 9th Nordic conference on human–computer interaction. ACM, New York, NY, USA, pp 100:1–100:6
3. Jochems N, Sengpiel M (2016) Introduction to the special issue on "Design for Aging". i-com 15:1–2. https://doi.org/10.1515/icom-2016-0013
4. Jochems N (2016) Designing tablet computers for the elderly a user-centered design approach. In: Zhou J, Salvendy G (eds) Human aspects of IT for the aged population. Design for aging. Springer, Cham, pp 42–51
5. Brewer R, Piper AM (2016) "Tell it like it really is": a case of online content creation and sharing among older adult bloggers. In: Proceedings of the 2016 CHI conference on human factors in computing systems. ACM, New York, NY, USA, pp 5529–5542
6. Yang YT, Chen B (2015) Web accessibility for older adults: a comparative analysis of disability laws. Gerontologist 55:854–864. https://doi.org/10.1093/geront/gnv057
7. Arfaa J, Wang Y (2014) A usability study on elder adults utilizing social networking sites. In: Marcus A (ed) Design, user experience, and usability. User experience design for diverse interaction platforms and environments. Springer, Cham, pp 50–61
8. Nahm E-S, Preece J, Resnick B, Mills ME (2004) Usability of health Web sites for older adults: a preliminary study. Comput Inform Nurs 22:326–334; quiz 335–336
9. Fisk AD (2009) Designing for older adults: principles and creative human factors approaches, 2nd edn. CRC Press, Boca Raton
10. Stevens MM, Abowd GD, Truong KN, Vollmer F (2003) Getting into the living memory box: family archives & holistic design. Pers Ubiquit Comput 7:210–216. https://doi.org/10.1007/s00779-003-0220-4
11. West D, Quigley A, Kay J (2007) MEMENTO: a digital-physical scrapbook for memory sharing. Pers Ubiquitous Comput 11:313–328. https://doi.org/10.1007/s00779-006-0090-7
12. Hawthorn D (2000) Possible implications of aging for interface designers. Interact Comput 12:507–528. https://doi.org/10.1016/S0953-5438(99)00021-1
13. Karahasanović A, Brandtzæg PB, Heim J et al (2009) Co-creation and user-generated content–elderly people's user requirements. Comput Hum Behav 25:655–678. https://doi.org/10.1016/j.chb.2008.08.012
14. Balabanović M, Chu LL, Wolff GJ (2000) Storytelling with digital photographs. ACM Press, New York, pp 564–571
15. de Barros AC, Leitão R, Ribeiro J (2014) Design and evaluation of a mobile user interface for older adults: navigation, interaction and visual design recommendations. Procedia Comput Sci 27:369–378. https://doi.org/10.1016/j.procs.2014.02.041
16. Motti LG, Vigouroux N, Gorce P (2014) Drag-and-drop for older adults using touchscreen devices: effects of screen sizes and interaction techniques on accuracy. In: Proceedings of the 26th conference on L'Interaction homme-machine. ACM, New York, pp 139–146
17. Motti LG, Vigouroux N, Gorce P (2015) Improving accessibility of tactile interaction for older users: lowering accuracy requirements to support drag-and-drop interaction. Procedia Comput Sci 67:366–375. https://doi.org/10.1016/j.procs.2015.09.281

18. Leavitt MO, Schneiderman B (2006) Research-based web design & usability guidelines. U. S. Dept. of Health and Human Services, Washington
19. Schorb B, Hartung A, Reißmann W (2009) Medien und höheres Lebensalter. VS Verlag für Sozialwissenschaften, Wiesbaden
20. Sengpiel M, Jochems N (2015) Validation of the computer literacy scale (CLS). In: Zhou J, Salvendy G (eds) Human aspects of IT for the aged population. Design for aging: first international conference, ITAP 2015, held as part of HCI international 2015, Los Angeles, CA, USA, 2–7 August 2015, Proceedings, Part I. Springer, Cham, pp 365–375

Perspectives of Older Adults and Informal Caregivers on Information Visualization for Smart Home Monitoring Systems: A Critical Review

Fangyuan Chang[(⊠)] and Britt Östlund[iD]

KTH Royal Institute of Technology, 141 57 Huddinge, Sweden
fancha@kth.se

Abstract. Although health monitoring systems in smart homes have been revealed as a significant tool to help people ageing in place, the density of data poses a challenge on the information visualization. This review aims to make contributions to find gaps in the field of information visualization regarding smart home monitoring for older people. Three kinds of information needs of older adults and their informal caregivers regarding smart home monitoring are categorized, including physical needs, emotional needs and cognitive needs. The research studies reflect that these needs are mainly used to discuss ideas of, design approaches for, the information visualization from ten aspects in the visceral level, behavioral level and reflective level. Results show that there is still a big gap existing in enabling older people and their informal caregivers to better understand smart home monitoring information. Some existing design recommendations can be improved while at the same time, some needs have not been manifested through information visualization. A wider understanding of older adults, informal caregivers and home living environment in all aspects are necessary.

Keywords: Older adults · Informal caregivers · Smart home · Visualization

1 Introduction

Due to the increasing aging population [1] and limited resources, the attention to how technology can support healthy and independent ageing has grown exponentially. People in the fourth age [2], a phase of growing disability and dependence, have particular needs for healthcare technology. When older adults become dependent on home help support, the home becomes both a private place and a working life environment for employed care givers or/and unemployed informal caregivers including family, friends or older spouses [3]. It is found that with ageing, human beings are more dependent on a home environment than previously, and are more willing to stay in their own home—age in place—despite facing the risk of ability declines [4–6].

One method to help people "age in place" is smart home monitoring [7, 8]. It is defined as a sensor system which can monitor parameters of users in the home unobtrusively and continuously [9, 10]. Examples of smart home monitoring for older

© Springer Nature Switzerland AG 2019
S. Bagnara et al. (Eds.): IEA 2018, AISC 822, pp. 681–690, 2019.
https://doi.org/10.1007/978-3-319-96077-7_74

people are technologies for functional monitoring, physiological monitoring, safety monitoring, sensory assistance or cognitive aids, social interaction support and security monitoring [11, 12]. Although research on smart homes has grown over the last decades, wide use of smart home technologies is far from reality [13]. Existing studies have revealed that older people have a high degree of acceptance of smart home monitoring when it provides personal utilities and values [14–16]. However, dense data is an obstacle for users to understand the information and perceive values of these systems.

Various research proves that the design of visualization exerts influence on the decisions made by both senior citizens and informal caregivers [17, 18]. As the bridge connecting the underlying digital technology with the real world, information visualization undertakes the responsibility of representing data abstractly, promoting insights and discovery as well as enhancing user experiences and interactions [19, 20]. It refers to the process of displaying data in a visual and meaningful way in order to enable users to understand the information regarding this data efficiently. According to existing studies, it is "using visual representations and interaction techniques which take advantage of the human eye's broad bandwidth pathway into the mind to allow users to see, explore, and understand large amounts of information at once" [21, 22]. A well-designed visualization considers the characteristics of older adults and their informal caregivers and is able to meet their information needs [23, 24]. Contrarily, an ill-designed visualization will result in ambiguous or unclear information and also cause misunderstanding. Information visualization from the interaction perspective has been proposed and proved as an effective method [25].

In this paper, we had a literature review about the information visualization regarding smart home monitoring for older people. Research questions are:

- What are the information needs of the older adults or their caregivers regarding smart home monitoring?
- What are the design recommendations that help manifest these needs in the design of visualization?
- What are the problems and gaps in this field?

2   Materials and Methods

2.1   Keywords

Several synonymous terms for senior citizens, informal caregivers, visual display systems and health smart homes were searched. (Terms related with senior citizens like aged or old or older or "older adult" or "older people" or "older person", etc.; terms related with informal caregivers like "homecare workers" or "care workers" or caregiving, etc.; terms related with information visualization like vision or "user interface" or visualization or display or "data visualization", etc.; terms related with smart home monitoring like "ubiquitous technology" or "ambient assisted living" or "home automation" or "electronic assistive technology" or "healthcare monitoring", etc.).

2.2   Inclusion Criteria

For the articles to be included, they are required to be written in English and published from 2008 to 2018. Conference abstracts and other reviews are excluded. Studies concentrating on older adults' or their families' information needs regarding visualization of smart home monitoring in home environments are included. Moreover, articles with respect to ethnographical accounts, sociological perspectives instead of software and hardware engineering are included. The other kinds of smart homes monitoring such as energy management are excluded.

The search was conducted in April 2018 and investigated the following databases: Web of Science, Scopus, PubMed, ScienceDirect and Google Scholar. These databases were selected because they cover a wide range of fields including aspects of human factors, human–computer interaction (HCI), ergonomics, gerontology, aging, as well as disability and homecare.

In total, 18 articles were screened from 3530 articles.

3   Result

Overall, from the aspect of older adults or their informal caregivers, there are three types of needs which are related to the smart home health monitoring information; design recommendations from ten aspects on three levels were identified to manifest users' expectations or perspectives on visualizations of the smart home monitoring systems. Because some factors are integrated with the others, the boundaries of them are not clear and entirely sharp, thus they have overlaps to a certain degree. The complete map is described in Table 1.

4   Analysis

To address the first research question in this paper, the information needs of older people and their informal caregivers were categorized as physical needs, emotional needs and cognitive needs. Physical needs are related to sensory-perceptual process; cognitive needs are relevant with information understanding; emotional needs are associated with personal emotion and feelings.

To solve the second research question in this paper, design recommendations from ten aspects were categorized in to three levels (the visceral level, the behavioral level and the reflective level) according to Norman's emotional design model.

4.1   Older People

As for the physical needs, even though older adults are different in their vitality and experience, aging is usually related to various changes in the sensory-perceptual process and mobility. Needs in this domain are mainly about impaired eyesight and motion declines, which have significant implications for visualization design [26, 27]. The visual decrement becomes a normal but severe problem for older people [17, 28].

Table 1. The map of reviewed articles.

Articles	Target groups	Needs			Design recommendations										
					Visceral level		Behavioral level				Reflective level			
		Physical needs	Cognitive needs	Emotional needs	Data abstraction	Presentation format	Reliability	Customizability	Sharing	Guiding	Safety and privacy	Self-identity	Assurance from others	context awareness
[29]	1	✓	✓		✓	✓		✓						
[30]				✓					✓		✓	✓	✓	
[31]			✓	✓			✓				✓	✓	✓	
[31]			✓	✓				✓				✓		
[33]		✓	✓	✓	✓		✓	✓			✓	✓	✓	✓
[26]		✓	✓	✓	✓		✓	✓					✓	
[27]		✓	✓		✓	✓						✓		
[17]		✓	✓	✓	✓	✓							✓	
[32]			✓	✓			✓				✓	✓		✓
[38]		✓	✓		✓			✓						
[37]		✓	✓		✓			✓						
[28]			✓					✓		✓				
[31]		✓	✓	✓	✓	✓		✓		✓				
[39]	2	✓		✓			✓	✓	✓		✓	✓		✓
[36]	2	✓	✓	✓			✓	✓	✓		✓		✓	
[35]	3	✓	✓	✓	✓		✓	✓	✓	✓	✓	✓	✓	✓
[34]		✓		✓		✓		✓	✓	✓	✓		✓	✓
[14]		✓	✓	✓	✓		✓	✓	✓		✓	✓	✓	✓

Tip: for the target group, 1 means older adults, 2 means informal caregivers, 3 means older adults and informal caregivers.

It mainly includes the decrease in visual acuity, declines in contrast sensitivity, a loss of dark adaptation as well as a reduction in the range of ocular accommodation [26]. These losses cause older people difficulty to locate crucial information from complex interfaces, to perceive small-scale components on a display, or to read small captions. Additionally, motor impairments, which include longer response times, declines in flexibility, less capacity in balance, may lead to challenges for older people to control small elements and moving targets [27, 29, 30].

The cognitive needs of older people are mainly about understanding specific info-graphics. It is not only related to the cognitive abilities of older adults themselves [18, 27], but also associated with their experience and cultural back-ground [31]. On one side, the cognitive decline can be seen as a big problem resulting from aging, including a diverse number of effects such as reductions in attention, memory, problem-solving, retrieval and reasoning [17, 18, 26]. It poses a big challenge to the older adults when they face dense information or complicate computer tasks (e.g. visual search). On the other side, older adults perform poorer than young people due to their lack of experience regarding the typical meanings of some elements in infographics (e.g. icon).

The emotional needs focus on older adults' feelings and emotions. Privacy, safety, isolation and being labeled are the main topics in this domain. There are conflicting attitudes towards the sense of safety and privacy. Some older people feel uncomfortable as they are "being watched" [29, 32] while others illustrated that the feeling of safety takes precedence over the sense of privacy [31, 33]. As for the feeling of being labelled vulnerable, older people express that it results in a negative emotion and they avoid being engaged in the care process [14]. The less engagement of older users leads to not only the sense of isolated, but also the shifting of heavy burdens onto those informal caregivers [34, 35].

4.2   Informal Caregivers

As some informal caregivers are the spouses or friends of those older care receivers, they are also old and have the same physical needs and cognitive needs. Namely, they have the same problems in vision, mobility and cognitive abilities and are lack of computer experience. However, due to some computer tasks, the requirement for them to deal with infographics is higher compared to that for older care receivers. Therefore, informal caregivers have more problems regarding cognitive needs [14, 34]. Firstly, they may encounter difficulties since they have to use the technology in a context which is not designed for them (e.g. complicated terminology) [36]. Moreover, it is difficult to ensure the quality and security of data collection and transfer (e.g. wrong operation), and the technical reliability is an issue (e.g. technology calibration and maintenance) [30, 36].

In terms of the emotional needs, informal caregivers think homecare work regarding smart home monitoring can cause alien feelings in familiar home sur-roundings, confuse the roles of family members and even lead to the appearances of detachment, mutual deception and resentment [14, 34, 35]. Some informal caregivers feel uncomfortable that they have to "spy" on older adults' information even if safety is the core issue [35, 36]. Meanwhile, incorrect machine usage and inappropriate caring can generate a sense of guilt [14, 35, 36].

4.3 Design Recommendations

Recommendations on the visceral level mainly focus on data abstractions and presentation formats. They refer to which data should be selected and in which form to show the selected data. Data abstractions focus on the graphical design of raw data (e.g. color, font, layout of graphical objects) while presentation formats are about efficient delivery of information derived from raw data (e.g. different types of charts, virtual reality or augmented reality). Several recommendations were proposed in this section: contrasting the color of background and text; changing the size of titles and content [27, 33]; using sufficient white space; using simple graphs rather than complicated graphs [27]; grouping elements according to their functions and displaying them with clear boundaries [29, 37]; adding the longitudinal data (e.g. timeline) [28]; changing the color or the shape instead of the size of graphic objects [17, 37, 38].

Recommendations on the behavioral level are mainly related to the experience of using the system with infographics. They can be categorized as reliability, customizability, sharing and guiding. Reliability is about the quality and security of information presentation. Customizability is associated with fulfilling the information needs of different target groups. Sharing refers to the information sharing and cooperation. Guiding is related to the operation instruction and the error prevention. Several recommendations were proposed: adding feedbacks after correctly collecting or transferring the information; increasing the transparency of collected data [27]; adding different visual version to support different users [29, 36]; ensuring all the users to understand the goal and guide of the homecare task [29]; showing the most used items [35, 36]; using display devices which users get used to; asking users questions about performing operations (e.g. error prevention and context help) [35, 36].

Recommendations on the reflective level are about reflections and elicitations of visualization, including safety and privacy, self-identity, assurance from others and context awareness (e.g. visual metaphors related to culture or experience). Unfortunately, studies in this aspect are few. Several recommendations arise: giving the options for older people to share their data to the specific person [29, 33]; avoiding ageism [39]; making the display system behave like a guest but not a host (e.g. asking rather than telling, supporting but not controlling); proving the options for learning how to cooperate well [32]; using more metaphors into visual design [35].

5   Discussion

5.1   Recommendations to Be Improved

As for the design recommendations on the visceral level, little is known about how to transfer raw data into information and how to present information in an efficient form. Whilst many publications concern data abstraction in this domain, there are few studies concerning presentation formats. That is to say, there is a lack of studies focusing on the practice to deliver complex information in different forms.

Problems regarding design recommendations on the behavioral level are lacking studies concerning details for sharing information and guides for dealing with error operation. Although the reliability and the customizability have attracted many attentions, sharing and guiding should also be emphasized. Firstly, smart home monitoring is normally passive and those collected data are sent to the others automatically. Therefore, older people don't have the autonomy to choose who can see the data and what kind of data can be shared. Recent studies begin to consider the importance of sharing information while problems like how to ensure the efficiency of sharing information or which information should be shared mandatorily still need to be solved. In addition, studies about guiding in the monitoring system often concentrate on making guidance tips clear and preventing errors. However, more attention should be paid on the robustness in operating the system especially when users have the wrong operation. Specifically, guides for dealing with wrong operation and error are also crucial.

Regarding design recommendations on reflective level, the issue concerns context awareness, namely, individuals' background regarding culture and experience. As older people and their informal caregivers are usually lack of digital experience while few studies considering context awareness, how to take advantages of individuals' culture for efficient information delivery in vision form becomes a big problem.

5.2 Needs to Be Emphasized

Concerning the identified needs of older people and informal caregivers, emotional needs including the sense of privacy and safety, self-identity and assurances from the others should be explored more.

Although the topic of privacy and safety has been discussed a lot, it is unknown about what affect the sense of privacy and safety most and which factor can affect their sense of privacy and safety most. It is found that the sense of privacy is related to not only the practice of sharing information visualization itself, but also the behaviors of informal caregivers (e.g. how informal caregivers express their suggestions according to the shared information) and the space where older people can relax without pressures on being monitored (e.g. relaxing shower time spent in the bathroom).

Regarding the self-identity and assurances from the others, the relationship between older people and informal caregivers while the importance of informal caregivers is ought to be considered. From the aspect of older people, they have few ideas about the goals of their informal caregivers' care work. The degree of communication about details of homecare between older people and their informal caregivers is low. The negative exchange of information leads to the isolation of older people and heavy burdens on informal caregivers, which results in negative attitudes towards smart home monitoring. From the aspect of informal caregivers, studies concerning the information needs of informal caregivers are much fewer than that regarding older adults. However, in the real living environment, informal caregivers have the direct impact on older people and might be the actual users of smart home monitoring particularly when older care receivers are less engaged in the care process. Their importance in the field of information visualization should be emphasized.

6 Conclusion

It is no doubt that smart home technologies have an enormous effect on home care. However, new challenges occur as numerous data need to be presented to older people and their families effectively and efficiently.

In this paper, we reviewed studies focusing on information visualization of smart home monitoring for older people. Information needs of older people and informal caregivers were classified into three types while existing design recommendations were categorized into ten aspects from three levels. Through analyzing identified information needs and design recommendations, existing design recommendations which need to be improved and some needs which have been less studied or ignored in this field were figured out.

Although recent studies start to maintain notions of helping older people and informal caregivers to better understand smart home monitoring information, there still exists room for extending this area. The information needs of older people regarding smart home monitoring should be considered comprehensively while at the same time, the meanings of home as well as the roles of informal caregivers should be considered, emphasizing factors such as cooperation, respect, privacy and security.

This paper has several limitations. Due to that the topic is multidisciplinary, related literatures are distributed in different scientific fields and journals and terminologies are different. For these reasons some results may be overlooked.

References

1. Chan A, Yasuhiko S, Robine J (2016) International perspectives on summary measures of population health in an aging world. J Aging Health 28(7):1119–1123
2. Peter L (1991) A fresh map of life: the emergence of the third age. Harvard University Press, Cambridge
3. Bratteteig T, Wagner I (2013) Moving healthcare to the home: the work to make homecare work. In: Proceedings of the 13th European conference on computer supported cooperative work. Springer, London, pp 143–162
4. Gram-Hanssen K, Darby S (2018) "Home is where the smart is"? Evaluating smart home research and approaches against the concept of home. Energy Res Soc Sci 37:94–101
5. Sixsmith J, Andrew J, Sixsmith J (1991) Transitions in home experience in later life. J Architect Plan Res 8(3):181–191
6. Fogel S (1992) Psychological aspects of staying at home. Generations 16(2):15–19
7. García-Herranz M, Haya P, Esquivel A, Montoro G, Alamán X (2008) Easing the smart home: semi-automatic adaptation in perceptive environments. J Univers Comput Sci 14 (9):1529–1544
8. Mshali H, Lemlouma T, Moloney M, Magoni D (2018) A survey on health monitoring systems for health smart homes. Int J Ind Ergon 66:26–56
9. Demiris G, Rantz J, Aud M, Marek K, Tyrer H, Skubic M, Hussam A (2004) Older adults' attitudes towards and perceptions of 'smart home' technologies: a pilot study. Med Inform Internet Med 29(2):87–94
10. Davey J (2006) "Ageing in place": the views of older homeowners on maintenance, renovation and adaptation. Soc Policy J NZ 27:128

11. Demiris G, Hensel BK (2008) Technologies for an aging society: a systematic review of "smart home" applications. Yearb Med Inform 3:33–40
12. Majumder S, Aghayi E, Noferesti M, Memarzadeh-Tehran H, Mondal T, Pang Z, Deen J (2017) Smart homes for elderly healthcare—recent advances and research challenges. Sensors 17(11):2496–2503
13. Liu L, Stroulia E, Nikolaidis I, Miguel-Cruz A, Rincon A (2016) Smart homes and home health monitoring technologies for older adults: a systematic review. Int J Med Inform 91:44–59
14. Cesta A, Cortellessa G, Fracasso F, Orlandini A, Turno M (2018) User needs and preferences on AAL systems that support older adults and their carers. J Ambient Intell Smart Environ 10(1):49–70
15. Golant S (2017) A theoretical model to explain the smart technology adoption behaviors of elder consumers (Elderadopt). J Aging Stud 42:56–73
16. Holden J, Carayon P, Gurses P, Hoonakker P, Hundt S, Ozok A, Rivera-Rodriguez J (2013) SEIPS 2.0: a human factors framework for studying and improving the work of healthcare professionals and patients. Ergonomics 56(11):1669–1686
17. Le T, Reeder B, Chung J, Thompson H, Demiris G (2014) Design of smart home sensor visualizations for older adults. Technol Health Care 22(4):657–666
18. Le T, Reeder B, Thompson H, Demiris G (2013) Health providers' perceptions of novel approaches to visualizing integrated health information. Methods Inf Med 52(3):250–258
19. Gross T (2003) Ambient interfaces: design challenges and recommendations. In: 10th proceedings on human computer interaction: theory and practice, pp 68–72. CRC Press, Florida
20. Hawthorn D (2000) Possible implications of aging for interface designers. Interact Comput 12(5):507–528
21. Thomas J, Cook A (2006) A visual analytics agenda. IEEE Comput Graphics Appl 26(1):10–13
22. Bederson B, Shneiderman B (2003) The craft of information visualization: readings and reflections. Morgan Kaufmann, San Francisco
23. Memon M, Wagner R, Pedersen F, Beevi A, Hansen O (2014) Ambient assisted living healthcare frameworks, platforms, standards, and quality attributes. Sensors 14(3):4312–4341
24. Byrne C, Collier R, O'Grady M, O'Hare M (2016) User interface design for ambient assisted living systems. In: International conference on distributed, ambient, and pervasive interactions. Springer, Cham, pp 35–45
25. Kosara R, Hauser H, Gresh L (2003) An interaction view on information visualization. State-of-the-art report. In: Proceedings of EUROGRAPHICS 2003, pp 123–137. Wiley, San Diego
26. Vermeulen J, Neyens C, Spreeuwenberg D, van Rossum E, Sipers W, Habets H, Hewson DJ, De Witte P (2013) User-centered development and testing of a monitoring system that provides feedback regarding physical functioning to elderly people. Patient Prefer Adher 7:843–854
27. Alexander L, Wakefield J, Rantz M, Skubic M, Aud MA, Erdelez S, Al Ghenaimi S (2011) Passive sensor technology interface to assess elder activity in independent living. Nurs Res 60(5):318–325
28. Lorenz A, Oppermann R (2009) Mobile health monitoring for the elderly: designing for diversity. Pervasive Mob Comput 5(5):478–495
29. Bock C, Demiris G, Choi Y, Le T, Thompson J, Samuel A, Huang D (2016) Engaging older adults in the visualization of sensor data facilitated by an open platform for connected devices. Technol Health Care 24(4):541–550

30. Berridge C (2015) Breathing room in monitored space: the impact of passive monitoring technology on privacy in independent living. Gerontologist 56(5):807–816
31. McNeill A, Briggs P, Pywell J, Coventry L (2017) Functional privacy concerns of older adults about pervasive health-monitoring systems. In: Proceedings of the 10th international conference on PErvasive technologies related to assistive environments. ACM, Rhodes, pp 96–102
32. Boise L, Wild K, Mattek N, Ruhl M, Dodge H, Kaye J (2013) Willing-ness of older adults to share data and privacy concerns after exposure to unobtrusive in-home monitoring. Gerontechnology 11(3):428–435
33. Coughlin F, D'Ambrosio A, Reimer B, Pratt R (2007) Older adult perceptions of smart home technologies: implications for research, policy & market innovations in healthcare. In: Engineering in medicine and biology society, 29th annual international conference of the IEEE. IEEE, Lyon, pp 1810–1815
34. Reeder B, Le T, Thompson HJ, Demiris G (2013) Comparing infor-mation needs of health care providers and older adults: findings from a wellness study. Stud Health Technol Inform 192:18–22
35. Jaschinski C, Allouch B (2017) Voices and views of informal caregivers: investigating ambient assisted living technologies. In: European conference on ambient intelligence. Springer, Cham, pp 110–123
36. Hwang S, Truong N, Mihailidis A (2012) Using participatory design to determine the needs of informal caregivers for smart home user interfaces. In: 6th International conference on pervasive computing technologies for healthcare. IEEE, San Diego, pp 41–48
37. Le T, Thompson J, Demiris G (2016) A comparison of health visualization evaluation techniques with older adults. IEEE Comput Graphics Appl 36(4):67–77
38. Le T, Wilamowska K, Demiris G, Thompson H (2012) Integrated data visualisation: an approach to capture older adults' wellness. Int J Electron Healthc 7(2):89–104
39. Wild K, Boise L, Lundell J, Foucek A (2008) Unobtrusive in-home monitoring of cognitive and physical health: reactions and perceptions of older adults. J Appl Gerontol 27(2):181–200

Usability in Electronic Judicial Process

Luís Olavo Melo Chaves$^{(\boxtimes)}$

Tribunal Regional Federal da 4ª Região, Porto Alegre, Brazil
olavoergonomista@gmail.com

Abstract. The first electronic procedural system of the Brazilian Federal Justice (eproc) began to be used in 2003. It was idealized by federal judicial officials. It currently has more than 5 million shares distributed. There were interviews with system users for improvements. Priority was the ergonomic criterion of workload and flexibility. As well as the usability factors associated with efficiency and user satisfaction in their task. There was investment in the area of Usability Engineering for improvements in the system. Usability measures were performed through a successful expert system. The development of a course for system developers was part of the intervention. The eproc system is now considered by the users an agile, functional and friendly system. It is an economy for public management and respect for the environment. No paper and print supplies are used. It is available online, 24 h a day, with Internet access. At the moment, it is expanding to other spheres of the Brazilian Judiciary.

Keywords: Usability · Participatory ergonomics · Federal public service

1 Introduction

This work has as context the implantation of the judicial acts in electronic means, within the scope of an institution of the Judiciary Power. The Brazilian Judiciary is one of the three powers of the Brazilian political system, called the tripartite system. He is responsible for judging by applying the law to concrete cases. This system is composed of common justice, federal justice, labor justice, military justice and electoral justice. In this institution, the Strategic Plan was established, which establishes the Vision of the Future—"To be a standard of excellence in the jurisdictional provision, in the service to the people and in the administrative management, with the consequent recognition of society". society, accessible, effective and qualified judicial service, "and to describe the following strategic objectives: to reduce the processing time of legal proceedings; improve external relationships; improve the image; optimize processes; encourage methodological and technological innovations; improve the quality of life and the satisfaction of the institution's people; strengthen personal and professional development; implement rational management focused on management indicators; improve internal communications. This is the locus of creation of the electronic judicial process. The electronic judicial process system was pioneered in Brazil in 2002, addressing, at first, only the issues of so-called federal small claims court involving small financial amounts. The system was developed with own information technology in open source —free software, with low cost to the public power. There was no employment of paid

© Springer Nature Switzerland AG 2019
S. Bagnara et al. (Eds.): IEA 2018, AISC 822, pp. 691–697, 2019.
https://doi.org/10.1007/978-3-319-96077-7_75

licenses, the people who worked on the development of the system were public servants. The system in its first version significantly impacted the modus operandi of those involved with a classic paper legal process. As a consequence there is a significant increase in access to justice in the first instance, since actions that took fifteen months to be attended now are being attended in approximately one hundred days. The workers who made direct use of the system, said "users" presented complaints. Registries of illness with this group of workers are now being investigated by health professionals and ergonomics. Using macroergonomic tools, in preliminary studies, in 2008, among several demands, the one that emerged was the one related to the interface of the new system. It was defined as an investment priority the solution to usability problems in the system, with reference to the listening of workers/direct users who suffered with the consequences of the design of the system. We qualitatively and quantitatively map the complaints, which we demonstrate in Table 1.

Table 1. Main complaints.

Frequency order	Complaints	%
1st	Improvement of shortcuts, tabs, attachment system, visualization, movement, identification, block sending, automation, etc. to reduce the number of clicks and facilitate the organization	30,15
2nd	Improvement in visual (including documents); more logical distribution of menu items; delete information, options, and spaces that are not needed	22,15
3rd	Alternatives that reduce the use of the mouse	12,78
4th	Include new options and new possible modules in e-proc	11,91

The percentage complement of 23.01% corresponded to questions of the order of machines, equipment and the organization as a whole that had no direct implications with the software. The workers' union at the time also pointed out that deficiencies in the system were causing occupational health problems. In this way ergonomic improvements were necessary and urgent. There was investment in the research, development and implementation of a usability policy.

2   Consulting in Usability Engineering

An external consultancy specialized in ergonomics and software usability was contracted, which had the tools to carry out, in partnership with the internal team of Institutional Ergonomics, a detailed analysis of the needs of changes in the system and then an ergonomic intervention aiming at the construction of software more appropriate to the work performed. User interviews, job observations and usability tests were performed. The users were selected through a representative sample of the following segments of the population: major and minor; greater and lesser work experience; equal distribution of proportion between male and female respondents. The results of these evaluations were translated by CmapTools, which were shared with the developers team. At this stage the definition of the map structure was composed of tasks; transactions; interface components; diagnoses and suggestions for improvement. The consultant defined as ergonomic criterion priority in evaluations:

a. Minimizing user workload (brevity of presentations and individual entries; minimum actions to perform tasks; informational density);
b. Management of errors (protection against errors; quality of error messages; correction of errors);
c. Compatibility (with the user; with your task; with the environment);
d. Consistency;
e. Meaning of codes and denominations;
f. User control;
g. Readability;
h. Feedback;
i. Grouping and distinction of information;

3   Monitoring Usability

To accurately diagnose the problems pointed out by the users and the degree of dedication given by the developers to the interaction problems. It was adopted a measurement instrument that monitored the users efficiency in 19 transactions (the most frequent and the most important ones) chosen from interviews with users, highly specialized in dealing with legal issues and with a high degree of interaction with the system, as well as interviews with developers of the same system. The activities carried out to implement the monitoring of the system transactions were:

(a) Installation and adaptation of the Usemonitor tool;
(b) Analysis and modeling of possible transactions to be monitored;
(c) Treatment of log data;
(d) Analysis of the results and assembly of the objective description about the frequency and efficiency of the users and their main strategies.

The occurrence of more than 40,000 of these transactions, totaling more than 660 h of work by a group of Federal Justice workers, allowed the construction of a base of quantitative measures on the efficiency of users in these 19 transactions.

The application of the usability-oriented analytical web technique and the Usemonitor tool allowed us to identify the problematic areas of the system interface. It was produced: a specification with 176 diagnoses of disconformities on the interfaces of this system, duly accompanied by respective suggestions for improvements on their screens and dialogues. Approximately half of these recommendations of this notebook, in a year were all implemented by the developers of this system. A new usability monitoring activity using the usemonitor tool in the 19 transactions considered in the previous year was carried out with the objective of re-applying the tool in order to assemble a second series of measures on the efficiency of users in these transactions.

The comparative analysis of the values of the first and second series of values showed that the changes made in the system were effective and that there is still a need to produce more modifications.

At the end of the Consulting activities, a set of diagnoses were distributed according to their level of severity and ease of review. Severe problems (36%) are those that manifest themselves in important and frequently used interfaces that, by their nature, can cause considerable time losses for users. The problems diagnosed as mild (27%) are those that manifest in secondary interfaces and rarely used, and that only disturb the beginners, mainly. The problems of medium severity (37%) are between these two extremes.

Regarding the ease of implementation of the suggested revisions 13% are considered difficult changes, 66% are considered easy changes and 21% are considered changes of average difficulty in the changes.

Most of those suggestions could not be implemented because they were not recognized by the developers as a problem, many because of lack of knowledge of ergonomics and usability on their part. A basic usability course was then developed for all the developers of the institution. Empowering the team of developers in user interface ergonomics has been established as a strategy for advancing the usability of the system.

4 System Developers Training in Usability

A course of ergonomics and usability was hired, especially adapted to the context of the personnel who were involved in the development of the system in question. The contents were directed to the aspects that concentrated the majority of interface problems detected in the evaluation. Also discussed were topics related to practices that could contribute to the improvement of other development processes.

The course was a customized initiative for professionals involved in the development of interactive systems to learn about the latest concepts, techniques and tools for Web application development, providing greater levels of acceptance, efficiency and user satisfaction.

The objectives of the training were:

- have a good understanding of basic concepts for the development of user interfaces, providing good experiences for the user;
- have a good understanding of the knowledge applied to the design of interfaces, in the form of ergonomic requirements, especially those proposed by ISO 9241—Interactive systems ergonomics;
- Know and be able to select the most appropriate methods, techniques and tools for the different contexts of analysis, design, evaluation, testing and monitoring of ergonomics and usability of interfaces;

The contents of the course were:

- Introduction and basic concepts: ergonomics, usability and user experience/sources of reference material;
- Requirements and recommendations for the interface development process: CCU-ISO9241-210, CMMU, Usability Engineering Cycle, User Experience Cycle;
- Standards, standards, criteria, requirements and ergonomic recommendations for interfaces: ISO 9241: 10 to 17;
- Methodological recommendations: approaches, techniques and tools for analysis and user-centered design, approaches, techniques and tools for the evaluation and monitoring of usability and user experience.

5 Adaptation of the System to the Use of Persons with Disabilities

The creation of a working group to adapt the electronic process to the use of people with visual impairment was successful, when the group included disabled computer programmers and a blind user who had a programming domain in open source systems. The feeling of empathy propelled quick and efficient solutions to the system.

6 User Training in Basic Ergonomics Tools

The main objective of the training was to use one staff per group of approximately twenty people in basic ergonomic intervention tools and secondary objectives to break with the resistance, still existing at the time, with the use of computerized systems and contribute to enrichment in the field of culture digital. These points were associated with the work station, so if the chair has adjustments, the adjustments must be explored and be at the service of increasing user comfort; the position of the video monitor should be seen and regulated in depth, height and inclination; the keyboard must ensure typing comfort; the mouse should be aligned with the keyboard and not cause positions of the wrists, fingers and hands to be left out of the so-called "neutral" axes; the levels of illumination, noise, vibration and temperature must be balanced according to the norms pertaining to the theme and levels of satisfaction of the working groups; respect for cognitive styles with differences in personality and the occupation of space in the workplace must be guaranteed, manifested in the disposition of objects on the table and in the interaction of subjects with their work object and in interpersonal relationships.

6.1 Training

It was divulged institutionally with the general objective of improving the working conditions in the institution and specific objective: to train a server, by work place, in basic knowledge of Ergonomics and basic functionalities of computer science to subsidize the improvement of the usability of the system. This training took place in the cities of Curitiba, Florianópolis and Porto Alegre (the three capitals of the Southern

Region of Brazil) in two stages, the second stage approximately 100 days after the first. Each step lasts 10 h/class. To be students of the course the student should be indicated by the local authority, considering the following items in order of priority:

(a) Take an interest in ergonomics;
(b) Familiarity with the use of computerized systems for local use;
(c) To be able to establish dialogue in its working group.

6.2 Didactic Strategies

Seven face-to-face courses were held, three in the city of Porto Alegre, two in the city of Florianópolis and two in the city of Curitiba.

In each of these courses scenarios were created that are absolutely identical to those used by the population, that is, a workstation composed of a chair, a computer with all its peripherals and the same software used in its original place of work.

Subjects were organized in pairs per workstation. The doubles had as a matter of fact that the two subjects had absolutely different statures. The goal was to provide monitor height adjustments, keyboard depth, chair adjustments; postural neutralization for a subject of a certain stature and then for the subject of different stature, at the same station.

E-mail accounts were created for anyone who did not already have a Facebook account.

The discussion of the use of a tool such as Facebook in the workplace served as an introduction to the classic theme of "digital culture vs. analog culture". The creation of this group in an informal network was studied as a strategic collaborative solution in the solution of basic ergonomic problems.

One of the characteristics of Facebook is the possibility of creating groups of members with well-defined rules for participation. In this way, a group of the class was created, where each subject could collaborate with an issue that was relevant to him or a contribution to solve an ergonomic problem pointed out in the universe of the group.

During these courses, in addition to the standardization of basic repertoire of ergonomics and computerized systems in implementation, contracts were established for the continuity and deepening of the issues, contracts of secrecy, contracts of respect and ethics in dealing with personal and professional issues.

6.3 Digital Culture as a New Culture at Work

The use of Facebook in the work environment for the participants of the course represented the concrete possibility of opening up a part of the hitherto insurmountable digital culture. For many, the computer represented the working tool of the moment, which only replaced typewriters and served as a support for a working text, or even represented a limited view of what the work is.

Being able to communicate with a co-worker who presents similar and common problems, but is physically distant, even in another city, is a possibility of reducing or even ending situations that have not been solved until then.

The intensive participation during two years within the group, contributed significantly in the ergonomic conditions of the institution.

The database generated in the Facebook group remains open in the workspace as a possibility of reviewing content and with each new day many members repost videos or texts and present the phrase: "we can not forget that:" as a clear manifestation of the learning that has occurred, occurs and will still occur. The use of Facebook, in addition to contributing to the demystification of the presence of a social network in the work environment, has brought a tool to learn ergonomics content in daily work in practice and finally brings us the reflection of the learning processes at work as effective work. To paraphrase Lhuilier (2013), "work is also to establish relationships with others, to engage in forms of cooperation and change, to enroll in a distribution of places and tasks, to confront points of view and practices, to experiment and give visibility to the capacities and resources of each one, transmit knowledge and know-how, validate the singular contributions. It is, in short, to be in a position to mark with its influence, its environment and the course of things. "The reflections carried out within this group of workers ask us about the place of work and the place of learning as something inseparable and therefore a true intersection between the culture of work and digital culture.

7   Final Considerations

In its latest version, in the three southern states of Brazil, 5.5 million lawsuits have only been filed electronically since its implementation. The speed of labor increases for workers, who previously operated on sheets of paper. The paper disappeared and with it were many tasks. Those who remained are of high cognitive load and high emotional load.

The question of usability has been present and the questions that remain to be investigated are: what mental models are associated with judicial decision-making? Are these mental models consistent with the mental models of system developers? What learning styles would be best suited for subjects interacting with systems in court environments (both developers and end users)?

References

Cybis WA (2006) UseMonitor: suivre l'évolution de l'utilisabilité des sites web à partir de l'analyse des fichiers de journalisation. In: 18eme Conférence Francophone sur l'Interaction Humain-Machine. ACM, Montréal, pp 295–296

Cybis WA (2009) Estudo de Validação de Abordagem de Monitoramento de Usabilidade baseada na Análise de Logs Orientada a Transações. In: Latin American conference on human–computer interaction, Mérida, México

Lhuilier, D (2013) Trabalho (F. S. Amador, Trad.). Psicol Soc 25(3):483–492

Finding User Preferences Designing the Innovative Interaction Device "BIRDY" for Intensive Care Patients

Jan Patrick Kopetz[✉], Svenja Burgsmüller, Ann-Kathrin Vandereike,
Michael Sengpiel, Daniel Wessel, and Nicole Jochems

University of Lübeck, Ratzeburger Allee 160, 23562 Lübeck, Germany
kopetz@imis.uni-luebeck.de

Abstract. The awakening process of artificially respirated patients on intensive care units from unconsciousness is called weaning. In this phase, patients experience difficulties to communicate their basic needs or to meaningfully contact staff and relatives. This means psychological distress for all affected persons - patients themselves, medical/nursing staff and relatives. One major goal of project ACTIVATE is developing and evaluating the innovative, ball-shaped input device BIRDY. It integrates recent technology in terms of sensors, actors, energy supply and wireless communication. Ventilated patients should use BIRDY to interact with the ACTIVATE system that is intended to support communication, provide relevant information and control smart appliances in the room. This quasi-experimental study is part of a larger requirements analysis and aims to show which physical characteristics of BIRDY are relevant for potential users and which values are preferred. In the study, subjects evaluated several everyday objects in a more or less handy form with characteristic values that could be eligible for the design of BIRDY. The subjects were divided into two peer groups: adults and senior adults. The latter was explicitly chosen due to the relatively high average age on intensive care units in Germany. The setting was created as realistic as in the laboratory possible. Participants conducted a pairwise comparison, ranked objects against fixed characteristics and chose a preferred object, which was used to interact with. Within this contribution, the results of the study and derived design proposals for the interaction device BIRDY are described in detail and discussed.

Keywords: Human centered design · Weaning · Intensive care patients
Interaction device · Paired comparison

1 Introduction

Intensive care units (ICU) are specific departments of hospitals, where the most critically ill patients are treated. Patients are often artificially respirated through an endo-tracheal tube and connected to a number of life-saving and monitoring devices through intravenous lines, feeding tubes, drains, catheters and more. When awaking from unconsciousness, many patients experience difficulties communicating even their basic needs to healthcare staff and relatives, causing distress for everyone involved.

© Springer Nature Switzerland AG 2019
S. Bagnara et al. (Eds.): IEA 2018, AISC 822, pp. 698–707, 2019.
https://doi.org/10.1007/978-3-319-96077-7_76

An important goal is to reduce the artificial respiration and to reinforce autonomous breathing, this process is called weaning. The research project ACTIVATE aims to address the problems of weaning patients with an interactive system that supports them in communication, (re-)orientation and autonomy. A key component of the system is the innovative, ball-shaped interactive rehabilitation device called BIRDY, which is developed and evaluated within the project. BIRDY is designed to assist patients in communication and to let them interact with the ambient ACTIVATE system using natural interaction patterns such as rolling and squeezing.

BIRDY is planned to integrate technology such as sensors, actuators, inductive energy supply and wireless communication to provide a helpful device that is easy to use. Equally important is to ensure patient safety and security as well as to meet hospital regulations regarding hygiene and technical devices. Furthermore, it is important that such a device has a very high level of acceptance, both among its users and the medical personnel. Besides its technical features, user acceptance will likely strongly depend on the look and feel of the device.

In this article, we describe a quasi-experiment as a part of the requirements analysis of BIRDY. The study aims to determine preferences of potential users regarding the properties of the device. To specify well-founded requirements regarding its physical characteristics, it is important to know (a) the size, weight, shape, surface properties and compressibility participants prefer and (b) if other characteristics influencing participants' preferences can be determined.

2 State of the Art

Since the 1960s, interaction devices have become relevant and much research in many application fields has been done. Yet, in the specific domain of supporting communication of artificially respirated ICU patients, only a few works have been published. One approach in this field, which has been published recently, is the work by Goldberg et al. [1] who aim to improve the communication between ICU patients and nursing staff with a technical solution. They analyzed the ICU context and specified requirements for an interactive system. Based on their findings, they developed a novel communication device allowing artificially respirated ICU patients to communicate with caregivers through a tablet-based system [2]. The design of their prototype device is minimalistic and focuses on ICU patients' needs, using both visual and auditory output. The system is still in development, but first tests were positively received by the patients who also provided helpful feedback. However, their prototype does not explicitly aim at supporting the reorientation of weaning patients, which is a crucial factor to prevent a prolonged healing process or even delirium. Furthermore, a tablet requires patients to be able to visually perceive the interface (they cannot explore it by touch). For patients who cannot see clearly (e.g., due to medication or missing glasses) or who are physically weak (even raising an arm would be too much), a tablet might not be the best solution. However, their findings provide valuable information about the ICU context.

Other related devices have been developed. For instance, the Roly-Poly Mouse [3] follows a ball-shaped approach as well. It is a wireless, rolling input device that combines advantages of 2D and 3D interactions. In a first user study, potential gesture range, hand postures and stability for upper forms were investigated. While the input method is very interesting for the ICU context, given the patients' situation, feedback, especially haptic feedback is needed. PALLA [4] is another spherical input device that provides simple feedback and is designed for games and leisure activities. Both Roly-Poly Mouse and PALLA are neither compressible nor suitable to the ICU context.

The described systems and devices provide insight into possible interaction methods but are in themselves not suited as an interaction device for ICU patients. Thus, a device specifically developed for the ICU context is necessary.

3 Methods

The goal of the study was to determine user preferences regarding the properties of an interaction device for in-bed-usage of ICU patients. Given the new form of interaction in this context, suitable properties were uncertain. User testing was needed to specify well-founded requirements.

3.1 Design

A within-subjects design was used to compare the objects with each other, rank them against fixed characteristics and determine the preferred objects.

3.2 Sample

Given the target group of ICU patients, a sample was selected that is similar in age and gender of ICU patients (mean age of ICU patients is approximately 64 years). The age of our 20 senior adults ranges from 58 to 84 years. They were recruited by means of existing contacts, flyers and asking potential participants in public places. Flyers were distributed at places frequented by seniors, such as pharmacies, homes for elderly, and welfare organizations.

To allow for generalizability, this sample was contrasted to a sample of 20 adults aged from 18 to 31 years. In order to conduct the study and gather the participants' evaluation of the interaction devices, inclusion criteria were sufficient health and mobility to participate in the study and being able to communicate verbally. The adult group was recruited by staff members of the Institute for Multimedia and Interactive Systems (IMIS) during student events.

Age, gender, hand sizes, computer literacy based on the Computer Literacy Scale (CLS, [5]) and Affinity for Technology Interaction (ATI, [6]) were recorded. Table 1 shows the participants demographics and data of the ATI and CLS scale. Just one participant was left-handed. Significant differences regarding the Computer Literacy (adults: $M_a = 24.65$, $SD_a = 1.87$; senior adults: $M_{sa} = 13.45$, $SD_{sa} = 7.45$; $t_{(38)} = 6.519$, $p < .001$, eta squared = 0.528, large effect) and Affinity for Technology Interaction ($M_a = 4.43$, $SD_a = 0.87$; $M_{sa} = 3.35$, $SD_{sa} = 1.22$, $t_{(36)} = 3.171$, $p < .001$, eta squared = 0.209, large effect) between the two peer groups could be found.

Table 1. Demographic sample data divided into adults, senior adults and total participants. Mean value (*M*) and standard deviation (*SD*) are displayed. ATI Score ranges from 1 to 6 (best), CLS score from 0 to 30 (best). Hand area is defined as length by width and is used as an approximated measure

Group	Age	ATI Score	CLS Score	Hand area (in cm^2)	Gender
	M (SD)	M (SD)	M (SD)	M (SD)	(Male/female)
Adults	23.45 (3.03)	4.43 (0.87)	24.65 (1.87)	172.08 (27.46)	9/11
Senior Adults	67.25 (6.60)	3.35 (1.22)	13.45 (7.45)	176.37 (29.49)	8/12
Total	45.35 (22.8)	3.92 (1.17)	19.05 (7.81)	174.58 (28.18)	17/23

3.3 Setting

In order to achieve high ecological validity, the study took place in a specially equipped laboratory. Participants lay in a hospital bed with the head-area adjusted at a 30° angle, a position ("upper-body position") usually applied to artificially respirated patients [7]. Additionally, the participants wore gloves to simulate the motor limitations of swollen hands, a common side-effect of ICU medication. The gloves are part of the age simulation suit GERT [8], developed to cause restricted hand mobility, grip ability, and tactile perception. Due to, e.g. physical weakness and connections to infusion cannulas, artificially respirated patients may have limited hand mobility. To simulate this limitation, participants of our study were instructed not to lift their elbows from the bed surface during the study.

3.4 Material

As starting point to select suitable objects for user testing, thirty commercially available small round objects with different characteristic values were acquired. All objects were potentially suitable for BIRDY. As objects have to be compared to each other, with increasing sample size, the amount of comparisons increases rapidly, so in a first step, a highly heterogeneous yet workable sample of objects for testing had to be selected.

A pretest with 12 participants was carried out at the health fair "Gesundheit Morgen" ["health tomorrow"] in Kiel, Germany 2017. Participants lay in a regular hospital bed and freely chose five objects out of a box filled with 30 objects to evaluate and finally rank them. Analysis of the data resulted in a ranked list of preferred objects. Based on these results, eight final objects were selected for the main study. Care was taken to get a high variety of the relevant characteristic values of size, weight, shape, surface texture and compressibility (how much force must be used to compress an object). The eight objects are shown in Fig. 1. A detailed description can be found in Table 2.

3.5 Procedure

After an introduction and laying down on the hospital bed, participants conducted a full paired comparison of the eight objects: they described their first impression of the object and, after picking up the comparison object, decided which ones they did prefer.

Fig. 1. Eight study objects with differences in size, weight, shape, surface properties and compressibility

Table 2. The eight study objects and their characteristic values (shape: ◯ = spherical, ▢ = egg-shaped; compressibility: I = hard, II = resistant, III = soft, IV = very soft; surface: A = rough with tight and tiny rubber nubs, B = even and crossed with rills, C = even and crossed with seams, D = even with rubber nubs, E = even, F = rough)

ID (#)	Object	Diameter (in mm)	Weight (in g)	Shape	Compressibility	Surface
1	Spike ball	64	45	◯	III	A
2	Bouleset ball	73	722	◯	I	B
3	Baseball	76	144	◯	II	C
4	Hedgehog ball	86	28	◯	III	D
5	Styrofoam egg	80 × 62	5	▢	I	E
6	Styrofoam egg	120 × 86	12	▢	I	E
7	Soft ball	120	44	◯	IV	E
8	Foam ball	89	30	◯	IV	F

Participants were asked to elaborate on their reasons for their preference, especially with regard to the objects' characteristics. The order of the objects was randomized to avoid order effects.

After the pairwise comparisons were completed, participants ranked the objects directly for preferred size, weight, shape, surface properties and compressibility. The top three choices for each attribute were recorded.

Participants then selected their preferred overall object and stated their reasons for their choice.

Finally, participants were instructed to freely interact with their preferred object in a given scenario. Figure 2 shows the procedure of the study.

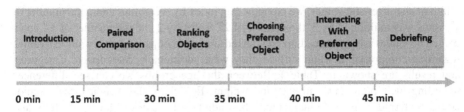

0 min   15 min   30 min   35 min   40 min   45 min

Fig. 2. Procedure of the study

4   Results

In this section, we present the results categorized by the three parts paired comparison, ranking objects and choosing a preferred object. Results concerning the interaction were published in [9].

4.1   Paired Comparison

The number of comparisons for a complete pairwise comparison results from choosing two objects (k) out of the number of objects (n). Following the equation below (1), for our eight objects, 28 comparisons were performed.

$$\text{Number of Comparisons} = \binom{n}{k} = \binom{8}{2} = 28 \tag{1}$$

Results of the paired comparison are displayed in Table 3. In total, the objects #1, #4 and #8 were preferred most often. Differences between adults' and senior adults' top choices can be seen. The adults more often chose #1 and #3 than # 4, while the choice of the senior adult group was more clearly towards #4.

4.2   Ranking Objects Against Fixed Characteristics

For the ranking of the objects regarding characteristic categories, the sum of mentions was recorded and weighted. As participants chose their top 3 objects regarding the

Table 3. Results of paired comparison; top 3 are marked bold; values represent, how often they were preferred in total; values range from 0 to 140 (total from 0 to 280); one missing value

Group	Object ID #							
	1	2	3	4	5	6	7	8
adults	**91**	34	**91**	86	87	46	31	**94**
senior adults	**113**	33	75	**114**	78	31	25	**90**
total	**204**	67	166	**200**	165	77	56	**184**

different characteristic values, we computed an index by weighting them, the number of 1st votes was multiplied by 3, the number of 2nd votes by 2 and the number of 3rd votes by 1. The values were summed up and ranked. As some objects had very similar values regarding compressibility and surface, they were additionally categorized. Categories and results are shown in Table 4. Between the age groups we can see differences regarding their preferences in size and weight. Especially the weight differs: senior adults preferred a much lighter object than adults. The index values for adults and senior adults were added up to calculate the value *total (t)*.

Table 4. Characteristic values of the objects ranked as the participants preferred them, grouped by adults (a), senior adults (s) and total sample (t); top 2 values of weight share the first place; compressibility: I = hard, II = resistant, III = soft, IV = very soft; surface: A = rough with tight and tiny rubber nubs, B = even and crossed with rills, C = even and crossed with seams, D = even with rubber nubs, E = even, F = rough)

Rank	Size (diameter in mm)			Weight (in g)			Compressibility			Surface		
	a	s	t	a	s	t	a	s	t	a	s	t
1	80 × 62	64	64	144	45	45[1]	III	III	III	D	D	D
2	64	86	76	30	28	144[1]	IV	IV	IV	A	A	A
3	76	76	80 × 62	45	144	30	II	II	II	C	F	F
4	73	80 × 62	86	44	5	28	I	I	I	F	C	C
5	86	73	73	28	30	5				B	E	B
6	120 × 86	89	89	5	722	44				E	B	E
7	89	120 × 86	120 × 86	12	44	722						
8	120	120	120	722	12	12						

Results for compressibility and surface are similar between the two age groups. Regarding the shape, the adults preferred an egg-shaped object while the senior adults preferred a spherical one. In total the spherical shape was preferred.

While female participants had smaller hands, the participants' hand size had no influence on size preference ($rho = -0.007$, $n = 40$, $p = 0.965$). Few participants named other characteristics they identified as important, namely subjectively perceived temperature (thermal conductivity; should not feel too cold or too warm) and that it should not be made of slippery material.

4.3   Choice of Preferred Object

The evaluation of the preferred object was determined by the frequency they were selected (see Table 5). The values show that in total the object #1, #4 and #8 were preferred most often. Differences between adults and senior adults can be seen regarding object #3 and #8.

Table 5. Choice of preferred object; values represent how often participants chose an object as their preferred one, broken down by age group; top 3 are marked bold; values range from 0 to 20 for each group (0 to 40 in total).

Group	Object ID							
adults	**5**	1	4	**5**	0	2	0	3
senior adults	**8**	1	0	**4**	3	0	0	**4**
total	**13**	2	4	**9**	3	2	0	**7**

5 Discussion

In this paper, we presented methods and results of a study to gather user preferences regarding characteristics of a new interaction device called BIRDY. To capture age-related differences, we recruited both adults and senior adults as participants. Based on the CLS and ATI Score values (see Table 1) the senior adults have significant less computer literacy and affinity for technology regarding interactions with digital computer systems. Additionally, the mean age of ICU patients is higher than 64 years, so we decided to follow preferences of the senior group if preferences between adults and senior adults differed.

We used two different methodic approaches to determine preferences of 40 participants resulting in similar results (objects #1, #4 and #8 were preferred). This seems to be an indicator for a stable preference rating. A possible limitation may be that the participants could only choose from our preselected objects (first 30, then eight).

The shape has been the most controversial characteristic. Senior adults preferred a spherical object, while adults preferred an egg-shaped one. We assume that adults would be more likely able to adapt how to interact with a spherical device than the other way around. Additionally, a spherical object is associated with playing/interacting, while an egg-shaped object might be negatively associated with fragility.

Regarding the size, objects with a diameter from 64 to 86 mm were preferred, sizes larger than 90 mm diameter were rarely preferred. We expected that the participants' hand sizes would affect their choice regarding BIRDY size, however, no such correlation was found. It is likely that the variety of possible interactions with the device was a reason here. If the ball was too large for the hand size, it could be rolled on the bed, otherwise, it could be held in the hand. Thus, the device should accommodate both interaction styles.

In terms of weight, objects between around 30 and 144 g were preferred. However, given the selection of everyday items, the objects' weight differed strongly and was not evenly spaced. By using objects evenly spaced on a weight distribution between 50 and 140 g, the results regarding weight preferences could be assessed more accurately. However, results of our study regarding preferred size and weight are supported by the results of the study of Perelman et al. [3].

Regarding compressibility, both groups of participants preferred easily or very easily compressible objects. Soft Materials seem to be best suited for BIRDY.

In terms of the surface and texture, participants clearly preferred objects with small rubber nubs or a rough surface over smooth surfaces.

The objects preferred most often (see Table 5) were again the objects #1, #4 and #8. This corresponds strongly to the results of the paired comparisons. These objects seem to have characteristics which lead to a high level of acceptance. Participants chose these objects due to their surface texture, softness, and size. For a more detailed analysis of the reasons why these objects were preferred see [10].

6   Conclusion

Based on the results, we defined several requirements relevant to the physical BIRDY device. Regarding size, a diameter between 64 mm to 90 mm seems to be most suitable. The preferred weight ranges from about 40 g to 150 g. Considering the technical lower bound of 70 g due to the required electronics, weight range should be between 70 g and 150 g. For interactions on a bed surface (rolling the ball over the sheet), weight seems less important, but it comes more important if patients hold the ball in their hand as lower weight requires less physical effort. The shape should be spherical as senior adults preferred it and it offers more degrees of freedom for possible interactions. For the surface, a rough texture seems to be more suited to tactile interaction. However, a rough, nubbly texture poses challenges for adherence to hygiene guidelines, as it would make the disinfection processes more difficult. Still, tactile feedback is important, not only because the majority of participants preferred a rough surface, but also because ICU patients' medication causes swollen hands and thus decreased tactile perception. A small number of nubs may be a reasonable compromise. Furthermore, we the material should be easy to compress, as it was the preferred choice by the participants and considering the decreased grip strength of ICU patients'.

Next steps include the development of BIRDY and studies to explore specific interactions between BIRDY and graphical user interfaces.

References

1. Goldberg MA, Hochberg LR, Carpenter D, Isenberger JL, Heard SO, Walz JM (2017) Principles of augmentative and alternative communication system design in the ICU setting. In: UMass Center for Clinical and Translational Science Research Retreat
2. Goldberg MA, Hochberg LR, Carpenter D (2017) Testing a novel manual communication system for mechanically ventilated ICU patients. In: UMass Center for Clinical and Translational Science Research Retreat
3. Perelman G, Serrano M, Raynal M, Picard C, Derras M, Dubois E (2015) The Roly-Poly Mouse: designing a rolling input device unifying 2D and 3D interaction. In: Proceedings of the 33rd annual ACM conference on human factors in computing systems. ACM, New York, pp 327–336

4. Varesano F, Vernero F (2012) Introducing PALLA, a novel input device for leisure activities: a case study on a tangible video game for seniors. In: Proceedings of the 4th international conference on fun and games. ACM, New York, pp 35–44

5. Sengpiel M, Dittberner D (2008) The computer literacy scale (CLS) for older adults—development and validation. In: Herczeg M, Kindsmüller MC (eds) Presented at Mensch & Computer 2008: Viel Mehr Interaktion. Oldenbourg Verlag, München, pp 7–16

6. Franke T, Attig C, Wessel D (2018) A personal resource for technology interaction: development and validation of the Affinity for Technology Interaction (ATI) scale. Int J Hum Comput Interact, 1-12

7. Wang L, Li X, Yang Z, Tang X, Yuan Q, Deng L, Sun X (2016) Semi-recumbent position versus supine position for the prevention of ventilator-associated pneumonia in adults requiring mechanical ventilation. Cochrane Database Syst Rev 1:CD009946

8. Groza HL, Sebesi SB, Mandru DS (2017) Age simulation suits for training, research and development. In: International conference on advancements of medicine and health care through technology; 12th–15th October 2016, Cluj-Napoca, Romania. Springer, Cham, pp 77–80

9. Vandereike A, Burgsmüller S, Kopetz JP, Sengpiel M, Jochems N (2018) Interaction paradigms of a ball-shaped input device for intensive care patients. In: Buzug TM, Handels H, Klein S (eds) Student conference proceedings 2018, Medical Engineering Science, Medical Informatics and Biomedical Engineering. Infinite Science Publishing, Lübeck

10. Burgsmüller S, Vandereike A, Kopetz JP, Sengpiel M, Jochems N (2018) Study of desirable characteristics of a communication device for intensive care patients. In: Buzug TM, Handels H, Klein S (eds) Student conference proceedings 2018, Medical Engineering Science, Medical Informatics and Biomedical Engineering. Infinite Science Publishing, Lübeck

Process Control

Case Studies Underrated – or the Value of Project Cases

Ruurd N. Pikaar[(⌧)]

ErgoS Human Factors Engineering,
P.O. Box 267, 7500 AG Enschede, Netherlands
Ruud.pikaar@ergos.nl

Abstract. In case studies a project is defined as the systematic design and implementation of a work system in the context of an investment project. It includes a system ergonomics design approach, an actual Human Factors (HF) intervention, feedback on project results as well as on methodology. HF activities are a small part of a project: other disciplines are involved and usually leading. HF Professionals interpret and integrate the results of scientific research. Feedback from practice could benefit researchers as well as practitioners. However, project results are not often published because it is not a part of the project scope, confidentiality, or simply a lack time and encouragement. It is not particularly helpful that the scientific community shows little interest in material presumably based on small sample sizes. To tackle this problem, IEA World Congresses and ODAM conferences since 2006 promoted company case study sessions.

A project case study is not only about looking back. New application areas and emerging technologies could benefit from HF knowledge. HF Professionals possess long time experiences, for instance regarding automation in process industries. They know the risks of high levels of automation and how to cope with them. This could be applied in autonomous shipping, traffic control, or remote operations. Transfer of proven HF knowledge requires published case studies. This paper presents an introduction in project case studies, gives a literature overview, and proposes a framework for the systematic reporting of cases. Some automation trends will be addressed, whilst showing possible benefits of knowledge transfer, or the need for this.

Keywords: Human factors engineering · Production ergonomics
Marketing ergonomics · Automation · Technology push · Project case study

1 Introduction

1.1 Human Factors Engineering

The International Ergonomics Association (IEA) defines *Human Factors* (HF) as: "the scientific discipline concerned with the understanding of the interactions among humans and other elements of a system, and the profession that applies theoretical principles, data and methods to design in order to optimize well-being and overall performance. HF Practitioners, contribute to the planning, design, implementation, evaluation, redesign

© Springer Nature Switzerland AG 2019
S. Bagnara et al. (Eds.): IEA 2018, AISC 822, pp. 711–720, 2019.
https://doi.org/10.1007/978-3-319-96077-7_77

and continuous improvement of tasks, jobs, products, technologies, processes, organizations, environments and systems in order to make them compatible with the needs, abilities and limitations of people". HF has a unique combination of three fundamental characteristics: (1) a systems approach, (2) it is design driven, and (3) it focuses on two related outcomes: performance and well-being [1]. HF Engineering is primarily a design and engineering activity. HF Professionals interpret and integrate the results of scientific research. They apply methodology developed by scientists, such as techniques for task analyses, workload assessment, or task allocation. Feedback on methodology would benefit researchers and practitioners.

A *project case study* is defined as the systematic design and implementation of a work system within the context of an investment project. A project case study paper should include a description of project steps in line with a systems approach, an actual HF intervention related to performance and wellbeing, as well as feedback on project results and methodology [2, 3]. Usually, HF Engineering is a minor part of a project; many other disciplines are involved and leading. Also note that a case study is not [3]: (a) a scientific experiment within a company setting; (b) a task analysis without the intention to intervene, or (c) a management program to reduce health & safety risks.

1.2   Why HF Practitioners Don't Write Articles

Did you lately read a Human Factors case study? Probably not, because there are few available (see Sect. 2). Why is that? For starters, it usually is not part of a project scope to write a paper about the project. Unlike scientists, HF practitioners are not rewarded in terms of funding or scientific status. Due to confidentiality, project results may not be published. This is particularly true for HF-related innovations which may bring a company ahead of its competitors. This happens relatively often, because developing new ideas is in the nature of HF inputs in projects.

New Idea – Introduction of Flat Screen Display Technology in Traffic Control. By the end of the 20th Century, large flat panel high resolution TFT-displays were introduced. Compared to the expensive and bulky projection cubes for video walls, the ergonomic advantages were obvious: improved sharpness, resolution, luminance (applicable in light environments), and easy mountable. However, it has been difficult to convince users (i.e. management) as well as video wall manufacturers of opportunities flat screen technology could, and indeed did, bring. For more information refer to DeGroot [4]. Published case studies, could have contributed to a transfer to this new technology. The author has several examples, of which two cases are summarized here.

Case 1 concerns the introduction of CCTV-supervision of two inland ship locks. Two operators work from a 270° glass walled control centre with an excellent outside view. A video wall consisting of projection cubes for CCTV-images of the locks and lock approaches was located in front of the operators, blocking the middle section of the windows. Operators complained about the bad resolution, low luminance and glare of the video wall, and about the blocked outside view. Human factors engineers developed a new design, based on flat screen technology, located overhead at each workstation (see Fig. 1). This redesign project could have been avoided!

Case 2. HF Engineers participated in an extension of a traffic supervision centre. The existing centre consisted of 3 multi-screen work desks, located in front of a large video wall.

Fig. 1. New technology applied for a lock control center.

A feasibility study (by an architect) concluded that the control center building would be too small for the new situation. Later it became clear that building regulations did not allow for a new building. The HF consultant introduced questions on why a large video wall was considered essential for the operator tasks. Is the information presented on the video wall relevant for all users? Are there alternative data presentation formats using less space on a video wall? Why not consider other display technologies? Design solutions were developed, including overview data presented at, and close to, each individual operator desk by means of high resolution flat screen technology. Thus, a new situation including 6 desks could easily fit in the existing center. Cost reduction was twofold: no new building, and significant reduction of costs for the display of overview information. Nowadays, this HF design is the standard solution for traffic control centers in the Netherlands [5].

Other reasons for not writing case studies are a lack of time, missing encouragement and assistance regarding scientific habits for writing a paper. Also, it is not helpful that, generally speaking, researchers don't show much interest in material with small sample sizes. This is of course the characteristic of projects in practice. Finally, HF Professionals are not driven by a need to publish, while researcher may not be allowed to participate in a conference without an accepted paper, no matter how thin the content may be. In an effort to change this, IEA world congresses since 2006 included company case study sessions, aiming at:

- Establishing a database of project case studies, thus increasing sample size with experiences from comparable projects.
- Enabling interaction across scientists and HF Professionals, in particular discussing need for, and practical relevance of, research.
- Assisting HF Professionals in getting case studies peer reviewed and published.
- Providing multi-disciplinary background to the scientific community. While researchers are experts in one or two scientific disciplines, the HF Professional has to apply many different HF disciplines, meanwhile understanding technical disciplines (construction, architecture, electrical, process engineering, etc.).

1.3   Technology Push

Emerging new technologies are another reason to publish project case studies. The 21th century introduced autonomous systems, real time remote control, and big data. For example, in the maritime area, HF knowledge on control and supervision of a vessel is limited compared to 50 years of applied HF in process industries or power generation. HF Professionals know about the effects of a high degree and high levels of automation. They know the risks, and how to cope with them. Automation experts of the 21th century should read the classic paper by cognitive psychologist Lisanne Bainbridge [6] on "the ironies of automation". Nowadays, within areas such as shipping, traffic control, or remote operations, the wheel is invented over and over again. The same mistakes are being made. Hence, transfer of proven HF knowledge and solutions will be beneficial and cutting project costs. The author believes, that for effective knowledge transfer, project case studies are essential.

2   Published Case Studies

2.1   Early Project Case Studies

In 1974, one of the first compilations of project case studies was "The human operator in process control" [7]. This book combines research and case studies, for example on the transition from local manual control to automated centralized control. It discusses questions on how many controls an operator can handle, and the application of new technologies, such as touch wire screens (yes: a first model of touch screens!).

Rijnsdorp [8] has been among the first to advocate a systems ergonomics approach in large scale automation projects. His book "Integrated Process Control and Automation" has been an engineering bible for process engineers, providing strategies and formulas for the design of chemical processes. A unique addition are 5 chapters dedicated to ergonomics: human operator tasks, alarm management, workplace design, etc. Rijnsdorp established "living" research for testing a systems ergonomics design approach. A major case has been an extensive participation of ergonomists in the Esso Rotterdam Refinery extension project. I had the pleasure to be part of this control room design project [8–10]. A few years after the refinery start-up, the new control room has been systematically reviewed [11]. After 20 operational years, HF Professionals have been part of the next generation redesign project.

The Esso Refinery case was reported at the 1989 IEA Conference "Marketing Ergonomics" (Netherlands). Other case studies by ergonomists at this conference and published in Ergonomics Volume 33 are: The design of a packaging workstation [12], The systematic development of a welding station [13], and cases from the steel industry [14]. Remarkable is the publisher's note (quote): "Readers will note that … some contributions are couched in informal, non-scientific language. Given the welcome participation of non-ergonomists and managers from industry who face ergonomic challenges …it was deemed worthwhile including this latter groups contributions to the debate …". A novelty at the conference were the on-site sessions. A session on control center design took place at the aforementioned Esso Refinery. Here, Moraal and Kragt [15], Pikaar et al. [9] presented papers on process control room design projects. Pikaar also included feedback by the engineering project manager, interior architect, and end user representatives. Extended versions of these papers have been published in the case study book "Enhancing Industrial Performance" [16]. Every chapter of this book includes feedback on the particular project, as well as a cost-benefit discussion. Other case studies in this excellent compilation are: (a) ship engine control room design [17], (b) a coal mining tunneling machine [18], and (c) manual handling in a discrete manufacturing process [19]. Although the cases mentioned in this section are >25 years old, they are still relevant for the HF profession. Technology has changed; human needs, abilities and limitations have not.

2.2 Company Case Studies

New initiatives on publishing project case studies can be attributed to the 2006 IEA World congress *Meeting Diversity in Ergonomics* [20]. Pikaar [21] introduced a standard case study format and summarized 10 cases accordingly, mainly related to control center design projects. This book [20] also includes chapters on hospital architecture [22], software development [23], and CCTV Ergonomics [24].

At the IEA2012 Congress (Recife, Brazil), special sessions on company case studies, i.e. applied ergonomics projects, were organized. The call for abstracts yielded 40 abstracts, and resulted in 18 full papers. Half of the contributions emphasized a systematic work task analysis, resulting in non-committal proposals for intervention. In 2012 this clearly represented the status and development of HF in Brazil. Cases from western countries related to manufacturing, steel industry, aviation, automotive, and health care. The contributions included: (1) Project based HF interventions, and (2) tools to manage HF programs in large organizations. Actually, most papers did not meet the definition of a project case study, because they lacked an intervention.

For IEA2015 (Melbourne, Australia), out of 68 abstracts, 45 proposals were qualified as relevant company case studies. The application areas represented logistics and warehousing, health care, product design for HF centered assembly (cars, aircraft), interface and web design, oil & gas, heavy industry, and rail. Half of the proposals originated in the congress hosting country, reflecting current interest in Safe Work Method Statement regulations. These papers present the development of reactive intervention programs, rather than pro-active design interventions. A pro-active approach was taken by Australian Post [25], where HF has been a part of the full systems design and implementation, including a formal HF review for approval by the Unions. Worth mentioning is the resemblance to an earlier case on airport baggage handling by Lenior [26], build by the same manufacturer. The manufacturer not only proposed a technical system, but also considered work organizational issues. We came across several more projects that resemble earlier projects/publications. For instance, Schreibers [27] reported in 2006 on job rotation, as a tool to reduce operator workload in parcel sorting/logistics. In 2015, the same issues could be found in Zuelch [28]. In conclusion: had the HF Professional known about the particular project, it could have been helpful for the more recent project.

2.3   Other Project Case Studies

More case studies can be found, or actually are hidden, in HF Handbooks on manufacturing, alarm management, process control, human reliability, CCTV, etc. An overview of HF aspects of CCTV (camera) systems design can be found in Pikaar [29]. Also, papers on process control graphics design projects are presented by Pikaar [30]. These topics will be addressed further in Sect. 4. New Technology.

3   Case Study Format

The idea to systematically publish cases may be attributed to Kragt [16]. Pikaar suggested to structure case study reports by using the following keywords [20].

Over the decades, HF professionals learned that a successful contribution to projects can only be achieved when the HF Professional adapts to the project phases and deadlines. The bottom right of Fig. 2 shows a generalized overview of engineering project phases based on ISO 11064-Part 1 *Principles for the design of control centers* [31]. Terminology may differ, depending on culture, country, and type of industry or organization. On a general level, there is evidence from reported case studies [32], that HF Engineering is not about additional efforts or higher project costs, probably on the contrary. It is important to stick to the systems ergonomics approach, originally by Singleton [33], always including a thorough situation analysis. The systems approach works well and is understood by the engineering community.

Project scope
- type of industry or organization
- project goal (example: build a power plant)
- investment level of the overall project
- realization period.

Structuring HF Engineering in the project
- project organization; position and responsibility of HF Professional(s)
- amount of HF work (hours), and % of investment influenced directly by ergonomics
- type and extent of user participation.

Main HF topics elaborated during the project.
- examples: job design, workplace design, interaction design.
- number of different jobs involved/to be designed
- number of different workplaces involved/to be designed.

Project phases - HF engineering steps
- project phases that included HF
- techniques applied per HF engineering step.

Lessons learned
- why did the project hire HF Engineers?
- did HF live up to the expectations
- typical results and cost/benefit of HF input
- lessons learned on the level of the project.
- lessons learned regarding HF methods.

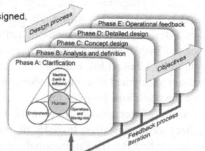

Fig. 2. Project case study format, first proposed by Pikaar in 2007 [20].

4 New Technology

The internet of things is everywhere. New technology is emerging fast. Automated production systems are being developed. Engineers work on autonomous systems, such as cars, trains, and ships. New technology pushes developments, management likes the new "toys". The aim may be full automation, however, the human factor will always be present. This section gives some examples of HF issues in new technology related projects.

Case 3 – Real Time Remote Control in Oil & Gas
Due to low oil prices, companies are looking to reduce production costs. For various reasons real time remote control could be a solution. It is proven technology in the Netherlands. According to HF case studies by Pikaar et al. [34, 35] successful remote control requires a thorough and detailed operator task analysis. Emphasis should be on communication between the on-shore control room and off-shore maintenance crews. UK and Norwegian industries are hesitant, and worker unions are against the idea. Knowledge transfer by HF evaluations of 4 Dutch on-shore control centers has been suggested. Unfortunately, no funding could be acquired. A missed opportunity to share knowledge: the wheel will be reinvented again.

Case 4 – Autonomous Ships
Several consortia are developing autonomous ships. However, these systems do need some kind of supervision from on-shore control centers. Knowledge transfer from process industries is evident, but is not happening. HF Professionals know that high levels of automation lead to less tasks and possibly boredom of supervising operators. Once something unexpected happens (and unexpected events will always take place), it is questionable whether the operator will be able to act adequately. Bainbridge [6] concluded so in 1983, Endsley [37] thoroughly elaborated on this in 2016. Project case studies from process industries are available, just need to be looked into.

Case 5 – Control Room of the Future
Vendors do a remarkable good job, selling as much screens and video walls as possible, which they might call "improving situational awareness". HF Engineers point in another direction, amongst others bases on extensive cognitive science research: less is better (see Bullemer [36], Pikaar [30]). Technology pushes companies towards showing all available data at the same time. What technology (intelligent systems) should do, is assisting the operator to select information needed for a specific process state, combining this information on a limited number of graphics. Cost reduction is evident: on a simple level, you don't need all the expensive instrumentation. On a complex level: it may very well be, that too much data presented doesn't provide an improved situational awareness. The cost of wrong actions or no action at all are high. There is no need for large screen displays: there is a need for process overview, either at the workstation, or at a shared location.

In conclusion: don't forget the lessons learnt in the past, when looking into the automated future. Learning from the past means searching for project case studies.

5 Conclusion and Discussion

Case studies are not yet considered science. However, if we publish cases in a validated format, and the number of cases grows, we surely are able to draw reliable conclusions. Published cases already indicate how useful and systematic they can be. Examples of useful knowledge transfer are given in the examples of Sects. 2 and 4.

HF Research and HF Engineering are different activities. The HF Professional interprets and integrates the results of research, often for rather complex man-machine systems. There is a one-way flow of articles from research to practice. Feedback from practice to research is limited, while researchers do not ask for it, and HF Professionals do not write about their projects. Both groups should be made aware of the need to improve marketing Human Factors. We need to prove the value of a HF input in new technology and new investment projects. HF in projects needs professionals, that are able to integrate all relevant aspects. We need to show the benefits. The best advertisement is delivering a good project, making the end users and the project owner happy with the results.

This paper started by mentioning the apparent gap between research and practice. Journals do not easily accept case study papers, because they don't fit a scientific format. Therefore, case studies require a different standard or format. The author [32] suggested the IEA to establish a database of systematically reviewed cases. A possible format has been suggested in Sect. 3. Intervention needs to be an important feature of a case study: the system or product has actually been implemented and is in operation. This enables us to draw "lessons learnt", and give feedback from practitioners to researchers regarding the usefulness and applicability of theories and methods.

Building a case study database should not be difficult in this age of online data. Content is available, for example considering project summaries HF Professionals need to provide to fulfil the requirements for registration. Also, looking at websites of major HF consultancy firms, case material is available, although in many different formats. The HF community needs a virtual environment to compile a database, such as Wikipedia, Research Gate, or Academia. Some flexible rules should be developed and accepted regarding the format of case reports, as well as for moderating the system.

The primary goal of HF is to contribute to human centered (re)design of systems [1]. It is not a goal to identify problems -things that others did wrong- and correct the problems within the limitations of an already implemented system. No, the focus shall be on business performance and investment projects: the design, redesign, or extension of production systems and other organizations [1]. Projects are run by teams of engineers, managers, and hopefully in many cases including Human Factors Engineers. So, are Human Factors Case studies underrated? Yes, they still are! But we can solve this by closing the gap between researchers and HF Professionals.

Acknowledgement. The author thanks his colleagues at ErgoS Human Factors Engineering for their contributions and giving an opportunity to work on this paper.

References

1. Dul J et al (2012) A strategy for human factors/ergonomics: developing the discipline and profession. Ergonomics 55(4):377–395
2. Pikaar RN (2012) Case studies—ergonomics in projects. Work 41:5892–5898
3. Pikaar RN (2015) Case studies—human factors in engineering projects. In: IEA2015 congress proceedings, Melbourne

4. DeGroot N, Pikaar RN (2006) Videowall information design: useless and useful applications. In: Proceedings IEA2006 congress meeting diversity in ergonomics
5. Schreibers KBJ, Bouchier JLA (2014) CCTV—case study traffic management highway tunnel. In: Broberg O et al (eds) Proceedings of the human factors in organizational design and management XI conference 2014
6. Bainbridge L (1983) Ironies of automation. Automatica 19:775–780
7. Edwards E, Lees FP (1974) The human operator in process control. Taylor & Francis, London
8. Rijnsdorp JE (1991) Integrated process control and automation. Elsevier, Amsterdam
9. Pikaar RN, Thomassen PAJ, Degeling P, VanAndel H (1990) Ergonomics in control room design. Ergonomics 33(4):589–600. https://doi.org/10.1080/00140.1390.0892.7168
10. Pikaar RN (1992) Control room design and systems ergonomics. In: Kragt H (ed) Enhancing industrial performance. Taylor & Francis, London, pp 145–164
11. Pikaar RN et al (1997) Ergonomics in process control rooms—part 3: the analyses. WIB report M 2657 X 97, International Instrument Users' Association WIB, The Hague
12. Mossink JCM (1990) Case history: design of a packaging workstation. Ergonomics 33(4):399–406
13. VanderVeen F, Regensburg RE (1990) Productivity, or the quality of work as the decisive factor in marketing ergonomics—design considerations for a new welding table. Ergonomics 33(4):407–411
14. Algera JA, Reitsma WD, Scholtens S, Vrins AAC, Wijnen CJD (1990) Ingredients of ergonomic intervention: how to get ergonomics applied. Ergonomics 33(5):557–578
15. Moraal J, Kragt H (1990) Macro-ergonomic design: the need for empirical research evidence. Ergonomics 33(5):605–612
16. Kragt H (ed) (1992) Enhancing industrial performance. Taylor & Francis, London
17. Döring B (1992) Determining human–machine interface requirements for a highly automated ship engine control centre. In: Kragt H (ed) Enhancing industrial performance. Taylor & Francis, London, pp 27–48
18. Verboven J, Vanwonterghem K, Verhagen P (1992) Ergonomic aspects of the development of a tunneling machine for coal mining. In: Kragt H (ed) Enhancing industrial performance. Taylor & Francis, London, pp 91–102
19. Steegs WL (1992) An ergonomic study of body complaints during the assembly of copy machines: implementation of a proposal for redesign. In: Kragt H (ed) Enhancing industrial performance. Taylor & Francis, London, pp 103–120
20. Pikaar RN, Koningsveld EAP, Settels PJM (eds) (2007) Meeting diversity in ergonomics. Elsevier, Amsterdam
21. Pikaar RN (2007) New challenges: ergonomics in engineering projects. In: Pikaar RN, Koningsveld EAP, Settels PJM (eds) Meeting diversity in ergonomics. Elsevier, Amsterdam, pp 29–64
22. Villeneuve J, Remijn SLM, Hignett JLS, Duffy AE (2007) Ergonomic intervention in hospital architecture. In: Pikaar RN, Koningsveld EAP, Settels PJM (eds) Meeting diversity in ergonomics. Elsevier, Amsterdam, pp 243–269
23. Mulder E (2007) An integral approach to make software work. In: Pikaar RN, Koningsveld EAP, Settels PJM (eds) Meeting diversity in ergonomics. Elsevier, Amsterdam, pp 211–228
24. Wood J (2007) CCTV ergonomics: case studies and practical guidance. In: Pikaar RN, Koningsveld EAP, Settels PJM (eds) Meeting diversity in ergonomics. Elsevier, Amsterdam, pp 271–287
25. Hehir S, Pikaar RN (2015) Design of a parcels sorting system. In: Proceedings 19th triennial congress of the IEA, August 2015, Melbourne

26. Lenior ONM (2012) Airport baggage handling—where do human factors fit in the challenges that airports put on a baggage system? Work 41(2012):5899–5904

27. Schreibers KBJ, DeGroot, VanTuil W (2006) How to design rotating tasks within jobs? In: Proceedings of 16th Triennial Congress of the IEA, Maastricht

28. Zuelch G (2015) Ergonomic evaluation of a U-shaped assembly system. In: Proceedings 19th Triennial Congress of the IEA, August 2015, Melbourne

29. Pikaar RN, Lenior TMJ, Schreibers KJB, DeBruijn D (2015) Human factors guidelines for CCTV system design. In: Proceedings 19th Triennial Congress of the IEA, August 2015, Melbourne

30. Pikaar RN (2012) HMI conventions for process control graphics. Work. https://doi.org/10.3233/WOR.2012.0533.2845

31. ISO11064-1 (2000) Ergonomic design of control centers—part 1: principles for the design of control centers. ISO, Geneva

32. Pikaar RN (2008). Ergonomics in engineering projects—how to achieve a booming business; In: Sznelwar L, Mascia F, Montedo U (eds), Human factors in organizational design and management—IX

33. Singleton WT (1967) Ergonomics in systems design. Ergonomics 10:541

34. Pikaar RN, Landman RB, DeGroot N, DeGraaf L (2012) On-shore supervision of off-shore gas production—human factors challenges. In: DeWaard D et al (eds) Human factors of systems and technology. Shaker Publishing, Maastricht, pp 293–305. https://doi.org/10.13140/2.1.2159.2963

35. Pikaar RN, DeGroot N, Mulder E, Remijn SLM (2016) Human factors in control room design & effective operator participation. In: Proceedings SPE intelligent energy international conference and exhibition, Aberdeen, Society of Petroleum Engineers

36. Bullemer P et al (2008) ASM Consortium guidelines—effective operator display design. Honeywell International Inc./ASM Consortium, Houston

37. Endsley MR (2017) From here to autonomy: lessons learned from human automation research. Hum Factors 59(1):5–27

The Life and Contributions of Neville Moray

P. A. Hancock[✉]

University of Central Florida, Orlando, FL 32826, USA
peter.hancock@ucf.edu

Abstract. This paper seeks, albeit incompletely, to provide a descriptive overview of the life and science of Neville Peter Moray. From his earliest contributions to the very foundations of cognitive psychology to his final works upon the ultimate purpose of humankind, he illuminated half a century of human knowledge with his erudition, his curiosity, and his manifest joie de vivre. The litany of his academic achievement is here leavened by some personal reflections of the character and influence of this unique scholar.

Keywords: Scientist · Mentor · Friend · Scholar · Neville Moray

1 His Early Career

Neville Moray was born on May 27th, 1935 and died on December 15th 2017 in the south of France. Moray attended a famous English, Catholic Public School; Ampleforth, before entering Worcester College, Oxford in 1953 where he initially read medicine. However, he soon found his passion for the expanding field of experimental psychology in which he obtained his doctorate degree in 1960. He spent one year as a visiting faculty member at the University of Hull in 1959 before proceeding to full-time positions over the next decade, first as Lecturer and then as Senior Lecturer at the University of Sheffield.

Moray's early work centered around the issue of selective attention, especially its auditory dimension as evaluated through dichotic listening methodology. Much associated with the famous Cherry 'cocktail party effect' Moray made this work the centerpiece of his research for more than twenty years. Some of his experimental and theoretical papers have become classics of the area [1]. But while continuing this line of 'pure' research, Moray was exposed to the applied challenges of his work on attention during a sabbatical year spent at Massachusetts Institute of Technology (MIT). In association with luminaries such as Tom Sheridan, and also under the informal tutelage of John Senders at Brandeis, Moray began to turn his attention to the issues of human-technology interaction. It was a direction that would occupy the next thirty-three years of his career up until his formal retirement in 2001.

Following his brief trans-Atlantic residence Moray (see Fig. 1), returned to Sheffield for a short interval but in 1970 took up an appointment first as an Associate Professor and then full Professor of Psychology at the University of Toronto in Canada. He was still exploring the experimental dimensions of selective attention as his various publications attest [2, 3]. However, he was, by his own admission, being led away from

© Springer Nature Switzerland AG 2019
S. Bagnara et al. (Eds.): IEA 2018, AISC 822, pp. 721–726, 2019.
https://doi.org/10.1007/978-3-319-96077-7_78

'pure' psychology and into human factors by the enticement of the practical and important challenges posed. These he encountered especially as the United Kingdom (UK's) NATO representative to the NATO Science Committee Special Panel on Human Factors in the later 1970's. For, by 1974, Moray had once more returned to the United Kingdom being first as Professor and then latterly Head of the Department of Psychology at the University of Stirling in Scotland. It was during this time that he developed a pioneering interest in the area of cognitive workload, for which his edited volume of 1979 remains a landmark text [4], triggering off any number of subsequent discussions and lines of development [5]. All the while, he integrated his extensive knowledge of human supervisory attention and nascent issues of operator workload into an integrated program of research on human interaction with automated and semi-automated technical systems [6]. As innovations such as driverless vehicles and autonomous drones are further developed in today's generation, Moray's foundational work on supervisory control and especially the dimension of trust in these systems assumes ever greater importance and exerts a strong and continuing contemporary impact [7].

Fig. 1. Professor Neville Peter Moray

2 The Human Factors Years

It may well have been his second sojourn at MIT in 1979 that finally persuaded Moray to formally change his appointment and research affiliation to human factors for, in 1981, he became a Professor in the field, of Industrial Engineering but now back at a familiar location, once again at the University of Toronto. Here, Moray became a prominent advocate for a fuller engineering understanding of the role of human operators in systems performance and systems design. Especially, he featured a concern for modeling, and also to a degree simulation, in making explicit predictions as to human response in complex situations. In practical terms he was much involved with activities in the nuclear power industry but also took part in National Academy work on the disposal of toxic waste. On one memorable occasion, in order to fulfill the safety requirements of an on-site breathing apparatus, Neville had to shave his beloved beard. It took some time for Neville to become Neville again!

He perhaps found his optimal professional niche at the University of Illinois at Urbana-Champaign (UIUC) where he moved to assume a permanent appointment in 1988, following one year as the Miller visiting Professor at UIUC. As a characteristic of his broad-ranging interests, Moray had formal positions in the Departments of Mechanical and Industrial Engineering, and Psychology, as well as the Institute of Aviation. In Mechanical and Industrial Engineering, he headed the Engineering Psychology Laboratory (EPL) for the next seven years. He continued to produce essential work in the domains of cognitive workload and supervisory control and trust [8]. In this latter arena, his pioneering efforts inspired many works [9]. However, he was also widening his vision to evaluate exactly how a full articulation of applied psychology could apply to the major issues facing the world [10]. In this, he anticipated the 'great challenges' vision of the National Academy of Engineering by more than a decade and a half. In particular, his 1994 Plenary Address at the triennial meeting of the International Ergonomics Association (IEA) and the paper which accompanied it, had tremendous impact [11]. Such effect were evident not only on the fields of Human Factors and Ergonomics, but on a number of futurists who featured his perspectives upon approaches to practical resolutions. These thoughts and insights remain vital today, especially as these global problems have grown worse over the intervening interval [12, 13].

Moray left Illinois in 1995 to take up a position at Valenciennes in France where he gave his engineering lectures in French. He was, by now, also collaborating with researchers from the University of Tsukuba in Japan where he was a visiting scientist on a number of occasions. In this latter work, he helped test and evaluate automated systems designed to assist in the 'go/no go' decision in aircraft take-off, among several other concerns for supervisory issues. The latter were emphasized in his 1995 visiting scientist position at the Japan Atomic Energy Research institute. The work produced reflects these respective interests but also features his growing concern for the 'big' problems that often task scientists as they approach their senior status.

His final, formal appointment was as a DERA Professor at the University of Surrey in Guilford back in England, a position he held from 1997 until his formal retirement in 2001. In 2000, a special, 'festschrift' session recounting and praising his life-long contributions was held at the joint Human Factors and Ergonomics Society and International Ergonomics Association Meeting in San Diego, California. It is no surprise that he was the recipient of the President's Awards of both of these entities; the highest honor that each of these societies confer.

3 Activities in Retirement

In his retirement years Moray became an accomplished artist (Fig. 2) and, as a resident of the south of France, he noted that to paint in that region one must simply love the riot of color and appreciate the magic of form. It is evident from his canvases that he did so, and with great élan. He continued to work actively, especially in areas related to Ergonomic theory [14] and even some very practical, case-study reporting almost up to the end of his life [15]. His final 'capstone' philosophical work, and arguably his "Magnum Opus," that summed up his overall perspective was the text entitled *Science, cells and souls: An introduction to human nature.* [16]. Both Moray and Sheridan [17] were writing in this same vein at the time and both works benefitted considerably from an on-going email interchange about content and conclusion. This final text of Moray's provides a fitting epitaph to his life in science.

Fig. 2. A representative canvas from Moray's brush as commissioned by the present author. The title of the piece is: 'Space × Space × Time × Time.'

4 Some Personal Reflections

On a personal level, Neville was supportive. He had an unquenchable intellectual curiosity about all that was around him and often, discussions with Neville would never touch on his professional work but range across much wider intellectual vistas to the great advantage of all present. Yet he was always generous with his time, his intelligence and his friendship. He was welcoming to all and was especially sensitive to beginning graduate students whose trepidation was calmed and ego boosted by his genuine interest and his virtually unparalleled mentoring skills. We have not simply lost a colleague and friend, we have lost a repository of knowledge, understanding and wisdom. It is hoped that Neville's own work can help us encode such wisdom so that it does not pass with the person but persists in forms much more intimate and empathic than the standard publication repositories. Further information as to fuller details on the life and contributions of Neville Moray can be readily accessed [18].

If, by any of the foregoing remarks I have implied that Neville Moray was a somber character, I wish to dispel that impression immediately and resolutely. Almost above all, Neville was fun. To be in his presence was to be energized. Not simply about science but about life. Meeting with Neville was always a pleasure. Though I am sure he was heir to the foibles and shortfalls of all of us, being able to talk and interact with Neville was always something one looked forward to. If the same could ever be said of me I would feel it would repay, if only in some small measure, his fellowship, mentorship, and friendship. From my own personal perspective, I shall always 'code' Neville as one part of the immortal triumvirate whose other members, John Senders and Tom Sheridan have now achieved almost legendary status in our corner of academe. Perhaps one of my fondest memories was the sight of Moray and Senders formally 'serenading' Sheridan on the occasion of his retirement. It was not the absolute height of musical achievement but in terms of friendship, I have rarely seen it equaled and have yet to see it surpassed. I miss Neville as a scientist, as a mentor, but above all as a friend. We shall not see his like again.

Acknowledgments. I am very grateful for the insightful observations of Professor Penelope Sanderson in helping to improve this work. The comments of Professor Thomas Sheridan were also most helpful in producing the final work.

References

1. Moray N (1967) Where is capacity limited? A survey and a model. Acta Physiol (Oxf) 27:84–92
2. Moray N (1969) Attention: selective processes in vision and hearing. Hutchinson, London
3. Moray N (1969) Listening and attention. Penguin Science of Behaviour, Harmondsworth
4. Moray N (ed) (1979) Mental workload: theory and measurement. Plenum Press, New York
5. Hancock PA, Weaver JL, Parasuraman R (2002) Sans subjectivity, ergonomics is engineering. Ergonomics 45(14):991–994

6. Moray N (1986) Monitoring behavior and supervisory control. In: Boff KR, Kaufman L, Thomas JP (eds) Handbook of perception and human performance, chapter 40, vol 2. Wiley, New York, pp 1–51
7. Hancock PA (2017) Imposing limits on autonomous systems. Ergonomics 60(2):284–291
8. Lee J, Moray N (1992) Trust, control strategies and the allocation of function in human–machine system. Ergonomics 35(10):1243–1270
9. Hancock PA, Billings DR, Olsen K, Chen JYC, de Visser EJ, Parasuraman R (2011) A meta-analysis of factors impacting trust in human–robot interaction. Hum Factors 53(5):517–527
10. Moray N (1993) Technosophy and humane factors. Ergon Des 1(4):33–39
11. Moray N (1995) Ergonomics and the global problems of the 21st century. Ergonomics 38 (8):1691–1707
12. Hancock PA (2018) In praise of civicide. Sustain Earth **(in press)**
13. Thatcher A, Waterson P, Todd A, Moray N (2018) State of science: ergonomics and global issues. Ergonomics 61(2):197–213
14. Moray N, Hancock PA (2009) Minkowski spaces as models of human–machine communication. Theor Issues Ergon Sci 10(4):315–334
15. Moray N, Groeger J, Stanton N (2016) Quantitative modelling in cognitive ergonomics: predicting signals passed at danger. Ergonomics 60(2):206–220
16. Moray N (2014) Science, cells and souls: an introduction to human nature. Authorhouse, Bloomington
17. Sheridan TB (2014) What is God? Can religion be modeled?. New Academia Publishing, Washington
18. https://en.wikipedia.org/wiki/Neville_Moray; http://moraysatmagagnosc.com/, http://cms.hfes.org/Cms/media/CmsImages/profile_moray.pdf

Ergonomics Analysis of Alarm Systems and Alarm Management in Process Industries

Martina Bockelmann[✉], Peter Nickel, and Friedhelm Nachreiner

Gesellschaft Für Arbeits-, Wirtschafts- Und Organisationspsychologische
Forschung (GAWO) e.V., Oldenburg, Germany
martina.bockelmann@gawo-ev.de

Abstract. A study dealing with human factors and ergonomics in the design of alarm systems and alarm management in the process industries was conducted at 15 workplaces for control room operators across different branches of industry in Germany. The results show that none of the systems under investigation fulfilled all the design recommendations derived from relevant literature, guidelines or standards. Thus, the results indicate an increased risk potential for incidents. Need for action to improve alarm systems and alarm management has, inter alia, been found particularly necessary with regard to alarm prioritization, consideration of operator performance limits, instructions on alarm handling and system support, continuous improvement processes as well as systematic training concepts and operator training concerning the alarm system and alarm handling. Results also suggest that there is still a lack of design improvements, as has already been documented in former research reports on critical incidents.

Keywords: Alarm system · Alarm management · Human system interaction
Process control operations · Ergonomics systems design · Process safety

1 Introduction

Investigations of critical incidents in the process industries with serious consequences for employees, companies, the environment and the general public reveal that much more importance should be linked to the application of ergonomics and human factors, especially in the design of alarm systems and alarm management. Among other factors, investigation reports on such events [e.g. 1–3] refer, inter alia, to the following inappropriate design conditions:

- high alarm rates/alarm floods
- poor prioritization of alarms
- non-ergonomic and/or inappropriate design of (alarm) displays
- non-response of critical alarms
- deactivated alarms
- lack of systematic training and training concepts for control room operators in dealing with critical situations

However, relevant guidelines [e.g. 4–7] as well as human factors and ergonomics literature [e.g. 8, 9] provide a broad range of design requirements and recommendations

© Springer Nature Switzerland AG 2019
S. Bagnara et al. (Eds.): IEA 2018, AISC 822, pp. 727–732, 2019.
https://doi.org/10.1007/978-3-319-96077-7_79

regarding the design of alarm management and alarm systems. As far as the available investigation reports are concerned, at least some of the knowledge and experience available has not yet been transferred into practice.

Therefore, this study was intended to clarify to what extent requirements and recommendations referring to human factors and ergonomics in process control systems design are being implemented in the design of alarm systems and alarm management in industrial control rooms in Germany.

2   Methods

A review of national and international guidelines, standards and human factors/ergonomics literature resulted in an update of a previously developed computer-based checklist [10] regarding the design quality of alarm systems and alarm management in the chemical industry.

A pilot study was conducted to test and modify the updated checklist. The final checklist version comprised 148 items, arranged in the following design areas:

1. alarm generation/alerting
2. alarm presentation
3. alarm prioritization
4. alarm system functionalities/technical measures
5. consideration of operator performance limitations
6. action guidelines and system interactions
7. control and feedback
8. alarm culture and philosophy
9. targets, performance and continuous improvement
10. documentation
11. training

This checklist was applied to different control rooms in three branches of industry; i.e. electrical power generation and distribution, food industry and chemical industry. Each alarm system, integrated in the process control system, as well as the alarm management were evaluated independently by two occupational psychologists, familiar with relevant human factors and ergonomics design requirements. Individual analyses usually lasted between 7 and 10 h. In some companies, experienced practitioners, such as technicians, system engineers and safety experts, also used the checklist to analyze the design quality. In total, 15 alarm systems in 14 control rooms were investigated in 12 companies throughout Germany. More detailed information regarding checklist development, suitability and application can be found in Bockelmann et al. [11, 12]. Critical deviations from normal operating procedures were not observed in any of the control rooms.

3 Results

From a human factors and ergonomics perspective, the results of this study indicate partly significant deviations from available requirements and recommendations for the design of alarm systems and alarm management. None of the alarm systems fulfilled all design requirements and recommendations from the checklist. Among other things, need for action has been documented with respect to alarm prioritization, consideration of operator performance limits, instructions and available system support, continuous improvement processes as well as systematic training concepts respective systematic operator training concerning the alarm system and the handling of alarms [12]. Selected results from the study will be presented briefly in this paper.

According to recorded alarm data, the average alarm rate during normal operations was frequently too high at the examined workplaces. Often, however, alarms which were presented to the control room operators were of no meaning to them, or at least not in the current process status. These were for example:

- old measuring points that no longer existed in the plant but in the software interface
- plant components that were down for different reasons
- information important to system engineers and technicians, but insignificant to the activities of control room operators
- alarm messages which control room operators did not know anything about
- alarm messages that did not require the attention or any actions by control room operators

Control room operators and supervisors reported that, during the first 10 min after a greater plant malfunction, alarm floods were the rule rather than the exception. Several control room operators said that they were sometimes "forced" to acknowledge alarms "blindly" due to high alarm rates.

In nearly two thirds of the alarm systems under investigation, prioritization according to importance and urgency was lacking for incoming alarms. Therefore, in these cases, the control room operators had to decide which alarms were to be given priority in each situation, based on their respective (and most probably restricted) experience. This deficiency seems to be especially important in critical incidences, when the alarm rate usually increases and the control room operator has to make appropriate decisions quickly.

For identifying an alarm and its causes, usually only single-line alarm messages, sometimes cryptically coded, were available to control room operators. Help systems with more detailed, alarm-related information were rarely available and then only for some selected alarms. Again, the operators had to rely on their personal knowledge and experience of dealing with these alarms, which can be especially critical, for example, in the case of rare events or process conditions and situations which have not yet occurred, at least not during their shifts or in training sessions, so that the operators are not familiar with them.

At workplaces where manual alarm shelving was possible by shift personnel, automatic reactivation of these alarms did not occur; e.g. after predefined time intervals.

There was also no reminder from the system. As a consequence, it cannot be ruled out that deactivated alarms will be forgotten and possibly permanently suppressed.

Control room operators usually received plant training and instructions on how to use the given process control system. However, systematic and specific training in dealing with alarms, high alarm rates and critical situations were quite rare. The investigation showed that useful process control system functions were sometimes unknown or unfamiliar to control room operators.

A comparison of the results of this study with factors contributing to incidents as mentioned in available investigation reports [cf. e.g. 1–3] still indicates similar problems in the design of alarm systems and alarm management (see Table 1). Present requirements and recommendations for an ergonomic design of alarm systems and alarm management have not been systematically implemented, at least not in many of the systems under investigation.

Table 1. Comparison of results from this study with contributing factors mentioned in available investigation reports (selection).

Study results (2016/2017)	Investigation reports
Average alarm rates in normal operations frequently too high (up to several hundred alarms per shift)	High alarm rates/alarm floods
Alarm messages bearing no meaning for the control room operators	
Reported alarm rates after greater plant malfunction mostly in the range from "likely over-demanding/difficult to manage" to "excessive"	
In nearly 2/3 of cases no prioritization	Poor prioritization of alarms
Usually only single-line alarm messages to identify causes	Inadequate information/misinformation
No automatic reactivation (or reminder) of alarms shelved by the shift personnel	Non-response of critical alarms
	Deactivated alarms
No (regular) systematic training and training concepts	Lack of adequate training/training concepts in dealing with critical situations

4 Discussion

The intention of the study presented was to check and inform about the implementation of given human factors and ergonomics requirements and recommendations for the design of process control systems with specific emphasis on alarm systems and alarm management. The investigation results revealed, at least in part, substantial design deficits and thus an increased risk for safe operations and occupational safety and health at work.

It should be considered that there is no statistical information available on the population of control rooms or process control systems in Germany, neither in general nor in relation to different sectors of industry and services. Therefore, it was and is not possible to draw a representative sample. This study refers to 14 control rooms only and, in addition, the participation in the investigation was voluntary. The results thus cannot be representative for control room design according to human factors and ergonomics requirements in Germany. Rather, it can be assumed that the sample should be positively biased. According to our experience from other projects, companies with rather problematic design conditions are much less willing to participate in such investigations. Based on the research strategy and the checklist available, however, it would always be possible to expand to more control rooms.

Be that as it may, despite design recommendations in standards and guidelines as well as scientific findings in the relevant literature, it appears that there is (still) need for information and a lack of implementation of information available or a lack of transformation of this knowledge into practice.

Therefore, the first thing to do is to engage in lobby activities. For the management of such companies it is important to learn about the relevance of human factors and ergonomics design requirements of alarm systems and alarm management and their potential to ensure or at least improve safe and effective plant operations as well as to help in avoiding fatal consequences that critical incidents might cause. Only then the necessary personnel, time and financial resources will be provided by the management.

Furthermore, accident insurance institutions, labor inspectorates and other safety and health experts should dedicate more attention to this topic (again) in order to identify design deficiencies, if there are any, make companies aware of them and offer support in dealing with these deficits to improve the design status and reduce the risk.

However, providing solutions in the design of alarm systems and alarm management according to human factors and ergonomics requirements involves sound and specific knowledge as well as a deep understanding of processes and problem areas; among manufacturers of process control systems, operating companies as well as labor inspectorates and workers' compensation boards.

Although many things in the design and operation of alarm systems are already well known and documented, others are not. The requirements and recommendations laid down in standards and guidelines are based on experience and good practice in this field as well as specific research and generalization of research gained from other areas. Specific, validated research approaches and results, however, are often missing. Therefore, research efforts are still needed to obtain validated scientific findings to prepare evidence-based recommendations for the design of alarm systems from a human factors and ergonomics point of view to operations in practice; i.e. the manufacturers of process control systems and companies who use these systems. But it will take time until research projects are instigated, completed and their results documented and made available. It will also require time to transform these findings into agreed guidelines to be available to the users concerned, i.e. manufacturers and operating companies. Therefore, research approaches and the transformation of their results should be intensified in order to make processes controllable for the operators, especially in the case of system malfunction. Then, a well-designed alarm system as well as a systematic and continuous alarm management supported by systematic and specific training concepts and procedures will become an essential element in the safety concept of process industry plants in order to keep them in a permanent safe state.

Acknowledgements. The research was partly funded by the Research Centre for Applied System Safety and Industrial Medicine (FSA) e.V., Germany (http://www.fsa.de/en/home/).

References

1. Kemeny JG, Babbitt B, Haggerty PE et al (1979) Report of the president's commission on the accident at Three Mile Island. http://large.stanford.edu/courses/2012/ph241/tran1/docs/188.pdf. Accessed 22 Mar 2018
2. Health & Safety Executive (HSE) (2000) Chemical information sheet 6. Better alarm handling. http://www.hse.gov.uk/pubns/chis6.pdf. Accessed 22 Mar 2018
3. U.S. Chemical Safety and Hazard Investigation Board (CSB): Investigation report: re-finery explosion and fire (15 killed, 180 injured). BP, Texas City, TX, 23 March 2005. Report no. 2005-04-I-TX, Mar March (2007). http://www.csb.gov/bp-america-refinery-explosion. Accessed 22 Mar 2018
4. ANSI/ISA 18.2 (2009) Management of alarm systems for the process industries. ISA, Research Triangle Park
5. DIN EN 62682 (2016) Alarmmanagement in der Prozessindustrie. Beuth, Berlin
6. EEMUA 191 (2013) Alarm systems. A guide to design, management and procurement. EEMUA, London
7. VDI/VDE 3699-5 (2014) Prozessführung mit Bildschirmen – Alarme/Meldungen. Beuth, Berlin
8. Bransby M, Jenkinson J (1998) The management of alarm systems. HSE Books, Sudbury
9. Hollifield B, Habibi E (2011) Alarm management. A comprehensive guide. ISA, Research Triangle Park
10. Bockelmann M (2009) Entwicklung und Überprüfung eines Prototyps eines Instrumentes zur Beurteilung und Optimierung des Gestaltungszustandes von Alarmsystemen – eine Machbarkeitsstudie (unpublished diploma thesis). Carl von Ossietzky Universität, Oldenburg
11. Bockelmann M, Nickel P, Nachreiner F (2017) Development of an online checklist for the assessment of alarm systems and alarm management in process control. In: Nah FF-H, Tan C-H (eds) HCI in Business, Government and Organizations. Supporting Business. Part 2. 19th international conference on human–computer interaction. Springer, Cham, pp 325–332. https://doi.org/10.1007/978-3-319-58484-3_25
12. Bockelmann M, Nickel P, Nachreiner F (2017) Alarmmanagement—Überprüfung des Gestaltungszustandes von Alarmsystemen mittels einer Checkliste (final report for the FSA e.V.). Oldenburg: GAWO e.V. (2017) https://www.fsa.de/fileadmin/user_upload/newsletter/newsletter_2017/Abschlussbericht_FSA_Alarm_2017-11-22_FDV-.pdf. Accessed 30 Apr 2018

Adaptive Human–Machine Interface Supporting Operator's Cognitive Activity in Process Control Systems

Alexey Anokhin$^{(\boxtimes)}$ ⓘ and Alexey Chernyaev

JSC Rusatom Automated Control Systems,
25-1, Ferganskaya Str., Moscow 109507, Russia
anokhin@obninsk. ru, ChernyaevAN@mail. ru

Abstract. The article covers the issue of information arrangement and presentation in a human–machine interface (HMI) for monitoring and control of complex process systems. Experience shows that there is no universal interface equally efficient in all cases. The article describes the concept of a multi-layered interface with three layers – system, analytical and functional. The same process information may be presented differently in these layers. Each layer is designed to support a certain type of an operator's cognitive activity: the system layer is meant for control support, the analytical layer – for assistance in the assessment and prediction of situation development, the functional layer – for the support of supervisory activities. Depending on the situation, each layer may be visible, semi-transparent or transparent (disabled). The multi-layered adaptive interface is described with the example of a VDU format for monitoring and control of a coolant circulation loop at a nuclear power plant.

Keywords: Multi-layered interface · Adaptability · Cognitive activity
Nuclear power plant · Operator

1 Introduction

While controlling complex process systems, operators solve various tasks. At a nuclear power plant, there are the following typical control tasks:

- switching – doing a sequence of process equipment switching actions (involving pumps, shut-off and control valves, etc.), and making sure that the actuation was completed properly;
- dynamic control – controlling the transients and making sure that the parameters are changing according to the intended pattern;
- checks and tests – performing checks, tests and equipment trial operation in localized process areas;
- supervisory control – overall supervisory control of a facility or a specific process area for early detection of deviations and threats to normal functioning.

Operators interact with the process system by means of a human–machine interface of several types. The modern nuclear power plants are monitored and controlled via a

© Springer Nature Switzerland AG 2019
S. Bagnara et al. (Eds.): IEA 2018, AISC 822, pp. 733–742, 2019.
https://doi.org/10.1007/978-3-319-96077-7_80

VDU-based interface. Each operator has several (from 4 to 10) VDUs at the workstation. Some of the VDUs permanently show the same VDU formats, e.g., an alarm log. Others may be used at the operator's own discretion when the operator wants to call the format of a particular process system, graphs, tables, or other information.

In order to ensure the common information environment and improve the group interaction in control rooms, a large video wall (large screen display) is installed, which usually shows a detailed diagram of the plant and the status of the main process parameters. In cases when safety assurance is needed, or if the computer system fails, operators use conventional panels and boards with electromechanical controls, buttons, switches, light panels, etc.

As digital control systems are implemented at nuclear power plants, the scope of information presented to operators is growing exponentially. Most process engineers or I&C designers are tempted to put as many sensors as possible and provide operators with all sorts of information. Moreover, operators in their turn are often asking for more and more data. Human–machine interface of a modern nuclear power plant already contains hundreds of VDU formats, and their number is growing. However, it will inevitably lead to the operators' information overload. This is especially true for complicated situations when the operator's activity involves high cognitive load, i.e. requires a great number of mental operations, like matching parameters with each other and with setpoints, recalling some rules, or doing certain arithmetic or logic operations. Another issue is that significant parameters and conditions appearing on displays and/or panels may remain unnoticed in the oversized information flow.

In this article, an approach is suggested to address this issue based on a combination of different ways of visualization of process information. A multi-layered interface resulting from such combination is more information-rich compared to the conventional one, with less VDU formats and better flexibility in information presentation with regard to the current conditions. The feature of such interface, when information may be presented in different scope and forms depending on the situation and the operator tasks, is called adaptability. The main principles of an adaptive interface are described in [1].

2 Organization of Adaptive Multi-layered HMI

2.1 Process Control HMI

The vast majority of VDU formats at the modern nuclear power plants are mimic diagrams of process systems with the dynamically changing process parameters presented in the digital format. The equipment state is shown with color filling or the changing shape of mnemonic symbols. Such process system display makes it easier to understand the process interrelations, to visualize the spread of a disturbance, and to monitor, select and control the equipment. On the other hand, the visualization in a process diagram is more equipment- rather than process-oriented. With this diagram, one cannot see the process dynamics. In order to evaluate the overall process state, the operator has to collect all necessary information (sometimes shown in various VDU formats), think it over and make a conclusion.

The latest two decades have brought the alternative ways to visualization of process information. The best-known and the most promising ones are ecological interface design, functional-oriented display and task-based display.

The *task-based display* provides information support to the operator for a certain task from the pre-defined list. One format shows the information which is relevant for one particular task only, even if the information is related to different process systems or uses different visual coding principles. Such an approach is especially useful when the tasks involves several process systems. The use of conventional system formats in such cases leads to information redundancy and requires extra navigation actions from the operator.

The *function-oriented display* is oriented on operator information support when a certain function defined on the basis of functional analysis has to be monitored or performed [2]. The information may be presented in the form of a conventional mimic diagram where each functional group of equipment (i.e. the equipment involved in the assurance of the particular process function) is highlighted visually. In other words, the process information is presented not independently, but in the context of the process function which has a certain state in each moment of time. Thus, the operator is focused not on the state of equipment, but on the state of the function. Such an approach has proven its efficiency in the monitoring and control of critical safety functions.

The *ecological interface design* is the image-based presentation of information aimed at easier comprehension and correct interpretation of data, especially in abnormal or unfamiliar situations [3, 4]. Information is presented to the operator in the form of an image displaying the physical and dynamic properties of the current processes and enabling the operator to understand the situation no matter how well it is known, even if there is no dedicated procedure in place. The ecological interface design is convenient when the operator has to monitor transients, assess deep disturbances rapidly, or match the process parameters and identify the process development trend.

This brief overview shows that there is no ideal universal interface which would be equally suitable for all kinds of tasks (see Table 1). One of the possible solutions is to provide the operator with different interfaces for different tasks [5]. However, this will lead to redundancy, the increase in the number of VDU formats and, potentially, to confusion in the selection of the best suitable format and navigation.

Table 1. Efficiency of different HMI types in support of typical operator tasks.

HMI type	Switching	Dynamic control	Inspections and tests	Supervision
Process system display	High	Low	High	Low
Task-based display	High	Medium	High	Low
Function-oriented display	Low	Low	Low	High
Ecological interface design	Medium	High	Low	High

2.2 Multi-layered Visualization Principle

Another option is to develop VDU formats which combine the features of some or all of the listed HMI types. However, such formats will inevitably be overloaded with information and therefore be incomprehensive. In order to tackle this issue, elements of different interface types may be joined in layers which appear or disappear depending on the situation or upon the operator's command [6]. In other words, the form, in which the same process information is displayed, may be adapted to the situation.

The multi-layered visualization principle may be seen on a simple example (see Fig. 1). The process system includes a water accumulator tank, an inlet pipeline with a control valve, and an outlet pipeline with a control valve supplying water to the consumer. The structure of the system can be easily seen in the system layer (Fig. 1A). In order to keep the required water level in the tank, material balance has to be maintained, which means the equal inlet and outlet water flows. Figure 1B shows the image – the analytical layer based on the principles of the ecological interface design. It enables one to match the inlet and outlet flow rates easily, and to see whether the current water level is equal to the set value (marked with a triangle). Knowing that the inlet flow is much lower than the outlet flow, one can make a conclusion that the water feed function is currently compromised. Figure 1C shows the analyzed system as two functions (the functional layer), one of which – "water feed" – is red.

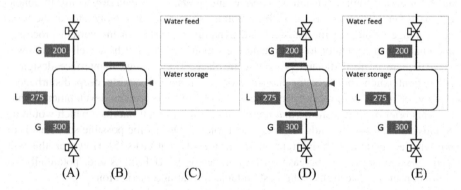

Fig. 1. HMI layers and their combinations: A – system layer, B – analytical layer, C – functional layer, D – system and analytical layers, E – system and functional layers.

The layers may be displayed independently or in combination with each other. Figures 1D, E show the analytical and the functional layers combined with the system layer. Such overlapping brings out the advantages of each information visualization method and avoids the operator overload.

In order to ensure the combination of layers, graphical objects of one layer shall not overrun the graphical objects of another layer. Objects of different layers may overlap if such overlapping causes no challenge for comprehension.

Each layer may be visible, semi-transparent, or transparent (invisible). The semi-transparent view is applied when several layers are combined. Making a layer semi-transparent makes a visual emphasis on the necessary layer, while the secondary one remains thin, and the operator's attention is not distracted.

2.3 HMI Adaptability

As it was noted above, adaptability of a multi-layered HMI means that a layer may be visible or invisible depending on the type and cognitive complexity of the task, the operator's role in the task, the knowledge of the situation and other factors. In order to select an optimal layer for the operators, one needs to have the relevant qualitative criteria.

The system layer is the basic one. It makes it easier for the operator to bind any additional information to the process equipment and the plant process areas. The system layer is a sort of a "map" for the operator, and its is advisable that it is shown in most cases. When the operator's attention has to be focused on the analytical or functional information, the system layer may be semi-transparent.

The analytical layer is needed in the following cases:

- the operators are unfamiliar with and unaccustomed to the situation, and the operational procedure is incomplete or inefficient; the operators have to find out themselves what happened and why;
- in order to understand the situation and make a decision, the operators have to do many mental operations: match several parameters with each other and with the setpoints, add and subtract values, etc.;
- the situation is dynamic, and while monitoring the situation, the operators have to decide quickly, whether the process is developing consistently in different parts of the system;
- in the current situation, it is important to monitor the material and power balances, and the levels of process media and energy.

The functional layer is efficient in the following cases:

- the operators have to watch whether the purpose of functioning of a process system or a group of equipment is achieved;
- in order to assess the performance of the equipment and systems, one needs to analyze the values of the process parameters and take several logical conditions into account, such as mode of operation and equipment state;
- there are alternative ways to fulfill the top-level process functions, which have to be taken into account by the operator when making a decision.

The functional and ecological approaches to information presentation are also efficient when one needs to assess the overall state of the plant, especially in case of a major abnormality involving deep material and energy imbalance.

The situations and tasks with the above factors are detected during the functional analysis and task analysis. The functional analysis also enables one to build the functional structure of the process facility and define the main criteria and logical conditions for the proper functioning of the facility.

3 Example of Multi-layered Adaptive HMI Implementation

3.1 Circulation Loop of RBMK Nuclear Power Plant

This section describes an example of implementation of a multi-layered adaptive HMI aimed at the monitoring and control of one process area at a nuclear power plant with RBMK reactor. The process area includes a reactor with coolant (water) pumped through it.

As water flows through the reactor, it is heated up and gets boiling. The resulting steam-water mixture comes to four steam drums where steam and water are separated from each other. Steam is forwarded to the turbine, and water is accumulated in the lower part of the drum and is forwarded back to the reactor. Coolant circulation between the reactor and the steam drums is ensured by the reactor coolant pumps (RCP). The reactor is conventionally divided into two parts – the left one and the right one. The steam drums, RCPs and the related pipelines form the coolant circulation loop which is also divided into left and right parts.

In order to preserve the material balance, steam forwarded from the steam drums shall be compensated by adding the same amount of water with the feed pumps via the feedwater units. The feedwater units include control valves which regulate the amount of water forwarded to the steam drums.

Some of the water fed to the reactor is forwarded to filters. The filtered water is returned to the steam drums. If the feedwater flow rate decreases dramatically, the emergency core cooling system is actuated. It supplies water from the emergency hydraulic accumulators directly to the reactor.

3.2 Functional Structure of Circulation Loop

The operation of the plant processes in question can be described in seven process functions:

1. Steam generation. The function involves steam and water separation in a steam drum. Performance of the function can be seen by the flow rate of steam delivered to the turbine and by the level of water in the steam drums (the level shall not exceed 200 mm). In the example, the function is performed normally.
2. Coolant accumulation. The function involves accumulation of water in the steam drum in the amount sufficient for reactor cooling. The function is considered degraded when the water level is below the setpoint (about – 500 mm). It can be seen in Fig. 2 that the water level in the left steam drum has nearly reached the lower setpoint which causes the reactor power decrease.
3. Coolant circulation. The function involves building up the pressure sufficient for coolant delivery from the steam drums to the reactor. The function is considered completed if the required pressure is ensured, the required number of RCPs are actuated (depending on the current reactor power level), and if the "RCP suction header temperature and pressure" pair is below the saturation line so that the RCP cavitation is avoided. In the example, only two RCPs are in operation on the left instead of three, and the pressure is insufficient.

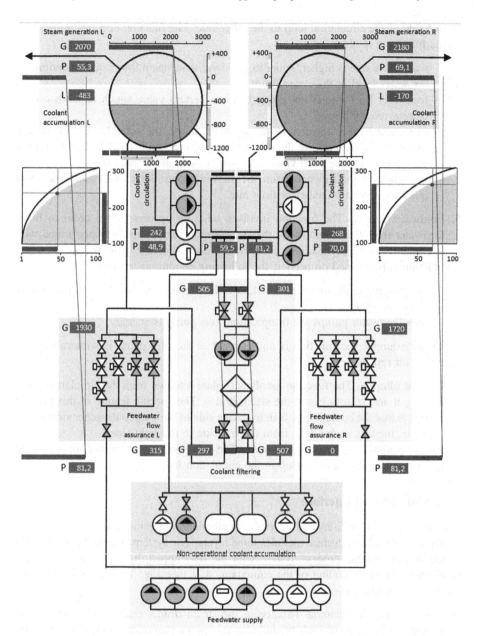

Fig. 2. Multi-layered interface for circulation loop monitoring and control at NPP with RBMK reactor.

4. Feedwater flow assurance. The function involves control of feedwater supply to the steam drums in the amount ensuring the material balance. The task is quite challenging as many flows are going through the steam drums, and water can leak from

one side of the circulation loop to the other side. The function operates normally if the following is observed:

- water flow rate is higher, equal to, or lower than steam flow rate, depending on the water level in the steam drum;
- feedwater flow rate is not going below a certain value which depends on the current reactor power;
- one control valve operates in the automatic mode, the second one – in standby mode; in order to ensure the control margin, the standby valve shall be within 30–70% range; the third valve is closed or is in the manual mode;
- the operator shall switch from the start-up controller to the main one upon reaching a certain feedwater flow rate.

5. Feedwater supply. The function involves building up pressure in order to bring feedwater from the deaerators to the steam drums. The function is performed with five main and three auxiliary feedwater pumps. For the case in question, the function is performed properly if the following conditions are met:

- the pressure in the discharge headers of the pumps is higher than the pressure in the steam drums;
- four feedwater pumps are in operation, one pump is standby.

In the example, the function is under threat as the fifth pump is withdrawn from operation for repair.

6. Coolant filtering. The function involves coolant letdown from the circulation loop, filtering it and returning it to the steam drums. The specific feature of this process system is that the coolant may leak from one side of the loop to the other side. In the example, there is such a leak from the left side to the right side.
7. Non-operational coolant accumulation. The function involves water supply to the reactor in case of pressure drop in the circulation loop.

3.3   Multi-layered Interface

The prototype of a multi-layered interface for circulation loop monitoring and control is shown in Fig. 2. The interface includes three layers. The system layer is a conventional mimic diagram of the process system with the overlapped digital values of process parameters and color coding of the equipment state (see Fig. 3). The analytical layer includes seven diagrams:

- two diagrams for material balance in the steam drums enabling the operator to monitor water level, identify the imbalance between steam and water flow rates, and to monitor whether the feedwater flow rate and level are approaching the setpoints for actuation of safeguards [7];
- two P-T diagrams to monitor the conditions for RCP cavitation;
- two pressure diagrams to match pressure in the steam drums, at the discharge of the feedwater pumps and in the RCP suction header;
- material balance diagram of the coolant filtering system enabling one to identify water leakage from one side of the circulation loop to another.

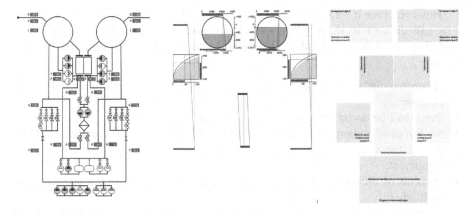

Fig. 3. Interface layers for circulation loop monitoring and control (from left to right): system layer, analytical layer, functional layer.

The functional layer includes eleven process function state indicators shown as boxes with the function names. Each indicator is colored in non-saturated red, yellow or green depending on the function state – whether the function is performed properly, is under threat (degraded), or is failing.

The importance of the functional layer can be illustrated with the following example. As it was noted earlier, the standby valve of the feedwater unit shall be in the 30–70% position. The 90% position is not a sign of malfunction; however, it narrows the control range. Informing the operator about this by an alarm is unfeasible as the event is insignificant, and the number of alarms is already high. In this case, the functional layer is the optimal way to draw the operator's attention to the event.

4 Discussion

HMI design is difficult to formalize. An idea may look simple when illustrated with a simple example. However, as soon as we come from the simple example (shown in Fig. 1) to the real prototype (see Fig. 2), we face a lot of problems which demand much creative effort and imagination to solve them. When developing the prototype of the VDU format described in this article, the authors faced the following challenges.

In order to build the functional layer, we need functional analysis focused not solely on the qualitative description of process functions, but also on exact quantitative results. In particular, the analysis shall give the carefully worded criteria enabling one to judge whether the function is performed properly. The criteria shall also be worded in terms of goals. E.g., the criterion for the heat removal function is the temperature of the medium being cooled down. It is important for the criteria to take into account the requirements for safety assurance of the systems, such as duplication and redundancy.

In order to build an analytical layer, operator tasks shall be analyzed. Such an analysis shall be based not only on process engineers' knowledge, but also on the real-life experience of operators gained at an existing plant, or at least with a simulator. For new build projects with no reference solutions in place, the task is quite challenging.

However, even if we have the necessary amount of input data to build all the layers, it is still difficult to shape the graphical composition of the VDU format. When designing the format, several factors shall be taken into account simultaneously, namely:

- parameters for visual matching shall be placed in front of each other as close as possible;
- there shall be enough free space to put diagrams and graphs;
- the layout of equipment functional groups shall be compact so that they could be put on the same template showing the state of the function performed by the equipment;
- elements of different layers, being animated and dynamic, shall not interfere with each other;
- a VDU format shall not be crowded, its composition shall be well-balanced, the mimic diagram shall have the minimal number of line intersections, the flows of process media shall be logical and easily traceable, etc.

It is easy to fulfill each of these requirements individually. But when it comes to meeting them altogether in a rich and complicated VDU format, it may end in failure because some of the requirements may be conflicting.

Information-rich VDU formats with information presented in unusual ways require thorough experimental testing. While testing, it is important that operators are trained and get accustomed to the new interface. In order to develop the adaptation mechanism, it is important to find out which layers, in which combinations, how often and in which situations will be used by operators.

References

1. Rothrock L, Koubek R, Fuchs F et al (2002) Review and reappraisal of adaptive interfaces: toward biologically inspired paradigms. Theor. Issues Ergon. Sci. 3(1):47–84
2. Pirus D (2004) Functional HSI for computerized operation. In: NPIC&HMIT 2004. ANS Inc., La Grange Park, pp. 1165–1172
3. Vicente KJ, Rasmussen J (1992) Ecological interface design: theoretical foundations. IEEE Trans. Syst. Man Cyber. 22(4):589–606
4. Burns CM, Hajdukiewicz JR (2004) Ecological Interface Design. CRC Press LLC, Boca Raton
5. Anokhin A, Marshall, E (2012) Adaptive human–system interface for control of complex systems (in application to nuclear power plant). In: 21st European meeting on cybernetics and system researches: EMCSR 2012. Bertalanffy Center for the Study of Systems Science, pp. 185–188
6. Anokhin A, Chernyaev A (2017) Multi-layered adaptive human–machine interface to support NPP operators. In: NPIC&HMIT 2017. ANS Inc., La Grange Park, pp. 26–35
7. Anokhin A, Ivkin A, Dorokhovich S (2018) Application of ecological interface design in nuclear power plant (NPP) operator support system. Nucl. Eng. Technol. https://doi.org/10.1016/j.net.2018.03.005

Implementation of Human–Machine Interface Design Principles to Prevent Errors Committed by NPP Operators

Alexey Anokhin$^{(\boxtimes)}$ ⓘ and Elena Alontseva

JSC "Rusatom Automated Control Systems",
25-1, Pherganskaya Street, Moscow 109507, Russia
anokhin@obninsk.ru, alenika-vega@mail.ru

Abstract. An approach to prevention of human errors committed by operators during monitoring and control of process systems is considered in the present paper. The purpose of the proposed approach is to eliminate the causes of errors and/or to reduce its impact on personnel activity when managing a complex process object. The reports on abnormal events occurred at nuclear power plants were used as a source of information about personnel errors. The events that took place at NPPs in 2009–2016 were analyzed. The events caused by deliberate violation of rules and regulations, lack of training and a poor safety culture were not included in this analysis. In order to categorize human errors, we used two criteria – sequence of operator task performance consisting of five stages and kind of error. These classification criteria form a two-dimensional matrix. Then, twenty-eight typical NPP operator errors extracted from the analysis and publications have been mapped on two-dimensional space. Based on analysis of standards and guidelines in the area of ergonomics and HMI design and taking into account the best practice, the seven principles and methods for error prevention have been proposed. It was shown which kinds of errors can be prevented by the use of particular principle. Thus, a set of rules are elaborated that allow HMI designer to choose the way to reduce the risk of human errors.

Keywords: Event analysis · Kind of error · Error prevention
Human machine interface

1 Introduction

When managing complex process systems, such as nuclear power plant (NPP), operator is required to perform various tasks including monitoring, decision-making and control activity. Insufficient consideration of human factors engineering (HFE) principles when designing instrumentation and control (I&C) system can lead to errors committed by operators. This can result in damage to process equipment, transition of the system to dangerous state, temporary interruption of production manufacturing and/or unsatisfactory quality of product [1].

Operators interact with NPP using several types of human–machine interface (HMI). The information on current operation conditions at NPP and all tools providing operator with possibility to control the plant are available via video display units

© Springer Nature Switzerland AG 2019
S. Bagnara et al. (Eds.): IEA 2018, AISC 822, pp. 743–753, 2019.
https://doi.org/10.1007/978-3-319-96077-7_81

(VDU) installed at operator workstation or via conventional (hardware) controls and displays, such as pointer or digital instruments, indicators, switches, push buttons, etc.

High complexity of process system and, therefore, complexity of HMI considerably increase amount of information and operator's cognitive load when performing numerous control and monitoring tasks. Such complexity contributes to the conditions in which the operator can make an error. Insufficiently trained or motivated operator, inappropriate process equipment design or unsatisfactory work conditions can be considered as root causes of error.

The following three strategies are proposed in [2], aimed at reduction of the risk of operator errors occurrence:

- replace or retrain the operator (compensation of insufficient personal peculiarities that contribute to the risk of errors occurrence);
- improve working conditions (improvement of inadequate organizational conditions, workplaces and HMI that contribute to the risk of errors);
- provide tools preventing an impact of errors on the process system (development of system that detects and corrects error).

It should be noted that the first two strategies are aimed at prevention of operator error using various approaches, while the third one is aimed at reduction of consequences of the error which have been committed.

All these strategies can be applied in case of NPP. Figure 1 illustrates them as virtual "safety barriers" preventing operator from negative impact on nuclear plant.

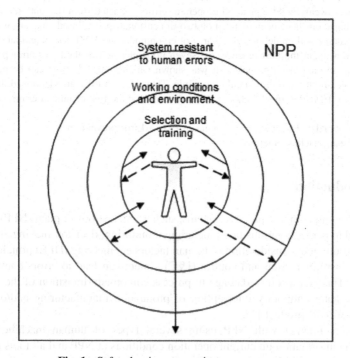

Fig. 1. Safety barriers preventing operator errors.

The first barrier consists in selection and training of operators. This barrier forbids the work of those operators who are not sufficiently trained to manage complicated situations, who don't have necessary skills and knowledge, and who have inappropriate physiological or mental capabilities. The first barrier can be implemented by many various methods, such as regular training, medical inspection before work, licensing of operators, etc.

The second barrier consists of various organizational procedures, hardware and software facilities that provide appropriate working conditions. These tools aim at development and keeping of safety culture at proper level, at formation of the required personal attitude towards the work and responsibility, at rational work and rest schedule, at provision of acceptable and comfortable environmental conditions, etc. The following measures can be considered as organizational procedures: maintenance of completeness and relevance of operational documentation, necessity to get permission before action, brief discussion before execution of the work. The second barrier also should ensure effective interaction between operator and process system. This includes the task of designing of such HMI which is harmonized with the physiological and cognitive capabilities of the operator as well as with the established professional stereotypes and expectations.

The third barrier ensures the tolerance of the system resistant to human errors. The third barrier consists in implementation of instrumentation and control system facilities which can detect the error and correct it or reduce its consequences. The following tool can be considered as examples of the third barrier: the use of automatic emergency protection programs, strengthening of resistance of the system to disturbances, providing operator with possibility to cancel operation and to return to the previous step of procedure, etc.

Properly designed HMI where the relevant capabilities and personal features of the operator are taken into account is a part of implementation of the second barrier. This can significantly contribute to reduction of risk of operator error when performing control actions at NPP.

When designing HMI for monitoring and control NPP power unit, it is necessary to take into account customer's requirements, recommendations of numerous regulations and guidances, operational experience and established professional stereotypes, as well as the characteristics inherent in population of the country where NPP is being constructed. In addition, there are a lot of ergonomics reference books and publications which contain many heterogeneous, sometimes contradictory to each other approaches and methods.

In this article, an attempt is made to analyze, summarize and systematize various approaches to HMI design aimed to prevention of operator errors.

2 Analysis and Categorization of Human Errors

2.1 Structure of Operator Task

There are many approaches to describe operator activity and to develop a model. Let's consider main phases and levels of operator activity [3].

Usually, structural models of activity are based on decomposition of operator activity to separate and relatively autonomous tasks. The task execution process, in turn, is divided into a previously fixed number of successively performed phases. Each phase can be decomposed to smaller elements of activity, such as actions and operations. Number of these elements depends on the task, while number of phases is fixed and depends on modelling philosophy.

The structural approach to modelling is based on some alphabet of typical actions (for example, search, movement in space, various kinds of perceptive and motoric actions), which constitute an elementary basis for description of operator's behavior. Actions are combined into elemental technologically significant step of activity – operation. The sequence of operations, implemented in accordance with a certain plan, forms technological task. Thus, technological task can be decomposed into separate operations.

A typical structure of process control task consisted of five phases performed sequentially by operator is as follows [4]:

- detection of signal – the time from the moment of appearance of the occurred event symptoms until the moment when the operator who previously was a passive observer detects these symptoms;
- diagnostics (identification) – identification of the occurred event, assessment of the situation and its potential threat to efficiency and safety of NPP operation;
- decision making – selection of the relevant operating procedure describing necessary actions to be performed by operating personnel in response to the event or detected symptoms;
- execution – execution of the actions prescribed by the operational procedures by the relevant operator(s);
- monitoring – check the effectiveness and the results of the performed actions, as well as adjust the actions as necessary.

Each phase of task performance to be decomposed to operations which can be categorized to four groups, namely [3]:

- perceptual operations – perception of information;
- cognitive operations – mental processing of information, memorization and decision making;
- motor operations – manipulation of controls;
- communication operations – interaction between operators.

2.2 Definition of Error

To establish relationship between operator activity and human errors, it would be useful to consider definition of a human (operator) error.

From the point of view of psychological theory, error is an inadequate reflection of objects and phenomena from the objective reality in the mind of a person. According to IEC 61513, human error or mistake is a separate action of personnel or procedure that leads to an unintended result. From the point of view of 'ergatic' systems theory, human error is some short-term human failure which is not associated with loss of ability to work (such kind of failure is called as a functional failure) [5]. Operator error can be defined as action or absence of action, which is prohibited by the operational regulation and results in deviation of the controlled process system parameters beyond the acceptable limits. Thus, operator error is any particular action performed or missed during operator activity, that is inadequate to current situation or results in overrunning the acceptable limits specified by the current process system operational mode [2].

2.3 Categorization of Errors

There are several ways to categorize human errors. The guidance [2] describes two categories of errors, namely

- omission – operator skips execution of the whole task or some element of activity;
- error committed during execution, such as erroneous selection, erroneous order of actions execution, untimely execution and poor-quality action.

Gubinsky in [5] interprets erroneous operator actions as a functional failure. He categorizes human errors to four groups, namely:

- the algorithmic functional failure – violation of logical order of actions performance prescribed by algorithm; there are three types of such kind of errors:
  - *omission* – the operations required by the algorithm have not been performed;
  - *unnecessary action* – the implemented operations are not required by the algorithm;
  - *broken sequence* – the operations required by the algorithm have been performed, but in wrong sequence (if this is significant);
- the parametric functional failure – *insufficient accuracy* of performing of action;
- the time functional failure – *untimely execution* of action;
- the wrong goal selection function failure – *incorrect choice of goal*.

Thus, the above described classification identifies six kinds of operator errors: (1) omission, (2) unnecessary action, (3) broken sequence, (4) insufficient accuracy, (5) untimely execution, (6) incorrect choice of goal.

2.4 Mapping of the Error Categories to the Task Performance Structure

Let's map the above specified error categories to the phases of task performance and to the kinds of elementary operations (see Fig. 2). This mapping reflects what kinds of errors are most probable at each phase and during execution of each type of operation. For example, omission of the signal, incorrect compare of current parameter value with setpoints, perception of unreliable data or late detection of signal can be committed during the event detection phase.

To reveal those phases of task performance where human errors are most probable, it is necessary to analyze the events and the accidents in which the operators committed errors. This analysis should identify typical errors and associate them with the phases of task performance. This make it possible to identify the most probable and most dangerous (critical) errors and to propose adequate measures that ensure compensation or prevention of these errors.

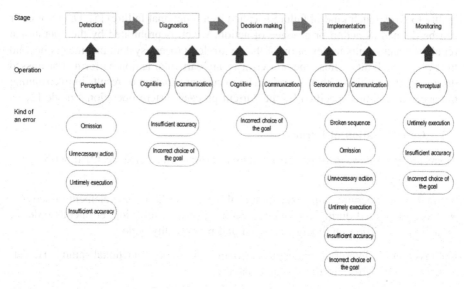

Fig. 2. Relationship between the phases of operator task performance, the types of operations and the categories of errors.

2.5   Analysis of the Events Occurred at NPPs

The reports on abnormal events occurred at NPPs were used as a source of information about human errors. The events that took place at NPPs in 2009–2016 were analyzed. The events caused by deliberate violation of rules and regulations, lack of training and poor safety culture were not included in this analysis. As a result of the analysis, the following typical and most frequent erroneous actions were identified:

- manipulation of correct control located at another power unit, another similar system, another loop instead of the required ones (incorrect choice): for example, the operator disconnected the 41BB18 switch instead of the 41BA18 switch;
- manipulation of wrong control (for example, neighboring) instead of the required one;
- setting of control to wrong position or setting wrong value: for example, setting the selection key to the 'Work' position when it was necessary to select another one; the operator typed value 123 instead of the value 0.123;
- performing an incorrect action: for example, the operator erroneously closed and confirmed the transition of the valve to the closed position;
- untimely implementation of action: for example, early locking a system or premature a pump stop;
- accidental (unintended) control action: for example, the operator accidentally touched the switch and turned it into incorrect position; the operator accidental touched the key managing the automatic power supply section for the turbine;

- erroneous space orientation (selection of wrong direction, wrong positioning, etc.): for example, the operator choses wrong direction of moving the trolley; the operator thought that he is located on one side of the refueling machine, while in fact he was located on the opposite side;
- error in detection and identification of information: for example, the operator made error when reading information from the display about the volume of water;
- omission of signal: for example, despite the presence of alarm that two shutoff valves are not open, this event was detected only after few hours.

The classification of human errors during control of NPPs as well as the causes and possible consequences of operator errors are considered in [6].

Based on the results from the analysis, it can be concluded that the most frequent and critical errors committed by NPP operators relate to the phases of decision making and implementation of actions. Confusion of unit or system and incorrect selection of control are typical errors.

It should be noted that the task performance phases, on which operator errors are most probably, depend on the features of process system and its nature. Usually, most frequent errors are associated with the most important and complicated phases, while the importance and the complexity of particular phase are different for various process systems. For example, in [7] it is noted that most of mistakes committed by pilots are related to the perception of information, i.e. they occur during the detection and monitoring phases.

2.6   Typical Operator Errors

Twenty-eight typical errors extracted from the analysis and publications [6, 7] have been mapped on two-dimensional space constituted by the categories of error (see the vertical axis in Fig. 3) and the task performance phases (the horizontal axis). This results in two-dimensional matrix of typical NPP operator errors. The empty cells of the matrix can be filled when new information on operator activity and NPPs operation will be gathered.

Prevention of operator errors aims to eliminate or reduce to a reasonable level the causes of typical errors, as well as to prevent or reduce the consequences of these errors. Let's formulate the basic principles, covering hardware and software methods and HMI facilities, which allow to compensate or to prevent errors.

3   Principles for Error Prevention

Based on analysis of standards and guidelines in the area of ergonomics and HMI design and taking into account the best practice, the following seven principles and methods for error prevention have been proposed (see Table 1). Each principle aims to eliminate the cause of one or more kinds of errors (see Fig. 4).

Management of human attention consists in development of tools supporting operators in timely detection of necessary information as well as reduction significantly the time required to search for the necessary HMI component. Implementation of this

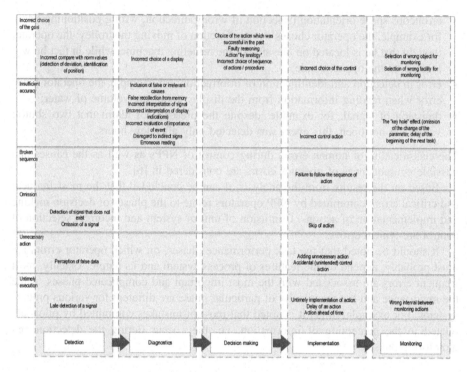

Fig. 3. Two-dimensional matrix of typical NPP operator errors.

Table 1. Human errors prevention principles and implementation methods.

Principle	Example of implementation method
Management of human attention	Add dynamics, brightness and contrast, use special layout
Implementation of redundancy and diversity	Use redundant coding and duplicate information
Providing feedback	Implement tactile, acoustic or visual human–system and system–human feedback
Complication of action	Add operation using hardware and software tools, and organizational procedures
Locking of erroneous action	Lock erroneous action and incorrect input
Unification and formation of steady skills	Unify actions and use the code consistent with stereotypes
Support operator with help and assistance	Use operator support systems (computerized procedures, diagnosis systems, timer, etc.)

principle prevents the errors of omission, untimely execution and incorrect choice of the goal at all the phases of task performance. Management of human attention is possible by means of incorporation of dynamics into any modality signal (flashing,

	Management of human attention	Implementation of redundancy and diversity	Providing feedback	Complication of action	Locking of erroneous action	Unification, steady skills	Support, help and assistance
Incorrect choice of the goal	V					V	V
Insufficient accuracy		V	V			V	V
Broken (wrong) sequence					V	V	V
Omission	V	V	V			V	V
Unnecessary action				V	V	V	V
Untimely execution	V	V				V	V

Fig. 4. The relevance of human errors prevention principles to categories of errors.

animation, etc.), display visual signal with high brightness or color contrast, use of high sound tone or volume contrast, use specially arranged layout of HMI.

Implementation of redundancy and diversity. This principle consists in redundant coding and/or duplication of information. Redundant coding is the use of several coding methods for representation of the same element of information. This provides operator with additional channels for self-control and allows to improve significantly the reliability of performing actions of various types. The duplication of information or controls prevents the errors of omission and untimely execution for all the phases of task performance.

Providing feedback is to provide operator with some confirmation that a certain event has occurred. A perceptible reaction of I&C system appearing in reply to operator's action can be considered as some kind of feedback. The second kind of such feedback is when operator provides some response in reply to occurred event, for example, alarm acknowledgment. Feedback can be tactile, acoustic or visual. This technique prevents, in general, errors of omission.

Complication of the action consists in the artificial introduction of additional operations necessary to perform the required action. The principle is used to protect against incorrect governing of controls and unnecessary actions on the phase of implementation of task. Generally, such errors are caused by inattention, negligence, deficit of time, combined activity, distraction, high workload, etc. Such errors are known in the practice of use of conventional NPP control rooms. For example, the button for increase of reactor power was pushed accidentally as a result of movement of the computer keyboard. Similar accident with another button was caused by the fallen helmet. Physical barriers, special software and hardware tool as well as organizational (procedural) methods can be used to complicate action for prevention of such unintentional errors.

Locking of erroneous action consists in creating a physical barrier which prevents operator to perform some action that is considered as erroneous in context of certain situation. The lock applies to both hardware and software controls. Locking helps to prevent incorrect action at the implementation phase of task performance caused by distraction, haste, rashness, negligence and other reasons. Input masks (format) preventing the typing of incorrect information or command can be used as a tool for locking of errors.

One of the most effective tools for errors prevention is *formation of steady skills* and formation of the workplace that is exactly consistent with these skills. These skills include both perceptual and mental skills, as well as skills to perform motor actions. The basic part of a human's skills is formed by a general cultural and professional environment and can be interpreted as a set of stereotypes (i.e., certain expectations). Another part of the skills, namely know-how skills and professional stereotypes are formed during accumulation of work experience. This approach to error prevention includes two mutually complementary activities: the formation of stereotypes and habits; and consideration of these stereotypes during HMI and workplace designing. The formation of steady skills and stereotypes ensures prevention of many kinds of errors at all phases of task performance. However, some stereotypes, for example stereotypes of diagnostics and decision making, are effective for simple tasks and can have a negative impact in complex situations. The situations are known when the operators tried to execute habitual actions in unfamiliar situation instead of implementation of other action strategy.

Support operator with help and assistance is a universal principle that aims to prevent all kind of errors. Help and support of operator during performance of task can be realized using various operator support systems (OSS). OSS is a wide range of tools aimed at helping operators at all phases of task performance – from detection of a signal to monitoring of process system reaction. OSS can be simple or complicated.

An example of a simple operator support system is a timer with forward count and countdown. The timer allows the operator to monitor duration of the process and the time remaining until a certain event should be occurred, including his own actions, such as next monitoring or control action, etc. Computerized procedures, support systems for analytical (cognitive) activities, diagnosis and forecasting systems and other special systems can be considered as a complicated OSS. For example, computerized procedures are designed to prevent many kinds of errors during the phases of decision making, implementation and monitoring of results.

One more important measure contributing to errors prevention is providing the operator with the opportunity to *cancel the action* before it will has become irreversible.

4 Conclusion

The theme of studying the causes of human errors and methods of preventing them is not new and is typical for complex and responsible areas of activity. However, information on it is numerous and fragmented. There are many publications which represent collection of human errors and propose various methods of their classification. However, almost always, the recommendations how to prevent new errors in the future are either very general or focused on some narrow task and on the particular conditions in which this error has been occurred in the past. The error prevention methods are scattered among numerous guidelines and standards. This does not allow the designer to effectively use them when designing HMI and workplace.

The main purpose of the present paper is to establish relationships between various categories of errors, various elements of human activity and general principles of error prevention. Human activity performed when managing process system is represented as a set of tasks. Performance of task consists of five sequentially executed phases. It was revealed that some kinds of errors are specific for some phases. Despite the fact that the proposed principles for errors prevention looks obviously, their formulation and systematization is the result of labor-consuming analysis. Seven basic principles were formulated, and it was shown which kinds of errors can be prevented by the use of particular principle. Thus, a set of rules are elaborated that allow HMI designer to choose the way to reduce the risk of human errors.

Let's consider an example of implementation of the proposed principles. For example, it is necessary to prevent some accidental control action. This is 'unnecessary action' which is usually committed during the phase of 'implementation' and caused by lack of time or distraction (see Fig. 3). One (or several) suitable principles can be applied, for example, 'complication of action' (see Fig. 4). The complication of action can be realized by introduction of additional operations (see Table 1), such as 'pushing' of soft (virtual) button in the computerized HMI, which confirms intention to perform action, or removing of the protective cover from control switch or push button in the conventional hardware interface.

References

1. Woods DD, O'Brien JF, Hanes LF (1987) Human factors challenges in process control: the case of nuclear power plants. In: Salvendy G (ed) Handbook of Human Factors. Wiley, New York
2. Miller DP, Swain AD (1987) Human error and human reliability. In: Salvendy G (ed) Handbook of Human Factors. Wiley, New York
3. Anokhin AN, Ostreykovskiy VA (2001) Ergonomics Issues in Nuclear Power Engineering. Energoatomizdat, Moscow
4. Rouse WB (1983) Models of human problem solving: detection, diagnosis and compensation for system failures. Automatica 19(6):613–625
5. Gubinsky AI (1982) Reliability and Quality of Ergatic Systems Operation. Nauka, Leningrad
6. Chachko SA (1992) Prevention of NPP Operator Errors. Energoatomizdat, Moscow
7. Bodrov VA, Orlov VJ (1998) Psychology and Reliability: A Man in Technical Control Systems. Institut psikhologii RAN, Moscow

Human Factors and Ergonomics' Contribution to the Definition of a New Concept of Operations: The Case of Innovative Small Modular Reactors

Stanislas Couix$^{(\boxtimes)}$ ⓘ and Julien Kahn

EDF R&D, Human Factors Group, 91120 Palaiseau, France
stanislas.couix@edf.fr

Abstract. Contributing to the earliest phases of design is an old challenge of Human Factors and Ergonomics (HF&E) experts. If HF&E contributions to Basic Design and Detailed Design of Nuclear Power Plants is acknowledge, HF&E experts' contributions to earlier phase of design like Conceptual Design or pre-conceptual design (PCD) are not frequent. Therefore, if participation of HF&E experts to the earliest phase of systems design is not new, up to our knowledge, there is no already described, validated and structured HF&E method to contribute to a pre-conceptual design (PCD) phase. The aim of the PCD phase is to prepare and address the scientific issues of a proposed new design. From an HF&E point of view, the scientific issues are related to the concept of operations envisioned for the new system. The paper presents the approach we are currently leading at EDF R&D during the PCD phase of the design of innovative small modular reactors (ISMR) in order to define its concept of operations. This approach is based on several methods we propose to articulate in order to contribute to fill the lack of described and validated HF&E method to contribute to PCD. The theoretical foundations of our approach are based on work analyses in reference work situations [1] and operational analysis [2]. The paper describes each step of the approach developed and the organizational conditions for the participation of HF&E experts into PCD.

Keywords: Human Factors and Ergonomics · Systems engineering
Conceptual design · Concept of operations

1 Introduction

One of the goals of ergonomics is to contribute to the design of man-machine systems. It is widely recognized that the involvement of ergonomics must take place as early as possible in the design process (e.g., [3]). Contributing to the earliest phases of design is thus an old challenge of Human Factors and Ergonomics (HF&E) experts.

Based on the four system design phases (e.g., [4]) nuclear plant design with pre conceptual design is divided in 4 phases: conceptual system design; preliminary (or basic) system design; detailed design and development; system test, evaluation and validation.

© Springer Nature Switzerland AG 2019
S. Bagnara et al. (Eds.): IEA 2018, AISC 822, pp. 754–763, 2019.
https://doi.org/10.1007/978-3-319-96077-7_82

HF&E contribution to Basic Design and Detailed Design is acknowledge in nuclear industry in general [5]. Nonetheless, and more generally, HF&E experts' contributions during earlier phase of design like Conceptual Design or pre-conceptual design (PCD) are not frequent. If participation of HF&E experts to the conceptual phase of systems design is not new (e.g., [2]), up to our knowledge, there is no already described and structured HF&E method to contribute to a pre-conceptual design (PCD) phase.

This communication is about the method developed by HF&E experts to contribute to the pre-conceptual design phase of a new system. This method is currently followed by the authors involved in a project of innovative small modular reactors (ISMR).

After a short presentation on the industrial context, and the underlying theoretical foundations, the method and the techniques used by HF&E experts in the PCD of ISMR will be described.

2 Industrial Context

In nuclear industry, the design of a new system is rare. New designs are not as frequent as in other majors industries involved in safety critical systems. Like most large systems, new nuclear reactors designs are mostly improvements or optimizations of past designs. It is not the case of the ISMR. This new kind of reactor is a major breakthrough with the designs of current large reactors designed, build and/or operated by EDF.

2.1 Innovative Small Modular Reactors

As the name implies, the first big difference compared to current large reactors is their size. ISMRs are multiple times smaller than current reactors. As a consequence, they produce less power. For example, current French reactor produce between 900 megawatts (MW) and 1600 MW. The ISMRs will typically produce from 25 to 300 MW.

The main reason for ISMRs is to tackle 3 major concerns in today's nuclear power industry [6]. Firstly, the small size of ISMR allows simple design with high level of safety, such as the use of passive systems. Secondly, another major benefit from ISMRs is that they will reduce design, construction and operation costs compared to current reactors. For example, their modularity will allow progressive investment as the modules can be built progressively. Plus, instead of assembling the reactor on the construction site, the reactors' parts can be sort of mass-produced and assembled at a central factory. The reactors are then moved to the construction site where they can be plugged to other big components also assembled in a central factory. Finally, ISMRs, thanks to their compactness and modularity, will be easier and cheaper to decommission.

The major technical breaks of ISMR project implies that the concept of operations of current large reactors won't be adapted to these new kind of reactors. In other words, the technical breaks imply socio-technical breaks. We need to revamp the way nuclear reactors are operated, otherwise, the objective of reduction of operation costs of ISMRs won't be reached. For instance, operation of French large reactors requires operating staff composed of at least 2 operators and a set of procedures for incidental or accidental conditions management. Actions of operators are controlled and verified by a crew member and a safety engineer. Their work is to ensure that the actions carried out by operators are the ones adapted to the state of the plant, and that operators don't miss any actions. In ISMR, thanks to passive safety systems, less actions from a human agent will be required to reach a safe state in accidental conditions. This technological breakthrough will have several impact on the sociotechnical system. For example, the need of two independent human agents verifying the actions carried out can be questioned. Thus, organisation of operating crews can evolve to match the new requirements to handle accidental conditions in ISMR.

2.2 Pre-conceptual Design Phase and the Definition of a Concept of Operations

At EDF, the aim of the Pre-Conceptual Design[1] (PCD) phase is to identify and solve the scientific and technical issues associated with the various design options identified.

At this stage, different options for the main systems, the main structures, and the concept of operations (ConOps) are discussed by expert designers and other stakeholders.

Many definition of ConOps co-exists. For example, according to [5], the ConOps is a description of how the design, systems and operational characteristics of a plant relate to the organisational structure, staffing, and management framework of the utility who will operate the system. More precisely, when defining a ConOps, several elements of the future socio-technical system have to be defined:

- the plant mission;
- the roles and responsibilities of humans and automation;
- the staffing, qualifications and training of the plant personnel;
- how plant personnel will manage normal, off-normal and emergency conditions (which includes the means they will use to control and monitor the plant, as well as the control room concept);
- how maintenance and modifications of the plant will be performed.

To sum up, a ConOps reflects the way the operations of a future plant is foreseen and imagined. It is a user-centric view of the system operation. In other words, the elements defined in the ConOps will have major impacts on Human Factors issues during operations. In this perspective, ISMR project managers have asked the authors to contribute to the definition of the ConOps of ISMR.

During the PCD of ISMR, the main purpose of HF&E experts is to highlight and identify ways to solve the human factors issues associated with the various options explored during the definition of the concept of operations. Thus, they first have to define the various options for the concept of operations.

There are strong links between the concept of operations and the design of main systems and structures. So, the HF&E analyses are strongly influenced by options taken and explored by systems architects and business specialists. In return, the analyses

---

[1] PCD is similar to NASA's «concept studies» [7].

carrying out by HF&E experts can strongly influence the design of main systems and structures as well as business options.

For example, as the ISMR will have a simpler design with less equipment, their operation will probably be easier. From an HF&E perspective, it means that we will probably be able to reduce the manpower associated with the operations. Thus, we need to define a new organisation of the operating crews. An output of this analysis can be to ask systems architects or I&C experts to automate some functions of the plant.

3   HF&E in the Definition of a ConOps

In our point of view, HF&E experts' participation to the definition of the ConOps (or any task in system design) must be theoretically and methodologically grounded.

3.1   A Model of Humans at Work

Our work comes within the scope of the "French-speaking framework" which primarily focuses on the analysis of human activity in situations [8]. In this approach, activity of operators depends on two major elements (Fig. 1): individual characteristics of the people and the elements of the situation in which they are working [9, 10].

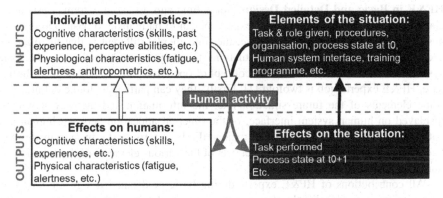

Fig. 1. A model of human activity, adapted from [11].

The individual characteristics of the people are either cognitive or physiological. For example, they refer to the skills of people, their perceptive abilities, their level of fatigue and alertness, etc. The elements of the situation refers to all the elements that are external to the people, that is the tasks and roles that people have, the operating procedures they use, the state of the process they are controlling, the human system interface they are working with, the training they were given, etc. In return, people's activity will have effects on both their characteristics and on the elements of the situation. For example, after performing activity, people will be more tired, their level of alertness will be reduced, the state of the process will be different, etc.

In this perspective, defining the concept of operations of the future socio-technical system means defining parts of the inputs of human activity.

3.2 Methodological Perspectives on HF&E in Phases of Design

Despite HF&E experts will to contribute as early as possible in the design of new socio-technical systems, it is rare. For example, it is the first time HF&E experts at EDF have contributed this early to the design of such a big system. In the nuclear domain, if international norms require the contribution of HF&E experts during design (ex: NUREG-711), the definition of the ConOps is not mentioned as an HF&E activity during design. This section reviews some of approaches to HF&E in design compatible with our theoretical background.

HF&E when the Design is Almost Over or the System is Already Commissioned. Until recently, Human Factors experts were often called at the end of the design, or even after design, when problems in human performance arise in a system, after the system was commissioned. In the "French-speaking framework", this kind of contribution is based on work analyses on the existing situation [1, 8]. Thus, the system to redesign or fix is perfectly known. The work analyses are easier, very powerful and reliable. The problem is that, at this stage, any modification will be extremely costly, even if the system is not already commissioned.

HF&E in Basic and Detailed Design. Some time ago, to avoid extreme costs associated with late involvement, HF&E experts began to participate to basic and detailed design. Compared to the previous case, the fundamental different is that there are no socio-technical systems and work situations to analyse yet. Nonetheless, the main systems and structures as well as the concept of operations are already defined. In this case, HF&E experts' job is twofold [1]. Firstly, they can participate to the definition of some elements of the future situation like, the information and the way it will be displayed on human system interface, or the function allocation between human and automation (see [2] for a review of major HF&E experts' tasks in system design). Secondly, they have to anticipate the effects of the various elements of the situation on human activity.

All contributions of HF&E experts during basic or detailed design are based on three sources (e.g. [1]). Firstly, they can perform simulations of human activity in the future system, with, for instance, wood and paper or virtual mockups, or even a full-scope simulator. Secondly, they also heavily rely on general knowledge in HF&E science. And finally, they can perform analyses of work in similar socio-technical systems, which means performing work analyses on an existing system that shares similar characteristics with the socio-technical system under design.

The rationale behind is the following. As human activity depends on the elements of the work situations, if HF&E experts analyse work in a system that share characteristics with the system under design, it is thus possible to anticipate the effects of some design features or characteristics on human activity. Then, it is possible to mitigate the negative effects or maintain the positive effects on human activity observed in the similar system for the future system. In other words, performing work analyses in similar systems means understanding what makes the similar system work or fail and transferring these results into the new design.

The problem is that during pre-conceptual design, the characteristics of the future socio-technical system are not known because defining them is actually the purpose of the pre-conceptual design phase. Thus, approach of HF&E in basic and detailed design is not directly applicable to the more conceptual phases of design.

HF&E in the Conceptual Phases of Design. Up to our knowledge, few elements in the scientific literature address specifically the contribution of HF&E experts in the definition of a ConOps, and more generally, in the early phases of system design.

A promising way for HF&E experts to contribute to conceptual design identified is the Scenario-Based Design approach [12]. Nonetheless, defining the ConOps is very similar to the definition of the "root concept" which is the basis for scenario definition that will contain the basic elements of the scenarios (stakes, actors, situations, etc.). In other words, defining the ConOps means to define the elements that will be put in a scenario. Thus, if scenarios can be used after this set of elements are defined, for example, to describe some elements of the ConOps to stakeholders of the project, another approach is needed to define this set of elements.

The contribution of HF&E experts to the conceptual phases of system design is explicitly mentioned by [2]. This approach is based on many techniques including operational analysis (projected analysis of the foreseen operational situations operators are likely to face with the new system), and the analysis of similar socio-technical system that are well suited for conceptual design. Nonetheless, few is said about the application of the techniques and their articulation in a structured and operational approach.

Consequently, there is a lack of HF&E engineering methods adapted to pre-conceptual design. Thus, a new approach, methodologically founded on the previous mentioned, has been developed and is currently followed during ISMR project to guide authors' action during the PCD phase.

4   Approach Currently Followed by Human Factors Experts During PCD of ISMR Project

This section details the four-step iterative process followed in the ISMR project to define the concept of operations and the organisational conditions of its application. This approach is mainly based on the analysis of work in reference situations [1], operational analysis and similar systems analysis [2].

4.1   Description of the Four Phase Approach

This approach is divided in four steps (Fig. 2):

- Defining options for the characteristics of the future system and operational situations;
- Identifying systems with similar characteristics and operational situations;

- Gaining knowledge on the similar socio-technical systems;
- Extrapolating the results to the system under design.

After a first loop in the process, it is possible to assess if the first options for the system characteristics will stand or any modification is needed. Finally, if the options seem positive on human activity and performance, they can be include in the concept of operations.

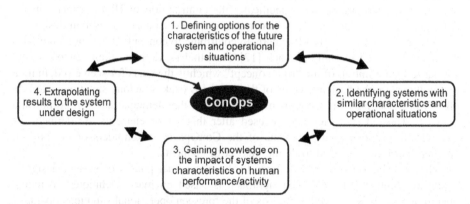

Fig. 2. HF&E approach during the pre-conceptual design of ISMR.

Phase 1: Defining Options for the Characteristics of the Future System and Operational Situations. The first step is to define the first options or hypotheses concerning the characteristics of the future systems (e.g., "operating several power units from a single control room", etc.) and the operational situations the socio-technical system will probably face[2] (e.g., "start-up of the plant", etc.). Of course, HF&E experts cannot do this by themselves. During ISMR project, it is done in cooperation with other experts involved in the project. For each design option discussed during these meetings, HF&E experts have to analyse if there is any major impact of human activity and performance.

This very first analysis is based on HF&E experts' general knowledge in HF&E science, and in work in similar systems if available. At this stage, this allows to identify if some options can be directly excluded, or if they should be investigated more deeply during the rest of the process. At the end of this phase, a first list of system characteristics (e.g., "extensive use of passive safety systems") should be available.

Phase 2: Identifying Systems with Similar Characteristics and Operational Situations. The aim of the second step is to identify systems that share similar characteristics and operational situations with the ones defined at the first step.

At the beginning, a first list based on authors' knowledge about other systems was established. Then, interviews with people that have valuable knowledge in these various systems should be carried out. This part is important HF&E experts could have

---

[2] This part of analysis is similar to what [2] calls «operational analysis».

been totally wrong about those systems or could have overlooked some characteristics that make the system very different to the point where the knowledge gained in these systems won't be transferable.

At the end of this step, a list of similar systems is available. For instance, during ISMR project, several systems that share characteristics with our future system has been identified (e.g. risky continuous process, multiple unit operated from a single control room): French hydro-electric power plants, coal and gas power plants, tele-intensive care units, offshore oil & gas production. Of course, other nuclear power plants have also been considered. Indeed, even if there are major technological differences, some aspects of the system will be the same.

Phase 3: Gaining Knowledge on the Similar Socio-Technical Systems. The third step in the process in to gain knowledge on the impact of the characteristics of the similar system on human activity in the similar operational situations by conducting a work analysis [1, 2]. In other words it means to identify what makes the similar socio-technical system work or fail in situations comparable to the anticipated ones for the future system. This analysis cannot be detailed: it should focus on important aspects of work, and the consequences of characteristics of the system on work. For example, in our study, characteristics like: control room layout, automation, task allocation, staffing, number of parameters to monitor, normal and off-normal situations management) have been investigated in similar work systems.

This could be done in two ways. Firstly, for all the systems identified, an analysis of scientific literature in HF&E issues in the similar socio-technical system should be performed. Secondly, if access to the people working (front-end operators, managers, etc.) in these systems if possible, some filed studies should be performed. Various techniques could be used according the authorisations obtained: interviews, observations, operational documents analysis, operational experience feedback analysis, etc. In this phase, HF&E experts' work is a "traditional" work analysis. The basic idea here is to understand workers their strategies to cope with systems characteristics and operational contexts and constraints. For instance, how operators in hydro-electric power generation face normal and off-normal conditions and determining how and at which cost they maintain high performance in the system.

Phase 4: Extrapolating Knowledge Gained to the System Under Design. The final phase is to extrapolate these results to the future socio-technical system. It is an important part of the analysis, because, if systems share some characteristics, they also have many differences. For example, if the analysis states that characteristics A and B in the similar system contribute heavily on the positive performance of humans, and only characteristics A will be possible in the future system, two options should be considered: the characteristic A should not be integrated into the ConOps, or a characteristic producing similar effect comparing to characteristics B should be sought. In this way, if some options for the characteristics are still unresolved, they should be analysed in the next looped of the approach.

A Non-Sequential Approach. In the early phases of design, some important aspects of the system can evolve rapidly. Thus, options investigated should be reanalyzed frequently. In this way, the approach cannot be sequential, even during the first loop.

If the design change, an analysis of the consequences of the change must be carried out to avoid analysis of outdated options for the characteristics of the future system.

4.2   Organisational Conditions of Our Approach

Finally, here are some organisational conditions for the participation of HF&E experts in the definition of the ConOps of the ISMR.

The definition of the ConCops is embedded with the instrumentation and control (I&C) work package as there are natural, historical and strong link between HF&E and I&C experts, especially when dealing with control room operations and definition of automation, which are two important parts of the ConOps. Nonetheless, alone, HF&E experts cannot define a ConOps by themselves. The fact that human activities depend on system characteristics implies to share information with other work packages like process design, safety, etc. HF&E experts cannot work isolated from other designers and other decision makers because, in a system, everything is in tight coupling and a modification in one part can affect many parts of it. For example, the size of the control room(s) will depend on the number of human agents needed to operate the plants and the way people are organised, which will have great impact on civil engineering. Thus, HF&E experts must cooperate tightly with experts from these other work packages and be involved in many major design and business decisions.

Concerning workload associated with the definition of the ConOps, in total, 4 HF&E experts are involved in this task, which represent 1 person-year. One is in charge of the definition of the concept of operations, which is me by the way, and I rely on the outputs from two others HF&E experts that are performing data collection in similar socio-technical systems, and one other HF&E expert who performed the analysis of scientific literature on human performance in similar systems.

5   Discussion

This paper presents the approach developed to define the ConOps of a new complex socio-technical system called ISMR. The approach is currently applied by the authors who are involved in the ISMR project. They are finishing a complete first loop of the approach. Nonetheless, some aspects of the approach can already be discussed.

A first lesson authors learned during the first loop is that there is a strong need to be flexible. As stated, during the conceptual phases of design, many elements of the system are modified. If some changes have no or minor consequences on the ConOps, others don't. Determining the consequences on the ConOps is really important as time could be lost analysing an option that is no longer possible, and don't analysing a new promising option made available as the design evolve. Thus, HF&E experts must ensure that they are aware of the last changes brought to the design. From our point of view, one good way is to have regular meetings with other stakeholders. To make this possible, a tight cooperation should be created and maintained. For this, HF&E experts must be legitimate in the design process: other designers must recognize their special knowledge and methods, and HF&E experts must give some innovative solution that work within the constraints brought by other designers. In other words, designing is always made of many trade-offs, and we must show how the solutions we promote can be beneficial for the future system.

A second lesson, is that it can be hard to get to people working at other complex socio-technical systems. Accessing other system and organising observations and interviews with them is a time consuming activity. One of the HF&E team member was almost dedicated to this task. Even if EDF owns some of the system we investigated to analyse some design options, it is difficult. People working in these system must be agreed to give you time even if there is no direct benefit for them. Nonetheless, this was made possible by the professional network the authors have and by giving people met the analysis the authors performed. This latter aspect is more important than it can appears first. Indeed, we learned that the managers and front-end operators met are always pleased to have an external look on their work.

References

1. Daniellou F (2004) L'ergonomie dans la conduite de projets de conception de systèmes de travail. In: Falzon P (ed) Ergonomie. Presses Universitaires de France, Paris, pp 359–373
2. Chapanis A (1996) Human Factors in Systems Engineering. Wiley, New York
3. Hendrick HW (2008) Applying ergonomics to systems: some documented "lessons learned". Appl. Ergon. 39:418–426
4. Blanchard BS, Fabrycky WJ (2010) Systems Engineering and Analysis, 5th edn. Upper Saddle River, New Jersey
5. O'Hara JM, Higgins JC, Fleger SA, Pieringer PA (2012) Human Factors Engineering Program Review Model (NUREG-0711, rev.3), U.S. NRC
6. Delbecq J (2012) Les Small Modular Reactors (SMR). Annales des mines – réalités industrielles (3):133–141. https://doi.org/10.3917/rindu.123.0133
7. NASA (2007) Systems engineering handbook (NASA/SP-2007-6105, rev 1). National Aeronautics and Space Industry, Washington, DC
8. Daniellou F, Rabardel P (2005) Activity-oriented approaches to ergonomics: some traditions and communities. Theor. Issues Ergon. Sci. 6:353–357
9. Leplat J (1990) Relations between task and activity: elements for elaborating a framework for error analysis. Ergonomics 33:1389–1402
10. Hacker W (1985) Activity: a fruitful concept in industrial psychology. In: Frese Michael, Sabini John (eds) Goal Directed Behavior: The Concept of Action in Psychology. L. Erlbaum Associates, Mahwah, pp 262–284
11. Leplat J, Cuny X (1974) Les Accidents du Travail. PUF, Paris
12. Carroll JM (2000) Making Use: Scenario-Based Design of Human–Computer Interactions. Massachusetts Institute of Technology, MA

Author Index

© Springer Nature Switzerland AG 2019
S. Bagnara et al. (Eds.): IEA 2018, AISC 822, pp. 765–768, 2019.
https://doi.org/10.1007/978-3-319-96077-7

Printed in the United States
By Bookmasters